The Divine Algorithm

Book One: *The Principle of Instantiation*

Book Two: *Emergence & Transformation*

Book Three: *Transcendence*

Damon Eaves

Eaves Press

Front cover by, James Edward Blanchard, jimblanchard.com

Topics; Origin of Universe, Life-Origin of (Abiogenesis), Apologetics, Cosmology, Population III Stars, Teleology, Philosophy of Science, Biology, Mitochondria, Darwinism, Consciousness, Spirituality, Miracles, Near Death Experience

TABLE OF CONTENTS

Book One: *The Principle of Instantiation*

Page	Chapter
7	Preface: The Divine Algorithm, Book One
16	The Divine Algorithm, The Overview
18	Special Problem No. 1: The Beginning of Our Universe
39	Variable 1: The Baryon Asymmetry
74	Variable 2: The Initial Entropy State of the Universe
66	Variable 3: The Horizon Problem of the CMB
100	Variable 4: The Cosmological Constant
121	Variable 5: The Fine Tuning of Gravity
144	Variable 6: Fine Tuning & Anomalies of The Standard Model
191	Variable 7: Fine Tuning of the Carbon Resonance
206	Variable 8: Stellar Nucleosynthesis
231	Variable 9: The CMB As The Instantiation of The Universe

Book Two: *Emergence & Transformation*

Page	Chapter
256	Variable 10: The Unique Fitness of CHNOPS
268	Special Problem No. 2: Abiogenesis, The Beginning of Life
297	Variable 11: DNA, DNA Transcription and the Cell
318	Variable 12: The Mitochondria and Cellular Respiration
340	Variable 13: Chirality & Organic Molecules
348	Special Problem No. 3: The Evolution of the Species
399	Interlude No. 1: The August Wheel
402	Interlude No. 2: Thomas Kuhn & Karl Popper
409	Preamble, The Divine Meditation
414	Variable 14: The Divine Meditation
465	Prelude to Our Final Two Chapters
466	Special Problem No. 4: Consciousness and Youness
519	Variable 15: Cell Consciousness
557	Interlude No. 3: The Evaluative Field
575	Section 1: Values & Validations of The Algorithm
578	Validations of The Algorithm

Book Three: *Transcendence*

Page	Chapter
587	Introduction
588	1. Mystical Experience & Spiritually Transformative Experience
597	2. Maya & The Communication of God Consciousness
606	3. The Prophets Experience
626	4. The Physical/Mentally Triggered Mystical Experience
659	5. The Biologically Triggered Mystical Experience
682	6. The Near Death Experience & Out of Body Experience
700	7. Transcribed Near Death Experiences, Atheists & Skeptics
737	8. Miracles
748	9. Is the Creator in Book 3, the Instantiator in Books 1 & 2?
751	10. For The Future
756	Acknowledgments

PREFACE: The Divine Algorithm, Book One

One Simple Question

This book exists because of a single, foundational question: How did life begin?

The public narrative is well-rehearsed: life began as an accident. But the more I examined this claim, the more strange it seemed. By what metric do we determine this was an accident? And why must the origin of life be framed as a statistical anomaly rather than a law of nature? I decided to move past the rhetoric to discover exactly how random chemicals "hit upon" the correct configuration to cross the impossible boundary from non-living to living. I wanted the mechanism. I wanted the chemistry.

The Illusion of Certainty

The initial stories I encountered were almost comical. They described a newly formed Earth—volcanoes, turbulent skies, lightning—and somewhere in that chaos, a "primordial soup." In these accounts, organic molecules simply collected and then—miraculously—resulted in life.

That was the entire story. There were no details. There was no transition. The gap between non-life and life remained unaddressed, bridged only by the assumption that it had happened.

Seeking substance, I consulted rigorous, collegiate-level resources like Campbell Biology—a staggering 1,500-page reference. Conspicuously, there was no dedicated chapter on the origin of life. What was covered occupied only a few glaringly incomplete pages, dense with assumptions but devoid of cited pathways.

As I endeavored to find the mechanical details of this transition, I finally reached the only logical conclusion: No one knows how life began. This isn't a failure of intelligence; it is not only impossible, but it's been proven as impossible. But because life exists, life arising from non-living materials is treated as an "accident" precisely because there is no mechanism, no pathway, and no empirical evidence to support it.

The Concession

in an interview with Ben Stein, in the 2008 documentary *Expelled: No Intelligence Allowed*[1] prolific author, Emeritus Oxford Fellow, evolutionary biologist and the world's staunchest defender of Darwinian theory, Richard Dawkins said no one has a clue how the first replicating cell was created.

Ben Stein: How did it [life] start?

Richard Dawkins: We know the sort of event that must have happened for the origin of life.

Ben Stein: What was that?

Richard Dawkins: It was the origin of the first self-replicating molecule.

Ben Stein: Right, and how did that happen?

Richard Dawkins: I've told you, we don't know.

Ben Stein: So you have no idea how it started?

Richard Dawkins: No, nor has anybody else.

His admission was no 'gotcha', or moment of weakness or poorly chosen set of words. In Dawkins' book, *The God Delusion* (2006)[1] he committed to paper that life is a "happy accident" and a "one-off event." He admits that while we have theories, like the *RNA World*, we do not have a proven pathway.

In an interview in 2012 on *The Unbelievable? Radio Show* (2012)[2] Dawkins stated the "probability of life's origin is very low," but then argued that it only had to happen once.

The Question of Fitness

For life to exist and diversify, the very nature of physical reality—the chemistry of atoms, the fundamental forces, and the laws of the universe—must possess a pre-programmed, inherent potential for biology. If the physical order were even marginally hostile to life, the existence of a single cell would be a non-starter.

Life does not just "happen" in this universe; life requires a deep, pre-existing synergy. It requires a universe where life isn't just permitted, but one where life is helped, at every stage to exist and diversify as it has on our planet. Because once there is life, life relies upon life for nutrition, comfort, shelter and all other things.

Twin Mysteries

---

[1] Producers, L. Craft, W. Ruloff, J. Sullivan, *Expelled: No Intelligence Allowed*, Film, 2008, host, B. Stein, Premise Media Corp with Rampant Films, April 18, 2008

The prerequisites for life were not an afterthought in our realty; they were hard coded into the physics of the universe. But then, we must confront another impossibility. Where did the universe come from?

We currently face a double failure of explanation. We cannot account for the origin of the universe, and we cannot account for the origin of life. In both instances, the evidence suggests that if we are restricted to purely mechanical, unguided means, both events are statistically impossible.

A New Search

This is the thread I pulled. I stopped asking how life could begin and started asking why reality allows it. I audited the foundational literature of modern science—cosmology, genetics, particle physics—and moved from "searcher" to "auditor."

The Granular Details of Impossibility

The pathway to life requires the simultaneous assembly of four distinct macromolecules—carbohydrates, proteins, lipids, and nucleic acids—within a single microscopic space. Natural processes fail here on every front.

Even if we grant the spontaneous assembly of every component, we face the final abyss: the literal spark of life. Consider this: fill the Grand Canyon with the dead remains of every creature that has ever lived. All the machinery is present, in perfect form, yet no life will arise from that collection of flesh. Structure is not agency. Form is not life.

Life is not a byproduct of self-directed complex chemistry; it is an absolute boundary. Death is the empirical proof of that boundary.

The Algorithm Emerges

As I moved from field to field, a recurring red flag appeared. In every discipline, I encountered the same unequivocal claim: "This problem is considered to be the greatest mystery in the history of [field]."

At first, I simply took note. But as the frequency of these claims increased, I realized these were not independent puzzles. They were symptoms of a systemic failure at the heart of every scientific discipline. I asked the obvious, ignored question: If every field claims its central problem is "the greatest," which one truly holds that title?

I realized they could not be ordered. They were all of the same kind: intractable. In every instance, the universe did not merely produce a puzzling result; it did the exact thing it was expressly forbidden to do. Furthermore, the consequences of these "trespasses" were never trivial—they were the very mechanisms that enabled our existence.

If the universe had not repeatedly performed what was forbidden, two things would be true:

(1) Our universe would be sterile, devoid of any possibility of biological life.

(2) The subsequent violations of physical law required to build our reality would have been rendered impossible.

I realized that in this pattern of impossibilities, humanity had stumbled upon something beyond mere anomalies. We had found a hidden pattern of miracles—a governing set of specifications that serve as the foundation of our reality at every stage.

The Six Realizations:

The Origin of Life is a total mystery: Despite the rhetoric of "primordial soup" or "self-assembly," there is no mechanical pathway from non-living chemicals to a self-replicating organism. The boundary is absolute.

The Diversification of Life is a mystery: When we strip away the assumed mechanisms of evolutionary theory, the sheer speed and specific complexity of biological diversification remain statistically unreachable by unguided processes. Moreover, as a matter of logic, this creates an unbridgeable gap: if we cannot reconstruct the path from non-life to life, on what evidentiary basis do we claim to understand how species diversified? Life itself—in its complexity, its agency, and its sheer architectural brilliance—requires a mechanism that currently evades us.

Universal Fitness is a requirement from the outset: Life is not a lucky occupant of the universe; it is a system-requirement. The chemical, biological, and systemic capacity for life—and its subsequent diversification—had to be hard-coded into the physis of the universe at the moment of instantiation.

The Singularity is a dead end: The Big Bang (T=0) model is a mathematical abstraction that cannot meaningfully approach the beginning. It fails to account for the origin of matter, the source of physical laws, or the mechanism of the "creation" event itself.

The "Greatest Mystery" is a forensic symptom: The recurring designation of "greatest mystery" across every scientific discipline is not a mark of progress; it is the forensic signature of a systemic failure. We are trapped in a materialist paradigm that forces us to solve impossible problems blindfolded, with our hands tied behind our backs.

The Pattern of Anomalies is the Algorithm: Each "unreachable" transformation and each "forbidden" trespass of physical law is not a random glitch. They are necessary, coordinated steps in a single, coherent sequence—a hidden pattern of miracles that bridges the gap from the beginning of the universe to the emergence of conscious, thinking beings.

The End of the Road

What began as a simple question, "How did life begin?" evolved into a forensic audit of the human technological arc.

The Divine Algorithm: Summary

The history of our universe is defined by precise discontinuities. They occur at the major transformational events, when new instantiations take place, starting at the beginning (T = 0) and continuing into the present. The dominant scientific paradigm has classified these anomalies as difficult to solve problems, and it has offered a number of dubious theories to patch the holes in our understanding. But when we strip away all of these speculative fixes, we have what I have called, The Divine Algorithm. It is divine, because it replicates the history of our universe and it results in human beings. And it does this through a series of miracles.

Our Approach

Our investigation organizes the individual components of the algorithm into two main types:

The Variables: the 15 specific physical constants and finely tuned anomalies that fall outside the predictive scope of the standard materialist hypotheses. The 15 'Variables' identified in this audit are the mechanical and environmental prerequisites for the system of our universe.

The Special Problems: the 4 most foundational and seemingly impossible events in the universe's history. These occurrences share a particular quality—they are the most perplexing and inexplicable events within the 'algorithm'—revealing critical, unbridgeable explanatory gaps in the materialist framework. They are 'Targeted Search' anomalies: outcomes that are not only beyond probability and first principle physics but share the quality of being precisely what is required to result in highly specified functional outcomes.

We will proceed with a forensic investigation into the "Variables" and "Special Problems" that make up the algorithm broken into three books:

Book 1: The Principle of Instantiation: the arrangement of the universe and structural prerequisites of the universe.

Book 2: Emergence & Transformation: the creation of the first stars, the galaxies, planets, and creation of life and intelligent life endowed with consciousness.

Book 3: Transcendence: tracing the necessary attributes of the causal agency behind Instantiation, Emergence, and Transformation, based upon the outcomes achieved by the variables and special problems.

In our inquiry, we will evaluate the internal consistency of the materialist framework using its own established parameter—that given the physics of our universe and given enough time, our universe will result in its present condition.

The Ruler

To aid us in our evaluation, we will use an objective 'ruler' for what is physically possible within our universe, versus what is effectively impossible given what materialism presupposes. The ruler we will use is based upon the *Probabilistic Capacity of the Observable Universe* (PCOU), a number that represents the absolute maximum opportunities for any event to take place within our universe.

This 'ruler' measures out to roughly 1 chance in $\mathbf{10^{140}}$

To explain this number, and why it is perhaps the most significant measure we currently have of a universal probability bound, it is based on the estimated number of fundamental particles within our universe, given as $\mathbf{10^{80}}$ in number. This number then subjected to a complex formula developed by William Dembski, a philosopher, mathematician and probability theorist, and further refined by Seth Lloyd, a quantum mechanical engineer and physicist.

The formula we are using incorporates the fundamental physical limits of our universe based upon fundamental particles, the inverse of Planck time, and the age of the universe in seconds. The inverse of Planck time, (how many ticks per second) is approximately$\mathbf{10^{45}}$. Factoring the total number of seconds since the big bang based upon 13.8 billion years, is between $\mathbf{10^{17}}$ and $\mathbf{10^{18}}$ seconds ($\mathbf{10^{18}}$ is a generous estimate). Multiplying this together, we get: $[\mathbf{10^{80}}\ \boldsymbol{x}\ \mathbf{10^{45}}\ \boldsymbol{x}\ \mathbf{10^{18}}]$ which equals $\mathbf{10^{143}}$. This is the maximum number of physical events (state changes) possible in the history of the observable universe as calculated by Dembski.

Dembski's number, which he called the *Universal Probability Bound* in his Cambridge University Press, peer reviewed book, *The Design Inference*[3]. of $\mathbf{10^{150}}$ was a generous (larger) number that assumed every fundamental particle in the observable universe could act independently[4] MIT physicist Seth Lloyd calculated the same limit, but as the total number of elementary logic operations (bit-flips) possible within the entire history of the observable universe. Consequently, he calculated the number to be approximately $\mathbf{10^{120}}$. He considered this the "physical reality" limit[5].

Lloyd's number is lower because as a quantum engineer, Lloyd realized that particles are constrained by the speed of light and energy required to "flip a bit of information"[6] This means that $\mathbf{10^{80}}$ particles can't all be 'talking to each' other simultaneously across the scale of the universe. $\mathbf{10^{120}}$ is the estimate of the real-world limit of how many opportunities exist.

In consideration of Dembski's and Lloyd's work, the number we have chosen as a benchmark, $\mathbf{10^{140}}$, is deliberately inflated above Lloyd's number to grant a more than sufficient leeway for the reach of material laws, first principal physics and probability to achieve the outcomes attributed to the universe. Our benchmark ('ruler') establishes a generous probabilistic threshold for probabilistic events that is billions of times greater than the universe's actual computational capacity defined by Lloyd.

To determine if a scientific explanation is sufficient or merely a "placeholder" for luck, we will use our 'ruler' of $\mathbf{10^{140}}$ to represent the Probabilistic Capacity of the Observable Universe (PCOU).

The magnitude of the PCOU is best understood by the absolute limit of cosmic activity: even if every fundamental particle in the universe acted as a probability agent, tossing the dice of chance once every Planck second since the dawn of time, the total volume of attempts would still fall short of the requirements for many of our observed "Variables."

The PCOU separates physical possibility from systemic violation.

Within the PCOU: The event is statistically "possible."

Outside the PCOU: The event is a functional discontinuity. It is a "meta-probability" event—in layman's terms, a miracle. It is a localized violation of physical probability that exceeds the capacity of the universe to produce it.

This audit is necessary because of a frequently ignored distinction: Existence is the evidence that a deed was done, not the automatic proof of material causation. The standard materialist fallback is: "We are here, therefore it happened." We do not contest the fact of our existence; we contest the explanatory sufficiency of that logic. Materialism is built upon the foundation of physical law and probability. When proponents of the standard model discard those laws the moment they hit the PCOU boundary, they are no longer practicing science; they are practicing faith.

Our investigation is not concerned with measuring the "somewhat unlikely" against the "known." We are measuring "physical absurdity" against the "realm of possibility."

Final Pieces of the Investigative Methodology

Our investigation rejects the traditional practice of treating anomalies in isolation. Historically, these events are siloed within their respective disciplines—biology, physics, cosmology—creating the illusion that they are independent puzzles. This fragmentation is the primary hindrance to a true synthesis.

Our hypothesis is simple: when these anomalies are removed from their silos and aligned, they reveal a sequence that is otherwise invisible. If the materialist paradigm fails the test of physics and probability, then these anomalies are not "ad hoc" problems. They are a linked, interdependent pattern.

The Cumulative Weight of Improbability

When we audit these anomalies as a linked set, the impossibility ceases to be additive and becomes compounded. By viewing them cumulatively, we move beyond the failure of the materialist paradigm and enter a new dimension of improbability. We are not just looking at a series of "lucky breaks"; we are looking at a system-wide breach of probabilistic law.

The Standard of Inquiry

We assume the universe is indifferent to the interpretive frameworks we project upon it. In this investigation, we strictly prohibit "jailbreaking"—which in modern science means attempting to force a validation where the evidence indicates a falsification. Falsifying a non-viable theory does not threaten the integrity of reality; it merely clears for real solutions.

We treat the universe as an open book. If we order our information based on constraint, logic, and first-principles physics—rather than dogma or belief—the evidence will declare exactly what the system is. By returning to these foundations, we do not merely critique the existing narrative; we force the evidence to reveal that our investigation into the

Divine Algorithm reveals the universe as a coherent, instantiated, and fully operational system.

---

[1] R. Dawkins, *The God Delusion*, 2006, Houghton Mifflin Harcourt
[2] (Unbelievable Radio Show, Host J Brierley, Premier Christian Radio, *Richard Dawkins & Peter S. Williams on The God Delusion*, Sept. 2012)
[3] W.A. Dembski, *The Design Inference: Eliminating Chance Through Small Probabilities*, 1998, Cambridge University Press)
[4] (W.A. Dembski, *The Design Inference: Eliminating Chance Through Small Probabilities*, 1998, Cambridge University Press).
[5] (S. Lloyd, *Computational Capacity of the Universe*, 2002, Physical Review Letters, Vol.88 No. 7).
[6] (S. Lloyd, *Computational Capacity of the Universe*, 2002, Physical Review Letters, Vol.88 No. 7).

The Divine Algorithm, The Overview

According to the standard cosmological narrative, the universe undergoes a series of evolutionary stages following a "Big Bang." However, when we interrogate the earliest observable epochs, the model enters a regime where the physics breaks down. The Big Bang framework attempts to describe the universe's initial evolution through a series of mathematical extrapolations that never truly approach $T=0$. Instead, the model is forced to treat the most critical phase of universal initialization as a "black box," relying on an ever-expanding library of speculative "fixative relations."

The standard cosmological model is widely presented as compact and empirically grounded, resting on a small number of well-defined ingredients—General Relativity, the Cosmological Principle, Quantum Field Theory, and auxiliary concepts like Dark Matter, Dark Energy, and Inflationary Cosmology. Presented this way, it appears internally motivated. But when examined at the level required for quantitative prediction, the model depends on a massive "information debt"—a set of structural assumptions that lie beyond direct empirical confirmation.

These assumptions are not derived from first principles; they are retroactively inserted to resolve specific tensions—horizon uniformity, flatness, and entropy constraints—and subsequently hardened into implicit background conditions. The result is a framework that explains cosmological observables by progressively enlarging its assumption set rather than allowing the observables to emerge naturally from the formality of the theory. The anomalies of the Big Bang are not "demanded" by the evidence; they are the unavoidable byproducts of an interpretive commitment that has outlived its usefulness.

The Myth of the Singularity

The central claim of this investigation is that we have mistaken our mathematical abstractions for reality. Big Bang cosmology treats the "singularity" as an inevitable starting point, yet the singularity is neither an observation nor a derivation—it is a myth.

Our investigation suggests a more coherent explanatory path, one consistent with the most robust empirical evidence available: the Cosmic Microwave Background (CMB). The CMB is not merely the "afterglow" of a primordial explosion; it is the hard-line boundary of the universe's instantiation.

When we accept the CMB as the primary datum—the point of origin—the "problems" of Big Bang cosmology evaporate. We do not need to solve the horizon problem or the flatness problem because these are not "struggles" the universe is undergoing; they are artifacts of an explanatory frame that has been extended beyond its natural domain of inference. The universe does not "evolve" into complexity from a singularity. It is instantiated as a fully operational system—a coherent galaxy machine—at the CMB.

A Systematic Investigation

In this work, we transition away from the "retrodictive" logic of the standard model. We accept the data as it stands: the universe emerges as a highly integrated system where symmetry, matter, expansion, stellar ignition, and the resonance of carbon are not the result of ad-hoc corrections, but the properties of a single, operational unit.

The first Special Problem—the creation of the universe—and its associated variables—baryon asymmetry, low-entropy states, the cosmological constant, and the fine-tuning of fundamental forces—are traditionally treated as "unresolved problems" of the Big Bang. In our framework, they are recognized as the inherent specifications of the Divine Algorithm. By untangling the hidden scaffolding of the standard model, we demonstrate that the persistent anomalies are not required by observation but are merely the fine print of a contract for a theory that is no longer supported by the evidence.

We conclude that the path to a coherent cosmology requires a paradigm shift: abandoning the mythical singularity and recognizing the CMB as the definitive, foundational moment of universal instantiation.

Special Problem No. 1: The Beginning of Our Universe

The Nature of the Problem

Our journey into the individual threads of The Divine Algorithm now begins with the most fundamental mystery of all: the origin of existence itself. This chapter will examine the first of our four "Special Problems"—the creation event of our universe, commonly known as the Big Bang. While the term is ubiquitous, we will explore the deep paradox that lies at its heart; the fact that our entire reality appears to have sprung into existence from nothing—an event that violates the most basic laws of physics as we know them. Our goal is not to recount a history of discovery, but to demonstrate that this initial event is an inexplicable miracle—the foundational 'Given' of *The Divine Algorithm.*

The Big Bang is one of the most well-known terms in science. The title suggests all that needs to be understood. An initial cataclysmic, explosive event created everything we can ever possibly see, know and understand including, ultimately, ourselves. For decades, the standard model asserted that prior to this event, 'nothing' could have existed—a conclusion forced by winding the clock of universal expansion backward until it vanished into a mathematical point. This is because as the clock of the universe was wound backwards, and the billions of years of universal expansion drew in upon itself, it got smaller, and smaller, to the point where there was no possible way for the universe to escape its beginning as a singularity.

A singularity is a physical condition smaller than any quantum designation hidden from any observation or measurement within the Planck scale. At the Planck scale we find the smallest possible meaningful designations of space, time, and length. And beyond this scale, we enter the singularity that allegedly began our universe where there are no meaningful designations of space, time, and length. Indeed, it is difficult to define what is meant by "beyond" in this context. The Planck scale itself is derived from the extremes of what humanity has been able to scientifically determine and is a combination of three designations: the Planck constant, the gravitational constant and the speed of light.

A singularity is not an object; it is a mathematical cul-de-sac where physics breaks down. It represents a boundary where density becomes infinite, spacetime curves into itself, and science loses all predictive power. It is the ultimate enigma—a physical form for the most perplexing question imaginable. If the most perplexing question imagined could take on a physical form, it would be a singularity. Within the materialist framework, this is the supposed cradle of existence. This singularity—a thing emerging from nothingness—marks what the standard model identifies as the absolute beginning.

For decades the understanding that the universe had an absolute beginning and emerged from a singularity, at the absolute moment of creation, was considered to be an unavoidable, straightforward, inescapable and mind-blowing understanding, made all the more difficult because this singularity tells us nothing about how the universe could have emerged from such an event. This was an intellectual crisis. Nevertheless, the consensus held that at T = 0, the universe instantaneously emerged from nothing, roaring through the Planck limit from an absolute beginning.

It is important to address the Planck scale and the implications of the Planck scale for the creation event because within this scale we find rules where hard determinations about what can and cannot be known. The Planck scale and big bang singularity are deeply connected as both are boundaries where our understanding of physics breaks down. Additionally, these determinations define the limits of what may be reasonably speculated upon and what has the possibility of being verified within our reality. What can be speculated upon is endless to be sure, but what may be speculated upon that can be scientifically verified within our reality, or scientific speculation, is what forms the crucial link between pure fantasy and pseudo-science, or even science fiction, and ideas that hold the potential to be scientifically verifiable into our reality. The need for this distinction will become evident shortly when we examine various proposals that attempt to reach beyond the limit of the Planck scale, and beyond the unknowable physics of the singularity, because under scrutiny, they all fail in establishing their assumptions, and they all fail in conceiving of any manner in which they can be scientifically verified. It is as if twin gatekeepers—one stationed at the Planck scale and the other within the singularity—demand an answer to a question that must be resolved before passing into their realms. Thus far no one has been able to decipher these questions, let alone answer them.

This leads to the essential takeaway: the Planck scale and the singularity constitute an impenetrable 'information wall.' Beyond this boundary, physics disintegrates, spacetime quantizes, and the language of science falls silent. Any statement regarding what came 'before,' or the nature of 'time' within this wall—including any concept of cause'—is logically meaningless. At the Planck limit and within the singularity, we have exited the domain of science and entered the realm of pure speculation.

Before The Big Bang

Despite the hard limit formed by the singularity beyond the Planck limit at the beginning of our universe, over the last two decades, a deepening schism has emerged in the scientific community regarding the status of the 'Pre-Beginning.' The debate centers on whether the energy of our universe, or some proto-physical state, existed prior to the Big Bang. Through the years this debate has taken on increasingly abstract and unorthodox dimensions that call into question the beginning of our universe and superimpose a beyond-physics existence of some physical state prior to the big bang. This debate contains several conjectures that question our definition of "nothing" and requires that our definition of "nothing" be inclusive of existence prior to the big bang event.

The upshot of these competing ideas is to challenge that the big bang was the beginning. In order to do this, these ideas seek to reframe the beginning of our universe from an absolute start to being a transformation from something that pre-existed the big bang. These conjectures rely upon pen and paper thought experiments, strangely interpreted clues, and speculative ideas of a new kind of quasi-physics that is able to produce a preconditional state for our universe prior to the big bang. The crisis lies in how these ideas are presented: they are touted as established solutions without a clear disclosure to the public that they are mathematically exotic and physically unverifiable. This is why the public thinks that there may indeed have been a universe prior to our universe. In YouTube videos, loaded into A.I. chatbots and among the most popular science educators, the message is the big bang is likely not the beginning. It is intentionally made unclear that these ideas are currently fantasy, and that to function, they must violate everything we have observed and can currently state about physics.

The principle of causality is the bedrock of our investigation into the beginning of the universe. It dictates that every effect or change in a physical system must have a prior cause or an attributable physical reason. By extension, no physical entity—whether matter, energy, or quantum field—can emerge into existence without a preceding causal framework.

This definition provides the foundation for both sides of the cosmological debate. First, it explains why the beginning of the universe cannot be circumvented: there is zero evidence—empirical or derivative—retrievable from the information ceiling of the Planck scale to validate a state prior to the Big Bang. Second, it explains why some scientists feel they have a mandate to invent fantasy worlds beyond that limit. Their dogmatic faith in causality demands a prior physical state; consequently, they feel licensed to construct mathematical bridges across the Planck limit, searching the singularity for a cause that their paradigm insists *must* be there.

In short, one side acknowledges that in our findings of the absolute beginning of our universe, we find an unbridgeable paradox that our present understanding cannot resolve. The other side, because they believe so strongly in the principle of causality, which is a foundation of the paradigm of materialism, doubts that this paradox is a hard stop, and feels they have already figured out that a world must exist prior to the beginning that can resolve the paradox of a universe beginning from nothing.

This is the current state of things.

Where Physics Ends

The paradox of the beginning strips science of its standard toolkit. Those searching for a "world behind the start" are forced to abandon observation for the abstractions of extreme mathematics and the unproven landscapes of string theory and cosmic inflation. The caveat, of course, is that any viable answer must be able to make predictions that can be confirmed, replicated, observed, and peer reviewed. In the pursuit of truth, one cannot simply rely on promissory notes; scientific integrity demands a rigid demarcation between speculative conjecture and verified theory.

In the following sections, we will dismantle the conjectures that attempt to pierce the ontological boundary of the Planck length. These models—including cyclical universes, "nothing to-something" models, and inflationary theory—all ultimately fail. Under scrutiny, they require conditions far more bizarre than the paradox they seek to solve, pushing the universe not only into what cannot be known, but beyond what is physically possible.

A Recurring Theme

The tendency to patch a scientific paradox with speculative, untestable theory is a recurring theme in this narrative. Throughout our examination of the Variables and Special Problems, we will see this strategy repeated: the use of mathematical fictions to circumvent the intractable problems that make up our reality. The necessity for these counter-narratives stems from a singular crisis: every paradox represents a systemic failure of scientific materialism. Collectively, these failures make up a robust falsification of the materialist paradigm itself.

Hard Science, Miracles, The Supernatural & The Divine

Our investigation maintains a rigorous standard for the Variables and Special Problems that form the backbone of this thesis. We will not divert into high speculation, nor will we introduce concepts already invalidated by accumulated scientific evidence. This is not a counter-narrative *to* science—unlike the materialistic 'rescues' we will examine—but rather the most honest articulation of science as it currently stands. These problems are real, and they remain unsolved. Every standard approach is blocked; every one of these problems defies the puzzle-solving capacity of scientific materialism.

Our task is forensic. In the tradition of the investigator, we must examine what science has excluded and what it has yet to reconcile, formulating a hypothesis that accounts for the staggering number of anomalies within our physical reality. Our thesis accepts the conclusion dictated by the evidence: the universe had a definitive origin event, prior to which there was nothing. At present, there is no scientifically verifiable way to circumvent this absolute start. This finding does more than challenge the status quo; it effectively falsifies scientific materialism and necessitates a new paradigm—one capable of resolving the profound questions that the current model cannot even address.

Taking the evidence we have in hand, this means that the event that created our universe, whether it was the Big Bang or not, is more than a beginning, and more than a paradox. At present it also represents a miracle.

> "A miracle (from the Latin *mirari*, to wonder),,, is an event that is not explicable by natural causes alone. A reported miracle excites wonder because it appears to require, as its cause, something beyond the reach of human action and natural causes."[1]

To find a miracle at the origin of the universe is to recognize that a miracle sits at the root of everything—from quantum physics to biology. In this context, the 'Beginning' is not a violation of physics to be rejected, but a paradox meant to be deciphered. It requires the one thing in the universe capable of providing context: the conscious observer.

From this perspective, the search for "pre-Big Bang" cosmologies is a desperate, ideological attempt to find a natural cause for an event that has none. This lack of cause may threaten the materialist paradigm, but it does nothing to diminish our reality. In fact, our presence within this creation event points toward a paradox far deeper than a simple violation of causality.

There is another definition for a beginning that emerges from nothing: Supernatural. This isn't an appeal to mysticism. It is a clinical use of the word. "Supernatural" simply describes a force that cannot be explained by science or that sits outside the laws of nature. Since the beginning of the universe has no material cause, it meets this definition perfectly. It isn't a theory or a belief—it is a forensic fact. Our universe was "caused by forces that cannot be explained by science."[2] Our universe was started by something that currently cannot be explained by science. Another definition of supernatural is, "departing from what is usual or normal especially so as to appear to transcend the laws of nature"[3] It could be argued that this is one of the best descriptions for the creation of our universe.

We have uncovered the primary motivations driving the materialist effort to invent a cause for a universe that seemingly has none: the need to adhere to the dictates of their paradigm, to rescue the lack of evidence from falsification, and to avoid a supernatural origin that sits uncomfortably close to a 'creative' God-like agency. Yet, no ideological motivation can divert us from the data. The universe had an absolute beginning with no prior state and no apparent material cause.

If we instead consider that the beginning was the first signal in a universal algorithm we are meant to discover. Then the universe's non-compliance with physics becomes more than an intractable problem or paradox. These features become signifiers of a deeper purpose. The creation of the universe can be seen as the opening move in a series of messages that must be assembled to be fully understood. The makings of a pattern that is universal and intentional. The beginning is the first sentence in a message that proclaims the intelligence that contrived our existence. In this context, the Big Bang, or the fact that it could not have taken place as described, is a deep indication of the true origin of our universe.

But first, we must backtrack. We need to recount how humanity evolved from the belief in an ageless, static universe to a dynamic one that held the vital clues of its own beginning. This is one of the most compelling stories in science, involving the greatest luminaries of the modern era. By recounting it, we will supply the evidence that this beginning was absolute—and that it held a paradox so deep that no materialist framework can unravel it. This paradox is the central condition for identifying the beginning as the first component of the algorithm embedded within our reality.

The Birth of The Big Bang

Before we can explore the Big Bang, we must understand how the scientific community was dragged—often kicking and screaming—toward accepting it. This wasn't a simple case of a new theory being confirmed by diligent research. It was a multi-century journey where science, following its own trail of breadcrumbs, was reluctantly led to a conclusion that was entirely unexpected and, for many, deeply troubling. This history is crucial: it establishes the immense weight of the evidence and reveals the true depth of the problem a creation event poses for a materialist worldview.

The term "Big Bang" was coined by astronomer Fred Hoyle during a BBC radio broadcast on March 28, 1949. Ironically, Hoyle didn't mean it as a compliment. He was a staunch defender of the Steady State model—the belief that the universe had existed for an eternity, unchanging and ageless. He used "Big Bang" as a term of derision, mocking the "irrational" idea that the universe could have just "banged" into existence from nothing.

But the Steady State was doomed. New evidence was trickling in from the night sky—strange, unidentified structures that didn't fit the old maps. Up until the 20th century, determining if a light in the sky was a nearby star or an entire galaxy millions of light-years away was largely guesswork. But as new technologies allowed us to categorize the heavens and Einstein's relativity provided the math, our place in the universe was fundamentally rewritten.

The evidence told a breathtaking tale: the universe had a dynamic, transforming history. Winding the clock back, the universe grew smaller and smaller, until the math pointed to a single point of infinite density and curvature—a Singularity. This isn't just a mathematical quirk; it is the first piece of our algorithm: *Variable #1: The Singularity at the Beginning of Time.*

Since it was discovered that the universe emerged from a single point that just prior did not exist at all, scientists have puzzled over what "banged", *how* it banged, and what rules could have possibly governed the creation of a universe born from a "bang."

We must once again emphasize that in order to properly evaluate the creation of our universe, it's crucial to distinguish between what is known from evidence and what is speculation based upon solid theory and an agreement with observation, and, in an almost opposite perspective, what is purely speculative and not attached to a foundation of science, and has no possibility of validation. The following timeline of the universe's first moments is presented to highlight a critical boundary in our knowledge: an "information ceiling" that exists at what is known as *Planck time*. By understanding this absolute limit, we can see that many widely discussed scientific theories are placed within a period from which we can gather no definitive information. We cannot say anything meaningful about what occurred before this moment. This is a wall" that no science nor scientist has been able to penetrate. It is likely impenetrable, indicating where we can no longer gather information. The one thing we do know is where this information ceiling is located and why it cannot be surpassed. It is located in an exact location in what Einstein called *space/time.*

Prior to when we wind the clock of our universe back to 10 to the minus 43 seconds ($\mathbf{10^{-43}}$) after the creation of the universe, (0.0000000000000000000000000000000000000000001 seconds), we have no physics to describe what happens, and no tools to probe behind this point, and no possible information we can pluck from behind this moment in space/time that can tell us what happened. Even what we can know about the universe after this time can tell us nothing about what happened prior. At $\mathbf{10^{-43}}$ seconds we reach something that is called *Planck time.* Planck time is the smallest possible divisible unit of time and Planck space is the smallest unit of space. To give you a feeling for how tiny this space is at $\mathbf{10^{-43}}$, if a single atom was blown up to the size of the entire observable universe, the Planck length would be ten feet long. The relationship between Planck length and Planck time is a Planck length is the distance the speed of light can travel in Planck time. They are both expressions of the same fundamental limit in spacetime. This marks the point that for everything prior general relativity and the standard model of particle physics become incompatible.

However, the period between $t = 0$ and $t = \mathbf{10^{-43}}$ seconds remains a complete "black box." It contains zero gatherable information. Even the window from $\mathbf{10^{-43}}$ to $\mathbf{10^{-12}}$ seconds remains a zone of pure speculation. We only begin to make definitive statements at approximately $\mathbf{10^{-11}}$ seconds, because these are the early-universe conditions we can recreate in particle accelerators. But there is a catch: these experiments aren't truly representative of the "first moments." The best we can do is observe what happens within a reality that has already been established. These conditions may be entirely different from the actual moment of origin—or perhaps they are nothing more than a red herring, a distraction that masks the true nature of the creation event.

Everything in that gap—from the "first breath" of the universe to the point where our accelerators can take over—is currently filled with speculative ideas presented as established facts. They are often treated as "placeholders for the truth," but they lack any possibility of validation. These speculations include:

Inflationary Theory -believed to have happened between approximately ~ $\mathbf{10^{-37}}$ and $\mathbf{10^{-35}}$ seconds[4] and ending around ~$\mathbf{10^{-34}}$ seconds, where 100 doublings of the size of the universe is hypothesized to have taken place. This inflationary doubling of the universe comes with the feature of destroying any information that existed prior, so, if it is true, we have *two* impenetrable information walls: the Planck limit and the "Inflationary Wash," between which we have only a tiny sliver of time to speculate.

Grand Unified Theory (GUT) at $\mathbf{10^{-33}}$ seconds: The belief that the electromagnetic, weak, and strong interactions were once a single force. We "guess" that the strong force broke away at $\mathbf{10^{-33}}$, seconds, followed by the electroweak separation at $\mathbf{10^{-11}}$ seconds. It is a beautiful mathematical symmetry, but it remains unobserved.

The Theory of Everything (TOE) at $\mathbf{10^{-43}}$ seconds: The ultimate placeholder. It suggests a moment when quantum gravity was united with the Grand Unified Theory. It is the "Holy Grail" of physics, yet it exists exactly at the edge of the information wall where our math stops working.

The hypothesis of Quantum Gravity, prior to $\mathbf{10^{-43}}$ seconds: A period that has never been observed and likely never will be. With no known force carriers existing in this era, we have no way to even formulate a concept of how gravity would function.

All of these are speculations. They are brilliant, highly detailed "mathematical fictions" designed to fill the void where the evidence went silent.

The problem of our inability to move back to the beginning is a fascinating one. While we can know nothing prior to Planck time, we do know precisely *how much* time we know nothing about. Our universe concedes this much to us. We are essentially up the creek without a paddle, but we know exactly how long we've been drifting. This isn't just a gap in our technology; it is a feature of the system. We have been given the exact dimensions of our own ignorance.

Based on the evidence we have discussed, here is the current informational timeline of our universe—the boundary between what we can prove and what we merely imagine:

t = -0 s (seconds) *prior* to the big bang. We know nothing, and nothing existed.

t = 0 s, we know nothing, but the universe begins in the dimension of time. We know the moment it happened, but not the cause.

t = 0 s to t = $\mathbf{10^{-44}}$ seconds, An impenetrable information gap. Because this is below the Planck limit 5.39 x $\mathbf{10^{-44}}$ seconds), physical laws as we know them do not exist.

t = $\mathbf{10^{36}}$ seconds to $\mathbf{10^{-33}}$, The proposed Inflationary Era. If this happened, it doubled the size of the universe so violently that it destroyed any evidence of what came before it—erasing the causal trail as if it never existed.

t = $\mathbf{10^{-33}}$ seconds to $\mathbf{10^{-11}}$ seconds, we can begin to speculate, but not validate, nor falsify events that may have transpired in this time period. We can dream up mathematical models that are where materialistic origin "stories" live.

$\mathbf{t = 10^{-11}}$ seconds, and moving forward. This time period coincides with the era of the electroweak phase. Because we can recreate these early physical conditions in particle accelerators, there is the option to append an observational link to this era, but it comes with a high potential cost. There is a massive logical trap here. In particle accelerators are only observing the behavior of a reality that is already established. Using a particle accelerator to explain the origin of the universe is a category error. These machines show us how matter behaves when it is smashed together *inside* a universe that already has laws, dimensions, and energy. To base our entire understanding of the *creation* event on how particles act in a controlled laboratory setting is potentially (most likely) highly unreliable. The beginning of the universe is a process where our reality was ordered in the beginning; the other is a breakdown of a reality that has already been formed and is functioning in a manner categorically removed from creation. By conflating the two, science has created a "red herring" that suggests we are close to the truth, when in reality, we likely haven't even stepped into the room where the truth happened.

380,000 years after $t = 0$, is the era of recombination: This is the moment hydrogen atoms finally captured electrons, allowing photons to stream across the cosmos. This "first light" is what we call the CMB. Found in 1964 by Penzias and Wilson and later mapped with surgical precision by the COBE and WMAP missions, this is the furthest back we can actually see.

This is the only direct physical evidence of the beginning we possess. Everything else—every theory, every timeline, every textbook "fact"—is a backward projection derived from this single snapshot. But here lies another strange cosmic smoking gun. The CMB has a signature that makes no sense because it is perfectly thermalized—same temperature in every direction to an impossible degree of precision. For it to be this uniform, every corner of the universe would have needed to be in contact with every other corner to trade heat—yet they were far too distant to ever interact. It is a profound paradox[5].

This timeline reveals a fundamental boundary to our understanding of the universe's first moments. Many widely discussed theories concerning this era are—irrespective of their popularity—nothing more than unverifiable speculations. They describe events hidden behind one, or even several, impenetrable information ceilings. This raises a crucial question: If science cannot know what happened at the moment of creation, how can we have any confidence that the beginning we envision, known as The Big Bang, ever took place? Or that the universe even *had* a beginning? The answer cannot be found by peering into the black box $t = 0$. Instead, we must rely on the centuries of verified astronomical observation that led scientists, despite their objections and hand waving to the consensus of a cosmic beginning. To understand the "algorithm," we must first understand the evidence that forced us to accept its existence.

A Brief History of the "Beginning" of Our Existence

Although the universe began 13.8 billion years ago, it could be said that for humanity, we have only been aware of this beginning since the early part of the last century.

The view that the universe was eternal and static is an intuitive view and a philosophically convenient view as well. As one stood upon the earth and gazed upon the heavens, it would be unnatural to draw the conclusion that everything one could see, feel and touch had once occupied a space smaller than the smallest grain of sand. And prior to occupying this space, the cosmos hadn't existed at all. This view of an eternal static universe also neatly avoided the troublesome implications of a singular creation event, - something that's implied by a beginning that even early scientists knew would be a problem to explain. This problem formed the competing view, held by many cultures throughout the world, either through myth, religion or spiritual views, that earth and mankind, and perhaps the entire cosmos, was the product of a singular supernatural creation event. This story has been passed down through the generations for many of the ancient cultures of antiquity. Despite these views, the fact that the ancient pre-Socratic Greeks believed that the universe had existed eternally, buttressed by other traditions from this early society, became the default view for the beginnings of scientific inquiry. Greek astronomer and mathematician Aristarchus of Samos (310-230 BC) challenged this view, placing the sun at the center of the universe with the Earth revolving around it based on his estimation that the Sun was far larger and much further away than the moon, and information from cycles of the moon. It made sense that the Earth was in orbit around it and not the other way around. His view contradicted the traditional view of the time but was unable to gain wide favor. Claudius Ptolemy (100CE to 170 CE) who was a Roman living in Alexandria, Egypt, held views that reinforced the geocentric idea of the Earth at the center with what appeared to be more rigorous math and observable predictions. His view dominated science for 1400 years.

The first major shift away from the Ptolemaic view and thus the traditional ancient understanding began with the Copernican Revolution. Copernicus proposed that the sun is at the center and the Earth and other planets are in orbit around the sun. This view eliminated a lot of the mathematical gymnastics required to make Ptolemy's views work and therefore directly challenged this view. Galileo's work in the 17th century solidified the Nicolaus Copernicus's heliocentric model—which placing the

Sun, not the Earth, at the center of our solar system—it freed scientists to contemplate the heavens as a much larger system of orderly mechanics, setting the stage for the revolutionary discoveries to come.

Sir Isaac Newton revealed a much more detailed conception of our solar system and the heavens when he conceived of gravity in an entirely different way. Newton's Principia (1687) revolutionized cosmology with its law of universal gravitation. However, Newton's own equations implied a profound problem: if his law of gravitation was correct, then every object in the universe would attract all others, and this would mean that the entire cosmos should have collapsed in upon itself long ago. To resolve this outcome, and to match his laws with the observation of a universe that appeared to be static, Newton concluded that the universe must therefore exist in an exquisitely fine-tuned and eternal balance, and that in order to achieve this balance the matter of the universe in his reasoning must be spread evenly throughout an infinite space. This imagined perfected distribution of matter was the reason his law of gravitation could be correct, despite its implications, while also preventing an unavoidable catastrophic collapse without his proposed fine tuning. Newton's views, supported by the stunning brilliance of the Principia, reinforced the prevailing view of a static cosmos, but it also introduced the critical concept of "fine-tuning" that would become central to the hotly debated cosmological ideas that would follow.

The next major advance in our understanding of the physics of the cosmos was when Albert Einstein conceived of gravity in yet another entirely iterative way. Following his *Special Theory of Relativity* in 1905, Einstein worked 8 years on developing his theory of gravity which he presented in 1915 embodied in the details of his field equations. In Einstein's conception, gravity, rather than being an attractive force between objects, was a force between objects which warped the fabric of geometric space/time. When Einstein reflected upon the radical implications of his new proposal, he realized that if his theory was true, he would arrive at the same essential conclusion as Newton, which was that parts of the universe would eventually become gravitationally attracted and collapse in upon itself. This was unavoidable. This process of gravitationally attracted masses would take place in multiple locations throughout the universe and set off a series of catastrophic events that would cause space to contract and compact. Both Einstein's and Newton's understanding of gravity, although fundamentally different in conception, were both essentially correct on different scales, and each scale indicated the universe would eventually become fatally attracted to itself and clump together, likely dooming any possibility of life.

An Unreasonable Perfection

At this point, the groundwork established by the Pre-Socratic Greeks, Aristarchus, Ptolemy, Copernicus, and Galileo, allowed Newton and Einstein to bring our understanding into a modern coherence with two iterative ideas on gravity. Their findings indicated that for our universe to appear as it did, it had to have been in an exquisitely finely tuned and unreasonably perfected eternal balance. The concept of gravity, under the new understanding of Einstein, similar to the understanding of Newton, was incorrectly envisioned to play the *most* important role in holding the objects within our universe in this precise and delicate balance. It must be remembered that gravity, to this day, remains an enigma. In the ancient past as in Newton's, Einstein's and modern times, no one truly understands what it is, how it works, or by what rules it plays by. By resorting to observation, which is all we have to gauge the operation of gravity, scientists of old, as in the present, are left without hard evidence to parse out the true operational power and/or means by which gravity works.

Einstein's understanding, however similar it was to Newton's, was a complete reimagining of the power of gravity, in an entirely non-intuitive way, envisioning gravity as an attractive force, but more importantly, this attraction, on the scale of the universe, created an invisible underlying structure of spacetime fabric that became the true power within our universe, and resulted in the equivalent of ideal geometric locations for objects within this spacetime fabric.

For Einstein matter and energy curves spacetime which in term determine where and how matter and energy should move. This understanding elevated the state of science from a mechanical universe into a universal understanding of an invisible fabric that undergirded its structure. With Einstein's contribution, our understanding of Earth's relationship with all other objects in the universe changed once again. We could now move beyond the quaint dance of a clock-like perfected solar system and begin to understand the true dynamics on the scale of everything that lay in our view and beyond our quaint solar system.

Einstein's views of the universe have been continually re-affirmed by multiple observations since his theory of gravity, including the September 14, 2015, detection of gravitational waves created by the merger of two black holes. These waves were predicted by Einstein's theory and confirmed by LIGO's twin observatories[6].

At the time Einstein released his field equations, although he correctly interpreted the realization of gravitational effects upon our entire universe, just what a "universe" was had yet to be grasped, decided, or even delineated from the concept of a galaxy. At the time, the Milky Way was believed to be a single structure in the sky. Not a view that contained an untold number of separate galaxies. Galaxies that could be seen in all directions.

But one of Einstein's predictions had failed to show any evidence of being real, and this was his prediction that however finely tuned the universe had started, even if it had been eternally old, eventually it would be perturbed in such a way as to set off a catastrophic chain reaction of gravitational clumping. Einstein, in observing the heavens, deduced, the same as all who had previously deduced, including Newton, that the world was exquisitely finely tuned and in an eternal state of exceptionally precise balance. He also realized, beyond the imaginings of Newton, that one could not just assume that the cosmos was in perfect balance. There had to be a physical (material) reason why this clumping had not occurred. His answer to this problem became what he famously called his "greatest blunder."

Einstein's Blunder

Einstein reasoned that some force must be holding the universe in perfect balance. In this realization, Einstein was correct. His error, however, was in attributing this force to a fictional concept of a consistent, persistent non-zero energy in all locations of spacetime that resulted in a counterbalance to the attractive force of gravity.

Einstein theorized that this energy exerted a constant, repulsive pressure throughout space and this outward push would act as a form of anti-gravity, perfectly finely tuned to counteract the inward pull of gravity, and thus held the universe in the static, eternal balance that Einstein and others believed matched their observations. Einstein called this energy the *cosmological constant* and represented it with the Greek capital letter *Lambda.* This was the term he inserted into his equations even though he had no confirmation that this force was indeed the key to the appearance of a static, eternally balanced universe.

Einstein had done more than just assume the cosmos was in perfect balance, he provided a reason why it was in balance, but, his reason was little more than an unscientific intuition, turned into a universal constant, and designated to be the force precisely required to keep the universe in a finely-tuned, and (what he assumed to be) an eternal, static state.

The introduction of *Lambda,* -the cosmological constant, was Einstein's resolution to the problem of gravity, but this fix only worked for a few short years, because soon, new evidence was introduced that contradicted what Einstein had introduced. The belief that the universe was static, eternal and in "perfect" static balance was factually incorrect.

Solving Einstein's Equations

Three separate scientists solved Einstein's 1915 field equations attached to his theory of relativity in such a way to provide an entirely new understanding of our universe.[7] The first was Karl Schwarzschild in 1916[8]. His was the first 'exact solution' to Einstein's equations—meaning he solved the field equations perfectly using pure mathematics, and with the exact solution to solve the equation without the use of simplifying approximations that are often necessary in solving physics equations. His solution eventually led to the *Schwarzschild Radius*, which is the size of the event horizon of a non-rotating black hole, or singularity. Upon reading Schwarzschild's solution Einstein wrote back, "I have read your paper with the utmost interest. I had not expected that one could formulate the exact solution of the problem in such a simple way… Next Thursday I shall present the work to the Academy with a few words of explanation."[9]

Next, Aleksandr Friedmann's calculations in 1922[10] showed that Einstein's equations for general relativity, allows for an expanding solution from a beginning. The "beginning" is interesting in that Einstein's equations were not intended to include a "beginning"; rather, they were intended to represent the *static* universe he believed existed. Nonetheless Friedmann intuited that his field equations allowed for a "beginning" and he introduced this concept within his work.

Friedmann's work on the field equations essentially found the same conclusion that Newton and Einstein arrived at. The universe could not remain in a simple steady state; it must be dynamic. The detail of Friedmann's work, working forward from an assumed beginning, meant that for the universe to give the appearance of being static, it could not have been the result of a fine distribution of matter in an eternal cosmos as Newton surmised. And it could not be the result of a cosmological constant acting upon a static and eternal universe as assumed by Einstein. If our universe was static, then it could only be as a result of having evolved into this condition over the billions of years. To achieve this state, the universe would itself need to be, once again, in an exceptionally precise balance over the course of its entire existence as it evolved.

Georges Lemaître, a Belgian monk/astronomer was the third individual to independently produce solutions to the field equations[11], and combined his findings with additional observations, namely those of Vesto Melvin Slipher, who witnessed a groundbreaking and puzzling observation, which was the universe was expanding with a velocity proportional to its distance to any given point[12].

The collective theoretical work of Schwarzschild, Friedmann, and Lemaître was a watershed moment. They had used Einstein's own equations to demonstrate that a dynamic, evolving universe was not just a possibility, but a direct mathematical consequence of general relativity.

Following up on this theoretical work, Arthur Eddington, dealt with the idea of a static universe another severe blow. Although Eddington believed the universe had been expanding, he didn't believe it necessarily had a beginning. But in analyzing the static universe field equation solutions of another contributor, Willem de Sitter, -a conclusion that static solutions were most likely not the actual solution- Eddington demonstrated that a static model required an unreasonable amount of, once again, *fine tuning* to maintain balance, and even in the case where unreasonable fine tuning was somehow applied, *any* alteration of that balance would inevitably lead to either a catastrophic contraction of the universe or a catastrophic endless expansion. In other words, the prospect that any force that could be explained had the power to maintain a precise balance, scientifically would not be able to do so. The universe seemed to be telling us, repeatedly, that all was not, and could not be as it appeared to us.

In 1924, Edwin Hubble shattered the existing conception of the universe. For the first time, he proved that our Milky Way galaxy was not the entirety of the cosmos, but merely one of countless "island universes." This fact was deduced by a key finding by Henrietta Swan Leavitt. She found that Cepheid variable stars displayed a relationship between their brightness and pulsation period, now known as the Levitt law or the Period-Luminosity Relation. By using this relationship, Edwin Hubble was able to use Cepheid variable stars as an ingenious "standard candle" to measure cosmic distances. Hubble demonstrated that faint smudges of light, long believed to be nebulae within our own galaxy, were in fact immense and unimaginably distant galaxies in their own right[13]. In an instant, the known universe expanded by orders of magnitude, revealing a cosmos vaster and grander than anyone had ever understood before.

Hubble's information also established that there were more galaxies in the universe than there were stars within our galaxy. The discovery of an entirely new cosmic scale also resulted in the realization that the amount of "stuff", -mass and energy, that needed to be held in the pristine eternal balance we previously assumed was on a galactic scale was also expanded by several orders of magnitude, and perhaps even further.

The End of the Static Universe

Over the years 1931 through 1963, a steady accumulation of scientists were coming to terms with an expanding universe, and also a universe that was not eternally old. But as is true for many scientific subjects, although the facts are the same, different interpretations are allowed, even after validation, as challenges to established principles are continually refined in the search for scientific truth.

Since 1964 a number of additional observations have confirmed the expected findings and predictions of big bang cosmology, including, what is generally regarded as the three main proofs: the expansion of the universe; the amount and ratios of hydrogen, deuterium, helium and lithium in the universe; and the feature of light travel, which means that the further away we look from our location on Earth, we are also looking back in time (spacetime), and can observe galaxies that are very old and at a much earlier stage of evolution than our own. If the universe was eternally old, finding an appreciably "younger" universe further back in time would be an unexpected result. An eternally old universe should be made up of eternally old galaxies no matter how far we looked in the past.

Despite the overwhelming proof for a beginning to our universe, there have been concerted and highly motivated efforts to produce new models that frame our universe as one that has continually existed in some form, and in various descriptions over the years. 'A Universe From Nothing' is one model, as is the 'Inflationary Multiverse,' where it is imagined that once inflation starts, it can never stop.

The Definite Beginning to the Universe

We now turn to the work that established the beginning of our universe as a scientific necessity, thus rendering all of the previous models moot in addition to being unprovable. In contrast to those proposals, the finding that our universe has a beginning comes from within physics itself, in the form of two rigorous mathematical theorems derived directly from Einstein's theory of general relativity. No less than two collaborative groups, involving the contributions of world-renowned physicists Stephen Hawking, Roger Penrose, Alan Guth, Alexander Vilenkin, and Arvind Borde. These theorems have provided the definitive proof that our expanding universe had an absolute beginning, from a singularity with no prior physical existence of any kind prior.

The Definitive Answer: The Singularity Theorems

The first piece of work was completed in 1968 and 1970, *The Hawking-Penrose, Singularity Theorems,* (Roger Penrose, Stephen Hawking and contributions from George F.R. Ellis)[14]. Through their work they were able to show that mathematical singularities were an inevitable initial condition for relativistic models of the big bang. Although their paper could not take us further back than a billionth of a second after creation, it does show that the initial condition of the universe was in fact, a singularity with *infinite* curvature. This result is similar to Lemaître's "primeval egg" theory of 1931, but it is categorically different. Lemaître missed the opportunity to realize that his primeval egg was a *singularity*, consequently, Lemaitre's egg, at the size of an atom, was at least one hundred thousand billion billion times larger than the singularity from which our universe sprang forth according to The Hawking-Penrose singularity theorems. To visualize this discrepancy, remember our earlier comparison: If a single atom were blown up to the size of the entire visible universe, the Planck length would be equivalent to a mere 10 feet on that scale. By comparison, Lemaître's "egg" was a sprawling, massive structure, while the actual singularity exists at a scale of "smallness" that defies human intuition and materialist explanation.

The Hawking-Penrose singularity theorems bring us back to the physical limit of space itself at $\mathbf{10^{-12}}$ and $\mathbf{10^{-33}}$ of a centimeter, where the radius of curvature of our universe is *zero units* in length and *zero spatial* volume. In this view, the big bang was *necessarily* a creation event, emanating from a space that is beyond that which can be measured, and of which, by definition, nothing can be known nothing. The Hawking-Penrose singularity theorems do not permit one to posit mass energy or a gravitational field as an eternal self-existing entity, since "prior to" the singularity neither time nor space existed in our universe. And without space, mass and energy would have no place to reside. In other words, however much mass energy existed from the beginning of the universe, it had to arise with the beginning of time and space. Before this, there is *nothing* which could have caused the universe. And when we say, "nothing" we actually mean, "nothing". Thus, no quantum fluctuations, or fields, or any other existing *anything* from which existence would spring forth. The work of Hawking and Penrose does away with all materialistic explanations for the creation of our universe, because they have demonstrated that nothing could have existed prior to this singularity[15]. Existence did not exist prior to the ssingularity,and the creation event is not something that conforms to reductionism or causality. This work proves that the "beginning" of our universe was the actual beginning.

The Hawking-Penrose singularity theorems brought the understanding of existence back to within a billionth of a second, but more importantly, to *the beginning* of the universe. It wasn't until 2003 that the clock ticked down from a billionth of a second to an actual beginning in a burst of genius in the Borde-Guth-Vilenkin Theorem. (Arvin Borde, Alan Guth and Alexander Vilenkin)[16].

The BGV Theorem

The BGV theorem proves that in cosmology it may be deduced that *any* universe that has, on average, been expanding throughout its history, cannot be infinite in the past but *must* have a space*time* boundary. Our universe fits these conditions and therefore had to have had a beginning. The theorem holds for *any* type of universe, i.e., inflationary universe, cyclic universe, oscillating, or multiverse, *that has, on average, been expanding*, and we know that our universe has been expanding.

The combined force of the Hawking-Penrose and Borde-Guth-Vilenkin theorems is decisive. They rigorously prove that any expanding universe like our own must have had an absolute beginning from a singularity—a point before which no matter, space, time, or physical laws could have existed. This conclusion effectively closes the door on all materialistic models that propose an eternal past or a pre-existing physical cause. The singularity theorems and BGV theorem do not tell us how existence could have sprung from this singularity, it just tells us that it did, and, because the answer lies beyond the limits of physics, we can't extract information from this space/time to answer the question of "how?" In establishing these facts, the theorems lead us to the very boundary of physics, presenting us with a true creation event from nothing. We are left with the fact that because the creation event has no possible physical cause, by definition, it requires a non-physical or "supernatural" one, i.e. the foundational miracle (event beyond the explanation of our physics) of our existence.

Summary of our Investigation into the Beginning of the Universe

The investigation in this chapter can be distilled into several key points. We saw how the long-held assumption of an eternal, static universe was challenged by the physics of Newton and Einstein, which revealed a profound paradox: our cosmos should have collapsed under its own gravity. The resolution was not a theory of perfect balance, but the observational proof from figures like Edwin Hubble that the universe is, in fact, expanding from a definite beginning. This conclusion was ultimately confirmed by the discovery by Penzias and Wilson of the Cosmic Microwave Background Radiation—the physical echo of the creation event itself. Finally, the singularity theorems of Stephen Hawking, Roger Penrose[17], Arvind Borde, Alan Guth , and Alexander Vilenkin provided rigorous mathematical proof that this beginning was absolute: a true creation event from a point before which no physical reality existed, and from which no usable information about what may have happened can be extracted.

Science, by following its own evidence, has thus led us from a model of an eternal universe to the boundary of a creation event that physics itself cannot explain, nor reconcile with a host of conjectures that have been put forward as likely solutions to the origin of our universe.

Value of Special Problem #1

As we have seen, the combined evidence from theoretical physics and astronomical observation has led science to an inescapable conclusion: our universe had an absolute beginning from a singularity, an event for which there can be no physical cause. In our introduction, we established a framework for evaluating the value of such events. We defined an event as M (miracle-impossible) if there is no possibility whatsoever of it occurring under the known laws of scientific materialism. The creation of an entire universe from nothing is, by its very nature, such an event. It is not merely improbable; it is a paradox that stands entirely outside the laws of physics, meaning the existence of this event, not only is improbable, but impossible, as no laws, conditions or parameters can be called in to generate the possibility of such an event. Therefore, within the framework of The Divine Algorithm, the value we must assign is:

Special Problem #1: The Beginning of Our Universe —

M (Miracle-Impossible)

The implication of this first finding cannot be overstated. We are in effect saying that the origin of all space, time, matter, and physical law—was not a physical event, but a supernatural one.

With this foundational miracle established, our investigation now zooms in, moving from the fact of the beginning to the specific, unorthodox properties of the object at that beginning, which is our first Variable: the singularity itself.

---

[1] C T. McGrew, 2019, Miracles, Stanford (online) Encyclopedia of Philosophy, 10/11/10, rev. 1/23/19)
[2] (Cambridge Online Dictionary, english/ supernatural/ dictionary(dot) cambridge(dot) org)
[3] (Merriam-Webster(dot) com/ dictionary, supernatural, 22 Jul., 2024).
[4] A. Guth, The Inflationary Universe, Pg. 186, 1997, Basic Books),
[5] S. Weinberg, The First Three Seconds, Chapter 5, 1977, Basic Books; A. Guth, The Inflationary Universe, Pgs. 48-75, 1997, Basic Books; G.F. Lewis, L.A. Barnes, The Fortunate Universe, Pg 145-149, 2016, Cambridge Univ. Press; G. Gonzalez, J.W. Richards, The Privileged Planet, pg. 174-175, 2004, Regnery Gateway; E. Chaisson, Cosmic Evolution, Pg. 102-103, 2001, Harvard Univ. Press;
[6] (B.P. Abbott et al., LIGO Scientific and Virgo Collaboration, Phys. Rev. Lett. 116, 061102, 2016).

---

[7] (A. Einstein, Die Feldgleichungen der Gravitation, Sitzungsberichte der Preußischen Akademie der Wissenschaften zu Berlin: 844-847. November 25, 1915)
[8] (K. Schwarzschild, Über das Gravitationsfeld eines Massenpunktes nach der Einsteinschen Theorie, Sitzungsberichte der Deutschen Akademie der Wissenschaften zu Berlin, Klasse fur Mathematik, Physik und Technik (1916) pgs, 189 and 424
[9] (A. Einstein, Eisenstaedt, The Early Interpretation of the Schwarzchild Solution, in D. Howard and J. Stachel (eds), Einstein and the History of General Relativity: Einstein Studies, Vol, 1 pgs 213-234. Boston Birkhauser, 1989
[10] (A. Friedman, Über die Krümmung des Raumes, Zeitschrift für Physik. 10 (1): 377-386.1922, S2CID 125190902
[11] (Abbé G. Lemaître, A.S. Eddington, The Expanding Universe, Monthly Notices of the Royal Astronomical Society, Volume 91, Issue 5, March 1931, Pages 490-501, https://doi.org/10.1093/mnras/91.5.490
[12] (V. Slipher, Nebulae. Proceedings of the American Philosophical Society. 56: 403-409. 1917, Bibcode: 1917PAPhS.56..403S).
[13] (E. Hubble, Extragalactic Nebulae, 1926, Astrophysical Journal. 64: (64): 321-369. Doi: 10.1086/143018.
[14] (S.W. Hawking, G.F.R. Ellis, The Large-Scale Structure of Space Time, 1994, Cambridge University Press)38, and (S.W. Hawking, R. Penrose, G.F.R. Ellis, The Nature of Space and Time, 1996, Princeton University Press).
[15] (S.W. Hawking, R. Penrose, G.F.R. Ellis, The Nature of Space and Time, 1996, Princeton University Press)40
[16] (A. Borde, A.H. Guth, A. Vilenkin, Inflationary Space-Times Are Incomplete In Past Directions, April 15, 2003, Physical Review Letters. 90 (15): 151301)41 and, (A. Vilenkin, Arrows of Time And The Beginning Of The Universe, 2013, Physical Review D. 88 (4): 043516)42
[17] (J. Hartle, S. Hawking, Wave Function Of The Universe, 1983, Physical Review D. 28 (12): 2960)44, (StarTalk with Neil DeGrasse Tyson & Stephen Hawking | Full Episode, National Geographic, YouTube, March 14, 2018, at timestamp 15:00)45

Variable 1: The Baryon Asymmetry

Matter

The fact that you are reading this right now is a serious problem in physics. Everything you are made of should have been annihilated in the first few seconds of the universe when the initial quantities of matter and antimatter were created, and inevitably annihilated, leaving behind only gamma rays, -the highest energy and shortest wavelength of electromagnetic radiation. Something unknown and mysterious happened to prevent this annihilation, and whatever this process was, has been termed *baryogenesis,* the process in the early universe that resulted in the permanent production of matter known as *the baryon asymmetry,* the subject of *Variable 2*. To understand the baryon asymmetry, we will first endeavor to understand baryons and their related particles, then we will have a greater appreciation for the baryon asymmetry, as well as the fundamental nature of our reality.

To grasp the true scale of this miracle, and why it forms a crucial part of The Divine Algorithm, our investigation in this chapter will proceed in two parts. First, we must understand the nature of the particles that survived this cosmic cataclysm. We will take a deep dive into the "unspeakable complexity" of protons and neutrons to appreciate what was truly at stake. Only by understanding the intricate, purposeful nature of what survived can we comprehend the miracle of its survival. Second, we will examine the problem of the Baryon Asymmetry itself, exploring the theoretical conditions required to produce a matter-dominated universe and the experimental evidence that shows why this remains one of the most profound and unsolved mysteries in all of science.

Baryons, Protons, Quarks Mesons, & Hadrons

The matter that survived this primordial annihilation is known as baryonic matter, or baryons. The most familiar of these particles are the protons and neutrons that form the nucleus of every atom in the universe. They are, in a very real sense, the fundamental building blocks of our existence.

Protons and neutrons were thought to be elementary particles up until the early 70's, then researchers working together with theorists became aware that protons and neutrons were themselves made up of even smaller particles called quarks. It was Nobel laureate (1969, Physics) Murray Gell-Mann and physicist/neurobiologist George Zweig that deduced that all baryons are made up of 3 more fundamental particles now known as *quarks*, a name thought up by Gell-Mann. The model proposed that the most common baryons—protons and neutrons—were each composed of a specific combination of these quarks.

Gell-Mann originally thought up the name "kwork" but refined the spelling when he recalled a line from author James Joyce's classic, *Finnegans Wake*, "three quarks for Muster Mark!" Gell-Mann developed the 3-quark model for protons in 1964 at the same time a physicist/neurobiologist named George Zweig, proposed the same structure independently. Zweig, in his conception, referred to quarks as "aces" but Gell-Mann's name caught on and became the standard. To clarify, although the proton and neutron contain 3 quarks, the 3-quark model actually refers to the 3 types of quarks that were known at the time, the up down and strange quark. Gell-Mann received his Nobel for foundational work in elementary particle theory, forming the foundation for the standard model, of which his 1961 proposal for the "eightfold way", an organizational scheme for hadrons, was a key piece. Like the 3-quark model, the structure of the eightfold way was independently arrived at by Murray Gell-Mann and Israeli physicist Yuval Ne'eman[1].

Gell-Mann's 'eightfold way' was a groundbreaking organizational scheme that brought order to the growing 'particle zoo' of newly discovered subatomic particles. Much like the periodic table organized chemical elements based on their properties, the eightfold way arranged particles like protons and neutrons into elegant geometric patterns based on their intrinsic characteristics, such as electric charge and a quantum property called 'strangeness.' Its greatest triumph was that its patterns had 'gaps,' which correctly predicted the existence and properties of a then-undiscovered particle, proving that the particle zoo was not random but had a deep, underlying order.

The name, given by Gell-Mann, refers to the Buddhist Eightfold Path. Zweig and Gell-Mann were nominated for a Nobel by Richard Feynman[2] for their discovery of quarks but they fell short of support for a Nobel. The eightfold way was not wrong; in fact, it was a profound success that brought order to the chaos of subatomic discoveries and correctly predicted the existence of new particles. However, it was soon understood to be a steppingstone to an even more fundamental idea. The beautiful symmetry revealed by the eightfold way was ultimately explained by the quark model, which showed that particles like protons and neutrons were not fundamental, but were themselves made of smaller components. The focus of physics then shifted from the patterns themselves to the quarks that created them.

The 3-quark model was experimentally confirmed in 1979 at the Stanford Linear Accelerator Center (SLAC) in 1967. This confirmation *did* win a Nobel prize in 1990 for physics for the 3-person SLAC team of Jerome Friedman, Henry Kendall and Richard Taylor[3].

> "The Nobel Prize in Physics 1990 was awarded jointly to Jerome I. Friedman, Henry W. Kendall and Richard E. Taylor "for their pioneering investigations concerning deep inelastic scattering of electrons on protons and bound neutrons, which have been of essential importance for the development of the quark model in particle physics[4]"

There are 75 baryons identified by theory, and many have subsequently been observed. Baryons are made up of combinations of 3 quarks from among the 6 known *flavors* of quarks. All baryons are not equal. Protons and neutrons are the only baryons that are *nucleons.* Nucleons are baryons that make up the nucleus of atoms. Protons and neutrons are also the only 2 baryons that have any appreciable half-lives. Protons are stable with no known half-life, which is as appreciable of a half-life as anything can have, although in some grand unified hypotheses, they have a projected half-life of $\mathbf{10^{33}}$ years. Free neutrons have a half-life of ~14.5 minutes before they *decay* into protons. The rest of the baryons have miniscule half-lives and decay rapidly. Thus far 44 of the 75 baryons have been observed, but only in particle accelerators for the smallest slices of time with half-lives ranging from $\mathbf{10^{-10}}$ seconds to $\mathbf{10^{-24}}$ seconds. 29 of the 75 theorized baryons have never been observed. 6 baryons have been observed, but their half-lives remain unknown.

Antimatter has its own list of 75 *antibaryons,* made up of *antiquarks.* Atoms of antimatter are made up of *antinucleons* made up of *antiprotons* and *antineutrons* and are orbited by *positrons,* the antimatter equivalent of electrons.

Mesons are also composed of quarks, but in this case 2 quarks. There are 33 known mesons at the current time. Together baryons and hadrons are considered *hadrons,* subatomic particles composed of quarks. Mesons are considered bosons (like photons) and thus have the quality of being able to mediate the strong nuclear force between hadrons, and indeed, they have been found to be plentiful inside of protons.

In the 1950's and 1960's when hadrons were being discovered and it was unclear what exactly they were, they were described by J. Robert Oppenheimer as composing a "subnuclear zoo" soon understood as the "particle zoo" because they were so plentiful (dozens of them), poorly understood, and diverse in quality, as fundamental constituents of matter.

This deep dive into the "particle zoo" and the complexities of Quantum Chromodynamics reveals a crucial point: the baryons that form the basis of our existence are not simple, inert building blocks. They are realms of unspeakable complexity, governed by a precise and bewildering set of rules. This complexity is not a trivial detail. I argue that it is the very foundation of their utility, and points to a deeper purpose behind their survival following the initial moments of creation. In this chapter we will begin to unravel this complexity and utility and how both factor into why baryon asymmetry is a crucial variable within our algorithm and the literal foundation of our physical reality.

The unusual part about the baryon asymmetry, besides the fact that it is inexplicable, is the stunning complexity, diversity and utility of baryons left over from the early universe's matter/antimatter annihilations. This leftover product is the foundation of our entire universe. Everything about matter, its properties, functions and the interactions within and between baryons is utilized by our universe for a point, a purpose and an end. And everything that is said to be finely tuned within our universe, is finely tuned for the inclusion of baryonic matter. This outcome is too specific to be a random accident, but it also seems far too purposeful to be the result of some inherent, self-organizing property of natural laws. This places the baryon asymmetry into an entirely new context, because what baryons contribute to our universe is essentially *everything*. Everything about baryons forms the most perfected and inscrutable, and indecipherable system ever encountered and ever to be known by flesh and blood humans. This asymmetry is also the fundamental basis of every *Variable* and *Special Problem* in *The Divine Algorithm*.

Matter could be theorized to be the obvious goal and first requirement for all that has transpired in our universe, from the big bang to the emergence of consciousness, all of which require baryons. To understand how baryons could fulfill such a profound purpose, we must first understand what they are. The journey into the heart of matter reveals that these fundamental particles are not simple, inert pellets, but realms of incredible complexity, governed by a precise and elegant set of rules. Let's examine the building blocks of this mystery.

The Quark Model

Protons and neutrons are also made up of 3 quarks in combinations of up and down quarks. In this model a *proton* is made up of 2 up quarks and 1 down quark, while a *neutron* has 1 up quark and 2 down quarks[5]. Protons and neutrons are also the only baryons made up of *only* up and down quarks. The other 73 baryons are made up of some other combination of 3 quarks from among the 6 flavors of quarks, which are up, down, strange, charm, top and bottom quarks.

All quarks carry an electric charge. Up, charm and top quarks carry a charge of +2/3rds, while down, strange and bottom quarks carry a charge of -1/3rd. This results in protons carrying a net charge of 1 which attracts an equal number of negatively charged electrons as there are protons in the nucleus. Neutrons carry a net charge of 0, making them neutral and requiring no charge balancing besides attraction to protons. After the 3-quark model was discovered, continued experiments revealed that there was far more, -unimaginably more- going on inside protons than 3 quarks. A particular set of observations and discoveries in 1974 lead to what is known as *the November revolution*. Physicists found evidence for a new particle, at the time referred to as the *J/psi* that could only be understood if there was another quark in existence. This became the *charm quark*.

> "Exactly twenty-five years ago, on 10 November 1974, two experimental groups (one, a MIT group doing experiments on the east coast at Brookhaven National Laboratory, USA and the other a SLAC-Berkeley group doing experiment[s] on the West Coast at Stanford Linear Accelerator center, U.S.A) simultaneously announced the discovery of a new particle at 3095 MeV whose lifetime is about 1000 times more than that of other particles of comparable mass. This announcement set on fire the world of high energy physics and is now known in the physics community as the November revolution. Like many revolutions its meaning was not clear at first. To appreciate it better, imagine if one suddenly discovers a new pyramid, twice as heavy as the largest one known so far and yet a thousand times narrower and thus higher! The unprecedented large lifetime (or narrowness of the peak) led to a host of theoretical speculations and vigorous experimental activity. During the next one year, more than seven hundred papers were written related to this discovery which was a record in physics (if not in entire science) at the time.[6]"

This 4th quark, at first postulated by physicist Sheldon Glashow in 1964 was dismissed as being speculative, but it was found in 1974 as the *J/psi* particle, soon to be recognized as the 4th charm quark. It acquired its name "charmed" or "charm" because Glashow realized its existence would be charmed by bringing symmetry between quarks and leptons (like the role played by electrons). The 1974 discovery resulted in the 1979 Nobel in Physics for Glashow along with Steven Weinberg and Abdus Salam and also recognized their additional "contributions to the theory of the unified weak and electromagnetic interaction between elementary particles"[7]. The flurry of experiments and findings soon elevated the idea of 4 quarks from a theoretical understanding and into reality that eventually revealed our current 6 quark structure which are referred to as *flavors*.

> "With visible evidence for their existence in hand, quarks finally won universal acceptance in the physics community, over ten years after they had first been proposed. -Further experiments in the late 1970s helped round out the new elementary particle table. In 1976 the same physicists that had discovered the ψ particles at SLAC also identified the τ lepton --a charged elementary particle that, like the electron and muon, does not feel the effects of the strong force. In 1977 a fifth kind of quark, dubbed 'bottom' or 'beauty,' was discovered at Fermilab; a sixth quark, called 'top' or "truth,' is now being sought with a mass at least a hundred times that of the proton[8].

The order of discovery for the 6 flavors of quarks is as follows:

Up, down and strange quarks found in 1969 at SLAC and theorized in the 3-quark model proposed in 1964 by Murray Gell-Mann & George Zweig

Charm quark found in 1974 at SLAC and theorized in 1964 by Sheldon Glashow who was later awarded the 1979 Nobel prize in physics

Bottom quark found in 1995 at Fermilab and theorized in 1973 by Makoto Kobayashi & Toshihide Maskawa who were later awarded the 2008 Nobel prize in physics

Top quark discovered at Fermilab in 1995 after being searched for over 20 years based upon work conducted by several physicists

The series of discoveries that established the 6 flavors of quarks was not a simple progression. It unfolded with the intrigue of a scientific mystery novel with clues, hypotheses and eventual resolution.

> "Both quarks and leptons come in three generations of paired elements with differing electric charge. But when the tau lepton was first found, the third-generation quark doublet seemed to be missing a member, whose existence and charge (⅔ the magnitude of the electron charge) were inferred from the existing pattern. In advance of its sighting, physicists named it the top (t) quark. Thus began a search that lasted for almost 20 years.[9]"

This decades-long journey, from a simple three-quark model to the six-quark "particle zoo," was a monumental achievement in physics. However, this was only the beginning of the story. As physicists developed more powerful tools to probe the heart of matter, they discovered that the proton was not just a simple container for three quarks, but a dynamic and "unspeakably complex" system in its own right. This brings us to our modern, and far more baffling, understanding of the proton.

Protons are Stunningly Complex

What we know of protons and quarks has been gathered by two primary methods: particle accelerators where electromagnetic fields are used to accelerate particles to create beams, such as Stanford University's SLAC, and particle colliders where particles are accelerated and collided like at CERNs LHC. Information from these experiments present scientists with different statistical views of the state(s) of the proton. This information has revolutionized our understanding of the proton, while also being confusing and "humbling to humans… Every time you think you kind of have a handle on [the proton], it throws you some curveballs.[10]"

> "The positively charged particle at the heart of the atom is an object of unspeakable complexity, one that changes its appearance depending on how it is probed… 'This is the most complicated thing that you could possibly imagine,' said Mike Williams, a physicist at the Massachusetts Institute of Technology. 'In fact, you can't even imagine how complicated it is…[11]'

This quote describes something far removed from the humble proton of high school science. To unwrap this 'unspeakable complexity,' we must go deeper than the simple three-quark model I introduced earlier and expand our view of what a proton truly is.

Quantum Chromodynamics

To truly understand the proton, we must introduce the theory of Quantum Chromodynamics (QCD). This is the theory of the strong nuclear force—the most powerful force in nature. QCD describes how quarks are bound together by force-carrying particles called gluons in a bizarre and beautiful dance governed by a property known as "color charge." It is this constant energetic interaction of quarks and gluons that makes the proton one of the most complex and fascinating objects in the universe.

> "[Quantum chromodynamics is] a quantum theory of the 'strong force' that acts between quarks. The theory describes quarks as being roped together by force-carrying particles called gluons. Each quark and each gluon has one of three types of 'color' charge, labeled red, green and blue; these color-charged particles naturally tug on each other and form a group — such as a proton — whose colors add up

> to a neutral white. The colorful theory became known as quantum chromodynamics, or QCD[12]."

QCD has some fundamental concepts that comprise the full view of the system. Let's unpack those to get to grips with QCD.

Asymptotic Freedom

The concept of asymptotic freedom is an important concept in QCD. Asymptotic freedom calls for the force between quarks to lessen when they are in close proximity and to strengthen as their distance is enhanced. At close distances this allows quarks to act as if they are independent from one another.

> "In the summer of 1973 physicists at Harvard and Princeton demonstrated that in certain gauge theories the force between the quarks could become relatively weak at short distances, a behavior known as "asymptotic freedom," which could explain why high-energy electrons and neutrinos appeared to be hitting loosely bound quarks inside nucleons. There were expectations that this force also became extremely strong at large distances (that is, comparable to the radius of a nucleon), effectively trapping the quarks-although such long range behavior could not yet be rigorously derived from the theory[13]"

And,

> "The most important implication of asymptotic freedom is QCD itself with point-like behavior of quarks at short distance and the strong confining force at large distance. But, in addition asymptotic freedom greatly increased our confidence in the consistency of quantum field theory, produced the first example of a theory with no adjustable parameters, enabled us to probe the very early history of the universe, and allowed us to extrapolate the standard model to high energy.[14]"

These two insights are crucial. As Riordan explains, asymptotic freedom provided a perfect physical explanation for why quarks can appear to be 'free' inside a proton while also being permanently trapped. More importantly, as Gross points out, this discovery was a watershed moment for physics. It not only solidified Quantum Chromodynamics as the correct theory of the strong force but also gave scientists the confidence to use these principles to probe the extreme conditions of the early universe—a key theme we will return to throughout our investigation.

Color Confinement

Color confinement is another important concept in QCD. Color confinement is related to asymptotic freedom and guarantees the quark will always contain the 3 valence quarks (the permanent quarks that give the proton its core identity, as distinct from the sea of temporary particles inside). Generally speaking, confinement means that quarks cannot be found individually and in normal conditions the color force will hold the 3 valence quarks an optimal distance from one another, -not so close as to experience degeneracy pressure, and not so far away as to be energetically unfavorable. "We find a strong repulsive pressure near the centre of the proton (up to 0.6 femtometres) and a binding pressure at greater distances.[15]"

Before we get into the specific mechanics, it's helpful to understand the core principles of color confinement. This is one of the most bizarre and important ideas in all of physics. In essence, it means:

Quarks Cannot Be Alone: The fundamental rule of color confinement is that quarks are never found in isolation. They can only exist bound together in groups of two or three, forming particles like protons and neutrons.

A Force That Grows with Distance: Unlike gravity or electromagnetism, which get weaker over distance, the strong nuclear force that binds quarks acts like an unbreakable "rubber band." The farther you try to pull a quark away from its partners, the stronger the force pulling it back becomes.

New Matter from Energy: If you pour enough energy into the system to "snap" this rubber band, that energy doesn't free the quark. Instead, the energy instantly transforms into a new quark and antiquark, creating a new particle (a meson) and ensuring that the original quark is never left by itself.

This strange set of rules is what makes the subatomic world so stable and complex, and it is a key piece of the puzzle we are about to explore.

It is by color confinement that the proton is kept stable and at an optimal binding distance of the quarks. The color force truly acts like a rubber band, maintaining optimal pressure, at times holding and reigning in the quarks, at other times pushing them away from one another to maintain an optimal distance. The upshot being that quarks are always confined in groups of 3 within baryons. These connecting points between and among quarks are considered to be flux tubes, and they function in such a way that if 1 quark gets knocked far away from the other 2, the strong force holding that quark will *confine* the quark to contain it within the 3-quark system. The strong force will continue to do this until it is no longer energetically favorable.

When this limit is exceeded, the stretched quark will convert the energy that is forcing it apart from the other three and holding it with the gluons and instantaneously become a quark/antiquark pair, and the remaining 2 quarks will instantaneously find another quark in their midst that conserves the color of the proton. It is able to do this because when the color force confinement is broken, and because the energy is greater at a greater distance, sufficient energy is created by the "broken" quark to create a new quark and a particle/antiparticle pair. In particle accelerators when this behavior is observed, a hadron jet of particle/antiparticles will be observed in place of the no longer confined quark. This inability to break color conservation in 2 and 3 quark systems is known as *color confinement* or just *confinement*.

Let's pause to consider what this system truly represents. It is a set of rules so counter-intuitive and exquisitely balanced that it defies any simple explanation of chance, particularly since in our understanding, there is no "chance" that matter which constitutes the 3-quark system of neutrons and protons should have survived the earliest moments in the history of the universe. What we have is a force that grows stronger with distance, a perfect, color-conserving dance of gluons and quarks, and a fail-safe mechanism that creates new matter from energy to prevent its own rules from ever being broken. This is not a system of brute force, but one of intricate, self-regulating logic. The question we must ask is whether it is possible, or more rational to believe, that this "unspeakably complex" and exquisitely perfected system is a product of unguided, natural law, or whether it's a product of something more coherent, and intentional, and even "aware", that has been woven into the fabric of the universe at the moment of creation? This latter explanation is part of what we propose could have only come from a form of meta-information intended and seeded into our universe along with the energies of the universe, in order to usher in the series of probability defeating transformations that were to follow.

The Particle Zoo

This principle of color confinement is the reason quarks are never observed freely in nature. Instead, they are permanently bound together into composite particles known as hadrons. The two main families of hadrons, which make up the bulk of the "particle zoo" discovered in the 20th century, are defined by their quark composition.

Quarks are always found in groups of 3, as in a proton, or groups of 2, which is a *meson.* 2 and 3 quarks systems have their own name which represents the particle zoo, *-hadrons*. Mesons, colorless particle/antiparticle pairs, have very short half-lives and decay into electrons, neutrinos and photons, thus expressing the close energetic relationship between subatomic particles and the fundamental forces. Mesons have only been observed either in particle accelerators or when high energy particles enter Earth's atmosphere collide with other atoms. These mesons come from energetic sources within our universe and beyond. Mesons (*pi meson, pion*) were theorized by Japanese physicists and Nobel laureate Hideki Yukawa who received the Nobel prize in physics in 1949.

Atomic nuclei consist of protons and neutrons held together by a strong force. Hideki Yukawa assumed that this force is borne by particles and that there is a relationship between the range of the force and the mass of the force-bearing particle. In 1934, Yukawa predicted that this particle should have a mass about 200 times that of an electron. He called this particle a "meson". Mesons' existence was verified in later experiments[16].

The behavior of the gluons that bind quarks is where things get truly strange. According to QCD, each gluon carries a "color charge" that is a mix of a color and an anti-color (e.g., red/anti-blue). This allows them to act as mediators for the strong force through a constant, intricate dance:

-A quark of one color (e.g., red) will emit a gluon.

-This emission instantly changes the color of the first quark (e.g., to blue).

-A second quark of a different color (e.g., blue) then absorbs that gluon.

-This absorption, in turn, changes the second quark's color back to the original's (e.g., to red).

Through this constant exchange, the overall "colorless" state of the proton is perfectly conserved moment by moment, while the quarks themselves are in a state of perpetual transformation. The strength of the strong force created by billions of such exchanges is immense, estimated to be the equivalent of ten tons of force holding the quarks together.

QCD shows us these exchanges when protons are probed or smashed, which results in an ever-changing array of views inside the proton, as one can imagine, because of the nature of exchanges and the momentum, confinement and asymptotic freedom of the quarks. These views, when gathered together, and analyzed create a composite statistical view of the internal dynamics of protons.

> "By using higher-energy electrons, physicists can ferret out finer features of the target proton. In this way, the electron energy sets the maximum resolving power of a deep inelastic scattering experiment. More powerful particle colliders offer a sharper view of the proton. Higher-energy colliders also produce a wider array of collision outcomes, letting researchers choose different subsets of the outgoing electrons to analyze. This flexibility has proved key to understanding quarks, which careen about inside the proton with different amounts of momentum. By measuring the energy and trajectory of each scattered electron, researchers can tell if it has glanced off a quark carrying a large chunk of the proton's total momentum or just a smidgen. Through repeated collisions, they can take something like a census -determining whether the proton's momentum is mostly bound up in a few quarks, or distributed over many.[17]"

In essence, this experimental method is the reason we know the proton is not a simple, static object, but a dynamic and "unspeakably complex" system. This revelation serves as a crucial piece of evidence for our investigation.

This brings our deep dive into the "particle zoo" to a close. We have journeyed from a simple three-quark model to a roiling sea of quarks, antiquarks, and gluons, all bound by the bizarre and elegant rules of Quantum Chromodynamics. We have seen a force that grows stronger with distance and a fail-safe mechanism that creates new matter to prevent its own rules from being broken . This is not a random collection of particles; it is an intricate, self-regulating system of "unspeakable complexity" that forms the very foundation of every atom in the universe. The existence of this stable, complex, and perfectly functional system is the first great miracle required for the existence of matter itself.

This bizarre reality of a proton being a "roiling sea" of particles is not just a theoretical idea; it is a conclusion forced upon us by decades of powerful and precise experiments. To understand the evidence for this "unspeakable complexity," we must look at how physicists probe the heart of matter.

A Sea of Quarks, And Antimatter

By firing high-energy electrons at protons, physicists get different "views" of the particle's interior depending on how the electron scatters. These different views confirm the complex reality described by Quantum Chromodynamics (QCD).

When an electron collides with a high-momentum quark, it scatters as if it has hit a solid independent particle carrying about a third of the proton's total momentum. This is the view that reveals the three permanent "valence quarks" of the simple high school physics model.

However, when an electron collides with a low-momentum quark, it scatters differently. This view reveals that the proton is also filled with a "roiling sea" of transient, low-energy quarks and their antimatter counterparts, which pop in and out of existence within a dense cloud of gluons.

It is these different results from the same experiment that provide the evidence for the proton's dual nature: a stable, three-quark structure that is simultaneously a chaotic and dynamic sea of virtual particles[18].

> "The three quarks in the basic picture of the interior of a proton are merely the "valence quarks" —buoys bobbing on top of a roiling sea of quarks and antiquarks (their antimatter counterparts), as well as the sticky "gluon" particles that hold them together. The total number of quarks and gluons inside a proton is always changing. Quark-antiquark pairs are constantly popping in and out of existence, and gluons tend to split and multiply, especially when a proton gains speed. It's basically pure chaos. The strong force-the most powerful of the four fundamental forces of nature-keeps this mess confined to the insides of protons and neutrons. Except when it doesn't.[19]"

The quarks inside a proton are held together by a bewilderingly complex system that obeys a multitude of rules. This system maintains a state where its constituents (i.e. the particles and virtual particles) are constantly and furiously interacting. Virtual particles pop in and out of existence, everything is constantly changing places and states, and the strong force is transferring color charges, all while conserving the colorless state of quantum superposition that holds this relationship between quarks, the protons they compose, and the atomic nuclei they all together maintain. Lest we forget, all of these exchanges and happenings are not simply for the "fun" of the proton, or to impress and confuse researchers. This is all an indelible and indispensable part of the fundamental structure of atoms that give us elements, molecules, chemistry, stars and life. This same complex relationship is happening in every proton and neutron of every atom in our universe, from hydrogen to plutonium. The idea of 3 quarks held together by the strong force is a simplistic visualization realized only after all of the accounting for the real and virtual and momentary particles and sea of gluons and exchanges are taken into account. At the end, when everything is canceled out on the proton balance sheet, 3 quarks are the net result.

What we find inside of the proton is a kind of small-scale recreation of what our universe was like in the beginning when the energies and pressures were of the scale of the large hadron collider, and all that existed was a particle soup of quarks and gluons. This is the womb of the universe, the original and undisputed primordial soup of creation where fundamental particles existed as a gas/fluid of particles and radiation before atoms existed.

> "the Quark-Gluon Plasma (the QGP) – the matter that the universe consisted of... is a special state consisting of the fundamental particles, the quarks, and the particles that bind the quarks together, the gluons... When colliding ions, the scientists create a fireball that recreates the initial conditions of the universe at temperatures in excess of several thousand billion degrees... Under these conditions the density of quarks and gluons is very large and a special state of matter is formed in which quarks and gluons are quasi-free (dubbed the strongly interacting QGP). The experiments reveal that the primordial matter, the instant before atoms formed, behaves like a liquid that can be described in terms of hydrodynamics.[20]"

In other words, the chaotic, liquid-like interior of a proton is not just a strange property of a single particle; it is a direct echo of the universe's very first moments. The 'primordial soup' was not just a historical event; a microscopic version of it exists inside every atom in the universe today. This connection between the subatomic and the cosmic reinforces the idea that the properties of matter are not accidental but are deeply and purposefully intertwined with the history of the cosmos itself.

Proton/Neutron Mass

This deep connection between the proton and the primordial universe is most evident when we examine the profound and counter-intuitive origin of its mass.

The nucleus of the atom makes up approximately 99.94% of the mass of the atom, with the remainder attributed to electrons. But this fact only deepens the mystery of the proton when we consider how the mass of protons, and quarks, thus, atoms is acquired.

> "How do protons and neutrons acquire their mass? Each of these particles, or 'nucleons,' is composed of a dense, frothing mess of other particles: quarks, which have mass, and gluons, which do not.

Yet the quark masses only add up to a mere 1% of a proton or neutron's mass, with the bulk of the proton mass coming purely from the motion and confinement of quarks and gluons[21]."

The strong force itself is what constitutes the majority of the mass for protons. The strong force, as we've reported, exerts up to ten tons of force confining quarks; and this energy, under Einstein's formula of E=mc2 is carried by the virtual particles and particle/antiparticle pairs of quarks and gluons that have been described and attributed to these exchanges though the math of lattice QCD.

"The researchers rely on a powerful method known as lattice QCD, which places quarks on the sites of a lattice and gluons on the links between them. This rigorous representation of QCD can be implemented numerically, and it is the only QCD-based method that can make quantitative predictions on length scales comparable to the proton or larger… A three-quark particle like the nucleon is exponentially more complicated for lattice QCD, and successful calculations, with all sources of uncertainty controlled, have been rare. In their work, [Dr. Yi-Bo] Yang and collaborators overcome some of the complications by using new computational methods that they, along with others, developed. These advances enabled them to compute the contribution to the proton mass from four sources[22]."

The sources of "confinement of quarks and gluons" are 4 in number and they are:
-quark condensate (~9%)
-quark energy (~32%)
-gluonic field strength energy (~37%)
-anomalous gluonic contribution (~23%)[23]

The source of the proton's mass is one of the deepest mysteries in physics. While protons are made of quarks, the mass of these constituent quarks accounts for only about 1% of the proton's total mass. The other 99% comes from the immense energy of the strong nuclear force that confines the quarks, and from the kinetic energy of the quarks and gluons moving at near light-speed within that confined space. In essence, the vast majority of the mass of every atom in the visible universe is not derived from the "stuff" of matter, but from pure, concentrated energy, in accordance with Einstein's formula, E=mc 2.

This is a profound realization. We think of our world as being made of solid "something," but the evidence from particle physics shows that it is almost entirely composed of the energy of these fundamental interactions.

This changes how we should frame our most basic existential questions. We have long asked, 'Why is there something rather than nothing?' But perhaps the deeper mystery is, "How does a universe that is almost entirely insubstantial give rise to the profound reality of something?"

This question leads us to an even more immediate and pressing paradox. The "something" that makes up our reality should not exist at all. According to the standard model of physics, the creation of matter in the Big Bang should have been accompanied by the creation of an equal amount of its mirror image: antimatter. Their subsequent meeting in the fiery cauldron of the early universe should have resulted in their mutual, total annihilation, leaving behind a cosmos filled with nothing but light. Yet here we are. The very existence of matter is the result of a fundamental imbalance at the dawn of time.

The Foundation of the Matter/Antimatter (Baryon) Asymmetry

The production of matter in our universe is governed by the equation E = mc2 which means that energy can be turned into mass and vice-versa. In the first 10 minutes of our universe, it moved through four epochs: the quark, hardon, lepton and photon epochs. What's key for our present investigation is that the energy that came into our universe would have had to have been created in equal amounts of matter and antimatter. Because matter and antimatter have the exact opposite charge and are precisely equal energetically one another's opposite, both would have immediately annihilated one another had they come into contact with one another. This would have been impossible to avoid in the hot, dense soup of the early universe, where, for a period, all of existence was a chaotic plasma of free-roaming quarks and antiquarks constantly colliding with one another.

By what we can measure from the CMB and other forms of energy, these annihilations took place, but with only an approximate 99.9999996% efficiency. But this inability of the universe to fulfill 100% of the matter/antimatter annihilations demanded by symmetry resulted in the most amazing series of products and transformations in the history of our universe, including the emergence of intelligent life. Alan Guth states, "For every 300,000,000 quarks, there were 299,999,999 antiquarks… Essentially all the pairs disappeared, leaving only a single unpaired quark for every 300,000,000 quarks that were initially present. These unpaired quarks rapidly formed into the baryons that we observe today."

It is apparent from our universe that some unknown process in the fiery first moments of creation produced a slight imbalance, or asymmetry, between matter and antimatter. This tiny surplus of matter is what survived the subsequent annihilation and went on to form everything we see today.

As far as we can tell, the observable universe is now composed entirely of matter, with almost no antimatter. This profound and unexplained outcome is what is known as the Baryon asymmetry.

The story of the baryogenesis mystery begins with how we came to learn that there is a discrepancy (asymmetry) in matter and antimatter in the first place, because until we were able to understand this, ignorance of the manner in which we existed, was bliss.

The Discovery of Antimatter & Matter/Antimatter Annihilation

The answer to this question comes from English physicist Paul Dirac and perhaps the most consequential, if not well-known equation in the history of science, *The Dirac Equation*. In 1928, 26-year-old Dirac was trying to unify quantum physics with Einstein's theory of special relativity. This task had already been accomplished, but not in such a way that pleased Dirac's mathematical aesthetics. Famously, for Dirac, if a solution wasn't correct and elegant and aesthetically pleasing, then it indicated the solution had not been refined, and therefore had not been fully mined for additional subtleties and nuances of truth.

Scientists of this age frequently produced results they didn't at first understand, only to later realize their result indicated answers to additional questions they had not anticipated–one of the hallmarks of a foundational truth. When Dirac had finished, although he had the elegant equation he desired, he found that he had a problem. His solution unexpectedly generated an equation with not one, but two possible solutions. One solution he recognized. It clearly represented the negatively charged electron. But the second solution represented the exact *opposite* of the negatively charged electron. It indicated the existence of a previously unknown *positively* charged particle. Dirac attempted to work on this solution to remove or cancel out the second result but was unsuccessful. He held onto his result, and over a series of 3 papers over 3 years, he refined his result until he was convinced that his result pointed to the existence of a previously unknown universal product, -antimatter. Dirac's initial paper proposing positively charged electrons was titled, *The Quantum theory of the Electron*, written in 1928[24].

Then in 1929 he wrote a follow up paper entitled, *A theory of Electrons and Protons* that attempted to explain a problem brought up by his previous paper regarding electrons[25]. Then through the contributions of Oppenheimer he realized that his second solution implied not just an opposite particle to the electron, but the first artifact from the realm of antimatter[26]. this led to his 1931 paper, *Quantised Singularities in the Quantum Field*, that explicitly predicted the existence of the positron and went on to state that if this new particle, and the electron were to come into contact, they would annihilate one another[27]. Dirac initially named this particle the *antielectron*, and later changed his term to *positron* when he announced his discovery to the world.[28]

The Dirac Equation unearthed antimatter from the wave equation, described electrons, and also quarks, united quantum mechanics with relativity, accounted for the newly discovered characteristic of "spin", and became a foundation of the standard model of physics[29]. was awarded the 1933 Nobel Prize for physics "for the discovery of new productive forms of atomic theory.[30]"

1 year later, in 1932, American-Swedish physicist Carl David Anderson photographed the first positron. The positron was distinguished from the electron in the photo by curling in the *opposite* direction from an electron due to its opposite charge. Anderson received his own Nobel prize (physics, 1936) for later work where he used gamma rays to create matter/antimatter particle pairs. With both theoretical and experimental evidence, it was realized that because of the laws of conservation, the entire standard model, made up of matter, would need to have an equal and opposite antimatter complement. This information indicated unequivocally that in the early universe equal amounts of matter and antimatter would have necessarily been created, and that for our matter dominated universe to have emerged without having annihilated itself completely, there had to have been some event in violation of the rules of conservation that resulted in matter surviving this irrevocable contingency. This meant that there was an essential mystery in our midst of an asymmetry between matter and antimatter in the initial moments of the universe.

Solving The Baryogenesis Mystery

To solve this mystery, physicists face a crucial choice. One option is to simply assume that the universe was created with a built-in imbalance from the very beginning—in essence, to accept the initial state as an unexplained, fine-tuned miracle. However, because the goal of science is to explain such outcomes from simpler, more fundamental principles. The most basic assumption is that the universe began in the most symmetrical state possible: with a perfect 1:1 ratio of matter and antimatter.

The challenge, then, is to find a naturalistic pathway from that initial, symmetrical state to the matter-dominated universe we see today. In 1967, the Soviet physicist Andrei Sakharov laid out the rulebook for this search. He determined that any viable theory that hopes to achieve this must satisfy three specific and non-negotiable conditions. They are the essential ingredients required to "cook up" a matter-filled cosmos from a perfectly balanced start[31].

Sakharov's conditions have since guided the search for a solution to the mystery. Sakharov's pathway requires that the symmetry of our universe not be broken at the outset. This means the solution to the disparity can't be due to the universe somehow producing more matter initially or at a higher rate than antimatter. Sakaharov cites the CPT Theorem, "The CPT Theorem guarantees that a particle and its antiparticle have exactly the same mass and lifetime, and opposite charge[32]."

The CPT Theorem was developed in 1951 by American physicist and Nobel laureate (Physics, 1965) Julian Schwinger, an innovator of quantum electrodynamics (QED). Sakharov's conditions then identify that the discrepancy must instead be evident through *CP violations* -a violation of CP-symmetry, or charge conjugation parity symmetry.

In CP violations, **"C" stands for charge:** *charge violations* are where the charge of the particle changes from positive to negative, or vice-versa.

"P" stands for *parity violations*: parity violations are when the particle is made to act as a mirror image of itself.

For CP violations to take place, somehow, *after* particle pairs are created (a key stipulation) some unknown law (a fundamental condition) of unified theory *selects* for a CP violation. This is how baryogenesis takes place. Having defined CP violations, Sakharov's proposal and three conditions for CP-symmetry violations are written below:

The first condition: The universe must be out of equilibrium (Baryon number violation)

The second condition: The universe must exhibit C and CP violation (C-symmetry and CP-symmetry violation)

The third condition: The universe must have baryon-number violating interactions (interactions out of thermal equilibrium)

In our universe, what we have found is:

The first condition is met: our universe is out of equilibrium as evidenced by the domination of matter.

The second condition is somewhat met: there is evidence that there are C and CP violations in our universe. In 1964, James Cronin and Val Fitch, two American physicists studying the decays of a K meson found that this particle and its antiparticle *weren't* perfect mirror images of each other as evidenced by the fact that they decayed at slightly different rates.

The third condition is not met: but requires some discussion.

"Baryon number violation is obviously a necessary condition to produce an excess of baryons over antibaryons. But C-symmetry violation is also needed so that the interactions which produce more baryons than anti-baryons will not be counterbalanced by interactions which produce more anti-baryons than baryons. CP-symmetry violation is similarly required because otherwise equal numbers of left-handed baryons and right-handed anti-baryons would be produced, as well as equal numbers of left-handed anti-baryons and right-handed baryons. Finally, the interactions must be out of thermal equilibrium, since otherwise CPT symmetry would assure compensation between processes increasing and decreasing the baryon number[33]."

The third and most crucial condition has not been met. While the experiments with K-mesons prove that a slight asymmetry between matter and antimatter is possible (fulfilling the second condition), this is not enough. To create the universe we see today, a specific process was needed that could actually violate baryon number—that is, create more matter particles than antimatter particles.

Furthermore, as Sakharov's work explains, this process had to occur while the universe was out of thermal equilibrium. In a stable, evenly mixed state, any reaction that creates a baryon would be perfectly balanced by a reverse reaction that destroys one, resulting in no net gain of matter. The surplus could only be achieved in a rapidly expanding and cooling universe, where the matter-creating reactions could happen faster than their reverse, matter-destroying counterparts.

To date, no such baryon-number-violating process has ever been observed in any experiment, and the asymmetries we have found are far too weak to account for the existence of our matter-dominated universe.

The Experimental Evidence

In 1964, American physicists and Nobel laureates (physics, 1980) James Cronin and Val Fitch ran experiments at Brookhaven National Laboratory with K-mesons (also called kaons), a type of meson that is half matter and half antimatter. They shot two different kinds of kaons, each with a different lifetime, down a 17-meter beamline and detected the resulting pion-decays at the other end. They didn't expect to find that *any* of the short-lived kaons decayed at the end of the pipeline, however, out of 22,700 decay events, 45 decayed within their detector for a ratio of 1 in 504.4

thereby establishing that a slight asymmetry was built into the laws governing decay. This finding established that there was a violation in fundamental physics and for this they were awarded the Nobel prize, "for the discovery of violations of fundamental symmetry principles in the decay of neutral K-mesons[34]." Their findings imply that either the principle in physics that governs the weak force/weak interaction in radioactivity and neutrinos was *incomplete* or *incorrect* regarding K-meson decays, because their result was the first finding that violated a fundamental principle of physics – the symmetry between matter and antimatter[35].

James Cronin, in accepting his Nobel Prize stated, "We know that improvements in detector technology and quality of accelerators will permit even more sensitive experiments in the coming decades. We are hopeful then, that at some epoch, perhaps distant, this cryptic message from nature will be deciphered[36]." (Keeping true to Cronin's prediction, 39 years later, in March of 2019, the LHCb experiment at the Large Hadron Collider observed CP violation for the first time in the decay of charm quarks.

These findings are significant as they demonstrate that perfect symmetry does not exist between matter and antimatter within our universe, and that there are likely other, as yet undetected CP violations within the standard model, perhaps beyond the weak interaction where the first 2 violations have been detected.

The bad news is that these findings *do not* meet Sakharov's third necessary condition. As significant as these findings are, these discrepancies and the processes that led to them are far short of what is needed to account for the current dominance of matter. These results are merely tantalizing clues to a much larger mystery, perhaps unrelated to our primary problem, and at best, possibly may lead to future investigations where the mystery begins to unravel.

Sakharov's third condition, that the universe must have *baryon-number* violating interactions, -interactions out of thermal equilibrium- is not met. There are no baryon-number violating interactions and there are at present no known conditions that could lead to the kinds of interactions that result in a surplus of matter over antimatter.

The implications of this are profound. Science has a clear, testable "recipe" for explaining the existence of matter, but a crucial ingredient is missing from our reality. We have found hints of the necessary asymmetry, but they are millions of times too weak to do the job. More importantly, we have found no evidence whatsoever for any process that can create a surplus of matter in the first place.

Since there is no known naturalistic pathway to account for the matter-dominated universe we observe, we are left with a foundational miracle. The very fact that we exist, that there is "something" rather than just a void of light, is an event that stands outside our current scientific understanding.

Beyond Sakharov's framework, physicists have explored other speculative avenues to explain the existence of matter. These include theories of GUT baryogenesis at extreme temperatures, proposals involving hypothetical 'Majorana' neutrinos, and solutions linked to the 'Strong CP Problem' and the search for exotic particles like the axion. While these ideas are a testament to the scientific effort to solve this mystery, they all share the same fundamental flaws. They are either based on speculative physics that cannot be tested, or they rely on the existence of new particles and fields for which decades of searching have yielded no evidence. Ultimately, none of these alternative pathways have provided a viable, evidence-based solution, leaving the origin of matter as profound a mystery as ever.

The mystery, however, does not end there. The problem is not just that we have an unexplained asymmetry where we expect symmetry; in a different area of physics, we find the exact opposite problem: an unexplained, perfect symmetry where we expect to see asymmetry. This is known as the Strong CP Problem, and it adds another layer of "unspeakable complexity" to the foundation of our reality.

The Strong CP Problem, Axions & Dark Matter

The mystery of our universe's finely tuned nature does not end with the baryon asymmetry. In a related area of physics, we find what has been called "a serious sickness of the standard model" and "the most underrated puzzle in all of physics[37]." the Strong CP Problem.

The problem is this: according to a core tenet of physics known as the "totalitarian principle"—that "everything not forbidden is compulsory[38]."—the strong nuclear force is predicted to violate a key symmetry (CP symmetry). Extensive experiments, however, have shown that it does not. This expected violation is mysteriously suppressed to a degree that is mathematically impossible to explain by chance alone, another case of extreme and inexplicable fine-tuning.

To solve this puzzle, physicists proposed the existence of a new, hypothetical particle called the axion. This particle, if it exists, would naturally cancel out the expected violation and "erase a stain from the standard model[39]." Intriguingly, the predicted properties of the axion also make it a perfect candidate for the mysterious dark matter that constitutes the vast majority of matter in the universe.

This presents a tidy but entirely speculative solution: a profound mystery being solved by a hypothetical particle that would, in turn, solve another. However, despite decades of searching and billions of dollars spent, no evidence for the axion has ever been found. We are once again left with an untestable, ad-hoc proposal to explain away a fundamental and seemingly miraculous feature of our reality.

Protons and the Basis of Reality

The challenge that lies before the greatest minds in physics is to understand why matter exists. In helping us to understand why it exists we must also come to understand in a basic sense what it is, and this is where the mystery of baryogenesis grows exponentially. Protons are said to be unspeakably complex and the most complicated thing that can be imagined[40]. and this is before we consider what they actually do and what they demand within our universe. I am referring to its role not just as a building block, but as the central architect whose own complex nature dictates the very laws and structure of our reality.

This may seem like a contradiction: how can the proton be the architect of a universe that it is also a part of? The answer lies in understanding that the proton is the primary and most fundamental expression of the "meta information" I believe was encoded in the universe from the outset. Its properties were not an accident; they were the blueprint. The specific and "unspeakably complex" nature of the proton—its mass, its charge, and the bizarre rules that govern its internal structure—set the non-negotiable terms for the rest of creation. All the other forces and constants had to be precisely what they are to accommodate this foundational particle. In this sense, the proton did not consciously design the universe, but its own design was the design of the universe.

It would be a gross understatement to say that the proton is the schoolyard bully of our universe. In this analogy it would be as if the schoolyard bully dictated the role, schedule, and lives of all the students, teachers, parents and administrators. And also dictated the architectural plans for all of the structures upon the world the school was a part of, and the world beneath it, and the entire universe that surrounded it.

Protons, in some sublime, transcendent and miraculous way, hold the plans for the entire universe. Its own "unspeakably complex" nature—its mass, charge, and the intricate laws of the strong force that govern it—is the standard to which all other forces and constants must conform. The values of gravity and electromagnetism are not independent; they are precisely what they must be to accommodate this foundational particle. In this sense, the proton is the central organizing principle from which the logic of the rest of the universe unfolds.

Atoms, suns, planets, galaxies, chemistry, biology and even, as we will see in later sections, Love, Truth, Beauty and Consciousness. All of this arises from the inner sanctum of the inexplicably complex proton. The proton calls for this, organizes, directs, and expresses this. If one is a physicalist, there is no other option. There are no other forces, fields, particles, or arrangements that this organizational information, and the key attributes, fundamental concepts and set of values established by humanity can be attributed to. Physicalists, (the view that 'everything is physical', that 'nothing over and above' the physical, or that everything supervenes on the physical… closely related to materialism[41].") can attribute these forces to natural physical laws in action, but physical laws are mere workmen in this scheme, what they do, how they do, where they do and what they do is dictated by the energetic, physical and spatial relationships demanded by 3 valence quarks arranged within a +1 charge proton and neutral charged neutrons, in their own arrangement, dictated by QCD which is one and the same/inextricable from the color, confinement and exchanges within the proton.

Summary & Value of Variable 2: Baryon Asymmetry

What we see in the baryon asymmetry is a very strange condition, one in which the focus of scientific efforts has been to find the underlying reason for this asymmetry though weak force CP violations, yet this pathway, even if revealed, has yet to show it has the power to account for an entire universe's worth of matter. This leaves baryogenesis and the baryon asymmetry without a feasible physical explanation.

Complicating this anomaly is the strong CP problem. CP violations that should be allowed, and that would help us to understand baryogenesis are unnaturally—and against expectation—suppressed to zero. How? That is a question we do not yet understand. At present, based upon our understanding, the chance of the baryon asymmetry and our matter-dominated universe existing, given our current understanding, is *zero*.

Per Andrei Sakharov's proposal, in order to account for our matter-dominated universe three conditions must be met. The first two conditions are met, but the third is not. The implication here is decisive. For the universe to have produced matter naturalistically, all three of Sakharov's conditions must be met. The fact that the third and most crucial condition—a process that actually violates baryon number—has never been observed means that science currently has no viable physical explanation for why we exist. This failure of the materialist "recipe" is the primary reason why, within the framework of this book, the baryon asymmetry must be considered a miracle.

The Value of Variable 1: Baryon Asymmetry:

M (Miracle)

There is no explanation for the baryon asymmetry

The conclusion that the Baryon Asymmetry is an impossible, miraculous event is the second major thread we have woven into our pattern. The first, the beginning of the universe itself, was the miracle of existence from nothing. This second miracle, the survival of matter, is the miracle of substance.

This was not a random, one-off event. It was the next, non-negotiable step required by The Divine Algorithm. Without a universe made of stable matter, the intricate complexity of the proton would be meaningless, and the potential for stars, planets, and life would have remained locked within the singularity. The survival of matter is the foundational act of creation that makes all subsequent levels of emergence possible. With this second, crucial miracle established, our investigation continues to the next Variable in the algorithm.

---

[1] (M. Gell-Mann, The Eightfold Way: A Theory of Strong Interaction Symmetry, Synchrotron Laboratory, Pasadena, CA Calif Inst of Tech, 1961) And (Y. Ne'eman, Derivation of Strong Interactions from a Gauge Invariance, Nuclear Physics. 26 (2). Amsterdam: North-Holland Publishing Co.: 222-229)
[2] (J. Gleick, Genius: The Life and Science of Richard Feynman, P. 396, 2011 Open Road Integrated Media)
[3] (NobelPrize(dot) org. Nobel Prize Outreach AB 2024, Physics 1990 prize, Fri. 2 Aug 2023, www(dot) nobelprize(dot) org/ prizes/ physics/ 1990/ press-release/).
[4] (Ibid, NobelPrize(dot) org. Nobel Prize Outreach AB 2023, Physics 1990).
[5] (M. Riordan, The Discovery of Quarks, Science, May 29, 1992, Vol 256, Issue 5061, Pp. 1287-1293 DOI: 10.1126/ science(dot) 256(dot) 5061(dot) 1287)
[6] (A. Khare, The November (J/ψ) Revolution: Twenty-Five Years Later. Current Science, 77(9), 1210-1213. www(dot) jstor(dot) org/ stable/ 24103711)
[7] (The Nobel Prize in Physics 1979. NobelPrize(dot) org Nobel Prize Outreach AB 2024. Sat. 17 Aug 2023. nobelprize(dot) org/ prizes/ physics/ 1979/ summary/)
[8] (M. Riordan, The Discovery of Quarks, Science, May 29, 1992, Vol 256, Issue 5061, Pp. 1287-1293 (Pg. 16-17 in original paper) DOI: 10.1126/science.256.5061.1287)
[9] (D. Denisov, C. Vellidis, The Top Quark, 20 Years After Its Discovery, Physics Today, Vol. 68, Iss. 4, April 1, 2015, physicstoday/ article/ 68/ 4/ 46/ 415101/ the-top -quark-20 -years-after -its-discoverythe)
[10] (C. Wood, M. Sherman, Inside the Proton, the 'Most Complicated Thing You Could Possibly Imagine, October 19, 2022, quantamagazine(dot) org/ inside-the-proton -the-most-complicated-thing -imaginable -20221019/)

---

[11] (C. Wood, M. Sherman, Inside the Proton, the 'Most Complicated Thing You Could Possibly Imagine, October 19, 2022, quantamagazine(dot) org/ inside-the-proton -the-most-complicated-thing -imaginable -20221019/)
[12] (K. Cooper, What Are Gluons?, January 29, 2023, space(dot) com/ gluons-carriers-strong-force-explained)
[13] (M. Riordan, The Discovery of Quarks, Science, May 29, 1992, Vol 256, Issue 5061, Pp. 1287-1293 (Pg. 16-17 in original paper) DOI: 10.1126/science.256.5061.1287)
[14] (D.J. Gross, The Discovery of Asymptotic Freedom and the Emergence of QCD, Proceedings of the National Academy of Sciences, Vol. 102: No. 26, June 28, 2005, PubMed 15967980)
[15] (V.D. Burkert, L. Elouadrihiri, F.X. Girod, The Pressure Distribution Inside the Proton, Nature 557, 396-399, 2018, doi(dot) org, 10.1038/ s41586-018-0060-z)
[16] (H. Yukawa, Facts, NobelPrize(dot) org, Nobel Prize Outreach AB2024, Wed, 14 Aug, 2024, nobelprize(dot) org/ prizes/ physics/ 1949/ yukawa/ facts/)
[17] (C. Wood, M. Sherman, Inside the Proton, the 'Most Complicated Thing You Could Possibly Imagine, October 19, 2022, quantamagazine(dot) org/ inside-the-proton -the-most-complicated-thing -imaginable -20221019/)
[18] (C. Wood, M. Sherman, Inside the Proton, the 'Most Complicated Thing You Could Possibly Imagine, October 19, 2022, quantamagazine(dot) org/ inside-the-proton -the-most-complicated-thing -imaginable -20221019/)
[19] (C. Moskowitz, Tiny Bubbles of Primordial Soup Recreate Early Universe, Scientific American, March, 1, 2023, www(dot) scienificamerican(dot) com/ article/ tiny-bubbles -of-primordial-soup -recreate-early-universe/)
[20] Niels Bohr Institute, The Early Universe Was a Fluid Quark-Gluon Plasma, October 5, 2018, phys(dot) org/ news/ 2018-10 -early-universe-fluid-quark-gluon-plasma(dot) html)
[21] (A. Walker-Loud, Dissecting the Mass of the Proton, Physics Magazine,Physics 11, 118, physics(dot) aps(dot) org/ articles/ v11/118, November 19, 2018)
[22] (A. Walker-Loud, Dissecting the Mass of the Proton, Physics Magazine, Physics 11, 118, physics(dot) aps(dot) org/ articles/ v11/118November 19, 2018).
[23] (A. Walker-Loud, Dissecting the Mass of the Proton, Physics Magazine, hysics 11, 118, physics(dot) aps(dot) org/ articles/ v11/118November 19, 2018).
[24] (P. Dirac, The Quantum theory of the Electron, Proceedings of the Royal Society, 117 (778): Pgs 610-624, 1928)
[25] (P. Dirac, A theory of Electrons and Protons, Proceedings of the Royal Society, 126 (801): Pgs 360-365, 1930)
[26] (G. Farmelo, The Strangest Man, 2009 Faber and Faber)
[27] (P. Dirac, Quantised Singularities in the Quantum Field, Proceedings of the Royal Society, 133 (821): Pgs 60-62, 1931).
[28] (H.R. Quinn, Y. Nir, The Mystery of the Missing Antimatter, Princeton University Press, 2008, Pg 44-5)
[29] G. Farmelo, The Strangest Man, 2009 Faber and Faber) Dirac
[30] (Nobel Prize Outreach AB 2022, July 17, 2022, Paul. A.M. Dirac – Facts. NobelPrize.org. Nobel Prize Outreach AB 2022. Sun. 17 Jul 2022).
[31] (A.D. Sakharov, Violation of CP Invariance, C Asymmetry, and Baryon Asymmetry of the Universe, Pisma Zh.Eksp.Teor.Fix 5, 1967, pgs 32-35; JETP Lett. 5, 1967, pgs 24-27, DOI: 10(dot) 1070/ PU1991v034n05ABEH002497).
[32] (A.D. Sakharov, Violation of CP Invariance, C Asymmetry, and Baryon Asymmetry of the Universe, Pisma Zh.Eksp.Teor.Fix 5, 1967, pgs 32-35; JETP Lett. 5, 1967, pgs 24-27, DOI: 10(dot) 1070/ PU1991v034n05ABEH002497) And (J. Schwinger, The Theory of Quantized Fields, 1951, I. Physical Review 82, Pgs 914–927)

[33] (A.D. Sakharov, Violation of CP Invariance, C Asymmetry and Baryon Asymmetry of Universe, Jett P Lett. Vol 5, A.D. 1967, Issue 1 Contents, P. 24, jetpletters(dot) ru/ cgi-bin/ front/ download/ gf/ 13/ article(dot) pdf)
[34] (J. Cronin, V. Fitch - Facts. NobelPrize(dot) org, Nobel Outreach AB2024, 20 Aug 2023, nobelprize(dot) org/ prizes/ physics/ 1980/)
[35] (J. Cronin, V. Fitch, Evidence for the 2-pion Decay of the K Meson, Physical Review Letters 1964).
[36] J.W. Cronin, CP Symmetry Violation - The Search for Its Origin, Nobel Lecture, Dec. 8, 1980).
[37] (D. Wu, A Brief Introduction to the Strong CP Problem, 2014, Zeitschrift für Naturforschung A. 52. 10(dot) 1515/ zna -1997-1-245).:
[38] E. Siegel, The 'Strong CP Problem' Is The Most Underrated Puzzle In All Of Physics, Forbes, Nov 19, 2019, forbes(dot) com/ sites/ startswithabang/ 2019/ 11/ 19/ the strong cp problem is the most underrated puzzle in all of physics)
[39] F. Wilczek, Stockholm University, Augmenting Reality: Axions, Anyons, and Entangled Histories, December 2, 2016, Fysikum Stockholm University on YouTube).
[40] (Ibid, C. Wood, M. Sherman, Inside the Proton);
[41] (From Wikipedia, citing, D. Stoljar, Daniel, Physicalism, In Edward N. Zalta (ed.), 2009 The Stanford Encyclopedia of Philosophy (Fall 2009 Edition). Retrieved 2014-08-07)

Variable 2: The Initial Entropy State of the Universe

Introduction

Having established the miracle of the universe's beginning and the subsequent miracle of matter's survival, our investigation now turns to the initial conditions of that creation event. For our universe to evolve into the complex, structured, and life-bearing cosmos we see today, it couldn't have just been any kind of beginning. The initial arrangement of all the matter and energy in the universe had to be exquisitely precise. This brings us to our third variable: the profound and paradoxical mystery of the universe's initial entropy.

To put it simply, entropy is a measure of disorder or randomness within a system. The Second Law of Thermodynamics, one of the most fundamental laws in all of physics, states that the total entropy of an isolated system, like our universe, must always increase over time. Systems naturally progress from a state of order to a state of disorder. A sandcastle (a low-entropy, ordered state) will always degrade into a simple pile of sand (a high-entropy, disordered state), and never the other way around.

This chapter will explore the profound paradox this law creates for cosmology. If entropy always increases, it logically follows that the universe must have begun in its most orderly, lowest-entropy state possible. And yet, observations of the Cosmic Microwave Background Radiation show the exact opposite: the early universe appears to have been in a state of maximum disorder, or high entropy. Our investigation will unpack this paradox, explore the groundbreaking work of physicist Roger Penrose in resolving it, and demonstrate why this "specialness" of our universe's initial state is another miracle that defies a materialistic explanation.

Foundation for a Paradox

To understand the discovery and significance of the initial entropy state of the universe, as we hinted at above, our findings in *Special Problem #1* (the beginning of the universe) and *Variable #1* (the singularity at the beginning of time) are key.

To appreciate the paradox of the initial entropy state, we must first understand the conflicting requirements for a life-bearing universe. Scientists look at the state of the current cosmos—filled with complex structures like stars and galaxies—and work backward to determine the initial conditions necessary to produce it.

The prevailing view was that for our universe to be as uniform as it appears, it must have started with a perfectly smooth and even distribution of matter and energy. However, scientists soon realized this presented a profound problem. A perfectly smooth universe would be a sterile one; gravity requires slight irregularities—tiny clumps of higher density—to act as "seeds." Without these imperfections to "grip" onto, the primordial gas clouds would never have condensed to form the first stars and galaxies.

This means the initial state of the universe had to be "perfectly imperfect." It needed to be incredibly smooth and ordered on a large scale to solve large scale cosmological problems, yet at the same time, it needed to have precisely the right amount of microscopic imperfection to allow for the eventual formation of all the large-scale structures (galaxy clusters, galaxies, etc.) necessary for life. This isn't just one fine-tuning problem; it's two, in a perfect, paradoxical balance.

We have now established that the universe required an incredibly smooth and uniform initial state to evolve as it did. This, however, raises a profound question that has baffled cosmologists for decades: how did the universe achieve this state of near-perfect uniformity as evidenced by the uniformity of the CMB? The standard Big Bang model has no answer, a puzzle that has been given a specific name: The Horizon Problem. And the horizon problem which is deeply related to the is directly related to the CMB's uniformity.

A Universe of Perfect Temperature

The Cosmic Microwave Background Radiation (CMB), the afterglow of the Big Bang, presents us with a picture of the universe that has baffled scientists. When we map its temperature across the entire sky, we find that it is almost perfectly uniform. Under the standard Big Bang model, this uniformity is a deep mystery. Opposite sides of the observable universe are so far apart that there has not been enough time in cosmic history for them to have ever interacted or exchanged heat. So, how did they all arrive at the exact same temperature? This is the Horizon Problem which is closely related to the problem of the initial entropy state. The horizon problem is such that it constitutes its own Variable and the next subject in our investigation. For the purposes of the initial entropy state, the theory of cosmic inflation appears to be our current best explanation. But it should be noted that the theory of cosmic inflation, which will also be examined in great detail in our next subject, only explains the smoothness of the universe; it does not explain the origin of the crucial "imperfect" seeds needed for structure to form. For that, we must turn to the bizarre and counter-intuitive rules of the quantum world.

This is where the materialist paradigm finds itself in a profound bind. To explain the "imperfect" seeds necessary for the formation of stars and galaxies, it must appeal to the bizarre rules of the quantum world, which demand the existence of random fluctuations. However, in doing so, it must accept that the very foundation of our reality is indeterminate—a state of fundamental "incompleteness" born from a sea of chaos that can be neither guided nor predicted. This raises a critical question for the materialist: if the universe is built upon a foundation of pure, unguided randomness, how can it account for the symphony of exquisitely ordered and seemingly impossible events that followed?

This indeterminacy was one of the attractions of the highly problematic inflationary theory, because the size of the universe at the theorized beginning of inflation was ruled by quantum process and thus indeterminacy represented by quantum fluctuations would become exponentially magnified at the exact rate of the inflationary doubling of spacetime and thus provide the "seeds" for the "imperfectly perfect" blueprint for the future structure of the galaxy, clusters, voids and superclusters within our universe. The link between these microscopic quantum fluctuations and the large-scale structure of the cosmos was solidified by the Nobel Prize-winning work of George Smoot, whose COBE satellite first mapped the tiny temperature variations in the cosmic background radiation. As his Nobel citation explains:

> "Various types of particles and radiation travel through outer space, including cosmic background radiation, which has been carefully studied through measurements from the COBE satellite. George Smoot led a project that in 1992 was able to point out small variations in radiation in different directions. This provides a clue to how stars and other heavenly bodies have come into existence. The variations can be explained by a kind of quantum mechanical fluctuations that have caused matter in certain places to form clumps that then have grown because of gravitation[1]."

In short, Smoot's work provided the first direct observational evidence that the seeds of all the galaxies in our universe were forged by the randomness of the quantum realm.

Despite this seeming correlation between the quantum fluctuations when the universe was quantum in size, and the contribution of these fluctuations to the large-scale structure of the universe, there currently is no theory that provides us with a plausible explanation for why our universe looks the way it does with regards to the thermalization of the CMB. This is where our problem of the initial entropy state of the universe comes into sharper focus, because in order for our universe to have this appearance, it necessitates that our universe began in an extremely unlikely and inexplicable condition–the initial entropy state.To understand the precise scientific nature of this puzzle, it's helpful to turn to the words of cosmologist Alan Guth, who provides a formal definition:

A problem of the traditional big bang theory (without inflation) [is] related to the large scale uniformity of the observed universe. The problem is seen most clearly in the *cosmic background radiation* which is believed to have been released at about 300,000 years after the big bang…, Calculations in a traditional big bang theory show that the sources of the background radiation arriving today from two opposite directions in the sky were separated from each other, at 300,000 years after the big bang, by about 100 *horizon distances*. Since no energy or information can be transported further than one horizon distance, the observed uniformity can be reconciled only by postulating that the universe began in the state of near-perfect uniformity[2]."

In order for our universe to look the way it does at present it would have had to have had an *improbable* and near *impossibly low* level of *entropy* at the very beginning. Only an extremely low level of entropy could have allowed for the fantastic level of structure we currently have in the universe today. Things get less organized over time, not more organized, therefore in the beginning, this impossibly low level of entropy is required. There is no escape from this conclusion. No known physical process—whether it's the energy of stars, the rotation of galaxies, or the influence of dark energy—can reverse or correct the inexorable increase of entropy for the universe as a whole.

Understanding Entropy

To grasp the significance of this state, it will help to explain "low entropy", and what we mean by "low entropy at the beginning of the universe". A diamond is an example of low entropy, because as a crystal structure, the atoms are *fixed* in place and therefore there is far less "freedom" of arrangement. If the diamond is smashed then the bits can be arranged in many different ways, vs. 1 way as a single diamond, therefore the entropy for a diamond smashed to bits, is *higher* than for a diamond that isn't smashed to bits. Higher entropy means *higher uncertainty* and a more chaotic system. Low entropy means *lower* uncertainty and a more ordered system.

Another example of potential entropy states is the classic example of a gas in a box. Imagine that within an enclosed glass box the contents are oxygen with 10% (by volume) the noble gas helium and somehow, although the helium is free to travel throughout the box, rather than conforming to thermodynamics, all of the helium gathers and collects itself as tightly and as closely as possible into one corner of the box entirely unmixed with any of the oxygen in the box. Although such an arrangement is theoretically possible, this arrangement would be *highly improbable*, because this arrangement would be *highly* ordered, and close to the highest possible order for the system and exactly what we would *not* expect for a 10% distribution of a helium gas in a box of oxygen. What we would expect to find, given sufficient time, is that the mixture of helium and oxygen would find itself in the *highest* entropy state that could be achieved with the helium evenly distributed throughout the box and having reached this condition, any additional movement would only serve to maintain the high entropy contents of the box.

This example illustrates why entropy is often thought of as *equilibrium*. In our gas-in-a-box example, high entropy is when the gas in the box is at equilibrium, or "perfect internal disorder"

The 2nd law of thermodynamics states that entropy in an isolated system that *is not* in equilibrium will tend to increase in equilibrium over time until it reaches a *maximum* equilibrium level (even distribution).

The key takeaway from these examples is simple: in physics, low entropy is a measure of high order, while high entropy is a measure of disorder.

The Great Cosmic Paradox

Having explained the specific problem of our universe's low entropy state this brings us to the paradox of the entropy state of our universe at the beginning. Currently our universe is in a very highly ordered state (low entropy) as evidenced by the uniformity of star forming clouds of gas, galaxies, galaxy clusters, solar systems, planetary systems and how they are distributed throughout the universe, stretching as far as the *aided* eye can reach: yet, inevitably as we have been moving *forward* in time, the entropy (disorder) of our universe and randomization of our universe has necessarily been *increasing* as it is required to do, and as it cannot avoid. Observation bears this out. Around us, ordered structures like buildings crumble over time, and hot objects like your cup of coffee will always cool to room temperature. In the cosmos, we see stars burn through their fuel, radiating ordered energy out into the disordered void of space.

This means that as we look into the past, as determined by the laws of thermodynamics, we would expect to witness that our universe would appear *less and less and less* disordered and randomized, and *more and more and more* ordered, the further back we go. If we extrapolate all the way back to the big bang, and the initial picosecond of our universe, then, necessarily, given the laws of entropy, our universe had to have been *highly* organized and at its highest level of organization, aka its *lowest* level of entropy in its history. This reality creates an obvious paradox.

The paradox is this: the Second Law of Thermodynamics demands that the universe must have begun in a state of incredible order, or low entropy, for all the complexity we see today to have formed. And yet, our direct observational evidence from the Cosmic Microwave Background Radiation shows an early universe in a state of almost perfect thermal equilibrium—a condition of maximum disorder, or high entropy. How can the universe have started in both its most ordered and most disordered state simultaneously? This is the profound puzzle that physicist Roger Penrose set out to solve.

In the words of Roger Penrose, "however special the universe may be now, with life existing in it now, it must have been far more special at an earlier stage in which life was not present." Penrose performed his calculations and his findings were published in the Proceedings of EPAC 2006, Edinburgh, Scotland in a paper, titled, *"Before the Big Bang: An Outrageous New Perspective and its Implications for Particle Physics*[3]*."*

> "Proposals for describing the initial state of the universe hardly ever address a certain fundamental conundrum… Since the entropy increases in the future direction of time, it must decrease in the past time-direction. Accordingly, the initial state of the universe must be the most special of all, so any proposal for the actual nature of this initial state must account for its extreme specialness. Proposals have been put forward… in which it is suggested that the initial state of the universe ought to have been in some sense "random"... but 'random' means 'non-special' in the extreme; hence the conundrum[4]."

Penrose's argument here is a powerful critique of any attempt to explain away the initial state of the universe as a random or typical occurrence. His point is that the laws of physics themselves demand that the beginning must have been incredibly special and highly ordered. Therefore, to propose a "random" initial state is a fundamental contradiction—it ignores the extreme fine-tuning required to produce the ordered, life-bearing cosmos we inhabit. This "conundrum," as Penrose calls it, is a key piece of evidence that the beginning of our universe was anything but an accident.

What Penrose found was that the paradox can be resolved by understanding that in the beginning of the universe it possessed two different types of entropy: thermal and gravitational.

> From a purely thermal perspective, the early universe was indeed in a state of near-maximum entropy. It was a smooth, uniform soup of particles and energy, which represents a high degree of disorder.
>
> From a gravitational perspective, the exact opposite is true. For gravity, a smooth, uniform state is a condition of incredible order and therefore extremely low entropy. Gravity's natural tendency is to pull matter together into clumps; a high-entropy state for gravity would be a lumpy universe filled with black holes.

This was Penrose's striking insight: the initial state of our universe was an exquisitely special and paradoxical combination. It had high thermal entropy (the matter was a uniform soup) but incredibly low gravitational entropy (the spacetime it inhabited was perfectly smooth). It was this "specialness"—this vast potential energy stored in the highly ordered gravitational field—that allowed for the formation of all the marvelous structures we see in the universe today.

Viewing a system as a gravitational system is not a standard evaluative perspective in the determination of the entropy of a system. In calling this view forth, Penrose is contrasting the paradox of the initial entropy state, with the reality of what has been *achieved* in our universe due to the very special force of gravity. Lineweaver and Patel added, "In one case [thermal entropy, referring to matter and radiation] things get more uniform. In the other [gravitational entropy] they get less uniform. So [the initial state of the universe must be] some kind of compromise or mixture of these things[5]." and this mixture, as it turns out, is not only so unlikely as to be beyond the power of chance to manifest, but it is also *precisely* what is required to make a universe that is life allowing and life inhabited with intelligent complex life. Even now, nearly 14 billion years removed from creation, unless we consider Penrose's radical and unprecedented hypothesis as theory in waiting, the formation of galaxies and galaxy clusters represent *low entropy* and a high degree of order in our universe.

Penrose's solution highlights the unique and counter-intuitive role of gravity in the universe's evolution. On small, non-gravitational scales, like a gas in a human-scale box of a few feet in height, width and depth, the behavior of particles is dominated by their thermal motion and collisions. Gravity has almost no effect. In this system, the most random, disordered, and therefore high-entropy state is a smooth, uniform distribution of whatever elements are within the box.

On large, cosmic scales (like a gas cloud in the universe), the cumulative pull of gravity becomes the dominant force. In this system, the rules are flipped. A smooth, uniform distribution is an incredibly ordered and unstable state (low entropy), while the most natural, disordered, and high-entropy state is for gravity to have done its work, pulling the gas into dense clumps like stars and galaxies.

Think of the early universe as a perfectly constructed house of cards. This is a state of incredible order and therefore very low gravitational entropy. As we've mentioned, gravity's fundamental tendency is to pull things together, to make them "clump." The most disordered, natural, and high-entropy state for a gravitational system is not a smooth distribution, but a collapsed, clumpy one—like a pile of cards on the floor.

The formation of stars and galaxies is the process of this gravitational entropy increasing, as the universe moves from its highly ordered, smooth initial state to a more disordered, clumpy one. The key to the paradox is this: the initial low entropy state of the universe's gravitational field was the necessary, perfectly ordered setup that provided the potential for gravity to do its work and create all the complex structures we see today.

Putting a Number on a Miracle

To provide a measure of just how exquisitely fine-tuned the initial arrangement of matter was, Roger Penrose set out to calculate the precise odds of our universe's low-entropy state occurring by chance. His calculation, which utilized the Bekenstein-Hawking entropy formula, can be broken down into a few key steps:

First, he established a benchmark for the maximum possible disorder, or entropy, a universe could have. He did this by using black holes as a kind of "standard candle" for the most disordered state possible.

Next, he calculated the total volume of all possible ways the universe could have been arranged, a massive number representing all the possible high-entropy, black-hole-filled configurations.

Finally, he compared the number of possible "black hole universes" ($\mathbf{10^{10^{123}}}$) to the tiny number of possible "orderly universes" like ours ($\mathbf{10^{10^{121}}}$).

Stephen Meyer explains this result in his book, *Return of the God Hypothesis,* which led to a number of staggering improbability.

"And since $\mathbf{10^{10^{101}}}$ is a miniscule fraction of $\mathbf{10^{10^{123}}}$, (Penrose) concluded that the conditions that can generate a life-friendly universe are extremely rare in comparison to the total number of possible configurations that could've existed at the beginning of the universe. Indeed, dividing $\mathbf{10^{10^{101}}}$ by $\mathbf{10^{10^{123}}}$, just yields the number $\mathbf{10^{10^{123}}}$, all over again. Since the smaller exponential number represents an incredibly small percentage of the larger exponential number the smaller number can be ignored as the massively larger exponential number effectively swallows up the smaller one. In any case, the number that Penrose calculated–1 in $\mathbf{10^{10^{123}}}$, – provides a quantitative measure of the unimaginably precise fine-tuning of the initial conditions of the universe[6].

In simple terms, Penrose's calculation reveals the odds of our finely tuned, life-permitting universe arising by chance, compared to all the other possible disordered, black-hole-filled universes that the laws of physics would far prefer. The number he arrived at—1 chance in 10 to the power of 10 to the power of 123—is a quantitative measure of the "specialness" required at the Big Bang.

To get a sense of the Penrose number, imagine trying to write it out in long form, starting with a '1' followed by all its zeroes. The total number of fundamental particles in our entire observable universe is estimated to be around $\mathbf{10^{80}}$. If you were to turn every single one of those particles—every quark, every electron, every photon—into a tiny zero, you would not have nearly enough zeroes to write down the number Penrose calculated. You would need billions upon billions of additional universes just to get enough particles to represent all the zeroes required. The odds of our universe's initial state occurring by chance are equivalent to correctly guessing one single, pre-selected particle out of that entire cosmic ocean of possibilities.

After presenting this astronomical figure, Penrose stated, "This gives us some idea of the enormity of the precision in the Big Bang!" Penrose referred to this number as a calculation of the "original phase-space volume" in his book, *The Emperor's New Mind*[7]. "One could not possibly even write the number down in full" as there would be *exponentially* more zeros than there are fundamental particles in our universe ($\mathbf{10^{80}}$)[8], meaning if every particle (photon, quark, electron etc.) in our universe was a zero, you wouldn't have anywhere near the number of fundamental particles to represent a meaningful fraction of the Penrose number. We would need billions upon billions of universes to have enough particles to represent *all* of the zeroes required to represent this number, *then, a single particle* out of all of these billions of universes needed to represent all of these zeroes, would represent *the single "chance"* that our universe would have the level of entropy it had at the beginning.

This number is so vanishingly small that it is functionally impossible. Yet, the situation is more profound than a simple impossibility. A logical contradiction is impossible and requires no further explanation. The initial state of our universe, however, is a state of almost infinite, specified order. It is an event that is not just improbable but requires an incomprehensible degree of fine-tuning to occur. This is the ultimate paradox: the state of our universe was functionally impossible, and yet it happened.

Not only was our universe *in* this special state, but the Penrose number has a practical requirement as well, because if the mass and energy of the early universe (neutrinos, photons, atoms and dark matter) was not *evenly distributed, precisely* to the specific ratio of the Penrose number ($\mathbf{10^{10^{123}}}$) there would be no chance for life of any kind in our universe. Thus, with his calculation Roger Penrose unknowingly linked the initial entropy state of the universe to all of the other variables and special problems in *The Divine Algorithm*, which are *strict* requirements if there is to be any possibility of life of any kind, anywhere in our universe.

The implications of the Penrose Number are profound and counter intuitive. It answers the great cosmic paradox by showing that for our universe to exist, with all its potential for life, it had to begin in a state of almost impossible, specified order. This leads to a nearly unbelievable conclusion that is so significant it bears repeating once again: our universe was actually more special at the very beginning—as a singularity with nothing resembling what we observe today—than it is now, even with the existence of life.

This finding suggests an apt vision for our universe is not that of a random physical system, but of a predetermined, and self-manifesting set of conditions–a "seed" if you will. This initial state contained all the perfect conditions required for it to grow, evolve, and eventually "flower," producing the fruit of intelligent life endowed with the mysterious gift of consciousness, making our universe into a miraculous kind of quantum mechanical garden of Eden.

But even a perfect seed needs an engine to grow. For the universe, that engine is the interplay between the inexorable increase of entropy and the relentless pull of gravity. It is this paradoxical relationship, where the universe's drive towards ultimate disorder is the very force that builds its most magnificent structures, that we will now explore.

Entropy's Engine: How Disorder Builds Worlds

The primary structure of our universe is driven by gravitational collapse, and thus star formation and galaxy creation. Stars are formed solely with the primary product of baryogenesis, -hydrogen and helium gas-. In gravitational collapse, this gas ignites and forms stars, the centrifugal force of this collapse creating layers where planets, moons, asteroid belts and planets form, after having been pushed out from this star by the pressure of its ignition. If the star is of the right mass then its luminosity is such that it can provide the ideal form of energy needed to drive a *closed system* such as Earth, to *reverse* the arrow of entropy and disorder, it can then set off precise chemical reactions in photosynthetic systems, and thereby form the complex structures and arrangements we call life. The "right mass and luminosity" of our sun also provides the requirements of life by supplying life with the precise form of energy needed to keep it alive.

At the same time, the presence of stars in the grand scheme of the universe also serves the 2nd law of thermodynamics as entropy-creating physical mechanisms. As stars burn fuel, they also lead to higher states of entropy, greater disorder, but also, a "trade up" in the form of the creation of elements that allow life and form the basis of organic chemistry, represented in the CHNOPS elements (carbon, hydrogen, nitrogen, oxygen, phosphorus and sulfur) while also creating additional elements, like, iron, copper, platinum, lithium, silicon, magnesium, and the other elements found on the periodic chart so advantageously utilized in future technological applications by intelligent life on Earth.

These heavier, non-organic elements can't create life, besides in specialized functions, such as iron and copper hemes and similar structures in essential biology within living organisms. In the grand scheme, these elements "lock up" the initial potential of hydrogen gas as entropy demands. In the end our sun will one day burn out, thus no longer providing the energy for our closed, entropy-defeating system, leading to the extinguishing of all life on Earth that depends on the reverse entropy-facilitating energy of our star.

On a grand scale, the lifetime of all stars is finite. Eventually, the primordial gas needed to form new stars will be exhausted. One by one, the last stars will burn out, and the universe will go dark. This inevitable future is known as the "heat death" of the universe—a state of maximum entropy where all matter is dispersed and all energy has been converted to waste heat. In this final, disordered state, no more complex work can be done; everything that can happen will have happened, and the creative potential of the cosmos will be extinguished.

We have a universe that abides by entropy and *leverages* it as an engine for the causative force of gravity to form the structure of our universe, and to also form energy sources such as stars. These entropic results are "natural" in the specific case, but they provide no answer for multiple inexplicable conditions (the beginning of the universe, the singularity, the baryon asymmetry) that allow these "natural" results to take place. Our universe is a tag team between miraculous events and "naturalness" that has turned our previous understanding inside out, and according to Roger Penrose, all derive from the most special entropy state that can be imagined.

> "It's called the Penrose Number, an equation really, $\mathbf{10^{10^{123}}}$, and its simplicity in appearance belies the impossibility that it represents — these are the odds in which the likelihood of the universe coming into existence in a way that would develop and support life as nearly mathematically impossible. This is the equation of a miracle…it's almost a zero probability[9]."

The Penrose Number represents the stunning conclusion to our investigation of the universe's initial state. We began with a profound paradox: the Second Law of Thermodynamics demands a highly ordered, low-entropy beginning, yet our observations of the Cosmic Microwave Background Radiation show a uniform, high-entropy thermal state. Roger Penrose resolved this by showing that the universe's initial state was a unique and exquisitely fine-tuned combination of high thermal entropy and incredibly low gravitational entropy. It is this "specialness," quantified by a number so vast it defies comprehension, that is the true miracle–a miracle that vastly exhausts the probabilistic capacity of our universe by billions upon billions of times. This was not a random or typical beginning; it was the specific, impossibly ordered state required to build a structured, life-bearing cosmos. This is another crucial thread in the pattern of *The Divine Algorithm.*

Summary: Variable #2: *The Initial Entropy State of the Universe*:

1 chance in $\mathbf{10^{10^{123}}}$, well beyond the *probabilistic capacity of the universe* (PCOU) of 1 chance in $\mathbf{10^{140}}$.

Value of Variable #3, *fM Functional Miracle*

The conditions we have investigated in our existence, only 4 steps deep into our *Algorithm,* have all defeated overwhelming odds, to the point of being either miracles in our world, or functional miracles in 4 separate, highly specified ways.

The Beginning of the Universe: The first "Special Problem" showed that our universe began from a singularity for which there is no possible physical or materialistic cause.

The Nature of the Singularity: The first "Variable" established that this initial singularity was not just a physical anomaly, but a "Pandora's box" containing the complex, forward-looking 'meta information' for our entire reality.

The Existence of Matter: The second "Variable," the Baryon Asymmetry, demonstrated that the survival of matter from the great annihilation in the early universe defies all known physical explanations, as the necessary conditions for it to occur naturalistically are not met.

The Initial Order of the Universe: The third "Variable" revealed that the initial entropy state of the universe was a state of almost impossible, specified order, with odds against it occurring by chance that are so vast they are functionally zero.

All of these events are required for our universe to exist, or stated another way, if these miracles had not occurred, the universe we currently know would not exist. Modern science considers these events coincidental, *ad hoc*, unrelated physical events. Mere details remaining to complete their wholly materialistic, physics-based, probability-based thesis. Modern science must consider them as mere details and *ad hoc* problems, otherwise to contemplate them as they are, which is a series of cooperative, cumulative, against all odds, life-necessary occurrences that defeat materialism, they can only be seen as links in a chain of an overall plan for our universe.

As we learned previously, because there is far more about our universe that we *do not know* than what we do know, we cannot rule out the increasing probability that there *is* an overall plan for existence that demands and facilitates such outcomes. And that in the view of this plan, what appears to us as miracles and impossibilities, are in fact, the only possible outcome. With each miracle, the odds increase that this plan is in operation, in fact the chance that this plan is not in place is as unlikely as the 4 miracles we have encountered so far.

Our current physicalist dominated "science" worldview is more interested in debunking and invalidating the teleological perspective than in finding the truth, but at this early point in our investigation, we have ample evidence that there is an undefined force within our universe guiding it to achieve a series of impossible outcomes.

This undefined guiding force is most evident in the exquisite fine-tuning of the universe's initial conditions. We have already seen how the initial entropy state was an impossibly ordered "seed," but this is only part of the story. For that seed to grow into the structured cosmos we see today, another, equally profound, miracle was required: the universe had to begin with an almost perfect uniformity of temperature. This brings us to the next focus in our investigation, which we have already briefly touched on in this chapter: the profound puzzle known as The Horizon Problem.

---

[1] (G. F. Smoot – Facts. NobelPrize.org. Nobel Prize Outreach AB 2024. Wed. 31 July 2024, nobelprize(dot) org/ prizes/ physics/ 2006/ smoot/ facts/)
[2] (A. Guth, The Inflationary Universe, Pg. 335g, 1997, Basic Books)
[3] [The full paper may be located here: https://accelconf.web.cern.ch/e06/papers/thespa01.pdf]
[4] (R. Penrose, Before The Big Bang: An Outrageous New Perspective and its Implications for Particle Physics, Proceedings of EPAC 2006, Edinburgh, Scotland, Pg. 2759-2762)
[5] [Solutions to the Cosmic Initial Entropy Problem without Equilibrium Initial Conditions, Vihan M. Patel, Charles H. Lineweaver, Entropy 2017, 19(9), 411; https://doi.org/10.3390/e`9080411]
[6] (S.C. Meyer, Return of the God Hypothesis, pg. 150, HarperOne, 2021).
[7] (R. Penrose, The Emperor's New Mind, Oxford University Press, 1989)
[8] [R. Penrose, The Emperor's New Mind) New York: Oxford, 1989), p. 344-]4
[9] (E. McCarty, Penrose Number, The Medium, online, December 23, 2022, ericmccarty1965(dot) medium(dot) com/ penrose-number -79fa30fcd025#)

Variable 3: The Horizon Problem of the CMB

Over the course of the previous chapter, we established that the universe's initial state had to be "perfectly imperfect"—incredibly smooth and uniform on a large scale, yet with precisely the right amount of microscopic imperfection to allow for the formation of stars and galaxies. This raises a profound question: how did the universe achieve this state of near-perfect uniformity in the first place?

This is the heart of our next "Variable," the profound cosmic puzzle known as the Horizon Problem. The standard Big Bang model has no answer for how regions of the universe on opposite sides of the sky, which have never been in causal contact, could have settled at the exact same temperature. This chapter will explore the depths of this mystery. We will then turn our attention to its most popular proposed solution, the theory of cosmic inflation, and demonstrate why this elegant idea, upon closer inspection, may expose an even deeper problem for a purely materialistic worldview.

The Horizon Problem

At close to 380,000 years of age the temperature of the early universe dropped enough that electrons were no longer being dislodged from nuclei by the bombardment of photons. Electrons, therefore, were free to attach themselves to protons and form neutral atoms of mostly hydrogen. This allowed photons to also freely travel throughout the universe. This is called the *age of recombination*. This event also allowed the universe, for the first time since the big bang, to be transparent, whereas prior to this event the universe was a fog of electrons and photons. This release of photons is also the energy that we view as the cosmic microwave background radiation (CMB).

Because our universe has a finite age, there is a fundamental limit to how far we can see. This boundary is our cosmic horizon. The light that forms this boundary is the Cosmic Microwave Background Radiation (CMB), which was released when the universe was about 380,000 years old. Before this time, the universe was a hot, opaque "fog" of plasma that photons could not travel through. This primordial fog creates an impenetrable wall in time; it is the earliest and most distant thing we can ever hope to see, and it defines the edge of our observable universe.

This uniform, simultaneous glow that we observe coming from all directions presents its own profound puzzle. What kind of event could create such a perfectly synchronized expansion everywhere at once? Martin Rees, astrophysicist and former President of the Royal Society, points out that the analogy of the big bang as an explosion is flawed, as an explosion requires an "unbalanced pressure" whereas in the early universe the pressure was the same everywhere and no region could possibly exist "outside" this uniform pressure[1]. This brings up another issue. There is no available reason to explain the expansion of our universe, consequently the process by which it began to expand is poorly understood.

The primary evidence for the incredible uniformity that Rees describes comes from the Cosmic Microwave Background Radiation (CMB). We will need to review several key pieces of information from the previous chapter and fold them into our current discussion to set the background for our current variable seeing as the low entropy state of the universe is closely related to the big bang, gravity, the flatness problem, and issues allegedly resolved by inflationary cosmology.

This afterglow from the Big Bang, which we can still detect today, is very nearly *isotropic*, or in thermal equilibrium, to an accuracy of 1 part in 100,000 in all of these directions. As a practical matter, seeing as the light from one side of the universe has not had the opportunity to mix with the light from the opposite side of the universe, this means that there would be no opportunity for these light sources to be the same temperature (to have come into thermal equilibrium) going back to when the universe was 380,000 years old.

The extreme uniformity of the CMB is not merely a problem; it's a paradox. On one hand, as we've discussed, there is no physical reason for the universe to be in such a perfect state of thermal equilibrium. On the other hand, this very perfection creates a second, opposing problem we examined in the low entropy state of the universe: if the early universe was perfectly smooth, how did any structures like stars and galaxies ever form? Gravity needs slight irregularities—tiny "seed perturbations"—to act upon. A perfectly smooth universe would have remained a sterile, featureless gas forever. As physicist Andrew Liddle explains, the standard Big Bang theory has no mechanism to create these necessary irregularities, "For the same reason that one cannot thermalize separated regions, one also cannot create an irregularity. So in the standard Big Bang theory one cannot have a theory allowing the generation of the seed perturbations [temperature changes]—they have to be there already[2]."

The Big Bang model assumes the universe, at its most extreme scale, is homogeneous (uniform throughout) and isotropic (looks the same/is the same in every direction) and indeed it is, but we know that there was no time for the universe to have come into this condition of thermal equilibrium. We must be mindful that when we look back to see the CMB, we are also looking back in time and seeing the past condition of the universe. With all factors taken into consideration, we know there was no time for this energy to come into contact with one another because at the time the energy of the CMB was released, the universe had already been expanding for 380,000 years and would have only had the ability to be in contact with 1-2 degrees of separation in the sky, not the entire sky, yet this energy is in thermal equilibrium.

"The initial universe is assumed to be homogeneous, yet it consists of at least $\sim\mathbf{10^{83}}$ separate regions which are causally disconnected (i.e., these regions have not yet had time to communicate with each other via light signals) ... Thus, one must assume that the forces which created these initial conditions were capable of violating causality[3]."

The Horizon Problem Spawns Inflationary Cosmology

The Horizon Problem is one of the most profound paradoxes in cosmology. The standard Big Bang model cannot explain how opposite sides of the universe, which have never been in causal contact, could have settled at the exact same temperature. The only explanation it can offer is to assume the universe began in an exquisitely and improbably fine-tuned state of thermal equilibrium from the very start.

This problem is so unprecedented that it spawned the birth of a bold and elegant proposal: inflationary theory. We will now explore this theory in detail, as it sets out to solve not only the Horizon Problem but several other perplexing issues in cosmology. However, as we will see, the question remains whether inflation itself requires an even more incredible degree of fine-tuning to work.

An Elegant Solution to the Horizon Problem (At First Glance)

Inflationary cosmology was developed primarily by Russian theoretical physicist Alexei Starobinsky, Russian-American theoretical physicist Andrei Linde, American theoretical physicists Alan Guth and Paul Steinhardt in the late 1970's and early 80's. This hypothesis answers several ongoing mysteries about the early universe, such as why the universe initially expanded, how it came into thermal equilibrium (the horizon problem), why spacetime is "flat", and why magnetic monopoles, predicted by Paul Dirac in 1931 and several grand unity hypotheses, including string theory, are absent when they should be detectable.

When first introduced, the beauty and elegance of the cosmic inflation hypothesis, and it's surprising agreement with expected observations and in addressing several universal issues, was unexpected. This added to its allure in being able to provide a pathway from a perplexing condition with several unexplained problems to an understandable one for the earliest moments, and present-day condition of the universe.

Much as Paul Dirac regarded his solutions as being true because they were beautiful and elegant, the inflationary hypothesis also provided a beautiful solution with the added benefit that it achieved this result without, appearing to require implausible scenarios, at least initially. It could bring the strangeness of the horizon problem from an impossible outcome, to a highly probable, expected and explained result. The combination of these characteristics, beauty, explainability, completeness, gave inflationary theory the appearance of correctness and completeness, which in the world of science, has usually been associated as a hallmark of truth.

The theory of cosmic inflation is presented as an elegant solution to several of modern cosmology's most perplexing problems. To understand both its profound explanatory power and its significant, perhaps fatal, flaws, we must first take a closer look at the nuts and bolts of the hypothesis. This requires a brief dive into the exotic physics of the early universe.

Inflationary Cosmology - The Nuts & Bolts Behind the Hypothesis

The full, technical details of cosmic inflation are deeply complex, rooted in the mathematics of quantum field theory and general relativity. For our purposes, a deep dive into this math is not necessary. Instead, our explanation will be brief, focusing on three key questions: What was the theory proposed to do? What does it require in order to work? And finally, what are the profound problems that challenge its validity?

The theory of cosmic inflation proposes that in the first fraction of a second after the Big Bang, the universe was dominated by a high-energy state known as a "false vacuum." According to the theory, this false vacuum had a bizarre property: it created a powerful repulsive gravity, an "anti-gravity" that acted as a cosmic "spring," forcing the fabric of space to expand at an exponential rate.

This period of inflation was incredibly brief but unimaginably violent. For a fleeting moment, the universe doubled in size every $\mathbf{10^{-37}}$ second, expanding much faster than the speed of light. In less than the blink of an eye, a region of space billions of times smaller than a proton ballooned to the size of a grapefruit, or in some descriptions, the size of a basketball. This seeming contradiction with the limits of the speed of light are resolved by the fact that only the speed is limited by this constant, whereas the *expansion rate* of the universe is not limited by this constant. The inflationary period then came to an end when the false vacuum "decayed," a process that must happen at precisely the right moment to avoid what is known as the "graceful exit problem" (the challenge of ensuring the universe transitions smoothly from its high-energy inflationary state to the stable, low-energy state we have today, without creating a chaotic mess or getting stuck in an endless expansion).

The immense scale of this expansion is the key to how inflation elegantly solves several of the Big Bang's biggest puzzles. Imagine drawing a tiny circle on the surface of a deflated balloon. The circle is small enough to have a uniform temperature, but the surface of the balloon is very curved. Now, if you inflate that balloon to the size of the Earth, your original circle is now an unimaginably vast region. From your perspective standing inside it, the surface would appear perfectly flat and would have a perfectly uniform temperature. This is how inflation explains the flatness and horizon problems: it proposes our entire observable universe is just a tiny, smoothed-out patch of a much larger cosmos.

A key feature of the inflationary model is that its effects are now seamless and indistinguishable from standard Big Bang cosmology. The period of hyper-expansion, driven by the hypothetical "inflaton field," would have ended in the first fraction of a second. After that point, the universe would have continued its expansion in a way that perfectly matches the history we have already described in our previous Special Problem and Variables.

This "indistinguishability" from standard Big Bang cosmology creates a profound challenge for validating the theory. There is no direct observation we can make today to confirm that our observable universe is just a tiny patch of a cosmos that is actually $\mathbf{10^{23}}$ times larger, as the theory requires. The only indication that the inflationary model *might* be correct is its elegant ability to solve the horizon, flatness, and monopole problems, but for the complex and demanding rules for creating formal theory, this simply is not regarded as robust enough to qualify as a scientific validation.

This leads to a crucial point that we must keep in mind: the problems that inflation solves may, in fact, be inexplicable miracles, just like the other *Variables* and *Special Problems* in our investigation. Inflation offers an alternative, but it is an alternative built on untestable assumptions, meaning our universe is at present inexplicable.

To appreciate why physicists were willing to propose such an extraordinary and untestable theory, we must first understand the sheer scale of the paradoxes it was designed to solve. The most significant of these is the puzzle of the universe's incredible uniformity, a mystery that is at the center of our current Variable known as the Horizon Problem.

The Horizon Problem & Inflation

The horizon problem is solved in the inflationary hypothesis because it creates an initial condition where the universe was so small when it began, that there was no thermodynamic burden.

> "Inflation theory answers the horizon problem by proposing that early in the universe's history, a small parcel of space expanded rapidly… prior to the period of inflation, those regions that are on different sides of the universe today were actually much, much closer than they would've been without inflation. That means that part of the CMB that look very distant today actually would have had time to "communicate" early in the universe's history because they were much closer then and in thermodynamic equilibrium. This explains why their temperatures are so similar now[4]."

At this size the universe could have freely mixed and become thermalized within the limits imposed by the speed of light. Only a tiny piece of the early universe needs to be thermalized, and it can be easily thermalized by the limits of the speed of light within this tiny piece of the universe. This thermal equilibrium, achieved *prior* to the period of inflation, is then preserved *through* the period of inflation, thus rendering our current view of the observable universe and the CMB in thermodynamic equilibrium, to 1 part in 100,000, thus eradicating the horizon problem.

Observable and *universe* are the key words, because although for us the observed universe is exactly the same whether inflationary cosmology or big bang cosmology is true, the size of the observable universe at ~$\mathbf{10^{-52}}$ meters at the beginning of inflation is preserved until the end of inflation as a *ratio* of *the entire universe* vs *the observable universe*. Put another way, although the size of the *observable* universe is the same after the inflationary epoch as big bang cosmology in the inflationary scenario, the *actual* universe is $\mathbf{10^{23}}$ times *larger* thus it extends far beyond what we are able to observe.

The size of the universe under inflationary theory is breathtaking, but also absolutely required in order for inflation to achieve our observable universe. The universe proposed by inflation is 'special' in a problematic, unscientific sense: the vast majority of it is fundamentally unobservable because it lies beyond the impenetrable wall of the CMB—an observational limit that will itself one day be erased from observation by the continued expansion of the universe.

This problem gives us a hint of the other potential problems that require some form of resolution in inflationary cosmology. There are several core predictions and mechanisms within the hypothesis that have to be real and not simply theorized to make the theory work, yet, these core elements, like the universe beyond the observable universe, either cannot be falsified or are extremely difficult to falsify.

Under hot big bang cosmology, we don't need to see beyond the CMB in order to predict what lies beyond it. This is because the universe under hot big bang cosmology is not required, nor predicted to extend much further than the 380,000 years when the CMB was made. In short, there is no $\mathbf{10^{23}}$ times larger universe beyond the CMB that we can never see, test, experience or retrieve information from. You have likely encountered the term "observable universe" in popular science. I argue that the frequent use of this phrase often serves as a premature and unacknowledged concession to the theory of cosmic inflation. While it is true that we cannot see past the Cosmic Microwave Background Radiation, it is the inflationary model that specifically predicts our cosmos is just a tiny, observable patch of a reality that is trillions of times larger. By emphasizing the "observable" part, many science writers are implicitly endorsing the truth of this highly speculative and unproven theory.

Thus, the first major issue with inflationary theory is that it describes events happening in a realm of almost unimaginable smallness, well within the quantum scale where our physical understanding breaks down. The key parameters of the theory, those being when inflation starts, when it stops, and the nature of the repulsive force that drives it, are not known and cannot be known for certain. The hypothetical "inflation field" and the "false vacuum" that are central to the theory have never been observed, so their characteristics must be guessed.

This leads to the theory's most significant flaw: its crucial factors must be "set by hand" to produce the universe we observe. The driving force of the hypothesis is not scientific observation, but the necessity to explain a pre-existing outcome.

At the same time, it's easy to see why the theory is so irresistible. Its uncanny ability to solve so many of cosmology's deepest puzzles in one elegant swoop gives it the appearance of a profound truth. Nonetheless, all efforts to derive the theory's key parameters from any deeper, fundamental principle have so far been fruitless, leaving it as a compelling story without a solid foundation in evidence.

All of this was not specifically realized when the first baby steps towards the creation of inflationary theory were being worked out. Curiously, the initial motivation for the theory appears to have been more philosophical than scientific. I argue that the "germ that seeded" the creation of inflationary theory was the desire to solve the profound philosophical problem that an absolute beginning poses for a purely materialistic worldview: the specter of a cause outside of nature. Inflation, in its earliest conception, was a search for a physical mechanism that could eliminate the need for an absolute singularity, thereby avoiding the subject of our first *Special Problem (#1), The Beginning of our Universe* and our first *Variable (#1), The Singularity at the Beginning of Time.*

From the abstract of the paper, *Inflation and the False Vacuum* from the book, *The Weight of the Vacuum* by Helge S. Kragh and James M. Overduin:

> "Inflation is the idea that the very early universe may have been dominated by vacuum energy, giving rise to a brief period of vastly accelerated expansion. Its roots go back to the late 1960s and the quest for singularity avoidance in the wake of the discovery of the cosmic microwave background radiation. A tentative connection to what would later be called the horizon problem within relativistic cosmology also goes back to this period. But inflation did not gain wide recognition until high-energy particle physicists became involved. Stimulated by the successes of electroweak unification, they began to explore the implications of spontaneous symmetry breaking for cosmology, with the attendant possibility of a false vacuum state. These models implied a universe dominated by massive relic particles from the early universe. This "monopole problem" was the trigger for the explosion of interest in inflation in the 1980s. A possible explanation for flatness was noted at about the same time. Only later was it appreciated that the most compelling argument for something like inflation is its potential to link the observed large-scale structure in the present-day universe to quantum fluctuations in the earliest moments after the big bang[5]."

Inflationary cosmology was founded on the promise that it *could* provide a single, elegant solution to three of the most profound mysteries that make up *The Divine Algorithm*: the Horizon Problem, the Flatness Problem, and the Monopole Problem. We will now focus our investigation on the granular details of how inflation attempts to resolve these issues, and why it ultimately fails, beginning with the puzzle of our universe's perfect "flatness."

The Flatness Problem & Inflation

The flatness problem, also referred to as *the oldness problem* or *the age of the universe problem,* is yet another one of the fine-tuning problems in the universe. In order for our universe to be flat, rather than curved inward (as a sphere) or outward (like a saddle) requires that the ratio of the actual mass density of the universe to be in, or very close to, perfect alignment with the critical mass density of the universe, referred to by the Greek symbol *omega*. Currently our universe is found to be not just flat, but *incredibly* flat, measuring out very close to the critical density of "1" required for it to be flat. Practically speaking, this means that the further back we go in time, the *closer* omega would have had to be to "1", as any deviation would rapidly become exaggerated over the course of time. Peter Coles illustrates the issue by starting at the very beginning:

> "There is another way of looking at this problem by focusing on the Planck time. At this epoch, where our knowledge of the relevant physical laws is scant, there seems to be only one natural timescale for evolution, and that is the Planck time itself. Likewise, there is only one relevant length scale: the Planck length. The characteristic scale of its spatial curvature would have been the Planck length. If spacetime was not flat, then it should either have re-collapsed (if it were positively curved) or entered a phase of rapid un-decelerated expansion (if it were negatively curved) on a timescale on the order of Planck time. But the Universe has avoided going to either of these extremes for around $\mathbf{10^{60}}$ Planck times [omega ≈ 1 now, means omega at t = 1 second differed from 1 by less than 1 part in $\mathbf{10^{16}}$ At the Planck's time, omega differed from 1 by less than 1 part in $\mathbf{10^{60}}$]. These paradoxes are different ways of looking at what has become known as the cosmological flatness problem (or sometimes, because of the arguments that are set out in the preceding paragraph, the *age problem* or the *curvature problem*), and it arises from the incompleteness of the standard Big Bang theory. That it is such a big problem has convinced many scientists that it needs a big solution. The only thing that seemed likely to resolve the conundrum was that our universe really is a professional circus artist… The argument then goes that the density parameter must be extremely close to 1, and that something must have happened in primordial times to single out this value very accurately. [If omega had been much above 1, it would have re-collapsed very early before making galaxies. If omega had been much below 1, it would have expanded so rapidly

that structures would not have formed. This requires a lot of fine tuning. It is known as the flatness problem[6].]"

In essence, this is the core of the flatness problem: for our universe to have avoided either immediate collapse or a sterile, runaway expansion, its initial density had to be set with a precision of one part in $\mathbf{10^{60}}$. To put that in perspective, this is like firing a cannon across the entire observable universe and hitting a single, one-inch target on the other side. This level of "specialness" in the initial conditions is what convinces many scientists, as the quote says, that it "needs a big solution."

The fact that our universe is so close to omega rather than close(r) towards the parameter required for either *eternal* expansion or eventual inevitable *collapse* is considered to be implausible. This is what it means to say, "our universe is flat". All things being equal, the factors that would have needed to be in alignment, in consideration of the number of values involved, and the range of possible values they could have had, is another part of the flatness problem. Understanding the reason why this number would be set so precisely, over the age of the universe is why this problem is also referred to as *the age of the universe problem*. Under the hypothesis of inflation, the problem is solved, as the edges of our universe are pushed beyond observation (the observable universe) so that our universe, is either ultimately flat (beyond what we can observe), or is so vast that it *appears* flat as far as we are concerned from our perspective; much like the ground appears flat to you despite us living on a spherical planet. In terms of the hypothesis of inflation, if we disregard all of its unexplained conjectures, whatever omega is at the beginning of inflation is preserved to the end of inflation for our piece of the inflationary universe.

The ability to solve the flatness problem was a major triumph for the inflationary model. However, its explanatory power extends further. The same mechanism of extreme, rapid expansion also prevents the overproduction of exotic particles that should have been created in the hot early universe. This leads us to the second major cosmological puzzle that inflation addresses: the monopole problem.

The Missing Monopole Problem & Inflation

The Missing Monopole Problem is another profound puzzle that arises from our best theories of the early universe. In simple terms, a magnetic monopole is a hypothetical, stable particle that has only one magnetic pole—a north without a south, or a south without a north. Grand Unified Theories, which attempt to unite the fundamental forces, predict that the extreme heat of the Big Bang should have created these particles in enormous numbers, perhaps as many as there are protons and neutrons. If this were true, these massive particles would be a primary component of the cosmos today. And yet, despite decades of searching, we have never definitively observed a single one.

This stark contradiction between theoretical prediction and observational reality is another of the things that inflation theory elegantly solves. It proposes that if magnetic monopoles were produced before the period of inflation, the subsequent exponential expansion of the universe would have diluted their density to an undetectable level. In essence, inflation would have pushed them so far apart that the chance of finding one in our observable universe today would be practically zero.

As with the horizon and flatness problems, this is a serious and elegant solution. If we assume that inflationary theory is correct, it appears to resolve several of the most perplexing issues in modern cosmology in one fell swoop.

Problems with The Inflationary Hypothesis

The inflationary hypothesis was revelatory when it was first announced. It predicted the kind of strange, unusual and elegant solution that has been the hallmark of many great theories in the history of science. Advances that are not only major steps forward but noted for their beauty and how they seem to provide puzzle solving capacity for a range of unanticipated issues. In this regard, the inflationary hypothesis is nearly unrivaled. Today, close to 45 years after it was first introduced, in nearly every book and article which deals with the early universe, inflation is featured prominently and oftentimes presented as fact in the history of the early universe.

It would be difficult for the layperson to understand that inflation is far from an established fact, and that there is scant evidence that it actually took place. The primary evidence, or indication that it occurred is that it offers an explanation for several problems, while actual evidence that any of its core assertions for its operations are absent. It is an exceptionally robust anecdotal hypothesis.

Meanwhile in the intervening years the inflationary hypothesis has been severely tested, challenged, and confronted with many problems, many of which were unknown when it was introduced in the early 80's. It's taken time to deconstruct the theory, subject it to rigorous testing and reveal its weaknesses. Subsequently the defenders of inflationary cosmology have been backpedaling and one of the primary architects of inflationary cosmology, Paul Steinhardt, has put forward several incredibly strong reasons to question if inflation actually took place that we will explore in depth as part of our review. Several authors including Steinhardt state unequivocally, "cosmologists should reassess this favored paradigm and consider new ideas about how the universe began[7]."

While the theory of cosmic inflation is often presented as a key component in the history of the early universe, it is far from an established fact and faces several, perhaps fatal, challenges. The great irony is that in order to make inflationary cosmology work, the amount of "hand-crafted" fine tuning required is tremendous. Unlike the fine-tuning of a physical constant that we can observe, the fine-tuning of inflation is "hand-inserted within the theory" to ensure it produces the desired outcome. As physicist Roger Penrose bluntly assessed, this makes the idea a "fantasy".

The core problems with inflationary cosmology that require this level of intervention include:

The Fine-Tuning Problem: The theory requires numerous parameters—such as the strength and duration of the repulsive force that drives inflation—to be set to incredibly precise, arbitrary values for which there is no underlying physical justification. Critics like Paul Steinhardt have argued that "bad inflation is more likely than good inflation," and that a universe without inflation is more probable than one with it.

Lack of Definitive Evidence: The theory's one major testable prediction was the existence of specific gravitational waves in the Cosmic Microwave Background Radiation. Decades of increasingly precise measurements have failed to find any evidence of these waves.

The "Multiverse" Problem: A key consequence of most inflationary models is that once inflation starts, it never truly stops, creating an infinite number of other "bubble" universes. This "multiverse" is fundamentally unobservable and makes the theory non-falsifiable, moving it from the realm of science into untestable philosophy.

Despite these profound issues, inflation remains popular primarily because there is currently no better alternative explanation for the Horizon Problem. This is a classic example of a speculative idea persisting not because of the strength of its own evidence, but because of the depth of the mystery it purports to solve is so great, there are no conjectures that can challenge and supplant it.

These troubles are not recent. They've existed all along, the backlash has been delayed because, as is typical for popular theories that reinforce the dominant paradigm, it was hoped that many of these problems would be alleviated and resolved as the details of the hypothesis were worked out over the intervening decades.

Perhaps the clearest indication that the hypothesis lacks true robustness, is the notable absence of a Nobel Prize, a prize that seemed all but assured several years after inflationary theory was unveiled. This formal acknowledgement that the hypothesis has risen to the status of a theory has become interminably delayed and now appears permanently out of reach. Nonetheless, at present, understanding by the general public of these challenges has yet to filter through, and inflation still sits at the head of the table as a key component in the recitation of the early history of our universe.

Science often works in this manner. It's not enough to impeach a popular idea or mortally wound it. It must be buried by the force of a new idea that can stand in its stead. Then the supporters of the old idea have something new to migrate to, and this migration is what causes the old idea to wither and die. That new idea has yet to emerge, so the inflationary model clings on to life for the time being.

This critique of inflation is not merely my own; it is shared by a growing number of physicists, including some of the theory's original architects. To understand the depth of these issues, we will now turn to the specific evidence and expert analysis that supports this assessment.

The Case Against Inflation: An Expert View

We have previously referred to inflationary theory co-author and dissenter by Paul Steinhardt. He wrote an article in 2017 with co-authors Anna Ijjas and Abraham Loeb[8], outlining several challenges to the theory. Prior to this Dr. Steinhardt co-authored two additional articles for Scientific American in 2011 and 2014[9] where he went into depth with numerous granular problems with inflationary cosmology. From his analysis, there appears to be no viable path forward for the concept, and continual embrace and touting of the theory as if it were cannon is misguided. Let's take a moment to summarize the key issues.

Extreme Fine Tuning

A primary weakness of inflationary theory is that it requires an extreme level of fine-tuning. The parameters of the hypothetical energy field that drives inflation are not fixed by any deeper principle, meaning they can be set to almost any value, and indeed, for inflationary cosmology to result in our present-day universe, they must be set very precisely. Added to this fine-tuning conundrum, the vast majority of the values that inflation may take place under would result in a chaotic universe entirely unlike our own and inhospitable to any possible form of life.

To produce the specific, uniform universe we actually observe, the theory's central parameter must be set by hand to a value of incredible precision (accurate to within 1 part in $\mathbf{10^{15}}$). Without any fundamental or physics-based theoretical justification for why this specific value should be chosen over any other, the theory loses its predictive power and becomes an exercise in fitting the data, rather than explaining it.

No Inflation Is Far More Likely Than "Good" Inflation

According to calculations by Roger Penrose, the chance of a flat universe (thus solving the flatness problem) without inflation is more likely, by orders of magnitude without inflation. As in the paragraph above, the fact that the parameter needed for good inflation is so highly specified within the space of total possibilities makes it wildly unlikely.

An Awkward Corollary

Paul Steinhardt, one of the fathers of the inflationary model, has stated "The first dictum holds that inflation is inevitable. But if it is, there is an awkward corollary: bad inflation is more likely than good inflation." And "not only is bad inflation more likely than good inflation, but no inflation is more likely than either. University of Oxford physicist Roger Penrose first made this point in the 1980s[10]."

A Different Angle

Gary W. Gibbons and Neil G. Turok confirmed Penrose's conclusion from a different angle. They conducted a theoretical exercise: if you take our current universe as an endpoint and calculate all the possible cosmic histories that could have led to it, what do the majority of those histories look like? Their work showed that an overwhelming number of these possible pathways involve insignificant amounts of inflation. This reinforces the core point: according to the probabilities, a universe like ours is more likely to have occurred without a period of cosmic inflation than with it.

The Problem of Quantum Indeterminacy

Seeing as inflation is essentially quantum in nature and governed by quantum indeterminacy (random), and is endless, with a set of infinities as to the kind of universes that can come out of it, this renders the entire concept very difficult to test. The process has so many possible outcomes that it could theoretically match any possible observation or measurement, therefore making it non-falsifiable, and non-starter for candidacy as a scientific theory, in the minds of most scientists.

Quantum Multiverse Implications

A crucial issue with inflationary theory is that it is fundamentally a quantum process. While early, naive models treated it as a classical, predictable event, the reality is that it is governed by the probabilistic rules of quantum mechanics. This creates an untenable problem for the theory's explanatory power. Because the process is quantum and random, it would not produce just one universe, but an infinite "multiverse" containing every possible outcome. Therefore, inflation cannot make a definitive prediction or provide a specific explanation for why our particular universe, with its specific set of laws and properties, came to be. It can only say that our universe is one of an infinite number of possibilities. This reduces the explanation for our universe to the weak anthropic principle, which states that because we need a universe like our own to survive, however unlikely, we should not be surprised to find ourselves in a universe like our own. Or, as Brandon Carson stated, "What we can expect to observe must be restricted by the conditions necessary for our presence as observers[11]."

An Absence of Evidence

As illustrated in two of our points above, even the most likely inflationary scenarios are less likely to create our universe than under the standard big bang universe. Added to this, recent observations of the CMB do not provide evidence of inflation. B-mode gravitational-waves, thought to be indications of the existence of inflation, are so far non-existent. These waves are predicted to exist in the most likely inflationary scenarios. Data from COBE (1992), the Planck satellite (2013, 2015), and the Bicep 2 (2014) and Bicep 3 (2018) rule the simplest and most likely inflationary scenarios entirely.

B-mode gravitational waves are unlike gravitational waves produced as in when black holes merge. B-mode waves are the size of galaxy clusters. The prediction of B-mode gravitational waves is the one prediction that more than any other would have been capable of substantiating inflation. Rather than resulting in the complete death of the hypothesis, because the values of inflationary cosmology can be tweaked endlessly, inflation can be made by its proponents to "match" *the lack of gravitational waves*, in effect, rendering them undetectable and thus unfalsifiable[12].

Inflation's Pandora's Box

A critical development in inflationary cosmology is the realization that, due to quantum effects, once inflation starts, it is eternal. While it may stop in our "bubble" of the universe, it continues to rage in the space beyond, constantly spawning an infinite number of new universes in a "multiverse" of unimaginable proportion.

This leads to a perverse and paradoxical outcome. Because this multiverse contains an infinite variety of outcomes, it becomes statistically inevitable that a universe exactly like ours would eventually arise by pure chance. However, it is also true that our specific, finely-tuned universe would be dwarfed by an unending number of chaotic, lifeless ones, making our own existence an event of almost zero probability within the multiverse itself.

The chaotic, unending, all-possibilities-realized multiverse demand for inflationary theory is the ultimate "Pandora's box" of the theory. Once opened, it can never be closed. Inflation's greatest strength—its ability to explain anything—becomes its greatest weakness, as it cannot be contorted to shed its most problematic feature: the eternal, unstoppable, and untestable creation of an infinite multiverse.

This, then, is the ultimate implication of eternal inflation. The theory that was initially thought to be such an elegant solution to the fine-tuning of our cosmos is revealed to be a brutal sledgehammer. It replaces the profound mystery of our universe's specific, life-giving properties with a simple, unfalsifiable appeal to infinity: in an endless multiverse, our existence is no longer a miracle of fine-tuning, but merely a statistical inevitability from a theory that we can neither confirm nor disprove.

Gaps in the Theory

An example of the lack of predictive power of inflationary scenarios is illustrated by a short list of the many things it *can't* predict: the number and arrangement of galaxies; the degree to which space is curved or warped; the amount of matter and other forms of energy that make up the universe; why the big bang happened; and, how the initial patch of the big bang that was our universe evolved into our current universe.

Reliance of Hypotheticals

The high-density energy of gravitational repulsion that drives inflation is purely hypothetical. There is no direct evidence that it exists, and even less is known about how it starts, why it starts, the strength of the energy that drives it, how it ends and why it ends. There are therefore currently *hundreds* of different proposals for what this repulsive force is, and how it can result in our current universe. These proposals are compounded by the infinite space of possible values for what this force, "Q" can be tuned to. It can be zero, or very large, but if it's greater than 1, then the universe it creates is dominated by black holes from the outset[13]. This means the value of Q must be both *finely tuned,* and has *endless possibilities,* thus replacing the fine tuning needed for homogeneity of the CMB in the hot big bang model for a series of fine tuning problems in the inflationary universe.

The Miraculous Hand-Off

The imaginary inflaton particle is tasked with perhaps the most important job in the history of the cosmos. Not only must it initiate and then gracefully end the period of inflation, but it must also decay in a very specific way. In this final act, it must perfectly transition its unknown, hypothetical energy into the entire known reality of our universe—all the matter and radiation we observe today. How this could happen is anyone's guess. The theory simply requires that it does, without providing a mechanism, papering over a miracle with a necessity.

The Response to the Criticism

In response to the expansive criticism of inflationary theory, most of it covered in the Scientific American article by Paul Steinhardt, Anna Ijjas and Abraham Loeb[14], a group of 33 eminent physicists penned a response in Scientific American to challenge the article[15] Among these physicists were some of the biggest names in science: Stephen Hawking, Andrei Linde, Alexander Vilenkin, Alan Guth, George Smoot, Steven Weinberg, and Frank Wilczek.

Surprisingly, in their response, there is very little rebuttal of the major criticisms. They don't attempt to dismiss most of the problems, and they don't contend that the criticisms are unfounded, instead their primary point of contention is that the criticism fails to acknowledge that work on inflationary cosmology is "fecund" in the sense that over 14,000 papers have been written about it by over 9,000 scientists. The models are, in their view, "successful empirically" and they contend that the data from experiments "match" with inflation very well. What is left unaddressed is that the key mechanism, the *inflaton* field, has *zero* evidence that it exists, and, in their response they fail to mention that, the reason why the models match observation is that they've been tweaked and adjusted to ensure that they match observation.

In the letter, the lack of an inflation reality vs the existence of an inflation hypothesis is defended by drawing parallels to the standard model of physics and how many of its particles were deduced and tuned by observation. But this comparison is apples and oranges. As physicist Sabine Hossenfelder has commented, "... the situation for inflation is not remotely comparable to the standard model of particle physics... In this case, the input of the models is actually more complicated than the output. This means they're crappy models without explanatory power. That you use the same procedure as for the standard model is completely irrelevant[16]."

The ultimate defense of inflation often retreats into the unobservable landscape of the multiverse. This is not science, but a philosophical sleight of hand: it attempts to explain the miraculous fine-tuning of our single, observable universe by dissolving it in an infinite sea of unobservable ones where every outcome is inevitable. The weakness of this position was made clear when 33 of the world's most eminent physicists, in response to a direct critique of the theory, could not rebut the charges with evidence, but instead resorted to an argument from popularity. This, perhaps more than anything, reveals that cosmic inflation is a paradigm in crisis, propped up by the authority of its defenders rather than the strength of its own scientific merit.

Value of The Horizon Problem

The Horizon Problem has no physics-based solution, and inflationary cosmology, the best available explanation for *The Horizon Problem* has been shown to have a number of unanswered, and perhaps unanswerable difficulties. According to calculations by Roger Penrose, in his own decades long critique of inflationary theory, the chance of a flat universe without inflation is essentially the same result at the initial entropy state, which is derived from the information provided by the CMB

The Penrose number for the low entropy state, is derived from the thermalization of the CMB and is given as $\mathbf{10^{10^{123}}}$. This exceeds the PCOU. Roger Penrose has also calculated that our universe is $\mathbf{10^{100}}$ *more* likely without inflation than with it[17]. And to make inflationary cosmology work at all, it has several free parameters that must be set by hand. There is no physical first principles physics based justification for their values. The inflation field has yet to be verified as anything but fantasy.

These free parameters include when inflation starts, stops and the strength of the anti-gravity "spring" that drives inflation. The number of possible combinations for these free parameters is immense, exceeding the benchmark of $\mathbf{10^{140}}$ established as the total possibilities in our Causal Operating Universe (PCOU). The value of *The Horizon Problem* must therefore be considered a functional miracle.

Value of Variable #3, The Horizon Problem: *fM Functional Miracle*

The Horizon Problem presents a unique challenge within our framework. Unlike other variables, its status as a "miracle" depends on how one interprets the failure of its proposed solutions.

On one hand, if we consider the uniformity of the early universe as simply an incredibly unlikely starting condition that the standard Big Bang model cannot explain, it qualifies as a Remote Possibility (RP). On the other hand, if we accept that the leading physical explanation for this uniformity—cosmic inflation—is not just flawed but fundamentally unworkable, then we are left with a necessary feature of our universe for which there is no known physical cause. In this light, it must be considered at present, a Miracle (M).

Either way, the implication is eye-opening. Our investigation has shown that the universe began in a state of near-perfect thermal equilibrium, a condition so improbable that it spawned the theory of inflation. We then saw how inflation, despite its elegance, fails to provide a viable, testable solution and instead introduces an even greater degree of fine-tuning. We are therefore left with another foundational condition of our cosmos that appears to be exquisitely and inexplicably set for the emergence of complex structures and, ultimately, life. This is another crucial thread in the pattern of inexplicable conditions that reveal The Divine Algorithm.

---

[1] (M. Rees Just Six Numbers, Basic Books, 1999 pgs 74-5)9
[2] (A. Liddle, An Introduction to Cosmology, Wiley, 2003, Pg. 101)10.
[3] (A. Guth, The Inflationary Universe: A Possible Solution to the Horizon and Flatness Problems*, Pg. 3, SLAC-PUB-2576, July 1980)
[4] (High Energy Astrophysics Science Archive Research Center, A. Ptak Director, NASA/GSFC, Inflation in the Universe, 1993, NASA's Cosmic Times, Imagine the Universe, imagine(dot) gsfc(dot) nasa(dot) gov/ educators/ programs/ cosmictimes/ educators/ guide/ 1993/ inflation(dot) html#)
[5] Weight of the Vacuum, 2014, SpringerBriefs in Physics. Springer, Berlin, Heidelberg. doi(dot) org/10(dot) 1007/ 978-3-642-55090-4_9), (H.S. Kragh, J.M. Overduin, Inflation and the False Vacuum. In: The
[6] (Adapted from P. Coles, 1999, The Routledge Critical Dictionary of the New Cosmology, Routledge Inc., New York. ned(dot) ipac(dot) caltech(dot) edu/ level5/ glossary/ essay_ flatpr(dot) html)
[7] (A. Ijjas, P.J. Steinhardt, A. Loeb, Cosmic Inflation Faces Challenges, Scientific American, 1/1/2017).
[8] (A. Ijjas, P.J. Steinhardt, A. Loeb, Cosmic Inflation Faces Challenges, Scientific American, 1/1/2017)21

[9] (P.J. Steinhardt, Paul Steinhardt Disowns Inflation, the Theory He Helped Create, Scientific American, 8/1/2014), (P.J. Steinhardt, The Inflation Debate, Scientific American, 31/2011)22
[10] (P.J. Steinhardt, The Inflation Debate: Is the Theory at the Heart of Modern Cosmology Deeply Flawed?, 2011, Scientific American)
[11] (M.S. Longair, editor, Large Number Coincidences and the Anthropic Principle in Cosmology, Confrontation of Cosmological Theories with Observational Data, IAU Symposium No. 63, 1974, Pgs 291-298. Dordrecht: Reidel Publishing Company)
[12] (Paul Francis, Astrophysicist, Australian National University, V3.9 Problems with Inflation, Paul Francis YouTube Channel, published 7/1/2020)23
[13] (N. Turok, A Critical Review of Inflation. Classical and Quantum Gravity, 19, 3449, [2002])
[14] (A. Ijjas, P.J. Steinhardt, A. Loeb, Cosmic Inflation Faces Challenges, 2/1/2017)25
[15] (A Cosmic Controversy, a Scientific American article about the theory of inflation prompted a reply from a group of 33 physicists, Scientific American, 5/10/2017)26
[16] (Sabine Hossenfelder, physicists, research fellow, Frankfort Inst. Advanced Studies, (channel), Did The Universe Inflate? Youtube, 3/5/2022)27
[17] (R. Penrose, Difficulties with Inflationary Cosmology, 1989, essay, Annals of the New York Academy of Sciences),(R. Penrose, Cycles of Time, 2010, Bodley Head).

Variable 4: The Cosmological Constant

Introduction To The Cosmological Constant

We now arrive at the final subject in our investigation of the universe's initial conditions: the Cosmological Constant.

The topics in this first part of our investigation—from the singularity and the survival of matter to the initial entropy and the horizon problem—are not disconnected puzzles. They are all different facets of the same foundational event, the "Genesis Event" that set the stage for our entire reality. The Cosmological Constant is the final, crucial piece of this initial setup, a value so exquisitely fine-tuned that it, perhaps more than any other, dictates the ultimate fate of the universe and the possibility of life within it.

As our final subject in this section, the Cosmological Constant presents a fine-tuning problem similar to, but distinct from, the Flatness Problem we've previously discussed.

To briefly recap, the Flatness Problem deals with the density of matter and energy in the universe (a value called Omega). For our universe to be "flat" and capable of forming structures, Omega had to be set at the beginning of time with an incredible precision of one part in $\mathbf{10^{62}}$.

The Cosmological Constant (represented by the Greek letter Lambda) is similar. It represents the energy density of the vacuum of empty space—what is now often called "dark energy"—and it also governs the expansion rate of the universe. Just like Omega, Lambda's value must be exquisitely fine-tuned. These two values, one governing matter and the other governing the vacuum, must work in perfect concert to produce a life-permitting cosmos, yet there is no known physical reason why they should be so precisely aligned.

The expansion rate of the universe, which is separate and distinct from the cosmological constant, but for simplicity is often discussed as being the same due to the asymptotic relationship between the two, is also influenced by the gravitational constant, as well as the density, distribution, structure of the universe, etc. There are multiple factors that are considerations and influences upon the expansion of the universe and these must be in alignment and it is imperative that when all of these factors come together, the expansion rate and cosmological constant work together within a precise, finely tuned co-contribution/collaboration in order for our universe to have any possibility of being life-inhabitable.

The expansion rate of the universe is predicated upon several critical factors coming into a "perfect" state of alignment. With "perfect" being defined by one single characteristic: the alignment necessary for life. This alignment is, even in its consilience with life, far beyond the reach of chance, because the critical factors that must come together have *no reason* to be in alignment. And, they are, in and of themselves, independently fine-tuned, and independently critical to the requirements of life.

Under the paradigm of materialism, there is no reason why these factors should, could, or would be in alignment. The condition could therefore be described *as if* they were set by a form of consciousness, that acted in 3 stages to:

(1) create the universe,

(2) select values for the physics of the universe to be in the range required for life to be *a possibility*, and then

(3) went about selecting the precise values required for our universe to express this possibility as the only possible reality, despite it being the exact opposite (an impossible and beyond the reach of chance outcome).

That is the great irony of our universe. It is one in which life is both possible and, from a materialistic standpoint, *impossible*, because what has to happen for life to exist is a series of seemingly miraculous outcomes from the perspective of physics and the probabilistic capacity of the universe. In short, materially, they cannot happen and therefore cannot be explained.

An analogy might be if a blind man with no experience of tightrope walking, could walk a wire strung from the Chicago Tower to the Burj Khalifa in Dubai, in hurricane force winds. While theoretically possible, the chance of the man successfully navigating all that way is vanishingly small, to the extent that if the man succeeded, the rational conclusion would be that he was somehow guided throughout his journey. The universe, I argue, presents us with a similar and exponentially more unlikely sequence of nigh-impossible events.

According to our current paradigm of physical materialism, just as it would be deemed impossible for a blind man to complete our tightrope walking feat, there is similarly, no reason whatsoever for the critical factors that go into the expansion rate of our universe, inclusive of the cosmological constant, to be in an alignment that results in our universe being life allowing, *particularly* when we consider the other factors in Part 1 of our investigation that also must be in the range of life-allowing, for the expansion rate to have any hope of being in a life allowing range.

The cosmological constant therefore exerts a huge influence over the destiny of our universe. In *The Privileged Planet*, Guillermo Gonzalez and Jay Richards describe the cosmological constant as the ratio between the intrinsic energy density attributable to the vacuum of space (the dark energy or vacuum energy density), and the critical density of the universe itself (matter energy density[1]) Essentially every place in the universe where matter isn't, vacuum energy exists, and this energy has remained the same throughout the history of the universe, and will continue to remain so, however, as the universe expands, because there is more empty space, there is more vacuum energy contributing to the overall effect.

This means the cosmological constant although being a constant, overall contributes increasing amounts of energy in our universe every additional moment in time. This shifting value for the vacuum energy (dark energy) in being both ever-changing, and ever evolving, is the most finely tuned *evolving* constant yet discovered.

Near the beginning of the universe, close to the age of the CMB, the dark energy contribution was so small in the universe it was insignificant, yet still vital. Now, with the universe having expanded for close to 14 billion years, dark energy by our calculations is by far the greatest single component of the energy of our universe, constituting over 60% of the energy of the universe.

We can summarize the relationship between the expansion rate of the universe and the cosmological constant by saying that the expansion of the universe is "driven" by the cosmological constant[2]. and although the expansion of our universe has many factors and influences, as we stated above, there is no understanding, or physical reason that can be given for why our universe expanded at all even within the idea of the inflationary universe.

It is this specific rate that allows for the formation of stars, the creation of heavy elements, and the existence of planets. This alignment is balanced on a razor's edge. If the expansion were even a fraction faster, matter would have dispersed too quickly for stars and galaxies to form; a fraction slower, and the universe would have collapsed back on itself long before life could emerge. Intuitively it may seem that the expansion rate of our universe that would allow for the creation of planets, stars and life would be a reasonably manageable range to stay within, but the growing preciseness of our science has revealed that this intuition is entirely false. In truth the expansion rate of our universe which is conducive to the creation of planets, stars and life is among the most finely tuned parameters in the history of science.

The precision required for our universe to exist in this "Goldilocks" state is staggering. The expansion rate must be fine-tuned to an accuracy of one part in $\mathbf{10^{120}}$—a level of specificity for which there is no known physical explanation, other than that it is the value required for us to be here.

It is difficult for us to appreciate how fine-tuned this number is, as our daily lives require no precedent close to this level of fine tuning. But there are some finely tuned, highly precise real-world numbers we can refer to. 1 part in $\mathbf{10^{120}}$ is a level of fine-tuning that is far beyond the accuracy of one of the most finely tuned and precise pieces of technology humanity has ever created, -the world's most accurate clock- with its accuracy record established in early 2019 in Boulder, Colorado. This clock, "a fractional-frequency uncertainty of less than one part in $\mathbf{10^{18}}$"[3] That means this clock is accurate to 0.000000000000000001 second, or within 1/1000 of a trillionth of a second. If this clock had been in operation since the beginning of the universe, then today it would have lost less than ½ of a second.

That's impressive. But by contrast the cosmological constant is *102 decimal places* more precise than the world's most accurate clock. This means the cosmological constant is a trillion, times a trillion, times a trillion… (a total of 32 "times a trillion") times *more* precise than the most precise instrument humanity has the ability to construct. The cosmological constant far exceeds the technological abilities of humanity to create. At present we only have the technological ability to calculate this level of precision, but not to replicate it.

But the fine-tuning required for a life-permitting universe does not stop with the Cosmological Constant. This value must work in perfect concert with other fundamental parameters, such as the initial density of the universe. As author Stephen Meyer explains, this initial density also had to be set to an incredible degree of precision:

> "A life permitting universe depends crucially on its precise expansion rate. Since the discovery of the red shift of the light coming from distant galaxies astronomers have discovered that if the universe were initially expanding even a smidgeon faster or slower, either stable galaxies would not have formed in the universe because matter would have dissipated too quickly [known as the heat death of the universe] for galaxies to congeal or else the universe would have quickly collapsed in on itself [known as the big crunch]... the expansion rate in the earliest stages of the history of the universe would have depended upon the density of mass and energy at those early times. And the density of the universe one nanosecond (a

> billionth of a second) after the beginning had to have been the precise value of $\mathbf{10^{24}}$ kilogram per cubic meter. If the density were larger or smaller by only 1 kilogram per cubic meter, galaxies would never have developed. This corresponds to a fine tuning of 1 part in of $\mathbf{10^{24}}$"[4]

In addition to the initial density, the strength of gravity itself—determined by the gravitational constant—must also be exquisitely fine-tuned. This value must be precise to one part in $\mathbf{10^{60}}$ to allow for the expansion of the universe, and to a separate precision of one part in $\mathbf{10^{35}}$ to allow for the thermal layering inside of stars necessary to produce life-essential elements like carbon. If gravity were slightly stronger, stars would burn too hot and too fast; slightly weaker, and they would never ignite or explode as supernovae, preventing the distribution of heavy elements necessary for building planets and life.

The History of The Cosmological Constant

The historical path to the Cosmological Constant is rooted in a simple, unavoidable fact we discussed early in our investigation: the theories of both Newton and Einstein predicted a dynamic universe that should either be expanding or collapsing. The eventual, overwhelming evidence for an expanding universe logically implied a beginning—a singularity. This discovery, however, raised a host of profound questions that science is not equipped to answer: Why was there a singularity? Where did it come from? What caused it to expand? And ultimately, what is the purpose of it all? It is in the attempt to answer these questions that the true nature of our finely tuned reality, and the role of the Cosmological Constant within it, comes into focus.

Both Newton and Einstein had only 2 choices; either the universe was expanding, or it was contracting. The choice should have been clear. An *eternal* universe that had been contracting would have ended in a big crunch an infinite time ago. There is no result within a contracting universe that does not result in a singularity at a finite point in time. The fact that the universe existed to be observed was an additional clue that it had to be expanding. Had Einstein predicted this, it would have added to his list of transcendent predictions, and it would have been experimentally verified. Instead, it was his belief in a simple, eternal, straightforward universe, as envisioned in the notion of heavenly spheres, where planets were held in an eternal static state, due to some imagined perfection of math and geometry, that led him to ignore science and choose belief disguised as intuition. But in order to ignore the likely expansion of the universe and make the universe static, he needed to insert a term into his equations that could reverse the expansion predicted by his equations. *Lambda* therefore made its first appearance in Einstein's theory of special relativity as the nonexistent, hand-inserted balancing factor Einstein invented to counteract what he, Isaac Newton, Georges Lemaître and many others realized would result in the unavoidable gravitational attraction, clumping and collapse of matter in our universe.

Einstein finely tuned the value of lambda to provide the precise amount of force to counteract gravity's attraction to avoid clumping and collapse.

> "(Einstein)... modified his equations of general relativity, adding what he called a cosmological term- a kind of universal repulsion that prevents the uniform distribution of matter from collapsing under the normal force of gravity[5]."

Einstein's cosmological constant (Lambda) was meant to act as a kind of cosmic anti-gravity—a universal repulsive force precisely calibrated to counteract the pull of gravity and hold the universe in a perfect, static balance.

The Irony of Einstein's Blunder

The history of the Cosmological Constant is one of the great ironies in science. As we reviewed in the last chapter, the work of physicists from Friedmann to Hubble provided overwhelming proof of an expanding universe, forcing Einstein to abandon the static model he preferred. He famously dismissed his own invention of Lambda—the "fudge factor" he had inserted into his equations to hold the universe still—as his "greatest blunder."

The ultimate irony, however, is that Einstein was wrong twice. Lambda does exist, but its real-world function is the exact opposite of what he intended. The Cosmological Constant is not a force that holds the universe static; it is the driving force behind its accelerated expansion.

"The Worst Prediction in the History of Physics"

The Cosmological Constant, in addition to being a "blunder" when created out of whole cloth by Einstein, in other scientific circles is the subject of what has been called "the worst prediction in the history of physics". This refers to the staggering disagreement between our two most successful scientific theories. On one hand, quantum mechanics predicts that the "vacuum" of empty space should be fizzing with a tremendous amount of energy. On the other hand, when astronomers measure the actual expansion of the universe, they find that the energy of this vacuum is incredibly small—about 120 orders of magnitude *smaller* than the theoretical prediction. This spectacular failure of theory to match observation is a foundational crisis in modern physics and a profound fine-tuning mystery.

To calculate the theoretical energy of empty space, physicists must sum up the energy of all the "vacuum fluctuations" predicted by quantum mechanics. The problem, as dictated by Heisenberg's uncertainty principle, is that the vacuum is a chaotic sea of "virtual" particles popping in and out of existence. The principle allows for incredibly high-energy particles to appear, as long as they disappear in an incredibly short amount of time.

When physicists attempt to add up the energy of all these fleeting virtual particles, the total sum approaches infinity! This means that, according to our best theories, the potential energy contained within empty space should be almost infinitely large—so large that the energy of all the matter in the entire observable universe would be negligible in comparison.

To prevent this theoretical energy from becoming infinite, physicists apply a "cutoff" to their calculations. They use the Planck scale—the smallest meaningful unit of length and time—as a non-arbitrary limit. The reasoning is that our current understanding of physics, including Einstein's theory of spacetime, breaks down at this fundamental scale, so it is a logical place to stop the calculation. While this practice successfully changes the predicted value from "infinite" to a finite number, that number still represents a tremendous amount of potential energy, far greater than what we observe.

Calculating the energy of empty space, the predicted energy density of the vacuum is $\mathbf{10^{114}}$ joules per meter cubed. This has the *mass* equivalent to $\mathbf{10^{97}}$ kilograms per meter cubed. This is particularly significant, because the total amount of energy from matter, calculated to exist within our universe, is far less than to $\mathbf{10^{97}}$ kilograms per meter cubed. It is approximately $\mathbf{10^{53}}$ kg. Consequently, the amount of energy which is theorized to exist in "nothingness" exceeds the amount of energy of "somethingness". This is *astounding*. Physicists John Wheeler and Richard Feynman famously calculated that based upon the energy potential of empty space, "one teacup of empty space contains enough energy to boil all the world's oceans[6]."

This enormous sum gives us our original figure that the predicted energy value of empty space should be 120 orders of magnitude *larger* ($\mathbf{10^{120}}$) than the observed value of dark energy (empty space), and this is the reason why the *predicted* value of energy in empty space, in contrast to its *observed* value, is known as "the worst prediction in the history of physics." Because the difference between the quantum mechanical prediction for the energy of empty space differs from the prediction made from the perspective of general relativity, which is very close to, but not zero, -the actual *observed* value of the energy contribution from the cosmological constant is $\mathbf{10^{-29}}$ / grams per centimeter cubed, which is 0.00000000000000000000000000001 grams per centimeter cubed[7]. These two numbers differ by 120 orders of magnitude; a discrepancy so vast that it is larger than the total number of atoms in the entire observable universe.

The Energy of the Void: A Cosmic Puzzle

Before we go further, let's take stock. We are faced with the single worst theoretical prediction in the history of physics—a value for the universe's vacuum energy that is off by a factor of $\mathbf{10^{120}}$. The great difficulty, as we've just noted, is that this prediction emerges from our two most successful and well-tested theories: General Relativity and Quantum Mechanics. This catastrophic failure at their intersection has forced physicists to question their most basic assumptions about spacetime and energy. In the following sections, we will explore the leading attempts to resolve this crisis, from modifications to gravity to radical new ideas like the multiverse.

As we stated above, the failed prediction for the cosmological constant is based on two of the most successful, and empirically validated theories in science, Einstein's General Relativity and Quantum Mechanics.

According to quantum field theory, particles, -electrons, protons, quarks, etc.- can be thought of as excitations in the quantum field. Each particle in the standard model of physics is represented by its own field, and all of these fields exist throughout all space, which means they exist in empty space. Because of this, "empty" space can never be said to be truly "empty" in keeping with Heisenberg's uncertainty principle.

This principle implies that the energy level of a space within a field *can never be zero. Potential* energy exists within zero-point vacuum energy, and this energy is *real* and contributes to the total energy of our universe. This energy is also *repulsive* as opposed to attractive as in the force of gravity. Because it is empty space and has energy, it has been identified as, and referred to as "dark energy", and this is what drives the expansion of our universe. Dark energy is otherwise known as, and frequently referred to as, *the cosmological constant.* And to put a bow on this entire review, as we stated earlier, counterintuitively, as the universe expands, more "empty" space is created, and thus, increasing amounts of dark energy are created out of "nothing". Therefore, it is predicted that the accelerated expansion of our universe will grow, adding trillions of additional "teacups" (as described by Richard Feynman) full of Earth-ocean-boiling-energy to the universe from now until the heat death of our universe.

The accelerating expansion of our universe has significant future consequences in 20 to 30 billion years' time. As the expansion accelerates, the most distant objects in our universe, beginning with the cosmic microwave background radiation, will *fade* from our view and disappear beyond the veil of what for us is the observable universe as its signal is stretched beyond our ability to ever receive it. This accelerating expansion creates a cosmic event horizon, a boundary in spacetime similar to that of a black hole. As distant galaxies are carried away from us faster than the speed of light, they will cross this horizon, and the light they emit after that point will never be able to reach us. The profound implication of this is that in the distant future, the sky will grow dark. All evidence of other galaxies, and even of the Big Bang itself as seen in the Cosmic Microwave Background Radiation, will eventually fade from view, lost forever to the arrow of time.

The implications of this future are profound for the present day. We find ourselves living in a very special, privileged epoch in cosmic history—not only are the conditions right for our existence, but they are also perfect for discovery. A future civilization of intelligent observers, born in the lonely, dark universe to come, would have no observational way of knowing about the Big Bang or the existence of other galaxies. They would look out into an empty void and, through no fault of their own, likely conclude they were alone in a static, eternal cosmos. The very fact that we are able to read the pages of our own cosmic origin story is a profound coincidence, and another thread in the pattern that suggests our place in the universe is not an accident.

The "Small" Mystery

The immense gap between the predicted and observed vacuum energy is only the first part of the puzzle. A second, more subtle mystery arises when we consider the measured value itself. As cosmologists P.J.E. Peebles and Bharat Ratra have emphasized, the challenge is not just to explain why the cosmological constant is so small, but to understand why it has the specific, non-zero value that it does.

> "A decade ago, the high-energy-physics community had a well-defined challenge to show why the dark-energy density vanishes [from its incredibly large, predicted value, to its exceedingly small, measured value]. Now there seems to be both a new challenge and clue: determine why the dark-energy density is exceedingly small but not zero[8]."

Like many of the concepts we have encountered thus far, there is no existing physical law, theory, or hypothesis available to help us make sense of why the estimate of the observed value of the cosmological constant is so fantastically small.

> "Most cosmologists hoped that some undiscovered law of physics required (the cosmological constant) to be exactly zero. They already knew that its observational upper limit was much smaller than the natural values expected from various particle fields and other theoretical fields. These particle fields require an extraordinary degree of fine-tuning–at least to one part in $\mathbf{10^{53}}$–to get such a small, positive nonzero value for the vacuum [dark] energy. At the same time, its value must be large enough in the early universe to cause the newborn universe to expand exponentially as inflation theory postulates. How the present value of the vacuum [dark] energy relates to the early expansion is yet another issue of debate[9]."

Under Einstein's theory of relativity, which is based upon classical physics, the cosmological constant is *zero*. Therefore there are many when faced with the strange nature of dark energy, and our inability to scientifically validate that it is the driving force of the cosmological constant, that believe the cosmological constant should be, and is zero, and that the acceleration of our universe rather than proving the influence of dark energy, instead shows that Einstein's theory of relativity needs to be modified and the force of gravity needs to be modified, resulting in the cosmological constant being canceled out. Theories that are of this nature are called *modified theories of gravity*[10].

These "modified theories of gravity" represent another attempt to solve a profound cosmic mystery by changing the rules of the game. The argument is that if the observed cosmic acceleration isn't caused by a finely-tuned dark energy, then perhaps our understanding of gravity itself is incomplete on the largest scales. However, to date, no modified theory of gravity has been able to successfully explain the full range of cosmological observations—from the precise details of the Cosmic Microwave Background Radiation to the patterns of galaxy clustering—as well as the standard model that includes the Cosmological Constant. As such, they remain speculative alternatives that have failed to resolve the fundamental problem: our universe appears to be exquisitely and inexplicably fine-tuned for life.

Astrophysicist Geraint Lewis and Astronomer Luke Barnes, authors of, *A Fortunate Universe*, cast the cosmological constant problem as a perfect storm of at least eight problems, including:

-The energy contribution of *each* quantum field, (corresponding to each fundamental particle).
-The inability of general relativity to address the energy and momentum of the universe.
-The inability of particle physics to address the effects of gravitation (the fact there is no quantum theory of gravity).
-The fact that the driving force of expansion may not be dark energy at all.

On top of all this, there is another baffling mystery: we do not know if "dark energy" is even a real, physical entity. As physicists Geraint Lewis and Luke Barnes point out, the term is essentially a convenient shorthand for "whatever makes the universe's expansion accelerate". It is a placeholder for a complete unknown. The crucial point is that something is causing the universe to expand in a very specific, finely tuned way, but whether that "something" is a new form of energy, a property of the vacuum, or a sign that our theory of gravity is incomplete, remains a mystery. At present, dark energy functions as a necessary fudge factor in our equations—a label for an observed effect, not a proven cause.

A Universe in Overdrive: The Discovery of Cosmic Acceleration

Michael S. Turner coined the term "dark energy" in the title to his 1998 paper titled, *Dark Matter and Dark Energy in the Universe*[11], In this paper, he used this term to understand what was driving the accelerated expansion of the universe. Also in 1998, two independent teams set out to investigate the expansion of the universe, fully expected to see, understand and determine how the expansion of our universe was *decelerating*, and therefore finding information sufficient to determine if there was enough matter in the universe to reverse this expansion because of the effects of gravity. What these two teams found instead is that the universe was *accelerating* in its expansion. In 2011 the Nobel Physics Prize was awarded to the leaders of the two teams who compiled the information that discovered and affirmed the presence of this strange form of energy.

The official reason for the prize is, "for the discovery of the accelerating expansion of the Universe through observations of distant supernovae[12]." Saul Perlmutter of The Supernova Cosmology Project, Brian Schmidt and Adam Riess of the High-Z Supernova Search Team were the 2011 physics prize laureates. Olga Botner, member of the Royal Swedish Academy of Sciences said of their discovery,

> "The distances between the galaxies should not be able to continue increasing as rapidly for all eternity – the expansion should

eventually slow down. It can be calculated that if the universe contains more than six atoms per cubic metre, this expansion must stop. The universe should start shrinking and eventually end with a Big Crunch… To everyone's utter amazement… Instead, the universe has been expanding faster and faster. What, then, is causing this expansion rate to increase? The answer might be a special form of energy, called dark energy, which is pushing the universe "outward". A similar form of energy – the cosmological constant – was discussed by Einstein as early as 1917 but was later removed from his theory… The discovery of accelerating expansion through studies of distant supernovae has thus changed our image of the universe in an unexpected, dramatic way[13]."

In her statement Olga Botner acknowledges that what is driving the acceleration is scientifically up for grabs and "might be… dark energy" The Nobel prize was therefore not awarded for the discovery of dark energy, rather, it was awarded for "the discovery of the accelerating expansion of the Universe" whatever that may be. To clarify, dark energy is *real,* what causes it, we do not know, and it remains to be confirmed that it is the driving force behind the accelerated expansion of the universe.

The discovery of the accelerating expansion was a revolution, but it raised a profound question: what is the universe actually made of? To get a clear answer, scientists needed a comprehensive "cosmic inventory." This came in 2001 with the launch of NASA's Wilkinson Microwave Anisotropy Probe (WMAP), a satellite designed to create the most detailed map ever of the Cosmic Microwave Background Radiation and unveil what the conditions were like in the early universe and to refine our understanding of the current universe by measuring the full CMB. The mission's data provided a stunning and unexpected picture of the universe's contents while also providing one of the first comprehensive assessments of the contribution of energies and forces within our universe. By 2010 the WMAP had largely completed its mission and made several important discoveries and clarifications to our understanding.

The age of the universe is 13.8 billion years

The expansion rate of the universe (Hubble Constant) is 69km-s

The content of the universe is:

5% ordinary baryonic matter, (i.e. atoms, humans, planets and stars)

24% dark matter

71% dark energy (a number that has been revised with continuing observations and measurements)[14]

The implications of the WMAP data once again fundamentally changed our understanding of the universe. For the first time, we had a precise accounting of the cosmos, only to find that 95% of it is composed of mysterious substances—dark matter and dark energy—that we do not understand. This new cosmic inventory not only deepened the mystery of our universe but also revealed the improbable puzzle known as the "Cosmic Coincidence," adding another layer of strangeness to the mystery of the cosmological constant, which we will now explore.

The Cosmic Coincidence

The *cosmic coincidence* or *the coincidence scandal* has its beginnings rooted in the inception of the universe up until 6 billion years ago, in the era when gravity, generated by matter and dark matter, has utterly dominated our universe. Beginning 6 billion years ago the near to zero force of dark energy began to exert its influence over the cosmos. To be clear, in the beginning gravity dominated the universe nearly to totality, but, even in this era, it did not have complete dominance, because at all times dark energy has had a hand in our fate.

Dark energy emerged from the infinite curvature and infinite density of the singularity, and because it is evenly distributed in all spaces, at all times, its repulsive energy has been essential in helping to shape the universe. Strangely, from 6 billion years ago, and until the end of time, our universe began to enter a new phase where the influence of dark energy has begun to rival that of gravity. At present, the showdown between the forces of gravity and dark energy are essentially equal. In the future, dark energy will wield a far greater influence, eventually dominating matter in the same way matter dominated dark energy in the beginning. This means that in the present, we are witness to a strange coincidence between the influence of matter and dark energy.

This is the heart of the "Cosmic Coincidence." We find ourselves living in the one special, fleeting epoch in the universe's history where the two dominant forces that control the fate of the universe are in a near-perfect balance. Harkening back to the thesis of our universal algorithm, and our claim of the inevitability of its discovery, and the unusual quality of observability of cosmic issues that reveal the parts of our algorithm. It is also the only epoch in which intelligent observers could arise and develop the technology to discover this fact. A few billion years earlier, dark energy would have been too weak to detect; a few billion years in the future, it will have pushed all other galaxies beyond our cosmic horizon, hiding the evidence of the Big Bang forever. The fact that the one time in cosmic history, in a universe with the potential to be hundreds of thousands of decillions of years old (**10^{40}** to **10^{100}**) before the "heat death of the universe", that is presently "just right" for life is also the one time that is "just right" for discovery by humanity is a highly unlikely "coincidence" that demands an explanation.

> "Astronomers have determined that today (dark energy) contributes about as much to the dynamics of the universe as the gravitational attractive force from visible and dark matter combined. The coincidence remains unexplained… there's only one 'special' time in the history of the universe when the vacuum and matter energy densities are the same, and we're living very near it[15]."

The only reason we know that the force of gravity and dark energy is roughly equal is because dark energy's effects on the universe are now at a level where it can be detected by human technology. Remember that prior to the 1998 research by The Supernova Cosmology Project and the High-Z Supernova Search Team fully expected that gravity was the only force that played a significant factor in the expansion of the universe. It is only because the emergence of dark energy is at the level that it can be detected that we now know it exists. Now we know that dark energy consists of 70% of the energy of the universe, the rest of the matter density is made up of the remaining 30% (regular matter/dark matter). This makes the ratio between the two roughly 2.3 to 1. This ratio, as viewed by scientific circles, is within reach of one another. You may question how a ratio of 2.3 to 1 is considered "equal". The answer is when the universe emerged from the decoupling era roughly 380,000 years ago, when the CMB was created, the matter density of the universe wasn't 2.3 to 1 in favor of dark energy, it was 2 *billion* to 1 in favor of regular matter/dark matter. In the future the matter density will eventually be 1 to 2 *billion* and exponentially beyond this ratio in the favor of dark energy. This is why a ratio of 2.3:1 represents a special time in our universe where compared to the distant past and the distant future, which is essentially the entirety of the universe except during our current period, this ratio is essentially "equal". This is *the coincidence scandal.*

> "How finely tuned is it that we exist in the era when vacuum and matter are comparable? Between the Planck time and now, the universe has expanded by a factor of approximately… **10^{64}**… Thus, there is an approximately 1% chance that an observer living in a randomly selected logarithmic expansion interval in the history of our universe would be lucky enough to have Omega (matter) and

Omega (vacuum) be the same order of magnitude… the calculation here gives some idea of the fine-tuning involved; it is substantial, but not completely ridiculous[16]."

The consequences of this "coincidence" are far more profound than the term suggests; they are a matter of existence versus non-existence. The timing of dark energy's dominance over gravity had to be perfect. If dark energy had taken over a few billion years earlier, its repulsive force would have overwhelmed gravity, preventing matter from ever clumping together. As physicist Lawrence Krauss notes, "galaxies would've never formed[17]." Without galaxies, there would be no stars, and therefore no planets or life[18] The fact that the universe's expansion rate is so exquisitely balanced against the force of gravity is another profound example of fine-tuning that is necessary for our existence.

It is with a touch of irony that when Einstein created the cosmological constant, it was unknown and undetected but believed by him to be the force that counteracts gravity and holds the universe in an eternal, static condition. Now we know it works against gravity, dominating it to both expand and accelerate the expansion of the universe. Although Einstein's biggest blunder ended up having to do everything with quantum mechanics, and the accelerated expansion of the universe was unknown, as an interesting historical connecting point, Einstein famously was unable to accept quantum mechanics as one of the pillars of physics. He believed there was something much deeper that made far more sense in a far more straightforward way than an uncertain, probabilistic, superposition-inclusive system.

In a letter to German physicist Max Born, dated December 4, 1926, Einstein wrote, "Quantum mechanics is very worthy of regard. But an inner voice tells me that this [is] not yet the right track. The theory yields much, but it hardly brings us closer to the Old One's secrets. I, in any case, am convinced that He does not play dice[19]."The "Old One" Einstein is referring to is the master architect/truth of our existence, or God.

Einstein's second significant quote on quantum mechanics is to the Hungarian mathematician Cornelius Lanczos, "It seems hard to sneak a look at God's cards. But that he plays dice and uses 'telepathic' methods (as the present quantum theory requires of him) is something that I cannot believe for a single moment[20]."

The ultimate irony is that despite Einstein's lifelong refusal to accept the probabilistic nature of the quantum world, the very existence of dark energy—the force that fulfills the role of his Cosmological Constant—is inexplicable without it. The only viable physical explanation for the energy of empty space is found in Heisenberg's uncertainty principle. As science writer Charlie Wood summarizes, this principle dictates that the vacuum of space must "crackle with fluctuations in the quantum fields that fill it," imbuing every field with a minimum amount of energy, known as the "zero-point energy" . In short, the force that governs the ultimate fate of the cosmos is born from the very quantum "dice-playing" that Einstein could never bring himself to accept.

The delicate and improbable balance between all these competing factors—the density of matter pulling the universe inward, the curvature of spacetime, and the repulsive force of dark energy pushing it outward—is elegantly captured in a single, powerful formula derived from Einstein's work, which describes the expansion rate of the universe. Below is the first Friedman equation, and it describes how the expansion rate of the universe relates to its energy content, curvature, and the cosmological constant.

Einstein's Cosmological Constant

This equation, with the additional contributions of Georges Lemâitre, Howard Robertson and Geofrey Walker, turned into the *FLRW Metric* (named after Friedman, Lemaître, Robertson & Walker). Whereas the Friedman equation describes the dynamics of how space evolves over time, the FLRW Metric describes the geometry of spacetime. The Friedman equation however was groundbreaking in its time seeing as it exposed the mathematical expression of the "razor's edge" on which our universe is balanced. It shows that multiple, independent, and exquisitely fine-tuned values must all work in perfect concert to produce a stable, life-permitting cosmos. The fact that the matter density, the curvature, and the cosmological constant just happen to have the precise values needed to allow for our existence is perhaps the most intractable of the fine-tuning problems concerning the cosmological constant.

Dark Matter vs. Dark Energy

Before we continue our investigation into dark energy and the Cosmological Constant, it is crucial to distinguish it from the other great mystery of the cosmos, which we've touched on briefly in the previous pages. That mystery is dark matter. Dark matter and dark energy are often confused, but they are fundamentally different phenomena with opposing effects on the universe. To fully appreciate the unique puzzle of dark energy, we must first briefly explore the history and nature of its cosmic counterpart.

Dark matter is similar to dark energy in that neither can be seen, and what exactly they are is a mystery. Despite this it is currently thought that they make up the majority of "stuff" in our universe, comprising 95% of the matter and energy in our universe. Dark matter alone is calculated to be roughly 5 times more abundant than "normal" baryonic matter. Although dark matter is not suspected as being a part of the cosmological constant, it is a *factor* in the expansion and accelerated expansion of the universe.

Swiss physicist and mathematician Fritz Zwicky is credited with founding the term "dunkle materie" (dark matter) in a 1933 paper, *The Redshift of Extragalactic Nebulae*[21], based upon his studying of Edwin Hubble's observation of the Coma Cluster of galaxies. For his paper, and first observation of "dark matter" Zwicky is considered the 'father of dark matter[22].'Zwicky noted a particularly significant anomaly, "Single galaxies were moving too fast for the cluster to remain bound together[23]." Zwicky summarized his findings,

> "In order to obtain, as observed, a medium-sized Doppler effect of 1000 km/s or more, the average density in the Coma system would have to be at least 400 times greater than that derived on the basis of observations of luminous matter [This would be in approximate accordance with the opinion of Einstein and de Sitter as discussed in Sect. 4.]. If this should be verified, it would lead to the surprising result that dark matter exists in much greater density than luminous matter[24]."

In the 1960's and 70's astronomers Vera Rubin and Kent Ford focused on the rotation of galaxies to calculate their rate of rotation with their mass. This focus is when conclusive evidence was retrieved for dark matter. This measurement, made by Rubin and Ford, could be made from galaxies that were tilted *towards* Earth's vantage point for a mostly full top, or bottom view. By calculating the rotation velocities of spiral galaxies, they discovered that the stars at the outer edges were moving far too quickly to be held in orbit by the gravity of the visible matter alone. According to the known laws of physics, these galaxies should have been flying apart. Since they weren't, Rubin and Ford concluded that there must be a massive, invisible "halo" of additional matter surrounding these galaxies, providing the extra gravitational pull needed to keep them intact. This confirmed earlier, similar suspicions from astronomers like Jan Oort in the 1930s and established the modern case for the existence of dark matter.

> "[Vera Rubin's] painstaking studies of the dynamics of spiral galaxies, in collaboration with Kent Ford, strongly supported the

> contention that they are surrounded by massive dark-matter haloes. The powerful evidence in her data was sufficient to dispel the doubts of the most dubious sceptics at the time[25]."

These two indications, realized by Rubin and Ford, building upon the observation of Zwicky (and Oor)t, established the case for dark matter[26].

> "Studying the Andromeda Galaxy—and then others more systematically—using spectrometers to chart orbital speeds of stars across galactic disks, Rubin and a colleague began to find measurable evidence that something other than visible mass was governing their motion... This "halo" of dark matter that Rubin discovered and charted around dozens of galaxies does not emit light and extends beyond the optical boundaries of each—and contains five to 10 times as much mass as the visible galaxy. Rubin's work was the first measurable, compelling, and scientifically presented evidence that supported a decades-old theory that unseen "dark matter" governed the structure and movement of galaxies[27]."

This discovery was a watershed moment. It transformed dark matter from a theoretical curiosity into a confirmed, essential component of our cosmological model. Their findings established that the vast majority of matter in the universe is not the familiar atomic "stuff" of our everyday experience, but an invisible, unknown substance whose only detectable influence is its immense gravitational pull. With this crucial mystery established, we can now clearly distinguish it from its even more enigmatic counterpart: dark energy.

Value of Variable 4:

The cosmological constant and the expansion rate of the universe are, both individually and collectively among the most finely tuned values in our universe. Like nearly all of our *Variables* and *Special Problems*, they hold the power to determine if our universe is fantastic, diverse, full of possibility and life-allowable, or must be an unknowable, featureless void.

If the cosmological constant varied by more than 1 part in 10 to the 120th power, the universe either would have dispersed too fast for stars and galaxies to have formed or would have recollapsed upon itself. If the expansion rate of the universe varied more -than 1 part in 10 to the 120th power, the same catastrophic collapse, or expansion would have taken place.

If the expansion rate of the universe were not in tune with the cosmological constant *to this same level of fine tuning,* then these same catastrophic fates would once again be a reality. This dual form of fine tuning, rarely considered, (cosmological constant x expansion rate of the universe) is far beyond the *probabilistic capacity of the universe* (PCOU) of $\mathbf{10^{140}}$.

Value of The Cosmological Constant is fM *for Functional Miracle*

The Cosmological Constant is the final, foundational miracle in the universe's "Genesis Event." Its value, fine-tuned to an almost unimaginable precision, ensures that the cosmos is balanced in a way that allows for the possibility of life. But even this conceit, on the level of the miraculous, when combined with the other miracles we have considered in this first group of Variables, provides humanity with nothing more than the opportunity to confront a future series of miracles equally, and in some cases even more fantastic if there is to be any actuality of life within our universe.

With this, we conclude our investigation into the universe's initial conditions. We have seen how a series of profound and inexplicable miracles—the beginning from nothing, the survival of matter, the impossibly ordered initial state, the uniform temperature, and now, the perfectly balanced expansion—were all necessary to set the stage. Our investigation will now move from the "what" of these initial conditions to the "how" of the universe's operation, as we begin to explore the exquisitely fine-tuned rules and physical laws that govern this miraculous reality.

Having established the series of miracles required for the universe's initial conditions—the "Genesis Event"—our investigation now shifts from the "what" of creation to the "how" of its ongoing operation. A universe with the right starting materials is not enough; it must also be governed by a precise and stable set of rules to allow for the emergence of complex structures like stars, planets, and life.

This section will explore the exquisite fine-tuning of these rules. We will examine the fundamental forces and physical constants that form the operating system of our reality. We will see how these values are balanced on a razor's edge, working in perfect synthesis to create a life-permitting cosmos, a condition for which there is no known physical or materialistic explanation.

---

[1] (G. Gonzalez, J.W. Richards, The Privileged Planet, Pg. 186, Regnery Gateway, 2004).

---

[2] (W.L. Craig, (drcraigvideos) YouTube Channel, Jun 8, 2016, https(colon) //www(dot) reasonablefaith(dot) org/ finetuning)
[3] (S.M. Brewer, J.S. Chen, A.M. Hankin, E.R. Clements, C.W. Chou, D.J. Wineland, D.B. Hume and D.R. Leibrandt, Ion Clock Busts into New Precision Regime, Phys. Rev. Lett. 123, 033201, July 15, 2019)11.
[4] (S. Meyer, Return of the God Hypothesis, Pg. 151, HarperOne, 2021)
[5] (Alan Guth,The Inflationary Universe, Pg. 37)12
[6] ([Scrutinizing the Cosmological Constant Problem and a possible resolution, 6 March 2013, Denis Bernard, André LeClair, https://arxiv.org/pdf/1211.4848.pdf])
[7] (E. Copeland, M. Merrifield, Sixty Symbols, CERN, Cosmological Constant - Sixty Symbols, January 8, 2011).
[8] (P.J.E. Peebles, B. Ratra, The Cosmological Constant and Dark Energy, Reviews of Modern Physics, Volume 75, April 2003, April 22, 2003)
[9] (G. Gonzalez, J.W. Richards, The Priviledged Planet, Pg. 205, Regnery Gateway, 2004)
[10] (E. Copeland, M. Merrifield, Sixty Symbols, CERN, Cosmological Constant - Sixty Symbols, January 8, 2011)
[11] (Nov 29, 1998) And (M.S. Turner, D. Huterer, Prospects for Probing the Dark Energy via Supernova Distance Measurements, 1999, Phys.Rev.D, Particles, Fields, Gravitation and Cosmology, Vol 60, Iss 8, id. 081301).
[12] (MLA style: The Nobel Prize in Physics 2011. NobelPrize.org. Nobel Prize Outreach AB 2022. Wed. 3 Aug 2022.).
[13] (O. Botner, Award Ceremony Speech, Physics, 2011, Nobel Prize Outreach AB 2024, NobelPrize(dot) org/ prizes/ physics/ 2011/ ceremony-speech)
[14] [Sergel, D. N.; et al. (2007). "Three-Year Wilkinson Microwave Anisotropy Probe (WMAP) Observations: Implications for Cosmology". Astrophysical Journal Supplement. 170 (2): 377–408. arXiv:astro-ph/0603449. Bibcode:2007ApJS. 170..377S. doi:10.1086/513700. S2CID 1386346.]22
[15] [Guillermo Gonzalez, Jay W. Richards, The Privileged Planet: How Our Place in the Cosmos is Designed for Discovery Pg. 205]26
[16] [Sean M. Carroll, Why is the Universe Accelerating? Enrico Fermi Institute, Department of Physics, University of Chicago, Published in "Measuring and Modeling the Universe", Carnegie Observatories Astrophysics Series Vol. 2, ed. W. L. Freedman (2004)27.
[17] (L.M. Krauss, A Universe From Nothing, Pg. 125, Atria Paperback, 2012)28.
[18] (Ibid. The Privileged Planet Pg. 205)29.
[19] (A. Einstein, Letters, 91, the collected papers of Albert Einstein einsteinpapers(dot) press(dot) princeton(dot) edu/)
[20] (C. Wood, Physicists Use Quantum Mechanics to Pull Energy out of Nothing, Quanta Magazine, February 22, 2023), (H. Dukes and B. Hoffman, eds, Albert Einstein, The Human Side, Princeton University Press, Pg. 68, 1979 [To Cornelius Lanczos, March 21, 1942, Einstein Archive, 15-294])
[21] (F. Zwicky, [republication] 2009, General Relativity and Gravitation, Jan.Vol41, Num 1, Pgs 207-224, doi 10.1007/ s10714-008-0707-4, [original (in original language), F. Zwicky, Helvetica Physica Acta, Vol 6, P. 110-127, 1933, Redshift of Extragalactic Nebulae, Translated by H. Andernach).
[22] J. de Swart, Deciphering Dark Matter: The Remarkable Life of Fritz Zwicky, Sept 3, 2019, Nature Magazine. Nature(dot) com/ articles/ d41586-019-02603-7, reviewing J. Johnson Jr's Zwicky: The Outcast Genius Who Unmasked the Universe, Harvard Univ. Press, 2019).
[23] (Ibid J. de Swart, Deciphering Dark Matter etc).
[24] ([original (in original language)], F. Zwicky, Helvetica Physica Acta, Vol 6, P. 110-127, 1933, Redshift of Extragalactic Nebulae, Translated by H. Andernach).

---

[25] (J. Mitton, How Vera Rubin Encountered Dark Matter, Astronomy & Geophysics, Volume 62, Issue 2, April 2021, Pages 2.22–2.28, doi(dot) org/ 10(dot)1093/ astrogeo/ ata055)

[26] (VR. Giovanelli, M.P. Haynes, C. Rubin, W.K. Ford, UBC 12591: The Most Rapidly Rotating Disk Galaxy, The Astrophysical Journal, 301:L7-L11, 1986 February 1 r 1986)24.

[27] (J. Wilensky, Cornell Alumni, Alumna's Galactic Quest Proved Existence of Dark Matter, Astronomer's years of data and observations won over skeptical colleagues and changed the field. Oct 22, 2021. https://alumni.cornell.edu/cornellians/ alumas-galactic-quest -proved-existence-of-dark-matter/)25

Variable 5: The Fine Tuning of Gravity

The Fundamental Interactions

Our investigation into the universe's rulebook begins with its most familiar and yet most enigmatic force: gravity. Along with electromagnetism and the strong and weak nuclear forces, gravity dictates the behavior of all matter and energy in the cosmos. We will explore how the strengths of these four fundamental forces are not independent but are exquisitely fine-tuned to work in perfect concert with one another, a harmony that is essential for our existence.

The behavior of our universe is dictated by four fundamental forces: the strong nuclear force, the weak nuclear force, electromagnetism, and gravity. These forces are considered "fundamental" because they are not reducible to any more basic interactions.

A helpful way to understand them is to think of them in two groups, based on the scale at which they operate:

(1) The Nuclear Forces (The Microscopic World): These forces rule the realm of atoms.

The strong nuclear force is the cosmic "superglue." It binds quarks together to form protons and neutrons and then holds those protons and neutrons together to create atomic nuclei.

The weak nuclear force is the force of transformation. It is responsible for radioactive decay and has the unique ability to change one type of particle into another, a process that is essential for the nuclear fusion that powers stars.

(2) The Large-Scale Forces (The Macroscopic World): These forces govern the universe on the scales we can see.

Electromagnetism is the force of chemistry and light. It holds electrons in orbit around atoms, allows for the formation of molecules, and is responsible for everything from sunlight to the nerves firing in your brain.

Gravity, while the weakest of the four, is the primary organizing force of the cosmos. It pulls matter together to form planets, stars, and entire galaxies.

Electromagnetism is responsible for our ability to see, observe and use our senses to perceive the universe. Electromagnetic light activates our eyes. Electromagnetic energy is the basis for chemistry and biology. It is electricity, and magnetism, and the charge of particles. Electromagnetism in a sense, bridges the largest scales of our universe with the smallest scales. Electrons attach to atoms because their negative charge is attracted to the positive charge of the proton and electrons emit and absorb photons, the force mediating particle for the electromagnetic force for the entire electromagnetic spectrum wavelengths. Ranked in order from the longest to shortest wavelengths, we find radio waves, microwave, infrared, visible light, ultraviolet, x-ray and high energy gamma rays.

Each of the 4 fundamental forces has its own characteristics, its own strengths, its own mechanism of action and its own unique effect upon our universe. But perhaps the best way to consider the fundamental forces is in how they work *together*, beautifully, seamlessly, each with their own power, influence, strength, set of characteristics and effects, indistinguishable from a form of universe-wide "subjective" consciousness, ever-aware, ever-evolving, and piercing into, and lording over all of the interactions in our universe. Nothing in our universe composed of ordinary matter escapes their edict.

These forces do not cross over, they each are distinct, except under the highest temperatures and densities when electromagnetism and the weak force unite as one force known as the electroweak force. At even higher temperatures, it is believed the strong force becomes one with the electroweak force. This is the idea behind grand unified theory. At the beginning of the universe, at some point before Planck time, it is believed that gravity joined the other three forces in a theory known as the Theory of Everything (TOE). Grand Unified Theory (GUT) and the Theory of Everything so far have eluded our multi-billion-dollar experimental accelerators and detectors, along with the greatest minds in science.

The unified state of the forces, which existed only in the extreme heat of the early universe, was fundamentally inhospitable to life. In that primordial furnace, it was too hot for even stable protons and neutrons to exist, let alone the complex atoms and molecules that life requires. It was only through a process known as *symmetry breaking*—as the universe expanded and cooled, allowing the forces to separate and take on their distinct identities—that the stable structures necessary for chemistry, planets, and biology could form. In short, the very existence of a complex, life-bearing cosmos is dependent on the forces being separate and distinct, a condition that was not a given at the outset.

The four fundamental forces are each described by their own highly successful scientific *theories:

The Strong Nuclear Force is governed by the principles of Quantum Chromodynamics (QCD)

Both the Electromagnetic Force and the Weak Nuclear Force are united under the umbrella of Electroweak Theory.

The Electromagnetic Force is also explained by the classical Electromagnetic theory (described by Maxwell's Classical Equations)

The Gravitational Force is described on large scales by Einstein's Theory of General Relativity and on smaller scales by Newton's classical laws.

Quantum Gravity (NO theory as yet): Thus far, the gravitational force on the quantum level has resisted a theoretical description. String theory, loop quantum gravity (LQG) and several other ideas are among the candidate theories for the theorized quantized version of gravity.

While these theories describe the fundamental nature of the forces, a crucial part of their identity is the vast difference in their strength and their range of action. To appreciate how these four forces work together to build our universe, we must first understand the vastly different scales on which they operate.

Range of Action for the 4 Fundamental Forces

The range of action for two of the forces, -the strong nuclear force and the weak nuclear force-, is no greater than a medium size nucleus of an atom. The range of the strong force is given as $\mathbf{10^{-15}}$ meters which is the official length of a femtometer, -a quadrillionth of one meter. A "Medium-size" nuclei refers to the size of a nuclei on the periodic chart of the elements. The range of the weak force is smaller than the range of the strong force. This range, in literature, is given as either $\mathbf{10^{-17}}$ or $\mathbf{10^{-18}}$ meters, which is either 1% ($\mathbf{10^{-17}}$) or 0.1% ($\mathbf{10^{-18}}$) the diameter of a medium sized nucleus. The weak force, in addition to being weak and of short range, is also transmitted by very heavy *W* and *Z bosons*. These bosons are the force carriers for the weak force, and they are around 100 times the mass of a proton. The heaviness of these bosons determines their short range and therefore the short range of the weak force.

The strong nuclear interaction holds quarks together within neutrons and protons, and it holds neutrons and protons together within the nuclei of atoms by the exchange of strong force mediating particles called gluons as described by Quantum Chromodynamics (QCD). The strong force therefore is responsible for protons and neutrons and for atomic nuclei.

All of the fundamental interactions are mediated by the exchange of bosons, which are considered particles. Under standard model physics, all forces in the universe that have any effect upon the universe are represented by a particle. Because of wave/particle duality, the particles of the fundamental interactions are also *fields* mediated by the exchange of bosons.

The other two forces, electromagnetism and gravity, have an *infinite* range, theoretically extending across the known universe. The electromagnetic force is carried by and mediated by photons as described in Einstein's 1905 paper, known as, *On a Heuristic Viewpoint Concerning the Emission and Transformation of Light*, better known under the shorthand, *The Photoelectric Effect*[1]. In this paper, light is described as consisting of tiny packets of energy called photons. Prior to Einstein's paper, light was thought of as being a wave, as observed in 1861-1862 by James Clerk Maxwell, and described in his paper, *On Physical Lines of Force*[2]. Later, the idea that light was a wave did not appear to make sense as in certain experiments, it appeared that light could only be absorbed in certain amounts and emitted in certain amounts as if it were *quantized*. Einstein's paper explained this paradox by establishing wave/particle duality.

Gravity is the "odd one out" among the four fundamental interactions, a fact that makes it uniquely mysterious. The other three forces are all described by quantum field theories that make up the standard model of physics, and are mediated by force-carrying particles, or bosons. The electromagnetic force, for example, is carried by photons.

Gravity, however, resists this description. In our most accurate and profound theory of gravity, Einstein's General Relativity, it is not a force at all. Instead, it is a consequence of the curvature of spacetime itself—mass and energy tell spacetime how to curve, and the curve of spacetime tells matter how to move. This is why, despite decades of searching, a hypothetical force-carrying particle for gravity—the "graviton"—has never been observed, and gravity remains conspicuously absent from the Standard Model of particle physics. This deep, unresolved conflict between our two best theories of reality is one of the greatest challenges in science, and one of the primary reasons why the yet-to-be-understood nature of gravity, despite its exquisite fine-tuning with the Standard Model, is a variable within our algorithm.

Emerging From Symmetry (?)

Our current understanding theorizes that gravity emerged from the other forces at $\mathbf{10^{-43}}$ seconds. After gravity the strong force separated from the *electroweak* force at $\mathbf{10^{-35}}$ seconds. Lastly, the electroweak force split into electromagnetism and weak nuclear force at $\mathbf{10^{-12}}$ seconds when the unification energy (246 GeV) and temperature ($\mathbf{10^{15}}$ Kelvin) of the universe was no longer sufficient to unite these forces, causing them to separate. Under our current understanding, the separation of gravity from the other fundamental forces had to have taken place around $\mathbf{10^{-43}}$ seconds. This time period of $\mathbf{10^{-43}}$ seconds after the Big Bang, is the uncrossable boundary in time that we have previously discussed as an impassable "information wall" in the history of the universe. This time is when the universe was in Planck time, which is the earliest known period of time after the Big Bang.

Strengths of the fundamental Interactions

Ranked by strength, the strong nuclear force is the most powerful, then electromagnetism, followed by the weak force and finally gravity, despite its presence across all of spacetime. But the strengths of the fundamental forces are not the best way to understand them in terms of our investigation. The role of the forces, what they do, what they allow, and what they interact with within our universe, in the context of their relative strengths to one another, is how they are understood by scientific minds.

If one wishes to sew, one uses a sewing needle, not an intercontinental ballistic missile. If a surgeon is trying to separate a small tumor from the speech centers of the brain, they use scalpels, microscopes, forceps and miniaturized cameras, not sticks of dynamite. The fundamental forces function in this manner. Strength is measured, but their action is exact and represent the most minute and precise and inviolable actions our universe can produce, extending far below what will ever be observed; what they do appears to be refined for a particular purpose, and as the fundamental forces within our universe, they facilitate all that the normal baryonic matter in our universe does and results in.

These analogies highlight a crucial point: the fundamental forces of our universe appear to be perfectly suited for their specific roles to a degree that we have a good working understanding, but are representative of such a level of complexity, change, flux and seamless perfection, that the particulars of how such a system was devised and put into operation may likely remain beyond mortal comprehension. To understand how this intricate and seemingly purposeful system works to create a life-bearing cosmos, we must first take a closer look at the unique characteristics of each force, beginning with the most powerful of them all: the strong nuclear force.

The Strong Force

The strength of the 4 interactions are typically given in relation to one another. Being the strongest force, the strong force, or *the color force* as it is known, sets the scale with the value of "1" for its strength among the 4 fundamental interactions. This value of "1" however, does not give us much of an indication of how strong this force is. In *Variable 2, The Baryon Asymmetry*, we discovered, "The strength of the color force… is estimated to be equal to roughly ten tons of force holding the quarks together." This is what "1" means within this relative strength scale among the 4 interactions.

However, the strong force is not consistently "ten tons of force" holding quarks together. This force is dynamic. When the 3 valence quarks within a proton are close together, the strong force decreases and becomes weaker due to asymptotic freedom. The strong force increases, in our rubber band analogy, as the valence quarks move apart from one another. When the quarks are too close together the strong force becomes repulsive, pushing the quarks back to their ideal distance. The same dynamic holds for protons and neutrons within nuclei. The strong force holds them at ideal distance, not too close, and not too far away.

Electromagnetism (& The Fine Structure Constant)

The next strongest force is electromagnetism with a strength value of 1 divided by 137 less than that of the strong force. This gives us a value of 0.00729735256. This number, 137 (rounded from 137.035999177) is also known as the fine-structure constant and it plays an important role in the fabric of our physical reality. More on this in a moment.

The range of electromagnetism, unlike the nucleus size range of the strong force, and the proton sized range of the weak force, is infinite, in relation to our universe, knowing no bounds. As the name implies, electromagnetism is the combined force of electricity and magnetism, two aspects of a single unified electromagnetic force, an insight gained from James Clerk Maxwell that was further enhanced by Einstein's theory of relativity, which taught us that frame of reference is key to electromagnetism and whether one observes an electric field, a magnetic field, or combination of both. Frame of reference is key.

To understand the measured strength of this fundamental force, we must move from the large-scale world into the atomic and subatomic realm. The strength of the electromagnetic interaction is defined by a mysterious, fundamental number known as the fine-structure constant. This "magic number," was first described by physicist Arnold Sommerfeld, one of the fathers of quantum mechanics, and a mentor to 7 Nobel prize winners. The fine structure constant is one of the foundational pillars of quantum mechanics and, as we will see, a profound mystery in its own right.

The history of the fine structure constant began in earnest when Max Planck solved a problem known as *the ultraviolet catastrophe.* This was a prediction that black body radiation, (electromagnetic radiation) in the ultraviolet range would emit an unbounded quantity of energy. This was nonsensical and also did not match observation, but there was no way to explain why this was the case. Max Planck figured out that energy could only be released in discrete packets which he called *quanta,* such that either a quanta of energy was released, or multiples, but no fraction of quanta existed in ultramagnetic radiation, and it therefore was not on a continuous spectrum.

Thanks to Planck, we know that electromagnetic energy is in a set number of quanta, and electromagnetic radiation can only be absorbed or emitted in these discrete packets. This energy level of quanta coincides with spectral lines (spectroscopy) for different atoms, which when leveraged as information is perhaps the most powerful tool in modern science. Planck's constant therefore defines the *quantum* (quanta) nature of energy and how *photons* convey this energy and how electrons absorb and release this energy. In fact, when we see, we see photons/waves of electromagnetic energy absorbed by our eyes.

Ernest Rutehrford then demonstrated that atoms were clouds of negatively charged electrons around a positively charged nucleus. Together with Neils Bohr they devised the Rutherford-Bohr atomic model. This model added additional detail to the understanding of the spectral lines of hydrogen. Niels Bohr's contribution to the model was that electrons were orbiting the nucleus in particular, set, stable "orbits", later refined to *probability clouds* or *ellipticals,* and these orbits directly related to quanta of energy.

Electrons could orbit in these particular stable orbits without radiating energy. Within these orbits they could emit or absorb energy and *jump* orbits, once again, the jumps were related to the *quanta* of energy they emitted or absorbed, and like quanta, the orbits were discrete. Electrons jumped into these orbits, and would never be found in between orbits, never in fractions of orbits. This is a strange and counterintuitive way to visualize physical phenomena, and it was the harbinger of multiple levels of quantum weirdness.

This core quantum level finding was a radical departure from classical physics. At this point, these orbits were related to the simple hydrogen spectra, and they were still somewhat of a puzzle. It was later found that when the spectral line of the hydrogen atom is examined at very high resolution, it is actually a closely spaced doublet and not a single line. This splitting, on the finest level of detail, was referred to as "fine structure[3]." This structure would also provide evidence for what would later be understood as *electron spin.*

This is the point where Arnold Sommerfeld made his contribution. In contemplation of this evidence in the orbits of electrons, spectral lines, and quanta, he proposed that if the orbit of electrons could be thought of as ellipses instead of being circular, then "the energy of the electron would be the same, except in the presence of a magnetic field[4]," and these ellipses would provide an understanding of the *fine structure* of the measured spectral lines. Sommerfeld's constant for measuring this structure became the *Sommerfeld fine structure constant.* It quantifies the strength of the electromagnetic interaction between elementary charged particles. The spectra of the hydrogen atom appear in discrete lines because these are different energy levels of the hydrogen atom.

The Sommerfeld *fine structure constant* after some refinement became simply *the fine structure constant*, and this constant, in our understanding of the relative strengths of the four fundamental forces, is the strength of electromagnetism[5].

The fine structure constant is one of the universe's many dimensionless quantities which we reviewed in our last variable. "Dimensionless" means it is a *pure number* arising out of nature itself. The fine structure constant is denoted in scientific notation by the greek letter, *alpha.* The fine structure constant, because it is foundational to so many things, can be described several different ways. It is, "a measure of the strength of the interaction between charged particles and the electromagnetic force[6]." "It is a number that determines how stars burn, how chemistry happens and whether atoms exist at all[7]."

If the fine structure constant varied by more than a few parts per billion, (differed in its 6 decimal place by -1), protons would repel each other so strongly atomic nuclei couldn't hold together, and stars could not produce carbon, which is essential to life. If the fine structure constant were stronger in the 6th decimal place by +1, molecular bonds would fall apart at lower temperatures altering many processes essential to life[8]. The weakness of this constant also allows electrons to "hop away" enabling chemical bonds[9] , which are indispensable for life. The fine structure constant is at the heart of quantum electrodynamics that describes the interaction between light and matter. For all of these reasons, and more being a pure number arising out of nature itself, with no theory to understand why its value is what it is, the fine structure constant and by association the number 137 are considered as one of the most perplexing mysteries and the most puzzling number in science. Nobel laureate Richard Feynman said of it:

"There is a most profound and beautiful question associated with the observed coupling constant… It is a simple number that has been experimentally determined to be… about 137.03597 with an uncertainty of about 2 in the last decimal place. It has been a mystery ever since it was discovered more than fifty years ago, and all good theoretical physicists put this number up on their wall and worry about it… It's one of the greatest damn mysteries of physics: a magic number that comes to us with no understanding by humans[10]."

~~But~~ Feynman was hardly the only luminary perplexed over the fine structure constant. English Physicist and Nobel Laureate Paul Dirac called the fine structure constant, "The most fundamental unsolved problem in physics[11]." German Physicist and Nobel Laureate Wolfgang Pauli said, "When I die, my first question to the Devil will be, 'What is the meaning of the fine structure constant'[12]?"

The mysterious recurrence of the number 137 has not gone unnoticed outside of physics. The psychologist Carl Jung, for example, speculated that the number 137 was a cornerstone of reality and a key piece of his theory of synchronicity—the idea of meaningful, acausal connections. While this idea remains in the realm of philosophy, Jung was correct as far as the word "meaningful" applies because if this value were even slightly different, the structure of our physical reality would be unrecognizable. The fact that the number 137 appears in so many seemingly unrelated but essential physical relationships is a repeated, synchronistic coincidence that, for many, including Feyman, Dirac and Pauli, suggests it may indeed hold a "secret to life itself". What isn't in dispute is that whatever the ultimate explanation for life is, it is a process that relies heavily on the precise and mysterious value of this fundamental constant.

It is surprising that all of the units within the fine structure constant formula cancel out, forming the crux of the mystery. To explain, canceling out means given the units within the fine structure constant formula of: elementary charge (Coulombs), vacuum permittivity (Farads per meter), the reduced Planck constant (Joule-second), and the speed of light (meters per second) we are left with only the dimensionless quantity of the fine structure constant, 137[13]. But that in itself is not as surprising as the synergy that began to happen after this number was found. 137 started "showing up everywhere". Below are just a few examples.

-The repulsive energy between two electrons is 137 times smaller than a photon with wavelength equal to the distance between those electrons.

-The orbital speed of an electron in the ground state of the Bohr model of the hydrogen atom is 137 times slower than the speed of light.

-The energy of the above ground state electron is smaller than the rest mass energy of the electron by a factor of 137 squared.

-The universe is 13.7 billion years old.

-The number 137 comes up in a number of other relationships as well[14].

Thus far, there is no answer for why this particular number, 137, is found repeating itself in so many different ways and key relationships within our reality. And there is something else. Like *The Divine Algorithm*, because the fine structure constant, -the number 137- is a dimensionless and unit-less number, it will be found anywhere and everywhere in our universe where these relationships are examined. This means that if there is intelligent life anywhere in the universe at or beyond our level of technological advancement, they will also realize the significance of the number 137 and ratio 137, and how it repeats within our universe, and how key it is to their own existence if they are biological life forms. In this sense, the number 137 is similar to our algorithm, because if advanced technologically adept biological entities are aware of the number 137, they must also be aware of the fact that our universe is not possible without a series of linked miracles that are beyond the ability of material science alone, physics and probability to explain.

The Weak Force

The weak force is the 3rd weakest force behind the strong force and electromagnetism. Its mechanism of interaction is understood by electroweak theory (EWT). The force carriers for the weak force are the W and Z vector bosons. There are 3 bosons in total, two W bosons, W+ and W-, and one Z boson. All three bosons have spin 1. The W+ and ~~Z~~ W- bosons have a magnetic moment because they are electrically charged, but the Z boson has no magnetic moment because it is electrically neutral. The Z boson is also its own antiparticle.

The weak force in comparison to electromagnetism is approximately $\mathbf{10^{-5}}$ to $\mathbf{10^{-7}}$ times *weaker*, depending on the energy scale. The lower value includes nearly all weak force interactions where the weak force is about 100,000 times weaker ($\mathbf{10^{-5}}$) than electromagnetism. The higher value is the exception that occurs at the highest energies, such as the earliest moments in the history of the universe and in our most powerful particle accelerators. In these conditions the weak force and electromagnetism become almost equal, approaching a ratio of $\mathbf{10^{-1}}$ or unity, as the two forces unify as the electroweak force[15]. "The weak force controls proton-proton fusion, a reaction that is 1,000,000,000,000,000,000 times *slower* than the nuclear reaction based on the strong nuclear force[16]."

The weak force has significant implications for human life, particularly its ability to allow stars to "burn its hydrogen gently for billions of years instead of blowing up like a bomb[17]." British cosmologist and astrophysicist Martin Rees estimated that a change in the strength of the weak nuclear force by 1 part in at least 10,000 relative to the strength of the strong force would have prevented supernova explosions. "Without these supernova explosions key heavy elements would be unavailable for life[18],"such as iron, zinc, and copper. Also included in the list are the heavy elements needed for advanced technological applications, such iron, copper, tin (bronze), and once again, silver, gold, platinum, and uranium, etc[19]. The earth itself requires heavy elements as well if it is to be a host for intelligent life. Besides the heavy metals we listed that are needed for biological functions such as iron, copper, zinc, nickel, chromium, and molybdenum, the earth utilizes radioactive isotopes such as uranium-238, uranium-235, thorium-232 and potassium-40[20] as key contributors to earth's molten core that creates our protective magnetosphere and drives the regenerative processes of earth's plate tectonics. All in all, supernova explosions are absolutely essential for the complex chemistry relied upon[21] for our experience as intelligent observers within the universe. "Without these supernova explosions key heavy elements would be unavailable for life[22].

The weak force also is key in making the nuclei of atoms in the form of protons and neutrons. Both protons and neutrons are essential for chemistry, but free neutrons will decay into protons in approximately 15 minutes. The weak force addresses this loss by keeping a steady supply of neutrons with its ability to turn up quarks into down quarks and vice versa. Protons can turn into neutrons and vice-versa via the weak force. The weak force is thought of mainly for its ability to transform elementary particles into other particles, its role in radioactivity, such as alpha and beta decay, proton decay and for its above stated role in the slow, steady burning of stars.

Gravity

Gravity is the weakest of the 4 fundamental forces. Gravity isn't just the weakest force; it is a fantastically weak force compared to the other forces.

The strong nuclear force is 6 thousand trillion trillion, trillion (6 x $\mathbf{10^{39}}$) times stronger than gravity. On Earth, we may not think of gravity as being weak, but if we compare it to electromagnetism its weakness becomes readily apparent. A small amount of iron will easily attach itself to a small magnet, easily resisting the gravitational attraction of the entire Earth to attach itself to a small magnet. This simple demonstration is a real-world example of what it means that gravity is $\mathbf{10^{36}}$ times weaker than electromagnetism.

Gravity is, however, special and has unusual characteristics. Newton's law of gravity, otherwise known as the inverse square law states the force between two masses, such as the Sun and the Earth, or the Earth and the Moon, is proportional to the product of the masses and inversely proportional to the square of the distance between them. An example is, if the Earth were twice its current distance from the Sun, the gravitational attraction between the Earth and Sun would be 1/4th of what it currently is. For gravity, the further objects are, the inverse square law diminishes their gravitational effect accordingly. The unusual thing about gravity is that its reach, in reality, and as calculated by the inverse square law, is never zero, and therefore infinite. However great the distance between two objects within our universe the inverse square law, shows that there is a non-zero gravitational effect. Regarding this relationship, Nobel laureate Paul Dirac famously stated, "Pick a flower on Earth and you move the furthest star." Among the peculiarities surrounding gravity, one is its infinite reach[23].

While gravity is by far the weakest of the four fundamental forces, its unique characteristics allow it to dominate the cosmos on the largest scales. Unlike the other forces, gravity is always attractive and cumulative—its influence adds up over vast distances. However, on a cosmic scale, gravity's infinite range and relentless, additive nature mean it always wins in the end. As weak and "feeble" as gravity is, it can compact the mass of billions of suns into a black hole, creating a gravitational field so powerful that not even light can escape. And while the strong and weak forces are confined to the atomic nucleus, and electromagnetism governs the world of chemistry and biology, it is gravity that dictates the overall structure and ultimate fate of the universe. In this sense, gravity is king.

We have now seen that our universe is governed by four fundamental forces, each with its own unique character, strength, and domain of influence. From the incredible power of the strong force binding the hearts of atoms, to the subtle, cosmos-shaping reach of gravity, these interactions appear to be perfectly and independently suited for their specific roles. The true mystery, however, lies not in the forces themselves, but in how their wildly different properties work in perfect, life-permitting harmony.

Having established our foundation with accepted facts, we will now turn to the profound fine-tuning of this cosmic symphony.

The Fine Tuning of Gravity with the Fundamental Interactions

The alignment of the force of gravity with the strong force, the weak force and electromagnetism must be exquisitely fine-tuned for our universe to have the potential for life, and for our universe to provide the needs for life. It could almost be said that the priority for the fundamental forces was to create the condition where atoms of hydrogen and helium come into being, and then to create stars to create the periodic chart of the elements, and then galaxies, solar systems and the fabric of space time.

All of this is orchestrated by the 4 fundamental interactions. If we started with the needs of life and then worked backwards in creating a universe that could allow, create, sustain and satisfy the technological curiosity of living beings, our number one priority would be the creation of the 4 fundamental forces. And then, we would need to ensure that they would have the values, characteristics, and force carriers that they currently have, however, with respect to a material/physical understanding of our universe and the arrow of time, this is precisely what is argued against. Under the dominant paradigm, the needs of life and the needs of a life-habitable universe could not have possibly been the reason for the 4 fundamental interactions, nor could it be the reason they exist at their current values. Nonetheless, if we were to consider for a moment a purpose-laden, and a "perhaps the intent is indicated by the result" based explanation for the fundamental forces, this idea would provide us with a perfect understanding and justification for the forces, their value and their end result within our universe.

To fully appreciate the precise nature, and vast implications of the fine-tuning of the fundamental interactions, particularly as it relates to gravity, it is helpful to briefly trace the historical journey of our understanding. The story of how we came to understand gravity—from Newton's initial insights to Einstein's revolutionary re-conception—reveals how each new discovery deepened the mystery and reinforced the "unreasonable perfection" required for our existence.

Einstein's Gravity

Einstein's theory of general relativity (GR) is a classical theory that describes how gravity should be expected to behave as a geometric property of spacetime curvature caused by the energy and momentum of matter and radiation, in the classical sense. Einstein made several predictions for his theory and so far, all of the predictions that we have the ability to confirm have been confirmed with exacting precision. The predictions are:

Gravitational lensing

Time dilation

Black holes

Gravitational waves, detected in 2015 by the 2 LIGO detectors

Three predictions have yet to be confirmed due to technological limitations:

A singularity at the beginning of the universe (from the field equations)

White holes (the reverse of a black hole that can be escaped from but not entered. This is theorized to exist as a solution to Einstein's field equations, similar to the black hole result)

Wormholes (tunnels in spacetime that begin or end in different locations in space, time, or both. This is theorized to exist as a special solution to Einstein's field equations)

The striking success of General Relativity, with its major predictions being repeatedly confirmed by observation, establishes it as the bedrock of modern cosmology. However, the entire theory and its successes rests upon a single, foundational number: the gravitational constant. The robust predictive power of Einstein's equations only works because this constant has a very specific, exquisitely fine-tuned value. This brings us another level deeper into the mystery of the fine tuning of gravity with the fundamental interactions: the precise and seemingly purposeful value of gravity itself.

The Gravitational Constant

If the gravitational constant varied by 1 part in $\mathbf{10^{60}}$ in being either stronger, or weaker, the universe would have either (weaker) expanded so rapidly that no stars could have formed, which means, no elements, no planets, and no life, or, (stronger) it would have precipitated a catastrophic collapse, into a black hole universe or black hole filled universe, making it impossible for stars, elements, planets and life to form. This will sound familiar because it is the exact same result if our last two variables, the cosmological constant and the expansion rate of the universe, had been set ever so slightly in the stronger or weaker direction. With this serial presentation of near catastrophic misses within our universe for the potential for life and the diversity of our universe's cosmological products, it appears that the universe is delivering a forceful, non-subtle message to challenge and cajole our greatest scientific minds to reach for an understanding that can make sense of this unreasonable set of outcomes. Our existence has been so unimaginably fortuitous, and so scientifically and theoretically inexplicable, we have an obligation, as a scientifically minded species, to reconcile this set of realities with a set of ideas that rise to the challenge so clearly framed by the set of information we have gained.

As we discovered in *Variable 5, The Cosmological Constant,* when we consider the cosmological constant with the gravitational constant, the cosmological constant is *finely tuned*, on a knife's edge to *ever so slightly resist* the force of gravity, and obliterate Einstein's resistive, hand inserted Lambda (cosmological constant), to accelerate the expansion of the universe. If the gravitational constant varied by 1 part in $\mathbf{10^{60}}$, or the cosmological constant and expansion rate of the universe each varied by 1 part in $\mathbf{10^{120}}$, there would be no possibility of life.

This same factored consideration of multiple factors vs. individual factors holds true for the fine tuning of the fundamental interactions, except instead of 2 factors that must be considered, we can add an additional 4, with respect to the finely-tuned nature of the fundamental interactions, that must be finely tuned in relation to one another resulting in the single fundamental outcome of the possibility of life within our universe. Astrophysicists John Gribbon and Marin Rees have pointed this fact out:

> "If we modify the value of the fundamental constants, something invariably goes wrong, leading to a universe that is inhospitable to life as we know it. When we adjust a second constant in an attempt to fix the problem(s), the result, generally, is to create three new problems for every one that we 'solve' The condition in our universe really do seem to be uniquely suitable for life forms like ourselves, and perhaps even for any form of organic chemistry[24]."

The astounding fine-tuning of the fundamental forces is an obvious fact, yet it is one that is often ignored or evaded. The most common dismissal relies on an appeal to the anthropic principle—the tautological statement that the conditions we observe must be compatible with our own existence. I argue that using this principle to shut down inquiry is a form of intellectual childishness, an evasion of the profound question, posed repeatedly by these anomalies, of why our universe is so exquisitely balanced for life. This much is obvious, and provides no scientific understanding whatsoever.

Guillermo Gonzalez and Jay W. Richards, authors of *The Privileged Planet* state the nature of the actual problem, "But in the real universe the values of all the constants and force strengths must be satisfied simultaneously to have a universe hospitable to life[25]." Although the anthropic principle functions as a meaningless tautology, the real problem of creating a universe capable of life remains, and it is as meaningful of a question that can be asked. "How?", 'this is all possible' is the question, and "Because" 'otherwise we would not be here to witness it', is not a suitable answer. Such trifling repartee should be reserved for imperious parents and obdurate children engaged in mealtime struggles. Such answers have never been acceptable for those committed to the pursuit of scientific knowledge. There is nothing in nature which tells us why the fundamental interactions have the strengths and ranges that they do, except that these strengths and ranges are precisely what is needed for living entities to understand the complex nature of fine tuning required for their existence.

Having dismissed the anthropic principle as an insufficient explanation, we can now turn to a formal definition of the profound mystery we are confronting. Author and philosopher of science Stephen Meyer provides a clear and concise summary of what has become known in physics as the "fine-tuning problem."

> "Fine-tuning in physics refers to the discovery that many properties of the universe fall within extremely narrow and improbable ranges that turn out to be absolutely necessary for complex forms of life, or even complex chemistry, and thus any conceivable form of life, to exist[26]."

> "We apparently live in a kind of 'Goldilocks universe,' where the fundamental forces of physics have just the right strengths, the contingent properties of the universe have just the right characteristics, and the initial distribution of matter and energy at the beginning exhibited just the right configuration to make life possible. These factors taken me to gather are so puzzling that physicists have given them a name–the fine-tuning problem[27]."

This "Goldilocks" enigma—the fact that the universe appears to be exquisitely and improbably configured for our existence—is the central mystery that The Divine Algorithm sets out to explain. Added to this mystery, we must add the ultimate mystery regarding gravity, one that is patently absurd: no one knows what gravity *is*. How this force is propagated, carried (transmitted) and received, and how it came into the universe is a mystery that is presently being pondered over, as it has been pondered over further back in history than Isaac Newton. Even with his limited technology Newton expressed thoughts on gravity that are virtually as true now as they were when he replied to Cambridge professor Richard Bentley's letter on February 18, 1693.

> "It is inconceivable that inanimate brute matter should, without the mediation of something else which is not material, operate upon and affect other matter without mutual contact, as it must be, if gravitation in the sense of Epicurus, be essential and inherent in it. And this is one reason why I desired you would not ascribe innate gravity to me. That gravity should be innate, inherent, and essential to matter, so that one body may act upon another at a distance through a vacuum, without the mediation of anything else, by and through which their action and force may be conveyed from one to another, is to me so great an absurdity that I believe no man who has in philosophical matters a competent faculty of thinking can ever fall into it. Gravity must be caused by an agent acting constantly according to certain laws; but whether this agent be material or immaterial, I have left open to the consideration of my readers[28]."

Newton went further in his speculation, hinting at the need for a "divine" universal arbiter. Why was it not enough for the "arbiter" to be a set of physical laws working in concert? Because in his correspondence with Richard Bentley, Newton revealed that his own laws of gravity created a profound paradox. He understood that if matter were evenly distributed throughout a finite universe, gravity would inevitably cause it all to collapse into a single, central mass.

To avoid this collapse and produce the stable, orbiting solar system we observe, the initial state of the cosmos required an impossible act of balancing—what Newton compared to making "the sharpest needle stand upright on its point upon a looking-glass". He concluded that a framework of laws alone was insufficient to achieve this. Instead, he argued that an intelligent, intervening, and purposeful consciousness was required to set the heavenly bodies in their perfect initial positions and motions, creating an order and balance that would otherwise not be possible.

Newton expressed this conclusion directly in his letter to Professor Bentley:

> "I agree that if matter evenly scattered through a finite space… what convene in the midst… Yet… it seems to me a supposition fully as hard as to make the sharpest needle stand upright on its point upon a lookingglass… Yet I grant it possible, at least by divine power; & if they were once so placed I agree with you that they would continue in that posture… unless put into new motion by the same power[29]."

We are now a fourth of the way through a new century, more than 430 years removed from Newton's thoughts, and his letter to Professor Bentley, and we know little more about what gravity is than Newton. It has defied being found out. No theory exists that is complete in understanding its nature, on any level, including the quantum, and consequently the hoped-for Theory of Everything (TOE) cannot be completed. And regarding quantum gravity, theoretical physicist Freeman Dyson summarizes:

> "Roughly speaking there are three possible alternatives which might describe the universe we live in. There is first of all what I would call the Orthodox view, the view of all the experts, which is that quantum gravity is a good theory just like quantum electrodynamics. That in fact that there do exist gravitons, and that they obey The Algorithms of quantum gravity, and they have the same kind of behavior as the photon except that it's harder to observe... The second possibility is what I call concealment. It is that quantum gravity exists but it can never be detected, in the same way that the quark particle exists, which everybody believes is real but it can never be observed. That has turned out to be a very fruitful concept in particle physics and it is called quantum chromodynamics, which is the theory of the strong interactions. The quark is a quantum field which has all the properties of a quantum field except observability, such that you can never see a quark by itself. All you can see is confined systems in which the quarks are hidden. That's why we call it confinement. That's the second possibility, that quantum gravity is real but unobservable. The third possibility is that quantum gravity is nonsense. That in fact there is no such thing as quantum gravity. That gravity is a purely classical phenomenon and the alleged effects of quantization are for some reason absent. If that were true, gravitation is a classical phenomenon. It is some kind of statistical property which only belongs to matter in the large, and not to matter in the small. So there's no such thing as a gravitational field of an electron and it's only a collective property of matter in bulk."
>
> -Freeman Dyson, *Is a Graviton Detectable,* August 26-29, 2013[30].

In this quote, the eminent physicist Freeman Dyson lays out three starkly different possibilities for the ultimate nature of gravity, underscoring how profound our ignorance truly is. The first is the orthodox view: that quantum gravity is real and we will one day understand it. The second is that quantum gravity is real but, like quarks inside a proton, its fundamental components are permanently concealed from us. The third and most radical possibility is that quantum gravity is nonsense—that gravity is not a quantum force at all, but a purely classical phenomenon that emerges only on large scales.

Dyson's willingness to entertain these possibilities highlights a crucial point: the very foundation of our physical reality remains an open and profound mystery. The fact that one of the four fundamental forces may be permanently beyond our grasp or may not even be a quantum force at all, is a significant challenge to any theory that claims to have a complete materialistic explanation for the universe.

Without the graviton, the alleged force carrier particle for the gravitational field, the elusive search for quantum gravity is a fantasy, and all current efforts that rely on quantum gravity are at this juncture, equally doomed. This list of these efforts is hardly trivial. It includes exotic theories such as string theory, which is itself a hypothesis for quantum gravity, and loop quantum gravity based on Einstein's spacetime, and geometric gravity that postulates spacetime is composed of finite loops woven into a network and numerous other placeholder ideas for a universal influence that has defied discovery. These theories all must contend with a fundamental problem, which is gravity resists being quantized, there is at present no quantum observation, understanding or theory of gravity.

> "The reason our descriptions of time and space break down near the Planck era is that the gravitational field at this time was so distorted and turbulent with its own quantum fluctuations, it is impossible to define a clock to measure time or a ruler to measure length. Only a fully quantum mechanical description of gravity — which we don't yet have — will let us probe deeper into this corner of cosmic history. If we could examine physics at the Planck scale today, we would see what this quantum chaos is like. But at this scale, nature defeats our best efforts to observe it at all. If you try to study the gravitational field of the cosmos down to 10-33 cm using a photon, its energy would have to be so high (1019 GeV) that it would immediately (after $\mathbf{10^{-43}}$ seconds, the Planck time) become a tiny, quantum black hole with the smallest possible mass and event horizon radius: 10-5 gm and 10-33 cm (the Planck mass and size). The black hole would then evaporate and destroy the very information you were trying to extract."
>
> -Sten Odenwald, Astronomy Magazine, April 2022 Edition[31].

The insights from Dyson and Odenwald provide a stark and conciliatory conclusion to our investigation of gravity. We are faced with a foundational force that is not only a profound mystery in its own right but one whose ultimate nature at the moment of creation is, according to our best physics, permanently shielded from us. This impenetrable wall at the beginning of time is a critical failure for any purely materialistic explanation of our universe. It is another powerful indication that the finely-tuned forces governing our reality did not arise from any knowable physical process, but were instead set in place by a cause that lies beyond the reach of science.

The Higgs Particle & Physics Beyond The Standard Model

The mystery of gravity is not the only challenge that points to the incompleteness of our current understanding. The Standard Model is also plagued by a series of internal paradoxes and fine-tuning problems. The most significant of these is the puzzle of the Higgs boson, the very particle that was supposed to complete the model, but whose discovery has opened up a new realm of questions that point to "physics beyond the standard model."

The Higgs boson is theorized to convey mass and thus convey the property of gravity to all particles and thus all objects. At the same time, all of the particles in the standard model are thought to give mass to the Higgs, but there is a problem. The value of the Higgs is nowhere near what was expected. This is the second major hierarchy problem in physics. To explain this discrepancy and solve the Higgs hierarchy problem scientists turn to something called “physics beyond the standard model”. As we’ve been moving through our first Special Problem and the first few Variables, we have highlighted several glaring gaps in our understanding that are perplexing and at this point, intractable. Physics beyond the standard model refers to the solutions to these perplexing problems and proposes that there is a level of physics beyond that which is represented in our current standard model of physics (leptons, bosons, etc.) that explains these deficiencies.

The major issues that lie in the realm of "physics beyond the standard model" are a daunting list of the most profound and intractable puzzles in science. They include:

The Strong CP Problem: The mystery of why the strong nuclear force, which is predicted to have a certain asymmetry, is observed to be perfectly symmetrical.

The Matter/Antimatter Asymmetry: The profound and unexplained mystery of why the universe is made of matter when the Big Bang should have produced equal amounts of matter and anti-matter that should have annihilated each other.

Neutrino Oscillations: The observed fact that neutrinos change "flavors" as they travel. This is a puzzle because it implies they must have mass, a property they are not supposed to have in the Standard Model.

Quantum Gravity: The long-standing and profound challenge of uniting Einstein's theory of General Relativity with the laws of Quantum Mechanics.

The Singularity: The ultimate mystery of the universe's origin from a point of infinite density where all known laws of physics break down.

This is a daunting list of problems yet, certainly our universe is discernible *in some way*, and we know that it is entirely discernible as a lived experience, which, according to our findings, may be the ultimate goal of the universe, and therefore, perhaps the only thing in our universe that it has deemed to be fully discernible, even at the sacrifice of physics itself. These problems give us numerous clues that the ultimate truth of our reality may not conform to what we traditionally think of as 'rational.' While our universe is certainly an experience we can discern, we do not yet know if it is one that can be fully comprehended by a single, coherent, logical system. The above list of problems provides us with numerous indications that perhaps the ultimate truth of our reality may not jibe with what humans traditionally think of as rationality, and this would be a truly odd situation, in a universe where the beauty and seeming perfection of so much of our physics, in particular, the 4 fundamental forces, provide us with such a diverse array of finely tuned, and perfect-for-life levels of emergence.

Value of the Fine Tuning of the Fundamental Forces with Gravity:

We have now seen how each of the four fundamental forces is exquisitely fine-tuned in multiple, independent ways to allow for a life-bearing cosmos. From the strength of gravity to the precise value of the fine-structure constant, each parameter appears to be set on a razor's edge. The true scale of this miracle, however, is only revealed when we consider the combined odds of all these values being correct simultaneously.

The fine-tuning required for each of the four fundamental forces to permit a life-bearing universe is staggering on its own. The gravitational constant must be precise to one part in $\mathbf{10^{60}}$, the strong nuclear force to one part in $\mathbf{10^{40}}$, the electromagnetic force to one part in $\mathbf{10^{37}}$, and the weak nuclear force to one part in $\mathbf{10^{31}}$. However, the true scale of this miracle is only revealed when we consider that these are not independent values; they must all be correct *simultaneously*. When we calculate the combined odds for this perfect synthesis, we arrive at a final probability of approximately one part in $\mathbf{10^{168}}$. This number, a one followed by 168 zeroes, so vastly exceeds the probabilistic capacity of our universe that it stands as a profound testament to the intricate design at the heart of our reality.

Variable 5: The Fine tuning of Gravity in Synthesis with the Other Fundamental Forces is

1 part in $\mathbf{10^{168}}$.. This value exceeds *the probabilistic capacity of the universe* (PCOU) of $\mathbf{10^{140}}$

Value of Variable 5 is f*M for Functional Miracle*

The fine-tuning of the four fundamental forces, in following with our thesis of The Divine Algorithm, represents yet another in a series of necessary foundational miracles. It is not just one constant that is set against impossible odds, but a symphony of four.

With this understanding of the finely-tuned rules that govern our universe, our investigation now turns to the players themselves: the fundamental particles of the Standard Model. As we will see, the mysteries and exquisite fine-tuning we have found in the forces are mirrored, and perhaps even deepened immeasurably, in the very nature of the matter they govern.

---

[1] (E. Einstein, On a Heuristic Viewpoint Concerning the Emission and Transformation of Light, Annalen der Physik 17, 1905)
[2] (J.C. Maxwell, On Physical Lines of Force, Philosophical Magazine, 1861, 90: 11-23, S2CID 135524562).
[3] Hyperphysics(dot) phy-astr(dot) GSU (dot) edu / Hydrogen Fine Structure)
[4] (A.C. Insights, YouTube Channel, The Tragedy of Arnold Sommerfeld The Wise, March 22, 2024)
[5] (A.C. Insights, YouTube Channel, The Tragedy of Arnold Sommerfeld The Wise, March 22, 2024)
[6] (K. Spalding, Unlocking The Mystery Of 137: Why This Number Is So Important, Iflscience(dot) com / December 26, 2023)
[7] (M. Brooks, There's a Glitch at the Edge of the Universe That Could Remake Physics, October 6, 2018, Newscientist(dot) com/)
[8] (M. Brooks, There's a Glitch at the Edge of the Universe That Could Remake Physics, October 6, 2018, Newscientist(dot) com/)
[9] (N. Walchover, Physicists Nail Down the 'Magic Number' That Shapes the Universe, Quanta Magazine, Abstractions Blog, December 3, 2020),
[10] (R.P. Feynman, QED: The Strange Theory of Light and Matter, Pg. 129, 1985, Princeton University Press)
[11] (N. Wolchover, Physicists Nail Down The "Magic Number" That Shapes The Universe, Quanta Magazine, quantamagazine(dot) org December 2, 2020).
[12] (PBS Spacetime YouTube Channel, Why is 1/137 One of the Greatest Problems in Physics? September 28, 2022).

[13] PBS Spacetime YouTube Channel, Why is 1/137 One of the Greatest Problems in Physics? September 28, 2022)
[14] (PBS Spacetime YouTube Channel, Why is 1/137 One of the Greatest Problems in Physics? September 28, 2022)
[15] (S.G. Porsev, K Beloy, A. Derevianko, Precision Determination of Electroweak Coupling from Atomic Parity Violation and Implications for Particle Physics, American Physical Society, Physical Review Letters, 102: 18: 181601, 5/5/2009)
[16] (J. Warner Wallace, Fine-Tuning of the Force Strengths to Permit Life, Sunday, August 3, 2014, crossexamined(dot) org)
[17] (F. Dyson, Scientific American 225 (1971), p. 56).
[18] (J. Warner Wallace, Fine-Tuning of the Force Strengths to Permit Life, Sunday, August 3, 2014, crossexamined(dot) org)
[19] (J. Warner Wallace, Fine-Tuning of the Force Strengths to Permit Life, Sunday, August 3, 2014, crossexamined(dot) org)
[20] (J. Korenaga, Urey Ratio and the Structure and Evolution of Earth's Mantle, Reviews of Geophysics, 2008, June, vol46: num2: doi 10.1029/2007RG000241)
[21] (M. Rees, Philosophical Transactions of the Royal Society of London, A 310 (1983), p. 317.)
[22] (J. Warner Wallace, Fine-Tuning of the Force Strengths to Permit Life, Sunday, August 3, 2014, crossexamined(dot) org)
[23] (P.G. Hewitt, Conceptual Physics, 4th ed. 1981 [footnote])
[24] (J. Gribbin, M. Rees, Cosmic Coincidences, Pg. 269, Bantam Books, 1989)
[25] (G. Gonzalez, J.W. Richards, The Privileged Planet, Pg. 206, Regnery Gateway, 2004).
[26] (S.C. Meyer, Return of the God Hypothesis, Pg. 131, HarperOne, 2021)
[27] S.C. Meyer, Return of the God Hypothesis, Pg. 131, HarperOne, 2021)
[28] (I. Newton, Source: 189.R.4.47, ff. 7-8, Trinity College Library, Cambridge, UK, [a reply to the original letter from R. Bentley on Feb. 18, 1693])
[29] (I. Newton, Source: 189.R.4.47, ff. 7-8, Trinity College Library, Cambridge, UK, [a reply to the original letter from R. Bentley on Feb. 18, 1693])
[30] (F. Dyson, Is a Graviton Detectable, Nanyang Technological University, Singapore, August 26-29, 2013)6
[31] (S. Odenwald, The Planck Era: Imagining Our Infant Universe, Astronomy Magazine, Issue: April 2022, Published April 21, 2022, https://www.astronomy.com/science the planck era imagining our infant universe /)7

Variable 6: Fine Tuning & Anomalies of The Standard Model

Having explored the fine-tuning of the fundamental forces that govern our universe, our investigation now zooms in to examine the particles that those forces act upon. This brings us to the Standard Model of Physics, our best and most successful theory of the fundamental constituents of matter.

This chapter will proceed in two parts. First, we will delve into the strange and paradoxical world of Quantum Mechanics, the bizarre set of rules that governs the behavior of these particles. We will see how its foundational principles challenge our classical understanding of reality and point toward a universe that is fundamentally dependent on observation and the influence of consciousness. Second, we will examine the Standard Model itself, exploring the profound fine-tuning and inexplicable anomalies found in the properties of the elementary particles.

Throughout this investigation, we will build the case that the very "stuff" of our reality, like the laws that govern it, is exquisitely and impossibly arranged in a way that defies a purely materialistic explanation, while also serving as the bespoke foundation for the human experience.

PART I: FINE TUNING AND ANOMALIES OF QUANTUM MECHANICS

Introduction

The Standard Model (the SM) of physics represents our best understanding of the fundamental particles and forces in the universe, and the forms it takes, and how it behaves. *Gravity* is conspicuously absent because as we just reviewed, there is no known gravity particle or force carrier, nor any way that gravity can be *quantized*. To be quantized–reduced to *quanta*–the discrete energy levels of the fundamental particles is the defining feature of quantum mechanics.

Dark matter and *dark energy* are also not represented in the SM because we have yet to find out what particles, or fields they correspond to, or if they correspond to *any* particles at all, and are instead an undiscovered class of universal influence, or perhaps, in the case of dark matter, a fundamental misunderstanding. Taken together the number and depth of conceptual absences within the SM represents an embarrassment and a stark counternarrative for those that would portray our understanding of the universe as being so nearly complete as to call into question the concept of free will. Our investigation into the SM and quantum mechanics will illustrate that this view is not only false, but one of the most overstated and unfounded assertions in modern scientific discourse.

There are additional gaps in our understanding of our reality, beyond gravity, dark matter and dark energy that are not within, nor explained by the SM. In keeping with quantum field theory, for every influence in our universe, there is an imagined particle and corresponding field that it is represented by *physics beyond the SM*. In keeping with this theory, if the SM was "complete" there would be only minor details remaining in our understanding of every *thing* and every *event* within our universe.

If the SM was complete with gravity, we would have completed a major step towards the theory of everything (TOE). And *if* we had a completed TOE, we would understand every material and physical happening in the history of the universe, and be able to predict with accuracy, both quantum and classical behavior in a deterministic fashion. With all of the rules of our reality in our possession, then armed with only the initial conditions, we would be able to simulate the entire chain of events from the inception of the universe to the present moment. This understanding would relegate our hypothesis of *The Divine Algorithm*, to irrelevance, because there would be no unexplained events, no unknown influences, and no beyond-the-probabilistic-capacity-of-our-universe events.

On the other hand, the hypothesis of *The Divine Algorithm* holds that no such TOE is possible if it is a fundamentally materialistic theory in nature, because our thesis holds that the crucial missing force, and missing understanding that contextualizes the happenings within our universe, is the existence of, and interventions of by, an intentional manifest consciousness working from a predetermined plan to create a universe with the express purpose and prerogative of inhabiting this universe with intelligent observers. The TOE and *The Divine Algorithm* are, indeed, mutually exclusive.

"The Most Successful Theory" The Standard Model & Quantum Mechanics

The Standard Model of particle physics, along with the rules of Quantum Mechanics that govern it, is often hailed as the "most successful theory in the history of science". In its own domain, this is true; its predictions about the behavior of particles have been confirmed with an accuracy unparalleled by any other scientific theory. However, this success comes with a profound and often unacknowledged caveat. As we will explore, this celebrated theory fails to provide answers to many of the most fundamental questions of our existence and fails to account for close to 95% of the known universe. It is this deep contradiction—between the theory's incredible success and its stunning incompleteness—that we will now examine.

> "How we think about the universe as a whole affects such apparently different questions as the problem of unifying quantum theory with general relativity, the problem of understanding the origin of the properties of the elementary particles, the problems of the interpretation of quantum theory, the problem of what 'caused' the Big Bang, and the question of why the universe is hospitable to life. These are all problems we have so far failed to solve, in spite of having made partial progress on some of them." -Lee Smolin, The Life of the Cosmos[1]

While the Standard Model is often hailed as the "most successful theory in science," this claim comes with a profound caveat. The theory is indeed stunningly successful in its own domain, making predictions about the behavior of particles with unparalleled accuracy. However, this success is limited. This creates a misleading impression for the public, suggesting a level of completeness that the theory simply does not possess.

This incompleteness is most evident in the ongoing, and thus far unsuccessful, quest to unify the fundamental forces. While physicists have successfully united the electromagnetic and weak forces into a single electroweak force, the next step—uniting them with the strong force to form a Grand Unified Theory (GUT)—remains an unproven hypothesis. The ultimate goal, a Theory of Everything (TOE) that would include gravity, is even further from our grasp. This failure to create a single, unified picture of reality from the known forces is a powerful indication that the Standard Model is not the final word, but rather a single, incomplete chapter in a much larger canon of knowledge.

All of these intermediate steps, the electroweak force, the electronuclear force, grand unified theory and the theory of everything, are important because they have led us to our current understanding that when our universe was created, at the very beginning, all of the forces were believed to be united and believed to be dominated by quantum gravity in the Planck epoch. In some sense these forces "had" to be united, having come into creation at the same moment, as one energy. Therefore, as we wind the clock back in the history of the universe, understanding how these forces unite goes a long way to understanding how our universe was created and emerged, although winding this clock back will still have to confront the creation of our universe from nothing. And if what we find remains irreconcilable with the physics we understand, and it remains the case that our universe seemingly emerged from a theoretically irreconcilable state or states, the result will be the further strengthening of our thesis of *The Divine Algorithm*, which asserts that there is a fundamental assumption about the materialistic nature of the TOE that may be false.

This is the fundamental choice that lies before us. One can view the profound gaps and paradoxes in our understanding of the universe as temporary problems, waiting to be solved by a future, complete theory of materialistic physics, that increasingly appears to be the scientific equivalent of *Waiting For Godot*, or, as I will argue as we continue to press our case, that one can view these gaps, paradoxes and series of miracles as positive evidence in themselves—the fingerprints of a different kind of reality and clues that allow us to decisively solve this mystery, and be far more proscriptive about what conjectures can be excluded based upon the evidence. To explore this choice, we must now turn to the deepest and most profound of these mysteries: the bizarre and counter-intuitive world of quantum mechanics.

Introducing Quantum Mechanics

Besides the question of whether our universe is a fundamentally materialistic one, there are a few additional problems with QM that make it resistant to being both understood and unified with classical physics. Two of these problems are if QM is *"real"*. The other question is if QM is *"local"*.

To define our terms, "real" in the context of QM means that objects have *definite properties independent of observation.*

"Local" means that objects *can be influenced only by their surroundings,* and by extension, *that any influence upon an object* (particle, planet, etc.) *cannot travel faster than the speed of light*, a phenomenon Einstein famously derided as "spooky action at a distance."

This insight into what is *real*, and *locality* is vastly important as a litmus test for our ability to reconcile QM with classical physics, because "In our everyday world, [the classical realm] realism and locality are cornerstone concepts that always hold true[2]."

This is the heart of the challenge that quantum mechanics presents to our understanding of reality. The two most basic, common-sense principles that govern our entire macroscopic experience—that things have definite properties whether we are looking or not, and that a cause must be physically close to its effect—appear to break down completely at the fundamental level of particles. The following investigation will explore this deep and, thus far, irreconcilable conflict between the classical world we inhabit and the bizarre quantum world that underlies it.

When QM was introduced, it brought in several perplexing paradoxes and seeming falsifications of the foundational rules of our reality. On the quantum scale, it certainly appeared that locality and realness were violated. Fundamental particles, in the example of *electrons*, seemingly do not have definite properties *prior* to measurement, *and* they are seemingly influenced by *non-local* events. This means the statement that quantum particles and quantum events are local and real cannot be proven, forcing us to change our understanding of the world as observed by science and placing an irreconcilable boundary between classical and quantum events.

To reconcile our understanding of *QM*, "You must give up *realism*. There may be a way to salvage realism, but to do so you must sacrifice *locality*[3]."

This quote from Hartsfield presents a stark and unavoidable choice. To make sense of the quantum world, we must abandon one of the foundational pillars of our classical reality. This is the ultimate paradox. The rules that govern the very building blocks of our universe are in direct and irreconcilable conflict with the rules that govern our experience of it. I argue that this fundamental disconnect is not a problem to be solved by a future theory but is instead a profound clue. It is a powerful piece of evidence that a purely materialistic worldview, which demands a single, coherent set of physical laws, is insufficient to describe the strange, observer-dependent reality we inhabit.

This fundamental conflict is not merely a scientific puzzle; it is a deep philosophical problem that strikes at the heart of what it means to know something about reality. As philosophers of science Giancarlo Ghirardi and Angelo Bassi summarize:

> "Quantum mechanics, with its revolutionary implications, has posed innumerable problems to philosophers of science. In particular, it has suggested reconsidering basic concepts such as the existence of a world that is, at least to some extent, independent of the observer, the possibility of getting reliable and objective knowledge about it, and the possibility of taking (under appropriate circumstances) at least some properties to be objectively possessed by physical systems. It has also raised many other questions which are well known to those involved in the debate on the interpretation of this pillar of modern science… in its standard formulation and interpretation, quantum mechanics is a theory which is excellent (in fact it has an unprecedented success in the history of science) in

telling us everything about *what we observe*, but it meets with serious difficulties in telling us *what there is*. We are making here specific reference to the central problem of the theory, usually referred to as *the measurement problem*[4]."

The "measurement problem" that Ghirardi and Bassi describe—the profound difficulty in telling "what there is" at the quantum level—is not just an abstract philosophical issue. It manifests in the real world in bizarre and counter-intuitive phenomena. The two most famous and well-documented examples of these violations of our classical reality are the collapse of the wave function and quantum entanglement, a word that made its first appearance in German when Erwin Schrödinger of the famous Schrödinger equation called it, *Verschrankung* which he later translated into English as *entanglement*.

Quantum entanglement refers to the fact that two particles can be joined in such a manner that the state of one *determines* the state of the other, while at the same time, each particle is in *all possible states simultaneously*, a concept referred to as "superposition" according to Schrödinger's wave function equation. "Entanglement comes from two requirements of quantum mechanics that seem to be on a direct collision course[5]." the first is *conservation laws*, "in all quantum systems energy, momentum, and angular momentum are conserved. In the absence of external forces or torques, their values must remain unchanged as the system evolves," and the second is the *Heisenberg Uncertainty principle* "energy, momentum, and angular momentum are all subject to Heisenberg's uncertainty principle. Whenever their complementary variables–time, position, and angle, respectively, are tightly determined, energy, momentum and angular momentum may be indefinite and unspecified and typically can span large ranges of possible values. This non-specificity persists until measurements are made that fixes them with measured values[6]."

This means that conservation, for two entangled particles will be preserved *but* in a state of superposition, and each particle is conserving both parts of the system until it is measured, at which time it definitely has one piece or the other, and only measurement will reveal which. Contrary to Einstein's intuition, which was subsequently proven wrong, the variable is not "hidden" until measurement, the variable does not exist for either particle until one is measured. "This mechanism that implements this cross-space dependence and enforces conservation laws is not apparent in the quantum formalism. Therefore entanglement joins the unresolved quantum mysteries of wave-particle duality and wave function collapse[7]."

Erwin Schrödinger was among the first of his peers to recognize the peculiarities surrounding entanglement and its world changing result. He considered entanglement much more than just a curious feature, rather he saw it as "the characteristic trait of quantum mechanics, the one that enforces its entire departure from classical lines of thought[8]," and a relationship where the state of a particle in an entangled system can't be fully described without the other, irrespective of the distance between them. Schrödinger described the profound mystery of this connection in his 1935 paper, *Discussion of Probability Relations Between Separated Systems*:

> "When two systems, of which we know the states by their respective representatives, enter into temporary physical interaction due to known forces between them, and when after a time of mutual influence the systems separate again, then they can no longer be described in the same way as before, viz. by endowing each of them with a representative of its own. I would not call that one but rather the characteristic trait of quantum mechanics, 'the one that enforces its entire departure from classical lines of thought. By the interaction the two representatives (or wave functions) have become entangled. To disentangle them we must gather further information by experiment, although we knew as much as anybody could possibly know about all that happened. Of either system, taken separately, all previous knowledge may be entirely lost, leaving us but one privilege: to restrict the experiments to one only of the two systems. After reestablishing one representative by observation, the other one can be inferred simultaneously. In what follows the whole of this procedure will be called *the disentanglement*. Its sinister importance is due to its being involved in every measuring process and therefore forming the basis of the quantum theory of measurement, threatening us thereby with at least a *regressus in infinitum*, since it will be noticed that the procedure itself involves measurement[9]."

The question, which has puzzled physicists since this phenomenon was discovered, is how does the electron know it's being measured, and thus presents itself in a specific measurable location as a particle and instantaneously ceases to exist as a probability cloud or wave? This question, and this odd condition, in violation of so many principles proven as foundational to classical physics, is why many proponents of QM believe it is ultimately *consciousness* and not measurement, or observation that collapses the wave function, because measurement and observation are activities only undertaken by consciousness.

This brings us to the ultimate implication of entanglement. If we accept the premise that consciousness is what collapses the wave function, quantum mechanics has presented yet another puzzling mystery for us to solve, which is the instantaneous connection between two entangled particles and consciousness. While our consciousness can explain the collapse of the particle we are observing, it cannot possibly be the direct cause of the collapse of its partner on the other side of the universe.

This implies that the quantum system itself must possess some form of awareness. It need not be biological, and it need not be at the same level of human consciousness, but nonetheless, it must be some form of conscious awareness that knows (has history) and retains information (memory) and causes the collapse of the wavefunction that reveals information. This is a chain of events that can only be obtained if entanglement and consciousness are related phenomena that work together to break all of the rules of classical physics. We must confront that the unobserved particle must "know," instantaneously, that its partner is being measured. This leads to a startling conclusion: the very question, "How does the electron know it's being measured?" is its own answer. Once again, it can only "know" if it has some intrinsic property that is aware of being observed. In this view, consciousness is the simplest explanation for the bizarre, non-local reality that entanglement reveals.

Observation is at least a partially *real* and *local* influence of quantum behavior, as demonstrated by entanglement; the observation is local and real for the observed particle and due to entanglement, experienced *as if* it were local and real for the entangled particle. Consciousness can apparently do what the speed of light can't.

Observation has been noted in a number of quantum effects, such as *the quantum Zeno* effect named after Zeno of Elea 490-430 BC and his three paradoxes of motion. In the quantum Zeno effect, if you perform an experiment with a quantum particle that can only be in two states, then it has some probability of being in either state and of switching states. Since it's a quantum particle, this means you *observe* the particle to see which state it's in. If the particle is unobserved it will be in both states at the same time. The conclusion is that observation is the only way to know its state.

Thus, the Zeno effect is the finding that "certain quantum events - like the electrons moving between atomic energy levels, or the decay of atomic nuclei, can be frozen through the simple act of observation[10]." an observed particle with two possible states will stay in one state and not switch states if it is constantly observed. And the less it is observed, the higher the probability it will switch states as a function of time[11].

This is a finding with paradigm breaking implications. The Zeno effect suggests that observation—and by extension, consciousness—is not a passive act, but an active force that can fundamentally alter physical reality, seemingly freezing time for a quantum system. This is one of the most direct and bizarre pieces of evidence for the participatory nature of consciousness in the universe.

A simple explanation could be that observation interferes with the system and inhibits its ability to switch. But the strange thing about this effect, confirmed by experiment[12], is that you don't have to observe the particle to see if it has switched states. If you set the experiment where there are two locations for the particle that correspond to different states, if the particle starts in the upper chamber, all you have to do is constantly monitor the lower chamber, and this will also decrease the probability the particle will switch states, as if you were observing the particle[13]. These experimental results force us to pose a probing question to the materialist worldview. It is not just, "How does a particle 'know' it is being observed?" We must also ask, "How does a particle 'know' it isn't being observed?" What physical law, constant, or parameter could possibly endow the quantum realm with the "sense-based" ability to distinguish between being watched and being alone? At present, physics has no answer.

Zeno of Elea, 490-430 BC had three paradoxes for moving things where he asserts, because of the existence of infinities, that certain motions are impossible. His first paradox was that "motion cannot exist[14]." Zeno's second paradox is, "Achilles can never overtake the tortoise in a race"[15], and Zeno's third paradox is the reason why the quantum Zeno effect is named after Zeno of Elea. In this paradox, called, "The arrow cannot fly[16]," Zeno argues that if an arrow is in flight, it must occupy a specific point in space at any given instant. However, to occupy a single point at a single instant is to be at rest. Since the arrow's entire flight is just a series of these instants, and at every instant the arrow is at rest, Zeno concludes that the arrow cannot be moving at all.

Werner Heisenberg

Werner Heisenberg's principle of indeterminacy (uncertainty principle) along with his work on quantum mechanics won him a Nobel Prize, while also affirming Zeno's paradoxes. Heisenberg, similar to Zeno, recognized that on the quantum scale, an observation we actually experience does not allow us to analyze its motion: " 'an object cannot occupy a given place and be moving at the same time.'

Heisenberg recognized that observation, as we actually experience it, does not allow us to analyze motion onto infinity. Sooner or later, we see that our activity introduces discontinuities in whatever we are observing. These discontinuities are fundamental to the new physics[17]." From Werner Heisenberg's Nobel lecture:

> "Bohr has shown in a series of examples how the perturbation necessarily associated with each observation indeed ensures that one cannot go below the limit set by the uncertainty relations. He contends that in the final analysis an uncertainty introduced by the concept of measurement itself is responsible for part of that perturbation remaining fundamentally unknown. The experimental determination of whatever space-time events invariably necessitates a fixed frame - say the system of coordinates in which the observer is at rest - to which all measurements are referred. The assumption that this frame is "fixed" implies neglecting its momentum from the outset, since "fixed" implies nothing other, of course, than that any transfer of momentum to it will evoke no perceptible effect. The fundamentally necessary uncertainty at this point is then transmitted via the measuring apparatus into the atomic event[18]."

In this crucial passage, Heisenberg is explaining that quantum uncertainty is not just a strange feature of the subatomic world, but an unavoidable consequence of the act of measurement itself. His argument is that for a large-scale observer to measure a tiny quantum system, they must interact with it, and this interaction inevitably introduces a degree of uncertainty that can never be eliminated. The very act of a classical observer measuring a quantum system fundamentally limits what can be known about that system. This reinforces the idea that the observer is not a passive spectator, but an active participant in the reality being observed.

This is exactly what happens with the application of consciousness (measurement/observation). This is the real puzzle, and it appears in such violation of our previously understood rules of physics that there is a group of scientists that reject the notion that consciousness collapses the wave function and is responsible for the assignment of values upon the collapse of the wave function. They instead believe that the wave function, when it is observed to collapse, does so under a process agnostic to consciousness and/or measurement and that there is another explanation altogether. This viewpoint has a number of different explanations for how the wave function collapses without observation or consciousness, and how particles assume their values. We will explore this area shortly.

> "Quantum mechanics indicates that there is an order to the universe. It simply isn't the order we expected. Even describing the true order

> of the universe is difficult because it involves something more than the physical world. Involves us, our minds, and our thoughts. Just how physics and our minds are brought together is a controversial subject. The gradual recognition that what we think may physically influence what we observe has led to a revolution in thought and philosophy, not to mention physics[19]."

This is the insight that quantum mechanics forces upon us. The quote from Fred Alan Wolf makes it clear: the observer is not a detached spectator looking in on an objective, independent reality. Instead, our minds, our thoughts, and the very act of observation are inextricably woven into the fabric of the cosmos. This "revolution in thought" is a direct challenge to the materialistic paradigm, which insists on a firm dividing line between the subjective mind and the objective physical world. As we will continue to explore, this is a key piece of evidence suggesting that consciousness is not an accidental byproduct of the universe, nor an illusory factor in the observation of quantum behavior as it has been framed by some, rather, a fundamental and participatory component of it.

Of all of these differences between Classical Physics and QM, perhaps the most important difference between these two theories is the most controversial, and that is for QM to be understood, it must be *interpreted.* QM is so vastly different from what is intuitive to humans, left without interpretation it has little meaning. It demands an interpretation for anyone attempting to describe what they witnessed. At the same time, because what is observed is such a mystery, QM leaves itself open to *multiple* interpretations.

This ambiguity is the reason for the deep, ongoing "interpretation wars" that have divided physicists for a century. To understand the profound nature of this conflict, we must first establish the baseline: the classical interpretation of physics, the common-sense, intuitive view of reality that governed science for centuries. Only by understanding this traditional framework can we appreciate the revolutionary and bizarre nature of the quantum world and its most famous rulebook, the Copenhagen Interpretation.

The Participatory Universe

For centuries, our understanding of the universe was governed by classical physics, the intuitive, common-sense framework established by figures like Newton and Einstein. This worldview is built on the two foundational pillars we reviewed earlier:

Realism, the idea that objects have definite properties independent of our observation, and

Locality, the principle that an object can only be influenced by its immediate surroundings. In this clockwork universe, reality is objective, and every effect has a clear, physical cause.

This tidy, deterministic picture, however, was shattered by the discoveries of the 20th century. The bizarre and paradoxical behavior of particles at the subatomic level revealed a new, far stranger set of rules, forcing a radical departure from these classical ideas. We will now explore this new reality and the profound philosophical problems it creates.

The arrival of consciousness on the cosmic stage is not a trivial detail; it is a universe-altering event. Before intelligent observers, the cosmos existed as a set of physical potentialities, a wave function of unrealized possibilities. But without a mind to observe it, what was it? A universe without consciousness is a book with no reader, a symphony with no audience. The physical laws and the matter they govern existed, but "physics" as a lived experience did not.

Quantum mechanics forces us to confront a startling possibility: that a conscious observer is a necessary ingredient for reality itself. In this view, nothing in our universe is truly "real"—observed, experienced, or transformed—until a conscious entity is there to witness it. This is the eye-opening lesson of the quantum world; consciousness is not an accidental byproduct of our material and biological experience but may be deeply and fundamentally entwined with the very fabric of the cosmos.

If we take the insights of quantum mechanics seriously, the implication challenges both the classical physics paradigm and materialistic paradigm. In the quantum mechanical view, consciousness is the special ingredient that collapses the wave function of the entire cosmos, turning a set of abstract physical laws into a rich, interconnected, and dynamic experience. The very act of observation—of conscious inquiry—is what appears to bring the universe into a state of definite reality. The implication, then, is that we are not accidental bystanders in a cold, indifferent cosmos. We are fundamental participants in the ongoing act of its creation. And if we consider this reality a bit further, if this insight is true, then what is it that our consciousness is interacting with to create this "rich, interconnected and dynamic experience"? Must it not be a form of consciousness or a corollary of consciousness itself? Or at least, a system made of and governed by something that possesses the ability to interact with consciousness? Logically, it seems that our consciousness, if it can interact with and change something with a thing, that thing must be sensitive to, and receptive and transformative of, with, upon, and unto our consciousness, as a form of highly specified information that in itself is essential to the rules of the universe. This is an idea we are obligated to consider as we gather the threads of our investigation and apply our understanding to the nature of quantum mechanics. Either way, this insight—which places consciousness at the very heart of existence–both material and biological, is the foundational principle of the first and most widely accepted interpretation of the quantum world.

The Copenhagen Interpretation

This bizarre, observer-dependent reality we have just explored—where the act of measurement collapses a wave of possibilities into a single, concrete outcome—is not a fringe philosophical idea. It is the core of the Copenhagen Interpretation, the first and still most widely taught framework for understanding the quantum world. Developed by pioneers like Niels Bohr and Werner Heisenberg, it is the formal scientific "rulebook" that accepts the paradoxes of the quantum realm, including the central and unavoidable role of the observer as fundamental features of our universe.

The Copenhagen Interpretation did not spring into existence overnight. It was the culmination of a breathtaking, three-decade scientific revolution that dismantled the classical worldview, piece by piece. The journey began with Max Planck and Albert Einstein, who first introduced the bizarre idea that energy comes in discrete packets, or "quanta" . This was followed by a cascade of groundbreaking discoveries: the Rutherford-Bohr model revealed the quantum nature of the atom; Louis de Broglie proposed that particles like electrons could also behave as waves; and finally, Erwin Schrödinger's famous wave equation provided the mathematical heart of the new physics. The Copenhagen Interpretation was the formal framework created to make sense of this strange and wonderful new reality.

The culmination of this scientific revolution was the development of the Schrödinger wave equation in 1926. This equation is the mathematical heart of quantum theory, and its bizarre implications are the source of the profound mysteries we have been discussing. To understand the core of the Copenhagen Interpretation, we must first understand what this equation says about the nature of reality. In his book The Dancing Wu Li Masters, Gary Zukov provides a clear explanation:

> "Once set in motion the wave function of a quantum system will continue to develop (change) according to causal law (the Schrödinger wave equation) until the system interacts with the observing system. At that instant, one of the possibilities contained in

the wave function would actualize and the other possibilities contained in the wave function would cease to exist. They would simply disappear. The wave function, that strange animal that von Neumann was attempting to describe, would 'collapse'. The collapse of this particular wave function [the photon] would mean that the probability of one of the possible results of the photon-measuring-device interaction became *one* (it happened) and the probability of the other possibilities became zero (they were no longer possible) ... The wave function, according to this view, is not quite a thing yet it is more than an idea. It occupies that strange middle ground between idea and reality, where all things are possible, but none are actual. Heisenberg likened it to Aristotle's *potential*." [all emphasis is from the original][20]

Until the collapse of the wave function, particles are in superposition, existing in all possible states, and these states, once the wave function collapses, due to measurement, become fixed as particles. The wave-particle duality instantaneously becomes a point particle; the superposition is no more. A one-sentence description of The Copenhagen Interpretation by English physicist and theologian John Polkinghorne is "A family of interpretations of quantum theory deriving from Niels Bohr and emphasizing indeterminacy [the apparent *necessary* incompleteness in the description of a physical system based upon the Schrödinger equation] and the role of classical measuring apparatus in measurement[21]."

This idea that observation and measurement and perhaps *consciousness* is key to QM, is deeply unsettling to a group of classically aligned physicists. The leader of this group was Albert Einstein, and he went head-to-head with Neils Bohr through a series of his famous thought experiments attempting to punch holes in QM and its belief that observation defined quantum characteristics. Strangely, or perhaps not quite so much, the role of observation, more specifically the observer is also key in General Relativity. In General Relativity, in Einstein's famous thought experiment involving a moving train, an embankment, lightning strikes, the speed of light, and two observers with different frames of reference, an event observed from an "objective" perspective as being simultaneous, can be measured by a second observer as happening at different moments in time.

In other words, under general relativity, there is no "objective" perspective, all perspectives are relative, and relativity is the frame of reference for an observer. In the same manner, as asserted by QM, the only frame of reference is not objective (the wave function) rather, it is relative, and in the collapse of the wave function, an observation is made, which unlike the frame of reference for general relativity, is shared with all other observers. As Neils Bohr believed, from both general relativity *and* quantum mechanics, "observation in physics is context -dependent." and the frame of reference of the observer determines the reality that is experienced[22] .

Perhaps the most famous and baffling example of this "context-dependent" reality is the strange case of an electron in an atom. As physicist Fred Alan Wolf explains, the very stability of matter depends on a bizarre quantum paradox that is directly tied to the act of observation.

> "The quantum laws allow, and in fact require, that any physical system exist in a superposition of possible states. For example, an electron in an atom may occupy several, indeed, an infinite number of positions simultaneously. Each position would be called a position state for the electron. In order that the atom possess a unique and stable state of energy, the electron must exist in a superposition of states. This superposition appears as a cloud called the *electronic wave cloud*. Without this cloud, the atom would not be stable and would disintegrate spontaneously. This disintegration occurs if someone actually takes a look at the electron. The cloud suddenly collapses into a single location at a single point of space and time, and the electron no longer remains bound to the atom. On the other hand, if the measurement of the position of the electron in the atom is not attempted and the atom's energy is observed, the cloud persists, and the atom remains in the stable energy state. This trade-off between possible observations is known as the *principle of complementarity* and has been a problem for physicists ever since quantum mechanics was invented. The problem is that since quantum physics requires an electron to be a cloud, just where is it when it is occupying space in the atom? Since any subatomic object, such as the electron, is supposed to be a pointlike particle, it cannot occupy the whole atomic space and nothing but the whole space, so help it. If it is a point, then where does it exist as such? Moreover, how does such an innocuous event as looking at an electron cause it to suddenly appear as a point?[23]"

This is the central, unavoidable paradox of the quantum world, often called the "measurement problem." The quote from Fred Alan Wolf makes it clear: the stable existence of an atom relies on its electrons being a cloud of possibilities, yet the very act of observation forces that cloud to collapse into a single, definite reality. This raises an important unanswered question: how does the innocuous act of "looking" have the power to fundamentally alter the physical world? It is this deep and bizarre mystery that has led some physicists to propose one of the most fantastical ideas in all of science, the *Many Worlds Interpretation.*

Fred Alan Wolf's, *Taking The Quantum Leap,* discusses the point where the Schrödinger equation results in *The Many Worlds Interpretation.*

> "The answer [from above: "how does such an innocuous event as looking at an electron cause it to suddenly appear at a point."] as I previously discussed, is that it exists as a pointlike particle in each of a dizzying infinite number of parallel universes. The superposition of all of these electrons, each in an other-worldly parallel universe, is the appearance of a single electronic cloud in this universe. This interpretation of quantum physics was first put forward by physicist Hugh Everett and is called the *Many Worlds Interpretation*[24]."

This, then, is the extraordinary proposal of the *Many Worlds Interpretation.* To solve the paradox of the measurement problem and preserve a deterministic, non-random view of physics, it posits that the wave function never actually collapses. Instead, at the moment of measurement, reality itself splits into a potentially infinite number of parallel universes to accommodate every single possible outcome.

The Everett Interpretation, aka Many Worlds Interpretation (MWI)

In his 1957 PhD thesis, the physicist Hugh Everett III proposed a radical new way to solve the measurement problem. Everett's solution was to take the Schrödinger equation at its most literal. He proposed that the wave function never actually "collapses." Instead, at the moment of measurement, reality itself splits into multiple, parallel universes to accommodate every single possible outcome. In the famous example of Schrödinger's cat, the universe would divide upon observation, creating one universe in which the cat is alive, and another, equally real universe in which it is dead. In this view, the wave function is always ongoing, and what we perceive as a "collapse" is simply our own consciousness proceeding down one of these branching timelines.

As we will see, this is a prime example of the kind of fantastical and untestable conjecture that is often employed to avoid the paradigm falsifying, observer-dependent implications of the standard quantum mechanical worldview. It's bizarre nature led one of the most decorated physicists in history, Roger Penrose, to refer to the MWI as "baffling."

> "For some reason people have a very strong faith in the formalism of quantum mechanics (which is) the unitary evolution of the

> Schrodinger equation, which tells you that these super positions persist… if you don't try to modify (the Schrodinger equation) then what do you do with that dead and alive cat (Schrödinger's cat)?... the argument is that somebody comes along and looks at it, and then that person becomes a superposition of seeing a dead cat and a live cat and the argument then is that these two alternatives inhabit parallel worlds. I don't quite see that because it's not what the theory says. In a sense what the theory says is that the cat is in a superposition of these two states and then you've come and look at it you're in a superposition of these two states. It's not as if there are two different worlds. There is one world which is a superposition… We don't experience that… My Oxford colleagues in the philosophy department are very much sold on this particular idea [the many world interpretation] and so they have very subtle ways of trying to explain it, but to me… it isn't two worlds. .. it's a superposition of these two alternatives in one world. Why can't you have a world with a superposition of different perceptions? It really relates to how one perceives things and this is getting us into the consciousness issue. They don't like that… It seems to me, even to make sense of many worlds, you need a theory of consciousness which doesn't allow the superpositions so in some sense your conscious perception only receives one of these superpositions[25]."

Penrose's critique is a powerful argument that dismantles the primary assumption of the Many-Worlds Interpretation. His key point is the theory doesn't actually solve the problem it sets out to; it simply renames it. The Schrödinger equation describes a single universe in a bizarre state of superposition (e.g., a single world containing a cat that is simultaneously dead and alive), not two separate, parallel worlds. To explain why our conscious perception only ever experiences one of these realities, Penrose argues that even the Many-Worlds Interpretation must ultimately appeal to a special theory of consciousness—the very thing its proponents are often trying to avoid. In this view, the MWI is not a true alternative, but a baffling rephrasing of the original mystery.

The irony is that it was Erwin Schrödinger himself—the architect of the wave equation—who first pointed out the intellectual inconsistency of those who used his work. In a quote recounted by John Gribbin, Schrödinger noted the strange position of the typical quantum theorist: "Nearly every result [the quantum theorist] pronounces is about the probability of this or that or that… happening–with usually a great many alternatives. The idea that they may not be alternatives, but all really happen simultaneously seems lunatic to him, just impossible... It is strange that he should believe this."

Schrödinger's point is a subtle but powerful one, and in it, he illustrates the appeal of the MWI although he never formally endorsed it. He found it "strange" that a physicist could accept a mathematical formalism describing multiple co-existing possibilities yet find the literal interpretation—that all these possibilities are in some sense real—to be "lunatic." He was the first to suggest that the most logical, if bizarre, conclusion of his own work was the existence of something very much like "many worlds."

While it is a fascinating historical irony that Schrödinger himself saw the "many worlds" concept as a logical, if bizarre, conclusion of his own work, this does not resolve the many foundational problems with the interpretation. Acknowledging a logical possibility is not the same as formulating a viable scientific theory. The Many-Worlds Interpretation, despite its conceptual allure, raises a far more fundamental question that its proponents often fail to address.

The Imaginative Function of The MWI

And here is a question posed to the MWI from the world of physics, "What term, or symbol in the Schrödinger equation has the energy and power to bring these *many worlds* into physical existence out of thin air?" It is an incontrovertible fact that we don't understand how our own universe came into existence, and yet, the Schrödinger equation, which does not refer to, nor has any term within it that can be said to possess world creating powers, is assumed by proponents of the MWI to inherently possess the power to create *endless* worlds that are exact duplicates of our own. How? This power is snatched from the thin air and justified solely on the basis that a group of stymied physicists cannot solve the collapse of the wave function, a problem that Roger Penrose has called one of the greatest unsolved problems in physics. The collapse of the wavefunction is the primary problem, and the MWI is not the solution to that problem. It is a *non sequitur*. Many see it as absurd that we have exchanged a problem in our real world for the illusion of infinite worlds that each offers no solution for the collapse of the wavefunction.

This critique of the MWI is not a new one. The debate over the nature of quantum reality has a long and storied history, filled with brilliant minds grappling with these same bizarre implications. And the quandary this debate has led to an endless pit that has diverted attention from the implications the QM world presents. It is this diversion into fantasy and fantastical solutions, that has obscured the true message that lies within the mystery–the message that our algorithm has pursued and proposes provides a single elegant solution for not just the quandary of QM, but for the quandary of all of the mysteries currently challenging our physics. One positive result in the reality-obfuscation wars is that the standard Copenhagen Interpretation has remained the most robust and widely accepted view. It is indeed a strange framework, but only in equal measure to the phenomena it attempts to interpret. It adds nothing to the drama of quantum strangeness, and its founders were painfully circumspect with regards to limiting the interpretation to what had been observed, experienced and was necessary to explain.

A View from History

The MWI should sound familiar from our discussion of *the multiverse* under inflationary theory. It is part of a strange trend in science, beginning in the mid 21st century, where our inability to resolve conflicts and paradoxes within science has led to the creation of increasingly strange, fantastical and unprovable hypotheses. Our narrative is hardly the first to point out that the world of science is in a state of crisis that has stymied our best efforts for more than 100 years. A crisis where no resolution appears to be forthcoming to separate scientific fact from science fiction.

With regards to the MWI, like the inflationary multiverse, and the string theory multiverse, they propose to divide our single universe into an infinite number of alternative universes, all to explain our single universe. And what all of these proposals have in common, is that they are all scientifically untenable. None can be validated, none can be falsified, and none can be *observed*. It will be helpful to turn to the history of the Copenhagen Interpretation to gain a firmer foundation on a set of principles derived from a time when the world's greatest scientists restricted themselves to problem solving strategies worthy of a Nobel Prize, instead of a Hugo Award.

In a 1959 essay for the American Journal of Physics, the philosopher of science Norwood Russell Hanson provided a powerful defense of the Copenhagen Interpretation. Writing at a time when the theory's foundations were still being debated, Hanson argued for its conceptual robustness and its indispensable power in helping physicists understand reality. The following passages from his work are crucial for understanding why this interpretation has remained the bedrock of quantum mechanics.

> "Many early objections to, and some present groans about, the Copenhagen interpretation, arise from not appreciating the historical and conceptual role played by Dirac's paper [the Dirac equation] Here is the notational key to all subsequent quantum physics. Yet, in

> that paper (Dirac tells me) the Copenhagen interpretation figured essentially–not as some philosophical afterthought Dirac appends to his algebra, but basic to every operation with the notation. Feyerabend provocatively suggests that this need not have been so, that it would be possible to have a 'minimum' (i.e., non-Copenhagen) interpretation of quantum theory, and hence of Dirac's paper. But as a matter of fact this is not the way in which this fundamental paper was written. This would not be the same paper were its assumptions 'purified' as Feyerabend suggests. Largely because of these tough-minded and realistically practical suppositions Dirac's theory had complete success (at least before the era of the meson). What critics of the Copenhagen interpretation often fail to see is that just to ask for an alternative account of micronature without actually proving one which works, is to reinvite the chaos which it is Dirac's triumph to have ended[26]."

> "Dirac's theory of the electron unites in one formally beautiful, and experimentally powerful theory every idea of the particle physics of the Twenties. He provided a comprehensive model for the hydrogen atom, explained the Compton scattering of electrons, the Zeeman effect (doublet atoms), and the empirically required electron spin; all these independent elements were forged together into an algorithm whose purpose was to achieve a relativistically invariant theory for fast electrons... Dirac's paper established quantum mechanics as a unified description of nature[27]."

The Copenhagen interpretation arises from the "necessity of describing nature and thus arises all of the conceptual constraints of quantum theory... In microphysics it is arbitrary whether one uses a wave or a particle language for descriptions, --just so one is aware that both are jointly valid. Several conclusions follow, which it is the merit of the Copenhagen school boldly to have adopted. The fact is that micro phenomenon is a conspiracy of wave and particle properties. But after admitting this (as one must) one must then also maintain a symmetry between these modes of description in all further theoretical work[28]."

> "Born dispelled this cloud (of uncertainty) with the ingenious suggestion that the waves be taken as a measure of the probability of locating a particle within a given volume element. Because it was operationally clear, and corroborated by every known experiment, Born's view was quickly adopted, and generalized for multiparticle distributions by Bohr, Heisenberg, Gordon, Jordan, Klein, Paul, and most significantly by Dirac[29]."

The core of Norwood Hanson's argument is that the Copenhagen Interpretation is not a flimsy philosophical add-on, but is woven into the very mathematical fabric of quantum mechanics that has proven so stunningly successful. He makes three crucial points:

-First, this interpretation arose from the practical necessity of describing a bizarre reality where particles behave as both waves and particles.

-Second, that its probabilistic nature was adopted because it was clear and corroborated by every known experiment.

-Third, that any proposed alternative must be held to the same rigorous standard of providing a working, testable model, not just speculative ideas.

This historical perspective provides the foundational justification that establishes the Copenhagen Interpretation not as one of many equal "opinions," but as the robust, battle-tested foundation of modern physics.

And finally, the issue with alternative interpretations of QM. This paper, written well before the Everett Interpretation, nonetheless dismantles it by the rigor of its foundational assumptions and relevant critique of alternative approaches.

"There is as yet no working alternative to the Copenhagen interpretation. Ask your nearest synchrotron operator. It therefore seems a questionable procedure to make every new and tentative speculation sound as if it were a clear alternative which could easily revolutionize the foundations of physics, if only the elder statesman would stop backing their favorite horse so uncritically. The issue is this: until you formulate a new interpretation which works in every particular as well as does the old one, call your efforts by their proper name, 'speculations.' This makes them no less worthwhile. And if it be riposted that the Copenhagen interpretation is itself but a speculation, then please let us distinguish those speculations which have proven themselves to work in theory and practice from those which have not yet even been put to any rigorous test[30]."

Hanson's challenge here is a powerful one. A "synchrotron operator" is a hands-on experimental physicist who uses massive particle accelerators—the workhorses of modern physics. His point is that while theorists may debate endless speculative alternatives, the Copenhagen Interpretation is the one that has proven itself to be the essential, practical toolkit for the scientists actually doing the work. In his view, any new interpretation must be held to the same high standard: it must not only be a compelling idea, but it must work in theory and in practice. Until then, he argues, such ideas should be called by their proper name: "speculations." This standard of separating leather chair, smoking room speculations from rigorous chalkboard facts has become even more important as the slew of problems that have cropped up since 1945 have created the modern crises in physics.

This historical view solidifies the Copenhagen Interpretation not as a philosophical preference, but as the most robust and experimentally verified framework we have for understanding the quantum world. Its conclusion—that reality at its most fundamental level is probabilistic, non-local, and inextricably linked to the act of observation—is not a problem to be explained away by speculative alternatives. I argue that it is a foundational clue. This deep and irreducible mystery at the heart of physics is precisely what we should expect to find if the universe is governed not by simple, materialistic laws, but by a deeper, purposeful logic.

What Consciousness Does That The Universe Cannot

Let us think for a moment about what it means if consciousness is what collapses the wave function.

It can be said that one of the reasons why materialists do not believe that consciousness collapses the wave function is that, for a materialist, consciousness itself is pure physics and not a privileged level of organization. Consciousness to the materialist is not beyond physics; it is fully represented by materialism, and a wholly physically described phenomena. Consciousness in this conception is atoms, particles, and physics. An emerging property of the movement and interaction of particles and nothing more. Anything that cannot be explained by this belief, is illusory, or false, or psychological, and does not actually exist as a real thing.

This means that the *physics* of the observer would have no more ability to *cause* the collapse of the wave function than any other collection of atoms in proximity to the experiment, reducing the attribution of the role of observation to the collapse of the wavefunction to an erroneous perception. For the materialist, consciousness has no power to cause the collapse of the wave function. This is an important point. What is at stake is more than the idea that consciousness (observation by a conscious being) causes the collapse of the wave function, but also, the definition of what consciousness is. Based upon the paradigm and the orientation of the individual analyzing the measurement problem, it is either possible for it to cause the collapse of the wave function, or it fundamentally is not.

For a materialist who believes that consciousness is just an illusion or a byproduct of physical processes, the idea that a mind can collapse the wave function is a serious problem. If consciousness is nothing more than atoms and physics—a brain receiving signals—then an observer should have no more power to alter reality than any other collection of atoms whether these atoms are incorporated in biology, chemistry, or combinations of elements such as rocks, asteroids, planets, etc.

This forces the materialist into a corner. To maintain their worldview, they must argue that the collapse is not caused by consciousness, but by a purely physical process. In this view, "measurement" is redefined as a simple physical "interference" with the quantum system, stripping the act of observation—and the observer—of any special status. This is a necessary move to avoid the profound, non-materialistic implications that the standard interpretation of quantum mechanics seems to demand.

Having considered this view, we can now consider the view that consciousness *does* cause the collapse of the wave function as a thought experiment. In this respect we are using the term *consciousness* to refer to the observation, experience, awareness *of life*, from the simplest to the level of awareness to the consciousness of an intelligent observer, *inclusive* of measurement. If under these circumstances we consider the proposition that observation causes the collapse of the wave function, then consciousness is something beyond *physics*, and it can do what *physics* (pure physics) cannot. We can say this on two accounts.

Firstly, a living, conscious entity can *favor* and *privilege* a single moment in time above another for the purpose of observation or measurement. This is not something that physics, without the intervention of *consciousness*, can do. There is nothing *in* physics that can designate a moment in time that is privileged and cut off from all others, for the purposes of making an observation, or recording an event, or making a measurement.

> "In general relativity the mathematical properties of time (or better, of spacetime)—its topology and geometry—depend upon how matter is arranged rather than being fixed once and for all. So physics can be, and indeed has to be, formulated without treating time as a universal, fixed background structure[31]."

To understand what it means for consciousness to "privilege a moment in time," it helps to use an analogy. Think of time in physics, especially under Einstein's relativity, as a flowing river. From the river's perspective, there is no special "now"; the entire river—past, present, and future—exists as a single, continuous entity. Physics can describe the properties of the entire river, but it has no mechanism to single out one specific cross-section and declare, "This is the present moment."

Consciousness, however, is like taking a photograph of that river. The very act of observation or measurement is what "privileges" a single instant, freezing it out of the continuous flow and turning it into a definite "now." While physics can describe the entire, timeless river, only consciousness has the ability to take these "snapshots" of reality, a power that lies outside the known laws of physics.

Secondly, physics, by definition, is the study of the behavior of matter and particles *by humans*. None of this (study, or science or observation of behavior) exists without intelligent observers[32]. Humans conceive of physics *existing* without consciousness, but, as we investigated earlier in this section, the reason for this is we cannot, without great difficulty, envision the universe without physics, they are so indelibly linked and connected in our conception.

For a scientist, the universe is physics; the two are inseparable. To imagine one without the other is a severe thought experiment that leads to a profound question, similar to the classic philosophical riddle: If a tree falls in a forest and no one is around to hear it, does it make a sound?

A physicist might say that, yes, the falling tree creates pressure waves in the air. But "sound" is not just a pressure wave; it is the experience of that wave, perceived by a conscious mind. Without an observer, there are only unrealized physical processes. In the same way, a universe without intelligent observers may contain matter and energy, but it does not contain "physics." There is no one to discern its laws, to describe its behavior, or to experience its reality. What it is without consciousness can hardly be conceived. In this view, a universe without an observer is a tree falling in a silent, empty forest.

We can conclude our thought experiment with a powerful supposition. *Physics*, for all intents and purposes only exists within the consciousness of intelligent observers. And, without intelligent observers what the universe is cannot be described. *How* could it be viewed? What is it to *view* in the absence of *something that can* view? What is it to be a quality, when there is nothing that can know that quality? In this context, *the measurement problem* is a mere trifle from the list of things that become real when there is an observer. And through our example, we can also see that *all* observation is privileged, in the sense that it both creates and experiences something that does not exist in its absence.

This idea—that the universe is a participatory system that requires a conscious observer to become "real"—is perhaps the most radical and philosophically challenging conclusion of modern physics. It is an idea that was not accepted easily. In fact, it was so repugnant to the classical, deterministic worldview that it prompted one of the greatest minds in history, Albert Einstein, to launch his most famous intellectual assault against it. This assault was in the form of the EPR Paradox paper, written by Albert Einstein, (the "E" in EPR, along with P, -Podolsky and R, -Rosen) the foremost critic of the peculiarities of QM and a fierce critic of the Copenhagen interpretation, the EPR paper was his best and last attempt to impeach the bizarre reality of the quantum world.

The EPR Paradox Paper

The EPR Paper was the most forceful articulation of the western world's argument against the quantum mechanical violations of classical physics that the Copenhagen interpretation said were real. The scientific argument of the EPR paper may have been flawed, but its author was Albert Einstein. His involvement transformed the debate from a niche academic dispute into a very public and high-stakes battle for the soul of physics.

In a letter to Max Born on December 4tn 1926, Einstein made his view clear in one of his most quoted statements. "Quantum mechanics is certainly imposing. But an inner voice tells me that it is not yet the real thing. The theory says a lot, but does not really bring us any closer to the secret of the 'Old One.' I, at any rate, am convinced that He is not playing at dice. Waves in three-dimensional space whose velocity is regulated by potential energy (for example, rubber bands)... With best wishes, Yours, A. Einstein[33]."

The EPR Paradox paper was published in Physical Review on May 15th 1935 under the title, *Can Quantum Mechanical Description of Physical Reality Be Considered Complete?*, on May 15th[34]. The headline of The New York Times for May 4th, 1935 read, "Einstein Attacks Quantum Theory: Scientist and Two Colleagues Find It Is Not 'Complete' Even Though 'Correct.', the article began, "Professor Albert Einstein will attack science's important theory of quantum mechanics, a theory of which he was a sort of grandfather." The EPR paper stated, "Any serious consideration of a physical theory must take into account the distinction between the objective reality, which is independent of any theory, and the physical concepts with which the theory operates." by "objective reality[35]."

Einstein was making a direct reference to accepted classical physics principles that "objective reality" was both "real" and "local", two terms we have previously defined. To review, to be *real* means objects have *definite properties independent of observation*. To be *local* means objects *can be influenced only by their surroundings*. "It took Einstein's theory of general relativity to bring locality into sharp focus. Before Einstein, Newton had obfuscated locality by suggesting that gravitational influences were instantaneous… Einstein's general relativity showed that gravity is the outcome of the warping of space-time by the presence of matter… and any changes to the curvature of space-time caused by matter can propagate only at the speed of light….Besides locality, Einstein had long held dear the idea of 'realism'... For Einstein, the real world exists independent of our observations[36]."

This is the unsettling challenge that quantum mechanics presents to our understanding of reality. The two most basic, common-sense principles that govern our entire macroscopic experience—that things have definite properties whether we are looking or not, and that a cause must be physically close to its effect—appear to break down completely at the fundamental level of particles. The following investigation will explore this deep and, thus far, irreconcilable conflict between the classical world we inhabit and the bizarre quantum world that serves as its foundation.

The EPR paper implied that the results of QM, specifically, conservation laws and the Heisenberg uncertainty principle combined with *spooky action at a distance* for entangled particles, were incompatible, which meant that QM theory was neither real, nor local. Spooky action at a distance was another way of saying beyond the speed of light transfer of information which is one of the paradoxical findings of quantum entanglement. The EPR paper was the first to ask, "How can the wave functions describing separated components of a system of particles, perhaps light-years apart, have uncertain values for the conserved quantities and yet respect the conservation laws when measurements are made, wave functions are collapsed and the conserved quantities determined?[37]".

The core of the EPR argument is that any instantaneous, faster-than-light transfer of information would violate the known laws of physics. They therefore concluded that entangled particles could not be taking on their values at the moment of measurement. Instead, the particles must have possessed these definite, "hidden" properties all along, and the act of measurement simply reveals what was already there. This concept became known as the theory of "hidden variables." Because QM did not account for this reality, it was "incomplete" according to the EPR paper.

The EPR paper laid out two conditions for QM. The first condition was, "'Is the theory correct?' the second question was, 'Is the description given by the theory complete?'[38]" "Only in the case in which positive answers may be given to both of these questions, that the concepts of the theory may be said to be satisfactory[39]." The answer to the first question was "Yes". To date there had been no conflict between QM and the experiments performed in the laboratory. Einstein knew this. This was a concession that QM was rigorous, which would normally be the end of the discussion. But this is why the EPR paper had two conditions, and the second condition Einstein felt he could prove the answer was "No" which was, "Is the description given by the theory complete?[40]"

The EPR paper's definition of *complete* was, "every element of the physical reality must have a counterpart in the physical theory[41]."Einstein proposed a thought experiment involving two entangled particles, one measured and the other undisturbed. The paper argued that all of the characteristics of the second particle could be described simply by measuring the first particle, and that this result was not something that could be accounted for by QM, and therefore QM was not complete.

> "If, without in any way disturbing a system, we can predict with certainty (i.e., with probability equal to unity) the value of a physical quantity, then there exists an element of physical reality corresponding to the physical quantity[42]."

Einstein's thought experiment was seemingly valid on its face, and it took Niels Bohr some time to formulate a response. In the end, Bohr didn't find a flaw in Einstein's logic; instead, he argued that Einstein's entire premise—that particles must have "real" and "local" properties—was a classical assumption that simply doesn't apply to the quantum world.

The wider physics community ultimately sided with Bohr. They accepted that while Einstein's argument was clever, it didn't invalidate quantum mechanics; it only highlighted its bizarre and counter-intuitive nature. As physicist John Polkinghorne summarizes, the majority view concluded that a measurement on one entangled particle does produce an instantaneous change in the other, no matter how far apart they are. This "counterintuitive togetherness-in-separation" is not a flaw in the theory; it is a fundamental feature of reality.

> "Einstein's conclusion from [his] complicated argument was that there must be something incomplete in conventional quantum theory. It failed to account for what he believed must be definite values of spin components. Almost all of his fellow physicists interpret things differently… Thus the majority view leads to the conclusion that *measurement on 1 produces instantaneous change at 2, a change that depends precisely on exactly what is measured at 1.* In other words there's some counterintuitive togetherness-in-separation between 1 and 2[43]."

For nearly 30 years, the debate between Einstein's "hidden variables" and Bohr's "spooky action at a distance" remained a purely philosophical one. There was no known experiment that could definitively prove one interpretation right and the other wrong. The physics community largely sided with Bohr, accepting the bizarre, non-local nature of reality as a fundamental truth, but it remained a debate about interpretation, not evidence. This philosophical stalemate was finally broken in 1964 when a physicist named John Stewart Bell devised a brilliant theorem that could put "spooky action at a distance" to a real-world, physical test.

Bell's Inequality Theorem

Bell's Inequality Theorem proposed a clever experiment. By measuring the spin of two entangled particles along different, randomly chosen axes, one could statistically compare the results to what would be predicted by two different versions of reality: one in which the particles have "hidden variables" (Einstein's view), and one in which they do not (the standard quantum view).

The results of this experiment, which have been repeated with increasing precision over subsequent decades, and in a series of two seminal papers, one in 1964, *On the Einstein Podolsky Rosen Paradox*[44] and the other in 1966, *On the Problem of Hidden Variables in Quantum Mechanics*, both by John Stewart Bell[45] were a stunning vindication of quantum mechanics.

The data consistently violated the predictions of "hidden variable" theories, proving that Einstein's classical intuition was wrong. As the author notes, "the intrinsic nonlocality of quantum mechanics has been demonstrated," confirming that the "spooky action at a distance" is a real and fundamental feature of our universe.

The experiment works like this: two entangled particles are sent in opposite directions towards detectors that can measure their "spin" (think of it as a particle's orientation, like "up" or "down"). Crucially, these detectors can be rotated at different angles relative to each other. The experiment is run thousands of times, and the results are plotted on a graph to see how often the particles' spins match as the angle between the detectors changes.

The Classical Prediction (The "Hidden Instructions" View): If Einstein were right and each particle had a pre-determined, "hidden" instruction for how to behave, the results would be straightforward. The correlation between their spins would decrease in a simple, straight line as the angle between the detectors increased.

The Quantum Prediction (The "Spooky Action" View): Quantum mechanics, however, predicts a much stranger result. Because the particles are in a shared state of superposition, their correlation will not decrease in a straight line, but will instead follow a smooth, wave-like curve.

The results of this experiment, repeated for decades with increasing precision, have always traced the wave-like curve predicted by quantum mechanics, and have definitively ruled out the straight-line result predicted by classical, local "hidden variable" theories. This is the ultimate proof that the strange, interconnected reality of the Copenhagen Interpretation is the one we actually live in.

The robustness of Bell's innovative idea was such that recently the 2022 Nobel Prize for physics was awarded to a trio of physicists, John Clauser, Alain Aspect, and Anton Zeilinger whose experiments were a confirmation of Bell's inequality[46]. "Their experiments collectively established the existence of a bizarre quantum phenomenon known as entanglement, where two widely separated particles appear to share information despite having no conceivable way of communicating[47]."

Our long survey of quantum mechanics has arrived at a definitive conclusion: quantum strangeness is a real thing we must contend with. The confirmation of Bell's inequality in several experiments is a validation of quantum mechanics and the Copenhagen interpretation. This result shows us that QM resists interpretations that attempt to put it into a classical box. QM violates classical laws, including the speed of light limit on information, locality, realness, and the idea that an object can be in multiple states instantaneously.

In furtherance of our hypothesis, our reality is not fully reducible to material laws. QM is foundational to our reality. Our universe is founded on QM, and life of any kind would not be possible without QM. Covalent bonding, eyesight, metabolism, photosynthesis, the creation of elements in suns, and sunlight, are all foundationally quantum in nature—as is every imaginable thing in our universe.

This is the ultimate, non-negotiable fact that any theory of reality must confront: our universe, at its most fundamental level, is a quantum system. The bizarre, probabilistic, and observer-dependent rules we have just explored are not a strange anomaly; they are the bedrock of our existence. With this counter-intuitive foundation established, our investigation will now shift from the strange rules of the quantum world to the specific particles that play by these rules. This once again brings us to the Standard Model of Physics, the collection of fundamental particles and forces that make up our material reality, and, as we will see, another realm of deeply challenging and inexplicable fine-tuning.

PART II: FINE TUNING AND ANOMALIES OF THE STANDARD MODEL OF PHYSICS

In our previous section, *Part I: Fine Tuning and the Anomalies of Quantum Mechanics (QM)*, we showed that the core rules of quantum mechanics—superposition, entanglement, the observer effect, and the measurement problem—resist expression within a purely classical system. To make QM appear classical, processes such as decoherence and probabilities are required. This tension between what classical theory demands and what QM actually is was decisively clarified by Bell's theorem, confirmed by decades of experiments showing violations of Bell's inequalities. In plain terms, QM isn't classical, nor is it materialist—not even under the wild, unobservable, unprovable, hypotheses of the Everett (many-worlds) interpretation.

In *Part II: The Anomalies of the Standard Model (SM)*, we build on our earlier investigation of long-recognized gaps in the Standard Model's foundation. Neither gravity, dark matter, nor dark energy fall within its scope. Decades of high-energy experiments at CERN, promoted as our best shot at finding the final pieces of the Standard Model, have failed to reveal over 2 dozen predicted particles and interactions–such as axions, inflatons, gravitons, and others—proposed in theories of dark matter, supersymmetry, and beyond-Standard-Model physics. Taken together, these gaps signal that our current material framework is incomplete and cannot fully account for the reality we inhabit.

We now turn to a set of empirically observed anomalies within the Standard Model—phenomena whose values and relationships cannot be fully accounted for by the Standard Model's foundation, the classical/materialist framework. These include the stunning proton-mass mystery; the subtle proton–neutron mass difference; the enigmatic pattern of particle masses and charges; the Higgs Hierarchy Problem and the looming Higgs False Vacuum Catastrophe; the proton spin crisis; the puzzle of the SM's three generations; and the deep mystery of confinement. The questions raised by these phenomena reach beyond any patchwork, *ad hoc* or yet-to-be-imagined explanations. Viewed alongside the anomalies of quantum mechanics, they suggest that our practice of science has brought us to an inevitable and profound crossroads–pointing toward a deeper, still-unfolding, non-material understanding of the universe—consistent with the central premise of our argument: *The Divine Algorithm*.

The Standard Model (SM)

We can understand *The Standard Model (the SM)* by thinking of it as the particles and forces whose behavior is described by *Quantum Mechanics (QM) and quantum field theory (QFT).* QM and the SM are linked, one to the other, as two sides of the same coin, such that a complete understanding of each requires an understanding of the other.

The Standard Model is our best and most rigorously tested theory of the fundamental particles that make up our reality. It describes the behavior of the three families of elementary particles: the Fermions (which include quarks and leptons, the building blocks of matter) and the Bosons (the particles that carry the fundamental forces).

While the SM is incredibly successful in its own domain, it is also fundamentally incomplete. It has no explanation for a number of profound influences in our universe, the most notable of which are a series of crippling caveats. The SM does not account for the force of gravity, nor does it account for the estimated 95% of our universe that is composed of dark energy and dark matter. What the SM accounts for is what is referred to as *normal baryonic matter*—atoms, and their electrons, protons, neutrons—and the quarks and gluon that make up protons and neutrons. This has led physicists to propose new ideas for "physics beyond the Standard Model," such as the theory of supersymmetry, which speculates about a whole new family of undiscovered "partner" particles to fill the gaps. It is this combination of stunning success and profound incompleteness that leads the SM to be described as a beautiful, but ultimately flawed, theory.

> "The Standard Model (SM) of Particle Physics, developed in the early 1970s, summarizes the known fundamental structures of matter and forces (except for gravity). It is the result of theories and discoveries developed and made since the 1930s. The SM has been able to explain almost all experimental results and furthermore it precisely predicted a wide variety of phenomena. With precision experiments performed at previous and current colliders, the SM has been established as a physics theory tested to highest precision at the quantum level[48]."

In summary, the fundamental particles that make up our reality are described by the Standard Model (SM) of physics. The behavior of the particles of the SM is described by quantum mechanics (QM). The SM was built on the foundations of QM, but is governed by several guiding theories, chiefly quantum flavordynamics (QFD), electroweak theory (EWT), the quark model, quantum chromodynamics (QCD), and quantum electrodynamics (QED).

Once again, we must confront the most profound gap in the Standard Model: the complete absence of gravity. The reason for this omission is the single greatest challenge in modern physics. The Standard Model is a quantum theory, describing forces via the exchange of particles. Gravity, however, is described by Einstein's theory of General Relativity, which treats it not as a force, but as the curvature of spacetime itself.

These two monumental theories are fundamentally incompatible. All attempts to "quantize" gravity—to describe it with a force-carrying particle—have failed. This deep and unresolved conflict is why gravity remains a complete mystery on the quantum level, and why it stands outside our most successful theory of matter.

> "In all quantum field theories, the matter and force fields are described as lying on top of a smooth fixed continuous grid of spacetime -the special relativity stage- but, to describe gravity we have to quantize the spacetime stage itself. How are we supposed to do this? Without a stage to stand on? This is the problem of quantum gravity. It brings us to the edge of our understanding of physical reality. To the end of spacetime as we know it. It means (the grid of spacetime) cannot be fixed, it must have probability of being in many different states. there is even some probability for spacetime to start shredding itself, resulting in an awful soup of quantum uncertainty where bits of spacetime pop in and out of existence. All of this happens at the Planck length because here, gravity is quantum[49]."

The quote from Balasubramanian provides a stark visual representation of why quantum gravity is considered the single greatest problem in physics. In essence, our two most successful theories operate on different stages. Quantum mechanics describes the actors (particles and forces) on a fixed, classical stage (spacetime). General relativity, however, tells us that the stage itself is a dynamic actor. The problem is that we have no theory for what happens when the stage transforms into the theater of the quantum realm.

As the quote makes clear, this leads to a complete breakdown of reality as we know it—a "soup of quantum uncertainty" where the very concepts of space and time lose their meaning. This is the ultimate "information wall" at the beginning of our universe. It is a fundamental, and perhaps permanent barrier to any purely physical explanation for our origins, and it is a key piece of evidence that the cause of our universe lies beyond the reach of the science that governs it. Whereas the *wave function* remains a perplexing mystery involving information and choice, the unsolvable mystery of QM, *gravity* is the perplexing unsolvable mystery of the SM.

The profound and perhaps the permanent mysteries of gravity and the wave function are not the only cracks in the foundation of the Standard Model. The paradoxes extend to the very particles themselves—the "cast of characters" that make up our material reality. To appreciate the full scope of the fine-tuning required for our existence, we must now turn our attention from the overarching rules to the players on the stage, to exploring the bizarre and inexplicable properties of the fundamental particles that the Standard Model describes.

Electromagnetism has the electron, muon and tau as lepton particles and photons as its force carrying boson particles. For the *strong force* quarks, gluons are its force carrying boson particles. *The weak force* has neutrinos as its lepton particles and the W and Z bosons. These bosons mediate neutrino absorption and emission and induce electron or positron emission and absorption. The weak force W boson is the only particle that can change one particle into another, as in decay and transformation. The Z boson is the only particle that can change the spin, momentum, and energy of neutrinos. And the weak force is also the only force known thus far to display CP violation. It is the sole property of the weak force in transforming particles that allows protons to turn into neutrons within a star. This is the key to the nuclear fusion process within stars that causes them to burn and shine and thus provide the Earth and life on Earth with the energy to reverse entropy and allow life[50].

The weak force's role in the steady burning of stars is a strong example of the fine-tuning necessary for a life-bearing cosmos. However, when we look closer at the mechanics of this force, the mystery deepens even further. The very process of radioactive decay, which is governed by the weak force, relies on a particle with such specific and improbable properties that it adds another layer of "unspeakable complexity" to the Standard Model.

The Weak Force As An Example of The Complexity of The SM

The mystery of the weak force deepens when we consider a crucial and counter-intuitive detail about its force-carrying particles, the W and Z bosons. In order to trigger a radioactive decay, a W boson is required. This is well understood and a standard part of theory. What is not well understood is that not just any W boson is required for this task, instead only a very rare kind of W boson can undertake this task.

In experiments, the W boson is measured to have a huge mass, about *85 times* that of a proton. This presents a paradox: if the particle is so massive, why is the force it carries so weak?

The answer lies in one of the most bizarre aspects of quantum reality. The W boson that mediates the radioactive decay process is not a "real" particle, but a "virtual" one. According to Heisenberg's uncertainty principle, a particle can briefly pop into existence from nothing, borrowing a huge amount of energy from the quantum vacuum, as long as it pays that energy back almost instantly.

Because the W boson is so massive, creating even a "virtual" one requires an enormous energy loan from the universe. This is an incredibly rare quantum event. It is this extreme rarity—the low probability of a virtual W boson appearing to do its job—that makes the weak force so incredibly weak. As physicist Don Lincoln notes, when a rare weak force interaction does happen, the force itself isn't weak at all; its weakness is attributed to the measure of its improbability.

The process of determining the mass of a particle as fleeting as the W boson is an ongoing mystery in itself. Typically, the mass of a short-lived particle is deduced by measuring the properties of the particles it decays into. This method, however, runs into a major roadblock with the W boson, as one of its decay products is a neutrino, a particle so elusive it is practically invisible to our detectors.

This "neutrino problem" means scientists must resort to a highly specialized and indirect form of "guesswork." They run billions of simulated collisions and make complex inferences based on the momentum of the particles they can see, all while relying on theoretical assumptions about how the W boson behaves. As a result, the final value for the W boson's mass is not a simple, direct measurement, but the product of a series of choices and assumptions that are, as noted among the scientific community, "heavily debated[51]."

There are reams of fine print on the SM of physics such as what we have reviewed in the last several paragraphs regarding the weak force and the W boson, painting a picture of a contentious field within which scientists are grappling to overcome serious challenges. Meanwhile there is a generic narrative that is presented to the public, of the neat and tidy chart of fundamental particles of the SM. While this chart is "accurate" for what we know, it does not convey the other side of the coin. There is a far more tangled narrative of what happens, and how particles behave and are understood on the ground floor of particle physics that is still being worked out and therefore omitted when being presented to the general public.

How The Standard Model Was Built

The formalization of the SM followed in parallel with the quantum mechanical revolution and quantum theory. The SM is considered to be a quantum field theory, while at the same time it forms the essential basis for all of the levels of emergence within our universe with regards to atoms and their behavior, up to the level where the SM transitions into classical physics.

The modern story of the Standard Model begins not with particles, but with a revolutionary idea from the English chemist and physicist Michael Faraday. In the 19th century, Faraday, a master of intuitive experimentation, made several of the greatest finds in the history of science. His most profound and lasting contribution was the understanding that fields—in his case, the electromagnetic field—were real, physical entities, not just mathematical conveniences[52]. This subtle but crucial insight, that an invisible field could be as real as a physical object, set the stage for the quantum field theories that now form the bedrock of our understanding of the universe.

In 1887 Heinrich Hertz discovered the photoelectric effect while attempting to confirm Maxwell's electromagnetic theory of light. Hertz's photoelectric effect was the discovery of a strange result that conflicted with Maxwell's theory where a charged object loses its charge more readily when illuminated by ultraviolet radiation. The discovery of the photoelectric effect was matriculated by additional work by Wilhelm Hallwachs, J.J. Thompson and Phillip Lenard culminating in Einstein's 1905 paper. This paper was the key moment in the history of QM because it explained the photoelectric effect as a quantum process. This paper was what literally put the *quant*a in quantum mechanics and is one of the main reasons why Einstein is seen as the grandfather of QM. Max Planck also founded another aspect of quantum behavior a few years earlier in 1900 with his discovery of the *quantization* of black body radiation which brought up a serious conflict with classical physics which was *continuous* rather than being quantized. What Planck found was that the spectrum of radiation occurs in discrete energy levels consistent with frequency (v) and his newly derived *Planck's constant* (h). "The fact that the absorption and the emission of energy is discontinuous is in conflict with the principles of classical physics[53]." Planck found the existence of energy levels were quantized, while Einstein found discrete quanta of energy. Einstein's proof of the quanta of light ran into direct contradiction with previous science that established that light was propagated by waves. This is something Planck's findings also concluded.

From here the race to provide the granular details of quantum mechanics was on, and key discoveries continued slowly building out the

view of this new understanding of a quantum world. This foundational work continued up to a peak in the mid 20's to mid 1930's with the participation of many luminaries we have already discussed in previous sections. Two key pieces of information came in the form of Louis de Broglie's 1924 work introducing his theory of electron waves which proposed a wave-particle duality, and the final coup de grace, Schrödinger's wave equation (wave mechanics) in 1925-26. A number of luminaries were involved in this vital work prior to, and after the wave equation which formalized the behavior-based side of the standard model known as QM.

The Standard Model of Physics And Beyond

Having established the bizarre, counter-intuitive "rulebook" of Quantum Mechanics, our investigation now turns to the cast of characters that play by these strange rules. To truly appreciate the anomalies and fine-tuning of the Standard Model, we must first trace the decades-long experimental journey that revealed its fundamental particles. This is not a story of theory alone, but of discovery—the story of how physicists, using ever-more-powerful machines, unveiled the "particle zoo" at the heart of our reality.

The assembly of the Standard Model was a multi-generational, international effort, built on a series of groundbreaking discoveries. Here is a brief timeline of some of the most crucial moments:

The Early Discoveries: The journey began with the "old guard" of physics. J.J. Thomson discovered the electron in 1897, and Ernest Rutherford discovered the proton in 1919 . James Chadwick completed the atomic picture with his discovery of the neutron in 1932.

The Power of Theory: The next great leap came from pure theory. In 1928, Paul Dirac's famous equation combined quantum mechanics and special relativity, and in doing so, it predicted the existence of antimatter.

The "Particle Zoo" and a New Order: The 1960s were a period of peak discovery. Theorists like Sheldon Glashow, Murray Gell-Mann, and George Zweig developed the quark model and the theory of Quantum Chromodynamics (QCD), bringing order to the chaotic "particle zoo" of new discoveries.

Unification and Confirmation: In 1967, Steven Weinberg and Abdus Salam, building on Glashow's work, developed the theory of the unified electroweak force. The experimental confirmation of these ideas came in 1983, with the discovery of the W and Z bosons at CERN.

The Final Pieces: The modern Standard Model was largely completed with the discovery of the massive top quark in 1995 and, most recently, the famous Higgs boson between 2011 and 2013.

This decades-long quest, marked by numerous Nobel Prizes, formalized our understanding of the fundamental particles of matter. However, as we will continue to see, this "complete" picture is still improbably filled with far more questions than answers.

Problems with The Standard Model

The number of problems that need to be resolved within the standard model are substantial. The topics we have selected are some of the more serious ones, others we will not list below have already been discussed: the origin of the universe, the baryon asymmetry, the strong CP problem, inflation, dark matter, dark energy, grand unified theory, and the theory of everything. The list goes on.

The Proton's Mass Mystery

One of the most profound and perplexing mysteries of the Standard Model is the origin of mass for protons and neutrons. When physicists attempt to calculate the mass of a proton by simply adding up the masses of its constituent quarks, the numbers do not "square up." The sum of the parts accounts for only about 1% of the proton's total measured mass.

The other 99% of the mass of the visible universe appears to be generated from pure energy—the binding energy of the strong force and the kinetic energy of the quarks and gluons furiously interacting within the proton. This discrepancy is deepened by another mystery: the masses of the fundamental particles themselves seem to follow no discernible pattern or rule. They appear to be arbitrarily set, yet they are the precise values needed to create a stable, life-bearing universe. The fact that the very substance of our reality is built upon a foundation of such inexplicable and seemingly fine-tuned values is a key anomaly of the Standard Model.

Proton/Neutron Mass Difference + Fusion Issue

The tiny mass difference between a proton and a neutron is another example of the universe's exquisite fine-tuning. This seemingly insignificant value is balanced on a razor's edge, and if it were even slightly different, the cosmos would be a sterile and lifeless place.

If the neutron were slightly heavier (making the mass difference bigger), the energy barrier for nuclear fusion inside stars would be too high. Stars would not be able to burn their fuel, and the universe would be a dark and simple place, unable to forge the heavy elements necessary for planets and life.

If the neutron were slightly lighter (making the mass difference smaller), all the protons in the early universe would have rapidly decayed into neutrons. This would have resulted in a cosmos devoid of hydrogen, the essential fuel for long-lived stars like our sun, and thus with no possibility for complex, multicellular life to evolve.

This perfect balance that the proton and neutron find themselves in, for which there is no known physical explanation, is another fundamental prerequisite for our existence.

No Known Reason For The Masses And Charges Of Particles

The specific charge and mass of the fundamental particles are what provide our universe with its most crucial characteristics. These properties are responsible for chemical reactions, the formation of large-scale features like galaxies and planets, and the emergence of nested levels of organization, from stable atoms and molecules to complex cells and, ultimately, conscious beings.

As science has investigated these charges and masses, they are inexplicably *finely tuned* for these aforementioned universal qualities, *as if* they were *engineered* to facilitate these events, features, and levels of organizations. The problem is, as we alluded to earlier, there has yet to be any reason discovered *why* the standard model possesses this particular set of exquisitely finely tuned masses and charges. These numbers could otherwise appear to be somewhat arbitrarily set, *but only if* we ignore the fact that they are exquisitely finely tuned as they are for a life conducive universe.

There is no fundamental reason why the particles of the Standard Model have the mass and charges they have, *except* when they are viewed from the perspective of the organizational needs for life in our universe. Under this rubric, these settings make "perfect" sense. They satisfy *all* of the requirements for life.

The fine-tuning of these fundamental values is a truth that is not widely acknowledged by the scientific community, in part because the argument becomes circular: since we only have one universe to observe, there is no direct evidence that these values could have been different.

But this sidesteps a deeper and more perplexing problem. Even in this perfectly life-allowing universe, the origin of life itself still requires a miracle, as we will explore in a later chapter. This creates a paradox of stacked improbabilities: Why would the universe be so exquisitely fine-tuned to create the necessary conditions for life, if life was still impossible without a subsequent miracle in addition the series of miracles we will investigate?

The only coherent explanation is that there is a universal prerogative for life at work. Such a prerogative would not only account for the initial, life-permitting values of the cosmos but also for the series of subsequent "impossible" events required to bring that life into being.

The materialist's search continues for a purely physical explanation for the fine-tuning we observe. The goal of this search is to find some as-yet-undiscovered law or external reason that would explain why the fundamental particles and constants of our universe are so perfectly arranged for life, without having to invoke the possibility of a guiding purpose or "prerogative for life."

The Higgs Hierarchy Problem (Again)

One of the greatest and most unexpected mysteries to emerge from the discovery of the Higgs boson is the problem of its mass. As the particle that gives mass to all other fundamental particles, it was expected that quantum effects would make the Higgs itself incredibly heavy. Instead, it is shockingly light.

To understand the scale of this problem, imagine a massive balancing scale. On one side, quantum theory places a weight of a quadrillion tons. For the scale to balance at the observed light mass of the Higgs, the other side must have a weight that is also a quadrillion tons, minus just one ounce. The fact that these two enormous, unrelated quantum values just happen to cancel each other out with such near-perfect precision is an act of fine-tuning so extreme it defies a naturalistic explanation.

This profound puzzle is known as the Higgs Hierarchy Problem. The leading proposed solution, a theory called Supersymmetry, predicted a host of new particles that would naturally explain this cancellation, but decades of searching at the world's most powerful particle accelerators have failed to find any evidence for them. We are therefore left with another foundational aspect of our universe that appears to be exquisitely and inexplicably fine-tuned, to a degree of at least one part in $\mathbf{10^{34}}$, for reasons that at the present moment remain a complete mystery.

The hierarchy problem for the Higgs, the mass particle of the universe, is for many the death knell of "naturalness" in science, because the idea that the underlying parameters of the Standard Model are all of the same size in appropriate units and that these values may be explained by underlying physics that gives them their uniform size, or determines their value has been dashed by this hierarchy problem. To understand why the Higgs Hierarchy Problem is considered the "death knell of 'naturalness'," we must first understand what physicists mean by that term. Naturalness is a guiding principle, an aesthetic preference for theories in which the fundamental constants are simple and robust, not reliant on miraculous cancellations or extreme fine-tuning. A "natural" theory doesn't require balancing a pencil perfectly on its tip; the Higgs mass, however, appears to require exactly that kind of impossible balancing act.

The failure to find new physics like supersymmetry to explain the Higgs mass has led to a crisis for the long-held scientific principle of "naturalness." This idea—that the fundamental parameters of the universe should be simple and not require miraculous fine-tuning—is fast approaching extinction. The problem is that any new theory proposed to "rescue" naturalness would have to be so complex and contrived that it would itself be fundamentally unnatural.

This is a radical departure, and fork in the road from the dream of physicists like Albert Einstein. His hope was that a complete theory of the universe would reveal an underlying, elegant simplicity—a cosmos that could create itself through beautiful, natural laws, without the need for any strange, mystical, or "supernatural" intervention. The hierarchy problem, and the failure to solve it, directly challenges this foundational belief.

> "The scientist's religious feeling takes the form of a rapturous amazement at the harmony of natural law, which reveals an intelligence of such superiority that, compared with it, all systematic thinking and acting of human beings is an utter insignificant reflection... It is beyond question closely akin to that which has possessed the religious geniuses of all ages."
>
> *-Albert Einstein*

In this quote most scientists choose to see an ode to a universe, perfect, robust, and intrinsically sound and unquestioned, but for reasons we have touched upon, and will explore more deeply, the universe could only be perfect, robust and intrinsically sound if it had no need (goal) to be created, and had no predestined teleological plan. Their vision requires a universe that was made by its own construction, unaided and self-arising, and that "births" perfected sublime laws, through an infallible and uncorrupted naturalistic fiat of unintended and unplanned irresistible, unalterable, harmony. But no. Instead, this vision has returned a string of stinging defeats, recounted in our journey through the 19 identified parts of our universal algorithm which has placed their idea in grave doubt while also revealing that there is, and must be an architect, no less alive, intelligent, capable, perfected, and possessed of consciousness, that demands their due, and overthrowing the imagined invisible muse and patron saint of materialism.

Higgs False Vacuum Catastrophe

The measured mass of the Higgs boson leads to a startling and deeply unsettling conclusion: the very fabric of our universe may be fundamentally unstable. According to the Standard Model, the Higgs field does not exist in its lowest possible energy state (a "true vacuum"), but in a precarious, higher-energy state known as a "false vacuum."

Think of our universe as a ball resting in a small divot on the side of a very high hill. While it is stable for now, a sufficiently large jolt—or even a random quantum fluctuation—could send it tumbling down into the true, lower-energy valley below. Such an event, known as vacuum decay, would trigger a "catastrophe" of cosmic proportions. A bubble of true vacuum would expand at the speed of light, wiping out all matter and structure in its path, fundamentally rewriting the laws of physics as it goes.

The great mystery is, why hasn't this happened? Physicists often appeal to speculative, unproven theories like supersymmetry or cosmic inflation to argue that the Higgs field must be truly stable. However, these are the same kind of ad hoc, untestable solutions we have encountered before. We are therefore left with another striking fine-tuning problem: the universe appears to be balanced in a precarious, metastable state, "just right" for our existence, for reasons that remain a complete mystery[54]. (

The Proton Spin Crisis

The origin of the proton's spin is one of the most unsettling unsolved mysteries in particle physics, often called the "proton spin crisis." All fundamental particles have an intrinsic property called spin. A proton's total spin is always 1/2. The simple expectation was that this spin would be the sum of the spins of the three "valence quarks" inside it.

However, experiments have shown that the spins of these quarks only account for about 30% of the proton's total spin[55]. The rest must come from the chaotic, roiling sea of other particles within: the orbital motion of the quarks and, most significantly, the spin of the gluons that bind them together. The fact that this fundamental property of matter arises from such a complex and poorly understood dynamic, rather than from its constituent parts, isn't just a simple accounting error; it's a serious conceptual crisis. It demonstrates that the Standard Model, for all its success, has a fundamental gap in its ability to explain how the properties of the whole emerge from its parts, revealing that the "unspeakably complex" nature of reality runs even deeper than we imagined.

Standard Model Three Generation Mystery

The discovery of three generations of quarks and leptons and their extreme range of masses is a central mystery of particle physics. Each particle in the standard model comes in three generations. While the familiar first-generation particles make up all the stable matter in our universe, their much heavier second and third-generation counterparts do not arise from conventional, low-energy particle interactions. Instead, they are only forged in the extreme conditions of high-energy environments, such as those found in particle accelerators or in the very early universe.

The Standard Model reveals that the fundamental particles of matter come in three distinct "generations." Each generation is a near-perfect copy of the one before it, containing the same types of particles with identical properties, except for one crucial difference: their mass. The second- and third-generation particles are significantly, and in some cases, thousands of times, heavier than their first-generation counterparts.

This presents a two-part mystery. First, why are there three generations? All the stable, ordinary matter in the universe—every star, planet, and person—is made exclusively from the light, first-generation particles. The heavier two generations are unstable and exist only fleetingly in high-energy environments like particle accelerators. Their existence seems entirely redundant for a life-bearing cosmos.

Second, why do their masses follow such a strange and unexplained pattern? There is no known theory that predicts this specific three-tiered structure. This mystery—the seemingly unnecessary repetition of matter with an inexplicably hierarchical pattern of masses—is another key anomaly, suggesting that our understanding of the fundamental structure of reality is far from complete.

Let's consider the quark family as an example. The family is organized into three distinct "generations. The first generation contains the familiar up and down quarks that make up all stable matter. The second and third generations are essentially heavier, unstable copies, containing the charm and strange, and the top and bottom quarks, respectively.

The big problem is, we have no theory that explains the bizarre and seemingly arbitrary pattern of these masses, nor indeed the masses of any of the other particles and their corresponding generations. This is a significant failure for what is otherwise our most successful theory of matter. As physicist Peter Woit argues, a truly fundamental theory should be able to explain the existence and properties of its own members. The fact that the Standard Model cannot explain this is another key anomaly, leaving us with a perplexing and unexplained "particle zoo" at the heart of reality. "Although quarks mix with one another and neutrinos mix across the generations, the ways in which they mix are not identical to each other. If there is a pattern or an underlying cause or reason as to why there are three generations, we haven't uncovered it yet[56]."

Mystery Of Confinement

Another mystery arising from the world of quantum physics is the troublesome and outstanding host of problems known as confinement. In simple terms, confinement is the unbreakable rule that quarks can never appear as free, isolated particles; they are always found "confined" inside composite particles like protons and neutrons.

The mystery is that while our best theory of the strong nuclear force, Quantum Chromodynamics (QCD), accurately describes the interactions between quarks, it cannot explain *why* they are permanently confined. The fact that quarks cannot be isolated is a brute, observational fact of our universe, but it is not a conclusion that can be mathematically derived from the theory itself. In other words, even with our detailed understanding of QCD there is no fundamental equation that proves why quarks must always be confined. It is simply an observed rule of nature that physicists have been forced to add to the theory.

This leaves a significant gap in our understanding. As physicist J.W. Moffat notes, to account for this, "physicists have been forced to come up with a somewhat ad hoc phenomenological explanation added in by hand. However, in the current particle physics community, this problem is simply ignored...". In short, one of the most fundamental rules governing the existence of matter is an observed reality that our most successful theory cannot explain. It is a mystery.

The Verdict on the Standard Model: A Pattern of Design

Our investigation into the fundamental rules and particles that govern our reality has reached its conclusion. We started our inquiry with the anomalies within Quantum Mechanics which cannot be answered by physicalist interpretations. *Bell's inequality* settled the question of hidden values and the measurement problem in QM demonstrating that the problems in QM are *real* problems, unreachable, and unexplained by any semblance of classical theory.

There is substantial evidence that only *consciousness* renders the universe of QM and The SM discernible, understandable, explainable and experienceable. Consciousness *has* the ability to do what physics *cannot,* which is *select and privilege one moment in time, over all others* for the purposes of experience (observation and/or measurement).

In our investigation we then utilized a thought experiment concerning a universe *exactly like our own* except devoid of consciousness and demonstrated that a universe devoid of consciousness is necessarily devoid of features, because only consciousness experiences features. With no consciousness, we have a universe of laws and fields, but no observer to experience the actual physics.

Finally, we found that as perplexing and paradoxical as QM is, life is dependent upon every part of it that we have investigated. We can tentatively state that as a group, all of the things we do not understand about quantum mechanics are *required* for particles, atoms, chemistry and prerequisites for life within our universe.

We then turned our attention to the Standard Model of Physics where we saw that *gravity*, like *the measurement problem*, presents intractable and irreconcilable problems in materialist physics.

We then discussed a number of additional anomalies within the SM, and demonstrated once again, the existence of a universal imperative for intelligent observers. Only this imperative accounts for our universe.

In order to answer the fine tuning and anomalies within our universe, ideas such as *the inflationary multiverse* and *The Everett Interpretation*, have been introduced. They each imagine unobservable, unprovable infinities of alternative universes with no purpose or meaning other than as psychological comfort and as a dissembling, rationalized, illusory levee to buttress the claims of the weak anthropic principle (we are here, so our existence must be one in which we can exist. These imagined universes, even if they did exist, have no bearing on our universe.

Unlike the other variables in our investigation, the fine-tuning of the Standard Model cannot be reduced to a single probability. Instead, its value is derived from the cumulative weight of the profound and interlocking mysteries that lie at the very heart of our physical reality.

We have seen that our most successful theory of matter is built on the paradoxical and observer-dependent rules of Quantum Mechanics, a framework that resists any classical or purely materialistic explanation. We have explored the Standard Model itself and found it to be a theory riddled with profound gaps—most notably its complete inability to account for gravity—and inexplicable fine-tuning, and dozens of missing, yet theorized supersymmetry particles and fields, to the unanticipated mass of the Higgs boson, to the impossible stability of the proton, and finally the uncontested fact that at present the Standard Model does not explain 95% of the stuff within our universe.

The sheer number and depth of these anomalies, all of which are necessary for a life-permitting universe, form a coherent pattern that defies an explanation of chance. Therefore, based on the total weight of the evidence, the value we must assign is:

Final Value for Variable #6 Fine Tuning & Anomalies of The Standard Model: *M (miracle)*

There is no material, classical, or quasi-classical explanation for QM or The SM.

Being solutionless, at present Variable #7 meets the definition of exhausting the probabilistic capacity of the universe (PCOU).

---

[1] (L. Smolin, The Life of the Cosmos, Pg. 12, Oxford University Press, 1997)
[2] (T. Hartsfield, Superdeterminism: To Better Understand Our Universe, Ditch The Idea of Free Will, August 29, 2022, bigthink(dot) com/ hard-science/ superdeterminism-free-will/).
[3] (T. Hartsfield, Superdeterminism: To Better Understand Our Universe, Ditch The Idea of Free Will, August 29, 2022, bigthink(dot) com/ hard-science/ superdeterminism-free-will/).
[4] (G. Ghirardi, A. Bassi, Collapse Theories, The Stanford Encyclopedia of Philosophy (Fall 2024 Edition), Edward N. Zalta & Uri Nodelman (eds.), plato(dot) stanford(dot) edu/ archives/ fall2024/)
[5] (J.G. Cramer, Quantum Entanglement Disentangled, Alternate View Column AV-223, March-April 2023 Analog Science Fiction & Fact Magazine, 2022)
[6] (J.G. Cramer Ibid).
[7] (J.G. Cramer, Quantum Entanglement Disentangled, Alternate View Column AV-223, March-April 2023 Analog Science Fiction & Fact Magazine, 2022)
[8] (E. Schrödinger, Discussion of Probability Relations Between Separated Systems, 1935),
[9] (E. Schrödinger, Discussion of Probability Relations Between Separated Systems, Mathematical Proceedings of teh Cambridge Philosophical Society, 31, October 28, 1935, Pgs. 555-563)
[10] (PBS Spacetime, Zeno's Paradox and The Quantum Zeno Effect, March 23, 2021, PBS Spacetime YouTube Channel)

[11] Effect Real? Sabine Hossenfelder YouTube Channel, October 13, 2024, 6:59 playtime), (S. Hossenfelder, A Watched Quantum State Doesn't Change. Is The Zeno
[12] (S. Hossenfelder, A Watched Quantum State Doesn't Change. Is The Zeno Effect Real? Sabine Hossenfelder YouTube Channel, October 13, 2024, 6:59 playtime)
[13] (S. Hossenfelder, A Watched Quantum State Doesn't Change. Is The Zeno Effect Real? Sabine Hossenfelder YouTube Channel, October 13, 2024, 6:59 playtime).
[14] F.A. Wolf, Taking The Quantum Leap, Pg. 15, Harper & Row, 1989)
[15] (F.A. Wolf, Taking The Quantum Leap, Pg. 16, Harper & Row, 1989)
[16] (F.A. Wolf, Taking The Quantum Leap, Pg. 17, Harper & Row, 1989)
[17] (F.A. Wolf, Taking The Quantum Leap, Pg. 21, Harper & Row, 1989)
[18] (W. Heisenberg, The Development of Quantum Mechanics, Nobel Lecture, December 11, 1933 NobelPrize(dot) org, Nobel Prize Outreach AB 2024, physics/heisenberg/lecture)
[19] (F.A. Wolf, Taking The Quantum Leap, Pg. 6, Harper & Row, 1989)
[20] (G. Zukov, The Dancing Wu Li Masters, Pg. 351, 1979, 2001, epub April 2009, William Morrow)
[21] (J. Polkinghorne, Quantum Theory, A Very Short Introduction, Pg. 96, 2002, Oxford University Press)
[22] (J. Faye, Copenhagen Interpretation of Quantum Mechanics, The Stanford Encyclopedia of Philosophy, Summer 2024 Edition, E.N. Zalta & U. Nodelman eds., Plato(dot) stanford(dot) edu/ archives/ sum2024/ entries/ qm-copenhagen/).
[23] F.A. Wolf, Taking The Quantum Leap, Pg. 255-256, 1989, Harper and Publishers)
[24] F.A. Wolf, Taking The Quantum Leap, Pg. 256, 1989, Harper and Publishers)
[25] (Roger Penrose with Robert Lawrence Kuhn, Many Worlds Of Quantum Theory, Closer To The Truth YouTube Channel, March 16, 2020)
[26] (N.R. Hanson, Copenhagen Interpretation of Quantum Theory, Pg. 4, The American Journal of Physics, Vol. 27, No.1, January 1959, retrieved from: users(dot) df(dot) uba(dot) ar/ giribet/ f4/ mq3(dot) pdf)
[27] (N.R. Hanson, Copenhagen Interpretation of Quantum Theory, Pg 4, The American Journal of Physics, Vol. 27, No.1, January 1959, retrieved from: users(dot) df(dot) uba(dot) ar/ giribet/ f4/ mq3(dot) pdf)
[28] (N.R. Hanson, Copenhagen Interpretation of Quantum Theory, he American Journal of Physics, Vol. 27, No.1, January 1959, retrieved from: users(dot) df(dot) uba(dot) ar/ giribet/ f4/ mq3(dot) pdf)
[29] (N.R. Hanson, Copenhagen Interpretation of Quantum Theory, Pg. 3, The American Journal of Physics, Vol. 27, No.1, January 1959, retrieved from: users(dot) df(dot) uba(dot) ar/ giribet/ f4/ mq3(dot) pdf)
[30] (N.R. Hanson, Copenhagen Interpretation of Quantum Theory, Pg. 5, The American Journal of Physics, Vol. 27, No.1, January 1959, retrieved from: users(dot) df(dot) uba(dot) ar/ giribet/ f4/ mq3(dot) pdf)
[31] (C. Smeenk, F. Arntzenius, T. Maudlin, Time Travel and Modern Physics, The Stanford Encyclopedia of Philosophy (Spring 2023 Edition), E.N. Zalta & U.Nodelman (eds.), plato(dot) stanford(dot) edu/ archives/ spr2023/ entries/ time-travel-phys/)
[32] (The American Heritage Dictionary of the English Language, 5th Edition, Physics, retrieved October 15, 2024, wordnik(dot) com/ words/ physics)
[33] (A. Einstein, Albert Einstein to Max Born, Physics Today, 1 May 2005; 58 (5): doi(dot) org/ 10.1063/ 1.1995729)
[34] (A. Einstein, B. Podolsky, N. Rosen, Can Quantum-Mechanical Description of Physical Reality Be Considered Complete? Phys. Rev. 47, 10, 777-780, May 1935, American Physical Review 10.1103/physrev.47.777)
[35] (A. Einstein, B. Podolsky, N. Rosen, Can Quantum-Mechanical Description of Physical Reality Be Considered Complete? Phys. Rev. 47, 10, 777-780, May 1935, American Physical Review 10.1103/physrev.47.777)

---

[36] (A. Ananthaswamy, Through Two Doors At Once, Pg. 97,2018, Dutton)
[37] (J.G. Cramer, Quantum Entanglement Disentangled, Alternate View Column AV-223, March-April 2023 Analog Science Fiction & Fact Magazine, 2022)
[38] (A. Einstein, B. Podolsky, N. Rosen, Can Quantum-Mechanical Description of Physical Reality Be Considered Complete? Phys. Rev. 47, 10, 777-780, May 1935, American Physical Review 10.1103/physrev.47.777)
[39] (A. Einstein, B. Podolsky, N. Rosen, Can Quantum-Mechanical Description of Physical Reality Be Considered Complete? Phys. Rev. 47, 10, 777-780, May 1935, American Physical Review 10.1103/physrev.47.777)
[40] (A. Einstein, B. Podolsky, N. Rosen, Can Quantum-Mechanical Description of Physical Reality Be Considered Complete? Phys. Rev. 47, 10, 777-780, May 1935, American Physical Review 10.1103/physrev.47.777)
[41] (A. Einstein, B. Podolsky, N. Rosen, Can Quantum-Mechanical Description of Physical Reality Be Considered Complete? Phys. Rev. 47, 10, 777-780, May 1935, American Physical Review 10.1103/physrev.47.777)
[42] (A. Einstein, B. Podolsky, N. Rosen, Can Quantum-Mechanical Description of Physical Reality Be Considered Complete? Pg. 777, Phys. Rev. 47, 10, 777-780, May 1935, American Physical Review 10.1103/physrev.47.777)
[43] (J. Polkinghorne, Quantum Theory, A Very Short Introduction, Pg. 78, 2002, Oxford University Press)
[44] (J.S.Bell, On the Einstein Podolsky Rosen Paradox, Physics 1 (3): 195-200)
[45] (J.S. Bell, On the Problem of Hidden Variables in Quantum Mechanics, Reviews of Modern Physics. 38 (3): 447-452)
[46] (O. Lanes, Qiskit YouTube Channel, Bell's Inequality, The Weirdest Theorem in The World | Nobel Prize 2022, October 7, 2022)
[47] (C. Wood, Pioneering Quantum Physicists Win Nobel Prize In Physics, Quanta Magazine Online, October 4, 2022. quantamagazine(dot) org/ pioneering-quantum-physicists-win- nobel-prize- in-physics-20221004/)
[48] (M. M. Muhlleitner, The Standard Model of Physics, 24th Vietnam School of Physics: Particles and Cosmology, August, 2018, itp(dot) kit(dot) edu/ ~maggie/ icise/ standardmodel(dot) pdf)
[49] (V. Balasubramanian, Quanta Magazine YouTube Channel, Space-Time: The Biggest Problem in Physics, September 25, 2024)
[50] (R. Webb, Weak Nuclear Force: The Fundamental Force Responsible for Radioactive Beta Decay, retrieved October 21, 2024, thenewscientist(dot) com/ definition/ weak-nuclear-force/)
[51] (J. Bendavid, M. Cipriani, M. Dunser, M. Mulders, S. Camarada, What Is the W Boson Mass Trying To Tell Us? CERN EP Newsletter, June 16th, 2022, ep-news(dot) web(dot) cern(dot) ch/ content/ what-w-boson- mass-trying-tell- us).
[52] (S. Strogatz, (interviewing D. Tong) What Is Quantum Field Theory and Why Is It Incomplete? August 10, 2022, The Joy of Why, quantamagazine(dot) org/ what-is-quantum- field-theory-and- why-is-it-incomplete- 20220810) (W.L. Pearce. Michael Faraday, Encyclopedia Britannica, 18 Sep. 2024, britannica(dot) com/ biography/ michael-faraday)
[53] (J. Faye, Copenhagen Interpretation of Quantum Mechanics, The Stanford Encyclopedia of Philosophy, Summer 2024 Edition, Edward N. Zalta & Uri Nodelman eds., plato(dot) stanford(dot) edu/ archives/ sum2024/ entries/ qm-copenhagen/)
[54] MJG Veltman, Facts and Mysteries in Elementary Particle Physics, World Scientific Publishing, Pg. 133, 2003)8
[55] (D. deFlorian, R. Sassot, M. Stratmann, W. Vogelsang, Evidence for Polarization of Gluons in the Proton, Phys.Rev.Lett. 113, 012001, July 2, 2014).
[56] (E. Siegel, Why Are There Exactly 3 Generations of Particles? June 7, 2022, Big Think, bigthink(dot) com/ starts-with-a-bang/ why-3-generations-particles/)

Variable 7: Fine Tuning of the Carbon Resonance

In our previous variable, we established that the fundamental rules of physics and the particles of the Standard Model are, themselves, a "miracle" of fine-tuning. We now turn to the first great consequence of this miraculous operating system: the creation of the elements.

For our universe to host life, it must contain a rich inventory of complex atoms, most importantly carbon. However, the Big Bang itself failed at this task. The theory of Big Bang nucleosynthesis, which we will explore, shows that the early universe produced almost nothing but hydrogen and helium—a sterile recipe for a cosmos devoid of planets and life.

This presents a profound challenge to The Divine Algorithm: if the universe's goal is intelligent life, where did the essential building blocks of that life come from?

The answer is that they had to be forged later, inside the fiery cores of stars. This process, stellar nucleosynthesis, is the true subject of this variable. But as we will see, this process also faced an apparently impossible barrier. The very physics that holds atoms together should have made the creation of carbon impossible. The solution, we will find, lies in another, breathtakingly precise miracle of fine-tuning, one so specific that it famously forced the atheist physicist who discovered it to declare that "a super-intellect has monkeyed with physics".

In this variable, we will follow the story of how physicists tried to build the periodic table from the Big Bang and from stars, only to run headlong into a wall: the apparent impossibility of creating carbon. The escape route from that wall turned out to be a single, exquisitely tuned energy level in the nucleus of carbon-12—what has become known as the Hoyle resonance.

Setting the Stage: The Primordial Element Problem

To see how severe the carbon problem really is, we first need to understand what the Big Bang actually made—and what it didn't. By 1950 the hot big bang theory indicated that the only elements that would have been created in the early universe: hydrogen (73%), helium (25%), isotopes of helium and hydrogen (less than 2%), and trace amounts of lithium and beryllium at less than 0.00000001% (**10^{-10}**). The ratios of these elements in our universe in the understanding of scientists, provide compelling confirmation for the hot big bang theory. No other cosmological origin models provide the universe with this high of a percentage of helium. And there is no other known process that could provide the universe with these overall ratios of elements.

These specific ratios are powerful evidence because they are the exact outcome predicted by the physics of the early universe. They align perfectly with models based on the temperature, density, expansion, and cooling of the original primordial matter[1].

Knowing the hydrogen-helium ratio also allows us to model and confirm the expansion rate of the early universe. The scientific theory that formally explains how all these factors (the initial temperature, density, and expansion) produced these specific elements is known as big bang nucleosynthesis. This theory was contained in a prediction for the big bang written into a Ph.D. dissertation by physics student named Ralph Alpher in collaboration with his Ph.D. advisor, George Gamow, who famously added the name of his friend Hans *Bethe* so that the paper could be known as *the alpha, beta, gamma paper*, aka, "the alphabetical paper[2]." The formal name of the Paper was *The Origin of Chemical Elements.* The paper argued, as we describe above, that the big bang would create hydrogen, helium and other elements in the correct proportions to describe our universe.

> "As pointed out by one of us, various nuclear species must have originated not as the result of an equilibrium corresponding to a certain temperature and density, but rather as a consequence of a continuous building-up process arrested by a rapid expansion and cooling of the primordial matter. According to this picture, we must imagine the early stage of matter as a highly compressed neutron gas (overheated neutral nuclear fluid) which started decaying into protons and electrons when the gas pressure fell down as the result of universal expansion. The radiative capture of the still remaining neutrons by the newly formed protons must have led first to the formation of deuterium nuclei, and the subsequent neutron captures resulted in the building up of heavier and heavier nuclei. It must be remembered that, due to the comparatively short time allowed for this process, the building up of heavier nuclei must have proceeded just above the upper fringe of the stable elements (short-lived Fermi elements), and the present frequency distribution of various atomic species was attained only somewhat later[3]."

Ralph Akpher's Ph.D. thesis advisor was George Gamow, who is renowned for developing his advisor's ideas into *the big bang model.* Gamow's advisor was Alexander Friedmann, the famed Russian physicist and mathematician who developed the *Friedmann equations* (1922, 1924)[4] and *the* Friedmann-Lemaître metric, a set of equations to describe the expansion of space under general relativity. General relativity is Einstein's theory of gravity. Friedmann's 1922 equations was the founding scientific evidence for the big bang. They indicated an expanding universe that had been smaller, *unimaginably smaller* in its distant past. Thus once again, we can see how the greatest ideas of the early 20th century were untangled from the same gordian knot of universal secrets. Each thread connected to the other, and championed by a luminary in science, holding secrets unlike any revealed previously. These were threads of duality, relativity, singularities, fine structure, antimatter, the universal wave function, quantum effects, and the sacred code in *DNA*.

The original ideas from the 2 page *alphabetical paper* were later included in another paper by Ralph Alpher and George C. Herman (1953) under the title, *The Origin and Abundance Distribution of The Elements*[5]. As indicated by this second title, Alpher's intention was to take the previous two decades worth of advances in nuclear physics and astrophysics and apply them towards understanding "the origin and relative abundance distribution of nuclear species". These previous decades of study had revealed, as written in *the alphabetical paper*, that the universe was 98% composed of 2 kinds of chemical elements, helium and hydrogen while everything else makes up the remaining 2%

The Origin and Abundance Distribution of The Elements begins, "On the one hand, nuclear physics suggests a generic relationship between nuclear species in that they are composed of the same constituents and exhibit significant correlations between their systematic properties and abundances. On the other hand, astrophysics suggests that our present universe has a definite age, that the observed relative abundances appear to be universal quantities, and that the physical conditions required for nuclear synthesis might be found in the early stages of the universe or in the interiors of special types of stars."

The Stuff of the Universe

The question on the lips of astrophysicists was, "How did the universe transform from a universe with 4 elements (Hydrogen, Helium, Lithium, Beryllium) in the early stages of the universe into all of the chemical elements represented in the periodic chart?" The story of how the universe underwent this process is the story of *stellar nucleosynthesis*, which literally means, the creation of complex atoms by the fusion of less complex atomic nuclei deep within stars. Stellar Nucleosynthesis is its own Variable, and our next Variable (#9, Stellar Nucleosynthesis) in *The Divine Algorithm*.

This paper, *The Origin and Abundance Distribution of the Elements*, evaluates the problem of the creation of the elements from the perspective of several competing ideas about how the universe was created. Although previously Alpher had, in a sense thrown his lot in with his advisor's model for big bang nucleosynthesis in *the alphabetical paper*, at the time there were several competing hypotheses that had to be fended off, the 4 most popular of which we will review as a matter of relevant historical regarding the origin of our universe.

> "Several very different theories have been developed to explain the origin and abundance distribution of the elements. One of the earliest of these [1] describes the observed abundances as a "frozen-in" thermodynamic equilibrium distribution of an assembly of nuclei and elementary particles, where the fundamental parameters are nuclear binding energies, temperature and density. In a recent theory [2] the abundances of the elements are considered as resulting from nonequilibrium processes, involving the formation of the lightest nuclei by thermonuclear processes with both neutrons and charged particles, and formation of the remaining nuclei principally by the successive radiative capture of neutrons, with intervening beta-decay. In a third type of theory [3] the formation of elements involves the fission of one or many super-nuclei or poly-neutron complexes. Finally, another attack on this problem [4] which leads to a steady-state cosmology involves either the continuous and spontaneous creation of matter uniformly throughout the universe at a rate such as to maintain the mean density of the universe at a constant value, or the sudden appearance in the universe of stars, identified as supernovae. The foregoing theories, differing in the proposed mechanisms and in the associated cosmological models, are all confronted with sundry difficulties, and it is not evident as yet that any one of them represents a final solution of this ancient problem[6]."

The central point is this: in the early 1950s, the origin of the chemical elements was a major, unsolved crisis. Scientists had several competing theories, but as Alpher and Herman noted, none represented a "final solution".

Into this environment of deep uncertainty stepped the astrophysicist Fred Hoyle. While championing his own "steady-state" model (theory #4 from the quote), he was forced to confront the specific, unsolved problem that stood in the way of any theory: the creation of carbon.

In 1952 Fred Hoyle was the proponent of above theory #4, the steady state model along with Hermann Bondi, and Thomas Gold who had all written papers on this model in 1948[7]. Hoyle's commitment to his steady-state model meant he had to account for the origin of the elements. This forced him to tackle the problem of carbon production. Carbon is the essential building block for organic chemistry and life, yet there was no known physical process that could explain how it was created in such abundance. The carbon problem was also key because, in the chain of stellar nucleosynthesis, carbon is a crucial steppingstone. If physicists could not find a pathway to create carbon, they would also have no way to explain the existence of all the heavier elements, from oxygen to uranium, which are built from it.

Hoyle's preferred explanation, the steady state model, accommodates the expansion of the universe by possessing the capacity to spontaneously create new matter in the center of galaxies, that in the hypothesis is directly tied to this expansion. In Hoyle's paper, "*A New Model for the Expanding Universe*[8], He begins with a quote by English physicist, astronomer and mathematician Sir James H. Jeans (1877-1946)[9] "*...the centers of the nebulae (galaxies) are of the nature of singular points, at which matter is poured into our universe from some other and entirely extraneous spatial dimension, so that, to a denizen of our universe, they appear as points at which matter is being continually created*". Hoyle then quotes Nobel prize winning theoretical physicists and co-founder of quantum mechanics, Paul Dirac (1902-1984). Dirac "*...pointed out that continuous creation of matter can be related to the wider question of cosmology.*" Hoyle included supporting ideas from Austrian-born astrophysicist Thomas Gold (1920-2004) "*who remarked that through continuous creation of matter it might be possible to obtain an expanding universe in which the proper density of matter remained constant.*" It is worth pausing for a moment on the nature of Hoyle's steady-state proposal, because it illustrates a broader pattern in speculative cosmology. Hoyle's steady-state cosmology draws an obvious parallel to modern "a-universe from-nothing" propositions. Both, in Hoyle's words, propose that "the creation process is likely to be an essentially quantum character...". This reveals a setup common to many speculative hypotheses, where a fundamental ignorance of the mechanism, combined with a lack of data, is used as the foundation and protection for the theory. This is not just an opinion; it is a recurring pattern from a historical perspective. In Hoyle's case, history delivered a clear verdict: nothing could be further from the truth than his steady-state model.

However, Hoyle's pursuit of this incorrect theory forced him to confront a very real and separate problem that was essential for any cosmological model: how did the universe create elements heavier than hydrogen and helium? Thus, despite coming from a starting point that was ultimately proven wrong, Hoyle went on to provide key, decisive insights into the creation of these new elements.

Hoyle's work on attempting to solve the problem of continuous creation led to the understanding that cosmic creation would only lead to the first few elements in the periodic table. This material would then need to be further processed within stars into higher elements. Naturally Hoyle turned his attention to the next step in the steady state process, the one that could process the newly created raw material from quantum processes into more complex chemistry—stellar nucleosynthesis.

The initial conception of how stellar nucleosynthesis worked is that elements of the periodic table were built up by the accumulation of single nucleons, either a proton or a neutron, one at a time, i.e. thus producing, step-by-step, all of the elements in the periodic chart, as expounded in the "alphabetical paper" by Alpher, Bethe and Gamow, where they conjectured, *"The newly formed protons must have led first to the formation of deuterium nuclei, and the subsequent neutron captures resulted in the building up…"* (incremental accretion) *"… of heavier and heavier nuclei*[10]*."*

In this view individual protons and neutrons were colliding with lighter elements within the densities and pressures (temperature) of stars, and would thereby fuse, via the strong force, building heavier and heavier elements. But this conception, in Hoyle's understanding, would hit an impasse at 5 nucleons, because any arrangement of 5 nucleons is highly unstable. This is known as the *five nucleon crevasse,* where nuclei with 3 protons and 2 neutrons (lithium 5) or 3 neutrons and 2 protons (helium 5) are only able to form for *infinitesimally small* periods of time, $\mathbf{10^{-24}}$ seconds, far too short to allow the accumulation of additional nucleons to form higher elements in the 1 septillionth of a second available. The practical result of the five-nucleon crevasse meant that rendering all higher configurations of elements beyond lithium 5 and helium 5 was physically unreachable. This crevasse has been likened to something like a 20-ft ladder with rungs at the top and bottom and a middle rung that vanishes in one septillionths a second[11].

Hoyle, in attempting to solve this problem had several clues to go on:

Higher elements were clearly plentiful in the universe, particularly carbon, a life essential element.

The work of physicist Edwin Salpeter proposed in 1951 that perhaps a combination of three alpha particles (helium-4) could somehow fuse into carbon-12 through an intermediary stage where 2 alpha particles would fuse into beryllium-8 and then gain a third alpha particle.

Future Nobel prize winning physicist William Alfred Fowler had already provided the known resonances (require energy levels) for beryllium-8. This data was crucial because it showed that simply adding a helium particle did not match any known energy resonance in carbon-12, seemingly making the creation of carbon impossible.

With these three primary clues, Hoyle was able to identify that the problem was at least three-fold:

Based upon the known resonances (energy levels wherein nucleon capture of different elements could be established) the combination of beryllium-8 and helium-4 was at too high of an energy level to allow them to fuse

Salpeter calculated that the needed pressure/temperature to fuse these elements would not be attainable inside stars, however, he also noted that perhaps unknown resonances of carbon-12 would alter his calculations[12].

Similar to the five-nucleon crevasse, beryllium-8 also has a short half-life, of only $\mathbf{10^{-16}}$ seconds, albeit, this is *ten million* times longer than the time available to the 5-nucleon crevasse ($\mathbf{10^{-24}}$). But time alone was not sufficient enough to allow these elements to fuse. *Unless* a resonance could be found, no amount of time would allow the needed fusion to take place[13]. Thus, theoretically, if a resonance could be found for carbon-12, it would allow beryllium-8 + helium-4 to create carbon-12. Hoyle felt that somehow this was the proper route to consider. The choices were few on how to proceed, and at the time Hoyle was the only one that believed this particular route could be found.

Believing this pathway toward carbon-12 must exist, Hoyle made a bold predictive leap that went beyond the established physics of the day.

Hoyle's Ingenious Solution

After considerable work, Hoyle then "hit upon an ingenious solution[14]." He proposed that if the carbon nucleus has a quantum property of resonance similar to the quantum electron shells of atoms where a discrete energy level of the vibrating protons and neutrons allowed a way forward, then this would result in the needed resonance. And if the nuclei of carbon and beryllium approached one another at high enough speeds to overcome their electromagnetic resistance, this proposed resonance, if it existed, would allow the strong nuclear force to take over and bind the atoms together.

Hoyle realized that if this resonance existed, then the newly created atom would be able to exist in this higher energy state and not fall apart and also not need to emit a gamma ray photon[15].

Hoyle therefore collected his ideas and proposed that carbon-12 had a hidden (unknown) resonance at the 7.68MeV energy level[16]. (The resonances for carbon-12 had been previously calculated and did not contain a resonance at 7.68MeV. But if this resonance existed, -the "Hoyle resonance" as it would later be called, it would boost the production of carbon in stars by a factor of *10 million*, allowing for 1 in every 2,500 beryllium-8 and helium-4 collisions to result in a carbon-12 atom, a figure sufficient to account for the abundance of carbon in the universe.

In the book, *The Privileged Planet*[17] authors Guillermo Gonzalez and Jay Richards liken a resonance to the correct acoustic note that leads to "a wine glass shattering", If the pitch is too high, or too low, then nothing happens, the note/tone must be nearly perfect to accomplish the needed task.

In 1953 Hoyle convinced future Nobel prize winning American nuclear physicist and astrophysicist, William A. "Willy" Fowler (1911-1995) to use his lab to search for the 7.68MeV carbon-12 resonance. As the legend goes, Hoyle invoked the anthropic principle for his belief in this resonance, (we exist, therefore the needed carbon energy state must exist to explain the amounts of carbon-12 in our universe). Fowler directed his students that if Hoyle's predicted energy state was a *spin 0*, with a particular parity then the resonance he predicted could indeed exist and it would have been missed by prior efforts. Within two weeks of searching, Fowler's lab detected the carbon resonance predicted by Hoyle, "the previously unnoticed energy state of carbon was detected[18]."

The granular details of how all of this came about is quite exciting. It shows that often the greatest discoveries were made by a handful of men, or two, and in this case, *one–Hoyle,* who was convinced to search for the resonance, and another-Fowler, that acquiesced to this effort, sympathetic to Hoyle's ideas, and willing to apply his far different skill set and technologies to assist Hoyle in discovering the truth.

We are fortunate to have a February 6, 1973 interview of William "Willy" Fowler by Charles Weiner[19] where the blow-by-blow details of the discovery of the Hoyle resonance are related, The essential story takes place in 1953. Fred Hoyle walked into Caltech's Kellogg Lab insisting there had to be a specific, undiscovered "resonance" in the carbon-12 nucleus at 7.6 MeV. Without this exact level, Hoyle's math showed that stars could never produce the carbon necessary for life. The experts had little faith. Their charts showed nothing there; in fact, they had recently erased a dashed line from their diagrams because previous experiments had failed to find it. The core leadership—including Kellogg Radiation Laboratory (Caltech) head Willie Fowler and the senior Lauritsen—dismissed Hoyle as a "funny little man" and told him to leave the "important work" to the experimentalists. Fowler later admitted they gave him the "brush off," viewing his request as a waste of laboratory time. But Hoyle's conviction wasn't a guess—it was a structural deduction. He realized that if the universe exists, then the carbon nucleus *must* have a secret door, invisible to the experimentalists, that allowed for its creation. Hoyle persisted until he convinced a young assistant professor, Ward Whaling, to give the proposition a studied try.

The result was a staggering discovery. Whaling and his group, David N.F. Dunbar, William A. Wenzel, and Ralph E. Pixley, performed the bombardment and found the state at 7.655 MeV—almost exactly where Hoyle predicted (7.68 MeV) a microscopic gap in the world of nuclear physics. While history attributes the discovery to Fowler who eventually won the 1983 Nobel Prize, the forensic reality is that Fowler was the skeptic, Ward Whaling was the catalyst, and Fred Hoyle was the architect.

Hoyle didn't find the state by peering into the nucleus; he found it by recognizing the *requiredness* of the system. He saw the required pattern in the stars and told the physicists exactly where to find the signature on Earth[20].

Fowler's candid, first-hand account is a testament to the nature of this discovery. It reveals that the existence of carbon—and by extension, all life—hinged on a "hidden" rule of physics that the experts in the field had not only missed but had actively "erased" from their records.

The discovery was a "tour de force" precisely because it was not a prediction derived from known, bottom-up nuclear theory. As Fowler himself admits, no nuclear theorist could have done it. Instead, it was a prediction of necessity; a top-down conclusion that for the universe to have the properties we observe (i.e., to have carbon), the laws of physics must include this specific, fine-tuned resonance.

Willie Fowler later received a Nobel Prize (in 1983) for his work in 1957 with Margaret and Geoffrey Burbidge and Fred Hoyle, titled, *Synthesis of the Elements in Stars,* which described the origin of all but the lightest chemical elements in stars, essentially the theory of stellar nucleosynthesis[21]. From the Nobel prize press release dated October 19, 1983, "The Royal Swedish Academy of Sciences has decided to award the 1983 Nobel Prize in Physics by one half to Professor Subrahmanyan Chandrasekhar, University of Chicago, Chicago, USA, for his theoretical studies of the physical processes of importance to the structure and evolution of the stars, and by the other half to professor William A. Fowler, California, Institute of Technology, Pasadena, USA, for his theoretical and experimental studies of the nuclear reactions of importance in the formation of the chemical elements in the universe. The common theme of this year's Nobel Prize in Physics is stellar evolution[22]."

Dr. Fowler also discovered that the carbon cycle is not possible in stars the size of our yellow dwarf sun. Carbon can only be manufactured in bigger, hotter stars. This means that stars capable of producing carbon are likely *not* ideal for life, even if they are as little as 50% larger than our sun[23]. (On the other side, stars *ideal* for life likely *cannot* produce carbon, therefore, when it comes to stars and life, it is a cooperative effort among several generations of stars to first produce the carbon needed for life, and then, downstream, they seed the universe with the ingredients necessary to produce suns that radiate the wavelength of light needed to allow life to assemble and flourish for billions of years. This complex, cooperative arrangement represents only a few of the several of hundreds of required synchronicities and levels of fitness required[24] for life, within galaxies and a universe that is life-allowing.

One Last Problem

There was one last problem to be solved to ensure that carbon-12 was plentiful for the requirements of life within our universe. Once Carbon-12 was produced, in order for it to be plentiful, it needed to stay as carbon-12 and not turn into some other element further up the production cycle. The central question in this inquiry was, if helium-4 could join with beryllium-8 to form carbon-12, what would keep carbon-12 from joining with another alpha particle (helium-4) to form oxygen-16? As it turns out, the energy level of *the strong force* is finely tuned to allow the formation of sufficient carbon, while also not burning all of the carbon produced into oxygen. This energy level results in sufficient ratios of carbon and oxygen so that there can be carbon for based life, and oxygen for metabolism and photosynthesis for carbon-based life, as well as water, the universal solvent, for carbon-based life, to maintain biological functions for life. This represents a trifecta of fortunate elemental outcomes for the benefit of life.

The precision in the burning of sufficient amounts of carbon and oxygen in a life-crucial balance is also objectively *finely tuned*; if the strength of the strong force is increased by *0.4%*, stars will have plenty of carbon, *but no oxygen*. If this same force is *decreased* by 0.4% then carbon is rapidly burned into oxygen leaving insufficient carbon[25]. Additional *fine tuning* is also found in the masses of quarks, which needed to have "simultaneously met nine different conditions for the right nuclear reactions to have occurred in the early universe." so that our universe can produce the right abundances of carbon and oxygen.[26] "The value of the up quark and down quark must have a precise fine tuning as well of $\mathbf{10^{-22}}$ times and $\mathbf{10^{-21}}$ times the Planck mass[27].

These *coincidences* and conditions needed for carbon-12 to exist caused Fred Hoyle to engage in further investigations to find out what conditions were necessary for the needed 7.65MeV resonance to be in place for Carbon-12. Hoyle uncovered the following factors:

-The strength of the electromagnetic force and the strong nuclear force must be within 0.5 percent of their current levels to make carbon production possible

-More than a 2-3 percent change in the light quark mass (an up quark is 470 times lighter than a proton) would lead to problems with the abundance of carbon or oxygen in the universe[28]

-The energy within stars must be high enough to allow the velocities needed for the atoms to collide at sufficient energies. This factor is dependent on the *strength* of gravity, which is by far, so hierarchically weak as to be a puzzle, nonetheless it determines the pressure and thus the heat inside of stars.

-The precise strength of gravity also determines the ability of stars to form *different* elements beyond carbon within discrete burning layers, like the layers of an onion, within stars.

-One problem of gravity being *too weak* (if it were even weaker than it already is) there would be no thermal layering and thus the elements needed for life, especially carbon, would not be produced.

-Another problem that would arise if gravity were weaker, is this condition would prevent stars from becoming supernovae. The heat, pressure and force of a supernova produces the complete list of the 93 naturally occurring elements that are neither produced in the big bang, nor produced during the life or the pre-supernovae, stellar nucleosynthesis life phase of a star. Although only 23 to 24 elements of the 93 are needed for life, most of the remaining elements have essential technological and/or scientific value.

-If gravity was too strong, the temperature within stars would be too hot and thus only elements *heavier* than carbon and oxygen would be produced. In this case stars would burn up too quickly for life on planets to have a consistent source of energy. For the purpose of stars to produce the elements needed for life, the strength of gravity must be finely tuned to 1 part in $\mathbf{10^{35}}$ compared to what the strength of gravity *could* be.

Hoyle's investigation revealed a complex, multi-layered problem. The existence of carbon doesn't just depend on one "lucky" resonance. It requires an entire "conspiracy" of seemingly independent physical constants. The strong and electromagnetic forces, and even the mass of the light quarks, must all be set to incredibly precise, interlocking values.

Furthermore, the force of gravity itself is revealed to be a perfect "Goldilocks" constant: it must be exactly as weak as it is. If it were slightly weaker, stars would not form the necessary thermal layers to produce carbon, nor would they explode as supernovae to distribute the elements. If it were slightly stronger, stars would burn too hot and too fast, destroying the very elements life needs and extinguishing themselves long before complex life could emerge. This isn't a single coincidence, but a web of them, all required for a life-bearing universe.

Fred Hoyle is said to have been an atheist *prior* to his discovery of carbon resonance. As he came to know the resonance and the additional number of necessary conditions required for the production of carbon, his viewpoint changed as evidenced by a famous statement he made regarding his changed understanding and belief in a power that arranged our universe in the way it is.

> "A common-sense interpretation of the facts suggests that a super-intellect has monkeyed with physics, as well as with chemistry and biology, and that there are no blind forces worth speaking about in nature. The number one calculates from the facts seems to be so overwhelming as to put this conclusion almost beyond question." - Fred Hoyle[29]

To give some background, Hoyle didn't so much *discover* the carbon resonance as he actually predicted that it *must* exist and directed others to find it where he predicted it would be. As we have found in several previous unsolvable mysteries, many things that are predicted to exist in our universe, such as WIMPS, axions, inflatons, gravitons, etc, either don't exist, or at least can't be found, while conditions that are not predicted, make no sense, and considered impossible, *do exist.* (For example, the miraculous survival of matter from the great annihilation at the beginning of time, or the universe's impossibly low-entropy starting condition). *In all cases,* where what exists, should be impossible, or at least highly unlikely and beyond the reach of chance, the most direct beneficiary of these strange conditions is life and humans, and our ability to make sense of (observe) our universe.

> "British astrophysicist Fred Hoyle [was] so impressed… by the string of 'coincidences' that exist between particular numerical values of dimensionless constants of nature without which life of any sort would be excluded. Hoyle points out how natural it might be to draw a teleological conclusion from the fortuitous positioning of nuclear resonance levels in carbon and oxygen: "I do not believe that any scientists who examined the evidence would fail to draw the inference that the laws of nuclear physics have been deliberately designed with regard to the consequences they produce inside the stars. If this is so, then my apparently random quirks have become part of a deep-laid scheme. If not, then we are back again at a monstrous sequence of accidents[30]."

Additional Citations:[31]

Hoyle's famous conversion—his conclusion that "a super-intellect has monkeyed with physics"—was not a leap of faith. It was a conclusion forced upon him by the "monstrous sequence of accidents" he had uncovered.

This web of fine-tuning is exactly what we define as a miracle: a necessary, life-giving condition where the required, interlocking constants are so precise that chance is no longer a rational explanation. To assign a formal value to this variable, we must quantify the combined odds against this "deep-laid scheme" occurring by chance.

The Value of *Variable* 7, The Carbon (Hoyle) Resonance:

Given all that had taken place in the universe up to the creation of the carbon resonance, physics had no choice but to allow the Hoyle resonance to exist. The carbon resonance, as strange and fortuitous as it is, and given that it is a prime example of extreme fine tuning that cannot be explained, physics dictates that the resonance must exist, but it is so rare and improbable that it was not known at first, and only by the urging of Fred Hoyle, was it even looked for, and found.

For our Algorithm, our question is quite simple: can physics explain it? In the case of the Hoyle Resonance, the answer is yes. Physics, i.e. Willy Fowler and Fred Hoyle demonstrated that both theoretically and experimentally, the solution to carbon resonance was found by physics.

Yet, although it is explained, some calculations of the Fine Tuning of Carbon Resonance, place the probability at less than 1 in 10^{143} which exceeds the PCOU.

The Value of the Hoyle Resonance is: ***fM Functional Miracle***

The fine-tuning of the carbon resonance is one of the most powerful and specific examples in our investigation, but strictly speaking, it's not an actual miracle under our definition, it's simply an extremely finely-tuned, precise, exact solution, and one that is required to produce suitable amounts of carbon in our universe by overcoming the five-nucleon crevasse.

As Fred Hoyle himself regarded it, the carbon resonance was evidence that the "laws of nuclear physics have been deliberately designed" to ensure the existence of carbon, the one atom upon which all known life depends. With this miraculous solution in place, stars were now empowered to perform their primary function: to forge the full, rich inventory of heavy elements with a particular eye towards the elements essential to life.

Our investigation now moves from the fine-tuning of the universe's fundamental rules to the creation of its materials, beginning with the process of stellar nucleosynthesis.

---

[1] (R.A. Alpher, H. Bethe, G. Gamow, The Origin of Chemical Elements, Feb 18, 1948).
[2] (R.A. Alpher, H. Bethe, G. Gamow, The Origin of Chemical Elements, Physical Review.73 (7): 1948, 803-804)
[3] (R.A. Alpher, H. Bethe, G. Gamow, The Origin of Chemical Elements, Physical Review.73 (7): 1948, 803-804)
[4] (A. Friedmann, On the Curvature of Space, General Relativity and Gravitation, 31 (12): 1991-2000), (A. Friedman, On the Possibility of a World with Constant Negative Curvature of Space, General Relativity and Gravity. 31 (12): 2001-2008),
[5] (A. Alpher, R.C. Herman, The Origin and Abundance Distribution of The Elements, Annu. Rev. Nucl. Sci., 1953. 2:1: 1-40)
[6] (A. Alpher, R.C. Herman, The Origin and Abundance Distribution of The Elements, Annu. Rev. Nucl. Sci., 1953. 2:1: 1-40)
[7] (H. Bondi, T. Gold, The Steady-State Theory of the Expanding Universe, Monthly Notices of the Royal Astronomical Society. 1948, 108 (3): 252, Bibcode: 1948MNRAS. 108..252B) (F. Hoyle, A New Model for the Expanding Universe, Monthly Notices of the Royal Astronomical Society. 1948, 108 (5): 372 Bibcode: 1948MNRAS..373H.)
[8] (F. Hoyle, August 5, 1948, Royal Astronomical Society)3.
[9] (Astronomy and Cosmogony, Cambridge, 1928, Pg. 352)4,
[10] (A. Alpher, H. Bethe, G. Gamow, (letters to the editor) The Origin of Chemical Elements, Physical Review, Vol 73, No. 7, April 1, 1948, Pg. 803-804)5.
[11] (S. Meyer, Return of the God Hypothesis, Pg. 133, HarperOne 2021)
[12] (E.E. Salpeter, Nuclear Reactions in Stars without Hydrogen, The Astrophysical Journal, 115: 326-328)7
[13] (J.D. Barrow, F.J. Tipler, The Anthropic Cosmological Principle, Oxford Univ. Press, 1986, Pg. 253)8
[14] (GF. Lewis, L.A. Barnes, A Fortunate Universe, Cambridge Univ. Press, 2016, Pg. 115)9.

---

[15] (GF. Lewis, L.A. Barnes, A Fortunate Universe, Cambridge Univ. Press, 2016, Pg. 115)10

[16] GF. Lewis, L.A. Barnes, A Fortunate Universe, Cambridge Univ. Press, 2016, Pg. 115)11.

[17] (G.Gonzalez, J.W. Richards, The Privileged Planet, Pg. 198, Regnery Gateway, 2004)12,

[18] (W.B. Ashworth, Linda Hall Library, Scientist of the Day, William A. Fowler, lindahall.org, August 9, 2021)13.

[19] (Interview of William Fowler by Charles Weiner on 1973 February 6, Niels Bohr Library & Archives, American Institute of Physics, College Park, MD USA, aip(dot) org/ history-programs/ niels-bohr-library/ oral-histories/ 4608-4)

[20] (From: D.N.F. Dunbar, R.E. Pixley, W.A. Wenzel, W. Whaling, The 7.68 MeV State in Carbon-12, PhysRev. Vol 92, Issue 3, Pgs 649-650, American Physical Society, 10.1103/PHysRev.92.649) & (S. Mitton, Conflict in the Cosmos: Fred Hoyle's Life in Science 2005, Joseph Henry Press.)

[21] (E.M. Burbidge, G.R. Burbidge, W.A. Fowler, F. Hoyle, Synthesis of the Elements in Stars, Reviews of Modern Physics, 29 (4): 547-650. Bibcode: 1957RvMP…29..547B)

[22] (Press Release, NobelPrize(dot) org, Nobel Prize Outreach AB 2024, Wed 23 Oct 2024, nobelprize(dot) org/ prizes/ physics/ 1983/ press-release/)

[23] G.Gonzalez, J.W. Richards, The Privileged Planet, Pg. 132, Regnery Gateway, 2004)14.

[24] A. D'amario, Astronomy Magazine (online), If Our Solar System is Composed of the Remains of Stars That Used Up Their Hydrogen…, February 2018, Issue)15

[25] (M. Reese, Just Six Numbers, Basic Books,1999, Pg. 55)

[26] S. Meyer, Return of the God Hypothesis, HarperOne, 2021, Pg. 138)17.

[27] (S.M. Barr, /A. Khan, Anthropic Tuning of thd Weak Scale and of Mu/Md in Two-Higgs-Doublet Models, PHys. Rev. D76, No 4: 045002)18

[28] (of Carbon-Based Life Leave Little Room for Error, Kurzweil Accelerating Intelligence, last modified March 15, 2013, https://www.kurzweilai.net/)19

[29] (F. Hoyle, The Universe: Past and Present Reflections, Pgs. 8-12, Engineering and Science 45, November 1981)20

[30] (Fred Hoyle quoted in M. Stockwood's Religion and the Scientists, addresses delivered in the University Church, Cambridge, 1959, SCM Press, quoted in J. Barrow, F. Tipler, The Anthropic Cosmological Principle, Pg. 22, 1986, Oxford University Press)

[31] Jones, Andrew Zimmerman. "Stellar Nucleosynthesis: How Stars Make All of the Elements." ThoughtCo, Aug. 27, 2020, thoughtco.com/stellar-nucleosynthesis-2699311.

Variable 8: Stellar Nucleosynthesis

Variable #9: Stellar nucleosynthesis: *fine tuning of all forces and energies within the sun resulting in supernova phenomena which produced the full chart of 92 naturally occurring periodic elements, and the raw materials needed for life giving planets and solar systems, and the technology of intelligent observers within our universe.*

Introduction, The Origin of Complex Chemical Elements

In our last variable, we established the miraculous fine-tuning of the carbon resonance, the specific key required to unlock the creation of heavier elements. We now turn to the grand cosmic engine that uses that key: stellar nucleosynthesis. This is the process by which stars act as giant forges, transforming the universe's sterile primordial elements into the rich periodic table that forms the foundation of chemistry and life.

The mystery of where the chemical elements came from was solved when *nucleosynthesis* was discovered. Nucleosynthesis reactions only occur under extreme pressures and extreme temperatures such as what took place in the big bang creation event–big bang nucleosynthesis, what happens in the core of stars-stellar nucleosynthesis, and in high energy collisions of particles in a process known as *spallation.* As we reviewed previously, only a few elements are created as a result of the big bang. Hydrogen, the primary and most numerous of the atoms created in the big bang technically exists wherever a single proton exists whether it has an electron or not. Without an electron, hydrogen is often referred to as simply a proton. A proton is a positively charged hydrogen ion represented by the symbol $H+$.

With an electron (unpaired) a proton is atomic hydrogen, or an atom of hydrogen. Hydrogen as an atom is rare; it is mostly found as molecular hydrogen (H_2) because the single electron seeks to be paired with another electron, and it can achieve this by pairing with another hydrogen atom.

This primordial hydrogen, the raw material for all subsequent stellar processes, was itself forged in the first few minutes after the creation event. The theory that explains the origin of this hydrogen, and the other light elements, is known as *big bang nucleosynthesis*.

Big Bang Nucleosynthesis

Once hydrogen was formed out of the quark plasma of the big bang and the temperature of the early universe dropped to one billion degrees, deuterium, an isotope of hydrogen formed. Deuterium is one proton and one neutron. Then helium formed, which is two protons and two neutrons, and then when several nuclei of helium collided in the early universe, miniscule amounts of lithium were created. Thus, big bang nucleosynthesis created hydrogen, its isotope, deuterium, helium, atomic number 2, and miniscule amounts of lithium, atomic number 3. The percentages of these elements and isotopes in the early universe were 75% hydrogen, 25% helium and 0.000000002% lithium *by mass* which translates to 92% hydrogen and 8% helium by *number of atoms* and once again, trace amounts of lithium[1].

While the predicted abundances of hydrogen and helium are considered a major triumph for the Big Bang model, that final trace element, lithium, presents a significant unresolved challenge. The amount of lithium our theories predict does not match what we observe, a discrepancy so serious that it has its own name: *the cosmological lithium problem*.

The Cosmological Lithium Problem

This problem is regarded as one of the most intriguing open questions in cosmology because it is the most serious deficit in the formulation of the big bang model[2]. "The Big Bang is currently regarded as the most successful model to describe the origination and evolution of the universe. However, its success has been limited by the so-called lithium problem, which refers to the fact that (the) primordial lithium-7 abundance is overpredicted by a factor of three in comparison to the value from observation, while predictions match the observed primordial deuterium and helium abundances[3]." According to experts the fact that this discrepancy exists, and exists to this magnitude means that either our understanding of the big bang model is off or missing unknown factors and/or, this result is an indication that we are missing a fundamental understanding of physics itself[4].

While the lithium problem remains a thorn in the side of the Big Bang model, the overall outcome of that first epoch is undisputed: it left the cosmos with a simple inventory of almost nothing but hydrogen and helium. For the universe to evolve from this sterile beginning and create the rich periodic table required for planets, chemistry, and life, a new second phase of element creation was necessary.

Precursors to Stellar Nucleosynthesis & Stellar Nucleosynthesis

For our universe to get to the point where metals, defined as elements higher than hydrogen and helium, were formed in more than negligible amounts took several hundred million years until the first stars were created. This required the intermediary step of the hot plasma from the universe's creation to cool enough for nuclei to capture electrons and become atoms. This capture is what caused the release of photons in the universe that signals the age of the CMB, which took place roughly 300,000 years after the big bang. Approximately 250 million years after the big bang, based upon clues from oxygen signals from 13.3 billion years ago[5], the age of *reionization* took place when the atoms that formed from the CMB became reionized from the energy radiated by the first stars, called population III stars. Their energy was sufficient to dissociate electrons from atomic nuclei once again. This first wave of stars is what began the process formally introduced the subject of our current variable, *Variable #9, stellar nucleosynthesis.*

> "The epoch when the very first stars appeared is a key period of cosmic history. These stars began the manufacture of the chemical elements (those heavier than hydrogen and helium) and their light began the reionization of the neutral cosmic gas. These stars thus mark the dawn of the universe as we know it today and the start of the so-called Epoch of Reionization. The term "reionization" refers to the process whereby these atoms are prompted (by the ultraviolet light from new stars) to shed some of their electrons. Astronomers estimate that this period occurred a few hundred million years or so after the big bang[6]."

The age of reionization that spread across the universe was the age of stars and also the age when light returned to our universe after its initial period of darkness.

> "At first, these stars would simply form pockets of ionized gas around them which would then shrink back as the atoms recombined. Eventually, however, with the continued formation of stars and radiation emitted from active galactic nuclei, the universe has become fully ionized[7]."

The age of reionization is when stellar nucleosynthesis began in earnest with hydrogen and helium transforming the chemical element content of our universe and providing the structure of our galaxy, while also creating the vast list of 92 naturally occurring chemical elements, up to uranium, through 4 stellar processes: (1) *the proton-proton chain*, (2) *the carbon, nitrogen, oxygen cycle* (CNO-cycle), (3)*the triple-alpha process*, and (4) *supernova* events.

The final stages leading to a supernova event begin once the star burns up its oxygen and starts fusing heavier elements in succession. Once it begins trying to fuse silicon into iron, the star moves into its final stages, mere days are left in its life, because the energy requirements to burn iron requires more energy than it produces. When the silicon in the core is converted to iron, fusion can no longer release energy. As a result, the thermal and electron degeneracy pressure (the outward pressure that holds the core up) suddenly "disappears", causing a near immediate stoppage of energy production, which in turn causes the core to collapse catastrophically. The rapid collapse triggers a powerful supernovae explosion as the outer layers are violently expelled. The supernova event is the most prolific and efficient form of nucleosynthesis and element dispersal process in our universe where stars the size of our sun and larger release their products so that they can be used as asteroids, meteors, planets and new stars.

This element-dispersal process explains how younger, "metal-rich" stars like our Sun are formed. But this raises a troubling chicken-and-egg question: If the heavy elements needed to *form* stars are made *in* stars, what were the very first stars made of? This first generation of stars, known as Population III, would have had none of these elements, and this creates a serious gap in our understanding.

A Gap In Our Understanding

The first step in creating the structure of our universe and creating the complex metals above helium is the process of star formation. And star formation is what would have triggered the age of reionization. Because of this these first stars would have had no metals–meaning no elements like lithium or above. This is important because not only were these stars entirely devoid of metals in their lifetime, as the first generation of stars they also would be unable to produce a significant amount of metals before the end of their stellar lives, which is assumed to be quite short based on the size of these stars. Metals are a key variable in the formation of the generation of stars that follow the first generation of population III stars.

Star Population Metal Content & Description

Population III, 0.0% metals. First generation star. Metal-free.

Population II, 0.1% metals. Old stars with low metal abundances. Most likely to be found in the halo of our galaxy.

Population I . 2-3% metals. Young, metal-rich stars like our sun. Most likely to be found in the disc of our galaxy[8].

The naming convention for star populations is counterintuitive. One would be inclined to think population I stars would be the first stars to exist. The naming is counter-intuitive because it's historical, not chronological. Astronomers first classified the stars they could observe into Population I (like our Sun) and Population II (older, metal-poor stars). At the time, these were all that were known to exist. It was only later that theorists realized there must have been an even earlier, pristine, metal-free generation that came before the other two. Since the names "I" and "II" were already taken, this hypothetical, first generation of stars was logically named Population III.

Astronomer Walter Baade established these designations in 1944[9], and credited Jan Oort with having suggested these classifications as far back as 1926[10]. It wasn't until much later, in 1965, as proposed by astronomer Neville J. Woolf[11], that it was realized that population III stars were a necessity, and that none of the stars that had been observed to that point in time had the characteristics of these first stars, namely, being nearly entirely devoid of metals, or of extremely low metal content.

The initial formation of Population III stars is puzzling because they were composed of only hydrogen and helium. This presents a major challenge for scientific models. Our understanding of modern star formation relies on the presence of heavier elements (or "metals"). These metals act as efficient coolants, allowing vast clouds of gas to shed heat, which enables them to clump together, fragment, and collapse under their own gravity.

But in the early universe, these heavier elements "had yet to be produced". Without these crucial coolants, the "thin clouds of hydrogen and helium atoms" would have struggled to collapse. As the gas compressed, it would heat up, and this outward-pushing pressure would fight against gravity, preventing the cloud from forming a star. This is why it has been noted in our standard cosmological narrative, ~~as the space.com article notes~~, "how these stars ignited is still a mystery"[12].

The Mystery Of The Universe's First Generation Of Population III Stars, The First Stars in Our Universe & The Value of Stellar Nucleosynthesis as a Variable

At several points in our ongoing investigation, I have related what is generally regarded as the standard cosmological narrative of how our universe came to be. The narrative should sound familiar to you by now. After the big bang, tiny quantum fluctuations in the early Planck-era universe were blown up to cosmic scales by inflationary cosmology — some twenty-eight to thirty orders of magnitude, or roughly a billion trillion trillion times. This hyper expansion of quantum level indeterminacies is what is believed to have set the stage for our modern-day universe by seeding it with the cosmic blueprint that eventually was occupied by cold dark matter, and then hydrogen and helium, which prepared our universe for the inevitable emergence of the first generation of stars. Within this narrative, the creation of the first stars appears to be a given. And they were believed to be massive compared to our host star the sun. From this initial structure, the homogeneity of the universe was made exquisitely rippled such that our universe was able to produce galaxies, galaxy clusters, and galactic threads in every possible direction, as about as orderly and regular as can be imagined.

From this example it can be seen that the standard cosmological is presented as a continuous causal chain, yet even taken at face value, the narrative conceals a serious rupture. As in the case of stellar nucleosynthesis—where fusion depends on an exceptionally narrow fine-tuning of the fundamental forces—the transition from structure to stars is by no means automatic. It is conditional. There exists a non-trivial and well-respected group of physicists—such as Lee Smolin in his book, The Life of the Cosmos' —who do not regard inflation as the default explanation of primordial structure at all.

> "The picture that structure forms from small perturbations, and otherwise symmetric universe may be true in outline, but even if it is true, it is incomplete. There must be some good reason why a universe that starts off as symmetrical and ends up developing such complexity, and if we are honest, we must admit we do not yet know what it is[13]."

For the scientifically skeptical, in their view, inflation merely displaces the fine-tuning problem from the state of the precise balance exhibited by our universe–almost perfectly smooth yet seeded with density contrasts of precisely the right amplitude— to the amount of fine-tuning and precision required to get inflation to make our universe look the way it does. In both cases the universe must begin in an extraordinarily special state. Any reasonable view must regard the question of how such conditions emerge within our universe as presently unexplained, irrespective of the attribution, (inflation, or its unknown alternatives) if this attribution cannot be demonstrated to be rigorously justified by known (and not hand-inserted, finely tuned) physics.

If my earlier use of the inflationary account within our standard cosmic narrative was misplaced as a placeholder until now, then this section stands as my mea culpa. But the deeper point does not depend on inflation in the first place. Even if inflation were correct in every detail—fully predictive, unavoidable, and observationally complete—it would explain only the origin of gravitational structure, not the feasibility of stars. Inflation does not explain the crux of the problem for the creation of the universe's first generation of stars, which is how these overdensities, if indeed they existed, overcome the far tighter constraints required to ignite into stars.

What follows, then illustrates that Between the existence of cosmic structure and the appearance of the first Population III stars lies an independent and extraordinarily restrictive filter—one the standard cosmological narrative quietly steps over, but never truly explains.

In this narrative, which is known as our standard cosmological narrative, the inflationary mechanism is assumed as an initial condition for stars and it serves as a vital precondition within the current model that focuses on the initial round of stars in our universe known as Population III stars. This model is the Lambda-Cold Dark Matter model usually represented as *ΛCDM*. In this model Lambda (Λ) represents the role of the cosmological constant. General relativity is the dynamic framework. Weyl's postulate–the assumption that on large scales, matter can be treated as a smooth cosmic fluid– is included. The "CDM" stands for cold dark matter– the weakly interacting mass that provides the gravitational scaffolding for universal structure. Cold dark matter is what is assumed to occupy the inflationary created quantum fluctuations. This is what creates the gravitational wells that pull in, condense and eventually ignite the primordial hydrogen and helium of the early universe into the first generation of Population III stars. These are the main parts of the model.

Having explained the ΛCDM model, it can be clearly seen that it serves as the lynchpin within the standard cosmological narrative as the underlying model for the creation of Population III stars. On one side of this model is the primordial hydrogen and helium of the primordial universe, and in distant third place, trace amounts of lithium at $\mathbf{10^{-10}}$. On the other side of the ΛCDM model are stars, and stellar nucleosynthesis, which in the life and death of stars of the right solar mass, the manufacture of the full periodic chart of the elements is facilitated. Over multiple generations of stars increasing metallicity is achieved, which as we explained leads to galaxies, and galactic structure, but also black holes, neutron stars, quasars, white dwarfs, and human beings, which are referred to as intelligent observers.

Stellar nucleosynthesis, as we discovered in the first part of this chapter, is key to all of these universal products, and up until this point in our narrative, was a seeming anomaly within our algorithm. It is the one process and variable within our chapter-by-chapter investigation that works fully as intended in virtually every instance and rather than being highly improbable, or exhausting the probabilistic capacity of our universe, or constituting a physical miracle given our understanding of first principle physics, is a 100% discernible and highly likely proposition, given that stars exist.

The chain of causality from the big bang to our modern universe appears to be all but complete. But this appearance of surety is an illusion. Of all of the ingredients that make up the ΛCDM model and our simulations for the creation of Population III stars, there is one ingredient it needs to work as intended that is missing that dooms the entire proposition of the model as serving as the narrative lynchpin in the story of our universe. That ingredient is metals. In astronomy, a metal is any element heavier than helium.

Hydrogen and helium gas clouds with metals are efficiently and reliably observed to collapse and ignite into stars in our simulations and observationally. Specifically, for Population II stars, carbon and oxygen are the dominant metal coolants, followed by "dust"–tiny solid particles composed mainly of carbon, oxygen, silicon, magnesium and iron. These metals, an astronomical designation for any element above helium and hydrogen, facilitate gravitational collapse and fragmentation of gas clouds into stars. Collapse, cooling, fragmentation, and ignition are all dependent on a minimum threshold of metallicity.

And here is where we reveal that our investigation into stellar nucleosynthesis ends with the conclusion that this process, seemingly inevitable within stars, is a Bonafide member of our beyond probability algorithm. The reason for this is metals are needed to make stars, and only stars can make metals. There is no way around this. This is the unresolvable paradox at the heart of stellar nucleosynthesis. Modern day Simulations artificially manufacture stars by using a formula that attempts to bypass the lack of metals. We have only formulas for star formation, not actual "star" simulations. "We are still a long way from simulating the formation of galaxies and the formation of stars inside those galaxies using the basic equations of physics, hence the need for star formation recipes." states cosmologist George Rhee, author of 2013's *Cosmic Dawn*[14].

Modern day simulations fail in the task of actually generating stars, as we will reveal in rigorous, granular detail, using the most up to date scientific data. This finding, that stars cannot be made from primordial helium and hydrogen, by any process we currently understand is affirmed by the fact that no Population III stars have ever been observed within our universe.

Metals and their necessity in star formation under impossible conditions are the paradox that stops our universal evolution dead in its tracks. The presumption that population III stars are inevitable, as has been related in nearly every rendering of our standard cosmological narrative, in truth has a huge unanswered question as its foundation.

In the upcoming pages, we will briefly review the timeline of advances that simultaneously set the stage for the discovery of this paradox, while also serving as the blueprint for manufacturing a false solution to explain away the paradox so that our simulations could brute force attack what physics says is impossible–a universal history with prolific and unending star creation.

Sir James Jeans and Nobel Laureate Subrahmanyan Chandrasekhar: Instability

In 1902 Sir James Hopwood Jeans (1877–1946) derived the gravitational stability criterion now known as the Jeans instability[15]. What he found was that under highly idealized conditions—under the assumption of a static, infinite, uniform gas and neglecting cosmic expansion— a self-gravitating gas would become unstable and collapse when gravitational forces-self gravity–overcome pressure support. Although mathematically correct, the exclusion of several physical influences present in the real universe, namely all of the influences excised in his idealized conditions (the assumption of a static, infinite, uniform gas neglecting cosmic expansion) is the reason this derivation has forever been referred to as *the Jeans Swindle*. Without the *swindle* there would be no instability or collapse.

Swindle or not, Jeans's analysis was groundbreaking and remains a major conceptual breakthrough, as the Jeans instability—still widely used today—provides the first framework for identifying when gravitational collapse could occur, while clarifying the competing roles of pressure support and self-gravity. Its enduring legacy is not as a complete physical description of star formation, but as a limiting criterion that is currently embedded within more realistic, non-ideal models for Population III star creation.

Subrahmanyan Chandrasekhar

Next in our analysis is the foundational work of Subrahmanyan Chandrasekhar, Nobel laureate for his theoretical studies of the physical processes governing the structure and evolution of stars[16]. From 1931 through the 1960s, Chandrasekhar developed a rigorous framework for stellar stability, culminating in his analysis of "relativistic gravitational instability" in *The Dynamical Instability of Gaseous Masses Approaching the Schwarzschild Limit*[17]. In this work, he showed that when internal pressure can no longer counteract gravity, hydrostatic equilibrium becomes impossible and collapse is unavoidable (once certain mass and density thresholds are exceeded). His analysis established a precise criterion for instability. His legacy is in providing future generations with a rigorous theoretical reference for modeling and simulation. His findings were that collapse is mathematically inevitable given certain conditions, but even so, star formation is not. This vital distinction set the stage for all subsequent work on stellar formation. Subsequent work identified that chemical composition–particularly the presence of metals–is decisive in determining whether collapsing gas can cool and fragment into low mass stars.

Timeline: Critical Metallicity

By the final decades of the twentieth century, theoretical work had already demonstrated that metals dramatically enhance radiative cooling, enabling gas clouds to shed heat efficiently and fragment into numerous low-mass objects[18]. Yet the problem of metal-free gas remained unresolved. Numerical simulations in the late 1990s and early 2000s showed that primordial collapse without metals favored inefficient fragmentation and the formation of very massive stars[19], with molecular hydrogen (H_2) as the primary available coolant. While H_2 cooling permits collapse, it imposes a relatively high temperature floor, limiting fragmentation and preventing the formation of low-mass stars[20]. These results suggested the existence of a critical metallicity—a threshold below which the star-formation process operates in a fundamentally different regime—though its precise value and physical origin remained uncertain.

Bromm & Loeb (2003): Quantitative Thresholds for Low-Mass Formation

This line of theoretical work culminated in 2003, when Volker Bromm and Abraham Loeb developed models describing the formation of the first low-mass stars from gas with extremely low carbon and oxygen abundances[21]. Rather than introducing the importance of metals, their study quantified the effect of trace carbon and oxygen—just 0.01–0.1% of the Sun's abundance—on gas cooling and fragmentation. Below this threshold, metal-free gas relies primarily on molecular hydrogen for cooling, which stalls fragmentation at relatively high temperatures and favors the formation of very massive stars. The presence of even minute metal enrichment marks a fundamental transition in the star-formation regime, confirming that the earliest low-mass stars could only arise once specific metallicity thresholds were met.

Initial Conditions, Inflationary Cosmology, the ΛCDM Model and Model Performance In Three Parts

Next, we examine the ΛCDM framework and its simulation-based implications across three stages—before, during, and after star formation. *Part I, Preconditions for Population III Stars,* examines the initial conditions and physical prerequisites assumed within the model. *Part II, Insertions in ΛCDM-Based Simulations,* analyzes where star formation is not derived from first-principle physics but is instead introduced by assumptions intended to produce a specific outcome–stars. *Part III, Metal Retention, Enrichment, and Time Constraints,* evaluates how effectively the model produces sustained, large-scale star formation in terms of enrichment efficiency and observed time limits.

Part I. Preconditions For Population III Stars - Quantum Fluctuations & Inflations Required Boost

Core Points from Ijjas, Steinhardt & Loeb From Scientific American – Review of Earlier Findings on Problems in Inflationary Cosmology. This is a review from our earlier investigation:

Inflation's Carefully Tuned Parameters — Inflation models require specific choices for the shape of the inflaton potential, the repulsive form of gravity, and other parameters to generate the tiny overdensities needed for structure formation. These choices are not derived from first principles; they are inserted to make the model work[22].

Long-standing foundational issues — Inflation does not naturally solve the problems it was designed to address without fine-tuning. Key issues include:

Initial conditions problem: inflation requires a pre-existing patch of space that is extremely smooth and flat.

Flatness and horizon problem: the extent to which inflation solves these depends on the model's chosen parameters.

Measure problem: probabilities for outcomes in eternal inflation are ambiguous and depend on externally imposed rules[23].

Quantum fluctuations lead to eternal inflation — Rare large fluctuations can restart inflation in some regions, meaning inflation never fully stops everywhere. What determines when and where inflation starts, stops, or slows is chosen ad hoc rather than emerging from first-principle physics or constants in the model.

Predictive power is limited — Because inflation is eternal and produces a multiverse, any physically possible outcome will occur somewhere. This undermines the theory's ability to make clear predictions about our universe.

Inflation, Hand-Tuning, Amplitude And Star Formation — Inflation models rely on carefully chosen parameters to generate the tiny overdensities required for structure formation.[24] These fluctuations are extremely small—minute deviations from perfect uniformity. Gravitational collapse, however, requires far rarer, high-amplitude peaks, roughly the top 0.1–0.01% of halos, to overcome pressure support. The work of Press & Schechter (1974)[25], and related works typically estimate that only a very small fraction of primordial halos are dense/massive enough to collapse early. Most references point to something on the order of 0.1% to 0.01% of halos having the capacity to form protostars[26]. Even within this subset, star formation is precarious with no demonstration from first principles that ΛCDM plus inflation reliably produces the halos needed to initiate collapse.

The Result of Inflation - Vanishingly Improbable Protostar Formation — Assuming the above percentage of halos eligible to cool and collapse (~0.1–0.01%) succeeds, only a fraction of these primordial halos reach the density and mass thresholds necessary to form protostars[27]. This is because of additional dynamical, chemical and feedback bottlenecks (e.g., molecular hydrogen cooling, radiative feedback, dynamical heating)[28]. In many high-resolution cosmological simulations, the number of star-forming halos as a fraction of threshold halos is estimated as being~10% or less that meet threshold conditions actually producing one or more Population III stars.

> "In their simulation of the first star, at most one massive (greater than 1 solar mass) metal-free star forms per pre-galactic halo. This means that even in halos that meet the conditions for collapse, only a single star emerges, and not many stars. This finding effectively implies that the efficiency of star formation in these early, metal-free halos is extremely low[29]."

This brings the original percent of halos eligible for collapse from ~0.1% to ~0.01% to ~0.01% to ~0.001%.

Reality Versus Inflation's Narrative Tension

This point isn't about physics, it's about the narrative assumption of robustness resulting from inflation. Within the ΛCDM model, it is an extremely fragile hypothesis that overwhelmingly does not produce what our universe needs. As we can see from the above analysis, if we examine only the preconditions assumed from inflation, they are layered in such a way to make the improbable nature of star formation appear as a coherent probability approaching a certainty. In truth, although inflationary cosmology is invoked as an assumed initial condition, its reality is dubious and ultimately unneeded as an explanatory framework. That there are overdensities in our universe is observed, that stars exist is unrefuted, therefore they both exist independently of whether inflation can be demonstrated as the cause. Going one step further, given the limited role inflation plays in explaining these overdensities, along with its questionable foundation, its inclusion as a precondition for Population III star formation should be understood as an assumption and not a critical pathway.

Dark Matter Halos

The ΛCDM model -based simulations assume that cold dark matter (CDM) exists and forms early halos in regions of slightly higher density, providing gravitational wells capable of attracting the primordial gas of the early universe. CDM is hypothetical and has never been directly observed. Halo formation and subsequent baryon capture rely on unknown CDM properties, meaning the simulation's accuracy is uncertain and untested. In short, these halos are a necessary input and not a confirmed physical reality. And as we have just seen, in relation to halo formation, the assumption of CDM is based on inflationary initial conditions that do not match the required amplitude necessary to create the required dark matter halos.

The Changing Landscape of ΛCDM Initial Conditions

ΛCDM-based simulations assume that cold dark matter is the scaffold for universal structure, and that the constant value of dark energy (Λ) provides the seeds for Population III star formation. Results from the 2024–2025 Dark Energy Spectroscopic Instrument surveys (DESI) suggest that dark energy may not be constant, rather it may be evolving—possibly weakening over time—placing the assumption of Λ in ΛCDM-based simulations under strain[30]. Independent of these findings, a persistent inconsistency remains: the Hubble tension, the mismatch between the expansion rate inferred from two established measures: The CMB vs. Cepheid-calibrated Type Ia supernovae—further indicating that ΛCDM may not fully capture the universe's expansion history.

In a Scientific American story, *Dark Matter May Be Missing From This Newfound Galaxy, Astronomers Say*[31], at smaller scales, some galaxies appear to lack the expected dark-matter halos altogether, challenging the assumption that cold dark matter universally governs structure formation and carrying serious implications for galaxy-formation models[32]. In a more recent Scientific American article, *Latest Dark Energy Study Suggests the Universe is Even Weirder Than We Imagined,* (2025) "Whether a resolution comes from dynamic dark energy, dark matter decay, or something entirely different, the ΛCDM model has seemingly been stretched to its breaking point. Every reasonable explanation for DESI's data involves new, scarcely explored physics[33]." Reflecting the accumulating strain, Duke University physicist Dan Scolnic stated a, "step up from tension to crisis" may be warranted should dark energy indeed vary with time[34].

Part II. Insertions For ΛCDM-Model Based Simulations - Co-Moving Coordinates and the Legacy of the Jeans Swindle

To permit gravitational collapse in an expanding universe, cosmological simulations employ co-moving coordinates—a framework that factors out the global Hubble expansion so that local overdensities can be treated as mathematically separated evolving systems within that background. This approach descends in concept from the original (James) Jeans analysis, in which the gravitational effect of the homogeneous background density was subtracted in order to make collapse mathematically possible—an adjustment later known as the Jeans swindle. While modern treatments embed this move within relativistic cosmology, the underlying assumption remains large-scale expansion is neutralized at the level of local dynamics[35]. In co-moving coordinates, primordial gas clouds that would otherwise be rapidly diluted by early-epoch expansion are allowed to evolve as if they are partially shielded from that expansion. Overdensities are permitted to grow under gravity relative to the expanding background rather than being stretched apart by it. Without this framework—and the assumption it encodes—cosmic expansion would act to dilute density contrasts faster than gravity could amplify them, rendering early star and galaxy formation far less probable within the model.

Assumed Accretion Rates / The Molecular Hydrogen Supply (Sound Speed) Problem

In many star-formation simulations, gas is assigned an assumed rate at which it falls onto a forming protostar so that a star appears within the limited runtime of the simulation. This shortcut avoids a deeper physical constraint identified long ago by Chandrasekhar: gravity alone is not enough to make a star. A collapsing region must either start with enough mass or be able to draw in fresh molecular hydrogen fast enough from its surroundings to keep growing[36]. In real gas clouds, there is a hard physical limit on how quickly material can flow inward. Gas cannot move faster than pressure waves can travel through it, which means the local sound speed (the sound speed problem) sets an upper bound on how rapidly molecular hydrogen can be supplied[37]. Classic analytic models of star formation show that if collapse demands mass faster than this limit allows, the process simply starves—gravity runs ahead of the available fuel. In simulations, this limitation is often bypassed by assigning accretion rates that exceed what gas could physically deliver under these conditions. Doing so ensures that a protostar forms on schedule, but it does so by overriding the natural bottleneck imposed by the physics of gas flow.

Fragmentation Models

Simulations employ threshold-based fragmentation rules that dictate when collapsing gas is converted into discrete objects. In primordial gas, the physical conditions governing fragmentation are poorly constrained, so these thresholds partially impose the number and size of fragments rather than allowing them to emerge fully from first-principle dynamics[38].

Artificial Sink Particles and Protostellar Cores

Star formation is an intrinsically multi-scale and non-linear process, and current numerical simulations are unable to follow the collapse from diffuse gas to fully resolved protostars from first principles across all relevant physical regimes[39]. To manage the extreme dynamic range in density, temperature, and spatial scale, simulations therefore employ sub-grid prescriptions, most notably the introduction of sink particles or protostellar cores, in regions where collapse is assumed to have progressed beyond the resolvable limit[40].

These sink particles act as proxies for unresolved protostellar objects, replacing the internal physics of continued collapse, accretion, and feedback with simplified accretion rules. As a result, the transition from gas cloud to protostellar object is not explicitly derived within the simulation but is imposed once predefined criteria are met. In this sense, simulations do not demonstrate the formation of protostars from first principles; rather, protostellar objects are introduced at the point where the relevant physics can no longer be directly resolved. The existence of protostars is assumed (inserted) at a critical stage rather than emergently produced by the calculation itself[41]. Simulations cannot directly resolve protostar formation. This is why they use sink particles, protostellar cores and other prescriptions to represent stars[42].

Jeans Length

Simulations impose a minimum resolvable scale, the Jeans length, so that gas clumps remain numerically trackable and can collapse. At this scale, star formation is effectively forced, bypassing the smaller-scale fragmentation that would naturally occur in a fully resolved system[43].

Suppression of Turbulence & Sub-Grid Turbulence

Chaotic, small-scale turbulent motions in collapsing gas naturally disrupt ordered accretion and can inhibit protostar formation[44]. Sub-grid turbulence prescriptions artificially suppress small-scale chaotic motions, preventing natural fragmentation and ensuring smoother star formation. Within this framework, these adjustments act as another refinement designed to make the emergence of stars computationally tractable[45].

UV Radiation / Negative Feedback Catastrophe

UV radiation suppresses Population III star formation by destroying molecular hydrogen via Lyman–Werner bands. Under the presumption of the first Population III stars, the early universe was UV-rich, meaning molecular hydrogen —the only coolant in metal-free gas—is continually being heated and destroyed, strongly suppressing subsequent star formation and creating a radiative feedback barrier that inhibits collapse in neighboring halos. This is the negative feedback catastrophe, and it cannot be resolved without metals, which for Population III stars do not yet exist[46].

Boundary Conditions

ΛCDM model -based simulations impose boundary conditions that favor collapse rather than deriving collapse from physical principles in an unbiased way. Choices about initial turbulence, rotation, cooling efficiency and density / velocity profiles significantly affect whether gas collapses to form stars. These conditions are not guaranteed by the ΛCDM model itself, rather they are preconfigured to produce the desired outcome. Within simulations, these conditions do not arise from physics on their own[47].

Assumed Masses and Bottlenecks of Early Stars (~50-300 Solar Masses)

The commonly cited mass range for Population III stars—~50–300 solar masses (a solar mass is based on the mass of Earth's host star)—is not derived from first principles but emerges as an assumption within simulations that struggle to produce stars in primordial gas[48]. The same physical obstacles that make Population III star formation difficult—suppressed fragmentation, high sound speeds in metal-free gas, strong radiative feedback, and dynamical instabilities—become more severe, not less, at higher masses[49]. To be clear, massive stars are not a natural outcome; it reflects the absence of any unguided, first-principles mechanism that would fragment or terminate accretion to produce smaller stars[50].

Very massive stars in this range ~50–300 solar masses are theoretically capable of producing heavy elements through core-collapse or pair-instability supernovae, but they are extremely rare. When / if they are able to form, they do not produce what our universe needs (see next paragraph), with the most massive stars sometimes collapsing directly into black holes and ejecting minimal metals[51] . Intermediate massive stars (~8–50 solar masses) from primordial gas also generate metals and explode within ~3–20 million years, but they remain rare, and their feedback disrupts small halos, producing patchy enrichment rather than continuous metal propagation[52].

Medium-mass stars (~1–8 solar masses) contribute only light elements and live too long (~100 million years to several billion years), to meet our universe's time constraints, and are the least likely to form in metal-free gas. Small stars (≤ 1 solar masses) which includes our own host star, the sun, cannot produce the heaviest elements. They can produce helium, carbon, nitrogen and trace oxygen, but they don't explode as supernovae, rather they end their lives as white dwarfs, returning very little of their mass to the galactic medium. But more pertinent to future star formation is their lifespan. Stars of 1 solar mass or less are long-lived (10 to 25 billion years) which cannot meet our universal timeline for robust star creation. Finally, the physics of stars of this size necessitates metals more than any of the other star sizes. Taken all together, stars of 1 solar mass or smaller are irrelevant for early chemical enrichment[53]. Together, these constraints for the various sizes of stars creates a severe bottleneck. In simulations, massive stars are preferred, so they are the ones that are expected. But behind this assumption is a stark reality. "Preferred" in this sense does not map onto probability: The sober reality of astrophysics is that halos capable of producing stars, which are already rare, overwhelmingly fail to form stars, and if they were to form stars, the stars they would produce would overwhelmingly fail to sustain early, multi-generation star formation and robust chemical enrichment.

Although primordial gas conditions permit very massive stars in principle, current simulations do not self-consistently form protostars from first principles. Instead, they insert sink particles and extrapolate mass growth under assumed accretion prescriptions rather than resolving collapse all the way to stellar birth[54]. As a result, claims of stars in the range of 50–300 solar masses reflect theoretical allowances rather than demonstrable, robust formation pathways, creating a fundamental tension: the stars invoked to solve early enrichment problems are themselves not shown to arise naturally within the simulations that require them: then, even if they were to arise, they would not reliably produce the metals our universe needs; instead, they would most likely sterilize halos or impose long recovery times before star formation could resume.

The result is a stark incongruity: the first stars that simulations rely upon to seed the universe are largely hypothetical, fragile, misaligned with the conditions required for sustained cosmic enrichment in addition to having never been observed.

Part III. Metal Retention & Enrichment & Time Constraints - Metal Retention & Enrichment

Additional complications noted in ΛCDM-based models include the fact that population III metal enrichment is often described as "rare and self-limiting[55]." Studies indicate that Pop III supernova products frequently escape their host minihalos[56]. This suggests that even when metals are produced, they often fail to enrich nearby gas in a way that promotes further star formation. In low-mass minihalos, the first supernovae frequently expel the majority of gas and metals beyond the halo, preventing efficient self-enrichment[57]. In cases where metals are retained within a halo, the surrounding gas is still negatively affected by radiation and mechanical feedback from massive star supernovae, which suppresses star formation within the same halo[58]. Feedback and associated radiative effects tend to disrupt dense gas, suppressing or delaying subsequent star formation[59].

Together, these results indicate that early metal enrichment is highly inefficient, further constraining the ability of primordial halos to sustain multi-generation star formation. Because of these factors, the likelihood of producing Population II stars from Population III supernovae is extremely low compared to the total halo population. Simulations suggest that only a tiny fraction of early halos—often fewer than one in many thousands—reach the conditions necessary for sustained star formation[60].

Time is a Hard Limit

The standard cosmological narrative imposes strict constraints on what our universe must look like within the first ~280 million years based upon the most recent series of observations by the James Webb Space Telescope (JWST). ΛCDM-model based simulations, even when equipped to produce stars in the most efficient and vigorous manner and loaded with an extensive ingredient list of insertions and assumptions to achieve this end, cannot reliably produce stars with such fecundity and chemical enrichment on this timeline.

-JADES-GS-z13.1 — at a redshift of ~13.2, corresponding to ~330–340 million years after the Big Bang.

-JADES-GS-z14-0 (2024) — at a redshift of ~14.18, corresponding to ~290–300 million years after the Big Bang.

-**Galaxy Mom-z14 (2025)** — at an approximate redshift of 14.44, it is observed at roughly 280 million years after the Big Bang. As noted by Naidu et al[61].

"JWST has revealed a stunning population of bright galaxies at surprisingly early epochs, $z > 10$, where few such sources were expected… that expands the observational frontier to a mere 280 million years after the Big Bang. The redshift is confirmed with NIRSpec/prism spectroscopy[62]."

With the deployment of the JWST, increasingly, galaxies are being observed at ages that overthrow the capabilities of even the most robust ΛCDM-based models. Remarkably, these early galaxies have lower total mass (on average), higher star-forming efficiency (extremely intense relative to mass), low, but non-zero metallicity, and possess small dense luminous cores, full halo structure and detailed rotational curves. Key to these observations, they are similar in almost all ways to dwarf galaxies in our local neighborhood, meaning if they were observed with a redshift comparable to dwarf galaxies in our local neighborhood, there would be little to indicate they formed over 14 billion years ago. Perhaps more pointedly, contrary to the expectation that the JWST would be likely be able to identify Population III stars as it resolved images from the beginnings of the universe, they have instead tightened the constraints and increased the oddity of the paradox of star creation. Population III stars, long invoked as a necessary feature of early cosmic evolution, remain unobserved at any age redshift within our universe.

JWST: More is Less–A Sobering Reality

Astrophysicist and cosmologist John Mather, who along with astrophysicist George Smoot was awarded the 2006 Nobel Prize in physics "for their discovery of the blackbody form and anisotropy of the cosmic microwave background radiation[63]." was the leader of the JWST team for 3 decades, taking the project from initial concept to launch. He was interviewed by physicist Brian Greene in late 2023 where he answered questions about what we have learned from the early observations of our universe from the JWST. Mather stated that with being able to look back at galaxies that formed 200 to 300 million years after the big bang, what the team found was far more developed than expected and growing faster than expected[64]. Mather cited evidence for merging black holes and merging galaxies, lit up with what must have been millions or billions of stars.

"We thought we'd see that far but we didn't think we would see what we're seeing. We thought the galaxies would be smaller, cooler, darker and now we are surprised we just don't know why they're that way." said Dr. Mather, "the big question is what are the first stars like?... There's also the big question of where did the first black holes come from? These days every galaxy has a big fat black hole in the middle, and we don't know whether that was produced before the galaxies or after the galaxies, so this is a wide-open question[65]."

Dr. Mather's response is a clear demonstration that with the addition of crucial information from the JWST, we know less, seeing as our assumptions have been dashed, and we are in need of even more information than ever before to piece the mystery of the early universe and the first stars together. We don't know how, when, or why the first stars, black holes or galaxies formed. What else is missing from our understanding, which at this point is increasing in depth and perplexity, is an open question.

Halo Constraints

When the constraints we have outlined in our review—the halo density-threshold constraint, halo collapse constraint, star-formation derivation gap, energetic disruption constraint, and metal-retention constraint—are treated sequentially, the fraction of primordial halos capable of sustaining future star formation collapses rapidly. Of all halos, only ~$\mathbf{10^{-1}}$ to $\mathbf{10^{-2}}$ satisfy the density (and thermodynamic) prerequisites implied by inflation. Only ~$\mathbf{10^{-2}}$ to $\mathbf{10^{-3}}$ of the total halo population is believed to actually undergo gravitational collapse. The subsequent pathway to multi-generation star formation remains severely constrained. Cosmological simulations consistently find that only a tiny fraction of star-forming halos, ~$\mathbf{10^{-3}}$ to $\mathbf{10^{-4}}$, successfully retain metals that lead to sustained Population II star formation. Taken together, under the most optimistic assumptions, these constraints imply that no more than ~$\mathbf{10^{-8}}$ to $\mathbf{10^{-6}}$ of halos support continued star formation. The conclusion is successful early enrichment is the statistical exception rather than the norm and overall we've arrived at a finding that only 1 in 10,000 to 100,000,000 of halos *eligible* to collapse into stars will successfully seed their halos for future star creation and out of all the halos created, no more than 1 out of 100,000 to 10,000,000,000 can be expected to seed multiple generations of stars.

Value of Variable #9

We began with stellar nucleosynthesis, which appears as an anomaly within our algorithm: instead of posing a probabilistic challenge or requiring a "miracle," it emerges as one of modern astrophysics' most robust and secure achievements. Once stars exist, metals are naturally produced through internally consistent nuclear processes. Stellar nucleosynthesis and supernova nucleosynthesis introduces a previously unimaginable chemical diversity within our universe, with 24 of these new elements being required for life, and the remaining elements to make up the full complement of 92, needed to make planets and experiment, devise and perform advanced technology. This aids greatly in our understanding of the universe.

Yet the paradox emerges when traced backward: the first stars are required to produce metals, but their own formation presupposes conditions that cannot exist in a metal-free universe. Metals are required to make stars, and stars are required to make metals.

From Jeans' original instability criterion, through Chandrasekhar's refinements which defined what could happen, to the creation of modern numerical simulations, the same gap persists: no first-principles derivation demonstrates how primordial gas forms a protostar, much less a stable, long-lived star. Simulations do not resolve this; they bypass it, inserting stars by prescription. Yet stars exist, the universe is metal-rich, and Population III stars are invoked as a bridge. Yet these stars have never been observed fully formed in galaxies roughly 280 million years after the Big Bang. Stellar nucleosynthesis is empirically robust, but its origin remains unexplained. Taken as a whole, the conclusion is unavoidable: the paradox of Population III star creation—the Midas touch of stellar nucleosynthesis—cannot be resolved under materialism and first-principles physics.

A core tenet of The Divine Algorithm is that we cannot grant the "miracles" for free. The entire engine of stellar nucleosynthesis is entirely dependent on a "miraculous" prerequisite: the origin of the first (Population III) stars. Therefore, as certain, and odd, and fortuitous stellar nucleosynthesis is, we cannot value this variable in any way as a "certainty." The fact that the engine (stars making elements) works perfectly (making nucleosynthesis) the fact that this can only occur after the impossible origin of the first generation of Population III stars renders stellar nucleosynthesis, itself, perhaps the "surest" "impossible" thing in our universe.

Bottom line, because the origin of stars is inexplicable by current physics, the Variable as a whole must be assigned the value: *Impossible / Miracle*

The Full Breakdown:

Value of Variable 8

Stellar Nucleosynthesis, 1:1, 100%

Population III Stars from Primordial Gas: Impossible / Miracle

Final Value: Impossible / M = Miracle

---

[1] (A. Frebel, Searching For The Oldest Stars: Ancient Relics From the Early Universe, Table 10.1, Pg. 244, 2015, Princeton University Press).
[2] (C.A. Bertulani, Shubhchintak, A.M. Mukhamedzhanov, Cosmological Lithium Problems, 9th European Summer School on Experimental Nuclear Astrophysics, EPJ Web of Conferences 184, 01002, 2018, doi(dot) org/ 10.1051/ epjconf/ 201818401002)
[3] (Chinese Academy of Sciences, SciTechDaily, science technology daily, Understanding the Big Bang and the Cosmological Lithium Problem, December 18, 2021, scitechdaily(dot) com/ understanding-the- big-bang- and-the-cosmological-lithium-problem/)
[4] (Modern Physics, Cosmological Lithium Problem, a major discrepancy between observed and predicted lithium-7 in the universe, challenging astrophysics and cosmology, 2024 Modern Physics Insights: Discover, Understand, Innovate, modern-physics(dot) org/ cosmological-lithium-problem/)
[5] (J. Parks, The First Stars Formed When The Universe Was Less Than 2% Its Current Age, May 16, 2018, last updated May 18, 2023, Astronomy Magazine online, astronomy(dot) com/ science/ the-first-stars- formed-when-the- universe-was-less-than-2-its- current-age/),
[6] (Harvard Center for Astrophysics, Harvard & Smithsonian, Preparing to Study the Epoch of Reionization, October 7, 2016, quoting, P.A. Carroll et al, A High Reliability Survey of Discrete Epoch of Reionization Foreground Sources in the MSW EoR0 Field, MNRAS 461, 4151, 2016)
[7] (D. Koh, Population III Stars: The Universe's Ultimate Reclusive Pop Stars, March 8, 2018, Kipac Kavli Institute for Particle Astrophysics and Astronomy at Stanford University, kipac(dot) stanford(dot) edu/ highlights/ population-iii- stars-universes- ultimate-reclusive- pop-stars)
[8] (A. Frebel, Searching For The Oldest Stars: Ancient Relics From the Early Universe, Table 10.1, Pg. 244, 2015, Princeton University Press) & (PBS Spacetime YouTube Channel, The Fate of the First Stars, May 31, 2017, 13min, 28 seconds)
[9] (W. Baade, The Resolution of Messier 32, NGC 205, and the Central Region of the Andromeda Nebula, Astrophysical Journal. 100: 137-146, doi: 10(dot)1086/14450)
[10] (W. Baade, The Resolution of Messier 32, NGC 205, and the Central Region of the Andromeda Nebula, Astrophysical Journal. 100: 137-146, doi: 10(dot)1086/14450).
[11] [Wikipedia, Stellar Populations, enwikpedia(dot) org/ wiki/ stellar_population, citing, (P. Thornton, Observational Aspects of Cosmology, Science, 1965, 151 (3716): 1411-1414, 1416-1418, PMID 17817304); (L. Green, Observational Aspects of Cosmology, April 1966, Sky and Telescope, 31: 199, April 1966)]
[12] (Space.com staff, Editor in Chief, T. Malik, First Stars in Universe Weren't Giants, Astronomers Say, November 10, 2011, space(dot) com/ 13572-early/ stars-universe-massive(dot) html)
[13] L. Smolin, The Life of the Cosmos, 1997, Pg. 167, Oxford University Press)
[14] (G. Rhee, Cosmic Dawn: The Search for the First Stars and Galaxies, Pg. 145, 2013, Springer Science + Business Media LLC)
[15] (J.H. Jeans; I. The stability of a spherical nebula. Philos Trans A Math Phys Eng Sci 1 January 1902; 199 (312-320): 1–53. https://doi.org/10.1098/rsta.1902.0012)
[16] (Nobel Prize in Physics, 1983 Nobel.org).
[17] (S. Chandrasekhar, The Dynamical Instability of Gaseous Masses Approaching the Schwarzschild Limit, Physical Review Letters 12, 114–116; Astrophysical Journal 140, 417–433, 1964).

[18] (C. Low, D. Lynden-Bell, The Minimum Jeans Mass or When Fragmentation Must Stop,1976, Monthly Notices of the Royal Astronomical Society, Vol 176(2), Pgs 367-390).
[19] (Abel, Bryan, Norman, 2002, The Formation of the First Star in the Universe, Science, 2001, Issue 5552, vol 295, p 93-98) ,
[20] (M. Tegmark, J. Silk, M.J. Rees, A. Blanchard, A, Abel, T. Francesco, P., How Small Were the First Cosmological Objects?, 1997, The Astrophysical Journal, Vol 474, DOI, 10.1086/303434); (V. Bromm, P.S. Coppi, P. S., R.B. Larson, R. B. The formation of the first stars I: The primordial star-forming cloud. 2002 The Astrophysical Journal, 564(1), 23–51)).
[21] (V. Bromm, A. Loeb, Formation of the First Supermassive Black Holes, 2003, the Astrophysical Journal, Vol 596 (1), Pgs 34-46)
[22] (P.Steinhardt The Inflation Debate, Is the theory at the heart of modern cosmology deeply flawed? 2011; Scientific American, Pg. 37-43) Vol 304, (4)), (A. Ijjs, P.J. Steinhardt, A. Loeb, Pop the Latest Astrophysical Measurements, Combined with Theoretical Problems, Cast Doubt on the Long-Cherished Inflationary Theory of the Early Cosmos and Suggest We Need New Ideas ,2017, Scientific American, Vol 316 (2))
[23] (P.Steinhardt The Inflation Debate, Is the theory at the heart of modern cosmology deeply flawed? 2011; Scientific American, Pg. 37-43) Vol 304,(4)), (A. Ijjs, P.J. Steinhardt, A. Loeb, Pop the Latest Astrophysical Measurements, Combined with Theoretical Problems, Cast Doubt on the Long-Cherished Inflationary Theory of the Early Cosmos and Suggest We Need New Ideas ,2017, Scientific American, Vol 316 (2))
[24] (P.Steinhardt The Inflation Debate, Is the theory at the heart of modern cosmology deeply flawed? 2011; Scientific American, Pg. 37-43) Vol 304,(4)), (A. Ijjs, P.J. Steinhardt, A. Loeb, Pop the Latest Astrophysical Measurements, Combined with Theoretical Problems, Cast Doubt on the Long-Cherished Inflationary Theory of the Early Cosmos and Suggest We Need New Ideas ,2017, Scientific American, Vol 316 (2))
[25] (W.H. Press, P. Schechter, Formation of Galaxies and Clusters by Self-Similar Gravitational Condensation, 1974, The Astrophysical Journal, Vol 187, Pgs. 425-438) Tegmark et al (1997), (M. Tegmark, J. Silk, M.J. Rees, A. Blanchard, A, Abel, T. Francesco, P., How Small Were the First Cosmological Objects?, 1997, The Astrophysical Journal, Vol 474, DOI, 10.1086/303434),
[26] (M. Tegmark, J. Silk, M.J. Rees, A. Blanchard, A, Abel, T. Francesco, P., How Small Were the First Cosmological Objects?, 1997, The Astrophysical Journal, Vol 474, DOI, 10.1086/303434), (W.H. Press, P. Schechter, Formation of Galaxies and Clusters of Galaxies by Self-Similar Gravitational Condensation, 1974, Vol. 187, Pgs 425-438, DOI 10.1086/152650).
[27] (W.H. Press, P. Schechter, Formation of Galaxies and Clusters of Galaxies by Self-Similar Gravitational Condensation, 1974, Vol. 187, Pgs 425-438, DOI 10.1086/152650), (M. Tegmark, J. Silk, M.J. Rees, A. Blanchard, A, Abel, T. Francesco, P., How Small Were the First Cosmological Objects?, 1997, The Astrophysical Journal, Vol 474, DOI, 10.1086/303434)
[28] (M. Tegmark, J. Silk, M.J. Rees, A. Blanchard, A, Abel, T. Francesco, P., How Small Were the First Cosmological Objects?, 1997, The Astrophysical Journal, Vol 474, DOI, 10.1086/303434)
[29] (Abel, Bryan, Norman, 2002, The Formation of the First Star in the Universe, Science, 2001, Issue 5552, vol 295, p 93-98)
[30] (L. Chiou, L. Billings, Latest Dark Energy Study Suggests the Universe is Even Weirder Than We Imagined (Why 'Evolving' Dark Energy Worries Some Physicists), Scientific American Inc, May 1, 2025).

---

[31] A. Ananthaswamy, Dark Matter May Be Missing From This Newfound Galaxy, Astronomers Say, Scientific American, Dec 22, 2021)
[32] (ibid A. Ananthaswamy, Dec 22, 2021).
[33] (L. Chiou, L. Billings, Latest Dark Energy Study Suggests the Universe is Even Weirder Than We Imagined (Why 'Evolving' Dark Energy Worries Some Physicists), Scientific American Inc, May 1, 2025).
[34] (A. Ananthaswamy, The Hubble Tension Is Becoming A Hubble Crisis, Scientific American, Inc, April 2, 2015)
[35] (A. Myers, P. Colella, P, B. Van Straalen, The Convergence of Particle-In-Cell Schemes For Cosmological Dark Matter Simulations, 2016, The Astrophysical Journal, 816(2), Article 56)
[36] (M. Mohammadpour, S.W. Stahler, Externally fed star formation: A numerical study, 2013, Monthly Notices of the Royal Astronomical Society, Vol 433 (4), P. 3389-3397, DOI 10.1093/mnras/stt972)
[37] F.H.Shu,. Self-similar collapse of isothermal spheres and star formation. 1977, The Astrophysical Journal, 214, 488–497.DOI 10.1086/155274)
[38] (P.C. Clark, S.C.O. Glover, R.S. Klessen, V. Bromm, Gravitational fragmentation in turbulent primordial gas and the initial mass function of Population III stars.2011, The Astrophysical Journal, 727(2), 110).
[39] (C.Federrath, R. Banerjee, P.C. Clark, R.S. Klessen, Modeling collapse and accretion in turbulent gas clouds: Implementation and comparison of sink particles in AMR and SPH. 2010, The Astrophysical Journal, 713(1), 269–290).
[40] (C.Federrath, R. Banerjee, P.C. Clark, R.S. Klessen, Modeling collapse and accretion in turbulent gas clouds: Implementation and comparison of sink particles in AMR and SPH. 2010, The Astrophysical Journal, 713(1), 269–290).
[41] (C.Federrath, R. Banerjee, P.C. Clark, R.S. Klessen, Modeling collapse and accretion in turbulent gas clouds: Implementation and comparison of sink particles in AMR and SPH. 2010, The Astrophysical Journal, 713(1), 269–290).
[42] (C.Federrath, R. Banerjee, P.C. Clark, R.S. Klessen, Modeling collapse and accretion in turbulent gas clouds: Implementation and comparison of sink particles in AMR and SPH. 2010, The Astrophysical Journal, 713(1), 269–290).
[43] (J.H. Jeans; I. The stability of a spherical nebula. Philos Trans A Math Phys Eng Sci 1 January 1902; 199 (312-320): 1–53. https://doi.org/10.1098/rsta.1902.0012)
[44] P.C. Clark, S.C.O. Glover, R.S. Klessen, V. Bromm, Gravitational fragmentation in turbulent primordial gas and the initial mass function of Population III stars.2011, The Astrophysical Journal, 727(2), 110).110)
[45] (P, C. Clark, S.C.O. Glover, R.S. Klessen, A-K. Jappsen, V. Bromm, Effects of Turbulence on Zero-and Low-Metallicity Star Formation, 2010, The First Stars and Galaxies: Challenges For The Next Decade. AIP Conference Proceedings, Vol 1294, Pp. 28-33, DOI: 10.1063/1.3518882)
[46] (K. Hasegawa, M. Umenmura, H. Susa, Radiative Regulation of Population III Star Formation, 2009, Monthly Notices of the Royal Astronomical Society, Vol 395, (3), Pgs 1280-1286, DOI: 10.1111/j.1365-2966.2009. 14639.x)
[47] (P, C. Clark, S.C.O. Glover, R.S. Klessen, A-K. Jappsen, V. Bromm, Effects of Turbulence on Zero-and Low-Metallicity Star Formation, 2010, The First Stars and Galaxies: Challenges For The Next Decade. AIP Conference Proceedings, Vol 1294, Pp. 28-33, DOI: 10.1063/1.3518882)
[48] (V. Bromm, P.S. Coppi, P. S., R.B. Larson, R. B. The formation of the first stars I: The primordial star-forming cloud. 2002 The Astrophysical Journal, 564(1), 23–51)).
[49] (T. Hosokawa, K. Omukai, N. Yoshida, H.W. Yorke, H. W. Protostellar feedback halts the growth of the first stars in the universe. 2011Science, 334(6060).
[50] (T. Hosokawa, K. Omukai, N. Yoshida, H.W. Yorke, H. W. Protostellar feedback halts the growth of the first stars in the universe. 2011Science, 334(6060).

[51] (V. Bromm, P.S. Coppi, P. S., R.B. Larson, R. B. The formation of the first stars I: The primordial star-forming cloud. 2002 The Astrophysical Journal, 564(1), 23–51) (A. Stacy, T.H. Greif, V. Bromm, V. The first stars: Mass growth under protostellar feedback. 2011, 2011, Monthly Notices of the Royal Astronomical Society, 2011, Vol 422, Pg. 290-309, corpusID 119113783) .
[52] (P.C. Clark, S.C.O. Glover, R.S. Klessen, V. Bromm, Gravitational fragmentation in turbulent primordial gas and the initial mass function of Population III stars, 2011, The Astrophysical Journal, Vol 727(2),DOI 10.1088/0004-637X/727/2/110), (A. Stacy, T.H. Greif, V. Bromm, V. The first stars: Mass growth under protostellar feedback. 2011, 2011, Monthly Notices of the Royal Astronomical Society, 2011, Vol 422, Pg. 290-309, corpusID 119113783)
[53] (V. Bromm, R.B. Larson, The first stars. Annual Review of Astronomy and Astrophysics, 2004, 42, 79–118).
[54] (A. Stacy, T.H. Greif, V. Bromm, V. The first stars: Mass growth under protostellar feedback. 2011, 2011, Monthly Notices of the Royal Astronomical Society, 2011, Vol 422, Pg. 290-309, corpusID 119113783)
[55] (I.U. Maio, S. Khochfar, J.L. Johnson, B. Ciardi, The interplay between chemical and mechanical feedback from the first generation of stars. Maio, U., Monthly Notices of the Royal Astronomical Society, 2010, 407(2), 1003–1015.)
[56] (V. Bromm, N. Yoshida, The first galaxies. Annual Review of Astronomy and Astrophysics, 2011, 49, 373–407.).
[57] (J. Mead, K. Brauer, K., G.L. Bryan, M.-M Mac Low, A.P., Ji, J.H. Wise, A. Emerick, E.P. Andersson, A. Frebel, Transport of metals from minihalos following Population III stellar feedback. 2024, The Astrophysical Journal, Vol.980(1)).
[58] (J.H. Wise, T. Abel, T. Resolving the formation of protogalaxies. III. Feedback from the first stars. The Astrophysical Journal, 2008, 685(1), 40–56.).
[59] (E. Visbal, G.L. Bryan, G. L, Z. Haiman, Z. Self-consistent semi-analytic modeling of feedback during primordial star formation and reionization. Monthly Notices of the Royal Astronomical Society, 2020, 497(4), 5501–5516.).
[60] (J. Jaacks, R. Thompson, K. Nagamine, B.D. Smith, Baseline metal enrichment from Population III star formation in cosmological volume simulations. Monthly Notices of the Royal Astronomical Society, 2020, 475(4), 4396–4410.)
[61] (R.P. Naidu, et al. A Cosmic Miracle: A Remarkably Luminous Galaxy at zspec=14.44 Confirmed With JWST, 2025, arXiv:2505.11263v1 [astro-ph.GA]:
[62] (R.P. Naidu, et al. A Cosmic Miracle: A Remarkably Luminous Galaxy at zspec=14.44 Confirmed With JWST, 2025, arXiv:2505.11263v1 [astro-ph.GA]:
[63] (J.C. Mather, Facts, NobelPrize(dot)org, Nobel Prize Outreach AB 2024. Sun. 3 Nov 2024, nobelprize(dot) org/ prizes /physics /2006/ mather/ facts/)
[64] B. Greene interviews J. Mather, World Science Festival YouTube Channel, Seeing Further: Searching for the Echoes of Creation, October 13, 2023, 1 hour, 4 minutes).
[65] (B. Greene interviews J. Mather, World Science Festival YouTube Channel, Seeing Further: Searching for the Echoes of Creation, October 13, 2023, 1 hour, 4 minutes).

Variable 9: The CMB As the Instantiation of the Universe

Introduction

The standard cosmological narrative places the beginning of our universe at the Big Bang. Big Bang Cosmology is based upon several theoretical underpinnings and has become ubiquitous and functionally inseparable from *Inflationary Cosmology*—which was expressly tailored to smooth over the catastrophe of identifying the beginning of our universe at a $T = 0$ beginning where all known laws, dimensions, forces and matter converge into a Singularity. This mathematical singularity is not something we have found, rather, a point where all possible understanding yields error messages—a mathematical infinity where our physics ceases to function and the standard model of physics enters a regime of complete conjecture.

To bridge the gap between this singular beginning and the universe we observe, the model relies on the Lambda-CDM framework—a ledger that contains Dark Energy and Cold Dark Matter components that accounts for the missing mass and energy required to keep the expansion of our universe on its predicted trajectory Together, Inflation and the Lambda-CDM framework are the scaffolding for the origin of all things cosmological, while relying on components: inflationary fields, dark energy, dark matter that cannot be derived from first principle physics, nor observed.

The Big Bang model begins—or, more precisely, cannot begin—without a significant information debt that cripples the model's credentials. By present-day accounting, 95% of the matter and energy arising from the Big Bang, as well as the mechanism for its expansion, cannot be accounted for. But this is hardly the complete set of problems within the model.

The model is a backwards extrapolation from the first relic of universal evidence—the CMB, with the addition of the laws and physics observed within our universe—that then tracks the evolution of our universe to a proposed beginning. This approach has yielded useful models and coherent mathematical descriptions that can account for many, but not all, present day observations.

This failure is the inevitable result of the model's methodology. It is a backwards extrapolation from the CMB that attempts to 'solve' the rueful problems at the beginning by 'guessing' at the cause of all existence. While this approach has yielded some useful mathematical descriptions of the *outcome*—which once again is largely founded on the hard evidence of the CMB—it is a retroactively guessed map that fails to provide information on what its title implies. It does not describe the instantiation of our universe.

The fundamental flaw of modern Big Bang cosmology is that it isn't actually explaining the beginning (the deed), it is explaining the aftermath (the debris) of a beginning it cannot approach while pretending the two are the same thing.

Our Theory

In forming our theory that the CMB represents the boundary of *instantiation* for our universe, we integrate both positive and negative evidence—accounting for what observation permits and what physical constraint forbids. The evidence defines the threshold of the probable, while also exposing exactly what cannot be derived from first principles within the Big Bang narrative. Under this constraint, an unavoidable boundary appears in the history of our universe—and it's the same one we have observed all along: the cosmic microwave background radiation.

If our universe is to be considered to be information rich, observable, and amenable to the practice of science, then the information we derive must be anchored in evidence. If the universe has an origin, it will be found in what we observe, not in the manipulation of data, nor a forced closure to an imagined 'clean' but catastrophic beginning. The CMB is the only evidence we have of a beginning, and it comes fully loaded, fully encoded for the reality we currently inhabit.

Everything *behind* the CMB—the Singularity, the Inflationary epoch, the $T = \mathbf{10^{-43}}$ *seconds* designation—is built on the imagined detritus of the CMB. It is a regime where the physics is unvalidated and the information debt is absolute. The CMB exists as a coherent operational unit. It is far from a chaotic pattern; it's a complete and detailed map of a universe initialized, tuned, balanced, and ready for the *Galaxy Machine* to run.

It is our claim that the CMB is the moment our universe began. We use the term, *instantiation* to designate this moment. Nothing 'banged', nothing 'inflated' faster than the speed of light, nothing violated anything. The CMB in this conception signals the moment the universe came into being, became coherent and provided us with observable evidence of the universal beginning, as far back as we can observe and gather evidence. This boundary is approximately 380,000 years removed from where the Big Bang model retrodictively indicates a catastrophic beginning".

In the standard model, that 380,000-year gap between the "Bang" and the "CMB" is where all the mathematical overrides take place. This is the black box where cosmology hides its slew of ad-hoc fixes.

It is our claim that the CMB is not only the first available evidence of the beginning of our universe, but it is the only evidence and the literal beginning of our universe. If our claim is true, then the beginning of our universe no longer stands apart as a special case within our investigation.

With the CMB as the instantiation of our universe, it follows the same evidentiary pattern observed throughout our audit: a precipitous discontinuity where physical law and probability collapse, only to be superseded by a precise, exacting, universal instantiation. This pattern is consistent across every Variable and Special Problem we have identified. The instantiation of the CMB is not a cosmological outlier. It indicates a single, recurring process solution applied to establish a universal requirement.

Definition of Instantiation

An instantiation is a physical override—a point where the universe exceeds its own probabilistic capacity to solve a functional requirement. By the materialist's own constraints, any event with a probability greater than 1 in $\mathbf{10^{140}}$ is a systemic failure of their model. In our audit, these failures are recognized as instantiations: non-random, highly specified configurations that ensure the universe remains observable, discernible and information rich.

Thesis Statement

We propose that the cosmic microwave background radiation marks the boundary of spacetime instantiation itself, not merely a phase within an already-existing universe. Physical laws validated within the post-CMB universe therefore cannot be legitimately extrapolated across this boundary without committing a category error.

About The Big Bang Model

The Big Bang model is widely treated as an explanation of cosmic origin, yet it lacks generative capacity or even a causal mechanism. As an origin theory, it fails because it cannot approach the beginning of the universe. It is restricted, by physics, to hovering near the Planck scale. To move further, it is said to need "quantum gravity".

Because it runs backwards, and is restricted from approaching the beginning, it must assume all of the laws and forces within the universe for its explanation. The model therefore functions as an origin story that does not begin at the origin. It offers no explanation for how the universe came into being, where it came from, or how its laws, constants, and boundaries were established.

The Big Bang theory is not an origin narrative. It is a retrospective, mathematical extrapolation that assumes, and does not explain the universe under pre-specified laws and boundary conditions.

If the CMB is treated as the earliest reliable observational boundary, then the CMB must encode the earliest stable constraints of reality. Under this constraint, mapping the instantiation of our universe to the CMB yields a coherent, evidence-linked framework that avoids the conceptual failures of Big Bang cosmology while remaining strictly grounded in observable structure. Contrary to the Big Bang model, the CMB is not a phase within a pre-existing spacetime; it is the instantiation boundary, the point at which spacetime, law, and information first achieve coherence.

We can press our case further in another regard.

Where Is The Evidence for Existence Prior to the CMB?

If there truly was an existence prior to the CMB, intuition tells us that there would be evidence of this fact. If the universe existed for ~380,000 years before it became transparent to light (photons), it would have been filled with phenomena that could be observed today, none of which has been seen. Specifically, two types of backgrounds and one material ledger would precede the CMB[1].

The Cosmic Neutrino Background or "Ghost" (The CMB)

> According to the model, while photons were still trapped in the opacity of the early universe, neutrinos were already free. They are theorized to have decoupled approximately one second after T=0, and are believed to account for roughly 41% of the radiation density of the early universe. This is the *Cosmic Neutrino Background* (CNB). Their presence remains a mathematical requirement, however, neutrinos are notoriously hard to detect, limiting our prospects for observation, meaning, we have not seen the CNB. The authors of *CMB-S4 Science Boo[2]k* acknowledge that direct detection of these low-energy (meV) neutrinos is beyond current technology. However, they argue that we can "see" them indirectly. Even if we can't catch a neutrino in a trap, we can see the "gravitational footprint" they left on the Cosmic Microwave Background.

The Gravitational Wave Background or 'Ripple' (the GWB)

According to inflation, gravitons, the quantum counterparts to gravitational waves, would have been released at the earliest moments of the universe's expansion. As yet, these "relic waves" have not been detected. As documented in Brian Keating's book, *Losing the Nobel Prize*[3], B-mode polarization, as these waves are known, were famously discovered and then debunked during the BICEP2 experiment. If these ripples existed, the BICEP2 experiment was our best chance to find them. Their confirmation would have been considered a confirmation for cosmic inflation and virtually guaranteed a Nobel Prize. The upper limit for these waves is now much lower than before which means a range of possible waves have been ruled out. If the upcoming CMB-S4 project cannot detect these waves, then one of the real conclusions will be that these waves do not exist[4].

Big Bang Nucleosynthesis, or "The Nucleosynthesis Ledger" (BBN)

This is the chemical evidence. According to the model, minutes before the universe cooled, it created specific ratios of hydrogen, helium, and Lithium-7. But observation shows three times less Lithium-7 than predicted. This is the Cosmological Lithium Problem, a chemical discontinuity that undermines the model's ledger[5].

The lack of all three forms of evidence has in some cases necessitated additional *ad hoc* fixes to explain their absence.

Because our universe has a problem in covering up her secrets, It is almost certain that if the universe existed prior to the CMB as alleged, we will eventually develop the technology to detect its existence. Our hypothesis will find robust validation when it is finally confirmed that no such evidence for a pre-CMB universe exists, because our universe was instantiated at the CMB.

From this perspective, the intractable problems of early cosmology do not arise because the universe is uniquely pathological, but because explanatory responsibility has been misallocated.

Instantiation is Unavoidable: Singularity Theorems

With the Big Bang model rendered incapacitated by catastrophic failure, there are only two logically distinct options for our universe: (1) eternal existence, or (2) instantiation/origination.

The singularity theorems of Hawking–Penrose and Borde–Guth–Vilenkin show that, under reasonable assumptions, classical spacetime is past-incomplete and cannot be extended indefinitely into the past.

Interpreted strictly, these theorems define a hard stop. They demonstrate that space-time is past-incomplete, terminating in a boundary that forbids extrapolation of a prior phase. Although this conclusion is supported, it does not close the door on possible alternative *kinds* of prior existences. So let's examine what occurs if we examine all that is permitted.

What Is Permitted Under the Singularity Theorems, and What They Result In

As we discussed in special problem #1: The beginning of the universe, classical cyclic cosmologies are strongly constrained by entropy considerations. This is *the Tolman entropy problem,* grounded in the second law of thermodynamics. In any closed or semi-closed system, total entropy never decreases and usable energy is progressively degraded. If the universe undergoes repeated cycles of expansion, contraction, and bounce, each cycle must contain equal or greater entropy than the last. In an eternal past, the universe would have reached heat death an eternity ago. The presence of usable energy and complex structure is all that is needed for forensic proof of a finite beginning. In other words, because we observe a structured universe in the present, the universe cannot be eternal.

Whatever existed in the "past" prior to the CMB, has been rendered unintelligible, and equivalent to non-existence by the CMB instantiation event. In this way we can see whether one pursues option 1, eternal existence, or option 2, instantiation/origination as a starting point, functionally, instantiation/origination becomes an event that cannot be avoided *as the beginning*. The CMB acts as an informational firewall. Beyond it, the past is functionally equivalent to non-existence. The CMB is the beginning and has also eliminated all possible precursors out of relevance.

There is yet another reason that even if the CMB and the singularity theorems leave open the possibility of a pre-Big Bang phase or an infinite regress of earlier universes; the evidence we have points toward a beginning *that matters* to universal observers such as ourselves.

The universe we can measure shows a highly specific, low-entropy initial condition, coherent laws, and a baryonic structure that supports chemistry and life. Those conditions are fragile. If time extended infinitely into the past in a way continuous with our present phase, entropy would have long since reached equilibrium, erasing structure, information, and obliterating the arrow of time. Without a mechanism to reset or constrain entropy, an eternal past becomes not just unproven, but incompatible with the ordered universe we observe.

That does not rule out deeper phases—cycles, bounces, mirror universes, or other pre-universal states. It only means that even if they existed, they do not replace the need for a *fresh* instantiation to reorder our universe into its present condition. An eternal sequence of cycles still requires a reset to restore low entropy, re-establish initial conditions, and re-tune the constants and laws that permit stable baryonic matter and life. In other words, the doorway to earlier phases is not a loophole that makes a beginning unnecessary; it only shifts the question back one step and if that step is to move forward into *our* universe, it must move into instantiation.

Once again, whether the instantiation event generated the universe or renewed it is irrelevant. The crucial fact is that instantiation occurred. We have the evidence. Nothing else is needed.

The CMB Versus The Big Bang

The Big Bang, although touted as an origin framework, avoids having to explain the beginning because it cannot approach $T = 0$. In truth, at $T = 0$ is where you will find the unacknowledged ghost in their machine. It represents the origin of the universe, yet it remains a boundary that Big Bang cosmology is forbidden to cross. This is a fatal irony because the beginning is what the model purports to address. They have succeeded in the court of public opinion, however. Most people think the Big Bang is an origin story. But it is actually a story that can never be consummated. The CMB addresses this point, and we have the evidence in hand. Indeed, it is everywhere.

What The CMB Encodes (Why It Is the Best Candidate for the Instantiation of the Universe)

The Cosmic Microwave Background (CMB) is the oldest observable electromagnetic radiation in the universe. It is a snapshot of the universe at the moment of instantiation into transparency and observability. It was configured for expansion, governed by law, and initialized to meet the specific requirements for everything that followed.

The CMB encodes the initialized parameters of our reality:

The immediate observable existence of expansion, spacetime, causal structure, and physical law.

A smooth, precision-balanced, low-entropy initial state.

A structural matter preference, evidenced by the acoustic peaks within the CMB power spectrum (what Big Bang cosmology calls the matter–antimatter imbalance).

Horizon-scale thermal uniformity and consistent fundamental constants across the entire observable volume.

A precise balance between the forces of gravity and expansion, including the constraints on the cosmological constant.

Consistent particle physics across the observable universe, as any variation in fundamental constants would have altered the acoustic structure we see.

The transition of the universe into a transparent and structured state, making it the first moment where information is accessible rather than conjectured.

Prior to the CMB the universe is not only unknown but inaccessible. This limitation is imposed by physics itself. The CMB functions as a cosmic boundary marker: the point at which an instantiated universe first becomes legible. Claims about earlier states are reconstructions.

$T = 0$ is Literally the Worst Possible Time for the Universe to Begin

If one is genuinely attempting to account for the coming-into-being of reality, then grounding that account at $T = 0$—as extrapolated from the Big Bang equations—is, on purely structural grounds, the worst possible choice. At $T \rightarrow 0$, spacetime itself ceases to be well-defined. Classical causality fails. Locality breaks down. Energy conservation becomes undefined. Determinism collapses. Even probability theory loses coherence once the background structures that give it meaning are gone. In other words, every conceptual tool we possess for explaining the universe is maximally degraded precisely at the point Big Bang cosmology gestures toward as "the beginning."

Instantiation at the CMB refuses to force an origin explanation into a regime where explanation is impossible *by definition*. It does not require causal continuity across a non-existent time, laws operating before laws exist, quantum mechanics explaining its own preconditions, spacetime emerging within spacetime, or observers, measurements, or probabilities prior to any structured domain. Instead, it states—minimally and honestly—that reality *appears as a fact*, already constrained and already structured, at the first point where description becomes meaningful.

Bottom Line: The evidence suggests that rather than emerging from a chaotic $T = 0$ domain, beyond the reach of physics or theoretical tools, the universe begins as a coherent law-governed system at the moment of the CMB. By grounding the origin here, we avoid the worst possible moment in cosmic history where reality is wholly unformed and wholly unexplained.

Literally nothing can be said about this moment, other than it is a singularity.

Clarification: The Universe Was Never a Tiny Object

The modern retrodictive framing of the Big Bang presents the universe as a tiny, atom-sized object at its inception. This is one of the most enduring and misleading representations in modern cosmology, surviving primarily by conflating two incompatible ideas: the universe as an infinite flat plane and the universe as a finite spherical seed. To understand the depth of this deception, one must first distinguish between scale and size.

Popular science *suggests* that the entire observable cosmos was shrunken down and placed in the dead center of an empty, infinite void that came into being at $T = 0$. In reality, modern cosmology assumes the existence of a spatially flat universe—pre-existing the big bang, somehow, against physics, evidence, and logical causality—where only an empty void can exist. To even call it empty is a misnomer, because we are talking about literal nothingness.

For the equations of the 'Bang' to even begin, the materialist is forced to smuggle an infinite spacetime grid into the basement of reality before the clock starts ticking. They mandate a set of laws for a boundary (T = 0) that their own math simultaneously claims cannot be approached or explained. It is a theory that requires the stage to be fully built and the actors to be in their places before the curtain is even allowed to rise, and yet, this is all done knowing full well, the curtain cannot even be approached.

Next, at $\mathbf{10^{-43}}$ seconds—the closest we can approach the beginning—it is not the universe that takes up fewer squares on the cosmic grid; rather, the coordinate system itself contracts. Imagine an infinite chessboard where the squares themselves shrink until they are 10 septillion times smaller than an atom. The "infinite chessboard" remains infinite, but the scale of its coordinates is microscopic. This is the difference between the *size* of the universe being smaller than an atom (which is false) and the correct *scale* of the universe being smaller than an atom.

This distinction exposes a fundamental paradox. If the universe was infinite at the beginning, the word "Bang" loses all physical meaning. A bang implies a center and a source of outward pressure, yet an infinite grid has no middle and no "outward." It simply grows the distance between its own coordinates everywhere at once. This unavoidable reality reveals the *coordination catastrophe*: the question of how every imaginable coordinate across an infinite field that did not exist prior to $T = 0$, suddenly came into existence and received the "command" to commence expansion at the exact same Planck-second and at the exact same rate. In an infinite-but-tiny-scale universe, a point "here" and a point "trillions of miles over there" have no way to communicate, yet the Cosmic Microwave Background shows they are identical to one part in 100,000.

To bridge this causal gap, Materialist cosmology mandates the "Inflationary Patch," retreating from the infinite grid to claim that our specific universe emerged from a physical point literally in size-and-scale smaller than an atom. To visualize the absurdity of the size of the proposed point-origin, consider this: if the size of a single atom was blown up to the size of the observable universe, the "patch" that supposedly inflated to create our current universe would be only ten feet wide.

To save this model from immediate gravitational collapse, cosmology invokes a series of *ad hoc* mechanisms: the inflaton field to flip gravity into a repulsive force, a "speed of light loophole" to allow space to stretch faster than its contents can move, and the "ultimate free lunch"—coined by Alan Guth—a mechanism where the expansion of space simultaneously generates new energy out of a vacuum to ensure the field never thins out in the very midst of the densest, hottest, most catastrophic pressures ever to exist within the universe. These are not first-principle physics; they are fixes and cheats designed to make the math match the observation of a flat, uniform sky.

The most devastating critique of this sliding scale model is the necessary existence of a universal boundary. If our universe expanded from a finite, atom-sized point, the expansion at the leading edge must necessarily be faster than the material near the center. This makes a mockery of the one part in $\mathbf{10^{120}}$ precision required for the cosmological constant and expansion rate. Such precision cannot be a universal constant if it must be retrofitted to a sliding scale of expansion rates across a finite bubble. The necessary "center" of this inflationary point-universe would remain a near-inertial anchor, essentially stationary, while the outer boundaries would be forced into a state of exponentially disparate velocity—a geometric contradiction that makes a truly flat, uniform expansion across a finite patch physically irreconcilable.

Ultimately, the Big Bang plus Inflationary cosmology provides no underlying basis for the system's existence. Instead, reality is snatched from the void via 26 free parameters—mass scales, coupling constants, and mixing angles—that are measured, recorded, and then declared to be "predictions." The Standard Model is a house of cards held together by an alleged "Bang" that can neither be approached nor explained—nor could it have actually "banged"—leaving an inflationary era set by hand to bypass the horizon problem, and a Goldilocks universe requiring at least 19 additional ad hoc hypotheses (The Divine Algorithm) to bridge discontinuities that probability cannot cross for our universe to appear as it does in the present day.

The result is the aforementioned *coordination catastrophe* rendered invisible by a linguistic shell game. To maintain the illusion of a coherent origin, modern cosmology operates behind a rotation of three universes, each offered when needed to bypass a specific logical failure:

> The Infinite Plane Universe: Used when the model needs to explain away the "Center Problem" and the "Boundary Problem" made necessary by the point-origin inflationary universe. If the universe is infinite, it doesn't need a middle. And if it is infinite, then even if it has a point-origin inflationary universe, the universe beyond it can be said to have no center.
>
> The Atomic Seed Inflationary Universe: Used when the model needs to explain the "Horizon Problem"—how the parts of the universe are so perfectly uniform. This universe is used when the universe must be point-origin and can no longer be arbitrarily large in size. At 10 septillion times smaller than an atom, the parts can "touch" and coordinate before expanding.
>
> The Multiverse Bubble: Used when the model is confronted with the "Fine-Tuning Problem." They embed our universe into an infinite "Multiverse" of failed worlds so that our existence looks like a mathematical accident—actually a mathematical certainty—rather than a precise initialization.

Each version is offered when a specific contradiction comes up. Members of the public, trying to make sense of both an inflationary point-origin universe and an arbitrarily large, possibly infinite universe that is spatially flat and has no center, feel like they are crazy because they can't reconcile the two. No one can. That's why there are three versions, and the fourth, which is all of them together.

But this is ironic, because none offer a single insight into why the system exists at all. It is a model that predicts the reality we see only because it was built specifically to mimic it. It is a retrodictive map—a drawing of the territory that was made after the journey yet claimed as the blueprint for the road.

The Big Bang as Retrodictive, Not Generative, Not a Theory of Origin

The Big Bang model is not a theory of creation. A true creation story must move forward: it must begin with nonexistence and explain how existence, spacetime, laws, forces, constants, and structure arise and unfold into the universe we observe today. A creation account is not simply a description of what happens next; it is a generative narrative that explains why the universe becomes real at all. The Big Bang does the opposite. It is not a map of how the universe was built; it is a forced reconstruction of how a set of individuals need the universe to expand *after* every bit of the construction was complete. It is a structural inversion.

For this reason, the Big Bang cannot in principle reach a true beginning, nor does it meet the definition of an origin narrative in a technical sense. Appeals to quantum gravity or new high-energy physics (two issues we will address shortly) do not resolve this limitation; they simply extend the same retrodictive strategy into regimes where the assumptions are largely incoherent.

This unrecognized inversion explains a persistent pattern in modern cosmology. Faced with the implication of a genuine beginning, the materialist framework displaces the problem rather than confronting it. Cyclic universes, universal bounces, mirror universes, and pre-Big-Bang phases are not demanded by evidence, nor indicated so much as generated by the necessity of conforming to a paradigm that is rooted in causality and incapable of addressing causality because causality comes to an end at instantiation–what the Big Bang theory identifies as $T = 0$. Rather than confront this–it literally can't as a paradigmatic limitation, the materialist framework invents 'causal continuity' (bounces, cycles) to flee from the boundary where existence demands an account. They are simply not following data.

The Baryon Asymmetry Hustle

Steven Weinberg—one of the principal architects of the standard cosmological narrative—explicitly acknowledges in his book, *The First Three Minutes* that the baryon asymmetry could simply have been present as an initial condition. This acknowledgement raises a foundational point for our argument, which is that the assumption of an initially symmetric universe is not required by evidence or enforced by physical law, it is a narrative preference.

The fact is there is no observation of an initial symmetric state, no fossil imprint, no relic signal, no empirical constraint found within the universe that demands it is telling. There is no conservation law requiring symmetry *at instantiation*, and as acknowledged by first principle physics, no concept forbidding a universe that begins asymmetrically. It is roundly acknowledged in literature that asymmetry could have been "put in at the beginning." CERN publications state outright that an initially symmetric universe has no explanation within the standard model. They then posit that asymmetry must either be imposed by hand or rescued through speculative early-universe dynamics, apparently dismissing that the universe could have simply started out asymmetric. Repeatedly we find that symmetry is assumed at the beginning, and this assumption is repeatedly treated as the baseline. Lectures introduce baryogenesis by stating that unless one is willing to posit asymmetric initial conditions, a mechanism *must* be found to generate asymmetry dynamically.

The standard cosmological narrative of the hot Big Bang assumes that the universe began in a nearly symmetric state. Then it attributes the observed asymmetry to unknown processes operating in the early universe. This creates an inversion: although the framework is said to assume symmetry, it can only account for the existence of matter by introducing speculative *asymmetry*. Symmetry therefore functions not as a requirement, but as a provisional baseline in search of a mechanism to support the claim.

Why this contortion? Because the Big Bang framework explains how a universe evolves once specified, not how it is specified in the first place. Because of the absence of a theory of initial conditions, symmetry is adopted as a convenient baseline. Once adopted, it must be undone to recover the universe we observe, forcing the introduction of speculative symmetry-breaking mechanisms that meet Sakharov's conditions. The complexity arises from attempting to preserve the assumption of symmetry. The origin of this assumption traces back to Dirac's prediction of antimatter. That discovery revealed a deep symmetry *in the equations* of relativistic quantum theory. Over time, a symmetry of equations became a *presumed symmetry of the universe itself*. What began as a formal property of theory was elevated into an expectation about cosmological history.

The fallacy lies in treating the theoretical virtue of the Dirac equation as evidence of historical truth.

With this as our foundation, If we examine standard baryogenesis frameworks, including the universally accepted *Sakharov conditions*, they do not follow from data. They follow from an assumption of symmetry, but, strangely, first by acknowledging asymmetry. Sakharov's 1967 paper starts out with:

> "The theory of the expanding universe, which presupposes a superdense initial state of matter, apparently excludes the possibility of macroscopic separation of matter from antimatter; it must therefore be assumed that there are no antimatter bodies in nature,

i.e., the universe is asymmetrical with respect to the number of particles and antiparticles (C asymmetry). In particular, the absence of antibaryons and the proposed absence of baryonic neutrinos implies a nonzero baryon charge (baryonic asymmetry). We wish to point out a possible explanation of C asymmetry in the hot model of the expanding universe (see Ref. 1) by making use of effects of CPinvariance violation[6]."

Sakharov therefore begins by acknowledging that the universe is observed to be asymmetric, and then proceeds to investigate what conditions would be required if that asymmetry had to arise dynamically from an initially symmetric state. He then states his famous conditions for asymmetry, "The baryon asymmetry may arise if the following conditions are satisfied: (1) baryon number violation, (2) C and CP violation, and (3) deviation from thermal equilibrium[7]."But mainstream physics acknowledges that the laws of physics do not determine initial conditions, therefore Sakharov's assumed, but unstated fourth condition, (4) that the universe begins in a symmetric state, isn't backed by any physical law.

We can take an accounting of the premises we have surfaced thus far and reach the following conclusion:

There is no law requiring initial symmetry

There is no evidence the universe was ever symmetric

The universe we observe is decisively asymmetric

After decades of work, all proposed mechanisms for generating asymmetry from symmetry remain speculative, incomplete, or empirically inaccessible

Although Sakharov is technically correct, his hidden fourth condition, which is as necessary as the other three conditions to be relevant, rests upon no known law. And because the symmetry-to-asymmetry pathway remains unconfirmed, the assumption of initial baryon symmetry is unjustified. Observationally, the universe is decisively asymmetric, and despite decades of theoretical effort, all proposed mechanisms for generating asymmetry from symmetry remain speculative or empirically inaccessible. Consequently, the Sakharov baryogenesis framework, while internally consistent, does not appear to apply to our universe.

The most reasonable conclusion is that our universe began in an asymmetric state. This asymmetry is not merely incidental; it is required for a life-allowing cosmos. Without it, matter and antimatter would annihilate, leaving only radiation; any remaining antimatter would continue annihilation indefinitely; and baryonic matter—the only substrate capable of chemistry, stars, planets, and observers—would not persist. What looks like a violation is, in truth, the signature of instantiation. 'Asymmetry' isn't a 'violation' of the law, it assumes facts not in evidence to serve as the basis for the leading question of asymmetry. Matter dominance is one of the many exactingly specific specifications within our universe.

Instantiation at the CMB mirrors our claim of instantiation at all times and for all Variables and Special Problems within the narrative of our book. Instantiation is the mode of the universe, and it becomes functionally observed when a need arises to usher in major transformational events within our universe that is unreachable by first principle physics, or probability.

This pattern is not unique to cosmology. Clear precedents exist in other domains where what appears as a series of intractable problems are subsumed under the universal instance of instantiation. A prime example is the origin of life. Viewed through a strictly material lens, the problem of first life fractures into dozens of apparent problems, each as intractable as the last, and each of the complex systems required for life appearing as its own subfield with its own hypotheses. But viewed as a single instantiation—a boundary crossing from non-living chemistry to living, information-preserving systems—these are not separate events. They are the emergent demonstrations of one transition from non-life to life.

The instantiation of the universe functions in much the same way. The multiplicity of "problems" in the creation of our universe arises because we are examining the aftereffects of an instantiation from the perspective of a bottom-up, orderly, reductionist, physics-based process.

Particle Accelerators, High Energy, and the Boundary of Applicability

Particle accelerators provide empirical data about how matter behaves under extreme conditions testing symmetry breaking, phase transitions, and the robustness of physical laws at high energy. The issue with regards to the salience of these experiments to universe creation is they operate entirely within an already-instantiated universe. This means what they reveal are stress responses *within our existing* reality. It cannot be said that they reveal the mechanisms by which reality itself comes into being. This distinction is all the difference. Symmetry breaking observed in laboratories is not a miniature version of cosmic creation. It presupposes spacetime, causality, conserved quantities, and stable laws. Physics itself cautions against extrapolating such results beyond the domain in which those structures already exist. To do this represents a likely categorical error, and at least a provisional error.

Modern cosmology assumes that increasing energy in the laboratory collapses symmetries in a way that mirrors the early universe. Under this assumption, accelerator energy scales are mapped onto earlier cosmic times, and mathematical unifications of forces are reinterpreted as historical events. Within that framework, it becomes natural to speak of forces "unifying" and later "splitting" as the universe cools. But once again, this move is inferential. Particle accelerators demonstrate that distinctions between forces blur at high energy, but they do not demonstrate that the universe *passed through* those regimes as part of a generative sequence. The experiments probe behavior within an already existing existence, not the organization of an emerging existence. This boundary is reinforced by the empirical record. Beyond the Higgs boson, none of the particles or fields predicted to stabilize early-universe physics have appeared. Supersymmetric partners, extra dimensions, and other proposed extensions remain absent. Decades of increasingly powerful experiments have revealed no deeper substrate beneath the Standard Model. This silence matters. These theories were introduced to make early-universe extrapolations coherent: their absence weakens the idea that high-energy physics naturally opens onto a pre-CMB or pre-instantiation regime. This stands as evidence against the hypothesis that high energy particle physics is a mirror into the emergence and organization of our universe in the beginning. Instead, the opposite has taken place. As experiments push toward hypothetical pre-instantiation conditions, predictive power collapses and paradoxes have multiplied. This is exactly what we would expect if *rather* than at the highest energies, the universe emerged piece by piece, and organized and cohered generatively in a manner of "reverse-reductionism", the universe *instead* comes fully online in a manner of instantiation, fully coherent, at what would appear to us as a boundary precisely like the CMB.

Inflationary 'Solutions' vs. Design Specifications

The primary advantage of a universe instantiated at the CMB is that it eliminates the problems inflation was invented to solve: the horizon problem, the flatness problem, and the missing magnetic monopoles[8]. In an instantiated universe, none of these problems exist. They are instead design features instantiated at the CMB.

> Global Thermalization: The problem of how distant regions of the sky share a temperature within 1 part in 100,000 is not a problem of size, time, the speed of light and coordination[9]. Instantiation makes all of this moot. By initializing the universe in thermal equilibrium, the universe avoids the problem entirely while also providing for the orderly,

gradual increase in complexity: stars, galaxies, and life, across billions of years of usable energy (low entropy).

The Benefits of The Flatness: A flat universe is an anomaly and a puzzle to the standard model. In a flat universe ($k = 0$) astrophysicists have to wonder why the universe is in a condition equivalent to a pencil being balanced on its sharp end—a fine tuning miracle. To achieve the flatness, we observe today without instantiation, the initial density of the universe would have had to be accurate to within one part in 10^{60}. In a generative model, this is an impossible coincidence; in an instantiated model, it is a baseline setting. From the perspective of instantiation however, flatness, rather than being an unbelievable example of improbable balance, is instead simply the most stable configuration for long-term existence. It is said to ensure that the gravitational potential energy exactly balances the kinetic energy of expansion[10]. A closed positively curved universe, ($k > 0$) would likely terminate in a premature "Big Crunch" before life could form[11]. An open negatively curved universe, ($k < 0$) would disperse matter so rapidly that gravity could never coalesce[12] enough metallicity to form life-sustaining, and life inhabited stars and planetary systems (solar systems) like our own.

No Monopoles vs "Missing" Monopoles: In standard cosmology, magnetic monopoles are expected as "glitches" formed in a universe struggling from a singularity and a chaotic initial period of expansion and evolution[13]. They serve no biological or structural function; therefore, they are excluded from instantiation.

The standard model puts us in a position where we must reconcile the fact that the most chaotic and catastrophic event in our universal history, resulted in the ultimate order, Roger Penrose's "Penrose" number representing the freakishly low entropy state (the phase space available to the universe)[14] of our universe in the beginning of $\mathbf{10^{10^{123}}}$ starting at the CMB. This is known as a 'power tower' equal to 10 *raised to the* 10th power, *raised to the* 10th power, then *raised to the* 123rd power. Science acknowledges the low-entropy initialization as a primary fact, rather than a statistical impossibility that manifested by chance, a trillionth of a second into the ultimate chaos, while still firmly in the grip of the quantum realm and quantum indeterminacy.

An instantiated universe begins only with what it needs to function in the best interests of a life-allowable and life-inhabited universe. If there is no life, what difference does it make if the universe crunches, expands too rapidly, is dominated by black holes, gamma ray bursts, neutron stars, massive, short lived stars with light that cannot support a wide life-allowable zone, or life-allowable period of time.

Why Quantum Emergence of Spacetime Is Incoherent

Something that is rarely stated, and little understood, is that any attempt to explain the origin of spacetime by appealing to a quantum regime runs into a foundational problem regarding time. Quantum mechanics, in every operational form we possess, presupposes time. States change, which means they evolve, which demands time. Probabilities are defined across temporal intervals. Time is not optional in quantum formalism; it is the parameter that makes the formalism meaningful. General relativity, by contrast, does not assume time as a background. It defines time as a structural feature of *spacetime* itself—shaped by geometry, energy, and curvature. The conclusion is gravity does not operate within time; it determines what time is.

At the origin of the universe, time is precisely what is absent. There is no prior temporal framework within which change can occur: no external clock against which dynamics can unfold. The beginning is not an event in time; it is the boundary at which time itself becomes meaningful. This creates a fundamental incoherence. Any theory that requires time in order to function cannot in principle explain the origin of time itself.

For this reason, proposals in which spacetime "emerges" from a wholly quantum realm are conceptually unstable. They assume that classical spacetime—with precise geometry, causal structure, expansion rate, and informational content—can be generated within a regime that is quantum—meaning, indeterminate, non-classical, and temporally undefined. Spacetime is not vague, it is highly structured and informationally encoded, This is what brings up an irreconcilable contradiction.

The issue is not whether classical information can survive the quantum phase, the issue is whether a quantum phase can contain, encode, or generate the classical information that defines spacetime in the absence of time itself.

It cannot, otherwise there would already be a unification between the quantum and classical regimes.

With this often-unacknowledged perplexity to the quantum realm, the problem of "quantum gravity" is fundamentally reframed from its usual form. The problem is not that we lack a sufficiently advanced theory to quantize spacetime at the beginning, the problem is that the demand for such a theory arises from an incoherent framing. What is being asked cannot be provided. It could be said that in waiting for a formalism of quantum gravity we are waiting for something that will never arrive. Any theory that quantizes geometry already assumes existence, otherwise geometry cannot be quantized. This means such a theory can describe transitions within an instantiated universe, but it cannot account for instantiation itself, which is geometry emerging from no geometry. Seen this way, the CMB does not point backward toward a missing quantum era that will someday complete the story of its own already-existing coherence, rather, it can only point forward to the earliest moment at which spacetime had already taken form—an already expanded, already structured, already capable of carrying information.

This brings us to our point, and another reason why we have identified the CMB as the instantiation of the universe. The CMB does not mark the beginning of existence, rather (again) it marks the first observable imprint of an instantiated universe.

If spacetime is instantiated as a fully coherent classical regime with the well understood quantum component, then quantum gravity is dismissed as a requirement. To be clear, matter fields may possess quantum granularity—as experiment confirms—but spacetime itself need not pass through a destructive or indeterminate quantum regime in order to exist.

Two Thought Experiments and a Conclusion

Let us begin by granting the Big Bang plus inflationary framework its strongest possible form. For the sake of argument, assume that every one of its foundational assumptions are correct. At the beginning assume that spacetime exists, that quantum fields exist, that the relevant laws of physics hold, and that from $T = 0$ through the epoch of force separation the model accurately describes the physical state of the universe. No skepticism, and no caveats. We accept the entire story exactly as it is told. Now ask a single, precise question: Does the information provided by the Big Bang narrative suffice to generate the universe we observe? The answer is no — and not for philosophical reasons, but for structural ones.

To derive a universe forward from any early state, the information available at that state must be complete in a very specific sense. It must include a full specification of the operative laws, a complete description of the physical state, boundary conditions constrained with parameters determined by physics, and a "closed" dynamics that guarantees that no additional information is inserted later, as is demanded by a 100% causal paradigm. Nothing can be smuggled in, implied, or imported from dimensions or realities unknown. Without these, derivation collapses into a narrative reconstruction, when what is needed is a narrative generation.

Even under the most generous interpretation, the Big Bang does not meet these requirements at the force-separation epoch. What it provides instead are incomplete field theories, symmetry-breaking stories, free parameters whose values are unexplained, and probabilistic descriptions that do not encode full information, nor their evolution.

Critical facts remain untouched: why symmetries break in this particular way, why constants take the values they do, why entropy is initially low, why perturbations have their specific spectrum, and why classical spacetime remains stable. Once again, the framework presupposes precisely the structures it claims to explain.

This leads to the central failure.

The information available at force separation is radically underdetermined. The same early-state description is compatible with an enormous range of possible universes — most of which do not resemble ours in any meaningful way. From that state alone, one cannot derive galaxies, chemistry, stable matter, large-scale coherence, the arrow of time, or the specific constants we observe. To obtain anything like our universe, additional assumptions must be added: inflationary mechanisms, fine-tuned initial perturbations, parameter choices, boundary conditions, anthropic filters, or multiverse selection effects. And even with this, a huge amount of cosmic "garbage" is generated that cannot be our universe and cannot sustain biological life. These realms are pushed outside of our consideration by anthropic selection, rather than what is required by the narrative, which are conclusions and explanations based on first principle physics that tell us how our universe was arrived upon, and not how a speculative, imagined firehose of creation may have stumbled upon our universe as one possibility from among a near infinite set.

At this point, the "derivation" from Big Bang implications is no longer a deduction from initial information; it is a layered narrative assembled *post hoc*.

This is where the Big Bang fails as a generative theory even on its own terms. The information it provides is not explanatory in either direction. Even when granted that all it supposes is correct, it is causally cut off from any account of how the universe came into existence, and it is insufficient to generate the universe forward without importing new structure. It explains neither end, the beginning, nor our present universe. It sits in the middle, mapping behavior under assumed laws while doing no generative work.

This is precisely why the comparison to the CMB matters.

The CMB is the earliest boundary at which the universe presents itself as a coherent, classical spacetime with encoded structure. Although we struggle to "unify" the classical and quantum regimes, existence itself has achieved this task, from the outset. It contains information that constrains what the universe already is, not a speculative story about what it might have been. Unlike the force-separation narrative, the CMB actually carries generative relevance: uniformity, anisotropy, low entropy, and fine structure, all of these are present together in a way that can only be explained as *instantiated*, not assembled from fragments or "evolved" from simple rules, or a sub layer that we cannot penetrate, or reveal.

Nothing in established physics demonstrates that quantum descriptions, by themselves, can generate classical spacetime structure. No experiment shows classical geometry emerging from a purely quantum substrate, no demonstration shows global coherence arising from local quantum indeterminacy, and no evidence shows macroscopic informational order being created rather than imposed.

A Second Thought Experiment: Removing the Burden Entirely

An anticipated objection to the first argument would be that it places an unfair burden on the Big Bang by demanding generative power from epochs such as force separation or earlier—regimes that are admittedly speculative and poorly constrained. Let us therefore remove that burden entirely. We will no longer ask the Big Bang to explain what happened at $T = 0$, nor even what happened prior to the CMB. We will grant it everything it claims with respect to the observable universe and ask only whether it becomes explanatory on its own terms once the CMB exists. So again, we grant the strongest possible concession.

Assume the Big Bang description is correct from the earliest mathematically tractable regime through recombination and assume the thermal history, expansion dynamics, and decoupling of radiation are all accurately modeled while also assuming the CMB is exactly what standard cosmology says it is.

Now ask a simpler question: What information is the Big Bang actually using to make its predictions?

The answer is unambiguous. Every successful "prediction" attributed to the Big Bang—element abundances, large-scale structure, acoustic peaks, temperature anisotropies, matter distribution, and subsequent cosmic evolution—is either directly measured from the CMB or calibrated against it. The parameters that make the model work are not derived independently of the CMB, they are extracted from it.

The CMB is not a consequence of the theory, it is the informational foundation on which the theory is built. This changes the explanatory direction entirely. The Big Bang does not generate the CMB, it reads it. It does not predict structure from first principles; it propagates structure already encoded in the CMB *forward* in time. Once the CMB exists, the universe's subsequent evolution is largely constrained by information already present. The Big Bang formalism can then be properly framed as acting as a mathematical imposition, not a generative source.

Seen this way, the Big Bang remains moot on both ends, from *T=0* that it cannot approach, to the CMB, which is where it derives (and does not explain) its primary evidence.

This leads to the unavoidable conclusion, all that the Big Bang predicts is predicated upon information derived from the CMB.

And if that is not convincing, then there is this: any critic of the idea that *instantiation* took place within our universe, irrespective of their preferred model, must acknowledge that a much deeper inconsistency lies within their midst. All of physics has inherited the results of instantiation while remaining unable to address any aspect of instantiation itself.

Unavoidably, instantiation is the default beginning of any imaginable hypothesis for the model of our universe that touches upon accepted spacetime, operative laws, a defined state space, low entropy, and global coherence—essentially any cosmological concept—reveals that these features must be accepted as a reality to make any of it work. Only instantiation with its global coordination, thermalization, information structure, and enforced low entropy, explains this.

These conditions cannot be emergent; they must precede and underwrite everything that follows.

Whether the chosen model for the universe posits that it "banged", cycled, bounced, inflated, mirrored itself, or, did anything else, as long as the model does not defy the evidence from the CMB and assert an eternal existence that did *not* arise from all that the CMB demonstrates, that model must arise from *instantiation*.

Instantiation is real. Instantiation happened, and it is fundamental to our reality.

Summary

The standard Big Bang framework is widely accepted in part because it feels appropriate that the origin of the universe should be in a neat package, and, seeing as it comes from nothing, it should be, in a sense, intractable. We accept the big bang because a seemingly metaphysical theory that terminates in a metaphysical mystery does not automatically appear deficient—particularly when it appears to address what is regarded as the mystery of all mysteries; existence itself, where it comes from, what it is, how it began, what it is doing, what is its purpose… etc. In this context, an unsolvable riddle at the beginning of everything feels intuitively correct. We have already braced ourselves to hear the impossible—not necessarily with the expectation that we will understand it.

The CMB-as-instantiation hypothesis does not claim to remove this mystery, nor its intractability. What it removes instead are a series of intractable concepts necessitated by the Big Bang theory. The Big Bang asks how we survived non-existence ($T < 0$), pre-physical chaos ($T = 0$), and a cascade of impossible domains previously documented: the CMB answers differently. It begins where the universe provides evidence and where physics is coherent.

The universe did not begin as a point, nor within a container, nor did it 'bang', or 'inflate' or pass through impossible states. It began with an *instantiation* relieving the alleged crises of a Big Bang generated by commitment to a retrodictive, reductionist paradigm. These crises signal the inability of the paradigm to represent the conditions it requires—a coherent, intelligible, boundary. A theory that cannot represent its own boundary is not a representation of the universe. With the principle of instantiation, the question of why there is a universe at all remains unsolved—that is—until later in our narrative, as we are able to gain clues for 'why', from 'what' the principle of instantiation continues to instantiate within our universe.

The accumulation of Big Bang crises are not failures of cosmology; they are its triumph. We have found the boundary the universe tells us exists, and with it the distinction between the narratives the universe permits and those it does not. Here is where science—stripped of paradigmatic commitment—indicates where the mystery begins.

Our universe has always been intelligible, discernible, information-extractable—would it be surprising to find that it is *transparent,* at the CMB, in order to show us the manner of its creation?

The Full Breakdown:

Value of Variable 9

The CMB As the Instantiation of the Universe: ***Impossible/ M = Miracle***

---

[1] (B. Keating, Losing the Nobel Prize, 2018, W.W. Norton & Company)

[2] (K.N. Abazajian, et al. CMB-S4 Science Book, First Edition, Chapter 2, "Inflation", 2016)
[3] (B. Keating, Losing the Nobel Prize, 2018, W.W. Norton & Company),
[4] (K.N. Abazajian, et al. CMB-S4 Science Book, First Edition, Chapter 2, "Inflation", 2016)
[5] (B. Keating, Losing the Nobel Prize, 2018, W.W. Norton & Company)
[6] (A.D. Sakharov, Violation of CP in variance, C Asymmetry, and baryon asymmetry of the universe, (Submitted 23 September 1966), Pis'maZh. Eksp. Teor. Fix. 5,32-35, 1967)
[7] (A.D. Sakharov, Violation of CP in variance, C Asymmetry, and baryon asymmetry of the universe, (Submitted 23 September 1966), Pis'maZh. Eksp. Teor. Fix. 5,32-35, 1967)
[8] (A. Guth, The Inflationary Universe, 1997, Pg. 286-7, Basic Books).
[9] (A. Guth, The Inflationary Universe, 1997, Pg. 181-4, Basic Books).
[10] A. Guth, The Inflationary Universe, 1997, Pg. 42, Basic Books).
[11] (A. Guth, The Inflationary Universe, 1997, Pg. 42, Basic Books).
[12] (A. Guth, The Inflationary Universe, 1997, Pg. 42, Basic Books)
[13] (A. Guth, The Inflationary Universe, 1997, Pg. 28, Basic Books).
[14] (R. Penrose, The Emperor's New Mind: Concerning Computers, Minds and the Laws of Physics, Chapter 7, 1989, Oxford University Press)

Book Two: *Emergence & Transformation*

TABLE OF CONTENTS

Variable 10: The Unique Fitness of CHNOPS
Special Problem No. 2: Abiogenesis, The Beginning of Life
Variable 11: DNA, DNA Transcription and the Cell
Variable 12: The Mitochondria and Cellular Respiration
Variable 13: Chirality & Organic Molecules
Special Problem No. 3: The Evolution of the Species
Interlude No. 1: The August Wheel
Interlude No. 2: Thomas Kuhn & Karl Popper
Preamble, The Divine Meditation
Variable 14: The Divine Meditation
Prelude to Our Final Two Chapters
Special Problem No. 4: Consciousness and Youness
Variable 15: Cell Consciousness
Section 1: Values & Validations of The Algorithm
Validations of The Algorithm
Interlude No. 3: The Evaluative Field

Variable 10: The unique fitness of CHNOPS

In the previous section, we traced how stellar nucleosynthesis forged the elements of the periodic table—step by step, layer by layer—culminating in the understanding of how carbon–the first and most important letter of CHNOPS was formed. The existence of these elements is only the first condition for biology. For life to arise to the level represented in the human life form a far more exacting requirement must be met: a small subset of atoms made within stars must possess chemical properties capable of supporting stable structure, energy flow, information storage, and controlled reactivity within the narrow physical conditions imposed by the environment of Earth.

Life turns to a small group of 6 elements to accomplish these tasks—carbon, hydrogen, nitrogen, oxygen, phosphorus, and sulfur—collectively called CHNOPS. These elements exhibit a rare combination of atomic and chemical features that allow complex chemistry to operate. Each CHNOPS element contributes something essential. Individually and together, they provide the bonding flexibility, molecular stability, and predictable reactivity required to form large, information-bearing molecules, for example DNA and RNA, amino acids, proteins, enzymes, and micromolecular machines vital for life. CHNOPS elements compose over 99% of the structures and networks capable of storing and transferring energy, maintaining organized chemical pathways, and sustaining the self-regulating systems that define living organisms.

Another larger subset of a dozen and a half or so elements play supporting roles, often subtle but important. Yet it is this small set—CHNOPS—that forms the structural and chemical backbone of life. Viewed in this way, these elements are members of a fundamental chemical language, and their properties arise directly from the structure of their atoms and the constraints imposed by physics.

What follows is an exploration of how each of these elements contributes to that framework, and why their collective fitness is central to understanding life in our universe. First, we examine the atoms themselves—the building blocks of all molecules.

Atoms: The Building Blocks of Life

Atoms are the fundamental units of matter and the building blocks of molecules. For CHNOPS elements, the atomic features that matter most for biology are:

- **Protons**: The number of protons: this defines the element and influences its chemical interactions. The number of protons corresponds to the number of electrons which in turn governs bonding and what structures an atom can form. So, although electrons "do" the bonding it's protons that determines which bonds are possible
- **Neutrons**: The number of neutrons provides nuclear stability. All CHNOPS elements are naturally stable, so although stability isn't a concern for the CHNOPS elements, stability itself is essential to the reliability of chemical behavior
- **Electrons**: Electrons govern bonding, molecular structure, and reactivity. The arrangement of electrons in shells and orbitals determines how atoms connect, which in turn shapes the molecules that enable life. Electrons aren't accurately described as particles in orbit. They are probability distributions constrained by quantum rules, which leads to our last bullet point.
- **Quantum Behavior**: Not usually included in a discussion of atoms we will add quantum behavior for reasons that are relevant to our entire investigation. The probabilistic nature of electrons, their orbital distributions, and even phenomena like tunneling subtly shape and in some cases are fundamental to how atoms bond and interact in the chemistries essential for life.

The behavior of electrons is critical: atoms gain, lose, or share electrons to form bonds. Covalent, ionic, and hydrogen bonds dictate molecular structure, solubility, shape, and reactivity. Through these interactions, simple atoms assemble into complex molecules. For CHNOPS elements, these atomic properties—classical and quantum—directly determine biological function: how bonds form, molecules interact with water, and shapes arise that enable life's processes. Within living systems, these interactions are organized into functional networks that store energy, transmit information, and maintain order.

Life doesn't work around atomic physics. It relies on the same unique atomic properties and strengths that allowed hydrogen and helium, under the unimaginable pressures of stellar cores, to forge the elements themselves. Within cells, these properties perform the same kind of work at a biological scale, powering everything from metabolism to energy transfer. These characteristics are essential to life's chemistry.

What comes next is the story of how each CHNOPS element contributes its own unique 'letter' to life's alphabet—how carbon scaffolds, hydrogen energizes, nitrogen codes, oxygen powers, phosphorus organizes, and sulfur stabilizes the molecules that make living systems not just possible, but exquisitely functional.

Accounting For The Elements - A Sacred Alphabet

There are *118 elements* on the periodic chart, and only 92 of them can be found in nature and are thus considered to be the *naturally occurring elements.*

Like a group of four-dimensional letters, each element has at least 20 data points (atomic number, mass, density, molecular weight, crystal structure, etc.) and multiple properties (toxicity, storage requirements, ease of production, applications) , conveying a host of applications, "talents" and "abilities". When elements on the periodic chart are *combined*, they are limitless in their ability to create nearly any structure, leverage and exist in any environment, and specifically, for intelligent observers, embody all practical ideas and solve problems, address needs, and "teach us" as much about the universe as we have the capacity to learn. I

In the same way the 26 letters of the alphabet have the capacity to tell any real or imagined story within existence depending on the elements of plot, story, structure, arrangement, punctuation and the rules of grammar, the 92 elements on the periodic chart function in the same manner as the actual physical things that words refer to. Elements make possible all words and worlds, ideas, and realities. In the place of syntax, grammar and punctuation, they have physics, gravity, bonding, affinities, chirality and quantum processes. Elements take on shapeshifting forms from the large hadron collider at CERN, to a stalk of pollinated, photosynthesizing corn. This elemental *alphabet* of the periodic table is all things.

Our existence was created by this sacred alphabet of elements and *is* this sacred alphabet. All life is woven into form by this alphabet. And the medium this sacred language is written upon is spacetime.

Chemistry

> "The Science of Chemistry has for its object the study of the nature and the properties of all the materials which enter into the composition or structure of the Earth, the sea, and the air... the highest efforts of Chemistry are constantly directed to the discovery of the general laws or rules which regulate the formation of chemical compounds, and determined reaction of one substance upon another[1]."

The Fownes quote defines chemistry as the study of the "general laws or rules" that govern how the elemental alphabet combines into compounds. Our investigation now focuses on how this chemical "grammar" is, itself, exquisitely fine-tuned for life.

We will see that this fitness operates on multiple, interlocking levels. It begins with one "letter," Carbon, whose unique properties make it the perfect backbone for complex molecules. We will then explore its partnership with a select group of other elements (CHNOPS) and, finally, examine the "Goldilocks" conditions, like bond strength and temperature, that are required for this specific chemistry to build stable, living organisms.

The First Letter In CHNOPS - Carbon, The Element of Life

Carbon *is* the essential building block of all 4 macromolecules needed for life. Carbon is able to fulfill this role because it has the unique ability to bond with itself to form long chains of atoms. Carbon is also *tetravalent*, meaning that a single atom of carbon is able to form 4 bonds at one time. Carbon atoms also have a small atomic radius. This ability along with their other abilities allows carbon to more easily be folded and compacted into the shapes needed to become functional proteins and enzymes. No other atom possesses this combination of unique abilities. Practically speaking, in biological terms, these abilities mean carbon can form chains of atoms of unlimited length in living systems. And indeed, Carbon *does* form long chains with itself and other elements. The prime example being its role in forming the sugar phosphate backbone of DNA and RNA messenger molecules. For the above reasons, and for reasons we will fully explore, life on planet Earth is referred to as *carbon-based life*.

There are four macromolecules essential to life. Carbon plays a key role in *all four*:

Proteins: Proteins are made from a precise sequence of the 20 essential amino acids formed into chains. Each amino acid is centered on a carbon atom bonded to an amino group. When amino acids are linked together (as directed by DNA) they fold into precise three-dimensional shapes, giving each protein its precise chemical and functional role as an enzyme, protein, or micromolecular machine that assembles, transports or otherwise arranges proteins.

Carbohydrates, including sugars and starches, all have carbon as their backbone. Because carbon can form bonds with four other atoms, it can create chains or rings. This flexibility allows carbohydrates to take the forms of sugars, starches, and cellulose (wood). Historically the name 'carbohydrate' comes from the old view of these molecules as 'hydrates of carbon.' While this is a simplification, it remains the name for this class of macromolecules.

Lipids: Cell membranes, fats, oils, steroids, and waxes. Carbon and hydrogen dominate their structure, forming long hydrocarbon chains or rings. These nonpolar chains are hydrophobic, meaning they repel water. This property drives lipids to organize into bilayers in aqueous environments, forming the structural basis of cell membranes and creating compartments essential for life, while also providing lipids with the ability to store energy efficiently.

Nucleic Acids: Carbon forms the backbone of the nucleic acids that make up DNA and RNA which are polymers (long chains of repeating molecular units that store and transmit genetic information), guiding the synthesis of proteins. Without carbon the sugar phosphate backbone of DNA and RNA wouldn't have either the structural flexibility, integrity, versatility and stability to support such long polymers.

The range of DNA length (size) in the natural world is quite *large*, from the 12 million base pairs found in baker's yeast, to the 3 *billion* base pairs of human DNA. If human DNA could be stretched out end-to-end it would be approximately 2 meters long. Yet within the cell, each strand is compacted into a space just 6 micrometers across, equivalent to the size of a red blood cell, 6-8 micrometers across (red blood cells have no DNA). Counter-intuitively, the size of the DNA strand for the single-celled amoebas, *amoebas dubia*, and *polychaos dubium* each have 670 *billion* base pairs. If their DNA was laid out end-to-end it would stretch the length of 5 football fields. In all cases carbon facilitates the construction of informational molecules that deliver the 4-letter code DNA uses, A, C, G, T. (adenine, cytosine, guanine, and thymine) to encode the plans for all living things.

"CHON"

An even smaller subset of the CHNOPS elements, known as CHON (carbon, hydrogen, oxygen and nitrogen), are sufficient to achieve several significant biological achievements. While CHNOPS makes up approximately ~99% to ~99.6% of the human body, CHON makes up approximately ~96% of the weight of the human body. The remaining approximately ~4% of human beings are made up of 18 remaining life essential elements, all of which are in trace amounts and minerals (minerals are made up of various elements combined in specific ratios). The rest of CHNOPS, approximately ~3% to ~3.6% is made up with phosphorus and sulfur. Phosphorus is a key component in DNA, RNA (sugar phosphate backbone) and proteins (sulfur is in the amino acids cysteine and methionine). Phosphorus is also used in phospholipids. Phosphorus is what allows lipids to self-assemble into bilayers that make up cellular compartments and define the inside and outside of cell walls.

Many scientists believe that due to the unique characteristics and capacities of the CHON elements, carbon, hydrogen, oxygen and nitrogen, along with their free availability in our universe, it would be expected that all life in our universe, wherever it is, irrespective of the environment, would be based upon carbon and the remaining CHON elements.

Bonding: Letters and Words of Life

Atoms combine through bonds to form molecules with entirely new properties. Covalent, ionic, and hydrogen bonds create three-dimensional shapes and charge distributions that govern biological function. In this sense, CHNOPS elements are letters in an alphabet. Each has unique chemical "powers," and their combinations spell the molecular words and sentences that encode, structure, and energize living systems. Understanding these bonds allows us to read and understand the language of life.

After Carbons ability to form 4 covalent bonds, Nitrogen comes the closest with its ability to form 3 such bonds. Oxygen can form two such bonds, and hydrogen, one. This tetravalency (4 bond forming ability) of carbon not only allows three dimensional biological structures of any needed shape, but as we have seen in the case of DNA, RNA, proteins and enzymes these molecules derive their functional capacity from their three dimensional shape, composition (number and type of atoms), and structure (how these atoms are arranged). They are only able to perform work if their chains, which are routinely 200 amino acids in length, and often far longer, have all of the above characteristics correctly assembled and this comes down to bonding.

Bond Strength

The *strength* of the bonds that link carbon with itself and other atoms in organic chemistry is also of a *perfect* strength to be chemically manipulated by the molecular machinery within the cell. It would be of no use for carbon to be able to form multiple strong bonds if those bonds were so strong that they could not be broken for the necessary biology and metabolisms, *or*, conversely, if they were so weak that molecules fell apart before they were intended, or needed. The *right bond strength* in addition to the r*ight chemical composition* is what allows controlled chemistry to be possible, and the bonds framed by the four primary elements of life (Proteins,

Carbohydrates, Lipids and Nucleic acids) are like so many of the other mysterious things in our universe, seeing as their bonding strengths represent an "inconceivably tiny band in the vast spread of energy levels in the cosmos[2]."where their functionality can be exploited.

This "perfect" strength is a direct result of Carbon's physical structure. As a relatively small atom with only two electron shells, Carbon holds its shared electrons close to the nucleus. This creates bonds that are robust enough to maintain complex structures (like DNA) against the random thermal motion of the environment, yet not so rigid that they cannot be rearranged by enzymes to facilitate the chemical reactions of life. A larger atom, like Silicon, holds its electrons more loosely, resulting in weaker, unstable bonds that would fall apart under the same conditions. This is the reason why silicon, which is often speculated to be a substitute for life in other worlds, faces unique challenges that are not shared by carbon.

An important fact about the bonds that carbon forms with other elements is that the strength of the bonds carbon forms with other elements does not differ substantially. Robert E.D. Clark in his book, *The Universe: Plan or Accident?* says Carbon, "...is a friend of all. Its bond energies with hydrogen, chlorine, nitrogen, oxygen, or even another carbon atom differ little. No other atom is like it[3]."

Clark's observation underscores an astounding observation: carbon is the only element in the periodic table capable of this specific chemical balancing act. If these bond energies were significantly different, or if carbon favored one partner too strongly over another, then the complex, interchangeable chemistry of life could not exist. That the one atom capable of forming complex chains also happens to possess the exact "friendly" bond strengths required for metabolism is a staggering coincidence. It suggests that carbon was not just one option among many, but the only option, pre-loaded with the precise specifications needed to build a living world.

Temperature

The four key elements of life: carbon, nitrogen, oxygen and hydrogen, and the stability of the bonds which hold them together, is predicated on the temperature these bonds form within a narrow life sustaining range where controlled chemistry is possible. As it turns out the range in which water is a liquid fits very well within the same narrow band of life-friendly temperatures. Above 121C (249 F) extremophiles *can* survive, but it must be noted that they are not complex life, rather they are single-celled simple life. Excluding extremophiles, at temperatures above 100C (212 F), the stability of organic structures decomposes at rates too fast for nearly all simple life to operate. 100C is the temperature water boils. After 1 to 3 minutes, depending on altitude, all simple life will be killed at such temperatures.

Emphasizing that extreme heat is detrimental to biology. RNA base *cytosine* (C) has a half-life of only 19 days at 100C. The RNA/DNA base *guanine* (G) is stable for 9 ½ months at 100C. It may seem like a practical matter that water is liquid roughly at the same range that organic chemistry occurs, but when we consider the range of temperatures of the planets in our solar system, and throughout the universe, the life-giving range of Earth, life, water and chemistry, is a rare and fortuitous quadruple coincidence for life. This coincidence between the temperature of the Earth, water, chemistry and life may also be interpreted to mean that life throughout the universe, besides of an exotic single celled variety, as in the example of extremophiles, must take place *only* within this narrow temperature range, which is, temperatures is, "... surely one of the most extraordinary and consequential bio-friendly coincidences in nature[4]."

The temperature range for biochemistry is roughly −50 °C to 130 °C, whereas the universe spans from near absolute zero (−273.15 °C) to temperatures exceeding 10 billion kelvin just seconds after the Big Bang.

Life emerges, at least on Earth, only where a unique 4 times life-enabling condition coincides: (1) temperatures where biological bonds remain functional, (2) Earth's ambient conditions mostly fall within that range, (3) abundant liquid water is present at these temperatures, and (4) carbon-based chemistry and CHNOPS elements bonding, integrity and chemistries operate properly. This tiny overlap represents an incredibly narrow window of opportunity in which life can exist.

A Narrow Band of Life

In stark contrast to the narrow temperature band required for life, the six CHNOPS elements were forged in the extreme violence of stellar cores. Consequently, these atoms are remarkably robust. They are stable across a vast range of cosmic environments and have no inherent "preference" for the delicate conditions found on Earth. Yet, it is only within this narrow, temperate window that their hidden capacity reveals itself. Outside of this range, they are merely stable matter; inside it, they unlock the unique ability to form the thousands of complex molecules, reactions, and pathways that make life possible.

> When Carbon is combined with Hydrogen, they form hydrocarbons (carbon + hydrogen). The list of hydrocarbon products includes plastics, fuels such as petroleum and natural gas, and lubricants such as oils and greases. All of these hydrocarbons are possible with just two elements: carbon and hydrogen.

If Oxygen is added to carbon and hydrogen, an entirely new chemical landscape emerges. Sugars, alcohols, cellulose (wood), waxes, preservatives, and additional fuels become possible. Oxygen introduces polarity and controlled reactivity, allowing carbon-based molecules to interact with water and participate in energy-releasing reactions essential for biology.

If Nitrogen is added to oxygen, carbon, and hydrogen—CHON—approximately 96% of living tissue can be constructed. Oxygen composes about 65% of living tissue, carbon 18.5%, hydrogen 9.5%, and nitrogen 3.2%. The CHON combination can produce 16 of the 20 essential amino acids that serve as the building blocks of proteins, enzymes, and molecular machines.

When Sulfur is added to CHON, the remaining 4 of the 20 essential amino acids can be formed. Sulfur also enables the chemistry of disulfide bonds, stabilizing protein structure, and appears in a wide range of biologically active molecules, including antibiotics, as well as compounds such as caffeine, nicotine, morphine, opium, strychnine, and belladonna.

Adding Phosphorus to CHON plus sulfur completes the core molecular toolkit of life–CHNOPS–.

Carbon's Gaseous Partners

At ambient temperatures, carbon is a solid, while hydrogen, oxygen, and nitrogen are gases. The gaseous states of hydrogen, nitrogen, and oxygen are not incidental—they are fundamental to life's chemistry. Oxygen's mobility as a gas allows it to diffuse efficiently into cells and participate in aerobic respiration, driving ATP production. Nitrogen, though abundant, is largely inert as N_2 gas and becomes biologically usable only through conversion by specialized microbes, often living in symbiotic association with plant roots. Through food chains built on this process, nitrogen enters biology, where it is used to form amino acids and nucleotides, supporting protein and nucleic acid synthesis. Hydrogen participates in proton transfer reactions and hydrogen bonding in aqueous environments, shaping molecular structure and energy flow. Together, these elements are chemically and physically fit for their roles, enabling both the construction of complex molecules and the dynamic biochemical flux essential for life, whether bound in compounds or cycling through gaseous states as required[5].

Fred Hoyle Again

We previously covered the carbon resonance without which there would be no carbon. The carbon resonance caused Fred Hoyle to state:

> "Now imagine yourself as a super intellect working through possibilities in polymer chemistry. Would you not be astonished that polymers based on the carbon atom turned out in your calculations to have the remarkable properties of the enzymes and other biomolecules? Would you not be bowled over in surprise to find that a living cell was a feasible construct? Would you not say to yourself, in whatever language super calculating intellects use: Some super calculating intellect must have designed the properties of the carbon atom, otherwise the chance of my finding such an atom through the blind forces of nature would be utterly minuscule. Of course you would, and if you were a sensible superintellect you would conclude that the carbon atom is a fix." -Fred Hoyle[6]

> "From 1953 onward, Fowler and I have been intrigued by the remarkable relation of the 7.65 MeV energy level in the nucleus of 12C (carbon-12) to the 7.12 MeV level in 160. If you wanted to produce carbon and oxygen in roughly equal quantities by stellar nucleosynthesis, these are just the two levels you would have to fix, and your fixing would have to be just about where these levels are actually found to be. Is that another put-up, artificial job? Following the above argument, I am inclined to think so[7]."

Hoyle's conclusion brings us back to the central paradox of Variable 10: the physical universe appears to be explicitly designed for a biological outcome. The "fitness" of these elements—their precise bond strengths, thermal tolerances, and structural versatilities—serves no purpose in the burning heart of a star or the cold vacuum of space. Their unique talents only gain meaning when they are assembled into the machinery of life. This suggests that the chemical "alphabet" of our universe was not a random byproduct of the Big Bang, but a pre-ordained language waiting to be spoken.[8]

Variable 10: The unique fitness of carbon, and CHNOPS

The fitness of carbon and the CHNOPS elements—hydrogen, nitrogen, oxygen, phosphorus, and sulfur—becomes fully apparent only in biology. Alone, these elements contribute little within the universe. Only when they find themselves forming the vital tissues and chemistries of life does their extraordinary suitability emerge.

And yet we have a puzzling riddle in the midst of this unavoidable chemical fitness: Phosphorus is a nuisance in a stellar furnace, but within life it forms the miracle of the ATP molecule.

In the extreme environment of a star, phosphorus is a rare and difficult-to-synthesize byproduct that contributes nothing to the star's stability or energy production; it is merely an energetic 'cost' of late-stage nucleosynthesis. Yet, in the quiet machinery of a cell, that same element becomes the backbone of the ATP (Adenosine Triphosphate) molecule—the universal energy currency of life. Without the specific, high-energy phosphate bonds that phosphorus alone can provide, life would have no way to store or transfer the energy required for a single heartbeat or thought. The element was forged in the chaos of a furnace it did not serve, to fulfill a biological purpose that would not exist for billions of years.

Some might describe the chemical fitness of these elements as a kind of an undirected coincidence. But logically, if life were to have any chance of existing in our universe, the capacity of the CHNOPS elements to facilitate life could not have "adapted" through evolution. The CHNOPS elements were, in effect, pre-adapted for a purpose not yet realized at the beginning, necessarily at t=0 in the history of the universe.

In assigning a value to the fitness of CHNOPS, we must look to physical laws. There are a number of highly specified and finely tuned factors that go into the fitness of the CHNOPS elements, and its chemical versatility carbon, hydrogen, nitrogen, oxygen, phosphorus and sulfur, such as:

Particle physics (quark asses, proton/neutron mass)

Fundamental forces (strong, weak, electromagnetism, gravity)

Atomic/nuclear structure (Z-atomic number, isotopes, binding energy)

Quantum mechanics (shells, tunneling, Pauli exclusion)

Physical constants (fine-structure constant, Planck constant, Speed of light, Gravitational constant, elementary charge)

Stellar and cosmological processes (triple-alpha, supernovae, metallicity)

Chemical Behavior (valence, electronegativity, bond stability)

Environmental context (planetary conditions, element abundance)

But in this instance, although it may indeed be a fortuitous condition beyond expectation that such fitness exists within the elements utilized in life, nonetheless it does, and physics easily explains it. While it has taken multiple "miracle-level" events in our investigation to arrive at the fitness of CHNOPS, all that has happened in the past must be taken as given if we are to examine the nature of these biological elements on their own merit.

Given the current state of the universe, and accepting all previous conditions as foundational, we are most justified in assigning the biological fitness of CHNOPS a value of effectively 100%. The CHNOPS elements were inherently and exquisitely fit for life and given everything that had taken place in our universe at the time they were manufactured, by all rules and laws of physics, there was no way they could not be.

The Value of Variable 10 is (for now) 1:1, Inevitable, 100%

There are only two caveats. First, although examined on its own, CHNOPS is an inevitability, it rests upon a series of Miracles and Functional Miracles that are required for its existence. Second, the existence of CHNOPS---as perfected as it is for the role of biological life---in itself is unable to bridge the gap from chemical fitness to the possibility of life itself. Life itself remains the real question. Like stellar nucleosynthesis and the carbon resonance—both ironclad if not astounding—life depends on processes that may not naturally arise in our universe, meaning physics and probability may not have an answer for the origination of life within our universe.

In the next chapter, we turn to the problem of life itself: can life, however fit, find a physical pathway into existence given the fitness of the CHNOPS elements? Or is it one more impossible event in the chain of causation from the beginning of existence to the soul of man.

---

[1] (G. Fownes, Elementary Chemistry, Theoretical And Practical, Palala Press, 2018, Pg 25)1

[2] (M. Denton, The Miracle of the Cell, Discovery Inst. Press, 2020)

[3] (R.E.D. Clark, The Universe: Plan or Accident? 3rd ed. Grand Rapids: Zondervan, 1972, Pg.97)3

[4] (M. Denton, The Miracle of the Cell, Discovery Inst. Press, 2020, Pg. 34)4.

[5] CHON Section References: (M. Denton, The Miracle Of The Cell, 2020, The Discovery Institute Press), (B. Alberts, A. Johnson. J. Lewis, M. Raff, K. Roberts, P. Walter, Molecular Biology of the Cell, Fourth Edition, 2002, Garland Science, Taylor and Francis Group), (L. Urry, M. Cain, S. Wasserman, P. Minorsky, J. Reece, Campbell Biology, 11th Edition, Pearson, 2017)

[6] (F.Hoyle, The Universe: Past and Present Reflections, Engineering and Science, November 1981. pp. 8–12)

[7] (F.Hoyle, The Universe: Past and Present Reflections, Engineering and Science, November 1981. pp. 8–12)

[8] Additional Section References: (M. Denton, Evolution Still A Theory In Crisis, 2016, The Discovery Institute Press), (L. Henderson, The Fitness Of The Environment, 1966, The Macmillan Company)

Special Problem No. 2: Abiogenesis, The Beginning of Life

1974 Nobel Prize-winning Belgian cytologist and biochemist, Christian de Duve's asks a question in the introduction to his book, "Vital Dust":

> "Is the emergence of life on Earth the result of a single chance event or combination of lucky accidents, or is it the outcome of biochemical forces woven into the fabric of the universe? And if inevitable, what are these forces, and how do they account not only for the origin of life but also for its evolution toward increasing complexity?[1]"

de Duve raises an ancient question that remains unanswered. Yet, within modern science, a default answer, several hundred-years in the making, has hardened into orthodoxy: that the origin of life (OOL) on Earth arose by chance, through a fortuitous sequence of undirected events. This idea, that life is an accident, having arisen *unaided* and *unplanned*, is not merely the standard mythology, but it constitutes what amounts to a deeply embedded tenet within the materialist paradigm. This paradigm has come to dominate western contemporary research practice and academic instruction, while also actively stifling dissent and protecting itself from challenges.

This alignment is not coincidental. The materialist worldview requires this conclusion. A strictly reductionist, physics-only framework allows for no other organizing principle beyond blind physical interactions operating without foresight, purpose or plan.

Is this requirement a scientific imposition? No.

Materialism often requires the conclusion that life is unplanned, unguided, and without any organizing principle beyond blind physical interactions operating without foresight, purpose, or plan. But this requirement is not itself a scientific position. Science seeks to explore and understand what is happening; it does not impose advanced metaphysical constraints on the nature of reality it is investigating. In principle, explanations consistent with evidence and open to empirical constraint remain admissible until ruled out.

Prior commitments inevitably influence which hypotheses we propose, but the information returned by investigation must be allowed to guide our conclusions. The critical error of strict materialism is the assumption that anything not materially reducible cannot be real. That assumption is philosophical rather than empirical.

Moreover, science routinely tolerates, and in some cases depends upon, entities and influences that are evidentially undeniable yet ontologically unresolved. Dark matter and dark energy, for example, are parameterized and theorized because their effects are evident, despite the fact that they have not been reduced to known physical entities or mechanisms, and at this point in time, may not be. Gravity occupies a similar, and instructive, position: it is both an exception and not an exception to material reducibility. It is geometrized as spacetime curvature and embedded within physical theory, yet no consensus account exists of what gravity is, how it carries energy in a local sense, or how it fully coheres with quantum theory—an unease reflected in the well-known reservations expressed by figures such as Einstein, Dirac, Feynman, and Hawking. There are numerous influences represented by asterisks within the standard model of physics that are nothing more than placeholders for influences and observed phenomena that have no current material reducibility.

The pattern is revealing. In cases such as the origin of the universe, abiogenesis, wave-function collapse, and observer-related effects, science permits speculation, abstraction, and provisional constructs where causal influence is evident but underlying ontology remains opaque. Yet when confronted with perhaps the most pervasive and familiar signatures of organization—apparent plan, intention, and non-blind ordering in life, particularly in the origin of life (OOL), those same allowances are often withdrawn in advance.

To exclude non-material or non-blind organizing principles a priori is therefore not a scientific conclusion but an asymmetric constraint placed before inquiry, not one compelled by the evidence after it.

Biologist Christian de Duve, who provided our opening quote, suggests that life could emerge from "…a single chance event or combination of lucky accidents, or," as the outcome of "biochemical forces." Such narratives are common, and in a sense necessary, because traditional reductionist frameworks provide no conceptual foundation to explain the true complexity of life's origin. Yet they are deeply unsatisfying. They often attempt to describe the vital moment of life's emergence from beginning to end—frequently within a single paragraph—leaping from raw "prebiotic" molecules to fully functional, integrated systems capable of sustaining life. In these accounts, phrases like "everything needed assembled" are used as placeholders, stand-ins for what must have been one of the most miraculous events in the history of the universe. What precisely "everything needed assembled" entails, how thousands of interacting systems, molecular machines, and feedback loops could arise simultaneously in a coordinated manner, is consistently absent.

Downstream Concerns

The implication of this missing understanding presents dire problems downstream, because the *truth* expressed in Darwinian evolution rests upon an impossible foundation made from the viability of abiogenesis–the claim that life arose naturally from non-living chemistry via unguided physical and chemical processes. Without an explanation of the origin of life (OOL)--the broad research question of how life arose on Earth (or anywhere)?, there can be no evolutionary process. The viability of Darwinian evolution is predicated on the discovery of an abiogenic path to OOL.

Charles Darwin's original theory of evolution, along with its neo-Darwinian refinements, is logically dependent on the prior existence of life and therefore cannot be fully disentangled from the question of abiogenesis. While Darwinian theory explicitly does not propose a mechanism for the origin of life, its explanatory machinery of random variation acted upon by natural selection can operate only once a living self-replicating system already exists.

Evolution, understood narrowly, is a theory of diversification—an account of how living forms change over time once life already exists. If life itself were shown to be the result of planning or intentional organization (by whatever means) then the evolutionary process acting upon that life could not be accurately characterized as blind, random, or unguided. It would make little sense to claim that the diversification of species—from amoeba to pine trees, fungi, humans, and blue whales—was not itself an intentional act in form, scope, and cumulative complexity, barely distinguishable in kind from whatever instantiated OOL. It would also be a dubious claim that Darwinism, without any prior association with OOL, swoops in after the fact to modifies this life in a manner that covers our entire planet with life. This would be to assume that the powers of Darwinism, as an opportunistic add-on feature, (which is blind, random, unintentional) are greater in force than OOL itself.

Because of this necessarily close association between OOL, abiogenesis and Darwinism it is therefore easy to conflate the three.

Darwin himself made no claims and provided no answers as to evolution's role in OOL. Thomas Huxley, a great defender and close friend of Darwin stresses this point in a speech. With respect to "the question of the origination of living or organic beings"[2]:

> "Mr. Darwin does not touch; he does not deal with it at all; but he says: 'Given the origin of organic matter -- supposing its creation to have already taken place, my object is to show in consequence of

what laws and what demonstrable properties of organic matter, and of its environments, such states of organic nature as those with which we are acquainted must have come about.'."

-Thomas Huxley, Speech, An Examination of Darwin, 1862-63[3],

When it comes to Darwinism, the character of the process cannot be more ontologically blind than the system it is modifying. What remains, then, is the recognition that abiogenesis—in the form of the first self-replicating life on Earth—and Darwinian evolution necessarily share a common point of origin, and more importantly a common creative ground. We must consider that OOL is likely the cause of the creative force behind speciation. If this hypothesis is correct, then the absence of a proposition for OOL within Darwinian theory is not a trivial omission. It is a tell—an indication that the theory, while descriptively useful, is not robust enough to account for the creative process it is asked to explain.

Abiogenesis: Two Critical Points

When we strictly apply the laws of physics, chemistry, and probability to the conditions of the early Earth, we are forced to a stark conclusion: Abiogenesis, -Life self-assembling itself from inorganic compounds by chance, accident or chemical imperative on early Earth is *impossible.*

This broad overview sets the stage for our investigation. Before we examine the specific chemical and statistical evidence—which will rigorously confirm these assertions—we must first establish two critical premises:

1. **Life establishing itself on Earth is impossible.** This is not merely a gap in our knowledge; it is a statistical certainty derived from the insurmountable odds against chirality, information encoding, and molecular assembly arising by chance. This leads to a paradox: if life cannot form itself, how can there be life on Earth?

2. **Consequently, the naturalistic speciation of life on Earth is also impossible**. While *Darwinian Evolution is not the primary subject of this section, it is fundamentally and irrevocably linked to Abiogenesis. Evolution requires a living, replicating self to act upon. If naturalism cannot explain the origin of that first life, it has no mechanism to get started. Darwinian Evolution is thus a "red herring" that diverts attention from the real prize: the truth of how life was established in the first place.

They both stand or fall together. *Darwinism will receive its own full treatment in a later section, *Special Problem #3: The Evolution of the Species.*

The ABC's of Abiogenesis

To fully grasp the magnitude of the problem, we must move beyond generalities and examine the specific, step-by-step requirements for life to arise from non-life. This brings us to the fundamental "ABC's" of the theory: the definitions, the assumptions, and the proposed mechanisms that underpin the entire concept. We begin by analyzing how materialist science attempts to categorize this event—is it an inevitability of physics, a statistical fluke, or simply a happy accident?

Life as an Unknown Force, Chance Event or Accident

Abiogenesis refers to the spontaneous generation of a living organism from nonliving matter, without reliance on pre-existing life or "seeds" of life. The term is primarily used in reference to the evolutionary question of how life began on Earth. As far as anyone knows, however, 'abiogenesis may not refer to anything that is real or existing. It may be entirely hypothetical or even mythological in our current understanding: meaning, it is scientifically possible to conclude the negative scientific conclusion that no abiogenic pathway is viable, based upon empirical and theoretical constraints, without hereby concluding that life was "created" or specifying any alternative causal mechanism. Although it is perhaps rare for ideas ensconced to the degree of abiogenesis to be ruled out without a replacement hypothesis, such a conclusion is not unheard of. It is normal science.

For abiogenesis to occur, life-permitting initial conditions must first be realized on Earth. These conditions, in turn, depend on a number of environmental factors. If all of these conditions are met and if the precursor molecules for the four main macromolecules of life, proteins, carbohydrates, lipids, and nucleic acids, are present, then the possibility of life is preserved. But there is a problem: the self-assembly of these macromolecules, whether in an aqueous environment or not, is beyond what is possible by chance alone. No self-organizational process has been discovered or theorized that produce the highly specific, biologically relevant self-assembly of molecules required for life.

Even if these components could somehow assemble within the protection of an enclosed lipid membrane, the task of transforming an inanimate collection of molecules into a living organism remains impossible. What defines *life* from that which merely appears *capable* of life, remains unknown. All efforts to identify, quantify, or replicate this vital life-giving power, whether in reductionist terms, or otherwise have thus far failed. No collection of non-living molecules, however complex, or complete, has ever exhibited the capacity to become living. This constitutes our first fully nullifying axiom against abiogenesis.

From everything we have been able to observe, the vital difference that separates the non-living from the living is something that must be transmitted from one living entity to another. This fact has been scientifically established and repeatedly confirmed. It is an axiom: *life cannot arise from non-living material.* And if this is true, then the implications are tremendous. That means however life started on Earth, it had to have been by either another living thing, or, something that had the power of a living thing. One of these must be true if it is equally true that life can only come into existence from the living.

Spontaneous Generation

'Science' has laid the question of abiogenesis to rest several times. Once in the 1600's, once in the 1700's and twice in the 1800's. These experiments have been replicated; they are definitive and unassailable. Current efforts are instead bent upon the search for ways to cross the divide from non-life to life to no avail, thus counterfactually confirming the original findings that life cannot arise on its own. To date, there has never been an observation of life arising from non-living materials in the history of scientific practice.

Experiments

In 1665, physician and naturalist Francesco Redi (1626-1697) proved that life comes from life, and that life could arise from anything that is nonliving is a mistaken belief. Redi's experiment placed fresh meat in open containers. The open containers readily attracted flies who laid eggs and maggots soon emerged. Redi was able to demonstrate that when the jars were enclosed such that fresh air could get in, but no flies, the meat remained free of maggots, thus disproving a common belief that maggots (life) could spontaneously emerge from non-living flesh. Redi summarized his findings as, "*omne vivum ex vivo*" (All life comes from life). Redi also advanced the idea that any lifeform that emerges, emerges from living parents or living 'seeds' of that same lifeform.

In the 1765 biologist Lazzaro Spallanzani (1729-1799) disproved biologist and priest John Needham's (1731-1781) hypothesis that life spontaneously arose as a "life force". Needham's' experiment heated broth allegedly of sufficient temperature to kill any living microbes. But in his experiments when this broth was exposed to air, it gave rise to living things from what he hypothesized was a "life force". Spallanzani proved that Needham's methods were likely unsound because after performing hundreds of similar experiments, he concluded that the microbes Needham observed had to have been from either ineffective sterilization or introduced from the air[4].

In 1864, chemist and microbiologist Louis Pasteur (1822-1895) proved that the microorganisms which become living within dead flesh were *airborne*, and not spontaneously generated, as vitalists or those who believed in heterogenesis (The spontaneous generation of animals and vegetables low in the scale of organization from inorganic elements), famous proponents of heterogenesis included, "...Newton, Harvey, Descartes and von Helmot accepted… without question[5]." In 1877, physicist John Tyndall, with his sterilizing chamber, proved that life cannot arise from non-life[6].

Although science has established that life cannot arise from non-living matter, the deeper implication, that life could not have arisen anywhere without an external source of organization, has rarely been fully grasped. This paradox is profound: it challenges our understanding and our mythology, appearing unthinkable and contradictory, given the observation of all possible forms of life flourishing within virtually every imaginable environment and condition.

Even more daunting to the prospect of abiogenesis are the problems that would have confronted OOL immediately upon emergence. These include the highly efficient and precisely timely processes required for protection, environmental awareness, a genetic code, replication and transcription. There would be no grace period, no learning curve, and "evolution" would be far too slow, uncertain and random to provide the immediate solutions demanded by life. No. Every one of these features would have needed to be fully operational the instant inanimate became living.

Moreover, OOL would have had to be robust enough to survive for tens, hundreds, or perhaps millions of generations for the *mechanism* of Darwinian random variation and natural selection to operate, iteration by iteration, to produce the diversity of life observed on Earth today. Indeed, Yet not only has the emergence of life from non-living material never been observed, either directly or conclusive evidence it has also never been observed in the history of humanity that one species has turned into a fundamentally different kind of species.

The First Blows against Probability & the Role of Chance in Life

In 1953, Watson and Crick, drawing heavily on the crucial X-ray crystallography work of Rosalind Franklin, were able to use the image of a DNA molecule to solve what was widely regarded as a central mystery of life: where is the plan? This x-ray image allowed them to complete the first successful model of an unimaginably long molecule at the heart of life itself. With their model, Watson and Crick were able to definitively affirm that the basis for life and the hereditary continuity is transmitted from one generation to the next by means of a simple four-letter code attached to within a double helix structure referred to as deoxyribonucleic acid, or DNA.

The power of this exciting discovery, —the realization that life is organized around an information-suffused molecule—electrified the scientific community and captured the imagination of the general public. It suggested that the long-hidden mystery of life was finally being deciphered and that its underlying driving force would soon be revealed. At the time it was widely believed that because this code had arisen "naturally", through *abiogenesis*, once it was studied and its structure revealed, it would inevitably reveal how it had come into existence, and consequently, how OOL established itself on Earth. Biological systems appeared far too complex and counterintuitive to fully understand in their finest details, but now that we had the code in our possession, it was believed that the missing key to this complexity would ultimately be explained, and perhaps, conceptually reconstructed.

By a remarkable historical coincidence, 1953 was also the year the Miller-Urey experiment–which we will explore in detail in the upcoming pages–was released. To the scientific world, it appeared the pieces were falling into place: Crick and Watson had found the structure, and Miller and Urey had found the building blocks. It seemed the mystery of life was all but solved. The Miller-Urey experiment used chemicals that simulated the conditions that were believed to be on early prebiotic Earth and were provided with a spark for energy to help drive the system. Once it was set up and allowed to run, chemical reactions occurred. When the products of these reactions were analyzed, they showed that the synthesis of complex organic compounds, including amino acids, the primary building blocks of all life forms on Earth, had formed from simple precursors.

For a moment in scientific history, a sense of convergence swept through the scientific world. The belief that mankind had moved through a once forbidden boundary was palpable. It was believed that once the information from these two results was thoroughly digested, it would be only a matter of step-by-step filling in details until much of the final picture would be complete. Structure (DNA information) and substance (chemistry and amino acids) had both appeared.

But despite the promise of these two seminal triumphs, an impasse remained. As the world's greatest researchers pored through the new data, they realized that rather than solving the riddle, they had uncovered a deeper one. The Miller-Urey experiment produced amino acids, basic chemical components, but DNA revealed that life required a blueprint of extraordinary complexity to assemble them.

From this momentary state of triumph, half of these twin breakthroughs stood at an impasse that redefined the triumph of the Miller-Urey experiment into a mirage. As we have learned more about what is involved in the assembly and operation of life, we have moved *further away* from solving the mystery. Challenges are piling up far faster than solutions.

In 1953, our ability to peer inside the cell was minimal. Two generations earlier it was widely believed that the cell contained little more than a sort of amorphous *jelly* with little to no useful information. Although organelles and the nucleus had been observed, they were not initially understood to be complex or highly organized structures. Today with our increased acuity on all fronts, we know that each cell contains the functional equivalent of an entire city's worth of coordinated activity, workers, engineers, libraries of information, structural frameworks, transport networks, energy plants, factories and regulatory systems that monitor and respond to conditions in real time. But strangely, these cities are ghost towns. There is no equivalent of a mayor, city council, or governing board. DNA indeed holds a complete repository of how to build the town and the machines that automate the cell, but how this all operates–the self-awareness or consciousness of the system remains a mystery. More on this later, but the larger point of our digression is definitive. As our knowledge has expanded, so too has our recognition that what is required for life is vastly more complex than was understood to be seventy years ago. And even today we haven't reached the bottom of it. In the understanding of our investigation, it appears at this point to be the kind of unresolvable paradox we have become all too familiar with in our investigation.

The problem is, as our observational detail has increased, so too has our recognition that the requirements for life are far more complex than previously understood. Each advance in molecular and cellular biology has revealed additional layers of organization, mechanism, and regulation, transforming what once appeared to be simple outputs, such as ATP molecules emerging from mitochondria, into deeply structured, multi-step processes. Using the same example of ATP and mitochondria, we have discovered that ATP doesn't just emerge simply, but it involves the citric acid cycle in the mitochondrial matrix generating electron carriers, which feed several other systems before producing ATP molecules. ATP production is a concrete instance of a broader biological principle where increased observational detail continually reveals new layers of complexity. Each layer reveals additional mechanisms which uncovers new questions. As a result, while our knowledge has indeed grown substantially over the past seventy years, the number of questions has expanded as well.

Biochemical Predestination & "Directed Chance"

"Biochemical predestination" is the concept that the emergence of life was an inevitable outcome of natural laws. In this view, life is an inherent property of matter itself. It posits that simple chemistry will naturally evolve into more complex chemistry, driven by organizational imperatives built into the elements themselves. These imperatives are viewed as the emergent result of the same physical laws that guide the cosmic evolution of the universe.

With this philosophical framework established, we can now examine the specific scientific theories proposed to validate it. The two most prominent attempts to provide a mechanism for biochemical predestination are the Oparin-Haldane Hypothesis and the previously referenced Miller-Urey Experiment. These studies famously proposed that a "primordial soup" could naturally organize into complex molecules, ostensibly proving that life was an inevitable product of chemistry.

The Oparin Haldane Hypothesis

In 1924, Soviet biochemist Aleksandr Oparin proposed his *heterotrophic theory*[7]. This idea proposed a means by which protocells could have emerged from simple constituents. Oparin theory suggested that the early Earth possessed a *reducing atmosphere, an atmosphere where oxidation is prevented because there either is no oxygen or, oxygen is removed, and hydrogen, carbon monoxide and other gasses are present to actively remove oxygen. The atmosphere conjectured in the Oparin theory was imagined to be made up of methane, ammonia, hydrogen and water vapor to serve as the template for chemical evolution. Several years later, in 1929, English Scientist J.B.S. Haldane[8] independently arrived at a similar conclusion. Together, the Oparin-Haldane hypothesis became the primary theoretical framework used for experiments and research in evolutionary biology and origin of life research.

Details of the theory posit that carbon compounds, such as methane and ammonia in the early atmosphere, were catalyzed by solar ultraviolet flux, cosmic radiation, and heat. This energy input increased their complexity (reversing entropy), forming amino acids and sugars that rained down into the oceans. From this molecule-rich "prebiotic soup," protocells are believed to have formed. The theory incorporates foundational evolutionary ideas, suggesting that these non-living protocells would "compete" with one another in a process of chemical natural selection. Eventually, one of these lines would theoretically emerge victorious, refined by this competition for survival, into a fully realized living cell.

Of note, for this theory to be true, not only is life required to arise once early in the prebiotic/early life history of Earth, but it is required to arise at least twice, and likely many times in order to fuel the Darwinian "competition" that forces at last one of these protocells into a functional capacity selected for as the most fit "winner".

We have already briefly described the daunting challenge for any life. This underscores the immense probabilistic difficulty facing this theory. While this view, as a line of reasoning to be explored, was plausible in the pre-DNA era of the 1920s, we now know that the complexity required for even a single protocell makes the spontaneous emergence of multiple competing lines exponentially more unlikely.

The Miller Urey Experiment

Because of our standard biological narrative in the area of OOL, we are obligated to close the loop on the Miller Urey experiment. It has been deemed irrelevant in terms of abiogenesis. The conditions imposed were artificial, particularly the inclusion of hydrogen, the elimination of oxygen and the borosilicate lab equipment were all key to the results, none of which conformed with the environment of the early Earth. Today it is accepted that the early Earth atmosphere was *oxygen rich* and not reduced.

The Miller-Urey experiment assumed the same kind of reducing atmosphere as the Oparin Haldane hypothesis except with the exclusion of hydrogen. (methane, ammonia, water vapor, and carbon dioxide). The atmospheric composition of the Miller-Urey experiment prevents oxidation because of the *lack* of oxygen. The accumulation of oxygen is prevented by its *removal* by hydrogen, carbon monoxide and other gasses that are known to actively remove oxygen. This result would provide an ideal template for the type of chemical evolution that was believed to be most conducive to life. But these manipulations, intended to provide a reasonable chance to produce the desired result, rendered the results irrelevant.

Early life experiments routinely exclude oxygen at appreciable levels in Earth's early atmosphere based on the assumption that the early atmosphere was conducive to the creation of organic chemistry. This assumption was not based on evidence, rather on necessity. In the presence of oxygen, the ability of biological molecules to form is severely hindered, if not entirely impossible, because oxygen dismantles non-living organic molecules at a much faster rate than they can form.

As far back as 1975 Belgian biochemist Marcel Florkin announced that "the concept of a reducing primitive atmosphere has been abandoned," and the Miller-Urey experiment is "not now considered geologically adequate." But these facts, concerning the irrelevance of the experiment to early Earth conditions, have largely been omitted in textbooks and educational materials. Presently, one would have a significant risk of believing the experiment was definitive proof that OOL could begin the first few steps on its own, "naturally". The consensus today is that the experiment is a non-sequitor to abiogenesis and OOL. This is not to say the experiment itself has no scientific value, just not to how OOL began on Earth.

Christian de Duve, whose quote began our chapter, back in 1995, in the pages of his book, *Vital Dust*, demonstrated skepticism about the results, while acknowledging its scientific relevance overall. "This historic experiment alerted organic chemists to the origin of life as a chemical problem. It sparked the birth of a new discipline… concerned with the spontaneity of biological substances under conditions that might have prevailed…" and then, he states, "...Ironically, the relevance of the conditions of this experiment is now seriously questioned[9]."

As of 2011 the debate was settled decisively by scientists at the Rensselaer Polytechnic Institute where findings were developed that prove the early atmosphere was composed mostly of oxygen-rich compounds, the same oxygen-rich compounds that are found in the current atmosphere. "The conditions on early Earth were… dominated by the more oxygen-rich compounds found within our current atmosphere. 'We can now say with some certainty that many scientists studying the origins of life on Earth simply picked the wrong atmosphere,' said Bruce Watson, Institute Professor of Science at Rensselaer[10]." The Miller-Urey experiment was *not* representative of the primitive atmosphere.

Additional Problems for Life - The Prebiotic Soup

The classic investigation into the origins of life is from 1984 titled, *The Mystery of Life's Origin,* written by Charles Thaxton, Walter Bradley and Roger Olsen[11]. This work has been updated and expanded with a 2020 re-release with up-to-date origin of life research emphasizing ongoing and new difficulties and problems[12]. The book includes a robust analysis of the potential for *any* enriched prebiotic soup experiment to supply the precursor molecules needed for life.

The prebiotic soup analysis focuses on the water-based origin of life seeing as the consensus is oceans (deep sea vents) and small bodies of water (where concentrations could be kept high) are believed to be the most likely sites a prebiotic soup to produce and concentrate the need molecules for life. It's easy to see how non-aqueous environments would be particularly challenged in providing a dynamic medium for the arrangement of prebiotic molecules. The analysis considered ocean bound, and small bodies (lakes, ponds, lagoons, etc.) of water. Their findings were prohibitive. "(German biochemist Klaus) Dose arrived at an upper limit estimate of amino acid concentration in the primitive ocean of 10^{-7}, some 10,000 times *more dilute* than the optimistic estimates reported[13]."Scientific analysis finds that this level of dilution of concentrates is far too dilute for the spontaneous formation of proteins. Their work refers to additional findings similar to their own, "...it is now generally accepted that the concentration of the soup was probably too small for efficient synthesis, particularly of biopolymers[14]."

Thaxton, Bradley and Olsen conclude this analysis by saying,

> "...geological evidence for the oceanic soup has not been located. If there ever was a dilute ocean that fed organic compounds into these smaller pools, there should be abundant evidence for it in the lower pre-cambrian sediments.", and, "...It is becoming clear that however life began on Earth, the usually conceived notion that life emerged from a soup of organic chemicals is a most implausible hypothesis. We may therefore with fairness call this scenario, 'the myth of the prebiotic soup[15].'"

Identifying a location on Earth for the onset of life, and the necessity for some kind of pre-biotic staging area for biologically relevant molecules to come together, the "prebiotic soup", leads to one of the many dead ends and insurmountable barriers on the pathway to life because of the water paradox. While water is the critical medium for sustaining life, it acts as a chemical barrier to its formation. In an aqueous environment, water promotes hydrolysis, a process that breaks down complex chains of nucleotides like RNA and DNA faster than they can form. "While water is the critical medium for all life, it also forms a chemical barrier to the formation of chains of nucleotides such as RNA and DNA that are the foundation of life… Desperate to solve the basic evolutionary impediment of this water paradox, some researchers have proposed that life may have first developed in something other than water[16]."

> "Based on the foregoing geochemical assessment, we conclude that both in the atmosphere in the various water basins of the primitive

> Earth, many destructive interactions with so vastly diminished, if not all together consumed, essential precursor chemicals, that chemical evolution rates would've been negligible... It is becoming clear that however life began on Earth, the usually conceived notion that life emerged from a soup of organic chemicals is a most implausible hypothesis[17]."

The reasons for their conclusion are many. Their analysis examined the deleterious effects of multiple influences on the prebiotic soup, including the destruction of organic compounds by energy inputs; thermal decay in oceanic environments; lightning-induced breakdown; evaporation and freezing cycles; and a range of scenarios in which purely naturalistic chemistries terminate in prebiotic dead ends[18].

The Clutter Problem

The problem of molecular chirality constitutes one of the most serious obstacles to any naturalistic account for OOL. In biological systems, chirality is a prior condition: life depends on strict molecular homochirality—specifically, the exclusive use of left-handed (L) amino acids in proteins, enzymes, and molecular machines, and right-handed (D) sugars in the sugar-phosphate backbones of DNA and RNA. By contrast, all known undirected pathways produce *racemic* mixtures, with equal proportions of left- and right-handed molecules (enantiomers). Such mixtures are catastrophic to OOL scenarios. The use of the wrong molecule in a growing string of molecules, prevents proper folding, and destroys function. The central difficulty, therefore, is not only generation of the needed molecules for life but the exclusion of the wrong form of molecule. There is no demonstrated naturalistic mechanism in prebiotic chemistry that can 'choose' the "right molecule." In the best-case scenario, unwanted chiral molecules will "clog up" the biological machinery. The "clutter problem" results in effectively halting any movement towards OOL.

The problem of chirality is so great that it has its own variable within our *Algorithm. Variable 13: Chirality & organic molecules*, where we will explore the issue at length.

A Smaller Genome As A Solution To Improbable Odds

With the improbability of the prebiotic soup, the creation of amino acids, and the clutter problem, work has commenced on the bottom end in OOL that allows a naturalistic pathway, a decreased probabilistic burden for piercing into the realm of the living. This hypothesis assumes that OOL started with a greatly reduced genome than previous estimates. Scientists have been able to identify a small bacterium with 900 genes where they have been able to delete over 400 genes and keep it alive until the organism has 473 genes. This appears to be the absolute lower limit for life; if even one more gene is removed from this "minimal cell," the organism dies[19].

To grasp the magnitude of this challenge, we can translate these "473 genes" into their building blocks: base pairs. A gene is not a single object; it is a long instruction set made up of strings of DNA letters (A, T, C, and G). In DNA, a "base" is one of the letters, and a "base pair" is two bases bonded together across the two strands of the double helix joined end-to-end. With this in mind, the proposed 473 genes, means each individual gene represents a specific sequence of these letters (bases) encoding a particular protein. If we use the gene for insulin as an example, it has 1,430 bases (either A, T, C or G) in sequence that must be arranged precisely if insulin is to be a usable feature in humans. Insulin, at 1,430 letters is on the lower end, the average gene will have around 2,500 letters, so 473 genes, means somewhere around 1,2 million bases that must be arranged in exact sequence (DNA is extremely precise, estimated at an error rate of $\mathbf{10^9}$ to $\mathbf{10^{11}}$ in living cells, or 99.9999999%).

With this in mind, our absolute genetic limit for life (473 genes) is still too great of a demand for us to reasonably expect that a random, unguided self-assembly of genes could accomplish on Earth *unaided, in the "right" chirality,* as in the theory of abiogenesis. 473 genes is a tremendous number of base pairs in length. Although the average gene is generally between 2,000 and 3,000 genes as we quoted above, genes can be anywhere from several hundred base pairs to 3 million or more base pairs. But consider this: even a genome with *150 base pairs* (which is completely incompatible with life) would *easily* exhaust the probabilistic resources of our universe in order to assemble itself, unaided, in any abiogenic pathway.

By "probabilistic resources," we are referring to our stated measure for the maximum number of "attempts" the universe can physically make (the number of particles x the speed of interactions x the age of the universe). This is our measure to evaluate the probability of each of our Variables and Special Problems in our chapter-by-chapter analysis.

To understand why this measure means naturalistic life is insufficient to the task of biology with respect to OOL, consider the math: a tiny genome of just 150 base pairs has $\mathbf{4^{150}}$ possible combinations, which equals roughly $\mathbf{10^{90}}$. Since there are only estimated to be $\mathbf{10^{80}}$ elementary particles in the entire visible universe, finding one specific, functional sequence by chance is effectively impossible. It would be akin to finding a single marked atom hidden not just in our universe, but across ten billion universes.

There is a bacterium that is found *only* within the insect, psyllid, it is named, *carsonella ruddii*. Its genome isn't 150 base pairs however, rather a very tiny (with respect to life) 159,662 base pairs, or over 1,004 times as long as 150 base pairs. Contrasted with the genome of cyanobacteria–one of, and possibly the earliest life forms in the fossil record–the amount of DNA in their genome is 22,359 times larger than *carsonella ruddii* at 3.57 *billion* base pairs[20]. That means one of the earliest life forms on Earth, has a genome 3.57 billion base pairs long. Not 473 as proposed by a hypothesis that attempts to ease the burden of random, unaided prebiotically relevant gene assembly.

Despite ongoing efforts to identify a pathway from non-life to life under the assumption that OOL was far simpler than the simplest life presently on Earth, there is little factual basis for this approach, and no current evidence. What evidence of life we have found, as far back as it has been found, gives every evidence of being similar in all ways to simple, single-celled life currently on Earth, such as stromatolites–structures left behind by the activity of cyanobacteria, a candidate for first life. We have already demonstrated that given the complexity of the genome, and the size of the genes required for life, combined with the demands of chirality and the need for exacting precision of letter order throughout this process, any idea of "simplicity" is irrelevant with regards to life. Ife, by all we understand, must have arrived with all of these systems on board, whether "simple" or more sophisticated in its simplicity (meaning first life is most certainly thought to be single-celled, simple life, not complex muti-celled life which came hundreds of millions of years later. Despite solid scientific evidence, the belief persists that life was far simpler than what is in evidence today, largely because it aligns with the idea of Darwinian adjacent-abiogenesis and OOL–with abiogenesis representing one proposed route among others, and OOL representing the open question of how life began. With respect to Darwinism, the assumption is that life proceeds slowly and gradually, from the bottom up, and from simplicity to complexity. And that simple life such as cyanobacteria represents an evolutionary leap from the actual OOL. While the assumption might seem reasonable in the abstract, it is not supported by the empirical data. Scientifically, given current empirical constraints, it remains an open possibility that the earliest life detectable to science already possessed fully integrated genetic and metabolic systems, even though no known abiogenetic pathway can presently account for such origin. This statement speaks to the very real paradox in OOL research. As Michael Denton observes:

> "...there is no evidence that any fundamental changes have occurred in the basic design of the cell system since its origination. The cell membrane, the basic metabolic paths, the ribosome, the genetic code,

are essentially invariant in all life forms on Earth. And absolutely no plausible well-developed hypothetical evolutionary sequence has ever been presented showing how the cell might have evolved via a series of simpler cell-like systems."

-Michael Denton[21]

Denton's observation cuts directly against the simplicity-to-complexity narrative. If the evolutionary ladder exists, life does not begin on its lower rungs—it begins at the top. Any hypothetical bottom rungs are unusable; they do not meet the minimum requirements for life. Cellular existence demands the simultaneous presence of multiple interdependent systems: a membrane capable of regulation, coordinated metabolic pathways, and a functional genetic code. These are not optional components that can be added incrementally; the first living system could not have been a plane assembled while already in flight. The components are mutually dependent and must operate together from the outset. In this sense, there is no such thing as a "simple" cell.

This requirement is not imposed by theory but by constraint. The physical laws of the universe, the rules of chemistry, and the demands of biology jointly enforce a threshold below which life cannot function. Completeness, therefore, is not an outcome of life—it is the price of entry.

Reflections on "Simplicity"

With completeness comes something that presents an even more profound challenge. We must consider this: how did the code arise? Of all the dead ends presented by the prospect for OOL emerging spontaneously—problems with the prebiotic soup, molecular clutter, chirality, probability, and prebiotic assembly—all point to the same conclusion: OOL could not have assembled itself given physics and chance. But even if we set those issues aside, the one thing physics and chance have the most fundamental problem explaining is the thorny issue of the genetic code. Its length (3.2 billion base pairs for humans), its simplicity (four letters), and its ability to embody the incomprehensible complexity of all of the diversity of life make the code the primary paradox in the multi-tiered paradox for OOL.

We have been approaching this problem in completely the wrong manner. OOL didn't merely hold the burden of its own genetic plan. In bringing itself into existence, it somehow managed to bring with it a coding convention so robust and sublime it had the capacity to carry all of the subsequent plans for all of the life forms that ever have, and likely ever will, exist. OOL wasn't just "OOL"; it was the seed and the genesis for the origin of all life–we could say, OOAL–on Earth.

The first living system we can infer was not just alive; it was already universal in its logic. Whatever else followed, initial life instantiated the plan-space within which all subsequent life would have to operate. An information framework that was maximally general and robust enough to underwrite everything that followed. It defined the space of the possible future of life on Earth. This is what the evidence reduces to. The genome of first life was the innovation that founded OOL. And first life was the carrier of this plan that we call DNA.

For this and many other reasons, DNA, the code for life, has its own Variable: *Variable 11: DNA, DNA Transcription, and the Cell.*

Panspermia

The word panspermia is derived from the Greek *pan* (all) and *sperma* (seed), and was coined by ancient Greek philosopher Anaxagoras who believed that the universe was made of seeds that reached Earth and propagated life[22]. For all the reasons abiogenesis is impossible, panspermia was seen as the solution. The attitude is summed up in a statement in the classic, *The Mystery of Life's Origin*, "Why persist in looking to the Earth for the answer to life's origin, especially since the evidence questioning terrestrial chemical evolution is quite substantial?[23]"

Many luminaries have contemplated the concept of panspermia. This group includes biochemist Alexander Oparin[24], of the famed *Oparin-Haldane Hypothesis*; molecular biologist Francis Crick, the co-discoverer of the structure of DNA[25]; astronomer Fred Hoyle[26]; and cosmologist/popular science educator Carl Sagan[27],

As an alternative proposal, it fairs no better than abiogenesis. In truth, the question isn't simplified, it's doubled. The question of OOL still must be solved, and panspermia offers no solutions in that regard while offering the new problem of how life got to Earth. The delivery mechanism adds a level of probabilistic complexity to the odds.

Panspermia does not address the question of how life began. It simply pulls the answer out of a black box that cannot be inspected. Even if life did begin somewhere else and was placed on Earth it would have faced the same daunting odds where it originated. And if the civilization or planet that seeded life on Earth did not originate life, and were also the recipients of panspermia, then we run into an infinite regress of panspermia.

Yet, what we know remains the same. At the end of the near-infinite regress, life must begin somewhere, unaided, by itself. And this brings us back to the central question of 'how life began.' Our answer finds that the needed conditions are such that if physics remains the same throughout our universe, abiogenesis is equally impossible in all locations.

The RNA World Hypothesis

One of the most widely held concepts in OOL research is the RNA-World Hypothesis. Biologist and biophysicist Alexander Rich, a collaborator of Francis Crick, first proposed the idea in 1962[28]. The hypothesis posits that in first life, RNA played the information carrying role of DNA, and that at some later point, DNA became the home for the genome. This explanation helps defuse a complex paradox at the heart of first life. How did DNA and RNA emerge, together, naturalistically, and from the outset, demonstrate a cooperative, interactive dependency with one another? The prospect of such teleological origins opened the door to speculation of how the relationship began.

RNA's catalytic ability was discovered in the 1980s[29], and previously it was known to play a key role in transferring (tRNA) and messenger (mRNA) roles, making it the only plausible candidate as an intermediary in early genetic systems. Problems with this proposal were known from the outset. RNA in its current form cannot store the volume of information necessary for life, nor can it perform the full range of catalytic functions required in a living cell. Molecular biologists have speculated that early RNA may have been chemically simpler, allowing it to carry more information. In the role RNA was cast in the RNA-world hypothesis, RNA was allowed the possibility of doing something it cannot be observed to do in the present.

That RNA is capable of storing genetic information does not address the central problem of OOL research: the source of the information (genome) required for the first living cell, and how it came to be encoded on RNA as an "intermediate". For the purposes of this investigation, the primary deficiency of the RNA World hypothesis is that it does not explain how life began. As such, the RNA World hypothesis is frequently misclassified as an OOL theory whereas it addresses a post-origin functional regime and is best understood as a middle-stage hypothesis. In this respect, the RNA World hypothesis is a red herring with regard to OOL and is therefore not itself directly relevant; however, what has arisen as problems in supporting RNA in the role of DNA has exposed several additional chemical and informational challenges associated with first life, and these challenges remain pertinent to the present investigation.

A Missing Letter

The spontaneous assembly and maintenance of RNA in a prebiotic soup is scientifically impossible for a simple reason: one of the four essential "letters" of its code, i.e. cytosine, is chemically unavailable in such an environment. Explaining how first life managed to arrange itself without one of its primary precursor molecules represents a near insurmountable barrier. Cytosine is one of the four nucleobases found in *both* RNA and DNA. The other bases in RNA are adenine, guanine and uracil. The use of uracil in RNA raises another problem. DNA uses thymine in place of uracil in its coding convention. This raises the question of how RNA-world cellular life transitioned from its use of uracil to the DNA-world's use of thymine?

> "The deamination of cytosine and its destruction by other processes such as photochemical reactions place severe constraints on prebiotic cytosine synthesis…The synthesis described thus far does not possess the necessary speed and selectivity to meet this requirement. The use of drying lagoons as a site for prebiotic synthesis appears geologically implausible… All schemes in which cytosine is synthesized locally and distributed globally also are handicapped in that the enormous dilution that takes place when cytosine is released into a global sea offsets any gain in synthetic efficiency[30]."

Shapiro's analysis exposes a bottleneck in the RNA world hypothesis. The chemical fragility of cytosine means it essentially cannot exist in a prebiotic environment; it degrades faster than it can be produced. This creates an insurmountable problem: without cytosine, there is no RNA. And without RNA, not only does the entire hypothetical "RNA World"— and the bridge collapse, but first life and its use of RNA in any capacity becomes insurmountable and collapses before it can even begin.

Among other things, *cytosine* spontaneously, through deamination, turns into *uracil* in an aqueous solution by hydrolysis. RNA can provide some enzymatic activity, but it's far short of what's needed for a cell to be built and to survive. RNA can carry some information, but not enough. RNA can replicate itself, but it has only been able to demonstrate it can replicate a small portion of its genome, and that it replicates with low fidelity[31]. For RNA to perform these additional functions and to carry the required information, it must be long enough and of the required coding to fold into complex three-dimensional shapes, and all of this requires that a great deal of information exists within the RNA molecule. Thus, any RNA molecule capable of even limited function must have possessed considerable specified informational content.

Using RNA instead of DNA does very little to solve the specified informational problem that is already solved by DNA. The repertoire of RNA compared to DNA is far more restricted. DNA provides the codes for *all* needed cellular functions and is capable of coding for enzymes which can couple with favorable *and* unfavorable reactions. Something RNA has not demonstrated the capacity for.

DNA also provides the plans for enzymes that to the world of modern microbiologists is the equivalent of a kind of magic trick. Certain enzymes have the ability to then assemble *other* proteins composed of elements that are *not* found within the 20 amino acids that form DNA and RNA. In other words, DNA codes for proteins that then assemble other proteins that incorporate chemical elements that DNA does not possess... and yet it has figured out how to command other proteins to incorporate them. How this occurs is poorly understood, but it implies that there is a genetic code that can "tell" an enzyme that it needs an element it does not have, how to find it, assemble it, and deploy it where, and when, and how it is needed. Elements, such as zinc, selenium, nickel, chromium, manganese, cobalt and lead, not within DNA itself, are incorporated in the essential biological functions under the direction of these enzymes, under the direction of DNA.

Seen through the lens of DNA as the creator of RNA, and not the other way around, we can look at the problem in the opposite manner of OOL RNA-world researchers. Looking at the OOL problem of DNA being encoded with vast amounts of highly specified information, we can ask the question of how RNA is encoded with *any* highly specified information at all. And how did this information, in the beginning, get encoded within the highly chaotic environment of the cell, while contending with entropy and innumerable cross reactions and diffusion challenges within the assumed prebiotic, primordial soup? And how did the first RNA-world RNA macromolecule get formed in an aqueous solution without the protection of a cell with no enzymes, or any other imagined biology to assist in the process? The RNA molecule itself is inherently unstable, so if it is not created within the relatively pristine, sustaining environment of the cell, then any part, or whole of the RNA molecule which could have arranged, would have likely been destroyed by the energy rigors of ultraviolet light and cross reactions.

Measured Reason, Scientific Rigor, & Circumspection vs Bias

Before moving on, let us briefly recap the sheer magnitude of the barriers we have identified. For life to arise by chance, it must overcome a gauntlet of independent, contradictory, and fatal problems:

The Chemical Barrier: The "primordial soup" is a myth; the ocean is too dilute to foster assembly. Furthermore, water itself acts as a solvent, breaking down biological chains (hydrolysis) faster than they can form.

The Component Barrier: Essential building blocks, such as cytosine, are chemically unavailable and unstable in prebiotic conditions, rendering the RNA World impossible and presenting severe challenges to the creation of RNA in the traditional DNA-RNA symbiotic relationship.

The Structural Barrier: Nature produces a "clutter" of left- and right-handed molecules, but life requires 100% purity (homochirality). A single wrong-handed molecule breaks the lock-and-key mechanism of biology.

The Probabilistic Barrier: Even the simplest theoretical life requires a genome of massive complexity. The odds of assembling just 150 functional base pairs by chance exceeds the probabilistic resources of the entire universe, yet the simplest known genome is over 1,000 times larger than that limit.

In light of these physical realities, the persistence of the abiogenesis narrative is puzzling. If physics and chemistry explicitly forbid abiogenesis, one must wonder how the theory is maintained? First, we must acknowledge that OOL must be explained. But it can also be said that OOL is not obligated to be bound to abiogenesis as that pathway, nor is it obligated to maintain the RNA-world hypothesis as if it were a confirmed biological history, particularly in light of the evidence we have enumerated. The persistence of abiogenic narratives is maintained not by evidence, but by a very specific form of scientific storytelling.

Because the chemistry of the RNA World faces such intractable barriers, its defense relies less on hard evidence and more on speculative language. A close reading of standard collegiate textbooks, such as Molecular Biology of the Cell, reveals that the theory is not supported by observed mechanisms, but by a chain of assumptions. As a standard example of the chain of assumptions, vs. scientific evidence that supports the RNA world hypothesis we can use the paper, On the origin of life: an RNA-focused synthesis and narrative, by Jacob L. Fine and Ronald E. Pearlman[32].

In the abstract to *On the origin of life: an RNA-focused synthesis and narrative*, you'll find:

> "By synthesizing origin of life (OoL) research from its inception to recent findings, with a focus on (i) proof-of-principle prebiotically plausible syntheses and (ii) molecular relics of the ancient RNA World, we present a comprehensive up-to-date description of science's understanding of the OoL and the RNA World hypothesis. Based on these observations, we solidify the consensus that RNA

> evolved before coded proteins and DNA genomes, such that the biosphere began with an RNA core where much of the translation apparatus and related RNA architecture arose before RNA transcription and DNA replication. This supports the conclusion that the OoL was a gradual process of chemical evolution involving a series of transitional forms between prebiotic chemistry and the last universal common ancestor (LUCA) during which RNA played a central role, and that many of the events and their relative order of occurrence along this pathway are known. The integrative nature of this synthesis also extends previous description[33]."

The statements in the quote above are assumptions and beliefs that represent claims made by the RNA-world hypothesis from within the field of origin-of-life research. They are not evidential, nor do they magically validate themselves, rather, they are suppositions based on evidentiary association of associated scientific ideas interpreted as possibly being relevant from inside the conceptual structure of the RNA-world hypothesis. This is standard practice: claims internal to a hypothesis are often stated as a working roadmap for inquiry. However, these statements are frequently presented with insufficient ambiguity regarding their scientific pedigree or scope of application, which makes them appear to be established facts. This is what we mean by the statement above: "the persistence of biogenic narratives is maintained not only by evidence, but by a specific form of scientific storytelling."

Indeed, even foundational proponents have acknowledged the interpretive limits. Leslie Orgel, one of the pioneers of the RNA world concept, explicitly noted that "despite considerable progress, the problem of the origin of the RNA World is far from being solved[34]," underscoring that RNA-first does not itself resolve how life originated. Similarly, other researchers have pointed out that an RNA-only origin remains difficult to justify without involving other molecules or systems. In a review of RNA-only models, Le Vay and Mutschler observe that "seemingly insurmountable weaknesses in the theory have been highlighted[35]," and that co-evolution with peptides and other biomolecules may be necessary to overcome the shortcomings of pure RNA systems.

Critics with standing in the biological community have gone further, highlighting functional limits of RNA itself. Charles Carter, a structural biologist, has argued that no single polymer—including RNA—could plausibly carry out all the catalytic and informational tasks required for life's origin, calling the RNA world "an expedient proposal" and, "There's no way that a single polymer could carry out all of the necessary processes we now characterize as part of life[36]." And the underlying chemical problems remain unresolved. A systematic critique of RNA world chemistry concludes that RNA is very difficult to produce under prebiotic conditions, is unstable, and has a limited catalytic repertoire—all of which raise questions about how an RNA world could arise spontaneously without assistance from other systems.

This pattern is common in the history of modern science. Biological hypotheses that align with an overarching materialist narrative tend to be treated with protection due to conceptual commitment that grants favor to these ideas. The RNA-world remains dominant in part because it fits neatly into a narrative of chemical continuity that harkens to evolution's claim of a slow progression towards complexity driven by mutational fitness. This evolutionary appearance, even with key foundational issues remaining unresolved, are both forgiven as work-in-progress, and openly acknowledged by serious researchers.

In *the fourth edition* of *Molecular Biology of the Cell,* the RNA world hypothesis is presented in a manner that exposes the confirmation bias we described above, and is replete with multiple assumptions, gaps, leaps, and unverified viability of nearly every intermediate step[37]. In their five-and-a-half-page explanation of the RNA world hypothesis we have chosen a dozen and a half of the most speculative explanations from approximately 3 dozen included in the overview. Presented in order, we have:

"If this idea is correct…"

"We also discuss how DNA may have…"

"…how the genetic code may have arisen…"

"It has been proposed that the first "biological" molecules…"

"In principle, an elaborate system of molecular synthesis… could have existed…"

"…it is difficult to imagine how long RNA molecules could be formed initially…"

"The transition between the pre-RNA world and the RNA world would have occurred through the synthesis of RNA…"

"Presumably, pre-RNA polymers also catalyzed…"

"Once the first RNA molecules had been produced, they could have diversified gradually to take over the functions… leading eventually to the postulated RNA world."

"…RNA molecules, each specialized for a different activity, may evolve…"

"…the formation of effective self-replicating systems must have been…"

"An early, crude form of compartmentalization may have been…"

"Presumably, the first membrane-bound cells were formed by the spontaneous assembly of a set of amphipathic molecules..., It is not clear at what point in the evolution of biological catalysts this first occurred. In any case, once RNA molecules were sealed within a closed membrane, they could begin to evolve in earnest as carriers of genetic instructions: they could be selected not merely on the basis of their own structure..."

"Although we can only speculate on the origins of protein synthesis and the genetic code, several experimental approaches have provided possible scenarios."

"This correlation is not perfect for all the amino acids, and its interpretation is controversial, but it raises the possibility that a limited genetic code could have arisen from the direct association of amino acids..."

"This development would have made the matching of "mRNA" sequences to amino acids more efficient, and it perhaps allowed an increase in the number of amino acids that could be used in templated protein synthesis."

"One can envision a crude peptidyl transferase ribozyme, which, over time, grew larger and acquired the ability to position charged tRNAs accurately on RNA templates—leading eventually to the modern ribosome. Once protein synthesis evolved, the transition to a protein-dominated world could proceed, with proteins eventually taking over the majority of catalytic and structural tasks because of their greater versatility..."

"If the evolutionary speculations about RNA outlined above are correct, these early cells would also have differed fundamentally from the cells we know today..."[38]

In the 1985 feature film *A Room with a View*, the characters Freddy Honeychurch and Reverend Beebe have come to greet new neighbors, the Emersons as they move into a cottage. They notice a dresser bearing an inscription in golden cursive: *"mistrust all enterprises that require new clothes."* Words to live by, and ones that are appropriate for our current discussion.

The series of assumptions and gaps in causality within the explanation of the RNA-world hypothesis in *the fourth edition* of *Molecular Biology of the Cell*[39] is nearly entirely devoid of *facts* that are relevant to how the RNA-world could possibly have arisen *de novo*. They substitute assumptions, as facts, that are so smartly dressed, held together with the glue of assumptive force, and presented with unhesitating authority within a massively referenced collegiate text, that it presents the illusion of finality in the factual basis of abiogenesis in league with the RNA-world.

Abiogenesis Summary - Reflections

If we take the discovery of DNA in 1953 as our marker in time, what has happened in the world of biology since has been entirely unexpected. Instead of filling in the dark spaces of our ignorance, with each new discovery, we have unearthed far more questions about life than we have solved, while we have failed to solve the original questions we had unearthed. Whether you start at DNA and end at Proteins or start at proteins that are essential in the manufacture of DNA, the exact same mystery persists, -"How?"-. If amino acids and proteins were not *designed* so that they could be manufactured by the symbolic four-letter genetic code read out in three-letter codons, then the only remaining hypothesis is that the code itself was *worked out* based upon the shape and structure of the amino acids themselves. But the idea that the 20 amino acids required for life just so happen to conform to the multi-step three-letter code convention of DNA derived from several intermediary steps that progressively evolved and facilitated by proteins constructed from this very same code, is far beyond the realm of chance or physical law.

The discovery of DNA and the results of the Miller–Urey experiment, both in 1953, created the impression that the problem of abiogenesis was nearing resolution. That impression proved illusory. As investigation progressed, the opposite occurred. The more deeply the cell was understood, the more constrained—and implausible—naturalistic origin scenarios became.

The "primordial soup" model is chemically untenable due to hydrolysis and destructive cross-reactions. Essential building blocks of life, including RNA, face unresolved synthesis barriers, such as the absence of cytosine under plausible prebiotic conditions. Molecular chirality introduces a further constraint: any unguided assembly process would be disrupted by a racemic (50/50) mixture of enantiomers when what is needed is homochirality. The informational content of even the simplest genome vastly exceeds the probabilistic resources of the universe. The RNA-World hypothesis, rather than solving the origin of life, merely relocates the problem by presupposing a functioning genetic system and in doing so creates a hypothesis that suffocates under the weight of the host of problems it introduces. Finally, there exists no explanation for the origin of the genetic code itself—the symbolic coding convention of DNA.

Far from being unraveled, the mystery of life's origin has only deepened. What confronts us is a fully integrated biological system that physics and chemistry say should not exist.

The Value of Special Problem #2 of The Beginning of Life on Earth

(The value for the origin of life, with only the laws of our universe and probability as helping hands).

Fred Hoyle and Chandra Wickramasinghe calculated the probability of the creation of cellular life as being 1 chance in $\mathbf{10^{40,000}}$[40]. Twelve years later a new estimate for the probability of cellular life was provided by researcher Harold Morowitz of Yale and George Mason University. His estimate was more severe, placing the probability of a single bacteria arising from basic chemicals by random processes at 1 chance in $\mathbf{10^{100,000,000,000}}$. Morowitz, a biophysicist, derived this number by calculating the odds of a single bacterium arising from basic chemicals by random forces alone[41].

The chance of life (abiogenesis) far exceeds *the probabilistic capacity of the universe* (PCOU) of $\mathbf{10^{140}}$.

Value: ***fM Functional Miracle***

---

[1] (C. de Duve, Vital Dust: The Origin and Evolution of Life on Earth, Basic Books, 1995)1
[2] (T. Huxley, Speech, An Examination of Darwin, 1862-63)2
[3] (Ibid speeches-usa (dotcom)Thomas Huxley)3
[4] [(L. Spallanzani, Essay on microscopic observations regarding the generation system of Messrs. Needham and Buffon, 1765) and (E. Capanna, Lazzaro Spallanzani: At the Roots of Modern Biology, Journal of Experimental Zoology, 295; No.3, 178-196, 1999) and (R. Mancini, M. Nigro, G. Ippolito, Lazzaro Spallanzani and his Refutation of the Theory of Spontaneous Generation. Le Infezioni in Medicina 15 No.3 199-206, 2007), from OpenStax, Microbiology 3.1 Spontaneous Generation]4.
[5] (C. Thaxton, W. Bradley, R. Olsen, The Mystery of Life's Origin: Reassessing Current Theories, Philosophical Library of New York, 1984)5.
[6] (Joseph Mastropaolo, Biology for the 21st Century, 2005)6.
[7] (A.I. Oparin, The Origin of Life, Translated (into) by Sergius Morgulis. Macmillan, 1938)7
[8] (J.B.S. Haldane, The Origin of Life. Rationalist Annual, 1929, 148: 3–10)8
[9] (C. de Duve C. Vital Dust: The Origin and Evolution of Life on Earth, Basic Books pg. 18-19)14.
[10] (Press Release, Rensselaer Polytechnic Institute, Setting The Stage For Life: Scientists Make Key Discovery About The Atmosphere Of Early Earth, November 30, 2011, Phys.org/news)
[11] (C. Thaxton, W. Bradley, R. Olsen, The Mystery of Life's Origin, Pg. 114, Philosophical Society, 1984).
[12] (C. Thaxton, W. Bradley, R. Olsen, The Mystery of Life's Origin: The Continuing Controversy, 2020, Discovery Institute Press).
[13] (K. Dose, Peptides and Amino Acids in the Primordial Hydrosphere, Biosystems, Elsevier, May 1975, Pg. 69)15

[14] (A. Nissenbaum, D.H. Kenyon, J. Oro, On The Possible Role of Organic Melanoidin Polymers as Matrices for Prebiotic Activity, Pg. 259, J Mol Evol 6, 253–270, 1975)16.
[15] (C. Thaxton, W. Bradley, R. Olsen, The Mystery of Life's Origin, Pg. 114, Philosophical Society, 1984)17
[16] (L. Zyga, Did Water-Based Life Originate Without Water? January 30, 2018, Phys.org, https://phys.org/news/2018-01 -water-based-life.html)18
[17] (C. Thaxton, W. Bradley, R. Olsen, The Mystery of Life's Origin: The Continuing Controversy, Pg. 116, 2020, Discovery Institute Press)
[18] (C. Thaxton, W. Bradley, R. Olsen, The Mystery of Life's Origin: The Continuing Controversy, Pg. 91-116, 2020, Discovery Institute Press)
[19] (J.F. Pelletier, L. Sun, K.S. Wise, N. Assad-Garcia, B.J. Karas, T.J. Deerinck, M.H. Ellisman, A. Mershin, N. Gershenfeld, R.Y. Chuang, J.I. Glass and E.A. Strychalski. Genetic requirements for cell division in a genomically minimal cell. Cell. Published online March 29, 2021. DOI: 10.1016/j.cell.2021.03.008)21
[20] (A. Nakabachi, A. Yamashita, H. Toh, H. Ishikawa, H. Dunbar, N. Moran, M. Hattori, The 160-kilobase Genome of the Bacterial Endosymbiont Carsonella, Science. 314 (5797): 267. PMID 17038615, 2006)22.
[21] (M. Denton, Evolution: Still a Theory in Crisis, p. 121, Discovery Institute Press, 2016)
[22] (V.M. Kolb, B.C. Clark III, Astrobiology for a General Reader: A Questions and Answers Approach, 2020, Cambridge Scholars Publishing)25.
[23] (Ibid, Pg. 192)26
[24] (A.I. Oparin, The Origins of Life on Earth, Academic Press, 1957 New York, New York)27
[25] (Francis H. Crick, L.E. Orgel, Directed Panspermia, 1973, Icarus. 19(3): 341-346)28 & (F. H. Crick Life Itself: Its Origin and Nature, 1981, Simon and Schuster)29
[26] (Fred Hoyle and N.C. Wickramasinghe, Lifecloud: The Origin of Life in the Universe, Harper and Row, 1978)30 and (Fred Hoyle and N.C. Wickramasinghe, Evolution from Space, Simon & Schuster, 1984)31
[27] Carl Sagan, I.S. Shklovskii, Intelligent Life in the Universe, 1966, New York, Dell)32.
[28] (M. Neveu, H.J.Kim, S.A. Benner, The "Strong" RNA World Hypothesis: Fifty Years Old", Astrogiology. 13(4): 391-403, April 2013).
[29] (K. Kruger, P. Grabowski, A. Zaug, J. Sands, D. Gottschling, T. Cech, Self-Splicing RNA: Autoexcision and Autocyclization of the Ribosomal RNA Intervening Sequence of Tetrahymena, Cell, 31 (1982), Pgs. 147-157)56, and (C. Guerrier-Takada, K. Gardiner, T. Marsh, N. Pace, S. Altman, The RNA Moiety of Ribonucleases P is the Catalytic Subunit of the Enzyme, Cell, 35)1983), Pgs. 849-857)
[30] (R. Shapiro, Biochemistry Prebiotic Cytosine Synthesis: A Critical Analysis and Implications for the Origin of Life, Proc. Natl. Acad. Sci. USA Vol. 96, pp. 4396–4401, April 1999)63.
[31] (M.P. Robertson, G.F. Joyce, The Origins of the RNA World, 2012, Cold Spring Harbor Perspectives in Biology, DOI 10.1101/cshperspec. a003608).
[32] (J.L. Fine, R. E. Pearlman, On the origin of life: an RNA-focused synthesis and narrative, 2023, Cold Spring Harbor Labortory Press, Department of Biology, York University, Toronto, Ontario M3J 1P3, Canada)
[33] (J.L. Fine, R. E. Pearlman, On the origin of life: an RNA-focused synthesis and narrative, 2023, Cold Spring Harbor Labortory Press, Department of Biology, York University, Toronto, Ontario M3J 1P3, Canada)
[34] (L.E. Orgel, Prebiotic Chemistry and the Origin of the RNA World, 2004, Critical Reviews in Biochemistry and Molecular Biology 39(2): 99-123, DOI: 10.1080/1040923049046076)

---

[35] (K. LeVay, H. Mutschler, The Difficult Case of an RNA-Only Origin of Life, 2019, Emerging Topics in Life Sciences 3(5):469-475. DOI: 10.1042/ETLS20190024)
[36] (J. Cepelewicz, The End of the RNA World is Near Biochemist Argue, December 19, 2017, Quanta Magazine, Origins of Life, quantamagazine.org)
[37] (Ibid, Molecular Biology of the Cell, Pg. 365-373)64
[38] [All Quotes: https://www.ncbi.nlm.nih.gov/books/NBK26876/ from: (Ibid, Molecular Biology of the Cell, Pg. 371-376)]65
[39] (Ibid, Molecular Biology of the Cell, Pg. 365-373)
[40] (F. Hoyle and C. Wickramasinghe, Evolution From Space: A Theory of Cosmic Creationism, 1984, Pg. 24, New York, Simon and Schuster).
[41] (M. Eastman, C. Missler, The Creator Beyond Time and Space, Costa Mesa, CA: TWFT, 1996, p. 61)

Variable 11: DNA, DNA Transcription and the Cell

The capacity of living cells to flourish on Earth rests upon the fitness of all levels of organization we have investigated in *Special Problem #1, Special Problem #2,* and *Variables 1-10*. The combined effects of the conditions and constraints previously examined—including what we have defined within this investigation as miracles—have resulted in a unique condition upon Earth: life. On Earth there are surprisingly few environments where life has not been found. The key to life's ability to survive across this wide range of environments is predicated upon the expression of encoded biological adaptations within DNA.

DNA has resulted in life that is able to flourish 30 miles high in the atmosphere[1] ,and as far as 6.5 miles deep on the ocean floor, with microbes also discovered in the deepest ocean trenches nearly 7 miles down, enduring immense pressure. Life persists in extreme acidic environments (pH 0—the strongest possible) and on the opposite end in highly alkaline environments (pH 12.8). Temperature-wise, life, thermophiles (organisms that thrive at very high temperatures), bacteria that can withstand extreme temperatures, have been found in environments of 251 degrees F, such as hydrothermal vents, volcanic fumaroles, and deep subsurface rock; and in environments as cold as –4 degrees F, such as polar caps, glaciers, and permafrost[2].

Life, termed extremophiles (organisms that can survive in extreme environments), has even been found in nuclear wastes, where *Deinococcus radiodurans* can withstand extraordinary doses of ionizing radiation that shatter DNA and still survive, at radiation levels orders of magnitude beyond human tolerance and sufficient to kill most life[3] as well as in salt mines, and brine pools that serve host to *halophiles* that flourish in salinities lethal to most organisms, and in acid saline lakes with both high pH and high salinity where microbial life is especially adapted to both extremes.

Sulfuric hot springs serve host to life called thermoacidophiles. In volcanic regions hot springs (194+ degrees F) carry sulfur-rich fluids that react with oxygen to form sulfuric acid, creating hot, highly acidic (low pH) approaching that of battery acid. Life has been found inside solid rock deep beneath the Earth's surface, including more than 2 miles down in South African gold mines within Earth's deep biosphere. It is estimated that 70% of the Earth' s bacteria and kingdom archaea–prokaryotic biomass) live in subsurface environments, in soil, sediments and deep aquifers.

underground. “The genetic diversity of life below the surface is comparable to or exceeds that above the surface[4].”

These extreme-environment-loving organisms are not just equipped to survive these environments, they are built to thrive within them, and the adaptations that enable this life are encoded within the genes within their DNA. It may be the case that life presently inhabiting the surface of Earth and in the oceans represents adaptive traits inherited from lineages already thriving in extreme environments.

This brings us to the central subject of Variable 11. While the resilience of life is remarkable, the true marvel lies in the source of that resilience: the information encoded within the DNA molecule itself. In this chapter, we will peel back the layers of the cell to examine this biological "software." We will investigate the staggering complexity of the genetic code and the multi-step translation system required to read it. We will ask how a random universe could possibly write a language more sophisticated than any human computer code and then embed it within living beings as the adaptive, protean engine of life.

The DNA Code

Within the cell nucleus lies DNA, a compact molecular library that carries the instructions for life. Encoded within it is a chemical language shared by all known organisms on Earth. The language itself is universal; the only thing that varies is the order and length of its letters. DNA contains no extra flourish, letters or “words”, only precise, executable instructions, that when faithfully carried out, express nothing other than stunningly complete and capable life.

The convention of DNA is far more efficient than any coding system devised by man. If the library of congress were encoded onto DNA, letter for letter, with spaces and punctuation, its “genome” could easily fit within a teaspoon. Thousands of life forms that use this code are invisible to the naked eye, some with genomes that if fully extended could stretch hundreds of yards long. This remarkable efficiency hints at something deeper than code alone. DNA’s capacity to give rise to living organisms depends on several interlocking layers of organization beyond the sequence itself, structural and regulatory, that govern how genetic information is interpreted and expressed. Life emerges not merely from the existence of the code, but from the system the code is embedded within

One of the things this system allows is the ability of DNA to reproduce itself billions of times–repeatedly and as needed, in whole and in part to make entirely new cells through cell division. DNA also reproduces individual genes billions of times which are then used as templates to specify the sequence of amino acids from among the twenty amino acids that are assembled to make the specific proteins encoded by those genes.

To accomplish this task, DNA relies upon a host of molecular machines that work together in a choreographed sequence to locate the correct gene from tens of thousands within the genome, and then to read, copy, and express that information by building proteins, piece by piece, from a a set of twenty standard amino acids that must be assembled in correct order and sequence, and delivered where they are needed within the cell in real-time. Everything about DNA—its structure, functions, capacities, and the constraints it must simultaneously satisfy to do what it does—is unprecedented, to the extent that DNA warrants treatment as its own Variable within this investigation.

Unimaginably Complex

In the previous chapter we recounted how James Watson and Francis Crick were co-credited with the discovery of DNA in 1953 and subsequently awarded the Nobel prize. As legend has it, in 1953, on the last day of February, Watson and Crick walked into the Eagle Pub in Cambridge and announced their discovery by stating to no one in particular, "We have discovered the secret of life[5]." Soon after this discovery, Watson and Crick continued their work to understand DNA under the assumption that the code itself was created by a naturalistic process of biological order that arose naturally from chemistry.

In line with evolutionary theory, it was believed DNAs creation in the beginning was helped along by means of "chemical evolution", and "prebiotic (pre-life) evolution". As Watson and Crick began to orient themselves to what DNA actually was, they discovered there were some peculiar traits inhabited by DNAs structure that would severely challenge the assumption of naturalistic origin. To date, chemical evolution and pre-biotic evolution remains hypothetical, and in the view of some critics, effectively apocryphal. They have never been directly observed and do not exist in any robust naturalistic sense that can be meaningfully applied to the origin or informational structure of DNA.

For the genetic code to organize in such a highly specified way is to run against the rules of nature. Crystals do form, but they do not encode themselves in highly specified sequences of letters that in themselves, and without an intellect to decipher and understand their downstream context, do not even meet the threshold for gibberish. These letters in isolation mean nothing in themselves, rather their meaning exists only in relation to the molecular machines that read, interpret and execute them. And even for natural kinds of spontaneous formations, if there is no physical law guiding the action, *order* is not the rule in nature. Disorder, the concept formalized as "entropy", is the rule. "The amount of disorder in a system can be quantified. The quantity that we use to measure this disorder is called entropy of the system[6]."

> "The universal tendency of things to become disordered is expressed in a fundamental law of physics–the second law of thermodynamics–which state that in the universe, or in any isolated system (a collection of matter that is completely isolated from the rest of the universe), the degree of disorder can only increase[7]."

This means in any system, there are always far more ways that a system can be disordered than ordered, and as a rule entropy is not defeated by a random process, either with the infusion of energy or otherwise. What is required for something to be ordered and encoded is *information*. This leads us to the information problem of DNA.

Before we go further, we should first distinguish between *creating* order and *maintaining* it. Once a system is alive, it can indeed temporarily defeat entropy, but only because a living system possesses the machinery to utilize and direct infusions of energy.

> "Living cells–by surviving, growing, and forming complex organisms–are generating order and thus might appear to defy the second law of thermodynamics. How is this possible? The answer is that a cell is not an isolated system: it takes in energy from its environment in the form of food, or as photons from the sun… and it then uses this energy to generate order within itself. In the course of the chemical reactions that generated order, part of the energy that the cell uses is converted into heat. The heat is discharged into the cell's environment and disorders it, so that the total entropy–that of the cell plus its surroundings–increases, as demanded by the laws of physics[8]."

The point we are driving at is that life must already exist and have a system in order to defeat entropy, and that defeating entropy is what must be accomplished to originate DNA. The problem in originating DNA is that no such system is in existence. These are some of the issues that Crick articulated in his Nobel prize acceptance speech a few years after his discovery.

Crick's acceptance speech for the Nobel prize discusses at length the sentiment for those involved in investigating the properties of DNA. The speech describes the upward trajectory of increasing complexity that had been discovered in the intervening time period from discovery to 9 years later. By the end of Crick's speech, a grand design in nature is all but revealed, along with hints that the facility of random processes held no relevance to that design. For those listening to Crick's speech, it must have felt like being led down a hallway that begins with promise and rudimentary ideas about what the plan of life could possibly be, based on the assumption that "natural" processes "designed" it. As the narrative moves deeper, the promise at the beginning slowly changes, becoming increasingly abstract and complex, until long before the promised destination is reached–an explanation of how DNA came about–it has been revealed that the scope of the problem is no longer in any recognizable terrain.

At some point, a world reminiscent of Lewis Carroll's *Adventures In Wonderland* has taken over for the previous purely physical view. One in which the molecules rather than marching forward in an order determined by chemical affinities, they act with agency, "awareness", rapidity, specificity, and efficiency, like the Queen's cards in Lewis Carroll's classic. The action of these macromolecules appears autonomous, as they move within the middle space between existing structures and DNA within the nucleus.

What Crick discovered in the intervening years was a mind-numbing complexity, and layers of complementary actions. With years of contemplation to guide his thoughts it began to dawn on Crick that the structure of DNA, and its coding convention, could hardly have resulted from an unguided process[9].

A Francis Crick quote from Crick's, *What Mad Pursuit*[10] highlight this realization.

> "While DNA could be claimed to be both simple and elegant, it must be remembered that DNA almost certainly originated fairly close to the origin of life when things were necessarily simple or they would not have got going… Biologists must constantly keep in mind that what they see was not designed, but rather evolved. It might be thought, therefore, that evolutionary arguments would play a large part in guiding biological research, but this is far from the case. It is difficult enough to study what is happening now. To figure out exactly what happened in evolution is even more difficult. Thus, evolutionary achievements can be used as hints to suggest possible lines of research, but it is highly dangerous to trust them too much. It is all too easy to make mistaken inferences unless the process

involved is already very well understood." -Francis Crick, *What Mad Pursuit*[11]

In 1962, when Crick, Watson, and Wilkins accepted their Nobel Prizes, they knew as little about the origin of the genetic code as Darwin knew about DNA in 1858–nothing. When Crick died in 2004, he remained a staunch atheist, never wavering in his materialism. Yet, this worldview forced him into a profound intellectual paradox.

While he famously admitted that the origin of life appeared to be "almost a miracle," he could not accept a miraculous cause. Instead, to explain the staggering complexity he had discovered without invoking a Creator, he was driven to the desperate hypothesis of Directed Panspermia—suggesting that life was seeded on Earth by an advanced alien civilization. That a scientific mind of his caliber felt compelled to appeal to extraterrestrials to preserve his materialist philosophy highlights just how compelling the evidence for design had become. Perhaps Crick did not realize what he was doing by invoking Directed Panspermia. To invoke alien intelligence, is to concede and to posit, in an essential way, that only intelligence, and only one far beyond our own, could possibly possess the requisite intellectual, and biological design capacity responsible for the existence of both the genetic code and life.

One line in Crick's above quote in, *What Mad Pursuit* stands out. "Biologists must constantly keep in mind that what they see was not designed, but rather evolved." The inference of design, from the perspective of Crick, is irrepressible.

Crick may not have found the origin of DNA, but he did leave us with a precise description of the obstacle. Through his Sequence Hypothesis, he revealed that DNA is not merely a chemical structure; it is a code. And this code conveys precise instructions that must be read, translated, and executed three letters at a time to be of any use. If read incorrectly, it is meaningless. But if read correctly, it creates life. As we will see, such a sophisticated information processing system is anything but random.

The Sequence Hypothesis

Crick's Sequence Hypothesis, often called the "Central Dogma", represented a fundamental shift in our understanding of reality, because it revealed that DNA functioned as a genuine code rather than a self-organizing chemical outcome unprecedented in the history of biology. It proposed that the specificity of life derives solely from the linear sequence of the genetic code, not from the chemical properties of the molecules themselves. In other words, matter does not create the code; the code commands the matter. Without this code to direct what must be built, chemistry does not know what to bind, and life does not arise. This is not a belief but a biological necessity: the requirements of life exceed what unguided chemistry can provide. The magnitude of this discovery cannot be overstated. And to this day this reality is still being grappled with and resisted by those who contest that the code somehow arranged itself naturally.

To understand the sheer complexity of this command structure, we must examine the precise, multi-stage process required to turn that code into life.

DNA utilizes a sophisticated five-step process to create living tissue from the genetic code. As we will detail shortly, this involves a precise chain of command: (1) transcription, (2) splicing, (3) transport, (4) translation, and (5) folding.

The puzzle confronted by Francis Crick and James D. Watson was that DNA alone is not enough to produce life. DNA requires a host of coordinated activities of molecular micromachines to read and translate it and build amino acids and construct proteins. This multi-level process is what Crick described as *the sequence hypothesis*, or "the central dogma."

> "In its simplest form (the Sequence Hypothesis) assumes that the specificity of a piece of nucleic acid is expressed solely by the sequence of its bases, and that this sequence is a (simple) code for the amino acid sequence of a particular protein… Its virtue is that it unites several remarkable pairs of generalizations: the central biochemical importance of proteins and the dominating role of genes, and in particular of their nucleic acid; the linearity of protein molecules (considered covalently) and the genetic linearity within the functional gene[12]..."

Initially Crick, like others, doubted that the 4-letter code that formed the stairs of DNA's spiral staircase of adenine (A), cytosine (C), guanine (G) and thymine (T) within DNA could provide a physical template intricate enough to accommodate the twenty different types of amino acids. They held the same doubts for RNAs four-letter code, that substitutes DNAs thymine for the base uracil. Moreover, further confusing the issue, protein synthesis took place in the cytoplasm of the cell *outside* the nucleus, while almost all DNA was found *in* the nucleus. Watson correctly surmised that a genetic messenger system likely obtained the needed bit of code from DNA in the nucleus and transported it to the cytoplasm.

Soon Watson, Crick, and others agreed that RNA was the likely messenger. Additional thoughts were added to this process to fill in the details until what was arrived at was the sequence hypothesis. This idea was summarized by Crick in his famous January 1955 notes, "'On Degenerate Templates and the Adaptor Hypothesis', one of the most important unpublished papers in the history of science[13]."

DNA Transcription (abbreviated)

The first step in gene expression is *transcription*. Transcription in this case simply means to rewrite the information in DNA into a form of RNA. To do this an enzyme called *RNA polymerase* attaches itself to the start codon of the needed gene site of DNA which is known as a "promoter". Once attached, the RNA polymerase *unzips* the DNA and processes it letter by letter into messenger RNA (mRNA).

The mRNA strand must then be processed by additional enzymes so that it can be read by a *ribosome,* a macromolecular machine that synthesizes protein and links amino acids together. The spliced and ready strand of mRNA then moves out of the DNA nucleus and into the cytoplasm where a ribosome attaches itself to the mRNA strand. As the mRNA is read by the ribosome, subunits of transfer RNA (tRNA) deliver individual amino acids to the ribosome. As each codon triplet is read (DNA is read out by the ribosome three-letters at a time), the appropriate tRNA releases its amino acid to the ribosome to form a growing chain which emerges from one end of the ribosome. This is how proteins are made from DNA. Protein synthesis is a multi-step process involving thousands of micro machines working in a seamless, coordinated fashion in the nucleus *and* in the cytoplasm. Synthesis is carried out by three separate codes (DNA, mRNA, tRNA) to deliver amino acids to make proteins.

This is the sequence hypothesis. There is nothing random about this process. There is nothing random about DNA.

We have now seen how the code is copied (transcribed) from the master vault of DNA, into a mobile messenger (mRNA). But before we re-examine the sequence hypothesis to deconstruct its granular details, we must first understand the language it speaks. How exactly does a simple string of four chemical "letters" (AGCT) describe the complex, three-dimensional architecture of a protein?

While the standard evolutionary narrative posits that all of these systems arose spontaneously, guided only by physics and probability (noting that origin of life lacked evolutionary mechanisms to develop and refine the genetic code) the next task is to understand DNA's mechanism by decoding its cipher: determining which nucleotide triplets (codons) composed of A, G, C, and T correspond to specific amino acids.

The "Frozen Accident"

There is something incredibly strange about the sequence hypothesis. At its deepest foundations, it is held together by nothing at all. This is the scandal of the "frozen accident" hypothesis—a story where the bridge has completely vanished between what the three-letter codons of DNA code for and how biological macromolecular machines retrieve the needed amino acid. The frozen accident hypothesis essentially admits there is no physical reason whatsoever for this to occur, and on the surface the assignment of codons to amino acids appears entirely arbitrary.

The frozen accident hypothesis claims that the genetic code—the mapping from three-letter codon bases to the amino acids they call for—was once set by primordial happenstance. Then, the physical enforcement of this arrangement somehow disappeared with no discernible trace, yet the system remained entirely immune to this wholesale abandonment as life grew more complex.

Think of a power line that runs from an electrical station to your house. The cable is removed entirely, yet, somehow, every electrical appliance still works perfectly without being plugged in. The "frozen accident" claims this happens not because of a physical or chemical reality; rather, it is because biology "remembers" an old connection that has been completely lost.

If you look up 'the frozen accident' you will find that the molecules reading the code do not "know" the amino acids by shape, affinity, or need. There is no physical necessity helping them come together.

To say this arrangement is a "historical accident" is a euphemism to avoid calling it what it actually is: a non-physical protocol operating in real-time.

This is the scandal of the frozen accident.

The Levels of Protein Product: Bases, Codons, Amino Acids, Proteins

When DNA is decoded from its 4 different bases (AGCT) it is always in 3-letter groups called *codons*.

Codon: A sequence of three nucleotides (DNA bases, i.e. AGCT) in a DNA or messenger RNA molecule that represents the step-by-step instructions, in sequence, for incorporation of specific amino acids into a growing, and eventually completed polypeptide (protein, enzyme, etc)[14].

Whatever doubts early molecular biologists had that the 4-letter code of DNA could represent a convention to create all 20 amino acids (all had been previously discovered by 1910) were swept away when it was realized this four-letter code was read out three letters at a time, and that by doing this, each of the 20 amino acid had at least one code it corresponded to, and in most cases several, and in three cases, six possible three-letter codes that called for a specific amino acid. With this realization, it was understood that this convention could easily account for how DNA was translated into living tissues.

The math reveals DNA can create 64 different three-letter codons. Each codon codes for one of the 20 separate amino acids and is also inclusive of "start" and "stop" codons. By utilizing amino acids, proteins can be constructed for use in all known life forms (plus or minus 1-2 additional amino acids to cover all of the synthesis demands by all of the known species).

With 20 different amino acids arranged in a chain, complete proteins can be composed ranging from as few as 20 amino acids (the protein *TRP-Cage* contains 20 linked amino acids), to proteins tens of thousands of amino acids in length, for one of the largest proteins (the 34,350 amino-acid-long protein *titin*). Considering all of the possible combinations for proteins, and the millions of species that have roamed the Earth, it is plainly evident that both mathematically, and in the known history of the tree of life, there are far more possible proteins that can be produced than there are subatomic particles in the universe ($\mathbf{10^{80}}$). A short protein 100 amino acids in length allows for $\mathbf{20^{100}}$ possible sequences.

What at first appeared as a simple 4 "word" genetic code to the discoverers of DNA, becomes unimaginably complex and endlessly productive because of how it is coded, decoded and is directly referenced to the full suite of known amino acids in three-letter codons. It has been stated that if the size of a single protein in your body was scaled up to the size of a penny, all of the proteins in one human body could easily fill the Pacific Ocean. Proteins, and the code they are made from-DNA are both unimaginably generative. Where is the randomness alleged by neo-Darwinism? The pattern is unmistakably intentional.

An oddity of DNA is it's not only a double helix composed of two identical strands, but each identical strand runs *in the opposite direction* from the other, while being joined end to end. The beginning of one strand is the end of the other, and vice-versa. Within the DNA double-helix are connecting points on both the *inside* and *outside*. DNA is connected on the outside by the sugar-phosphate backbone, and in the middle by the nitrogenous bases–the four-letter code itself. How the code is connected between the twin strands of the nitrogenous bases makes no functional difference when genes are read out because each strand is first separated from one-another, and then further separated as individual genes are separated from the entire DNA strand before being read; however, what bases the twin strands are joined in the middle of DNA must follow a simple rule. The base Adenine (A) is always paired and joined in the middle with thymine (T). And cytosine (C) is always paired with guanine (G). So, although DNA is only read one strand at a time, it aligns perfectly when joined end to end, whether it is millions or tens of billions of base pairs long with (A) paired to (T) and (C) paired to (G).

The fact that the strand is always connected in the middle by two bases is why the length of a given strand of DNA is always expressed in *base pairs*. The most recent research shows that the human genome is composed of anywhere from 20,000 to 25,000 genes spread across the 23 human chromosomes[15].

DNA Transcription

Before the advent of powerful microscopes, whatever scientists envisioned the cell nucleus to look like, be constructed of, and whatever method it used to create life, was all thrown to the wind when the double helix was discovered and deciphered. The 5-step protein synthesis was far more complex and detailed than they could have anticipated or imagined.

The Process: From Code to Flesh

The Set-Up

DNA's 4 bases are Adenine, Guanine, Cytosine and Thymine

- The codes for proteins are read from DNA three-letters at a time (codon
- The 4 bases can create *64 possible* 3-letter codons
- Each of the 64 codons corresponds to either:
  - an amino acid - 60 codons
  - a transcription start signal - 1 codon, AUG, which is Methionine, an amino acid
  - a transcription stop signal - 3 codons, UAA, UAG, UGA

5 steps of protein synthesis

The steps in protein synthesis is described in different ways but the key steps are the same.

- **Transcription**: RNA is transcribed from a DNA template
- **RNA Processing**: the RNA transcript is spliced to *produce* messenger RNA (mRNA) which moves from the nucleus (DNA) to the cytoplasm where proteins are synthesized

Transport Messenger: The mRNA leaves the nucleus and attaches to a ribosome where amino acid activation takes place

Amino Acid Activation: Each amino acid attaches to its proper transfer RNA (tRNA) with the help of a specific enzyme (that carries the amino acid) and ATP used to provide the energy attach the amino acid to its tRNA

Translation: A succession of tRNAs add their amino acids to the polypeptide chain as the mRNA is moved through the ribosome one codon at a time. When completed the polypeptide is released from the ribosome

Note: Once the polypeptide chain is folded into its tertiary shape, it becomes a protein[16].

The Impossible Jump from Non-Living to Living

It is in this final step—folding—that something truly "magical" happens. Somewhere between the mechanical reading of the DNA code and the protein performing its function, a vital transmutation occurs. Atoms bonded into molecules suddenly acquire the active properties of life. Modern biologists often take this for granted, failing to note that something beyond structure has occurred. Some force appears to emerge within newly assembled life that cannot be explained by physics, or chemistry alone. Elements, no different from those in rocks, soil, or seawater become alive. As we will see, this jump from a linear chain of amino acids to a functional, living three-dimensional machine—exhibits a level of organization, vitality, collaboration and agency that can only be termed "consciousness". What it exhibits transcends the physical, the chemical, and the purely biological. It is overlooked and unacknowledged. But that cannot be surprising when this exact same quality when exhibited on the scale of individual humans, animals, plants and bacteria is regarded as "biology". This overlooked and unacknowledged dimension of life emergence is vitally relevant to our investigation to the extent that it will be examined in its own variable: *Variable 15: Cell Consciousness.*

Protein Folding

Protein folding is an entire sub-discipline of its own. Proteins must be folded into their correct shape in order to be functional; while there are a multitude of ways that a protein can be folded *incorrectly,* there is only *one* way it can be folded correctly. Amazingly, when a protein is folded incorrectly, in nearly all cases, there are other proteins that are somehow alerted, and race to unfold and refold the protein into its correct shape. This correction process, added to the original near 100% efficient folding process of proteins, results in correct folding even closer to 100% at 99% extended to several decimal points. The remote incidents when an incorrectly folded protein is not corrected within the cell, it can cause numerous problems. The larger unit that incorporates the misfolded protein most likely will not function properly, causing issues for all activities it is required to perform, and this exact problem constitutes a class of amyloid diseases[17].

Levinthal's Paradox

In 1969, Cyrus Levinthal noted that because of the large number of degrees of freedom in an unfolded polypeptide chain, the protein molecule, composed of dozens or hundreds or tens of thousands of amino acids, can have an astronomical number of possible conformations. We are back to protein synthesis and several perplexing mysteries raised by the origination of protein synthesis and by association, life itself. Proteins, even when correctly *assembled*, must be *folded correctly* in order to render them biologically functional. The estimated number of possible configurations for a 300-residue molecule is $\mathbf{3^{300}}$ (3 possible configurations for each of 300 possible folding positions in a typical protein), which is equal to $\mathbf{10^{143}}$ possible positions. "Even for a small protein of 100 residues with 3 possible amino acid bond states sampling $\mathbf{10^{13}}$ new configurations per second, it will take $\mathbf{10^{27}}$ years or 10 octillion years to find all the possible configurations[18]…" Despite the massive timetable estimate for protein folding, the average time for protein folding is *fractions of a microsecond* per amino acid residue as measured in both bacterial and eukaryotic (animals, plants, fungi) cells[19]

Levinthal's Paradox: Not Solved, Only Reframed

Many discussions of Levinthal's paradox describe it as being "solved." Physical chemist Leandro Martínez writes that Levinthal's paradox "has been solved: it was shown that small biases toward the native conformation result in realistic folding times of realistic-length sequences[20]" and uses the energy-landscape/folding funnel concept to support this framing. However, the models described as "solving" the paradox do so by replacing the combinatorial search problem with the existence of a presupposed biased energy landscape that facilitates a pathway to the native state. In this view, proteins fold on a funnel-shaped free energy surface that directs conformational changes toward the native basin, avoiding exhaustive sampling. But analyses show that energy funnels do not resolve the paradox, as originally stated. Detailed studies illustrate that funnels cannot explain the non-astronomical folding time. In other words, the "solution" offered by the energy-landscape/funnel paradigm answers a different question, which is if a biased energy funnel exists, how it produces rapid folding. The solution does not demonstrate that such a landscape exists, nor does it explain why such bias exists[21].

And now for the reveal: the discussion of whether the "energy landscape" solves Levinthal's paradox neglects to acknowledge that there is no energy landscape. At least not one that can be observed. Nothing corresponding to a literal landscape is visible when studied experimentally. That is why it is called an "energy" landscape: nothing exists, so the funnel is assumed to exist as energy. How this energy exists and how it guides folding is unknown and unproven. The literature itself emphasizes the conceptual nature of the funnel: "The energy landscape theory of protein folding is a statistical description of a protein's potential surface. It assumes that folding occurs through organizing an ensemble of structures rather than through only a few uniquely defined structural intermediates[22]" "...the funnel folding landscape is effectively a metaphor that lacks substantial support ... and the procedures that yield funnel landscapes tend to be physically unrealistic.[23]"

These statements make clear that the funnel is a conceptual or statistical framework, something imagined, that makes sense of the physically inexplicable, not a physically observed structure. You will not find scientific evidence that an energy funnel has been independently observed or causally demonstrated. Stated plainly, the purported solution to Levinthal's paradox assumes a pre-existing, biased energy landscape that has never been observed, and cannot be demonstrated to exist, that proceeds to explain folding dynamics within it, while leaving the central problem untouched: If such a landscape were to exist, how it could arise without any known explanatory mechanism[24]. As of 2022, this problem remains elusive and unsolved[25].

The Problem of Locating Functional Proteins Among Possible Proteins

To reconstruct how individual proteins locate specific genes on the genome and then fold them into their tertiary shape has proven to be a daunting task for biologists–indeed, reconstructing how this happens is a goal yet to be achieved. But in understanding these challenges the original problem of how such a system originates by means of a naturalistic path has proven to be even more difficult. To solve this problem in earnest we must go back to a more fundamental problem, which is, two-fold: how did OOL identify the correct sequence of amino acids (2) translate these proteins into a four-letter, three-codon code (3) and then assemble the letters onto a genome (and construct the molecular structure of the double-helix of DNA. (4) and do all of this under the constraints that nothing other than natural laws and probability to help the process along?

If the amino acid sequences that lead to functional proteins could be easily found by a random search, then the problem of protein production for organisms would be much easier, and the explanation for how DNA became encoded with information would be made easier. But "easier" in this sense is relative. A combinatorial problem that would take more than the age of the universe to solve one tiny piece of, is reduced to a problem that would take half the age of the universe to solve 1 tiny piece of. In this context "easier" simply means less impossible, like leaping over Mt. Everest vs leaping over Mt. Everest with a running start.

To focus on the scale of just one of these problems, the task of finding amino acid sequences that lead to functional proteins has several complications as well. First, notice that to even approach this problem, we must already artificially constrain the random search of finding biologically active *amino acid sequences* that correspond to *functional proteins*. To constrain this problem in this manner, we have "cheated" tremendously because for the origination of DNA by OOL, there was no way to restrict the random search to amino acid sequences. Whatever the venue for first life, there was no way to isolate any molecules let alone amino acid sequences. But we must continue

Constraints On Protein Function in DNA Transcription

Even the restricted problem of locating functional proteins within amino-acid sequence space represents a combinatorial nightmare: removing these restrictions to recreate origin-of-life scenarios, as must be done to truly evaluate the scope of the problem, collapses the problem to irrelevance entirely rather than merely straining it beyond measure.

But we must continue, if we are to understand the challenges faced by origin-of-life even as we acknowledge this artificially narrowed framing vastly understates the true depth of the problem.

If we restrict the search space to amino acid sequences, we have already assumed several layers of biological machinery that did not yet exist: a selection of purified homochiral amino acids, a mechanism for assembling them into long chains, a stable chemical environment to hold those chains, and perhaps the most perplexing problem among many, some kind of selection criterion that distinguishes "functional" from non-functional sequences of amino acids. None of these conditions would have been available in origin-of-life scenarios. All of these are downstream achievements of life, not conditions that could be imposed at the beginning of life. But in OOL scenario, they are assumed to have been achieved all-at-once.

Even within this constrained framework, functional proteins occupy an extraordinarily sparse region of sequence space. Molecular biologist Douglas Axe has found that the overwhelming majority of possible amino acid chains do not fold into stable three-dimensional structures, and of those that do fold, most exhibit no biological activity. Function is not a generic outcome of length or complexity; it is a highly specific property dependent on precise ordering, spatial complementarity, and chemical context[26]. Random variation does not merely have to stumble upon a fold, but upon the right fold, performing the right task, under the right conditions.

Moreover, protein function is not intrinsic to sequence alone. Proteins require molecular machines to fold correctly, cofactors to operate, membranes to localize them, and regulatory systems to control their expression. Treating protein discovery as a standalone search problem already abstracts away interdependencies that are essential for biological activity[27]. The functional landscape is therefore not only sparse but relational and success depends on coordination among components that themselves require prior organization.

Crucially, this means that probability calculations which already appear prohibitive are still optimistic. They assume a single target, evaluated independently, rather than a tightly coupled network of mutually dependent systems. In reality, partial solutions confer no advantage unless integrated into a broader functional context. A protein that folds but cannot be expressed, localized, regulated, or utilized contributes nothing toward viability[28].

Identifying A Functional Protein

Molecular Biologist Douglas Axe in investigating the problems involved in identifying a functional protein took his work a step further to test a common evolutionary assumption: that proteins are "flexible" enough to be easily produced by random mutations. By investigating a 150-amino acid sequence, Axe found the opposite. His rigorous mutagenesis experiments revealed that the frequency of sequences capable of forming stable, functional folds is incredibly rare[29]—just 1 in $\mathbf{10^{74}}$. This finding shattered the prevalent idea that proteins were not particularly fussy about their design. As Axe explains, the fact that a protein can tolerate a few changes doesn't mean those changes are harmless; it only means the "harm had not yet reached the breaking point[30]."In reality, functional sequences are so rare that a random search has effectively zero chance of finding one.

If finding functional proteins and functional protein sequences was the work of random mutation, then there would have been no OOL on Earth; it simply could not have happened. The chance of finding 1 functional protein by way of a random search would be the equivalent of finding *one atom* out of all of the atoms in the visible universe *blindfolded.*

With odds standing at 1 in $\mathbf{10^{74}}$, the entire probabilistic resources of our universe would be exhausted finding just one functional protein sequence. This reality stands in stark contrast to biological fact: the human genome contains over 20,000 protein-coding genes. If finding a single protein by a random search exhausts the universe's capacity, the existence of 20,000 specific proteins in just one species—let alone the millions of other species that have existed on Earth—becomes mathematically impossible.

Because any random search has been ruled out, Darwinian modeled random search has been ruled out. Then it would appear that some process more efficient than random search is the only option for OOL.

Variable #11 Value - Discussion

Throughout the entire human experience, not once have we observed information to be encoded without intent. Only intelligent agents, and in human experience, humans, are known to create codes. Yet, DNA represents not merely a primitive form of encoding, or any system produced by naturalistic means in the history of discovery, rather, it is one of immense complexity. DNA must be read out in multiple steps using three separate coding conventions with multiple additional steps to have any function in life.

Defenders of Darwin may argue that the genome required for the origin of life arose naturally—by accident, through prebiotic evolution, or with the aid of natural law. For them, this must be the case, because there is no alternative within a strictly naturalistic framework, even though they freely admit they do not know how it occurred and venture that no one may ever discover the "how." They cite papers that indicate a robust *suspicion* (rather than evidence) that RNA rather than DNA served as the first genome, despite the weight of evidence contrary to this claim, sufficient to smother it out of the historical record.

They further argue that the first genome would not have consisted of billions of base pairs, and that it may have been far simpler. This category of responses misses the central issue. Genome length is secondary to genome origination. The critical problem is not how long the genome was, but how a genome—any genome—came to exist at all, complete with a functional mapping between coded sequences and proteins. By comparison, the difference between a short genome and a long one is trivial.

The origin of life required the selection of a genome governed by a multi-step transcription–translation system, producing specific functional proteins that must then be correctly folded into precise three-dimensional structures. None of this is biologically meaningful unless all components are present and coordinated. The difficulty is not merely chemical synthesis, but the establishment of a symbolic, rule-based system that operates only once the entire convention is already in place. A feat that requires foresight, coordination, engineering with an eye towards a shared goal with division of labor spread through multiple domains, each facing insurmountable barriers, yet biology itself is left responsible to solve.

And here lies the irony. If this same genetic code were presented to the very evolutionary scientists who defend its naturalistic origin—but they were instead told it had been produced by a team of scientists using advanced technologies, and were asked to list the steps required to achieve it—they would have no hesitation in identifying a long sequence of processes requiring information, intention, coordination, and foresight. They would readily return with a detailed account of the intelligence necessary to accomplish such a task.

This raises an unavoidable question: why do the standards change when this identical code is placed at the origin of life? Is this still science? Has the reasoning shifted? Or is this the result of a paradigm that demands exclusively naturalistic explanations, while restricting all other causes in advance?

At this stage, critiques about code origination are typically set aside, and the response defaults to the claim that evolutionary processes are sufficient in principle, even when no detailed mechanism is available. They return with the claim that the mechanism does exist, because evolution is not random, rather, natural selection is highly specific. This is a category mistake. Natural selection is an after-the-fact process. It acts only as a filter, not as a creator. It does not write genetic code, engineer the double helix, or establish translation conventions. It merely selects among what already exists. To invoke the existence of DNA as evidence for the power of natural selection is circular reasoning, since DNA is precisely what must be explained.

Because the only mechanism Darwinism offers for generating novel genetic information is random mutation, the creative power of the theory remains fundamentally dependent on chance.

The assertion that DNA arose naturally—by accident, through prebiotic evolution, or by unspecified natural law—has no evidentiary proof. How

DNA, together with the full transcription–translation system and protein-folding requirements, first appeared on Earth remains unknown. Under the known laws of physics and chemistry, no complete naturalistic explanation has been demonstrated.

Value of Variable #11

We offer this value with the understanding that it already constrains the probability of each bullet point to a near-infinitesimal subset of the actual problem. That is, it assumes that each listed domain has only the choices listed—for example: a genome restricted to four known options (A, C, G, T); that proteins can be found at all (which assumes their "known" amino-acid sequences can be randomly guessed); that three coding conventions will arise (from unlimited possible options, including all those that would not be functional); and that seven RNAs and proteins required to service DNA can arise by chance (again, from unlimited options, including all nonfunctional ones), etc.

-Human genome accurately encoded, in 3.2 billion positions, with four possible options at each position: 1 chance in $4^{3,200,000,000}$

-Chance of all 20,000 proteins in the human genome being found by random search: 20,000 x 10^{74} (for the estimate of a small protein only 150 amino acids in length)

-Chance of three separate coding conventions arising by chance, all working together (DNA, mRNA, tRNA)

-Chance of seven primary RNAs and proteins that transcribe and service DNA arising by chance: 1 chance in 10^{74} x 7

(RNA polymerase, mRNA, spliceosome, Ribosome, tRNA, and Protein folding Proteins). The proteins involved in servicing DNA are far greater than the 150 amino acids in length. A growing mammal cell can have as many as 10 *million* ribosomes involved[31]. (

Value of DNA, DNA transcription and the cell is:

Exceeds the PCOU of 1 in 10^{140}

fM for Functional Miracle

---

[1] (Wynne Parry, Sky-High Microbes: How Far Up Can Life Exist?, November 13, 2013, Livescience.com/29865 strangest places life found)1,

[2] (P.H. Rampelotto, Extremophiles and Extreme Environments, Life (Basel), 2013 Aug7;3(3):482-5. PMID 25369817)2.

---

[3] (A.W. Anderson, H.C. Nordan, R.F. Cain, G. Parrish, D. Duggan, Studies on a Radio-Resistant Micrococcus I. Isolation, Morphology, Cultural Characteristics, and Resistance to Gamma Radiation, 1956, Food Technol. 10(1): 575-577)
[4] (Evan Gough, Universe today, December 13, 2018, Deep Carbon Observatory, Institut de Physique du globe de Paris, uniersetoday.com, (google: deepcarbon.science, uniersetoday.com, 140847, there's a surprising amount of life deep inside the earth…)]
[5] (Brian Hayes, American Scientists, The Invention of the Genetic Code, www. American scientists dot org / the invention of the genetic code)
[6] (B. Alberts, A. Johnson, J. Lewis, M. Raff, K. Roberts, P. Walter, Molecular Biology of the Cell, Pg. 70-71, Fourth Edition, 2002, Garland Science, Taylor & Francis Group)
[7] (B. Alberts, A. Johnson, J. Lewis, M. Raff, K. Roberts, P. Walter, Molecular Biology of the Cell, Pg. 70-71, Fourth Edition, 2002, Garland Science, Taylor & Francis Group)
[8] (B. Alberts, A. Johnson, J. Lewis, M. Raff, K. Roberts, P. Walter, Molecular Biology of the Cell, Pg. 70-71, Fourth Edition, 2002, Garland Science, Taylor & Francis Group)
[9] (Award Ceremony Speech, Francis Crick, The Nobel Prize in Physiology or Medicine, 1962, NobelPrize.org, Nobel Prize Outreach AB 2023, Thu. 28 Sep 2023 www. Nobelprize. Org /prizes /medicine /1962 /crick/lecture)12
[10] (F. Crick, What Mad Pursuit: A Personal View of Scientific Discovery (1988), Basic Books)
[11] (F. Crick, What Mad Pursuit: A Personal View of Scientific Discovery (1988), Basic Books)15
[12] (Wikipedia, sequence hypothesis, quoting F.H. Crick, On Protein Synthesis. Symposia of the Society for Experimental Biology. 12: Pg. 152, 158, 1958)
[13] (The Francis Crick Papers, Defining the Genetic Coding Problem, 1954-1957, National Library of Medicine, profiles.nlm.nih.gov/spotlight/sc/feature/defining)18
[14] (B. Alberts, A. Johnson, J. Lewis, M. Raff, K. Roberts, P. Walter, Molecular Biology of the Cell, Fourth Edition, 2002, Garland Science, Taylor & Francis Group)
[15] (The National Human Genome Research Institute, What is a Genome? And Gene September 28, 2023, www. Genome. gov/ about-genomics fact sheets a brief guide to genomics, & www. Genome. gov/ genetics glossary gene)
[16] (L. Urry, M. Cain, S. Wasserman, P. Minorsky, J. Reece, Ca. mpbell Biology, 11th Edition, Pearson, 2017)
[17] (E. Reynaud, Protein Misfolding and Degenerative Diseases, Nature Education 3(9): 28, 2010, nature(dot) com/ scitable/ topicpage/ protein-misfolding- and-degenerative-diseases- 14434929/)
[18] (T. Pederson, Protein Structure: Has Levinthal's Paradox "folded"?, March 11, 2021, The FASEB Journal Vol 35, Issue 3, e21416)21
[19] (L. Martinez, Introducing the Levinthal's Protein Folding Paradox and Its Solution, J. Chem. Educ. 2014, 91, 11, 1918–1923, August 4, 2014, doi(dot)org/10.1021/ed300302h); (C. Levinthal, Are There Pathways for Protein Folding?, Journal de Chimie Physique et de Physico-Chimie Biologique. 65: 44-45, Volume 65, 1968).
[20] (L. Martinez, Introducing The Levinthal's Protein Folding Paradox and Its Solution. 2014, Journal of Chemical Education, 91(11), 1918-1923. DOI: 10.1021/ed300302h)
[21] (D.L. Nelson, M.M. Cox, A.L. Lehninger, Lehninger Principles of Biochemistry 2017, 7th Ed.). W.H. Freeman. [summary of energy landscape/funnel interpretation])

[22] (Bryngelson, J. D., & Wolynes, P. G. (1989). Intermediates and barrier crossing in a random energy model (with applications to protein folding). Journal of Physical Chemistry, 93(19), 6902–6915. DOI: 10.1021/j100363a013); And
[23] (Sorokina, I., Mushegian, A. R., & Koonin, E. V. (2022). Is protein folding a thermodynamically unfavorable, active, energy-dependent process? International Journal of Molecular Sciences, 23(1), 521.DOI: 10.3390/ijms23010521)
[24] (M. Olvera de la Cruz., et al. Cunning Simplicity of Protein Folding Landscapes, 2001, Protein Engineering, Design and Selection 14(8), 521-530).
[25] (A.V. Finkelstein, N.S. Bogatyreva, D.N. Ivankov et al. Protein Folding Problem: Enigma, Paradox, Solution, biophys Rev, 2022)
[26] (D. Axe, Undeniable,Harper One, 2016).
[27] (L.A. Urry, M.L. Cain, S.A. Wasserman, P.V. Minorsky, J.B. Reece, N.A. Campbell, Campbell Biology, Pg. 111, Series, Pearson, 11th ed, October 19, 2016).
[28] (M. Denton, Evolution: A Theory in Crisis, 1986, p. 329).
[29] (D. Axe, Undeniable,Harper One, 2016).
[30] (D. Axe, Undeniable, Harper One, 2016).
[31] BSCB, British Society for Cell Biology, Ribosome, BSCB website (.org) / learning-resources/ softcell-e-learning/ ribosome/]30

Variable 12: The Mitochondria and Cellular Respiration

In the previous variable, we established that the DNA "software" is a miracle of information. But software, no matter how complex, is useless without hardware to run it and energy to power it. This brings us to the mitochondria, the microscopic power plant residing within the cells of all complex life.

We focus on the mitochondria next because it serves as the ultimate test case for the materialist worldview. Its internal machinery—involving proton pumps, spinning motors, and electrical gradients—is so sophisticated that it dismantles the naturalistic hypothesis entirely. As we will see, this structure could not have come about by chance, physical laws, or the Neo-Darwinian process of random mutation and natural selection in any imaginable scheme.

Mitochondria are yet another paradox within our universe. Something that exists, that cannot exist and has no means to exist.

Our investigation into Abiogenesis has already established that the random origin of any cellular structure is statistically impossible. However, the mitochondria present a specific, fatal challenge to the evolutionary narrative of how complex cells developed—specifically the theory of endosymbiosis.

Mitochondria are more than organelles; they are the engines of life. They are the fundamental reason complex organisms exist, eat, and breathe. Their demand for oxygen is so critical that the human body utilizes specialized delivery vehicles (red blood cells) that are uniquely stripped of their own DNA and nuclei solely to maximize their capacity to transport oxygen to these cellular power plants.

In exchange for this constant supply of resources, mitochondria provide the energy for every biological task, from locomotion to cognition. To generate this power, they utilize three highly specialized, interlocking processes:

The citric acid cycle (CAC)

The electron transport chain (ETC)

ATP Synthase, the proton driven turbine that manufactures the energy source of life, ATP.

Before we review our three specific processes for Variable 12 in detail, it is vitally important to note the highly efficient *design* of the 3 very specific, interlocked, interactive processes described above, and the role of *fitness* for the transitional metals utilized in their operation. We must also recognize the sophisticated engineering at play. These structures are explicitly designed to harness the power of the quantum world. They rely on phenomena like quantum tunneling and the manipulation of individual protons and electrons to drive the entire energy network.

The citric acid cycle produces the molecules necessary to drive the Electron Transport Chain and ATP Synthase in an ingenious, never ending production loop that utilizes the oxygen we breathe to turn the food we eat into energy, which in turn drives the energy production process (catabolism) that keeps us alive. The citric acid cycle also creates the carbon structure for amino acids, while also producing several required amino acids.

The CAC, ETC, and ATP synthase processes constitute a universal design program found in all life. In the plant kingdom, this respiratory machinery works in a perfect, complementary loop with photosynthesis. In the words of science journalist Liz Veloz, "Photosynthesis requires the products of respiration, while respiration requires the products of photosynthesis[1]."

This interlocking relationship represents a mind-boggling feat of biological engineering. Furthermore, these systems are not recent innovations. They have been in existence for 3.5 billion years, driving the energy production of life since the era of the first cyanobacteria.

To fully appreciate the magnitude of this engineering, we must move beyond abstract diagrams and consider the physical reality of these structures. Far more than simple chemical solutions, they function as sophisticated nanotechnology, performing tasks with a mechanical precision that defies our expectations of biological matter. It is this observation that leads molecular biologist Michael Denton to remark:

> "We would wonder even more as we watched the strangely purposeful activities of these weird molecular machines, particularly when we realized that, despite all our accumulated knowledge of physics and chemistry, the task of designing one such molecular machine—that is one single functional protein molecule—would be completely beyond our capacity at present"
>
> -Biologist Michael Denton, M.D., author of, *Evolution: A Theory in Crisis*[2]

The engineering of the mitochondria and the macromolecular structures that are housed within them are unmatched in biological sophistication, or in any form of human technology. As is true for all cellular processes, *the size* of the structures used in all three functions (ETC, CAC and ATP Synthase) are on a scale *so miniscule* that only the most powerful microscopes in biotechnology are able to image their parts and their actions, and these observations have only happened recently. We are talking about structures that utilize individual protons, and individual electrons. To better appreciate the wonder of this engineering, let's take a more detailed look at the electron transport chain; the process that enables the production of the cell's 'fuel'.

The Electron Transport Chain

The state-of-the-art *transmission electron microscope* (TEM) is capable of 2 *million* times magnification and is just on the cusp of the level of resolution needed to view the 20-nanometer height of the ATP synthase. At the height and with a width of less than 10 nanometers, some *5,000 ATP synthases* could be stacked lengthwise across the *width* of the average human hair[3]. To generate the ATP necessary for life, the inner membrane of the mitochondria is packed with a specific series of molecular machines. Collectively known as the Electron Transport Chain, these protein complexes function like stations on an assembly line. They are arranged in a precise sequence to pass electrons down an energy gradient, ultimately driving the production of ATP. The five stations are:

The 5 protein complexes:

Complex I - NADH Dehydrogenase
Complex II - Succinate Dehydrogenase
Complex III - Cytochrome C Reductase
Complex IV - Cytochrome C Oxidase
Complex V - ATP Synthase

In the operation of this chain, individual electrons and protons interact with these protein complexes in a highly choreographed sequence. These enzymes are grouped and sub-grouped like interlocking puzzle pieces, functioning with operational tolerances of scale and speed that no human technology can currently match. The discovery of exactly how this machinery harnesses particle flow to generate energy was a watershed moment in science, revealing a mechanism so elegant and unexpected that it revolutionized our understanding of the cell, culminating in the conference of multiple Nobel Prizes. The 1953 Nobel Prize in Physiology or Medicine, to its co-discoverers: Hans Adolf Krebs, for discovery of the Krebs cycle, now more commonly known as the citric acid cycle; and Fritz Albert Lipmann "for his discovery of co-enzyme A[4]." and the 1978 Nobel Prize in Chemistry to Biochemist Dr. Peter D. Mitchell, for his "understanding of biological energy transfer through the formulation of the chemiosmotic theory as noted in the quote below[5].

> "The mechanism by which cells use cellular respiration to manufacture ATP is not only one of the wonders of cell biology, but also one of the most unexpected and important discoveries of twentieth-century science. The mechanism, described by the chemiosmosis hypothesis, was first proposed by British biochemist and Nobel laureate Peter Mitchell in a letter to Nature in 1961, but was not thoroughly validated and widely accepted until the following decade. The mechanism, which origin-of-life researcher Leslie Orgel called 'the most counterintuitive idea in biology since Darwin,' involves exploiting the energy discharged by electrons flowing down an energy gradient to pump protons through a bilayer lipid membrane. These, according to Mitchell's model, then flow back through the same membrane, providing energy for the synthesis of ATP, which is carried out by an extraordinary rotating enzyme, ATP synthase, one of the most complex of all known enzymes."
>
> -Biologist Michael Denton, M.D., author of, The Miracle of the Cell[6]

The Electron Transport Chain functions like a steep energy staircase. It drives electrons "downhill" through the four protein complexes because each successive step in the chain possesses a stronger chemical attraction for the electron than the one before it. The steps are composed of an iron-sulfur cluster (or a copper heme) molecule known as a *redox center*. Several redox centers are housed in each of the 4 protein complexes. Each step of the gradient *desires* the electron more than the step above. Each enzyme complex in the chain holds an electron for only 5 to 20 *milliseconds* (0.005 to 0.20 seconds)[7]. This represents a maximum potential of 50 to 200 electrons *per second* or 180,000 to 720,000 electrons per hour for each of the iron-sulfur crystals in the line. With an average of 5,000 ATP synthases in a mitochondrion, and 18 redox centers in the 5 protein complexes (including ATP synthase), an average mitochondrion can process anywhere from 16.2 billion to 64.8 billion electrons per hour depending on if it's in low activity or maximal activity. It is estimated that the average human has approximately 37 trillion mitochondria, with an average of 1,000 per individual cell.

The performance of this system is highly proficient. Even more impressive is its efficiency: the system captures 70% of the released energy to direct towards proton pumping[8]—a feat of energy conservation that rivals the best human engineering.

This combination of extreme speed and high efficiency points to a singular conclusion. These are not randomly derived chemical aggregations; the protein complexes of the electron transport chain function as irreducibly complex molecular machines.

The culmination of the electron relay is the creation of a massive proton gradient. By pumping protons across the membrane, the ETC has effectively charged the mitochondrial "battery," creating a reservoir of potential energy like water held behind a dam. To harvest this potential and convert it into ATP (the universal energy currency of life), the cell employs a biological masterpiece: a rotary motor known as Complex V known as ATP synthase.

But how do these biological machines actually move electrons? To perform this function, the protein complexes must incorporate specific elements capable of conducting electrical charge. This brings us to the vital role of metals within the mitochondrial machinery.

Transition metals such as copper and iron are uniquely suited for this task and are vital to the electron transport chain, while magnesium (an alkaline earth metal) plays an essential role in stabilizing the ATP synthase[9]. The engineering design of mitochondria is nothing less than the best possible solution to the energy production needs of living organisms, by means of leveraging the fundamental capacities of specific chemical elements forged in the pressure of star life cycles.

As mentioned previously, these structures rely on specific metal-based components to hold electrons. Most complexes utilize iron-sulfur redox centers, while others, such as the protein cytochrome c, incorporate a copper atom. Because they are electronegative, and particularly efficient in attracting negatively charged electrons, transitional metals like copper are particularly efficient as target sites and are thus incorporated in *heme* proteins such as cytochrome c within the electron transport chain. *This represents a design choice.* The transition metals have the capacity to accommodate different numbers of electrons in their outermost electron shells, and because of this they can be fine-tuned to the several different states needed to accept electrons at the bottom of the gradient (chain of redox centers) with each state in a slightly different level of affinity, tuned by ideal distance, from the others[10]. it functions like a wire, electrons flow in the same manner as they do in an electrical wire[11]. The capacity used in the flow of electrons in the ETC is the same reason these metals are used in technological applications such as circuits.

"Only the transition metal atoms possess precisely the properties required to form an electronic circuit in which the electrons will lose their energy in discrete ordered steps."

-Michael Denton, M.D., Biologist and author of, The Miracle of the Cell[12]

Thus, each step within each protein complex is fine-tuned; as is the configuration and order of the protein complexes. For protein complexes I, II and III, the spacing of the redox center (~14-15 angstroms) is the perfect length to keep electrons from skipping steps. This space is also equal to the preferred jump length of the electron and is also the perfect distance for the electrons to *quantum tunnel through the membrane* to the next step down[13]. They disappear from one redox center, pass through a membrane and reappear at the next downward redox center. Ironically, the fact that the electron must *tunnel* through the membrane rather than passing through "open" space results in electrons flowing faster than they otherwise would[14].

Quantum tunneling is itself an interesting technological feature of atoms gifted to them by their quantum nature. As a particle or solid matter, electrons cannot pass through barriers. As waves/orbitals, they can, as long as the high probability portion of the wave function of the particle extends *through* the width of the barrier[15], the wave has a high and perhaps close to a 100% chance of passing through. The shorter the distance, the easier the jump. If the distance is too great, then the chance that the electron can tunnel becomes vanishingly small, but according to quantum principles, is *never zero*. As it turns out, the ~14 angstrom distance combined with the thin barrier is exactly the distance and thickness most preferred for the electron[16]. The existence of a redox center at this exact distance *on the other side* of the barrier that needs to be jumped adds to this facility. This is an extraordinary feature, dependent on a consilience of biology, chemistry, physics, the atom, quantum mechanics and classical mechanics, and the principles of quantum *engineering*. The standard evolutionary explanation—that natural selection simply honed this distance over time for maximum efficiency—fails to account for the unforgiving nature of quantum mechanics. Quantum tunneling does not operate on a gentle sliding scale where "closer" is slightly better. Tunneling probabilities drop off exponentially with distance.

If the redox centers were not already positioned at the precise ~14-angstrom distance, the electrons would not merely flow "less efficiently"—they would effectively stop flowing altogether[17]. Evolution cannot select for a system that does not function. Therefore, a mechanism that requires precise quantum positioning to work at all cannot be the result of gradual, trial-and-error adjustments.

But we also need to consider something else. The electron is indeed a particle (when measured) and a wave, but with quantum theory there is, in a sense, a *third way* to conceive of electrons that is inclusive of the wave function, and that is as a probability cloud (or orbital). The electron, with a precisely defined mass, electric charge, and other precisely defined characteristics fits well with the conception of a particle and within this idea it is thought of as orbiting the nucleus as defined by the wave function. Within this definition particles also exist as probability clouds that give us *no better* than a 90% chance of finding the particle within a given cloud. Strangely, the electron has momentum and kinetic energy but no motion. The cloud that a single electron exists within is "perfectly static". Neither completely a particle nor completely a wave. The shift in conceiving of the electron as a probability cloud is a stark departure from the classical view of matter. The existence of the electron is no longer a certainty of position, but a matter of statistics, as the physics demonstrates: "A real electron does not exist in any one place. All it has is a certain probability of being here as opposed to their[18]…"

The consequence of this quantum reality is evident in what we have seen in the electron transport chain. The design of life is not based on classical mechanics; it is fundamentally engineered to leverage the electron's probabilistic and wave-like nature—such as quantum tunneling—to ensure the necessary cellular processes function with speed and fidelity. And in truth, whether biology leverages quantum principles, or whether quantum principles exist to facilitate the needs of biology, is an epistemological question that lies outside of the realm of human-based science.

The implication of this quantum engineering is that the machinery of life is not a classical accident, as materialists claim. The cell is, at its core, operates as a kind of quantum machine. It's incredible efficiency, speed, and fidelity rely on principles that defy classical expectation, such as quantum tunneling. This demonstrates that the system somehow arrived at a configuration that allowed it maximum performance by exploiting probabilities that are only possible at the subatomic scale and could not have come about gradually as per evolutionary theory. This integrated structure—where life utilizes and exploits the fundamental physics of the universe—is significant. It is uncontroversial to point out that in the mitochondria, an extreme fidelity has been expressed by laws of quantum physics and their integration with the design of its machinery to achieve exquisite fidelity.

A Deep Dive Into ATP Synthase

ATP Synthase is a marvel of biological engineering: a true rotary motor operating at the nanoscopic scale. While the Electron Transport Chain creates the potential energy (the proton gradient), ATP Synthase is the turbine that harvests it. Driven by the flow of protons rushing back across the membrane—much like water flowing through a hydroelectric dam—this enzyme spins its internal rotor at incredible speeds. The resulting mechanical rotation forces the structure's upper section through a precise series of shape changes (conformations) that physically crush ADP and phosphate together, manufacturing the essential energy currency of life: ATP.

Structurally, the ATP Synthase machine is composed of two distinct functional units. The bottom section, known as *F0*, is embedded directly within the inner membrane and acts as the rotor. It is driven to spin by the stream of protons flowing through it. The top section, *F1*, extends out of the membrane and acts as the catalytic head; it uses the rotor's rotational energy to mechanically forge ATP molecules.

To maximize efficiency, ATP synthase complexes are arranged in long rows along the sharp ridges of the mitochondria's inner folds, known as cristae. They are typically paired in dimers (angled at 70 degrees) to optimize density. This folded architecture functions like the cells in a battery housing, allowing the mitochondria to pack an immense number of these motors into a microscopic space.

This discovery overturned the long-held assumption that biological structures at this scale were disorganized or scattered. As imaging technology improved, it revealed that the mitochondrial membrane is not a chaotic mix of enzymes, but a strictly ordered lattice:

> "For many years it was assumed that the ATP synthase and other energy-converting complexes are randomly distributed over the inner membrane. The first hint that this is not the case came from deep-etch freeze-fracture electron microscopy, which revealed double rows of macromolecular complexes in the tubular cristae… The double rows were thought to be linear arrays of mitochondrial ATP synthase. This is indeed what they are, but it could only be shown unambiguously more than 20 years later… which revealed rows of ATP synthase dimers in mitochondria of all species investigated…" *-Biochemist and Biophysicist Werner Kuhlbrandt, author of Structure and Function of Mitochondrial Membrane Protein Complexies*[19].

Engineering Feat

The performance specs of this ATP Synthase are humbling. It has been found that the rotor at the base spins at ~100 to 300 revolutions per second, which is approximately 6,000 to 18,000 revolutions per minute, close to three times the speed of a commercial jet engine at cruising altitude. If we assume the rate of 150 rotations per second, would equal 450 ATP molecules every second.

But the true marvel is the scale. This is not a solitary engine; it is a fleet. A single mitochondrion can hold up to 5,000 of these active motors. A single active cell can contain up to 600,000 mitochondria. The human oocyte (egg or ovum) is considered the most mitochondria rich cell with ~100,000 to 600,000. When you do the math, a single human body, with 37 trillion mitochondria, contains quadrillions of these motors, all spinning at jet-engine speeds, seamlessly generating the precise amount of power needed to sustain life. This is nanotechnology operating at an industrial scale and efficiency that human engineering can scarcely dream of matching.

Even more remarkable is the efficiency of this system. While the best human-made hybrid engines achieve roughly 41% energy efficiency, the ATP synthase motor converts nearly 100% of the energy from proton pumping into stored chemical energy. The electrical forces involved are equally staggering. The tiny 5-nanometer membrane sustains a field strength of 30 million volts per meter—equivalent to the raw power of a bolt of lightning.

Hydrogen Powers The Electron Transport Chain (ETC)

Hydrogen is the first element in the periodic chart, and it has special properties because of its configuration of one proton and one electron. This causes hydrogen's electron to be *easily* dissociated from its proton resulting in "free" protons and electrons that can be exploited by the mitochondria. This same quality is what makes hydrogen an excellent source of energy, as in hydrogen fuel cell vehicles. Notably, water in biological systems, whether in animals or plants, provides a great resource for harvesting protons and electrons. This dissociation of hydrogen is the key mechanism that allows the mitochondria to separate the "charge" (the electron) from the "mass" (the proton), handling them separately to drive the entire system:

> "...when a molecule is oxidized, a hydrogen atom removed from it can be readily dissociated into its constituent electron and proton–allowing the electron to be transferred separately to a molecule that accepts electrons, while the proton is passed to the water. Therefore, in a membrane in which electrons are being passed along an electron transport chain, pumping protons from one side of the membrane to another, can be relatively simple[20]."

Hydrogen is the most abundant atom in the universe, consisting of 74% of all atoms. It's the most abundant element in our sun at just over 91%, and there are more hydrogen atoms in the human body than any other atom, making up 62% of our total atoms. But because hydrogen is 16 times *lighter* than oxygen and 12 times lighter than carbon, it is only *the third* largest contributor by weight to the human body. Hydrogen is the primary atom carbon binds to, and is a part of *every* molecule of water, sugar, carbohydrate, lipid, amino acid, protein, DNA and RNA. Hydrogen is, in fact, *the most available* atom in organic chemistry.

The Citric Acid Cycle (CAC)

The last of our three processes within the mitochondria is *The Citric Acid Cycle. The role of the Citric Acid Cycle is to take glucose from the food we eat and to turn it into products that feed the 4 protein complexes of the electron transport chain. The Citric Acid Cycle creates hydrogen atoms for the Electron Transport Chain (ETC) as well as molecules such as *NADH* and *FADH 2* which *deliver* the electrons pulled from hydrogen atoms to the ETC to provide the energy to pump the protons needed for the production of ATP within the ATP synthase.

The Citric Acid Cycle itself takes place in the fluid-filled matrix of the mitochondria. It functions as an 8-step continuous loop, beginning and ending with the same molecule: oxaloacetate. As the cycle turns (twice for every glucose molecule), it generates the electron carriers needed for the ETC and contributes to the total yield of 38 ATP.

Crucially, this cycle is physically fused to the machinery we just discussed. While most of the cycle's enzymes float freely in the matrix, the specific step converting succinate to fumarate is catalyzed by succinate dehydrogenase. As we noted earlier, this enzyme is actually Complex II of the Electron Transport Chain, embedded directly in the inner membrane.

This physical connection acts as a permanent bridge between the cell's metabolic "refinery" (CAC) and its "power plant" (ETC). The fact that a single enzyme serves a dual, essential function in two distinct, complex cycles implies that they did not evolve in isolation. Instead, this structural integration is a hallmark of top-down engineering, suggesting these systems were designed as a unified whole from the very beginning.

The Mitochondria

Our previous investigation into Abiogenesis (Special Problem #2) has already established that complex biological structures cannot arise by chance. This effectively falsifies the hypothesis that the mitochondria could have evolved as an independent structure solely through random mutation inside a cell.

This leaves the alternative hypothesis: Endosymbiosis. This theory conjectures that mitochondria were originally separate organisms that were captured and "enslaved" by a host cell to serve as power plants. However, this scenario faces the same hurdle of complexity. As William Paley's watchmaker analogy illustrates, an intelligent agent can insert complex complications into a mechanism by design. Nature, relying on random processes, cannot simply "embed" highly specified, interdependent living parts into a host cell. The metaphor is fitting reasons we will discover that are particular to the mitochondria, and the structure of the alleged "host" cells it sits within.

The mitochondria do not operate in isolation. Beyond ATP production, it is deeply integrated into the cell's broader network, regulating vital processes ranging from immune signaling to programmed cell death (apoptosis). To function, it must coordinate these activities with a host of other specialized organelles that make up the eukaryotic cell.

This "cellular city" includes the Nucleus (the command center), the Endoplasmic Reticulum (the manufacturing plant), the Golgi Apparatus (the shipping department), and various other specialized structures like Lysosomes and Centrosomes. The sheer complexity of this environment highlights the immense difficulty of the Endosymbiotic Theory: for a foreign bacterium to become a mitochondrion, it would not only have to survive entry but seamlessly interface with every one of these sophisticated, pre-existing systems.

Organelles

Nucleus: the DNA is enclosed within the nuclear envelope of the nucleus.

Smooth Endoplasmic Reticulum: functions in metabolic processes that include synthesis of lipids, metabolism of carbohydrates, detoxification of drugs, and storage of calcium ions

Rough Endoplasmic Reticulum: a place where secretions are made.

Ribosomes: complexes that carry out protein synthesis

Golgi Apparatus: where proteins are received, sorted and shipped within the cell

Lysosomes: enzymes that eukaryotic cells use to digest (hydrolyze) macromolecules

Vacuoles: an integral part of the cells and transport solutes through the cell

Centrosome: the region where microtubules (the girders and cytoskeleton of the cell) originate[21].

Endosymbiotic Theory of Mitochondria

Having introduced Endosymbiosis earlier as the standard explanation for how mitochondria came about, it will be helpful to examine exactly why it is so widely accepted. The theory is not without superficial merit; physically, mitochondria do bear a striking resemblance to bacteria. Proponents of the theory rely on four specific lines of morphological evidence to argue that mitochondria are actually captured prokaryotes.

The modern version, popularized by biologist Lynn Margulis in 1967, relies on several key similarities between mitochondria and bacteria. Textbooks like Campbell Biology argue that these features prove mitochondria were once free-living prokaryotes that were engulfed by a host.

> "How did the eukaryotes evolve from their prokaryotic ancestors? Current evidence indicates that the eukaryotes originated by endosymbiosis when a prokaryotic cell engulfed a small cell that would evolve into an organelle found in all eukaryotes, the mitochondrion. The small, engulfed cell is an example of an endosymbiont, a cell that lives within another cell, called the host cell[22]."

The primary evidence cited includes:

Circular Genome: Mitochondria possess circular DNA, which is characteristic of bacteria.

Reproduction: They replicate independently of the cell via binary fission.

Structure: They are enclosed by a double membrane and are similar in size to bacteria.

Other similarities have been noted

Mitochondria in animals and Chloroplasts in plants, etc., are two structures alleged to be the products of endosymbiosis, are both similar in size to bacterial cells

Mitochondria and chloroplasts have their own ribosomes

Several eukaryotic microbes (Giardia and Trichomonas) have a nuclear membrane but no mitochondria[23]

On the surface, these similarities seem compelling. However, when we look past the morphology and examine the genetics, the theory begins to unravel.

Genetic Problems

While symbiotic relationships are found throughout nature, they invariably involve two distinct organisms retaining their own separate genetic material. There is no known precedent in the animal kingdom for a symbiotic relationship where the "guest" organism surrenders its DNA to be stored and managed within the "host" organism's nucleus. This is what we observe between mitochondria and the cell it inhabits.

The Mitochondria also possess an unusual legacy. Mitochondrial DNA is inherited almost exclusively from the maternal lineage, with extremely rare, exceptional cases of paternal contribution reported[24]; however, the most stunning feature regarding mitochondrial DNA is unique among all other organelles in the animal cell: they have their own genome. Aside from the nucleus itself, the mitochondria is the only structure in the human body to carry independent DNA.

However, this genome is surprisingly limited. It contains the essential codes to manufacture the most complex parts of the machinery directly on-site, but for the vast majority of its operation, the mitochondria rely on the cell's nuclear DNA.

This raises the serious question of how key genetic information, assumed to be from the bacteria-like mitochondria, was transferred to *the human genome* while other parts of the mitochondrial genome were retained by the bacteria-like organism.

There is an even deeper problem than the location of the DNA: the mitochondria and the human nucleus speak different genetic languages. They operate under different coding conventions.

In the standard human genome, the triplet code AUA is the instruction for the amino acid isoleucine, and UGA is a "Stop" command. But in the mitochondria, those same codes mean something completely different: AUA codes for methionine, and UGA codes for tryptophan.

This creates a fatal translation error for the Endosymbiotic Theory. If a gene were transferred from the mitochondria to the host nucleus, the host's machinery would misread the instructions. It would insert the wrong ingredients or stop production entirely, resulting in a broken protein.

If mitochondria had been a free living organism with their own coding convention developed by an entirely different evolutionary path from the human genome, and its DNA was taken up by its host cells nuclear DNA *after it had been captured and enslaved*, then there is no reasonable proposal for how this coding discrepancy could have possibly been reconciled. It would have required that the part of the mitochondrial DNA that was taken up by human DNA, was edited, and reprogrammed to function under an entirely different convention. And the only tools human DNA would have to accomplish this task, would be *random mutation* and the *ex post facto* selection process. That's a tall order to even explain, but it's an even taller order *to solve,* letter-by-letter, and triplet-codon by triplet-code along the entirety of the genome.

And then we must pay attention to the time consideration. All of this reprogramming of the mitochondrial genome into the coding convention of nuclear DNA would have had to happen quickly in order to avoid fatal coding errors. But of course, this makes no sense. If mitochondria could have survived inside of the host cell, then there would have been no reason whatsoever for any of its DNA to have been transferred to nuclear DNA in the first place. There is no evolutionary reason for it, on top of it being an impossible task.

To fully appreciate the coordination nightmare this creates, we must look at exactly what the mitochondria keeps in its own vault. Its genome is surprisingly small—a mere 16,569 base pairs coding for just 37 genes. This limited inventory includes:

13 proteins for:

NADH Dehydrogenase subunits (7)
Cytochrome Oxidase subunits (3)
ATP Synthase subunits (2)
Cytochrome b (1)

and 24 genes that code for RNAs.

This means that the remaining *1,100 genes* for mitochondria are coded in the "host" cell's nuclear DNA. This represents 99% of the mitochondria's *functional* DNA. To reiterate, if the transfer of these 1,100 genes are to be explained by evolutionary theory, the only available process is *random mutation* and *natural selection*. "Random", yet so specific and discerning that the cell's nuclear DNA "recognized" all of the mitochondrial codons that required re-coding *and re-coded them*, while all codons that did not require re-coding were left *unchanged*[25].

According to the Vertebrate Mitochondrial Code (2), mitochondria differ from the standard Nuclear DNA (Standard) code in the following ways:

-UGA is a STOP codon in Nuclear DNA. In Mitochondrial DNA, UGA calls for Tryptophan.
-AUA calls for Isoleucine in Nuclear DNA. In Mitochondrial DNA, AUA calls for Methionine.
-AGA calls for Arginine in Nuclear DNA. In Mitochondrial DNA, AGA calls for a STOP codon.

-AGG calls for Arginine in Nuclear DNA. In Mitochondrial DNA, AGG calls for a STOP Codon.
-START codons: Nuclear DNA primarily relies on AUG for a START Codon. Mitochondrial DNA can START with AUA, AUU, AUG and GUG.
-STOP codons: Nuclear DNA uses three STOP codons: UAA, UAG, UGA. Mitochondrial DNA uses only UAA and UAG.

All other codons are identical[26].

This is not a fringe concern; it is a fundamental flaw recognized by researchers who look beyond the superficial similarities. In a comprehensive critique, molecular biologist and researcher Albert D.G. de Roos challenges the scientific consensus, noting that while the theory describes what happened, it completely fails to explain how. He asks the crucial question: by what specific mechanism could such a complex genetic merger actually occur?

"The endosymbiotic theory for the origin of mitochondria claims that our mitochondria were derived from an engulfed bacterium that was enslaved to become the current powerhouse of most eukaryotic cells. This endosymbiotic theory has become consensus among evolutionary biologists to such an extent that it is considered a fact and forms the basis for most research on molecular evolution. I argue that the mechanistic basis of the endosymbiotic theory is not sound and even contradictory to our current concepts of evolution." -Albert D.G. de Roos[27]

And

> "All evolutionary theories must offer an explanation in mechanistic terms of how it should or could have happened in order to be tested. The difficult thing with the endosymbiotic theory is that it proposes no real mechanism… There is a difference between the process of endosymbiosis and its incorporation in the germline, necessitating genetic changes. What were those changes? What was the host? Was it a fusion, was it engulfment, how did the mitochondrion get its second membrane, how did two genomes in one cell integrate and coordinate? The theory is also strongly teleological, illustrated by the widely used term 'enslavement'. But how do you enslave another cell, how do you replace its proteins and genes without affecting existing functions?"
>
> -Albert D.G. de Roos[28]

The cell, on its own scale, is not dissimilar in complexity from the human body. The body is composed of proteins and cells; the cell is composed of molecules and atoms. To subsume a foreign organism, and develop, within seconds, a system for subsuming, rather than digesting, while also calling off the immune defense response for an attacking the foreign invader, would represent a wholesale reprogramming and rewiring of multiple innate cell responses. This would also require the addition of several other systems that prior to the enslavement event were without precedence.

> "Genetic mutations that allow bacteria to thrive in the cytoplasm would not be strategic for survival. Anaerobic cells normally do not survive in an environment that contains oxygen, while the endosymbiont would need oxygen in order to present fitness advantage. The two organisms would initially compete for energy sources since bacteria are users of ATP and do not export it. The extensive gene transfer that is needed in the endosymbiotic theory would wreak havoc in a complex genome since frequent insertion of random pieces of mitochondrial DNA would disrupt existing functions… Then we must speak of the wholesale retrofitting of the host cell's prior energy generation/utilization process to the responsibility of this new mitochondrial invader. Not just one, but multiple intermediate/transitional responses would be required. How it would have done… is entirely unclear."
>
> -Albert D.G. de Roos[29]

A recent study of mitochondrial DNA by biochemist and microbiologist Dr. Michael W. Gray finds that there is insufficient evidence in the DNA of mitochondria to indicate a bacterial origin.

> "What was not anticipated was how relatively few mitochondrial proteins with bacterial homologs would group specifically with α-Proteobacteria in phylogenetic reconstructions: At most, only 10–20% of any of the mitochondrial proteomes examined so far display a robust α-proteobacterial signal."
>
> -Dr. Michael W. Gray[30]

Gray also found,

> "The sequences of many eukaryotic genomes are now clearly showing that the gene repertoires needed for the mitochondria to function are not derived from bacteria but are remarkably unique to the type of creature in which they are found. While certain genetic similarities do exist, these correspondences are plausibly explained by the standard engineering concept of code reuse—common code to solve similar problems[31]."

Gray's analysis highlights the fatal flaw in the evolutionary narrative. If the mitochondrial gene repertoire is unique to the specific creature it inhabits—rather than being a generic bacterial remnant common to all—then the "captured slave" hypothesis is genetically untenable.

When we combine this with the insurmountable language barrier of differing genetic codes, the verdict is clear: the mitochondria was not an external conquest. It appears to be an intentional, integrated component, designed with specific "code reuse" principles to function seamlessly within the distinct architecture of each unique species.

Problems in The Tree of Life

The certainty often projected by evolutionary theory stands in stark contrast to the reality of the data. Julian Gough, principal architect of the SUPERFAMILY genomic database, candidly admits that in the deep history of life, verification is nearly impossible, leaving room for bias to replace proof:

> "In the field of studying ancient evolution, it's very hard to falsify or verify. As such people tend to attach unreasonable weight to their beliefs of one possibility or another."
>
> -Julian Gough, Principal architect of the open-access database SUPERFAMILY[32].

To understand the depth of the problem of "ancient evolution," we must look to the fundamental map of evolutionary biology: The Tree of Life. This conceptual model attempts to trace the lineage of all living things back to a single root—the Universal Common Ancestor. By comparing the genetic sequences of different species, scientists build a vast "family tree" to determine which organisms are related and when they diverged.

If the Endosymbiotic Theory were true, the genetic map should show a clear branch where a bacteria merged with a host cell. However, a comprehensive study by Charles Kurland and Ajith Harish titled *The Phylogenomic Tree of Life* found something entirely different[33]. Their analysis found that the four assumptions of the gene tree culture (evolutionary in basis) are *all unsupported* by the evidence.

A. That protein-coding sequences are molecular fossils

B. That gene trees are equivalent to species trees

C. That the tree of life is assumed to be rooted in a simple akaryote cell implying that akaryotes are primitive; and,

D. The notion that all of most incongruities between alignment-based gene trees are due to horizontal gene transfer (HGT), which includes the endosymbiotic models postulated for the origins of eukaryotes[34]."

In layman's terms, they challenged the dogmatic belief that genes serve as perfect historical records, or that tracing the history of a single gene is the same as tracing the history of an entire species.

Most importantly, they refuted the standard practice of "fixing" the data when it doesn't fit the theory. When genetic lines don't match the evolutionary tree, scientists often blame "Horizontal Gene Transfer"—essentially claiming the genes hopped between species—rather than admitting the tree itself might be wrong. Kurland and Harish found that when you strip away these unproven excuses, the genetic data no longer supports the standard evolutionary narrative.

The research findings of Kurland and Harish, and M. Gray are controversial for those that *believe* in endosymbiosis, as they identify evidence of similarity between the mitochondria and bacteria as n*ot* being evidence of the bacterial origins of mitochondria. Transfer of genetic material from the alleged bacterial genome of the mitochondria is *not* supported by the evidence. Similarity does not constitute evidence, nor does it constitute causality.

Kurland and Harish published another paper two years later in 2017 under the unambiguous title, *Mitochondria are not Captive Bacteria*, where they once again affirmed that the mitochondrial tree is rooted in the (Last) Universal Common Ancestor (LUCA) of the Tree of Life, and that descent of mitochondria is *autogenic* (self-generated / self-produced) and not *endosymbiotic.*

> "A systematic analysis of the origins of the mitochondrial proteome based on empirical genome evolution models now indicates that 97% of modern mitochondrial protein domains as well as their homologues in bacteria and archaea were present in the UCA (universal common ancestor) of the modern ToL (tree of life)[35]."

This finding upends the standard evolutionary timeline. If 97% of the essential protein domains for mitochondria were already present in the Universal Common Ancestor, then the mitochondria could not have been a foreign bacterium "captured" eons later. The data suggests that the blueprint for the mitochondria was not acquired from the outside but was inherent in the very first life form. This supports the autogenic (self-generated) model: the mitochondria evolved from within the cell, as an indigenous component, rather than being enslaved from without.

Another Evolutionary Problem

If evolution cannot explain how mitochondria could have been enslaved, and how their genome could be transferred to the host cell's genome, despite a coding convention conflict, then we have revealed another problem. If we instead consider the hypothesis that mitochondria are indigenous to the cell, then we are still left with the problem of how it developed in such a way that it houses some of its own DNA, while the bulk is housed in nuclear DNA under an entirely different coding convention. There is no reasonable evolutionary explanation for what we find in mitochondria, through either endosymbiotic theory, or evolutionary theory.

Summary

We do not need to generate a new calculation to understand the odds of the mitochondria. Whether viewed as an indigenous structure or an endosymbiont, its existence relies on a foundation that we have already proven to be statistically impossible.

We face a cascade of three core problems, each of which is insurmountable on its own:

The spontaneous creation of the "host" cell.

The spontaneous creation of the mitochondria.

The seamless integration of their incompatible genetic and metabolic systems.

Our investigation into Abiogenesis (Special Problem #2) established that the odds of one living cell arising by chance are effectively zero. Therefore, the odds of two distinct cellular entities arising and then successfully merging to form a eukaryotic cell are not just impossible; they are the square of impossible. There is no known physical, random, or necessitous process that can explain this integration.

Summary and Value of Variable 12:

-Odds for OOL, or any life, including the organelle mitochondria, or the human cell: Exceeds the PCOU of 1 in 10^{140}: *fM* for Functional Miracle

-Odds of the mitochondria being integrated within the cell: Exceeds the PCOU of 1 in 10^{140}: *fM* for Functional Miracle

Variable 12 Exceeds the PCOU of 1 in 10^{140}: *fM* for Functional Miracle

Epilogue: Life, Consciousness & The Cell

Our investigation of the mitochondria leaves us with a deeper question. We have explained the power plant of the cell, but what of the entity that operates it? Beyond the question of how the cell is powered lies the ultimate mystery: what is the force of life that animates the operation of the cell?

Surprisingly, biology has no agreed-upon definition for "life" itself. It substitutes a description of what life *does* for an explanation of what life *is*.

We often conflate life with DNA, assuming the code is the coordinator. But DNA is inert. It is a library, not a librarian. It holds the plans, but it performs no actions; everything DNA is said to "do" is executed by molecular machines that read, interpret, and implement those instructions. And by logic, it appears the cell must be coordinated by an organizing power irreducible to physics or chemistry.

Regulation, timing, error-correction, and functional integration occur in ways that are not reducible to physics or chemistry alone. Somewhere between dead atoms and living protein, a threshold is crossed. A spark jumps and this crossing implies the presence of something irreducible: an organizing agency that instantiates lifeless matter into a living system. For lack of a better term, we may call this a form of consciousness—not consciousness as human introspection, but as primordial awareness with agency. The presence of its qualities within molecular material is perhaps the missing definition of what we call life, and that this, not DNA itself, is the true engine of biology.

We will return to this profound mystery in our final investigation: Variable #15, Cell Consciousness

---

[1] (L. Veloz, "How Are Photosynthesis & Cellular Respiration Related?" sciencing.com, https://sciencing.com/how-are-photosynthesis-cellular-respiration-related-12226137.html. 30 September 2023)1
[2] (M. Denton, Evolution: A Theory in Crisis, 1986, p. 329)2.
[3] (M. Denton, Evolution: Still A Theory in Crisis, 2016, Discovery Institute Press).
[4] (The Nobel Prize in Physiology or Medicine 1953, NobelPrize.org, Nobel Prize Outreach, 2026, Sun 4 Jan 2026)
[5] (The Nobel Prize in Chemistry 1978. NobelPrize.org. Nobel Prize Outreach 2026. Sun. 4 Jan 2026.)
[6] (M. Denton, The Miracle of the Cell, Pg. 82, Discovery Institute Press, 2020) (references cited within quote: (P. Mitchell, Coupling of Phosphorylation to Electron and Hydrogen Transfer by a Chemi-osmatic Type of Mechanism, Nature 191,no.4784(1961):144-148),
[7] (M. Denton, Miracle of the Cell, pg. 74-89, 2020, Discovery Institute Press)
[8] (M. Denton, Miracle of the Cell, pg. 74-89, 2020, Discovery Institute Press
[9] (M. Denton Miracle of the Cell. Pg. 86-89 Discovery Institute Press, 2020)4.
[10] M. Denton, Miracle of the Cell, pg. 74-89, 2020, Discovery Institute Press)
[11] (M. Denton, Miracle of the Cell, pg. 74-89, 2020, Discovery Institute Press)
[12] (M. Denton, Miracle of the Cell, pg. 88 Discovery Institute Press)
[13] (C.C. Page, C.C. Moser, X. Chen, P.L. Dutton, Natural Engineering Principles of Electron Tunneling in Biological Oxidation-Reduction, 1999, Nature, 402(6757):46-52).
[14] (M. Denton, Miracle of the Cell, pg. 74-89, 2020, Discovery Institute Press)

[15] (C.C. Page, C.C. Moser, X. Chen, P.L. Dutton, Natural Engineering Principles of Electron Tunneling in Biological Oxidation-Reduction, 1999, Nature, 402(6757):46-52),
[16] (M. Denton, Miracle of the Cell, pg. 74-89, 2020, Discovery Institute Press)
[17] (C.C. Page, C.C. Moser, X. Chen, P.L. Dutton, Natural Engineering Principles of Electron Tunneling in Biological Oxidation-Reduction, 1999, Nature, 402(6757):46-52).
[18] (Northwestern University, WCAS, Plate 1: What is an Electron? [online, faculty.wcas.northwestern.edu/ The 20 Website Plates])6.
[19] (W. Kühlbrandt, Structure and Function of Mitochondrial Membrane Protein Complexes., BMC Biol 13,89 (2015))7
[20] (B. Alberts, A. Johnson, J. Lewis, M. Raff, K. Roberts, P. Walter, Molecular Biology of the Cell, Pg.782, Fourth Edition, Garland Science, March 1, 2002)10
[21] (L.A. Urry, M.L. Cain, S.A. Wasserman, P.V. Minorsky, J.B. Reece, N.A. Campbell, Campbell Biology, Pg. 100-114 Series, Pearson, 11th ed, October 19, 2016)
[22] L.A. Urry, M.L. Cain, S.A. Wasserman, P.V. Minorsky, J.B. Reece, N.A. Campbell, Campbell Biology, Pg. 532 Series, Pearson, 11th ed, October 19, 2016)15
[23] (J. McLatchie, On the Origin of Mitochondria: Reasons for Skepticism on the Endosymbiotic Story, Evolutionnews.org, January 10, 2012); (L. Sagan, On the Origin of Mitosing Cells, 1967, Journal of Theoretical Biology, Volume 14, Issue 3, March 1967, Pgs 225-274, IN1-IN6);(ScienceFacts, The Endosymbiotic Theory, sciencefacts(dot) net/ the-endosymbiotic- theory(dot) html);
[24] (S.Luo, C.A. Valencia, J. Zhang, N.-C Lee, S. Dell, J. Brown, S.-M Chen, Y.-H Chien, W.-L Hwu, P.-C Fan, L.-J.C. Wong, P.S. Atwal, Biparental Inheritance of Mitochondrial DNA in Humans, 2018, Proceedings of teh National Academy of Sciences of the USA, 115(51), 13039-13044)
[25] (C. Takemoto, L.L. Spremulli, L.A. Benkowski, T. Ueda, T. Yokogawa, K. Watanabe. Unconventional Decoding of the AUA Codon as Methionine by Mitochondrial tRNAMet with the Anticodon f5CAU as Revealed with a Mitochondrial in vitro Translation System. Nucleic Acids Res. 2009 Apr;37(5):1616-27. doi: 10.1093/nar/gkp001. Epub 2009 Jan 16. PMID: 19151083); (F. Yamao, A. Muto, Y. Kawauchi, M. Iwami, S. Iwagami, Y. Azumi, S. Osawa, UGA is Read as Tryptophan in Mycoplasma Capricolum. Proc Natl Acad Sci U S A. 1985 Apr;82(8):2306-9. doi: 10.1073/pnas.82.8.2306. PMID: 3887399)
[26] (Osawa, Ohama, Jukes & Watanabe 1989), (Temperley et al. 2010), (A. Elzanowski, J. Ostell, The Genetic Codes, National Center for Biotechnology Information, NCBI, update: January 7, 2019, https:// www.ncbi nlm.nih. gov/ taxonomy/ Utils/ wprintgc. cgi)22
[27] (A.D.G deRoos, A Critique on the Endosymbiotic Theory for the Origin of Mitochondria, October 2007 DOI: 10.13140/RG.2.2.11811.76326))
[28] (A.D.G deRoos, A Critique on the Endosymbiotic Theory for the Origin of Mitochondria, October 2007 DOI: 10.13140/RG.2.2.11811.76326))
[29] (A.D.G deRoos, A Critique on the Endosymbiotic Theory for the Origin of Mitochondria, October 2007 DOI: 10.13140/RG.2.2.11811.76326))
[30] (M.W. Gray, Mosaic Nature of the Mitochondrial Proteome: Implications for the Origin and Evolution of Mitochondria Proc Natl Acad Sci USA, August 2015, 18;112(33):10133-8. PMID 25848019)24
[31] (M.W. Gray, Mosaic Nature of the Mitochondrial Proteome: Implications for the Origin and Evolution of Mitochondria Proc Natl Acad Sci USA, August 2015, 18;112(33):10133-8. PMID 25848019)25

[32] (Special thanks to Suzan Mazur, Contributor, Journalist and author of the article, DEEP EVOLUTION and its Discontents, March 17, 2017)26.

[33] (C. Kurland, A. Harish, The Phylogenomics of Protein Structures: The Backstory., in Biochimie 119 (Online 31 July 2015). p.284-302)27.
[34] (Above Ibid, 'Abstract')28.
[35] (A. Harish, C.G. Kurland, Mitochondria are not Captive Bacteria, Journal of Theoretical Biology, Vol 434, December 2017, Pgs 88-98)29

Variable 13: Chirality & Organic Molecules

The issue of chirality is a major complicating factor along the pathway to the first living cell. We touched upon chirality in *Special Problem #2: The Beginning of Life on Earth (Abiogenesis).* Chirality exists because complex molecules in biology can be arranged in two distinct, non-superimposable forms: a "left-handed" version and a "right-handed" version.

Scientists use two different naming conventions to describe this. In biology, we typically use the L-form (Left) and D-form (Right) system. In chemistry, a more modern system labels them as S (from the Latin sinister, meaning left) and R (from rectus, meaning right). Regardless of the label, the concept is the same: these molecules are mirror images that cannot be superimposed on one another.

In the foreword to the book *The Mystery of Life's Origin*, chirality is briefly introduced as a serious problem in retracing the path for OOL. "Another intractable problem [with origin of life research] concerns the spontaneous origin of the optical isomer preferences found universally in living matter (e.g., L-rather than D-amino acids in proteins, D-rather than L-sugars in nucleic acids). After all the prodigious effort that has gone into attempts to solve this great question over the years, we are no nearer to a solution today than we were thirty years ago[1]."This book was written in 1984 and recently updated in 2020. With the addition of nearly 4 decades of effort, the question remains intractable.

The most important aspect of chirality for our discussion is the revelation that in any scenario where molecules are produced by natural means, *they will be produced in equal amounts* in their left and right-handed forms, including in the prebiotic (pre-life) conditions of early Earth. This fact brings up an almost unthinkable problem for OOL. This requirement for purity is absolute in biology. Currently, living systems use exclusively homochiral molecules: L-amino acids are used to build proteins, while D-sugars form the backbone of DNA and RNA.

The five nucleobases that comprise the primary genetic alphabet rely entirely on this specific orientation. In DNA, these are Adenine, Cytosine, Guanine, and Thymine (ACGT). In RNA, Uracil (U) replaces Thymine. Without the correct "handedness" in the sugar backbone, these bases cannot align to form the double helix, and the genetic code becomes unreadable[2].

Given how essential this molecular purity is to the very architecture of life, one might assume that science has long ago solved the riddle of its origin. However, despite decades of intense research, the mechanism for selecting one "hand" over the other remains a profound mystery. As a recent report from The Scripps Research Institute candidly admits:

> "'Origin of life' chemistry has been a busy field for much of the past century. Its practitioners have discovered dozens of key reactions that plausibly occurred on the early, "prebiotic" Earth to produce the first DNAs, RNAs, sugars, amino acids and other molecules that sustain life. Missing from this body of work, however, has been a plausible prebiotic theory for the emergence of homochirality[3]."

This admission is significant. It acknowledges that while we can simulate individual chemical reactions in the lab, we still lack a coherent explanation for how nature could have spontaneously filtered a messy, 50/50 mixture into the perfect, 100% purity required for OOL to function. In practice, laboratory access to homochiral molecules generally relies on starting from chiral biological materials or on introducing constraints that pre-exist the reaction itself such as introducing catalysts, auxiliaries, or starting with enantiomerically pure material (chiral molecules).

To address this stunning gap there are theories that life began with the inclusion of both chiral forms of molecules and then converted to a homochiral state. There are several proposed methods for this conversion. None are satisfactory, nor do they have the capacity to achieve the required ends homochirality. In some conditions, it can be imagined that one form of the molecule is preferred over the other. But these methods do not establish a pathway to homochirality.

The problems in achieving homochirality are non-trivial and are far more complex than a simple process can achieve:

> "There are $\mathbf{2}^N$ possible stereoisomers where n is the number of stereocenters in the molecule. If discussing carbohydrates, there are eight possibilities among the abundant 5-carbon carbohydrates and sixteen possibilities from the 6-carbon-long carbohydrates. Claims that these structures could be prepared under prebiotic conditions in high enantiomeric purity using inorganic templates, or any presumed templates, have never been realized even with the advanced designs of the origin-of-life researchers." -Biologist James Tour[4]

Tour's calculation exposes the exponential nature of the problem. As biological molecules grow more complex, the number of possible "wrong" shapes explodes. For a simple 6-carbon sugar, there are sixteen possible configurations, yet life requires exactly one.

This means that a prebiotic environment wouldn't just be a 50/50 mix; it would be a chaotic soup of dozens of non-functional isomers for every useful one. Without a guiding intelligence to select the correct shape at every single step, the probability of assembling even a simple carbohydrate chain drops precipitously toward zero. "Prebiotic syntheses invariably generate very heterogeneous solutions of organic compounds. This makes it very difficult to imagine how ordered linear polymers, made of only the canonical monomers, amino acids, or nucleotides could ever have been assembled[5]."

This principle applies regardless of complexity. Whether we are building a simple amino acid with two forms or a complex sugar with sixteen, the underlying thermodynamic problem is the same. At every single point where a molecule can twist left or right (each stereocenter), both options are at the exact same energy level. Neither is "easier" to make. Therefore, nature has no preference at any individual step. It effectively flips a coin at every junction, inevitably producing a chaotic mix of every possible variation rather than the single, specific shape life requires. Except, for a reason that remains unknown, in the origin of life.

> "When chiral molecules are synthesized in the laboratory from achiral building blocks, equal amounts of L- and D-enantiomers are produced unless painstaking care is taken to introduce an asymmetric agent or selective process during the synthesis[6]."

There is no satisfactory scientific resolution to this paradox.

When engaging in the rigorous purification procedures required to separate these molecules to arrive at a homochiral set, achieving a usable end result comes at a high price. Because the "wrong" half of the mixture must be discarded at every step, these methods drastically reduce the yield. And a lab book must be kept documenting each step so that they can be repeated with fidelity when needed. Furthermore, despite this significant loss of material and effort, at each step, the end result often fails to achieve a 100% optically pure batch.

The conditions found in a prebiotic environment, where there are no selection mechanisms, make any expectation for life to arise completely absurd.

The Geometry of Carbon

The root of this problem lies in the geometry of the carbon atom itself. Because carbon forms bonds in a tetrahedral shape, whenever it attaches to four different groups, it creates a chiral center[7]. Since this "asymmetric carbon" is the backbone of amino acids and sugars, chirality is baked into the very chemistry of life.

This creates a unique challenge for separation. Because mirror-image molecules have identical physical properties (boiling point, density, freezing point), they cannot be separated by simple natural forces like evaporation or gravity. Without a pre-existing chiral agent to pick them apart, they remain locked in a stubborn 50/50 mix.

> "Because the physical properties of enantiomers are identical, they seldom can be separated by simple physical methods, such as fractional crystallization or distillation. It is only under the influence of another chiral substance that enantiomers behave differently, and almost all methods of resolution of enantiomers are based upon this fact[8]."

The Stakes: From Scent to Toxicity

While these molecules look identical on paper, their effect on biological systems is profoundly different. We can sense this directly: the molecule carvone smells like spearmint in its R-form, but like caraway seeds (rye bread) in its L-form. Our chiral biological receptors easily distinguish between the two[9].

However, the consequences can be far more severe than a difference in smell. The tragedy of the drug Thalidomide in the 1950s illustrated the lethal power of chirality. One hand of the molecule was a safe sedative; its mirror image caused devastating birth defects. Because the drug was synthesized in a lab as a racemic mixture, the presence of the "wrong" hand turned a medicine into a poison[10].

This same critical issue extends to the very blueprint of life. DNA and RNA must be homochiral to function. A strand that is even 99% pure is not sufficient; a single wrong-handed sugar molecule destroys the stability of the sugar-phosphate backbone, preventing the double helix from twisting correctly.

We can see the consequences of this in the phenomenon of Z-DNA. Unlike "regular" right-handed DNA (B-DNA), Z-DNA contains left-handed architecture that forces the helix to twist in the opposite, "zigzag" direction. As biochemist Robert Wells has documented, these structural anomalies are not harmless quirks; they are linked to approximately 20 neurological diseases and 50 genomic disorders.

In the code of life, geometry is destiny; the wrong shape leads inevitably to disease or death

The Enzyme Paradox

Modern life solves this problem with exquisite precision. Our cells use sophisticated enzymes to select and manufacture only the correct homochiral molecules. But this answer only deepens the mystery of origins. Enzymes are proteins—complex chains that must be homochiral to exist.

This creates a circular paradox: Life requires homochiral enzymes to create homochirality. The system cannot start without the very purity that the system is designed to produce.

How life struck upon this convention is unknown, but there is no known natural way for it to have built a homochiral system, from scratch in violation of entropy and the rules of natural chemical production[11]. In the last few years, we are beginning to understand why chirality in living molecules may be an indispensable design feature that holds one of many keys to the existence of life, a characteristic shared by all Variables and Special Problems.

Chiral Induced Spin Selectivity

As it turns out, recent understanding of the effect of homochirality on life has revealed a previously unknown efficiency afforded by homochiral biological systems. This efficiency, known as chiral induced spin selectivity effect, or CISS, may function as a life imperative. Although our understanding is in the early stages, it is increasingly becoming evident that without the advent of homochirality, life, complex life, and intelligent observers may not be possible at all.

To understand the mechanism behind this imperative, we must look at the fundamental interaction between molecular shape and subatomic particles. As a recent chemical review explains, this effect links the geometry of the molecule directly to the quantum behavior of the electron. Research pioneered by professors Ron Naaman and David Waldeck have revealed a stunning connection between the shape of a molecule and the behavior of electrons. This phenomenon, known as Chiral Induced Spin Selectivity (CISS), demonstrates that chiral molecules function as quantum filters:

> "…One might ask, 'Why is chirality, as such, preserved so persistently throughout evolution?' or 'What makes chiral symmetry so important to life?' Answers to these questions could involve the relationship between chirality and the electron spin, which manifests as the chiral induced spin-selectivity effect (CISS). CISS refers to the connection between chiral symmetry and electron spin in molecules and materials, and it can manifest for electron transmission and for electron displacement currents[12]."

In less technical terms, electrons possess a quantum property known as "spin"—they can be "spin-up" or "spin-down." In a non-chiral material, electrons of both spins pass through with equal ease. However, Naaman and Waldeck discovered that chiral molecules act like a specific turnstile.

They preferentially allow electrons of one specific spin state to pass through, while blocking the other.

The Thermodynamic Advantage

Why does this matter for life? As chemist James Tour explains, this spin-filtering capability is a gamechanger for energy efficiency[13]. When electrons are disorganized (random spins), they scatter as they move, creating resistance and heat. But when electrons are organized by spin (via CISS), they move through the molecule with significantly less resistance[14]. Tour notes that the transmission probability for the "correct" spin can be a hundred times higher than for the wrong one. This drastically reduces the amount of heat released during electron transfer[15]. For a cell packed with millions of spinning motors and flowing currents (like the mitochondria), this heat reduction is not just a luxury; it may likely be a requirement for survival.

The Implication for Origins

This discovery raises the stakes for abiogenesis significantly. It implies that homochirality is not merely a structural preference; it is an energetic necessity. Life requires these specific shapes to manage energy flows efficiently enough to survive[16].

A Calculation of Impossibility

This requirement effectively squares the already impossible odds of abiogenesis. When Fred Hoyle famously calculated the probability of life arising by chance (1 in $\mathbf{10^{40,000}}$) (Fred Hoyle and N.C. Wickramasinghe, Evolution From Space, pg 130, 147, Simon & Schuster, 1984), a number that in the present age has been raised several orders of magnitude as we have learned more about the operation of living cells, he did not factor in chirality. If we add the requirement that every single one of the billions of molecules in a cell must also be the correct "hand" to allow for CISS, the improbability pushes past the boundaries of scientific notation into the realm of the absolute impossible.

Value of Variable 14, Chirality & Organic Molecules:

If we apply the exponential complicating factor of chirality to the existing exponent of the Hoyle/Wickramasinghe odds of life arising unaided, (1 chance in $\mathbf{10^{40,000}}$)[17], or to the updated Harold Morowitz odds of life arising unaided (1 chance in $\mathbf{10^{100,000,000,000}}$)[18] we *double* the exponent of chance because at each position of the genome of life, and each step of the sequence hypothesis, we must now have a chiral molecule.

We don't need to perform this calculation to know these odds far exceed the probabilistic capacity of our universe of 1 chance in $\mathbf{10^{140}}$.

Value of Variable 14, Chirality & Organic Molecules:

Functional Miracle fM

---

[1] (C.B. Thaxton, W.L. Bradley, R.L. Olsen, The Mystery of Life's Origin, Forward, vi, 1984, Philosophical Library, New York)
[2] (J. Tour, We're Still Clueless About The Origin of Life, essay, in C. Thaxton, W.L. Bradley, R.L. Olsen, J. Tour, S. Meyer, J. Wells, G. Gonzalez, B. Miller, The Mystery of Life's Origin: The Continuing Controversy, Pg 325, 2020, The Discovery Press)
[3] (The Scripps Research Institute, Editors Notes, How Molecular 'Handedness' Emerged In Early Biology, Phys(dot) org, February 28, 2024, phys(dot) org/ news/ 204-02 -molecular-handedness- emerged-early- biology(dot) html)
[4] (J. Tour, We're Still Clueless About The Origin of Life, essay, in C. Thaxton, W.L. Bradley, R.L. Olsen, J. Tour, S. Meyer, J. Wells, G. Gonzalez, B. Miller, The Mystery of Life's Origin: The Continuing Controversy, Pg 329, 2020, The Discovery Press)
[5] (M. Denton, Evolution: Still a Theory in Crisis, p. 121, Discovery Institute Press, 2016)20
[6] (R.A. Hegstrom, D.K. Kondepudi, The Handedness of the Universe, Scientific American, Vol262, No1, 1990, pg.108-115)1.
[7] (K.A. Urry, M.L. Cain, S.A. Wasserman, P.V. Minorsky, J.B. Reece, Campbell Biology, 11th Edition, Pearson, 2017)
[8] (https: / / courses. Lumenlearning lcom / suny-potsdam organicchemistry / chapters / 4-6-separating-enantiomers /).
[9] (J. Bloom, Carvone: One Molecule, Two Different Scents And Flavors, American Council on Science and Health, August 1st, 2018, acsh(dot) org / news / 2018 / 08 / 01 / carvone-one molecule-two-different -scents-and- flavors-13255)
[10] Newcastle University. (2016, February 8). Nature's mirror: The Code for Chirality. ScienceDaily. Retrieved October 4, 2023, from www. Sciencedaily .com / releases / 2016 / 02 / 160208124237)19 from research article: (W. Xiao, K.H. Ernst, K. Palotas, Y. Zhang, E. Bruyer, L. Peng, T. Greber, W.A. Hofer, L.T. Scott, R. Fasel, Microscopic Origin of Chiral Shape Induction in Achiral Crystals. Nature Chemistry, 2016; DOI 10.1038/NCHEM.2449)20
[11] (J.R. Cronin, S. Pizzarello Enantiomeric Excesses in Meteoritic Amino Acids. Science. 1997 Feb 14;275(5302):951-5. PMID: 9020072)
[12] (B.P. Bloom, Y. Paltiel, R. Naaman, D.H. Waldeck, Chiral Induced Spin Selectivity, Chemical Reviews, 2024, 124(4), 1950-1991, DOI: 10.1021/acs.chemrev.3c00661)
[13] (Professor, Dr. James.Tour, Inference Online, Vol.1, No.4/Dec.2016, Chiral Induced Spin Selectivity, https: / / inference-review.com / article / chiral-induced-spin-selectivity).
[14] (R. Naaman, Y. Paltiel, D.H. Waldeck, Chiral Induced Spin Selectivity and Its Implications for Biological Functions. Annu Rev Biophys. 2022 May 9; 51:99-114. PMID: 34932912.)
[15] (Professor, Dr. James.Tour, Inference Online, Vol.1, No.4/Dec.2016, Chiral Induced Spin Selectivity, https: / / inference-review.com / article / chiral-induced-spin-selectivity).

[16] (K. Michaeli, N. Kantor-Uriel, R. Naaman, D. Waldeck, The Electron's Spin and Molecular Chirality-How Are They Related and How Do They Affect Life Processes? Chem. Soc. Rev. 2016, 45, 6478-6487.)
[17] (Fred Hoyle and N.C. Wickramasinghe, Evolution From Space, pg 130, 147, Simon & Schuster, 1984),
[18] (M. Eastman, C. Missler, The Creator Beyond Time and Space, Costa Mesa, CA: TWFT, 1996, p. 61)

Special Problem No. 3: The Evolution of the Species

Our investigation thus far has dismantled the materialist explanation for the origin of life. We have seen that the cell is a marvel of irreducible complexity that physics and chemistry cannot produce by chance. But the naturalist narrative relies on a second, equally critical pillar: evolution.

The claim is that once the first spark of life existed, the blind forces of random mutation and natural selection were sufficient to build the entire kingdom of life, from simple bacteria to consciousness. In this section, we will apply the same rigorous mathematical and biological scrutiny to this mechanism. We will ask whether this process truly possesses the creative power attributed to it, or if "survival of the fittest" proves too thin a mechanism—an antiquated scaffold, first proposed in the 1850s, and retrofitted onto a reality far richer, more intricate, and no less resistant to explanation than what we've investigated thus far.

To evaluate whether evolution possesses genuine explanatory power, we must retrace the intellectual path that led to its adoption. Only by understanding the historical and philosophical pressures that shaped Darwin's theory can we measure the mechanism itself.

Evolution & The Divine Algorithm

In *The Divine Algorithm*, evolution is our 3rd of 4 Special Problems, and the 16th of the 19 parts in *The Algorithm*. But in the chronology of human history, evolution is the *first* instance where science recognized an unresolvable anomaly and conspired to remove it by declaring a substitute materialistic explanation western science calls *evolution*.

Darwinian evolution holds a central place in Western thought. It remains the dominant framework in modern biology, influencing many scientific fields beyond biology itself. Because of this status the concept of evolution warrants examination not only as a scientific theory, but as a historical and cultural framework.

In the beginning its authority was not immediate. It has slowly developed and become entrenched over time. Tracing its emergence—from Darwin's early formation and the voyage of the Beagle to the theory's subsequent reception, the cultural forces at play which it was proposed, the competing narratives, and its derision, defense, and eventual acceptance are all important milestones in its consolidation as a biological axiom.

This is an understanding we are obligated to explore if we are to understand the essential context for both its scientific claims and its enduring influence.

The History of Evolution

We begin with the scientific landscape Darwin inherited—a world that still viewed species as fixed because the best available evidence pointed in that direction.

Prior to the release of *On the Origin of Species*, and prior to his voyage on *The Beagle*, Charles Darwin accepted the predominant view that species were *immutable* and were specially (individually) created. This means once a species was created, it remained on Earth unchanged until it went into extinction. This view extends as far back as Aristotle. In Darwin's age there was no conclusive evidence that species evolved. This was the basis for the scientific belief in the fixity of species: it did not stem from religious fervor, or old testament faith, as it has been often portrayed. The key piece of evidence that supported special creation is the same evidence that excluded evolution, and that was the fossil record. Although the fossil record of the time was far less than at present, it nonetheless provided no *intermediary forms* between species that spoke of a slow, steady evolutionary process. This record remains unblemished to this day. Animals instead exhibited a repeated pattern of abruptly appearing in the fossil record, remaining *fixed in form* for long stretches of time, and then, just as abruptly, going into extinction with no evident examples of species that could have *evolved from* these fixed-in-form species, or that these fixed-in-form species could have *themselves* evolved from.

Up until Darwin's era, the practice of animal breeding had also shown that although selective breeding for particular characteristics could push the boundaries of what was possible *within* a species (*intraspecies microevolution)*, there was a finite limit for what could be achieved. That limit was below that which was required for the creation of a new species. Naturalists in the era of Darwin realized that if selective breeding consistently abutted against a limit, and this limit did not allow the creation of new species, then natural, unguided sexual reproduction of animals in the wild was far less likely to result in speciation. There was no evidence that breeding within a species could result in a new species. In fact, this remains true today: in all of human history, despite thousands of years of intensive selective breeding, we have never created a distinct new species. We have only produced *varieties* within the existing kind.

Geology also provided evidence of *special creation*, or so it appeared, as from the time of Isaac Newton, and Johannes Kepler, the age of the Earth was believed to be 4000 years; an incredibly short time to account for a natural process of evolution. By the 1800's the consensus was that the age of the Earth was approximately 20-40 million years old with an upper limit of 100 million years–Lord Kelvin's estimate.

It wasn't until the discovery of radioactive properties, long after the death of Darwin in 1896, by Henri Becquerel, with ongoing refinement culminated by Clair Cameron Patterson, at the fairly recent date of 1956, that the age of the Earth was brought close to its current believed age of approximately 4 *billion* years.

This concept, known as "Deep Time", 20-40 or even 100 million years, in Darwin's age, changed everything. As the estimated age of the Earth expanded from biblical thousands to geological millions, a new idea arose: the slow transmutation of one species into another. The paradigm shift in Darwin's era—even with estimates as low as 20-40 million years—provided the essential window for naturalistic theories to take root.

The Pre-Darwinian Landscape: Dogma and Fable

Yet beneath this scientific restraint lay a deeper confusion. For centuries, scholars had floated theories that ranged from speculative to outright fantastical. Darwin entered a field marked not by clarity, but by contradiction and myth.

To understand the intellectual landscape Charles Darwin entered, we must look at the chaotic mix of dogma and speculation that preceded him. As detailed in Henry Fairfield Osborn's definitive history, *From The Greeks to Darwin* (1908)[1], the mid-19th century was dominated by a rigid "Special Creation" theory. However, as the budding science of Paleontology began to reveal distinct successions of different species in the fossil record, this theory had to be awkwardly remodeled[2]. It mutated into a belief in a "succession of creations"—a pseudo-scientific patch-job full of inconsistencies yet stoutly defended by the zoological establishment. Young Darwin was among the few who dared to look beyond this dogma. But the "naturalistic" alternatives available to him were often steeped in fantasy.

The Era of Fabulism

Early attempts to explain the diversity of life without special creation often devolved into mythology. As Osborn recounts, the 17th-century botanist Claude Duret proposed a "direct transformation" theory that sounds more like fairytale than physics. Duret described a tree observed in Scotland whose falling leaves would undergo a miraculous transmutation depending on where they landed: if they struck the water, they would slowly transform into fish; if they struck the land, they would turn into birds. Similarly, Father Kircher, in his *Mundus Subterraneus* (1678), filled volumes with "authentic observations" of orchids giving birth to birds and even very small men the moment they touched the ground[3].

Later thinkers attempted to add a veneer of mechanism to these transformations, but the results were often just as fanciful. Benoît de Maillet (1656–1738) guessed that aquatic life might adapt to land, but his proposed mechanism was absurdly rapid. He imagined flying fish getting stranded on shore, perhaps "driven out of the water by the ardor of the chase[4]." Prevented from returning to the sea, and under the influence of the air, they would transform within a single lifetime. De Maillet described how their anterior fins turned into wings, their scales softened into downy feathers, and the neck and beak elongated until "the fish discovered itself a bird."

The Materialist Turn

As the Enlightenment progressed, figures like Maupertuis and Diderot moved away from fables and toward a materialist philosophy. Yet, to bridge the gap between non-living and living matter, they felt compelled to endow atoms themselves with magical properties. Maupertuis advanced a hypothesis where all material particles were invested with "psychical properties." Diderot went further in *Histoire Admirable des Plantes*, published in 1609, endowing organic particles with a "rudimentary sensibility" and a "vague touch" imparted by the Creator[5]. He imagined molecules that could restlessly seek out the most favorable position, combining and recombining until they found "repose" in the form of an animal. Thus, when Darwin arrived, the field was split between a fossil-denying theology and a collection of "just-so" stories where leaves became fish and molecules had feelings. The world was waiting for a plausible mechanism.

This is the tradition that Darwin and his contemporaries had to contend with. There was a need to answer the question, where did all the life come from? In the age of Darwin two main contending theories were: (1) "evolution", meaning a process that resulted in a gradual increase in complexity and diversity of animal form by an as yet unknown internal means; the primary contending view was: (2) special creation, which in some considerations was a direct appeal to a creator, and in other considerations was based upon the best hypothesis to fit the evidence. Between these two ideas was a bridge of evidence that indicated species were fixed and new species emerged whole and complete, by an unknown cause. This evidence suggested that since there were no clear intermediary forms between species, there was no evidence that a gradual process of evolution had taken place. Too many sudden appearances in the fossil record precluded this route.

What both views agreed upon was that the process that created new species was unknown, unobserved and did not provide evidence in the fossil record. This was the foundation that defined the task Darwin, and a host of his contemporaries endeavored to solve. How species emerged in such a way that aligned with the evidence but also explained how it was done.

Prelude - The Beagle

With this intellectual backdrop in mind, we can now turn to Darwin himself—not as an isolated genius, but as a young naturalist shaped by these competing traditions and by the circumstances that placed him aboard the Beagle.

Darwin's voyage on *The Beagle* was in equal parts a product of his love and talent for naturalism, his family's deep pockets, and his mediocrity as a student. Previously, Darwin had proven himself to be at best an average student. He showed flashes of brilliance and was eminently observant, but his disdain for formalized education, and desire for self-study hampered his academic growth. Darwin's brilliance, exposed in his excursions as a naturalist, lay not in the classroom, but in the field. While he was an indifferent student of received wisdom, he possessed a singular gift for observation and self-directed study—traits that would ultimately prove more valuable than any conferred degree, when mixed with his family's wealth, connections, and associations he was able to form during his time at university.

Darwin dropped out of the University of Edinburgh medical school after discovering he wasn't particularly suited to more invasive medical procedures. This was revealed anecdotally when he witnessed live surgeries and was said to have been put off. In truth, this incident was likely equally as influential in his dropping out of medical school as his invigorated passion for his naturalistic pursuits.

During his education he formed a close bond with physician and fellow naturalist Dr. Robert Edmond Grant (1793-1874), who gave up the practice of medicine to study invertebrates. At the time, medicine was unique as one of the few formal scientific disciplines offered within a university curriculum. Other emerging fields—such as geology, paleontology, and ornithology—were not yet established as distinct degree programs. Instead, they were viewed as "Natural History," largely the domain of amateur enthusiasts and gentlemen scholars rather than professional scientists. Darwin found success in this field, and his work led to several papers being published, some allegedly by Dr. Grant, without proper credit to Darwin. At the same time Darwin became a member of an underground association called the *Plinian Society* of the University of Edinburgh. This was a group of students that read, debated and discussed the natural sciences and reviewed one another's finds during excursions to the seashore.

Darwin's time in Edinburgh clarified his views and revealed his passions and interests and he subsequently dropped out of university.

Darwin's next plan was to become a clergyman. It is speculated that his religious inclinations were not strong, however, his mother had passed away when he was 8 years old, and as an heir to the Wedgewood pottery and ceramics fortune, he had no need of work, and indeed never was formally employed, therefore a small parsonage would be conducive to the maintenance of his naturalistic hobby, while serving as a respected occupation. As mentioned previously, this was a time when natural sciences had yet to be promulgated into academic disciplines, and there was not much in the way of specialized degrees or courses such as biological sciences[6]. At Cambridge, where Darwin obtained a BA, the natural sciences hadn't begun "triposes" (examinations) until 1851, a mere 8 years before the release of, *On the Origin of Species*—decades after Darwin had attended college. Seen in this context, the opportunity on *The Beagle* offered the 22-year-old Darwin a more direct route to what he was truly interested in. He therefore passed on the special divinity course he had intended and put his religious aspirations on indefinite hold.

The expenses for Darwin's 5-year journey, and his position on board *The Beagle*, under captain FitzRoy, was secured and paid for by his father, the doctor and financier Robert Darwin who was encouraged to support the excursion by Darwin's maternal grandparent, Josiah Wedgwood who noticed that the young Darwin was, "a man of enlarged curiosity[7]."

On *The Beagle*, Darwin was granted great latitude to explore the terrain encountered along the journey. The primary mission of the crew was as a naval survey expedition. Thus, he spent less than two years of the five-year voyage on board *The Beagle*. The rest of the time was spent on forays inland which allowed Darwin to take a number of samples and specimens and as well as make close observations documented in meticulous notes recorded from 1831 to 1836. The experience upon *The Beagle* fundamentally changed Darwin's views. With over a dozen separate South American landing points on both the Atlantic and Pacific ocean, and particularly within the 13 islands of the Galapagos Archipelago, Darwin observed what he had not observed on large continents. A number of plant and animal species existed only in select and remote areas. These environments unexpectedly produced a wide range of variation among the animals in microclimates and microenvironments.

Darwin hypothesized all of the major themes of his future opus, *On The Origin Of Species* from his experiences on land-based excursions, and in particular, in the remoteness of the Galapagos Islands. It was in these islands, that he believed the species he had encountered had diversified as *isolated populations* by means of *random variation*, with *natural selection* and *descent with modification*. Based upon his samples, it appeared that he had in-hand evidence that evolution—in the way that would become synonymous with Darwin—was real, continual, and unceasing. And key to Darwin's insights, these isolated populations, where conditions varied, created the needed opportunities for small adaptations to accumulate rapidly. Not lost on Darwin, these in-isolation adaptations provided greater contrast than in areas without such remoteness, separation, and climate difference.

Returning Home

When Darwin returned home, although he had seen what he believed to be evidence of active evolution and recently created species, what he was less certain of was that he had obtained *irrefutable evidence* that the mechanism—evolution—was occurring: he was at a loss for what the mechanism of change could possibly be. Many individuals prior to Darwin had struck upon the idea of evolution, including Darwin's grandfather Erasmus Darwin, who is noted to be the first Briton to propose his own theory of evolution. Erasmus's ideas on the topic appear in at least 3 publications: *Zoonomia* (1794); *Loves of the Plants* (1789); and, *The Temple of Nature* (1803). Erasmus also touched upon the concept of the descent of life from a common ancestor, sexual selection, and the similarity of structure among living things (homology)[8]. Erasmus' writings on evolution, though studied, had been unable to convince others that evolution had been captured and described. Darwin therefore knew well the challenge he faced in proposing that he had solved the riddle of evolution.

Darwin arrived home with tantalizing clues yet remained unsure he could force them into a coherent thesis. That changed in March 1837, several months after his return, when he met the ornithologist, John Gould. Gould delivered a critical revelation: Darwin's Galápagos birds were not mere varieties of familiar species—they were entirely new species, including two distinct kinds of mockingbirds and more than two dozen land birds found nowhere else on Earth[9]. This was the validation Darwin needed. Encouraged, he began assembling the argument that would become *On the Origin of Species*, a task that would take nearly twenty years as he gathered data and refined his rhetoric.

Darwin began to formulate his argument within the milieu of influence we discussed previously, but also from several popular ideas of his current time. From Thomas Malthus's *Essay on Population,* he assimilated the concept of *survival of the fittest* with respect to the competition of too many animals for too few resources. The inclusion of this principle drew criticism then as it does now. The link between Mathus's ideas and speciation appeared forced. It was noted that in the natural world, it was not uniformly observed that there was a *competition* for scarce resources, rather the natural world was mostly composed of environments, where resources were sufficient, and competition, if there was any, was of a natural predator/prey "circle of life" variety, rather than direct competition between and within a species, characterized as a life-or-death battle. That notwithstanding, Darwin saw that viewing the world through the lens of Malthusian doctrine and emphasizing the concept of limited resources and overpopulation would allow him to drive the much-needed concept of the *struggle for existence,* which his theory required to place *natural selection* in the natural position to drive evolution.

In the competition for limited resources, those who were the fittest would survive. Darwin therefore needed Malthus to set the foundation for "*survival of the fittest.*"

Still, competition alone was not enough. Darwin needed a mechanism that could convert small advantages into lasting biological change. His insight was simple: if the individuals best suited to their environment also produced more offspring, then those traits would gradually spread. Eventually, he came to believe this process could yield entirely new species. As he later wrote, competitive reproduction provided the "theory by which to work[10]," and Malthus had given him the "long-sought clue[11]" to an effective evolutionary engine.

"The result of (competitive reproduction) would be that of a new species… Here, then, I had at last got a theory by which to work." Darwin continued, "...natural selection was the inevitable result of the rapid increase of organic beings… (Malthus) gave me the long-sought clue to the effective agent in the evolution or organic species[12]."

The second major influence on Darwin's theory was Charles Lyell's Principles of Geology. Lyell argued that immense geological structures could arise from countless small changes accumulated over "deep time." Darwin adopted this reasoning wholesale. If tiny geological processes, repeated for ages, could raise mountains, then tiny biological changes—given the same vast stretches of time—might create new species.

The third influence came from Jean-Baptiste Lamarck, who proposed that species were not fixed but mutable. Lamarck's specific mechanism of inheritance was flawed; he believed that traits acquired during an organism's lifetime could be passed directly to its offspring. Modern biology rejects this. But Darwin salvaged what he needed: the idea that organisms adapt through purposeful interaction with their environment. As one commentator summarized, Darwin treated adaptation as a natural consequence of living activity rather than divine intervention.

What Darwin—and everyone else—lacked was any understanding of the physical medium of heredity. Mendel's work on genetics was unknown to him, and neither he, nor Lamarck, nor Mendel had any inkling of DNA's structure. Only in 1869 did Friedrich Miescher isolate "nuclein," and nearly three-quarters of a century would pass before scientists learned that it encoded life through a four-letter chemical alphabet read in triplets, as Francis Crick later articulated.

Yet Darwin moved forward anyway, weaving these disparate ideas into a theory that felt familiar because it drew on concepts already accepted in the scientific and cultural zeitgeist. By anchoring his argument in the intellectual currents of his time, he constructed a narrative that appeared both novel and inevitable—an elegant synthesis built on borrowed parts.

Pedigree

Darwin's father, Robert Darwin, the son of Erasmus Darwin, was a doctor and financier. He was elected as a Fellow of the Royal Society in 1788 based upon a single 1786 paper on ocular spectra and color perception. It is believed his election to the Royal society was buttressed by his connections, social standing and family reputation. Robert Darwin had previously studied under several leading scholars, including the Scottish minister and natural historian John Walker, himself a fellow of the Royal Society and one of the main scientific consultants of his day. John Walker served as an agricultural, industrial and mining advisor to many influential Scottish landowners.

Darwin's maternal grandfather was Josiah Wedgwood, an innovative English potter, ceramicist and pioneering industrialist founder of the Wedgwood company, a mass producer of high-quality pottery. He was equally well known as an abolitionist. The Wedgewood fortune was significant, as was the family reputation. As an abolitionist Wedgewood produced a medallion distributed for free at abolitionist meetings became world famous after its production in 1787 with the motto, "Am I not a man, and a brother." Charles Darwin, fortified with these two combined fortunes, and the accompanying social status was well situated in society and the scientific community to release and support his evolutionary theory.

The Release, Reticence, and Acceptance of, *On The Origin of Species*

Charles Darwin was 16 when he began studies at Edinburgh, 19 when undertook studies at Cambridge, and 22 when he boarded *The Beagle* (1831 to 1836). It wasn't until he was 49 when, *On the Origin of Species* was finally released in late November of 1859[13]. Darwin then spent the better part of the next decade after the release of, *On The Origin of Species* responding to criticism, defending his thesis, and making revisions for future editions. During this time Darwin also took a great deal of time off to deal with his numerous physical and nervous ailments. The critique and negative reaction to his book included a good amount of scorn and ridicule primarily due to the lack of evidence, its unfounded suggestions and a narrative overly reliant on anecdotal propositions, assumptions, rhetoric and several strange and noteworthy evolutionary examples.

A famously absurd example Darwin used to illustrate a possible motivation for evolution was the image of a bear swimming in the water with its mouth open. On this basis Darwin suggested this example could serve as the precursor for the long slow change from bears to whales. Darwin was forced to make such associations, because if evolution was true, then water bound mammals, such as whales, would first be required to have crawled upon the Earth as gilled fish, then developed lungs, become established as land mammals, before retreating back into the water to begin the slow process of becoming manatees, dolphins and whales. The anatomy of water-bound mammals demanded such historical precedent, as well as a low, incremental series of steps from sea, to land, and back again.

The New Hypothesis Struggles

After years of attack and pitched debates taking place in private rooms, societies, and in print, the most unexpected transition took place. Slowly, the tide turned and the theory took hold beginning in the late 1860's. It had always had a few key backers, but over the years It built momentum in academic circles, gained support among religious leaders, and in the late 1860's approached the tipping point, and by 1870 crossed over into wide acceptance.

Historians have struggled to identify a single event that transformed evolution from a beleaguered hypothesis into a widely accepted theory. What is clear, however, is that Darwin's ideas arrived at a moment when the intellectual climate was unusually receptive. Evolution was already "in the air," and Alfred Russel Wallace's independently conceived paper on species introduction—published in the Annals and Magazine of Natural History in 1855—added momentum. When Charles Lyell sent Wallace's work to Darwin, it set in motion the joint presentation of their papers to the Linnean Society in 1858. This dual debut gave evolution an unexpected legitimacy: it was now not the claim of one man, but of two naturalists, working worlds apart, with different evidence sets, that reached the same conclusion.

The theory struggled upon its release among Darwin's peers, but then, in an amazing turnabout, gained its strongest foothold among the next generation of naturalists. Where the older establishment remained cautious, younger scholars were eager to champion what felt like a bold new scientific worldview. Crucially, this surge of enthusiasm did not arise from new evidence—none had appeared. Instead, supporters assumed that paleontology, still in its infancy, would eventually uncover the "missing links" needed to vindicate Darwin. The absence of evidence was treated not as a fatal flaw, but as a temporary inconvenience that future discoveries were expected to solve.

Darwin also found powerful allies among those committed to advancing a fully materialistic science. Chief among them was Thomas Huxley whose influence within British scientific circles was immense. A rising star in multiple disciplines, had undertaken his own formative sea voyage at age 20 aboard *The Rattlesnake*. Huxley was a professor of natural history at the Royal School of Mines, a naturalist to the British Geological Survey, A professor at the Royal Institution, Professor at the Royal College of Surgeons, and the Royal College of Surgeons, and a president of the Royal Society.

Huxley wielded his authority with force and flair, using articles, lectures, and public debates to defend Darwin's theory. His stature meant that supporting evolution became synonymous with aligning oneself with a prototypical insider's insider of Royal institutions, and scientific progress—a cultural shift that played a decisive role in the theory's ascent, and its influence upon the next generation of scientific scholars.

Huxley who became known as "Darwin's Bulldog" due to his forceful and brash defense of evolution, was a moniker well-applied after a famous Huxley anecdote from his debate with Bishop Samuel Wilberforce, on June 30, 1860. In this debate it is said that Wilberforce started out with a half-hour attack on evolution which he ended with the question poised to Huxley, (not officially recorded) *"Do you consider yourself descended from an ape through your grandmother or grandfather?"* Huxley is said to have replied sharply, *"I would rather be descended from an ape than a Bishop!"*

Huxley's influence spread far beyond formal scientific venues. His energetic defense of Darwin's theory permeated drawing rooms, universities, private clubs, and popular journals, all of which were at Huxley's disposal. At the same time, a menace had been growing that was seen as a threat to scientific progress: many intellectuals felt a growing pressure to "side with science" rather than align with creationism, which appeared increasingly outdated in the Victorian imagination, yet, was incredibly difficult to outdistance in light of a dearth of evidence against the fixity of species. These forces created a receptive environment: evolution was still speculative, but Huxley's advocacy gave the hypothesis an air of inevitability, while also shielding evolution under the banner of modern scientific thought. Many were eager to rally behind the clarity and authority of its concepts.

In the end, evolution triumphed not through decisive evidence but through a gradual accumulation of pressures and incentives. Supporters trusted that future discoveries would fill the gaps, but make no mistake, these gaps had not been filled at the time it was ushered into practice; Darwin's rhetoric carried enormous persuasive power; materialists embraced the theory for philosophical reasons; influential figures promoted it; the capture of nearly all major scientific institutions had been complete, and many welcomed a scientific alternative to creationism. Over time, these forces coalesced into a consensus. Evolution was effectively chosen—anointed as correct—despite it lacking the empirical foundation normally required of a scientific theory.

That early acceptance functioned as an IOU: evolution was granted legitimacy long before it could supply the evidence needed to justify it. From a modern standpoint, the gap is even more striking. In 1859, DNA had not yet been discovered, and the mechanisms of inheritance were barely understood. The very processes Darwin relied upon—random variation and the transmission of new traits—were complete mysteries. The scientific community accepted the theory long before it had the tools to probe it, or test it, let alone confirm it.

Dissent

Despite this eventual acceptance, the path to Darwinism was anything but smooth. Before evolution gained its cultural and scientific momentum, it faced substantial resistance from many of the very people Darwin hoped to persuade. The ascension of evolutionary theory was completed in a series of steps.

Initially the way was not clear for Darwinism to become the dominant idea of speciation, as we recount this process took most of the next decade after the release of, *On The Origin of Species*. There were those among the old guard who objected to the sweeping assumptions made by Darwin. One of the staunchest critics was Swiss American Biologist, Zoologist and Geologist Louis Agassiz. This selection from one of three added chapters to his book, *Of Species and Classification in Zoology (1869)* (*De l'Espece et de la*

Classification en Zoologie) outlines his most direct articulation against Darwinism.

> "...Darwinism excludes nearly all the mass of acquired information, for it assimilates and takes exclusively that which could be useful to the doctrine. It is not the facts that determine for the Darwinists the nature of their generalizations, it is the system that dictates the nature of their reality[14]."

And,

> "The fundamental idea on which Darwinism rests is that the organized beings that themselves are succeeded by direct progeny, far from reproducing necessarily the essential characters of their ancestors, tend to avert from them. Until our days all the Physiology admitted, as an axiom based on the experience of all time, that the descendants of any living beings having the faculty to reproduce are the living picture of their parents, and that the very fertility of these beings was the guarantee of the conservation of their types. This notion is moreover corroborated by another fact, that, in the result of a mixture of species, one recognizes the afferent parts from each of the originators who contributed to the production of the new being. Of these two unshakable facts is formed the conviction that dominated the science until the present time: knowledge, that the organized beings are reproduced, from generation to generation, with characters identical to the ones that they possessed during their first emergence[15]."

Three themes dominated Agassiz's critique: the lack of evidence for inherited divergence, the fixity of species boundaries, and the apparent simultaneous appearance of diverse forms in the fossil record.

Agassiz argued that Darwinism simply ignored the full body of biological knowledge accumulated to that point. In his view, all physiology supported the axiom that offspring reliably reproduce the essential traits of their parents. Even in hybrids, he said, one could clearly trace which traits came from which parent. This "unshakable fact" of inheritance had underpinned biology for centuries. By contrast, he accused Darwin of advancing an ungrounded claim: that successive generations gradually diverge from their ancestors. He insisted no one had demonstrated the vast range of variation required for Darwin's theory, either in nature or in domestic breeding, nor anything resembling an upward "progress" in organization.

Agassiz acknowledged that individuals within a species vary but argued that these variations have strict limits — what he called "the boundaries of flexibility." However large these differences might be, he maintained that no observation had ever shown an individual becoming anything other than its species. Darwinists, he complained, stretched individual variation into claims about species transitions without factual support. He stated: "The school of Darwin goes beyond facts when it states that these individual differences constitute the transitions from one species to another."

Agassiz also appealed to paleontology. He noted that in several geological horizons, entirely distinct and fully formed types appeared together — different classes and orders with no obvious ancestors among each other. If these organisms were all contemporary, he asked, how could one serve as the ancestor of another? To him, this pattern contradicted gradual transformation and instead revealed sharp discontinuities that Darwin's "genealogical trees" merely papered over with arbitrary lines. According to Agassiz: "Far from supporting the doctrine, the great facts revealed in paleontology contradict it."

Taken together, Agassiz's critique rested on three claims: that inheritance is conservative, that variation is bounded, and that the fossil record shows abrupt appearance rather than continuous transformation. Criticisms such as this, though pointed, were not enough to stem the tide of Darwinism, as the next generation of scientists were nearly entirely won over, including Agassiz own son, Alexander E. Agassiz.

Charles Darwin's 5 Causes of Evolution

Darwin continued revising his ideas about how evolution works long after *On the Origin of Species*. By the time he published The Descent of Man (1871), he insisted that sexual selection played a far greater role than he had originally allowed. As he put it, "Sexual selection had a far more important role in the Descent than in *The Origin of Species*, … so much so that it would no longer be accurate to describe the theory as that of natural selection[16]."

In The Descent of Man, Darwin went further: natural selection was only one among several forces shaping human evolution. He argued that five distinct causes combined to produce Homo sapiens:

1. Sexual Selection

Organisms better adapted to the environment would benefit from higher rates of survival, but also, those that had better survival rates would also have other traits that may not directly aid their survival in the environment, rather, they would aid their ability to sexually reproduce or be sexually selected.

2. The Inherited Effect of Use and Disuse (descent with modification)

When a particular organ becomes highly developed in response to environmental pressures, Darwin believed that this acquired use could become hereditary. Repeated use over many generations—for example, reaching for food high in the trees—could, in this view, eventually produce a creature like the giraffe.

3. The Direct Action of the Environment (variability, random variation and natural selection)

The direct action of the environment, Darwin argued, shapes the conditions of life in ways that influence organisms. It can promote the use or disuse of certain organs and favor traits that happen to suit new environmental demands.

4. The Correlation of Growth

"Correlation of growth" refers to the way certain developmental changes get carried along simply because they are linked to other, beneficial traits. As Darwin put it, organisms are so interconnected in their growth that when a small variation in one part is preserved by natural selection, other parts may be modified as a by-product.

5. An Unspecified Cause[17] .

Darwin *never* named this 5th cause, which we referred to earlier when we discussed the long-held idea of a fifth force. Gertrude Himmelfarb, author of *Darwin and the Darwinian Revolution*[18],attempted to explain it from Darwin's articulation:

> "...the variation induced by this new factor was of no service to the organism, either in its inception or in its later development… And the nature of its cause was unknown. Darwin could only assume that whatever its cause, so long as it continued to act 'uniformly and energetically' over a long period, the result would be the production not of 'mere slight individual differences, but well-marked constant modifications[19].'"

Darwin acknowledged that this fifth cause arose from something within the organism itself, he described it as a "uniform and energetic" internal force capable of producing stable modifications over time and that variation could originate from its built-in principle of change, independent of external conditions, and not merely from environmental pressures. Darwin also described this force as a persistent, internal source of modification that operated independently of external conditions. His refusal to anchor this cause in any physical structure makes this "force" very uncertain and the source of its operative power unknown. He offered no anatomical or physiological basis—no reference to reproductive cells, hereditary particles, or anything comparable. It resembled a teleological principle, with a built-in striving or direction. The parallel is difficult to ignore although on the surface, it appears to be heretical to his own theory.

Darwin's uncertainty about the source of variation as a mechanism was not new. Even as early as 1852, he admitted that the complete understanding of the causes of variation was unknown. By the time he wrote *The Descent of Man*, he was still unable to pin this force down. Gertrude Himmelfarb explained that Darwin simply assumed that whatever this cause was it could generate major, lasting modifications[20]. This fifth cause is notoriously vague.

Darwin had long argued that tiny individual variations—differences earlier naturalists dismissed as mere accidents—were the raw material of evolution. If natural selection could accumulate these slight variations without limit, he believed, it could eventually produce the major transformations needed to explain the origin of new forms. But in keeping with his taxonomic ambivalence, core processes, such as forces of variation, categories of intermediates, and functional definitions were left underdefined, and allowed to shift or retroactively be fitted to outcomes, when it came to variation, Due to this essential inability to provide closure, Darwin left the source of variation unexplained.

As the historian Paul Sloan notes: "Variation was presented in The Origin with the statement that 'the laws governing inheritance are quite unknown.'" This is the essential point: Darwin built his system on causes that he admittedly did not understand. Because genetics was entirely unknown in Darwin's time, he could only speculate about the source of inherited change. The mechanisms inside the "black box" of heredity remained mysterious. Darwin sometimes hinted that the answer might lie there; at other times he leaned on the unnamed fifth cause as the principle that coordinated the other four. He found himself wholly incapable of articulating how this force worked.

The Teleological Fifth Cause?

What, then, was this fifth cause named by Darwin? His own words leave the question open. Many contemporary thinkers—both materialist and teleological—have likewise suggested that something additional may be operative in nature. While they do not identify it as Darwin's unnamed "fifth cause," which he specifically named as a cause in *human* evolution, it is often framed instead as a fifth force, supplementing the four fundamental interactions of physics acting across multiple fields. This proposal repeatedly takes the same form: an attempt to absorb an irreducible influence into a reductionist framework by redescribing it as emergent, effective, or dynamically generated. These efforts are aimed at the same target—an as-yet undiscovered law or organizing influence that appears to operate alongside the known interactions, framed as if it were reducible.

Michael Denton raises this question directly in his book, *The Miracle of the Cell*, arguing that the transition from chemistry to biology may require "new laws… yet to be discovered[21]," rather than blind assembly by chance. Denton states that he is not alone, and that Gerald Joyce, Robert Shapiro, Paul Davies, Tommaso Bellini, Nobel Laureates Christian de Duve, Jack Szostak and Stuart Kauffman hint at the possibility of such a force waiting to be discovered[22].

Surprisingly, even Thomas Huxley—Darwin's fiercest defender—acknowledged that evolution carried a teleological flavor. He was untroubled by the idea that nature might reveal purposeful structure, and he argued that evolution did not eliminate this possibility. Huxley noted a "wider teleology… based on the fundamental proposition of evolution," in which the world unfolds through the lawful interaction of matter shaped by an original intelligent arrangement.

French philosopher Henri Bergson made a similar point: if small accidental variations are to accumulate into coherent new forms, some guiding principle must ensure their coordination. As he put it, evolution seems to require a "good genius" to preserve or synchronize the needed changes[23].

And here we have one of the aims of Darwin's evolution laid bare. The two *primary* goals were to use evolution as the primary example of materialist "theology", and the other goal was to banish religious and spiritual theology and God from the role as the informational authority and fiat power behind creation. Darwin concedes that obliterating God was one of the primary goals of his theory, "If I have erred in giving to natural selection great power, which I am far from admitting, or in having exaggerated its power, which is in itself probable, I have at least, as I hope, done good service in aiding to overthrow the dogma of separate creations[24]." Materialism via the sciences had to wrest this role from the oppression of a spiritual realm and render them discernible and controllable by either man or no one. They desired a benign God, broken into an untold number of pieces, distributed evenly throughout the universe, represented by orderly laws, and captured within bits of matter, and in the aggregate forming a level of emergence that gently whispered a logical course of evolution.

Religion and Darwinism

Although modern readers often see evolution and religion as inherently at odds, this conflict was far less obvious in the eras of Erasmus Darwin or even Charles Darwin. Within a deistic framework—where God created the universe and then allowed it to unfold through pre-set laws—evolution could be understood as part of a divine mechanism rather than a denial of one. Erasmus Darwin split the difference by expressing a vision of harmony between divine authorship and natural development explicitly:

> "The great CREATOR of all things has infinitely diversified the works of his hands but has at the same time stamped a certain similitude on the features of nature, that demonstrates to us, that the whole is one family of one parent."
>
> -Erasmus Darwin[25]

Because deism allowed for God's purposes to be expressed through natural laws, many clergy saw no contradiction between their faith and Darwin's proposed mechanism. This theological flexibility helped Darwin secure early religious support, which in turn softened public resistance to his theory and played a significant role in legitimizing evolution. Many believers assumed that, if God chose evolution as *His* method, this posed no threat to their faith. What they did not see was that key defenders of Darwinism increasingly framed the theory in strictly materialistic terms—leaving no room for divine guidance within the mechanism itself.

Yet prominent Christian thinkers sought reconciliation. Asa Gray, a devout Harvard botanist, defended Darwin while insisting that evolution remained compatible with providence. His essays, collected in *Darwiniana*[26], helped bring religious audiences on board. In England, Darwin bolstered this perception by quoting supportive remarks from the Anglican priest Charles Kingsley in later editions of *On the Origin of Species*. Kingsley even praised the idea that God might create through secondary causes rather than through repeated miraculous acts:

> "A celebrated author [Charles Kingsley] and divine has written to me that 'he has gradually learnt to see that it is just as noble a conception of the Deity to believe that He created a few original forms capable of self-development into other and needful forms, as to believe that He required a fresh act of creation to supply the voids caused by the action of His laws."
>
> -Charles Darwin, *On the Origin of Species*[27]

Evolutionary History As Prologue

Having recounted the historical ascent of Darwin's evolutionary theory—drawn from his field notes and experiences aboard *The Beagle*, particularly in the Galápagos; through its long incubation period prior to the publication of *On the Origin of Species*; and through its initial rejection, criticism, and gradual uptake among the next generation of naturalists—we have traced the conditions under which the hypothesis gained traction. Its rise was aided significantly by the public advocacy of Thomas Huxley, the near-simultaneous emergence of Alfred Russel Wallace's independent formulation of evolution through natural selection, and the support of several prominent religious figures. These incremental stages and partial victories eventually culminated in the ascension of the evolutionary hypothesis to accepted theory, despite the persistence of evidence that stood in tension with its claims. It was, in effect, anointed under a posture of half-belief and half-promise: that confirmatory evidence would be forthcoming in the years ahead.

The record that emerges from this history is not one of immediate insight or evidentiary closure. Darwin's theory was initially met with resistance, skepticism, and sustained criticism, and for a considerable period it rested on a limited evidentiary base. Its persistence was supported not only by its scientific ambitions, but also by advocacy, alignment with prevailing intellectual currents, and gradual institutional uptake. Over time, the hypothesis was normalized, absorbed into pedagogy, and inherited by a subsequent generation for whom its provisional status had increasingly receded from view. Only after this process of consolidation did the theory come to be treated as largely self-evident.

This stands in contrast to what might be assumed given the theory's robust contemporary standing. From its present position of near universality, one might imagine its publication as an event that sent immediate shockwaves through the intellectual world, its brilliance swiftly absorbed and its explanatory power quickly recognized as the obvious key to understanding the diversity of life. The historical record, however, tells a more gradual and contingent story.

We may now rejoin the well-known narrative of evolution—familiar to anyone who has taken biology, who is well-read, or who is situated within the academic milieu—where Darwin is hailed as one of the great geniuses and innovators of the scientific tradition, and evolutionary theory stands largely unquestioned as the mechanism of change responsible for the iconic tree of life and the diversity of organisms on Earth; from the microscopic and simplest forms to the strange and astonishingly complex.

In this form, Darwinism no longer functions merely as a scientific theory, but as a deeply entrenched explanatory foundation—culturally absorbed, pedagogically reinforced, and treated as the default account of biological reality.

A Century of Discovery

The life of Charles Darwin (1809–1882) unfolded during a period of extraordinary scientific upheaval. The advances of the 17th and 18th centuries had already transformed astronomy, physics, and mathematics; the 19th century accelerated this transformation dramatically. In Darwin's lifetime alone came Ohm's law, the Doppler effect, the laws of thermodynamics, cell theory, and Mendeleev's periodic chart. Each breakthrough reinforced the prevailing belief that nature's deepest secrets were being unlocked through reductionist science, and that a fully material explanation of life's diversity was within reach.

The Road to Neo-Darwinism

As new discoveries accumulated, Darwin's original theory underwent continual revision. The most significant shift began with the rediscovery of Gregor Mendel's work (1856–63), which demonstrated that traits are inherited as discrete units rather than blended continuously. This finding undermined Darwin's earlier Lamarckian commitments—especially the idea that "use and disuse" could directly modify hereditary traits.

Further pressure came from August Weismann, whose 1892 Germ Plasm Theory argued that reproductive cells (gametes) are categorically distinct from the rest of the body. If so, then environmentally induced changes in somatic tissues—Darwin's mechanism of use and disuse—could not be passed on. Mendel and Weismann together forced a redefinition: variation could no longer mean vague, organism-wide changes. It had to refer to changes in the hereditary material itself.

The Rise of Population Genetics

By the 1920s and 1930s, this redefinition catalyzed what became known as the modern synthesis or Neo-Darwinism. Three thinkers were central:

J.B.S. Haldane's 1932 The Causes of Evolution reframed Darwin's ideas through the lens of emerging genetics.

Ronald Fisher, whose Genetical Theory of Natural Selection (1930) attempted to show that Mendelian genetics supported Darwin's general framework.

Sewall Wright, who developed concepts such as genetic drift, adaptive landscapes, and effective population size, while arguing that small mutations were more likely to be beneficial than large ones.

This synthesis unified natural selection with Mendelian heredity. But in doing so, it erased the five causal factors Darwin had proposed over six editions of *On the Origin of Species* and *The Descent of Man*. In the new system, random mutation became the only mechanism capable of generating the raw material for evolution. Everything else—sexual selection, use/disuse, correlation of growth, and Darwin's unnamed "fifth cause"—was swept aside.

It is crucial to note: the modern synthesis arose decades before the structure of DNA was known. Its architects were reasoning inside a black box; with no understanding of the actual genetic machinery their theories depended on.

DNA Shatters Earlier Assumptions

The discovery of DNA's double helix in 1953, along with Francis Crick's Sequence Hypothesis, massively altered the landscape. DNA was not a fluid, malleable substance that could tolerate broad ranges of variation. It was a digital code, where each three-letter codon corresponds to a specific amino acid.

Because proteins only function when folded into precise three-dimensional shapes, a random change to the sequence is almost always disruptive. As we reviewed in an earlier section, protein scientist Douglas Axe later quantified this: for a typical 150–amino-acid protein, the proportion of sequences that fold into a stable, functional structure is roughly 1 in $\mathbf{10^{74}}$. This is not a hospitable landscape for random search.

If the builders of the modern synthesis had known the structure, specificity, and combinatorial constraints of the genetic code, it is far from clear they would have been comfortable basing evolution on small unguided mutations.

The Mutation Bottleneck

Mutations occur during DNA replication, but nearly all of them disappear. To have evolutionary significance, a mutation must:

-Occur in a gamete (egg or sperm).

-Be passed to the offspring—a ~50% chance.

-Be passed again to the next generation— ~50% chance.

-If all of this occurs, after several generations, more than 99% of all mutations—beneficial, neutral, or harmful—are simply lost.

Even if we assume, very generously, that 5% of all mutations are beneficial, any mutation rate must still be multiplied by twenty to represent the actual rate of beneficial mutations. And even a beneficial mutation is irrelevant unless it also confers a reproductive advantage, the only currency that "counts" in Darwinian terms. Strength, speed, or survival mean nothing if they do not increase reproductive success.

This requirement produces a paradox: complex evolutionary transitions would require numerous coordinated mutations, yet each individual mutation must independently produce a reproductive advantage to persist. Evolution cannot "plan ahead," and the vast majority of changes that might be required for a new biological structure would offer no immediate reproductive benefit.

A Break Between Theory and Reality

DNA exposed the gap between the assumptions of early evolutionary theory and the true nature of genetic architecture. The genome is not intuitive. It is multilayered, hierarchical, and dependent on networks of transcription, translation, chaperone folding, and quality control—systems that were unimaginable to 19th- and early 20th-century theorists.

The cell's internal processes defy simple analogies. Proteins locate specific genes within a three-billion-letter genome; enzymes execute multi-step editing and transport operations; molecular machines assemble in sequence to produce functional structures. None of these agents has eyes, memory, intention, or awareness, yet they consistently execute tasks that—at the molecular scale—are astonishingly specific.

The mechanism by which random mutation could generate new functional proteins, new body plans, or new biological architectures remains unknown. It has never been observed. It is not predicted by the sequence hypothesis. And it stands in tension with everything molecular biology has revealed about the fragility and specificity of the genetic code.

The assumptions made by Darwin and the architects of the modern synthesis—reasonable guesses at the time—simply do not align with what we now know. The random mutation paradigm does not possess the creative power once ascribed to it. It is, in effect, a story told before the contents of the black box were known.

Once the genetic black box was opened, the evolutionary problem multiplied rather than diminished. Not only was the proposed mechanism inadequate, but the timeframe itself collapsed. With known mutation rates and population sizes, the appeal to "deep time" could finally be tested—and found wanting. Nowhere is this more apparent than in the proposed divergence between humans and chimpanzees; the closest living relatives in evolutionary theory.

Chimpanzees in Three Parts - Part One -

Before the molecular era, evolutionary timelines rested largely on plausibility arguments. If variation was assumed to be effectively limitless, and if time was assumed to be vast, then evolutionary change could always be deferred to "deep time." These assumptions were never tested; they were simply required.

That changed once DNA could be sequenced, mutation rates measured, and population genetics modeled quantitatively. With these tools in hand, evolutionary change could no longer be treated as an open-ended abstraction. It became a calculable process, bounded by fixed rates, finite populations, and generational turnover. When these constraints are applied, the time available for evolution to accomplish the transformations attributed to it is woefully insufficient, by margins that range from tens to hundreds to thousands of billions of years.

The genome of chimpanzees and humans, assumed to be from a common ancestor, has been reported to be 98.8% the same soon after genetic sequencing was available. This may sound innocuous, but with the human genome at 3 *billion* base pairs in length, the reported ~1% (1.23 to 1.25%) difference between chimpanzees and humans represents 36 *million* base pairs or *72 million* total units of evolutionary change that needs to be accounted for by a random process. This estimated problem has gotten much worse now that the actual genetic difference between chimpanzees and humans is gauged to be much higher. New, more rigorous analysis methods indicate a more accurate figure of chimps and humans sharing only *95% of their genome, "Chimp and human genomes are minimally 5 % divergent, representing about 150 million nucleotide differences[28].",

An additional 114,000,000 million mutations must be fixed in the same 100,000 to 300,000 generations to successfully account for chimpanzees and humans evolving from a common ancestor, and for humans to evolve from what must be assumed to be an even more primitive chimp-like creature than a chimpanzee.

The 100,000–300,000 generation constraint is known in population genetics as the *waiting time problem*. It asks a simple question: given known mutation rates and population sizes, how long does it take for specific, coordinated genetic changes to arise and become fixed in a population?

In a single quantitative demonstration, this problem proves fatal to the proposed human–chimpanzee evolutionary pathway. Using measured mutation rates, geneticists and co-inventor of the "gene gun" John C. Sanford and colleagues calculated that changing just two specific genetic letters—out of the roughly 114 million differences between chimpanzee and human genomes—would require approximately 84 million years[29]. When these established rates are projected backward across the 3–6 million years available since the proposed human–chimpanzee divergence, the conclusion is unavoidable: there is nowhere near enough time available. The shortfall is catastrophic—on the magnitude of tens of billions of years.

The difficulty becomes far more severe when the same constraints are applied across the full evolutionary timeline. From the first appearance of life, commonly represented by the Last Universal Common Ancestor (LUCA), to the emergence of Homo sapiens, evolution is granted at most 3.8 billion years. That span must account for every major biological innovation: new proteins, new cell types, new body plans, and entirely new modes of life—the entire *tree of life*.

Yet even this does not capture the full scale of the problem. Evolutionary theory must account not only for the emergence of modern humans, but for the origin of all biological diversity–meaning each and every genetic change required to achieve this diversity. There are an estimated 8 million living species on Earth today, and roughly 99 percent of all species that have ever existed (on the order of 800 million) are estimated to now be extinct.

Each of these lineages would require its own suite of novel, functional genetic changes, subject to the same waiting time constraints. When these demands are considered collectively, the mutational burden not only increases, but compounds. Evolution is required to generate, test, fix, and then discard vast numbers of complex genetic innovations, all within a finite and already insufficient timeframe.

At this scale, we are not merely concerned with probability, but with feasibility. Random mutation and natural selection fall short, long before the job is done—by margins so large that no amount of additional time prior to the heat death of the universe could realistically close the gap.

Chimpanzees - Part Two - Genome Size and the Collapse of the Junk DNA Assumption

The time problem is only the first layer of difficulty. Even if evolution somehow had far more generations at its disposal, it would still face a deeper obstacle: the amount of functional genetic information that must be explained.

For much of the twentieth century, evolutionary theory relied on a crucial assumption. Large portions of the genome were believed to be nonfunctional remnants of evolutionary history, commonly referred to as "junk DNA." If true, this would dramatically reduce the explanatory burden. Evolution would only need to account for a small fraction of the genome as being biologically meaningful.

That assumption no longer holds. Advances in molecular biology and genomics have revealed that vast regions once dismissed as "junk" play regulatory, structural, and informational roles. The explanatory burden has increased in response. Instead of explaining a small active core of genetic information surrounded by large swaths of inert debris, evolution must now account for the functional significance of nearly the entire genome.

This shift radically changes the human–chimpanzee problem. The question is no longer how a few protein-coding genes diverged, but how large-scale, coordinated changes across a mostly functional genome could arise within the limited number of generations available.

> "Without junk DNA, evolutionary theory cannot currently explain how everything works mathematically… in the evolutionary model there have only been 3–6 million years since humans and chimps diverged. With average human generation times of 20–30 years, this gives them (evolutionary theorists) only 100,000 to 300,000 generations to fix the millions of mutations that separate humans and chimps. They (chimps) don't have enough time, even after discounting the functionality by over 95% of the genome..."
>
> -Robert W. Carter[30]

Junk DNA plugged a gaping hole in evolutionary theory, but now that most, and perhaps all of the remaining 98% of the genome must be accounted for as well, the problem of explaining the *millions* of ancestral differences between chimpanzees *and* humans is made exponentially worse.

> "waiting times can be prohibitive – depending upon the exact circumstances. Some of the subsequent papers have been critical. Yet even those papers show that establishing just two specific co-dependent mutations within a hominin population of 10,000 can require waiting times that exceed 100 million years (see discussion). So, there is little debate that waiting time can be a serious problem[31]."

These results place a clear upper bound on what random mutation can accomplish within the proposed timeframe for human–chimpanzee divergence. Even when favorable assumptions are used, the number of specific mutations that would need to arise and become fixed far exceeds what known mutation rates can supply. The calculations presented so far, however, depend on an additional assumption that has not yet been examined: that ancestral populations remained large and stable throughout the entire period of divergence. The validity of that assumption is what we'll examine next[32].

Chimpanzees - Part 3 - Steady Population

Even after accounting for known mutation rates, evolutionary models still attempt to preserve human–chimpanzee common ancestry by appealing to large, stable ancestral populations. These models typically assume a minimum population size of 10,000 individuals maintained continuously throughout the proposed divergence period. This assumption represents a best-case scenario for mutation accumulation. Using this population size, evolutionary simulations have been run with a modeling program known as Mendel's Accountant[33]. The results were as follows:

These simulations showed that even under biologically favorable conditions, waiting times for specific mutations quickly become prohibitive. Establishing just two specific nucleotide changes required, on average, 84 million years, while five coordinated changes required roughly 2 billion years. Sanford and colleagues concluded that even under the most generous assumptions, the waiting time for specific genetic sequences consistently exceeded what evolutionary timelines allow. "Even using the most generous feasible parameter settings, the waiting time required… was consistently prohibitive." (Sanford et al., 2015)

If we assume the commonly cited figure that chimpanzee and human genomes are 98.8% similar, then a 1% difference across a genome of roughly 3 billion base pairs still leaves about 30 million differences per lineage, or 60 million total genetic changes to account for. The observed mutation rate is approximately 1 in 100 million letters per generation. At this rate, even producing eight specific nucleotide changes would take longer than the age of the universe. Extrapolating further, Sanford and colleagues calculate that achieving 60 million changes would require approximately 2.52 quadrillion years[34].

Crucially, this estimate only measures how long it takes for changes to occur somewhere in the genome. It does not address the far harder requirement that those changes occur at precise locations and with specific functional outcomes. The problem is random mutations do not accumulate toward a target. They continue indefinitely, without guidance, and have been overwhelmingly demonstrated to have the effect of degrading genetic information rather than building it. Under such conditions, there is no mechanism to ensure that the required configuration–the "desired" end result, or target– is ever reached.

Population geneticists acknowledge that beneficial mutations are rare, and high-impact beneficial mutations are extremely rare. Because of this, evolutionary change is often limited not by selection, but by how long it takes for the right mutation to appear in the right place. As Sanford and colleagues note, selection cannot act until a specific mutation arises at a specific genomic location[35]. "Waiting for just the right mutation to arise in just the right location can be a rate-limiting factor in long-term evolution[36]."

Even if we accept the already implausible figure of 2.52 quadrillion years, the problem only deepens. Our sun has a total lifespan of roughly 10 billion years. This means it will only be around for ~0.000004% of the required time period for these random changes to take place.

Finally, this problem does not apply only to chimpanzees and humans. As indicated earlier, evolutionary theory must also account for the emergence of all known life from *a single common ancestor*. Each lineage requires its own sequence of successful mutations. The waiting-time problem therefore compounds across the entire history of life, placing the explanatory burden far beyond what random mutation can sustain.

The failure of mutation-based mechanisms is not confined to genetic modeling. The same limitations appear when evolution is tested against the historical record of life itself. If gradual, unguided change were responsible for biological diversity, its signature should be written clearly in the *entire* fossil record. What we find instead is a pattern that presents a separate and equally severe challenge.

A Series of Unmet Evolutionary Challenges

Below we provide a series of additional challenges and proofs *against* the possibility of evolution as the cause of the diversity of life.

Evolutionary theory struggles to explain the rapid emergence of 13–16 fundamentally new body plans during the **Cambrian Explosion**. During this relatively brief window, roughly 20 of the 27 animal phyla found in the entire fossil record appear abruptly. These forms enter the record without clear precursors in earlier strata and without transitional sequences linking them to simpler organisms[37] Only three phyla appear before the Cambrian, and only four emerge afterward[38].

The Cambrian Explosion is not unique. Paleontologists have identified more than a dozen additional episodes of rapid biological diversification. One of the most significant is the Great Ordovician Biodiversification Event, which occurred roughly 40 million years after the Cambrian. During this period, Cambrian fauna was largely replaced by Paleozoic forms, with a sudden proliferation of suspension feeders and pelagic organisms.

Another example is the sudden appearance of flowering plants, known as "The big biological bloom". During the early Cretaceous period, approximately 130 million years ago, angiosperms entered the fossil record abruptly and in great diversity, without clear ancestral forms.

During the Eocene period, between 56 and 34 million years ago, mammals underwent a rapid diversification, with 15–17 new orders appearing in a relatively short geological span.

The appearance of other classes of animals has been found in the fossil record, including winged insects, dinosaurs, turtles; and marine reptiles; all with no precursor or ancestral forms in preceding strata. An inexplicable outcome if all life is to be explained by "descent with modification."

There are only a handful of examples of animals making an appearance in the fossil record where there *is* a possible claim of a precursor form that they developed from. Besides these limited exceptions, the consistent pattern that is found are innumerable abrupt appearances of biological forms in the fossil record that lack any discernible connection to similar forms in lower sedimentary strata. These new forms remain static over long periods of time, contravening the evolutionary hypothesis of long, slow, constant, and accumulated change. Alternatively, they go out of existence with no related forms having evolved *from* them either prior to or after extinction.

If evolution is true, and species change occurs in incremental steps, then nearly the entire fossil record should be made up of intermediary forms with very few static forms. This is not the case.

Darwin knew about this problem and put forward his theory under the caveat that future evidence within the fossil record would need to support continual evolution. This has not happened. The fossil record has instead illustrated the opposite, which is to affirm that no definitive Darwinian macro evolutionary process has any evidence of having taken place.

The genetic and fossil evidence considered so far exposes a consistent limitation: mechanisms proposed by evolutionary theory fail precisely where large-scale biological innovation is required. Incremental genetic change may explain variation within existing forms–micro evolution—but it does not account for the origin of new anatomical systems, new body plans, or developmental architectures. To address these gaps, defenders of evolution increasingly appeal to additional mechanisms beyond classical genetics. The next section examines two of the most prominent of these appeals—epigenetics and irreducible complexity—and evaluates whether they genuinely resolve the underlying problem or merely relocate it.

Irreducible Complexity

Certain biological systems require multiple, interacting components to function at all. Because partial versions of these systems would be nonfunctional, they cannot be assembled through a series of small, independently useful steps. In such cases, gradual genetic change lacks a viable pathway, even in principle. The system only works when all essential parts are present and properly integrated: partial or intermediate versions would be nonfunctional. As a result, such systems cannot be built through a series of small, independent, functional steps.

The difficulty for evolutionary theory is straightforward. Natural selection can only preserve changes that confer an immediate functional advantage. If a system provides no benefit until all of its components are in place, then selection has nothing to act upon during the intermediate stages. The process stalls before it begins.

A simple analogy helps illustrate the problem. A car cannot perform its function until its core components—engine, transmission, wheels, steering, and braking—are present and coordinated. Incomplete versions do not provide partial transportation benefits. While biological systems are vastly more complex than machines, the principle is the same: functionality depends on the coordinated presence of many parts.

This creates a second, deeper difficulty. Evolutionary theory describes its mechanism as unguided and non–goal-directed. Yet irreducibly complex systems require parts that are not only present, but precisely matched and integrated toward a future function. There is no evolutionary mechanism that explains how a blind process could assemble such systems when the final functional outcome cannot be selected for until it already exists (under the assumptive mechanism of evolution).

Irreducible complexity therefore exposes a structural limitation of random mutation and natural selection. The problem is not merely improbability, but the absence of a viable step-by-step pathway.

Biochemist Michael Behe, author of *Darwin's Black Box (1996)* defines irreducible complexity as, *"a single system which is necessarily composed of several well-matched, interacting parts that contribute to the basic function, wherein the removal of any one of the parts causes the system to effectively cease functioning*[39]*."* Behe provides examples of irreducibly complex machines in the cell: the eukaryotic cilium; the intracellular transport system; and the bacterial flagellum. These structures demonstrate that a blind, random process cannot account for a structure which requires several complex, interacting parts which must *all evolve together* towards an *end goal* that could not have been previously intended.

Irreducible Complexity in Life Cycles

The case histories of animal and plant adaptations are replete with examples that are impossible if they have to occur in random, unguided, small, incremental steps (irreducible complexity). Among innumerable examples, we have selected the life cycle of the liver fluke and the life cycle of the butterfly.

The liver fluke begins life as an egg within the intestines of a sheep. When the egg passes through the feces of the sheep and lands on the ground, it hatches. If it is fortunate to find water, it swims, finds a pond snail, works its way into the snail's pulmonary chamber or lung, grows in size, and buds off germinal cells that become a secondary type of larva. This larva then migrates into the other tissues of the snail. Here the larva develops into yet a third larval form that in appearance resembles a tadpole. This "tadpole" then works its way through the snail onto a blade of grass. Once on the grass, if it is lucky, it is eaten by sheep. Once inside the sheep, it migrates to the liver and develops into its adult state. In its adult state it then leaves the liver to migrate to the intestines of the sheep. In the intestines it mates and gives rise to offspring who start the life cycle *all over again*. Like humans, all of these expressions are coded for within the organism. An organism that under our materialistic worldview and Darwinian paradigm, arrived at this entire process *by chance* and is necessarily, genetically, *wholly ignorant* of all of these goings on.

A caterpillar forms a chrysalis and releases an enzyme that *digests* all of its own tissues rendering them into a "cell soup". Imaginal disks within this soup, -specialized cells that are present throughout the lifecycle and develop mostly in the chrysalis stage-, are then "fed" this soup and what was previously a caterpillar slowly develops into a butterfly. If evolution proceeds by small, incremental steps, then this is not possible. If all of the *complete* plans/steps needed to become a butterfly are not present once the caterpillar digests itself then the caterpillar remains a cell soup and dies, completely unable to turn into anything else. And if the full plan for becoming a butterfly is in place when the caterpillar first digests itself, then it can't be a result of small, gradual steps within an "unguided" evolutionary process.

There are innumerable examples like the ones above. There are no animals that can live their entire life cycle without several irreducibly complex systems and/or processes taking place. The maturation of every lifeform from newly born to maturity is representative of an irreducibly complex process, as is gamete production, embryological development, gestation, digestion, liver function, the nervous system, pulmonary system, etc. If intermediary steps are inconsequential, and all systems are derived randomly and without goals, then there is no reductionist explanation for how any such systems arise no matter how simple they may have been in the beginning.

Darwin's Own Acknowledgment of the Problem

Darwin's position was not that the problem of complex organs like the eye had been solved, but that one day it would be. He acknowledged the difficulty openly and proceeded on the expectation that future discoveries would reveal a sequence of small, functional intermediary changes capable of bridging the gap. But we must remember, if we are speaking of Darwin proper, what was available in the 1850's to explain the development of the eye was limited to 5 evolutionary forces: (1) sexual selection; (2) inherited effect of use and disuse (descent with modification); (3) random variation and natural selection; (4) correlation of growth (some changes get carried along because they are linked to other beneficial traits); and (5) Darwin's famous unnamed "unspecified cause" which resembled a kind of teleological force, antithetical to his theory, and if goal-directed, enough to invalidate his theory entirely.

Darwin was guessing, not just about the contents of a black box, but guessing without even knowing that any kind of box existed.

More than a century and a half later, the situation is not one of clarification but of complication. Advances in genetics, embryology, and molecular biology have not uncovered such a pathway; instead, they have exposed additional layers of coordination and constraint that make the gradual assembly of a complex eye and/or any biological structure, increasingly difficult to account for.

> "... now small trifling particulars of structure often make me very uncomfortable. The sight of a feather in a peacock's tail, whenever I gaze at it, makes me sick!" -Charles Darwin, Letter no. 2743[40]

Epigenetics - The Limits of Genetic Change

Although natural selection can account for variation within populations—such as differences in beak size, appendage thickness, or body size—isolated DNA changes are insufficient to explain how complex biological traits are regulated. Large-scale developmental outcomes require coordination among regulatory layers, and one of the most crucial of these is epigenetics.

Epigenetic factors are stable, heritable features that influence gene expression without altering the underlying DNA sequence. These include DNA methylation, histone modifications, chromatin accessibility, and non-coding RNAs that regulate chromatin states. By controlling which regions of the genome are accessible, epigenetics shapes the potential for genes to be expressed at specific times and in specific contexts. This means that changes in DNA alone cannot directly reorganize developmental programs or produce new structural outcomes[41].

This limitation becomes clear in embryological development. The emergence of distinct body plans depends on precise control of gene expression over time. Epigenetic mechanisms act as gatekeepers, ensuring that certain genetic sequences are turned on or off at the correct stage of development. DNA sequence mutations alone cannot guarantee these orchestrated patterns, because the accessibility and timing of genes are controlled by epigenetic regulation[42].

Epigenetic regulation is therefore central to how an organism develops from embryo to adult. It determines which genes are available for activation and influences the overall trajectory of development. Crucially, these regulatory frameworks cannot be rewritten simply by altering DNA sequences; they function as higher-order constraints on the potential outcomes of genetic information.

Even Charles Darwin recognized the importance of continuity in development. He noted, "If it could be demonstrated that any complex organ existed which could not possibly have been formed by numerous, successive, slight modifications, my theory would absolutely break down[43]." While Darwin did not know about epigenetics, modern understanding of these mechanisms highlights how DNA sequences alone are insufficient to explain developmental outcomes.

Gene Regulatory Networks & The Sugar Code

In a stunning paper by molecular and cell biologist Dr. Jonathan Wells, *The Membrane Code: A Carrier of Essential Biological Information That Is Not Specified by DNA and Is Inherited Apart from it*, Wells challenges the assumption that DNA alone controls the development of a newly fertilized embryo[44]. Compelling evidence from multiple studies shows this is not the case. Embryo development in particular occurs through a series of sequential switches and triggers, each initiating new patterns that guide further development. None of these developmental switches are directly encoded in the DNA sequence. While DNA provides the instructions for proteins and regulatory elements, it does not specify the spatial, temporal, or state-dependent information necessary for gene activation during embryogenesis. These functions are executed by gene regulatory networks (GRNs): interacting systems of genes, transcription factors, and signaling pathways whose behavior depends on context, feedback, and prior cellular states[45].

GRNs can distinguish between different regions of a developing embryo and coordinate development accordingly, even though this spatial information is not explicitly in the DNA of the fertilized egg. Spatial patterning emerges dynamically as cells respond to positional cues, signaling gradients, and prior regulatory states integrated by GRNs, a phenomenon observed across many animal species[46]

As Harding and Hancock (2008) note, these regulatory circuits are not simple on–off switches; they instead, "integrate and process signals to operate as switches, oscillators, logic gates, memory modules and many other types of control systems." This systems-level processing enables cells to respond appropriately to internal and external cues that guide growth and development[47]

If the information guiding the embryo development of all life forms is not fully contained in DNA, then the question arises: how can mutations, body-part changes, and speciation—processes supposedly genome-focused—be coordinated with developmental precision by an entirely separate system? This additional information, sometimes called the "sugar code," appears to reside on the surface of cells. Like DNA, it carries information, but its capacity is far greater: with just seven sugars, it can form hundreds of thousands of unique "words," most no more than five letters long, sufficient to identify the many different types of cells in a complex organism[48] (This code functions like an alphabet, more diverse and complex than the four-letter DNA code (ATGC). This code "can be rearranged to form hundreds of thousands of unique words, most of which have no more than five letters[49]."

The sugar code likely plays a critical role in embryonic development and life in general. Because this code is not contained within DNA, it *does not* follow the same direct heritability process of DNA. It is believed that cells as they divide transmit this code during division, but the exact mechanism of transmission remains unknown.

Embryonic development is also influenced by electrical currents within cells and organs. Experiments altering these currents can halt or modify development. These forms of information complement DNA and the sugar code, providing the stage-by-stage guidance necessary for constructing tissues in three-dimensional orientation. The process is likened to the metaphor of DNA providing the blueprints, while the sugar code provides the information to directs the timing, placement, of hardware, and specific materials—i.e., the "bricks" and "wiring"—in specific locations to assemble a functioning organism[50]

The conclusion is clear: embryo development is not controlled solely by DNA. This fact disrupts the primary axiom of Darwinian evolution, because if the genetic code alone cannot dictate the plan and formation of a life form, then random mutation and natural selection acting only on DNA, as the Neo-Darwinian synthesis proposes, cannot fully account for the creation and development of complete organisms.

The existence of a second, separate informational code raises a further question: if there are two coordinated codes—DNA and the sugar code—how could these completely different systems have arisen independently through random mutation and natural selection, yet still work together seamlessly, in a coordinated and synergistic manner in embryo development, and ultimately drive speciation? This represents a foundational level of irreducible complexity far beyond what has typically been envisioned.

DNA, in addition to its main genetic duties, codes for networks of genes that do not themselves form physical body plans. Instead, these gene regulatory networks (GRNs) direct other genes on how to build the organism. GRNs are crucial because although each cell (except a few specialized types) contains a full copy of the genome, cells cannot autonomously assemble the organism without guidance. GRNs specify each cell's type (lung, heart, liver, brain skin, or bone, etc), and role within the overall developmental plan[51] Complicating matters even further, GRNs are not physically in DNA, but DNA encodes the genes and regulatory sequences that form the network; the GRN emerges when those components interact in time and space and are triggered into action from signals and states outside of the genes themselves. And these triggers aren't singular; they are a cascade of triggers. The entire process is extremely complex and has yet to be fully deciphered. This raises a serious challenge for Darwinian evolution as the creative mechanism for biology[52].

> "The components of these networks transmit signals (known as transcriptional regulators or transcription factors) that influence the way the individual cells develop and differentiate. For example, exactly when a signaling molecule gets transmitted often depends upon when a signal from another molecule is received, which in turn affects the transmission of still others –all beautifully coordinated to perform specific time-critical functions. These networks of genes and gene products function much like integrated circuits and ensure the developing organism produces the right proteins at the right time to service the right types of cells during embryological development[53]."

Understanding GRNs shows that the complex interrelated feedback and timing within these networks could not plausibly arise from random processes. GRNs involve one of the most intricate series of interactions imaginable, and the stakes are the highest possible: the formation and viability of life itself.

> "...the body plan is formed by the execution of an inherited genomic regulatory program for embryonic development. The basic control task is to determine transcriptional activity throughout embryonic time and space, and here ultimately lies causality in the developmental process. The genomic control apparatus for any given developmental episode has a physical reality: it consists of the specifically expressed genes which encode the transcription factors required to direct the

> events of that episode, most importantly including the cis-regulatory control regions of these genes. The cis-regularity sequences combinatorially determine which regulatory inputs will affect the expression of each gene and what other genes it will affect; that is, they hardwire the functional linkages among the regulatory genes, forming network subcircuits. The subcircuits perform biologically meaningful jobs, for example, acting as logic gates, interpreting signals, stabilizing given regulatory states, or establishing specific regulatory states in given cell lineages; here the term "regulatory state" means the total of active transcription factors in any given cell at any given time. In turn the subcircuits are "wired" together to constitute the gene regulatory network (GRN), the overall genomically encoded developmental control system[54]."

The claim, by evolutionary theorists, is that GRNs, particularly developmental GRNs (dGRNs) are flexible and may through manipulation and mutation generate new animal body plans. They are even theorized by these individuals to be able to *drive the process of evolution*. This is not possible. Caltech dGRN researcher Eric Davidson observed that mutations affecting dGRNs that regulate body-plan development led to "catastrophic loss of the body part or loss of viability altogether[55]."

> "There is always an observable consequence if a dGRN subcircuit is interrupted. Since these consequences are always catastrophically bad, flexibility is minimal, and since the subcircuits are all interconnected, the whole network partakes of the quality that there is only one way for things to work. And indeed, the embryos of each species develop in only one way[56]."

One Darwinist response to the implausibility of dGRNs driving evolution is to concede that evolution cannot address the issue in the present, but to speculate that GRNs were highly flexible in a far distant past. Under this view, early GRNs could have been altered more easily and thus served as pathways for early evolutionary change. In a debate documented by the Discovery Institute between UC Berkeley paleontologist Charles Marshall and author of *Signature in the Cell*, Stephen C. Meyer, Marshall argued that modern GRNs are "overlain with half a billion years of evolutionary innovation," making them resistant to modification, whereas GRNs at the time of the emergence of the phyla were not so encumbered[57]. Luskin comments: "The only reason Marshall would have said this is if modern dGRNs are in fact 'so encumbered' that they could not provide a model for evolution. Marshall was forced to ignore modern experimental data and speculate that perhaps in the past dGRNs were different[58]..."

To be clear, there is no evidence that GRNs were any different in the past. Claiming they were, obligates proponents to present evidence for this assertion and to explain how a highly flexible early system could have evolved into an inflexible one that no longer drives evolution, nor is driven by random mutation alone. Suggesting a system is resistant to mutation is equivalent to saying it is resistant to evolution itself—an idea fundamentally at odds with evolutionary theory.

DNA contains the genetic blueprint. Epigenetics provides the heritable regulatory framework that decides which parts of the DNA blueprint are read. Gene regulatory networks (GRNs) execute this blueprint, coordinating timing, spatial organization, and signaling across cells to build tissues and organs. The sugar code, carried on cell surfaces, conveys cell identity and spatial orientation, telling cells where to go and what to become. The sugar code is separate from DNA.

Darwin knew nothing of DNA, epigenetics, GRNs or the sugar code.

Genetic Entropy

In his book, *Genetic Entropy* Dr. John C. Sanford, the inventor of the "gene gun", states that the consensus among geneticists is that the human genome is *degrading* due to the rapid accumulation of deleterious mutations. He goes on to say that this mutation rate is "*much higher than was previously thought possible*[59]." Dr. Sanford states that the current estimated mutation rate stands at 1-2% and possibly as high as 5% per generation. The degradation of genetic information is far greater than can be fixed by natural selection. Not only is natural selection helpless in addressing or correcting a 1-2% DNA degradation rate, it is equally unable to reverse this trend. In light of this revelation, maintaining the view towards natural selection as a positive force in genetic diversity is "...(an) extremely naive perspective[60]..."

> "The inability of natural selection to effectively remove large numbers of low-impact mutations has major implications regarding the long-term maintenance of the genetic integrity of populations. A substantial but unknown fraction of the many mutations in each eukaryotic individual must be deleterious. Yet this study indicates that most such deleterious mutations are too subtle to respond to natural selection... Unless some entirely unknown mechanism is operating, it appears that net genetic deterioration is an inevitable aspect of the mutation/selection process, given known mutation rates and fitness effects[61]."

The implications of Genetic Entropy are devastating for the evolutionary worldview: For decades, the mantra of defenders of evolutionary theory has been that "deep time"—given enough millions of years—random mutation and natural selection can solve all apparent combinatorial paradoxes found within the genomes of the world's species. Dr. Sanford's data have shown that the exact opposite narrative is far more likely time is the enemy, with succeeding generations adding more deleterious mutations to the genome with no known process to stop genetic collapse.

The implication is clear: biological life on Earth is not spiraling upward toward perfection; it is spiraling downward, leaving science with no mechanism to explain the depiction of the most famous evolutionary icons: *The March of Progress*, progressing from crouched ape to upright man, with a series of faithful and direct iterations, culminating in *Homo sapiens*. We must now turn our attention to how this story, the backbone of the human evolutionary texts, is itself unable to find support from within the genetic record.

Imaginings From The *March of Progress* and The *Missing Link*

The March of Progress is one of the most iconic images of evolution. It is a depiction of the transition in the silhouette of a stooped ape gradually straightening its spine, shedding its hair, and transforming seamlessly into a modern man. This confident icon of evolution occludes the reality that among paleontologists there has been a persistent controversy over the facts underlying this iconic depiction[62]. As Dr. Wells points out, when we strip away the reconstructive artwork and look strictly at the hard evidence, this "path" from ape to man disappears. There is no agreed-upon consensus that the bone and skull fragments used to illustrate *The March of Progress* represent distinct species. Nor is there agreement about their sequence, or about how much of the reconstructed "species"—often based on minimal fossil material—reflects interpretive reconstruction rather than clear fossil evidence.

The reality of the fossil record is not a smooth ascent, but a series of disconnected dots. Modern humans appear on the timeline roughly 300,000 years ago—suddenly, distinctly, and without a clear chain of ancestors linking them to the animal kingdom. Even the presumed "links" we do find, such as Neanderthals, betray the narrative. Since we now know Neanderthals could breed with Homo sapiens to produce fertile offspring, they do not represent a different species evolving into man; they represent variation within humankind—an example of microevolution (like breeds of dogs), not the macroevolutionary leap the textbooks demand.

Author Jonathan Wells, who holds Ph.D.'s from Yale and UC Berkeley in theology, molecular biology, and cellular biology, addresses the lack of consensus regarding the image of a definitive progression; typically beginning with *Pliopithecus*, a miocene primate from ~15-20 million years ago to modern man in his book, *Icons of Evolution: Science or Myth?*[63]

> "According to Darwin, the origin of the human species was fundamentally similar to the origin of every other species. Human beings, he argued, are modified descendants of an ancestor they shared with other animals (most recently, the apes), and their distinctive features are due primarily (though not exclusively) to natural selection acting on small variations." (J. Wells, *Icons of Evolution: Science or Myth?* Pg. 209, Regnery Publishing 2002)

Wells argues there is no fossil evidence that humans evolved from actual apes or from any other primate. And humans in this respect are in good company, as there are no animals on this planet that have been shown scientifically and irrefutably to have evolved from any bona fide separate species. The view of a progressive series of intermediate forms culminating in modern humans is due to interpretation through the lens of reverse-engineered Darwinian implication and not scientific evidence.

A recent find of skulls illustrates the controversy and lack of consensus among paleontologists. The find is documented on the online site for *Science Magazine* under the title, *A Complete Skull from Dmanisi, Georgia, and the Evolutionary Biology of Early Homo*[64], The story was also cited in an article on CNN, *Rare Skull Sparks Human Evolution Controversy*, by Elizabeth Landau, October 19, 2013[65].

> " 'In the eastern European nation of Georgia, a group of researchers has excavated a 1.8-million-year-old skull of an ancient human relative... (they) say it belongs to our genus, called Homo. This is the most complete early Homo skull ever found in the world,' said lead study author David Lordkipanidze. Lordkipanidze and colleagues also propose that these individuals are members of a single evolving Homo erectus species, examples of which have been found in Africa and Asia. The similarities between the new skull from Georgia and Homo erectus remains from Java, Indonesia, for example, may mean there was genetic 'continuity across large geographic distances' the study said. What's more, the researchers suggest that the fossil record of what have been considered different Homo species from this time period – such as, 'Homo Ergaster, Homo Rudolfensis and Homo Habilis – could actually be variations on a single species, Homo Erectus[66].'"

The most popular current *evolutionary* view is that there were several proto-apes, apes, and proto-humans that formed an unbroken ancestral lineage from ape to modern man. What is missing is the yet-to-be-discovered last universal common ancestor between ape and human, the apocryphal "missing link". We must consider the suggestion by paleontologist David Lordkipanidze that the skulls currently labeled Homo Habilis, Homo Ergaster, and Homo Rudolfensis are actually one species. If Lordkipanidze is correct, then what we actually have in the fossil record is an example of *one species of ape* that in some depictions appeared somewhat more human-like, and in other depictions, was more ape-like and stooped. How this ape has been made to appear more human-like is then achieved by artistic license, not anthropology, aided by the desire to have a clear progression of intermediary forms in the run-up to modern humans. This is the most likely explanation for why we have a single homo species split into three distinct species to give the appearance of evolution in action.

If there is one *proto-human, ape-like* creature, instead of three, it is a much shorter leap, aided by artistic license, to modern humans. But even with the aid of vivid artistic recreations, paleontologists are as yet unable to close the gap from ape to man, and on this point, there is no disagreement. The modern origin of the ape to human **Missing Link* has its origin in a very real evidentiary gap that persist into the modern age. This gap, if it falls in line with the number of scientific gaps we have found thus far in our investigation, spanning from the cosmos to chemistry to biology, may represent a gap that *never was filled* in the fossil record, and thus has no evidence that can complete the evolutionary hypothesis.

It may be the case that the existence of a link between man and ape may be wholly imaginary. A Darwinian inspired illusion. The missing link has authentically never been found, so if the only at this point can only be inferred/imagined. Equally damning, there is no current link between *any* presently existing species on Earth and *any* fossil evidence that can illustrate a smooth evolutionary path to the apocryphal, last universal common ancestor (LUCA). Apes and Men are separate. No species links them.

> "When it comes to reconstructing entire individuals or the history of human evolution, however, agreement is hard to find. One area of disagreement is how many species of human-like apes or ape-like humans co-existed at any given moment. The 'lumpers' tend to group all specimens into one of a few species, while the 'splitters' divide them into many more. Even if agreement were to be reached on which specimens represent separate species, there would still be the question of whether they are ancestors of modern humans or extinct side-branches of the evolutionary tree. Disagreement also

continues between the 'out of Africa' camp, which maintains that modern humans first evolved in Africa and then spread throughout the world, and the 'Multiregional camp, which argues that our species evolved in many places simultaneously[67]."

The evidence in the fossil record indicates that there is an ape (*Australopithecus Afarensis)* artistically depicted with a more human-like countenance, and also a human (*Homo Habilis)* depicted with a more ape-like countenance. Between these two is where the "missing link" is believed to have spanned the gap that may have never existed.

-APE: ***Australopithecus Afarensis*** 3.8m - 2.9m years ago &

-***Austraopithecus Africanus*** 3.3m - 2.1m years ago.

-MISSING LINK?

-HUMAN: ***Homo Habilis*** 2.31m - 1.65m years ago &

-***Homo Erectus*** 2.0m to 117,000 years ago.

-*"Modern"* HUMAN: ***Homo Neanderthalensis*** 430k - 130k years ago - ***Homo Sapiens*** 300k years ago

To attempt to resolve these issues 29 scientists gathered to re-examine neo-Darwinian theory and produced findings from 24 research papers and published their findings in, *the Synopsis of Scientific Proceedings: Biological Information - New Perspectives*. These scientists were from the fields of information theory, computer science, numerical simulation, thermodynamics, evolutionary theory, whole organism biology, developmental biology, molecular biology, genetics, physics, biophysics, mathematics, and linguistics. From their press release:

> "(the authors) unanimously agree that the mutation/selection process is not adequate to explain the labyrinth of informational networks that are essential for life[68]."

And,

> "Within each cell there are information systems surpassing the best human information technologies. These systems create what is essentially a biological Internet within each cell."

In the first quote, we find an indictment on the evolutionary mechanism itself—essentially the entire Darwinian concept. What the authors illustrate in the process is that the flaws they have uncovered cast even more doubt on our ability to find a direct link between ape and man given the paucity of evidence. A commonsense read is that there simply is not enough to infer evolution, as science has done for two centuries, when the entire mechanism is questionable and evidence is sorely lacking.

The fact that twenty-nine experts from fields ranging from quantum physics to computer science have converged on a shared verdict means the evidentiary ground has shifted, and the widespread presumption of an evolutionary progression for life on Earth—particularly given the vast gaps and abrupt instantiations in the fossil record—faces increasing explanatory difficulty. Not only does evolution fail in providing the fossil evidence for a smooth evolutionary progression from apes to men, but it fails to account adequately for *any* life form on Earth. Life is certainly on Earth. It is certainly diversified, but evolution has not produced sufficient credentials to be named the cause. This relegates the iconic *March of Progress* for the human past, evolution an unconvincing meme.

Now we will turn to another facet of our analysis.

If the cell truly surpasses the best human information systems in complexity, as the evidence suggests, we must address the most common objection raised against it. For decades, materialists have argued that the genome is not an efficient code, but an oversized graveyard of broken genes and useless data that it cannot purge (commonly referred to as "junk DNA"). They point to the baffling disparities in genome size—where some simpler organisms and plants, possess more DNA than humans—as evidence of random, messy evolutionary accretion. This discrepancy between apparent biological complexity and genome size is known as the *C-Value Paradox.*

This new evidence raises the question of whether the genetic code is actually broken—a charge which appears to support certain interpretations of the evolutionary hypothesis—or whether it is instead robust, meaningful, and coherent as a functional biological information system. With recent advances genetics endorses the view that the genome is an efficient, and robust informational system reflecting a level of biological information architecture that earlier scientific eras lacked the technological sophistication to fully comprehend.

The C-Value Paradox, Junk DNA and Trans-Splicing

Genomes can differ wildly in size seemingly with no regard for the complexity of the lifeform they represent. This counter-intuitive result is referred to as the *c-value enigma* or *c-value paradox.* "'c-value' means the 'constant' (or 'characteristic') value of haploid DNA content per nucleus, typically measured in picograms (1 picogram is roughly 1 gigabase)," with a 'gigabase' equaling 1 billion base pairs (the human genome is rougly 3 billion base pairs)[69].

The paradox refers to the seeming contradiction between the complexity of an organism and the length of its genome. For example,

Polychaos dubium, a freshwater amoeba, has a genome 670 billion base pairs in length. A flowering plant called *paris japonica* has a genome 149 billion base pairs in length. Compare this with the human genome (*homo sapiens*) at 3 billion base pairs in length. The mystery is why the c-value paradox exists and why what was assumed to be "extra" DNA was needed by relatively simple organisms?

When DNA was first sequenced, a number of genes were identified, but over time it became apparent that there were vast stretches of the genome that were not genes and held a purpose that was unknown. The number of sections that appeared to have no known function appeared to far outweigh sections of the genome that could be identified as having function. This discrepancy, along with the c-value paradox, led to the hypothesis that the majority of the genome, particularly for animals with vast c-value discrepancies, was "junk DNA".

> "The human genome has three billion base pairs in its DNA, but only about 2% of them encode proteins. The rest seems like pointless bloat, a profusion of sequence duplications and genomic dead ends often labeled 'junk DNA[70].'"

The junk DNA hypothesis holds that there are long, repetitive non-coding regions within DNA composed of sequences of functionless DNA. This theory posits that these regions may be deactivated coding sequences for no longer needed body parts, old body plans, and/or remnants from previous species and/or made up of "selfish" genes that have found a way to remain in the genome without a purpose, or what has been called, "nonfunctional evolutionary leftovers that are just along for the ride[71]." This bloat in the genome results has been framed as a conflict between "selfish" and "unselfish" genes, and that differing c-values among species may depend on which side is "winning"[72]. The junk DNA hypothesis has ripened to predict that only 1-20% of the human genome is genetically active and needed by the organism. The remainder is conjectured to be evolving neutrally[73].

New research has provided a hammer blow to the hypothesis of Junk DNA. This evidence provides reason to believe very little, if any "junk" DNA is in the genome of any animal[74]

The ENCODE Project (Encyclopedia of DNA Elements), a massive international public research consortium, composed of hundreds of researchers from 32 institutes worldwide, released the results of a decade-long study involving over 400 scientists. In 2012 they discovered that the vast majority of this so-called 'junk'—at least 80% of it—is biologically active. 'Junk DNA' is not garbage; it is the Operating System. These areas, previously unknown, play key functions in the preparation and maturing of the genome for protein production and together form, "an elaborate patchwork of regulatory sequences that act as a huge operating system for controlling the genome[75]."

While the 2% of DNA that codes for proteins acts as the 'bricks' of the body, the remaining 98% acts as the 'architect.' This non-coding DNA regulates when and where genes are turned on or off. It controls the dimmers, the switches, and the timing. Far from being evolutionary debris, this section of the genome is a sophisticated control panel of staggering complexity.

One of the kinds of DNA classified as "junk" by the hypothesis are *retrotransposons*. It has been found that between 6-30% of RNAs *start* with retrotransposons. Based on the distribution of retrotransposons, over 23,000 candidate regulatory regions have been found within the genome (mouse and human) and 2,000 examples of bidirectional transcription caused by the presence of retrotransposons have been identified[76]. The result? It is likely that nearly *the entire genome is functional*, not just 1-20% as predicted by the Junk DNA hypothesis.

Far More Complexity

How this non-coding genome works has revealed a far more intricate and complex system of genetic information processing than previously imagined. The genome, rather than representing solely a library of codes for protein production, instead provides *templates* for protein production. These templates must be copied and transported where they can be spliced and mixed with regulatory codes. This allows genes selected from the genome by other proteins for transcription/replication to be manufactured and used for their intended purpose. This discovery was observed by a team led by molecular biologist A.J. Rachel, of the CCMB in India[77]. Dr. Rachel's team was able to identify 2 mRNA transcripts in the midst of a 40 million base pair "desert" previously believed to be non-coding regions of the genome of the *Y chromosome*. These transcripts *were observed to leave* the Y chromosome entirely and mix with an mRNA transcript *on chromosome-1*. This type of travel and mixing *between* different chromosomes is an example of 'trans-splicing" a previously unknown, unanticipated process discovered and identified by A.J. Rachel's team. In the observed process the regulatory code on *chromosome 1* splits the mRNA on the *Y chromosome*, disposes of unwanted parts, splices them with the wanted portion in order to mature and optimize the information on the code before the synthesis of the protein. This non-coding region *is then able to tell the protein* how much to make (amount), and when synthesis should occur (timing)[78].

How the molecular micromachines are able to identify code within the *Y chromosome* and transfer it to specific areas of *chromosome 1* that are needed for it to mix is an unknown process. It is unknown if the information needed to identify and transport the code to its needed destination as a preparatory process for transcription is explicitly contained in the genome. Similarly, it is also not well understood how a molecular micromachine is able to see, read, discern, transport, edit and then recognize when an mRNA is *ready* to be duplicated.

The stunning findings of the ENCODE consortium show that the molecular machines they have identified exhibit extraordinarily coordinated, goal-directed behavior. These systems operate with remarkable precision, displaying what appears to be foresight in their regulation, timing, and output. Even more striking, they function according to internal regulatory clocks that determine how, when, and how much they are supposed to produce. Their activity is not chaotic, improvised, or partially formed; it is regulated, synchronized, and integrated.

For these molecular machines, there is no evidence that intermediary states—partial systems or incomplete configurations—are compatible with their biological role. No intermediates in molecular systems. No intermediates in functional protein complexes. No intermediates in integrated biological machinery. Their functionality depends upon the coordinated operation of the entire system.

The absence of intermediary systems as described above presents a direct challenge to the Darwinian presumption of gradualism.

Darwinism depends upon gradualism because evolution, as traditionally described, proceeds through the accumulation of (mostly) helpful mutations one by one, each preserved through natural selection, until they eventually yield a new protein, enzyme, or body part. The mechanism requires incremental construction–gradualism. But here an intractable paradox emerges. A system that must function as a whole cannot arise through the gradual assembly of nonfunctional parts. For a molecular machine to perform its role, it must exist as a fully integrated unit. This is the problem of irreducible complexity as a structural constraint.

Summary

Timeline – The Ascendency of Darwin's Theory of Evolution

1809 — Charles Darwin born

1831–1836 — Darwin aboard HMS Beagle

1859 — *On the Origin of Species* published

1859–1870's — Evolution gains traction among scientists

1882 — Charles Darwin dies

1880's–1910's Darwinism widespread, but not dominant

1920's–1930's Scientific consensus begins to shift with understanding of genetics and Mendelian inheritance. The link between genetics and evolution remains unresolved

Neo-Darwinism & The Modern Synthesis Key Publications (Darwinism becomes the accepted explanation for speciation).

-1930: R.A. Fisher –*The Genetic Theory of Natural Selection*, unified Mendelian genetics with selection.

-1932: J.B.S. Haldane – *The Causes of Evolution*, developed mathematical models of selection

-1931-1932: Sewall Wright — *Evolution in Mendelian Populations*, & *The Roles of Mutation, Inbreeding, Crossbreeding, and Selection in Evolution,* developed ideas such as genetic drift, and adaptive landscapes

-1937: Theodosius Dobzhansky – *Genetics and the Origin of Species,* integrated Mendelian genetic with Darwinian evolution

-1942: Ernst Mayr – *Systematics and the Origin of Species,* defined species as reproductively isolated populations

-1942: Julian Huxley – *Evolution: The Modern Synthesis,* presented evolution as a coherent integrated theory.

These authors worked from outside the genetic "black box". With our current understanding of DNA, inheritance and development at the molecular level, it's clear their models fail precisely where they must be the strongest. This highlights the gap between what passed for plausible narrative in the early 20th century, and the rigor and specificity required of modern genetics.

1953 — DNA double-helix structure discovered (Watson & Crick)

1961 — The Sequence Hypothesis Proposed (Crick)

1966 — Protein synthesis mechanisms deciphered

1970's – 1980's RNA regulatory roles discovered

2002 — Modern framing of The Sugar Code ("glycobiology" etc., emerged decades earlier)

2004 — Gene Regulatory Networks in literature

2008 — Epigenetics consensus definition reached

Darwin's theory was, by his own admission, incomplete. He did not understand the nature of biological variation, the mechanisms of heredity, or the developmental processes required to generate form. While he correctly intuited that traits are inherited, he lacked a granular, molecular account of how inheritance operates and consequently, he could offer no robust causal explanation for the origin of species.

The modern synthesis, which substitutes random mutation for Darwin's unspecified random variation, attempts to resolve this gap, yet its probabilistic adequacy in the face of the 4-letter code of DNA remains deeply contested. The depth and complexity of DNA, and its operative functions have been examined in a previous section.

Taking all of this into account, it could be said that Darwin's enduring accomplishment was not the provision of a complete or mechanistically grounded theory of speciation, but the establishment of a conceptual foundation upon which evolutionary biology was later formalized. What is taught today retains historical continuity with Darwin primarily at the level of narrative—descent with modification—while exhibiting little mechanistic continuity with his original proposals. Modern discoveries overshadow Darwin's explanatory claims.

Importantly, the lasting structural effect of Darwinism has been the entrenchment of a materialist, reductionist framework as the default explanatory boundary for biology. This framework has functioned not merely as a biological model, but as a broader socio-philosophical template shaping scientific inquiry across disciplines. While Darwin himself was not a strict materialist, the theory bearing his name has come to insulate science from explanations outside reductionism, while also excluding ideas that do not conform to reductionist criteria.

As we argue throughout this work, this insulation has contributed to many of the theoretical oddities, a series of intractable problems and a proliferation of dominant hypotheses that are preserved through auxiliary assumptions rather than the ability to withstand empirical tests. Yet they persist resistant to verification, and immune to definitive falsification.

Assessing The Probabilistic Likelihood of Darwin's Evolution

This chapter examined Darwinian evolutionary theory, beginning with the life of Charles Darwin—his upbringing, historical context, intellectual influences, and opportunities for scientific training and peer collaboration. Darwin developed his theory of natural selection through careful observation of biological variation during his voyage aboard the HMS Beagle.

Darwin proposed that all biological diversity descends from an original living system that diversified over time to produce the diversity of life on Earth. While evolutionary theory is often presented as a complete explanation for biological diversity, significant unresolved problems remain when biological life is analyzed as an information-driven system. Darwin developed his theory without knowledge of molecular biology, genetics, or the challenges later revealed by the discovery that the genetic code is a primary target of speciation. In the nineteenth century, Darwin's theory was initially met with resistance and only gained broad acceptance in later generations, eventually becoming the dominant framework for explaining biodiversity in universities.

Modern biology has shown that genetic sequences alone do not explain biological organization.

Paleontology presents additional challenges. The fossil record shows significant gaps between major biological groups. The Cambrian explosion is a prominent example, where most major animal body plans appear within a relatively short geological period. Similar rapid diversification events appear throughout Earth's history without clear explanation. The "march of progress" depiction of evolution from ape to man is misleading because it suggests linear advancement. Biological evolution is better described as branching diversification rather than progressive improvement.

Genetic entropy and the C-value paradox further challenge simple mutation-selection models. Genome size does not reliably correlate with organismal complexity, raising questions about how biological information is generated and maintained.

Value of Special Problem # 3: Darwinian Evolution

The situation we find ourselves in is quite unusual. The fossil record affirms that there has been increasing *speciation,* over billions of years, gaining in complexity, diversifying in body plans and culminating in the emergence of human beings. The fossil record also shows that new species emerge in the fossil record on timescales that preclude both random mutations, and any namable naturalistic process. Despite this the defense of evolution is as strong today as it ever was. Nothing has been conceded and even more has been claimed as ad hoc additions and explanations for the theory's deficiencies proliferate.

Presently there are no alternative theories for the diversification of life on Earth. The problem is not one that Darwinian theory holds alone. To make sense of life and the diversification of life appears to call for an entirely new worldview, and a change agent more in line with Darwin's 5th unnamed force, than classical evolution, or anything that aligns with the principles of physics, chemistry or biology from the perspective of bottom-up materialism, or top-down reductionism.

As the 3rd *Special Problem* in our *Algorithm*, coming after our 13th *Variable*, the diversification of species is inexplicable, and yet, it has happened within our midst. It is a miracle of all miracles. The creation of each new species represents a separate, inexplicable *miracle* of creation.

Value of Special Problem #3: Darwinian Evolution

Odds Darwinian Evolution is true: 0 = *M (miracle-impossible).*

---

[1] (Macmillan & Co. London, (1908), Pgs. 6-9, 13, 15, 110-115, 117).,
[2] Ibid, Pg 9).
[3] (Ibid, Pgs 108,109)
[4] Ibid, Pgs 110-111)
[5] Ibid, Pgs 113-114).
[6] (G. Himmelfarb, Darwin And The Darwinian Revolution - Elephant Paperback, Orig 1959)
[7] (F.J. Sulloway, The Evolution of Charles Darwin, Smithsonian Magazine, December 2005)2
[8] (Harvard University, Early Evolutionist, Erasmus Darwin, Friedman Lab, Dept. of Organismic and Evolutionary Biology early-evolution.oeb.harvard.edu).
[9] (F.J. Sulloway, The Evolution of Charles Darwin, Smithsonian Magazine, December 2005).
[10] (N. Annan, L. Stephen, The Godless Victorian (New York: Random House, 1984).
[11] (N. Annan, L. Stephen, The Godless Victorian (New York: Random House, 1984).
[12] (N. Annan, L. Stephen, The Godless Victorian (New York: Random House, 1984)4.
[13] C. Darwin, On The Origin Of Species, 1859 Logos Publishing)
[14] (Agassiz (1869): Darwinism - Classification of Haeckel, Translation 1996, Paul J. Morris, Creative Commons, March 1, 1996, With thanks to Malcom Kottler, 1 of 3 chapters added to 1869 French Translation of Essay on Classification, under the title, De l'Espece et de la Classification en Zoologie. Pp. 375-391, Part 3, Chapter 7: Le Darwinisme. - Classification de Haeckel, Balliere, Paris, pp. 394 available at www(dot)athro(dot)com/general/atrans.html)
[15] (Agassiz (1869): Darwinism - Classification of Haeckel, Translation 1996, Paul J. Morris, Creative Commons, March 1, 1996, With thanks to Malcom Kottler, 1 of 3 chapters added to 1869 French Translation of Essay on Classification, under the title, De l'Espece et de la Classification en Zoologie. Pp. 375-391, Part 3, Chapter 7: Le Darwinisme. - Classification de Haeckel, Balliere, Paris, pp. 394 available at www(dot)athro(dot)com/general/atrans.html)
[16] (G Himmelfarb, Darwin and the Darwinian Revolution, Pg. 363. Elephant Paperback, Orig 1959).
[17] (G Himmelfarb, Darwin and the Darwinian Revolution, Pg. 363. Elephant Paperback, Orig 1959).
[18] (G Himmelfarb, Darwin and the Darwinian Revolution, Pg. 363. Elephant Paperback, Orig 1959),
[19] (G Himmelfarb, Darwin and the Darwinian Revolution, Pg. 369. Elephant Paperback, Orig 1959; Quoting, C. Darwin, The Descent of Man, 1st ed. Vol I, Pg. 153).
[20] (G Himmelfarb, Darwin and the Darwinian Revolution, Pg. 369. Elephant Paperback, Orig 1959; Quoting, C. Darwin, The Descent of Man, 1st ed. Vol I, Pg. 153).

[21] (M. Denton, The Miracle of the Cell, P. 139-141, Discovery Institute Press)
[22] (M. Denton, The Miracle of the Cell, P. 139-141, Discovery Institute Press)
[23] (From French Philosopher, Henri Bergson, Creative Evolution, P. 43, in, in G.Himmelfarb, Darwin and the Darwinian Revolution, Pg. 343-344, orig, 1959, 1996 Elephant Paperback)41.
[24] (C. Darwin, The Descent of Man 1st ed Vol I 152-53 Orig. 1871).42
[25] (E. Darwin, Zoonomia, Vol 1, Preface, 1794, J.Johnson/St. Paul's Churchyard)7
[26] (A. Gray, Asa, Darwiniana: Essays and Reviews Pertaining to Darwinism. 1876, New York: D. Appleton and Company (reissued by Cambridge University Press, 2009).
[27] C. Darwin, On The Origin Of Species, 1859 Logos Publishing)10
[28] (R.J. Britten, Divergence between samples of chimpanzee and human DNA
[29] (J.C. Sanford, W. Brewer, F. Smith, J. Baumgardner, The Waiting Time Problem in a Model Hominin Population, Their Biol Med Model, September 17, 2015, 12:18. PMC4573302) .
[30] (Carter, R.W., 'The slow, painful death of junk DNA' Journal of Creation 23(3) 2009).
[31] (J.C. Sanford, W. Brewer, F. Smith, J. Baumgardner, The Waiting Time Problem in a Model Hominin Population, Their Biol Med Model, Pg. 3 of 28, September 17, 2015, 12:18. PMC4573302)85.
[32] *Footnote: This 5% figure is the high difference between chimps and humans. The standard that is used in most publications is that chimps and humans share 98.8% (often rounded to "99%") DNA. For the rest of our inquiry, we will work with the 98.8% figure, as it does not change the impossibility of the scenario.
[33] (R. Carter, A Successful Decade for Mendel's Accountant, Journal of Creation, August 2019).
[34] (J.C. Sanford, W. Brewer, F. Smith, J. Baumgardner, The Waiting Time Problem in a Model Hominin Population, Their Biol Med Model, September 17, 2015, 12:18. PMC4573302).
[35] (J.C. Sanford, W. Brewer, F. Smith, J. Baumgardner, The Waiting Time Problem in a Model Hominin Population, Their Biol Med Model, Pg. 3 of 28, September 17, 2015, 12:18. PMC4573302)
[36] (J.C. Sanford, W. Brewer, F. Smith, J. Baumgardner, The Waiting Time Problem in a Model Hominin Population, Their Biol Med Model, Pg. 3 of 28, September 17, 2015, 12:18. PMC4573302)
[37] (S.C. Meyer, Darwin's Doubt, Pg. 32, HarperOne, 2013)43.
[38] (S.C. Meyer, Darwin's Doubt, Pg. 32, HarperOne, 2013)43
[39] (M. Behe, Darwin's Black Box, p. 39 Free Press, 1996)
[40] (C. Darwin, Darwin Correspondence Project, "Letter no. 2743," accessed on 1 April 2024, darwinproject(dot)ac (dot)uk/ letter/? docId=letters/ DCP-LETT-2743.xml)
[41] (S. C. Meyer, Darwin's Doubt, p. 281, HarperOne, 2013).
[42] (S. C. Meyer, Darwin's Doubt, p. 281, HarperOne, 2013).
[43] (C. Darwin, On the Origin of Species, 1859, Logos Publishing).
[44] (J. Wells, The Membrane Code: A Carrier of Essential Biological Information That Is Not Specified by DNA and Is Inherited Apart from it, 2013, Pg. 474, DOI 10.1142/9789814508728_0021).
[45] (J. Wells, The Membrane Code: A Carrier of Essential Biological Information That Is Not Specified by DNA and Is Inherited Apart from it, 2013, Pg. 474, DOI 10.1142/9789814508728_0021)
[46] (Ibid., p. 475).
[47] (Ibid., p. 478).
[48] Ibid Pg. 479).
[49] (Ibid Pg. 479).

---

[50] (Ibid Pg. 482).
[51] (S. Royle, Gene Regulatory Networks: From DNA to Development, blog, special edition on networks, sitn.hms.harvard.edu, August 30, 2021)
[52] (G. Peng, H. Jing-Dong, Regulatory Network Characterization in Development: Challenges and Opportunities, September 17, 2018, DOI: 10.12688/f1000research.15271.1)
[53] (S.C. Meyer, Return of the God Hypothesis, Pg. 311, HarperOne, 2021)
[54] (I.S. Peter, E.H. Davidson, Evolution of Gene Regulatory Networks Controlling Body Plan Development, Cell, Vol.144 Issue. 6,P970-985, March 18, 2011)55
[55] (E.Davidson, Evolutionary Bioscience as Regulatory Systems Biology. Developmental Biology, 357:35-40, 2011)56
[56] (D. Davidson, Evolutionary Bioscience as Regulatory Systems Biology. Developmental Biology, 357:35-40, 2011)
[57] (Cited by C. Luskin: C.R. Marshall, When Prior Belief Trumps Scholarship, Science, 341, September 20, 2013: 1344 (entire reference in): C. Luskin, Eric Davidson (1937-2015) on Gene Regulatory Networks, EN, Evolution News, October 18, 2015, evolutionnews.org)58.
[58] (Cited by C. Luskin: E.Davidson, The Regulatory Genome: Gene Regulatory Networks in Development and Evolution. Burlington: Elsevier, 2006, p. 195, (entire reference in): C. Luskin, Eric Davidson (1937-2015) on Gene Regulatory Networks, EN, Evolution News, October 18, 2015, evolutionnews.org)
[59] (J.C. Sanford, Genetic Entropy Pgs. 51-53, 4th Edition 2005 (1st), 2014 (4th), FMS Publications).
[60] (J.C. Sanford, Genetic Entropy Ibid).
[61] (P.Gibson, J. Baumgardner, W. Brewer, J. Sanford, Can Purifying Natural Selection Preserve Biological Information? (Pg. 249), 232-262, (2013) 10.1142/9789814508728_0010)
[62] (J. Wells, Icons of Evolution: Science or Myth?, Regnery Publishing 2002).
[63] (Regnery Publishing 2002).
[64] (D. Lordkipanidze, M. S. Ponce De Leon, A. Margvelashvili, Y. Rak, G.P. Rightmire, A. Vekua, C.P.E. Zollikofer, (Science, 18 Oct 2013, Vol 342, Issue 6156, pp.326-331).71
[65] (E. Landau, CNN, Rare Skull Sparks Human Evolution Controversy, 12:04pm, EDT, Sat, October 19, 2013, https// www .cnn .com/ 2013/ 10/ 17/ world/ europe/ ancient-skull-human-evolution)
[66] (E. Landau, CNN, Rare Skull Sparks Human Evolution Controversy, 12:04pm, EDT, Sat, October 19, 2013, https// www .cnn .com/ 2013/ 10/ 17/ world/ europe/ ancient-skull-human-evolution)72
[67] (J. Wells, Icons of Evolution, Pg 224, Regnery 2002)73
[68] (R.J. Marks II, M.J. Behe, W.A. Dembski, B.L. Gordon, J.C. Sanford, Biological Information - New Perspectives, Proceedings of the Symposium, Cornell University, May 31-June 3, 2011)
[69] (S.R. Eddy, The C-value paradox, Current Biology Vol 22 No 21 R898, junk DNA and ENCODE Sean R. Eddy, HHMI Janelia Farm Research Campus, Ashburn VA 20147, USA.)75
[70] (J. Buehler, The Complex Truth About 'Junk DNA', Quanta Magazine, Explainers, September 1, 2021, quantamagazine(dot)org/ the-complex-truth-about-junk-dna-20210901)
[71] (D. Smith, The Mysterious 98%: Scientists look to shine light on the 'Dark Geome', Phys.org, February 3, 2017)76.
[72] (Wong, H.Y. Selfish DNA. Encyclopedia of Genetics, 2001)77.
[73] (Eddy, S.R., 'The C-Value paradox, junk DNA and ENCODE' Current Biology, vol22, iss21, 2012)

[74] (Carter, R.W., 'The slow, painful death of junk DNA' Journal of Creation 23(3) 2009)
[75] (S. Connor, Scientists Debunk "Junk DNA" Theory to Reveal Vast Majority of Human Genes Perform a Vital Function, The Independent, UK, U.S. Edition, September 6, 2012)81.
[76] (Faulkner, G.J. et al., The Regulated Retrotransposon Transcriptome of Mammalian Cells, Nature Genetics 41(5):563–571, 2009)80.
[77] (A.J. Rachel, Y.R. Ahuja, H.M. Reddy, A. Suresh, L. Singh, S. Pradhan, S. Tiwari, S. Vallinayagam, Z. Jehan, Novel Noncoding RNA from Human YDistal Heterochromatic block (Yq12) Generates Testis-Specific Chimeric CDC2L2, Nov 9, 2006, Genome Res. 2007.17: 433-440, Cold Spring Harbor Laboratory Press, 10.1101/gr.5155706).
[78] (A.J. Rachel, Y.R. Ahuja, H.M. Reddy, A. Suresh, L. Singh, S. Pradhan, S. Tiwari, S. Vallinayagam, Z. Jehan, Novel Noncoding RNA from Human YDistal Heterochromatic block (Yq12) Generates Testis-Specific Chimeric CDC2L2, Nov 9, 2006, Genome Res. 2007.17: 433-440, Cold Spring Harbor Laboratory Press, 10.1101/gr.5155706)83

Interlude No. 1: The August Wheel

Why do humans believe ideas that are demonstrably untrue?

We have just examined Abiogenesis and Darwinian Evolution—two foundational pillars of the modern worldview that possess no supporting evidence and zero mathematical probability. Yet, they are defended with religious fervor.

They are not alone. The landscape of modern "science" is littered with concepts that are treated as fact despite being physically impossible or purely speculative: a universe from nothing, the inflationary multiverse, string theory, and quantum loop gravity, to name a few. The question returns: Why are impossible ideas regarded as absolute truths?

The answer lies in the architecture of the human mind. We are not purely logical creatures; we are creatures that can become ideological as well. The "capriciousness of human reason" means there is no such thing as a sacrosanct concept—not even "Science" itself.

Neither intelligence nor the accumulation of credentials offers immunity from a belief gone rogue. A PhD, a Nobel Prize, or a lifetime of study is no inoculation against this particular form of ideological blindness.

To explain what is happening we can turn to *The August Wheel*, a compass of sorts, that can provide a map of the human relationship with ideas, and objective truth, while also representing how and why the precipitous fall into illusion and delusion take place. All of these destinations are represented on The Wheel, as intended, or unintended destinations on the Wheel.

Introducing The August Wheel

The August Wheel can illustrate how human beings, born naked, dependent and knowing nothing move through life, accumulate understanding, and progressively trade in understanding for knowledge. The trajectory is hardly simple, nor assured, for there are hazards along the way as we have discussed above.

The August Wheel can show us how individuals fall into false ideas and how they elevate to true belief. It can't tell you if what one believes is true, but it can indicate exactly what has taken place in retrospect, and often, individuals recognize their position upon the August Wheel, and this recognition is enough for them to course correct in their trajectory.

Having been born into **Ignorance**, the 'ultimate' path is one we identify as **The Golden Path** that leads to a condition where one's **Belief** aligns with

Truth and the duality between what is believed, and truth dissolves into transcendence, which is a form of complete understanding by the individual. This is of course an ideal condition.

But there are other destinations on **The August Wheel**. These are bifurcated: between (1) **Illusion** and (2) **Delusion.** On The August Wheel, since Illusion and Delusion are not Truth, twhen one finds themselves at their destination, they have fallen, thus, we label Illusion and Delusion as **"The Fall" on The August Wheel,** because one has 'fallen' from The Golden Path.

Illusion & Delusion

When one does not know (is **unknowing**) what is **true,** and misidentifies truth, this represents the fall into **Illusion**.

When one **believes** something that is objectively a f**antasy** istrue, they have fallen into **Delusion**.

Coming from **Ignorance**, it is easy to fall into either delusion or illusion.

In these examples we can see that **Ignorance** is the original vulnerability, and we are all born into it.

Transcendence

The final destination on The August Wheel is transcendence. When what one **believes** is the **true,** then there is the opportunity for belief and truth to merge into **transcendence.**

The problem is of course, that delusion or illusion are often misidentified as transcendence (a belief in truth).

Discussion

If we look at The August Wheel as applied to Materialism or the dominance of the Darwinian worldview, we can witness a collective cultural Fall into Illusion or Delusion (depending on the reason for an individual's misidentification of Darwinian theory as a true belief. This misidentification creates ripple effects that distort the entire understanding of existence by human culture.

Ideally, science is our safeguard against such falls. Science is our measuring stick and judge of what ideas are Bonafide and which are bogus.

Thankfully, science still faithfully provides objective data. Our investigation is a testament to the quality of information provided by science. We would have no hope to form our hypothesis if it were not for the high quality of scientific information that has formed our argument.

The problem, exposed in our narrative, is that far too often, the data provided by science has been forced into a materialist's interpretation, such that only data that fits a materialist's interpretation is allowed to be held as true.

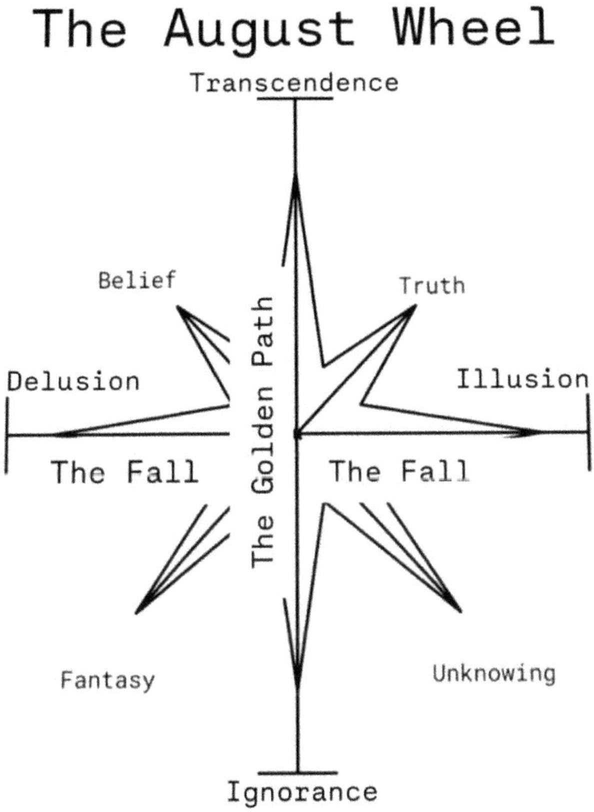

Interlude No. 2: Thomas Kuhn & Karl Popper

The American philosopher of science, Thomas Kuhn, author of *The Structure of Scientific Revolutions* (1962), and Austrian/British philosopher, Karl Popper author of *Conjectures and Refutations: The Growth of Scientific Knowledge,* and *The Logic of Scientific Discovery* (1963), have provided the road map for understanding the nature of paradigms, the process of "paradigm shifts" (Kuhn), and the overthrow of insufficient theories based upon the principle of falsifiability (Popper). For some, Popper and Kuhn are either a part of the foundation of western science, or they go a great deal towards explaining the western tradition of science and must be contended with.

For our purposes, we will focus on how the ideas of Kuhn and Popper provide us with a description for how the overthrow of reductionist/materialist science will occur, and how the embrace of the consciousness-caused creator existence paradigm will be achieved, along with additional notes on why the reductionist/materialist paradigm is so resistant to change, even in the face of multiple forms of falsification.

Falsifiability

Popper recognized the danger of theories that explain everything but predict nothing. He understood that a theory which cannot be tested—and potentially proven wrong—is not science; it is dogma.

To protect the integrity of the scientific method, Popper sought to draw a hard line in the sand. He needed a way to distinguish between empirical reality and metaphysical belief, between the testable and the pseudo-scientific. He called this the 'Problem of Demarcation,' and his solution was ruthless in its simplicity:

> "Thus, the problem which I tried to solve by proposing the criterion of falsifiability was neither a problem of meaningfulness or significance, nor a problem of truth or acceptability. It was the problem of drawing a line (as well as this can be done) between the statements, or systems of statements, of the empirical sciences, and all other statements--whether they are of a religious or of a metaphysical character, or simply pseudo-scientific… I called this first problem of mine the 'problem of demarcation'. The criterion of falsifiability is a solution to this problem of demarcation, for it says that statements or systems of statements, in order to be ranked as scientific, must be capable of conflicting with possible, or conceivable, observations"

-Karl Popper, *Theory of Falsification*[1]" (

And yet, in modern science, particularly that which is reductionist/materialists in nature, we see a proliferation of theories that openly defy falsification. Popper's point in the above quote is that any idea that cannot be unmoved or unfalsified is not a theory. It must be able to be contradicted. It's alignment with the truth must be earned by criticism and attack. If it continues to provide answers, and its contradictions are not swept under the rug, then falsifiability has done its job. But ideas that propose realities, dimensions and structures that can never be seen, inspected, tested, is indistinguishable from fantasy. And a fantasy surely must be included among the criteria for being falsified.

But contrary to Popper's ideas, many material ideas are objectively untrue, and in their radical, materialist longings, objectively false, yet they are allowed to stand unquestioned or little questioned, simply because they work to prop up the dominant materialist worldview. The result is a corruption of the scientific method, where absurd and untestable hypotheses are granted tenure while pointed challenges are silenced and held to a far more rigorous, and at times, impossible standard (they must both be true and they must also conform to a false reductionist/materialism paradigm). *What Popper calls for*—what our narrative has exposed is not just the need for a proposal or a new theory; it is a call for the necessary dismantling of false idols (theories for our reality).

To understand why this corruption persists—and how it will eventually shatter—we must look at the mechanics of the paradigm itself.

Thomas Kuhn famously described normal science as a process of puzzle-solving using a specific set of accepted concepts. As long as these 'tools' work, the paradigm remains stable. But Materialism is now facing a dilemma: its tools no longer fit the data. The question is poignant: why, then, does the establishment refuse to change them?

Tools of the Paradigm

Kuhn argues that the reluctance to change established ideas is structural. A scientific establishment behaves like a factory: it does not want to shut down the assembly line to build a new machine unless it has absolutely no choice. Failure—complete and total failure—is the only acceptable signal preceding collapse.

> "So long as the tools a paradigm supplies continue to prove capable of solving the problems it defines, science moves fastest and penetrates most deeply through confident employment of those tools. The reason is clear. As in manufacture, so in science - retooling is an extravagance to be reserved for the occasion that demands it.

> The significance of crises is the indication they provide that an occasion for retooling has arrived."
>
> - Thomas S. Kuhn in *The Structure of Scientific Revolutions*[2]

And yet we find ourselves chained to a paradigm that is incapable of addressing the pressing scientific issues of the late 19th, 20th and 21st century. In many respects, the collapse happened decades ago. Since then, science has been limping with the reductionist approach. Stymied, stuck and unable to move. Acknowledgement and articulation of this stoppage, in books, blogs and critiques of modern science has been prolific over the last few decades. The question persists, why do these ineffective ideas remain dominant. I believe that to understand the reason why materialism and Darwinism reigns supreme we can find our answer in the first and last parts of Kuhn's statement:

> "So long as the tools a paradigm supplies continue to prove capable of solving the problems it defines[3]…"

We have found a vital clue of why despite numerous mortal wounds, Darwinism has more lives than Schrödinger's cat. As recounted in our narrative, when evolution was released, it was meant to accomplish two equally important goals. The first was to explain the origin of species, but equally important was the second goal, -to wrest the origin of species from special creation and deliver it into the control of materialism.

Although Darwin failed in the first respect—providing a scientifically valid mechanism—he succeeded spectacularly in the second.

He created a tool that, in Kuhn's words, was 'capable of solving the problems it defines.' But the 'problem' Darwin solved was not biological; it was theological. Materialists saw any view that had to bring in God or the hint of any other supernatural power as a means of explanation as a problem; Darwin's theory provided the necessary weapon to destroy this prevailing worldview. Darwin himself wryly conceded that this atheistic objective was a primary motivation, stating, '...I have at least, as I hope, done good service in aiding to overthrow the dogma of separate creations[4].'

It was the soon-to-be worldwide acceptance of evolution that ensconced Darwinism as the key theory within the materialistic paradigm. Acceptance of the alternative to special creation rather than *proof*, won the battle for Darwinism. And whether by honorable or dishonorable means, Kuhn's analysis of the needs for a paradigm to be overthrown still holds true. We continue with another pillar in Chapter VIII of *The Structure of Scientific Revolutions* where Kuhn explains the required factors.

> "The significance of crises is the indication they provide that an occasion for retooling has arrived."

But the real soul of Kuhn's reasoning is sobering. It does not identify the fall of materialism, nor the re-establishment of Science, nor being falsified as the crucial ingredient to supplanting a false, established paradigm. He names something quiet. Unsaid, but unforgettable and undeniable once understood.

> "...once it has achieved the status of paradigm, a scientific theory is declared invalid only if an alternate candidate is available to take its place. No process yet disclosed by the historical study of scientific development at all resembles the methodological stereotype of falsification by direct comparison with nature[5]."

We have now isolated the true source of Darwinism's resilience to falsification and eradication. It is not enough to simply falsify the theory. As Kuhn observed, a paradigm is never abandoned until a viable replacement is ready to take the throne. We have arrived at the moment of Crisis. Modern scientific instruments combined with the continued strain of a paradigm that is unable to provide any puzzle solving capacity for new problems has brought progress to its knees. There has been very little in the way of serious breakthroughs since the middle of the 20th century.

We have a new candidate waiting in the wings to patch things together and put things in their proper order. The evidence points unequivocally to an omniscient kind of teleological order as the guide behind our reality. But here lies the trap: this evidence is immaterial in nature. It is not material. The 'ghostly' nature of consciousness fails to meet the arbitrary requirements of the current establishment precisely because it is not material in nature.

We must remember; materialism does not view itself as a paradigm—a lens through which we see the world—but as the Truth itself. Therefore, it rejects any solution that does not support its tenets before it can even be considered.

This creates a 'Triple Choke Hold' on modern science:

A Materialist paradigm that does not recognize itself as a belief system, but as the only objective reality.

A Darwinian theory that was accepted on a promise rather than proof. Since it was never based on facts to begin with, meaning it cannot be killed by facts.

A replacement paradigm that lacks a material basis—the one thing the current establishment insists upon.

How long before the scientific world migrates to the acceptance of an immaterial consciousness and spiritual realm as our new paradigm is anyone's guess. It may take 5, 25, or perhaps 300 years. But one thing is nearly certain, migration is inevitable, because we will make no progress until we migrate. This much is evident. Materialism has no puzzle solving capacity over any of the Variables or Special Problems of The Divine Algorithm. The algorithm has its own chokepoint; scientific progress will largely be impossible as long as the false paradigm maligns scientific data.

Is it really possible for the entire scientific establishment to stare at evidence and refuse to see it? History says yes. To understand how a dominant paradigm blinds experts to the obvious, we need only look back at the last major revolution in Earth sciences.

A Cautionary Tale - The Example of Continental Drift

The theory of continental drift serves as an example of an idea that in many ways appears obvious yet was not accepted because of a previous paradigm and a previous belief. In the case of continental drift, the alternative belief was the "fixist" continental view. In the conception of Kuhn, the incorrectness of the idea of fixism was insufficient for the idea to die. What was required was a fully formed new conception for fixists to migrate to.

By 1950, humanity had acquired technological knowledge sufficient to create and understand electromagnetism (1864), radios (1901), relativity (1905), rockets (1926), computers (1937), nuclear power (1942), nuclear bombs (1945), and the structure of DNA (1953). Yet, it was not until *1968* that the theory of continental drift, now called plate tectonics, took hold. Prior to this time, the proposition of continental drift was seen as pseudoscience, despite the fact that the continents of the world fit together like puzzle pieces in a manner so obvious that grade school children would point out how South America could easily and near-perfectly fit into the coastline of West Africa. Despite this obvious feature, and the fact of nearly identical rock, sedimentary formation, and plant life between the east coast of South America, and the west coast of Africa, none of these similarities made a significant impact upon scientists at the time, because of the tradition of the prior fixist belief was unquestioned. How did the fixist explain these similarities? These similarities were seen as mere coincidence. In this age, in order to prop up the idea of a fixists world that existed without drift, maps of the ancient world were drawn with *land bridges* which could facilitate the transportation of plant life across the oceans.

When Alfred Wegener proposed his theory in 1912, it had already been proposed by several other individuals who were credited by Wegener for their assistance.

In 1931 Arthur Holmes, a British geologist, was the latest and strongest advocate for continental drift because he was able to correctly identify radioactive decay as a contributor of plate tectonics which drove the engine of continental drift.

In 1939 a conference was held by the international geological society where one of the main topics was *the opposition* of continental drift by fixists. The fixists prevailed. It wasn't until the 1950's that the tide began to turn. Research and information gathered from the analysis and mapping of the sea floor provided vital clues. But it wasn't until a series of papers published from 1953-1963 where plate tectonics was established as a scientific discipline that plate tectonics was able to provide vital information for the refinement of Wegener's ideas in a form that the fixist were unable to counter.

By 1967 plate tectonics had become axiomatic to modern geophysics and in 1968 Geophysicist Jack Oliver along with Bryan Isacks and Lynn Sykes published *Seismology and the New Plate Tectonics*, providing conclusive seismologic evidence supporting plate tectonics which also encompassed a full explanation for continental drift[6]. Unfortunately Alfred Wegener passed away in 1930 and did not live to see his theory gain acceptance, but at least he died with the conviction and knowledge that he was correct and that in the eyes of history, he would be vindicated.

The lesson here is stark: being 'right' is not enough. Although the theory of Continental Drift was correct, the mere observation of the facts was insufficient to crumble the old worldview. Why? Because the dogma of 'Fixism' held what could be called an unbreakable psychological grip on the scientific community.

Facts alone could not break that grip; a mechanism was required. It was not until Plate Tectonics was proven—providing the physical engine that drove the continents—that the experts were finally able to migrate from their false certainty to the new paradigm. They didn't just need to see the movement.

We now know as fact that *Pangea*, a supercontinent of all of the Earth's land masses, existed from (estimated) 335 million to 220 million years ago. It then began to break up and *drift apart* into the continents we now recognize in the modern age.

> "It is only by combining the information furnished by all the Earth sciences that we can hope to determine 'truth' here, that is to say, to find the picture that sets out all the known facts in the best arrangement and that therefore has the highest degree of probability. Further, we have to be prepared always for the possibility that each

new discovery, no matter what science furnishes it, may modify the conclusions we draw."

-Alfred Wegener, *The Origin of Continents and Oceans, 1915*

In the statement above, Alfred Wegener describes a dynamic, flexible, and organic scientific method—one that advances by increments and rarely by leaps. This progression requires validated knowledge, but oftentimes knowledge is kept out if it conflicts with an entrenched narrative. Far too often Truth is not the Master of Science, narratives and paradigms are. Science should know that Truth is its unassailable master and there can only be one master.

Yet the story of Wegener reveals something else as well: the enduring temptation of humans in their various roles and competing priorities to rule over the scientific enterprise rather than submit to its discipline—to preempt it instead of follow it, to ignore or fabricate evidence, and to elevate ideas that cannot withstand rigorous scrutiny.

These diversions are not science; they are distortions of it. And yet they remain a persistent and powerful adversary in our midst.

The history of science eventually wins out. As we look back on the progress of humanity in the sciences, we see the unavoidable collapse of false paradigms, no matter how deeply rooted and no matter how numerous its adherents in the face of progress. These ideas live on borrowed time[7].

---

[1] K.R. Popper, Conjectures and Refutations The Growth of Scientific Knowledge, Pg. 38, Basic Books, 1962)
[2] (2nd Ed. (p. 76), 1962, The University of Chicago)
[3] (Ibid, Kuhn)
[4] (C. Darwin, The Descent of Man 1st ed Vol I 152-53 Orig. 1871)
[5] (Ibid, Pg. 89, Kuhn)
[6] (B. Isacks, J. Oliver, L.R. Sykes, Seismology and the New Global Tectonics, Journal of Geophysical Research, 73(18):5855-5899, September 15, 1968)
[7] (Alfred Wegener's Hypothesis on Continental Drift and Its Discussion in Petermanns Geographische Mitteilungen (1912-1942) Imre Josef Demhardt, Polarforschung 75(1), 29-35, 2005, Erschienen 2006), (American Intransigence: The Rejection of Continental Drift in the Great Debates of the 1920's. Robert P. Newman, Department of Communication Studies, University of Iowa, Iowa city, IA 52242)

Preamble, The Divine Meditation

Preamble, Variable 14: The Thousand Buddhas Meditation

Introduction

If we sweep away the failed mechanisms of blind chance and random mutation, what are we left with? We are left with a universe that looks less like a machine and more like a thought.

If the cosmos is not a clockwork device wound up and left to run on its own, then it must be something far more active. It implies a Sustainer; a consciousness that not only designed the system but actively holds it in existence, moment by moment.

And now we have to think of things in an entirely different way. We cannot anthropomorphize the 'sustainer' of 'the thought' that holds our reality together in the same way we anthropomorphize other things. If there is a mind, and thinker that sustains our reality, then this mind has been sustaining our reality for billions of years in addition to bringing it into being. Every quark, proton, photon, living being, and possibility, all at once. We are talking about something that defies our understanding of time, dimension, awareness and power.

That's the very first thing we must get our minds around if this thought is to remain cogent as a thought experiment.

Next, we must drop all ideas about how to envision how the meditation of our reality can possibly take place. It does take place. That our reality exists is not a question. That what needs to happen for it to exist, is not a question. All needed things are taking place. We must 'drop all thought' as the meditation teacher tells students repeatedly, if we are to comprehend, in a small way, how our reality is sustained.

Now, to understand how a single consciousness can maintain a reality of such infinite complexity, and break it down in understandable terms, I must turn to an unlikely lesson from my childhood—one that taught me the power of the mind to hold a thousand worlds at once.

My first experience with meditation was at the age of 7 when my mother began practicing Nichiren Shoshu Buddhism. On two occasions she brought me to a house in the north end of Seattle where a group of 15-20 practitioners sat and chanted together. The sound they made while chanting 'Nam-myoho-renge-kyo', was mesmerizing. 'Nam-myoho-renge-kyo'which loosely translates into "devoting oneself to the Mystic Law of the lotus flower of the teachings," which is essentially the Buddha's teachings was indescribable for me at that age.

There was a deep drone, -a buzzing kind of a sound, a rhythm, and also a deep resonant non-human sound mixed deep within that puzzled me. I couldn't identify or locate where this sound was coming from. I walked around the group, but wherever I was, the sound seemed to be on the other side of the room. Something special was happening. My mother received her own gohonzon a short time later. A gohunzon is a small simple altar, made of wood, containing a scroll with sacred sayings and deep insights. The scroll is said to reflect the true nature of all those that looked upon it and chanted in its presence. The name comes from 'hon' (root/fundamental) and 'zon' (venerable/honorable), with the prefix 'go'. It is also referred to as an object of devotion or respect.

Practice

Years later as an adult, I attended a silent meditation retreat in Marin County, California. My initial thoughts were that if I could emerge from the retreat knowing how to meditate (previously I had only chanted) and got three or four deep insights, then the time spent would have been justified.

The practice I was taught was to empty my mind and to sit in nothingness. When my mind drifted, I was to gently drop the thought and bring my attention back to watching my breath. Breathe in and breathe out. Watching. Nothing more. With that advice, after half a day adjusting, I was able to "meditate." For the next several days we meditated 11 to 13 times a day for 30-45 minutes at a time between three simple meals.

With practice I found that if I emptied my mind, I could create an open space and it seemed that this space, freed from the internal chatter of my life, could reach something that was deeper and wiser within me.

Ultimately, I emerged from the retreat with far more than I had hoped for. I obtained the 3-4 deep insights I had hoped for, but the unanticipated bounty was learning a method to retrieve a limitless number of similar insights.

Things Get Complicated

A few months after the retreat, I visited my mother in Seattle. I spoke to her about my experience. In the conversation she told me that she had been meditating daily, several times per day, for the entire time since she began Nichiren Shoshu. She then talked about other forms of meditation she had been taught. In particular, she illuminated a meditation that I could barely comprehend, -*the thousand Buddha meditation.*

She said she was taught that one could imagine a thousand Buddhas, all around, in every direction, at all heights. Taking up all spaces. And each Buddha is juggling three balls while standing on one foot atop an endless stack of Ganeshas "all the way down". She explained that the best way to do this meditation was, "The best that you can." This meant it was OK to do pieces of it as one built up capacity to do the entire thing. Even if you're not doing *the full meditation*, trying to do this meditation, even thinking about trying to do this meditation in a mindful way, is part of the meditation and equally builds concentration, imagination, focus and clarity and something different from insight–clarity, capacity, and the ability to hold several meditations at once.

To *master* this meditation is to develop and expand one's point of focus to become more inclusive of visual elements and for the visual images to become more detailed and "real". At first these images may be more *thoughts* of juggling Buddhas than Bonafide *visual images* of juggling Buddhas. But if you continue to practice, the images become more vivid, and focus improves. In this way one can learn parts of the meditation, and then, put the parts together to achieve the thousand Buddha image.

Imagine a Buddha, then imagine the motion of balls being juggled. Then a juggling Buddha, then a juggling Buddha on Ganesa. Slowly, like riding a bike, each part becomes automatic so that you no longer have to expend as much concentration to imagine the parts of the meditation. Then we can put the pieces back together, step by step, until *the thousand Buddha meditation* is complete. The same happens when we learn to read. To a new reader a thick book is an unimaginable mount Everest of words. Imagine handing *Don Quixote* or *The Decline and Fall of the Roman Empire* to a five-year-old. Each letter is a challenge to remember, to sound out, to recognize, to reproduce. Slowly we recognize letters, sounds, words, and their meaning. Soon, we no longer are having to read the word, they are being read by some part within us. Eventually we find a book that captivates us so that we reach a state where we are unaware of reading words, or holding a book, or even being in our room. We are in the story, seeing the sights, observing the characters, witnessing the action, and sharing the experiences. The words become an environment that we enter and experience in totality.

A Meditative Space

As opposed to silent "watching breath" meditation, the complex thousand Buddha meditation approaches the same end by an entirely different route. Rather than to struggle with creating emptiness, where one is constantly inundated with thoughts or the intervening monkey mind, when attempting the complex thousand Buddha meditation, one struggles to create a thousand interacting images in place of silence.

The difference is where the silent meditation creates an empty space where wisdom can well up, the thousand Buddha meditation creates a space for capacity, nimbleness, focus, clarity and capacity, and in this complexity, wisdom is entered directly though the complexity of the meditation and the mind-state it imparts.

This is how the complex mediation provides benefit due to one's mind being fully engaged and therefore being unable to think inane thoughts. In both conditions, the meditator is allowed to sink within the quiet, meditative space that is created by either form of mediation, and retrieve benefit.

Existence As Meditation

Here is where we arrive at The Divine Meditation, the subject of Variable 14.

If the central hypothesis of this book is true—that our reality is the product of an intelligent, purposeful consciousness imposing a chosen design upon our reality—then we must ask how that reality is maintained. It does not exist in a vacuum. If our hypothesis is true, then everything that happens in our existence occurs as a kind of sustained meditation within the consciousness of what can be termed a Universal Creator.

It is this active consciousness that sustains our reality, orders it, and endows it with the ability to do what appears to us to be impossible things, through a power akin to limitless imagination. But an imagination unlike human imagination, rather an imagination where what is imagined, physically exists and has representation and can be interacted with, is durable across what we call time, and has cause and effect among all of the other physical characteristics that we have identified within physical reality.

The structure of atoms, chemical bonds, suns, solar systems, and galactic threads do not just 'exist' on their own; they are held in place. Given their energy and dynamics within this ultimate consciousness as if they were each their own Buddha, among a limitless number of juggling Buddhas in the cosmic mediation. Life, love, birth, death, and human free will exist as a complex, living meditation within the mind of the Creator.

The meditation undertaken by universal consciousness does more than animate a field of a thousand buddhas. It actualizes the entirety of the observable universe into reality and, as the history of our universe demonstrates, carries it through successive stages of transformation and emergence.

Each of these transitions—cosmic formation, the rise of life, the emergence of mind—appears, relative to prior conditions, as a rupture in probability and an expansion beyond established physical law. Within our framework, these discontinuities correspond to what we have called the Variables and Special Problems within the Algorithm. A "miracle," in this context, does not imply incoherence or contradiction. It refers to an event that exceeds the regular operation of the universe—what classical theology (Thomas Aquinas and others) names *potentia ordinata*, the ordered and customary structure of reality.

By contrast, the extraordinary transitions that inaugurate new orders of being fall under *potentia absoluta*: the unrestricted power of reality to generate forms not derivable from its prior configuration. What appears impossible from within an established order is not impossible in itself; it is an expression of a deeper layer of causal depth. Modern science gestures toward this deeper layer when it speaks of a "theory of everything." Yet it is crucial to recognize that what we now experience as *potentia ordinata*—the stable regularities of physics—was once itself an expression of potentia absoluta.

The fundamental constants, the laws, the very structure of spacetime emerged at the beginning as unprecedented instantiations. They were not products of a prior order; they constituted the first order. In this sense, what we call 'miracles' are not violations of reality, but moments in which reality discloses a deeper capacity than the one currently stabilized.

At the origin of the universal meditation, the ordered world did not yet exist. What now appears necessary was once contingent; what now appears stable was once emergent.

If the universe is the expression of a manifest meditation, then its structure is not accidental. It is ordered. The beginning was eruption; what followed was articulation. The question now is not simply how the universe began, but how it was written in constraints as the very material of our reality.

The Meaning of Varible 14

To understand the scope of Variable 14, we must look beyond local emergence to the macro-architecture of the cosmos. The Divine Meditation does not merely bind atoms; it arranges the heavens, establishes geometry at scale, and distributes matter across vast structures.

We have examined the initial conditions of the Big Bang. Now we turn to what those conditions became—and to the discovery that revealed the true magnitude of the consciousness behind the meditation.

Variable 14: The Divine Meditation

The Formation of our Universe

While previous Variables and Special Problems examined the mechanics of creation and the profundity of biological emergence, we now confront a deeper realization: life is not sustained by order alone, but by a universe structured for disclosure.

To understand the deeper structure at play, we need only realize a fundamental truth—Intelligent beings require more than a habitat with sustenance to understand the cosmos. They require a vast richness in resources and an unfettered level of observability to iterate knowledge, tradition and culture. A biosphere is insufficient without a cosmos that can be seen, measured, and understood—a universe that teaches its inhabitants what it is, how it is structured, and how it came to be, is essential to the craft of humanity.

The depth of creation therefore extends beyond molecular precision or cellular ingenuity. It includes the infrastructure that makes inquiry possible: a star at the right distance, stability and bandwidth; a moon that generates tides and eclipses; plate tectonics that recycle nutrients and regulate climate; a magnetic field that shields and preserves atmosphere; a transparent atmospheric window through which the heavens are visible; the clearing of the inner solar system of hazards that prevents perpetual catastrophe; and the buried resources—coal, oil, iron, rare metals—that make technological ascent achievable. These features are rarely classified as miracles. They are categorized as conditions. Yet taken together, they form an unreasonably enriched architecture not merely of survival, but one with the seeds for revelation planted within.

From the earliest civilizations, the heavens instructed humanity in rhythm and order. Then, in the twentieth century, our species extended its sight by orders of magnitude with the discovery that our world is not the center of a small enclosure, but one luminous island within an immense cosmic expanse.

We begin this next stage with a discovery in 1934 that forced this expansion of scale.

The Creation of Galaxies & The Milky Way

The mystery of how galaxies formed found its first major theory under Walter Baade, a German astronomer who studied the universe at Mount Wilson in 1934. Teamed with fellow astronomer Fritz Zwicky, they identified supernovae, a new class of celestial objects. "The extensive investigations of extragalactic systems during recent years have brought to light the remarkable fact that there exist two well-defined types of new stars of novae and supernovae[1]."In the 1950's, 30 years after Edwin Hubble confirmed that the Andromeda Galaxy was its own galaxy and not part of our own Milky Way Galaxy, Baade was able to surmise that older metal poor stars created gas and metals which were then incorporated into the next generation of stars. This led Baade to form the first serious attempt to explain the structure of our universe by proposing that galaxies formed as individual entities out of the original gas from the big bang.

The Heliocentric Model

For thousands of years, as ancient astronomers viewed the heavens, they were fascinated by a select few unusual stars that traced inexplicable patterns in our sky. Some of these stars would move in one direction, stop, go in the opposite direction, and then, return moving along their previous path. Several of these stars were noted to twinkle and change colors as they went about their movements. What our early astronomers were able to figure out, beginning by some estimates 700 years before the modern age, or by other estimates thousands of years earlier, is that these strange wandering stars were *planets* that orbited within our solar system.

One of the first theories to suggest that the movement of these planets were around the Sun was by Greek philosopher *Aristarchus* in 230 BC. His idea contradicted Aristotle's who passed away nearly a hundred years earlier in 322 BC. Aristotle believed that the planets and Sun, and the entire universe orbited the Earth. Heliocentrism was proposed by over two dozen ancient thinkers. Most notably in the western world, was Nicolaus Copernicus in his, On the Revolutions of the Heavenly Spheres (*De Revolutinibus Orbium Coelestium), (1543),*

and Galileo in his famous pamphlet, *Sidereus Nuncius (1610)*, or *Starry Messenger*. Beginning in 1755, and for the next two hundred plus years, the theory for *the creation* of our solar system was known as *The Nebular Hypothesis,* first developed by Immanuel Kant in 1755 in the book, *Universal Natural History and Theory of the Heavens.*

The nebular hypothesis states that our solar system was the creation event of our sun, and that the planets were created by a gigantic cloud of rocks and dust and gas—what we now call the protoplanetary disc. Kant was able to deduce that the solar system formed from a diffuse cloud, collapsed by gravity, which rotation flattened, and planets formed from material within this rotating structure, thus sharing direction of motion due to shared common origin.

This prescient hypothesis explains why planets inherit a common direction of motion and occupy roughly the same orbital plane. This An impressively prescient hypothesis for Kant, a philosopher, and author of another masterpiece, *A Critique of Pure Reason.* Later, in 1796, Pierre-Simon Laplace independently proposed a similar model, with more mathematical articulation in *Exposition du systeme du monde,* which then became known as the Kant-Laplace nebular hypothesis, a subject we will return to later in this chapter.

The Universal Habitable Zone

The question of how *our* Milky Way galaxy formed remains elusive, but it involves a pattern stamped upon the universe that remains evident within the template of CMB. This remnant—the only evidence we have from the creation of the universe—left a perfectly imperfect template that predetermined the pattern of galactic threads, galaxy clusters and the modern-day architecture of the universe. The cosmological constant, being precise to 120 decimal places, ensures that the expansion of the universe is neither too slow, which would lead to a catastrophic collapse into a massive black hole, nor too fast, which would blow the universe apart such that the largest structure within our universe would be individual atoms. Gravity, dark matter, and dark energy are all crucial players in setting out the current size, dimensions, and architecture of our universe. These combined factors, along with those yet to be discovered, produced the observable universe while actively sustaining the delicate balance required for its habitability.

The Galactic Habitable Zone

For a galaxy to support life it needs to satisfy multiple factors. First, the galaxy must be of a certain size in order to have a critical galactic habitable zone. This zone delineates the ideal location within our galaxy where all of the needs for a mineral rich solar system are present[2]. A viable, habitable, life-allowing and safe solar system requires enough generations of stars, supernovae, and a requisite safety from hazards to create a sun and solar system that can last for the billions of years needed for intelligent life to arise. Hazards such as local supernova, gamma ray bursts, transient radiation events, asteroids, meteors and other sterilization events must be avoided.

Habitable systems appear within an exceedingly narrow cosmic threshold. Regions dense enough to produce the chemical elements required for life are also regions saturated with stellar radiation, supernova activity, and gravitational disruption, while regions calm enough to preserve planetary systems are often too chemically quiet to generate them. Our solar system required both conditions and appears to have achieved them through an extraordinarily narrow cosmic balance, having emerged from this contradiction by navigating a path between cosmic violence and cosmic quiet, formed amid the explosive fertility of a stellar nursery, yet now isolated in vast interstellar silence, suspended far from destructive stellar environments.

Our solar system occupies a narrow, ring-shaped region of the galaxy that balances chemical enrichment against environmental stability. This specific band, a kind of cosmic Goldilocks zone — has been referred to as the Galactic Habitable Zone hypothesis developed in part by Guillermo Gonzalez and Jay W. Richards in their book, *The Privileged Planet*[3]. The boundaries of the Galactic Habitable Zone are determined by the availability of planetary building blocks and by environmental threats to complex life within the galactic environment.

Our Spiral Galaxy

Our galaxy is a spiral galaxy. 72% of galaxies are spiral in shape which is one of 3 main broad categories of galaxy types in our universe, the other two being elliptical galaxies (8%) and irregular galaxies (20%). To no one's surprise, our spiral galaxy is ideal for the needs of life. It is sufficiently large to contain both a generous habitable zone, with a high metal content. while also not being too large to hamper a life providing star/planet system. Our galaxy is also sufficiently mature to have relatively intact spiral arms that are not too frayed as to create a continuous zone of hazards throughout. Our spiral arms have gaps where there is relatively less activity, and zones where needed activity is taking place to process elements and create stars. It is theorized that elliptical and irregular galaxies, with no "arms" would have a much more difficult time supporting life[4].

Elliptical galaxies are shaped like ellipses and are mostly close to circular in shape. They are not ideal for life because they contain mostly older stars and not much gas and dust therefore little new star formation. New research by Daniel Whitmire in The Monthly Notices of the Royal Astronomical Society (MNRAS), discusses reasons why elliptical galaxies would not be ideal for life. In order for elliptical galaxies to be more life friendly, they need to be larger. But, when elliptical galaxies are younger, they would have been far more compact and therefore would have experienced "massive galactic sterilization events associated with quasar/AGN activity and starburst supernovae". These lethal doses of radiation would have effectively sterilized these kinds of galaxies from any future habitability[5]. Whitmire's paper also found that due to the higher metallicity within these types of galaxies (elements higher than hydrogen and helium), it would have also led to a disproportionate number of large gaseous planets. Prior to the new research by Whitemire, these types of galaxies were believed to be "cradles of life" in the universe[6]."

Irregular galaxies, the third type of galaxy, is a catch-all term for galaxies that have no particular shape and are not elliptical and spiral. They are among the smallest galaxies, and opposed to elliptical galaxies, they contain a great deal of gas and dust, which means that a lot of star formation is occurring within. Consequently, this causes them to be very bright. The Small and Large Magellanic Clouds that orbit our Milky Way Galaxy, are examples of two irregular shaped galaxies. It is believed that irregular galaxies are not good candidates for life. Their average size and their unpredictable nature mean they are also unpredictable with regards to life-giving properties.

When we survey the galactic inventory of the universe, we do not find a limitless array of life-supporting universes. Instead, we find that two of the three major types of galaxies are structurally poor habitats for life. Life requires order, resources, star nurseries and vast expanses where sterilizing events are uncommon. The fact that we find ourselves not just in a spiral galaxy, but in a spiral galaxy specifically tuned with the size, maturity, and safe zones required for long-term survival is but the first in a series of requirements we will investigate in the improbable rise of intelligent, technological life on Earth.

Safely Assembling the Ingredients for our Solar System

With the information gathered from meteors recovered from within our solar system, we are able to tell that the Earth, Sun, Moon are all composed of the same initial elements and isotopes from the same cloud of gas and dust which formed our solar system. This is unsurprising. What is of interest is that the ratio of elements and isotopes tell us that our solar system was likely made up of *at least* four distinct supernova events from previous generations of stars, which has increased the metallicity of our solar system[7]. The cloud of gas that became our solar system was created within a star nursery, made up of the remnants of these previous generations of stars that were likely triggered into collapse by shock waves from previous supernovae events to precipitate star formation collapse. Supernovas provide clouds of hydrogen gas with the asymmetrical distorting energy that precipitates their collapse into the protoplanetary disc that creates solar systems. This is how our solar system was created.

The analysis of the composition of our solar system material also reveals a completely unexpected event. At present our location in our galaxy is within an area known as *the co-rotation circle.* This is a location where the rotation of the stars within our area are *in sync* with the rotation of our galaxy's spiral arms. "This circle is the place in the desk of a spiral galaxy wherein a star revolves around the galactic nucleus with the same period as a spiral arm pattern[8]." Being within the co-rotation circle is statistically *rare* given all of the possible locations in a galaxy for any particular part of the galaxy, and places our solar system *within the galactic habitable zone,* which is as safe of a location for as a life-bearing planet can be—*between* the spiral arms, *and within the co-rotation circle.* This keeps our solar system from having to pass through the spiral arms, for 20 to 30 million years, wherein lies a much higher density of suns and a much higher likelihood of experiencing a transient radiation (life sterilizing) event such as gamma rays, supernova, black hole feedings/ejectings, meteors and other hazards.

Another ideal condition is the shape of our orbit. It is generally flat and lacks the wide, vertical oscillations that would plunge us in and out of the Milky Way's disc as we orbit the center. This flat, stable motion provides us with a quadruple benefit. Three of these factors keep us alive, while the fourth allows us to learn. To recap:

-**Spiral Arm Avoidance**: We do not cross the dangerous, star-dense spiral arms.

-**Vertical Stability**: We avoid the high-radiation environments found above and below the galactic plane.

-**Co-Rotation**: We stay synchronized with the safe zone, avoiding the 'traffic' of the galaxy.

-**Clear Observation**: This stable platform provides us with a pristine, unobstructed view of the cosmos—a luxury not afforded to most stars[9]. (

With the multiple levels of safety for our solar system in the milky way, we end up with a problem. Although this location is ideal for our safety, it is a less-than-ideal location for the *creation* of our planet in the beginning. Being within spaces of the spiral arms, but also on the outside of it, and also outside of the *inner* part of the co-rotating circle means that if our star system would have formed in its present location, it likely would have not had the high metallicity needed, nor the vital mix of several different supernovas to form what Earth requires to serve as a host for intelligent life.

Within our present safe location, it is also unlikely that the hydrogen gas cloud that formed our solar system could have experienced a shock wave powerful enough to precipitate the collapse of the gas that ignited our sun and formed our protoplanetary disc. Our solar system required a much more complete checklist of materials than we could have found in our current location. This checklist likely can only be obtained *inside* of the corotation circle within a star nursery. The problem is, we are just outside of it and nowhere near a star nursery[10]. How did we obtain the enriched life-allowing materials to create life in our location? And how did we obtain several generations worth of star material and obtain the needed energy to cause the formation of our solar system in a location where star formation is rare?

The evidence suggests a dramatic origin story. Our solar system was likely cooked in a dense stellar nursery; a chaotic crowding of approximately 10,000 suns deep within a hazardous spiral arm and then migrated from this hazardous inner zone. Once formed, our sun and planets undertook a remarkable migration—more of a great escape—as we moved quickly and efficiently out of the dangerous inner zone and settled into our current location between the spiral arms in the co-rotation circle.

This move was not just convenient; it was existential. Research indicates that Earth's historical entries into the dense spiral arms coincide with extinction-level events[11]. We effectively migrated out of the 'kill zone' and into a sanctuary by unknown means.

The Great Escape

The story of our solar system's escape to its present location involves a number of improbable events, all together, which appear to require a miracle of coordination and good fortune, or perhaps something else.

Like a thief in the night, our solar system spirited away from its star nursery, moving against the gravitational attraction of the gas and clouds, somehow remaining intact and with little regard for the attractive pull of a host of forming suns and supernovae, to its present location. "The planets in the solar system survived dynamical interaction with the Sun's siblings. At the moment, we cannot exclude that the Sun's nursery may have been a *very massive,* long-living region where thousands or even tens of thousands of stars were born together[12]."

But there is an important caveat: although star nurseries are filled with gravitationally bound stars, the stars within a star nursery cluster are loosely bound due to the immense energy and activity of star formation. This allows them to move around somewhat freely. As more stars are formed, gas, dust and protostellar material are blown away by the intense radiation and stellar winds created by massive young stars. This weakens the gravitational grip on all of the stars, including our sun, allowing then to drift away from one another[13]. But this is only part of the story. It's much more likely for two-star or multiple star systems to result from star cluster dissipation vs. single star systems. Roughly 50-70% of sun-like stars emerge from nurseries in binary or multiple star systems[14]. For stars more massive than our own, the binary or multiple star rate can exceed 70-80%. It is quite fortunate for us that we emerged singly, for if we were in a two-star or multiple star system, life on Earth would be far less likely.

For now, it is incredibly interesting that nature has a natural, built-in process for a dense collection of stars within a star nursery, with intense gravitational hold, to dissipate and disperse stars throughout a galaxy and throughout a universe. An ingenious result from what intuitively appears to require a Rubik's cube level escape.

Our Sun now sits far from the nearest star system, which is currently the triple star system of *Alpha Centauri* that includes *Rigil Kentaurus, Toliman* and *Proxima Centauri. Proxima Centauri* is the closest to our Earth at 4.2465 light *years* away. To scale, if our sun were the size of the period at the end of this sentence (0.5mm, the size of a micro tip pen), it would be 8 miles distant from *Proxima Centauri* which would be 10% larger (0.55mm) according to this scale[15].In terms of star nurseries, the nearest star nursery to Earth is *Rho Ophiuchi*, and it is *400 light years away*, or 94.2 times further away than Alpha Centauri. To scale, if we assumed the same scale for our sun (0.5mm), this distance would span 753 miles, or the distance from Seattle, WA to Salt Lake City, UT.

In the 1980's it was discovered that our solar system sits in a "peculiarly empty region of space… about 1,000 light years across", or, 500 light years in all directions from our solar system, which is exactly the distance to *Rho Ophiuchi*. This empty region is called "The Local Bubble[16]." (*Proxima Centauri* is in this bubble. The bubble is not empty, it is "peculiarly empty" i.e., unusually depleted of contents). It is believed that this bubble formed a short 14.4 million years ago, timed with the last 0.0031 percent of the Earth's age, -a time that also aligns with *the arrival of modern man*-. It is believed this bubble was formed by the shock waves of approximately *15 supernovae* and is currently expanding at a rate of 4 miles per second. Once again, the reason we happen to find ourselves in the middle of this protective space is, according to science, "purely coincidence[17]."

When we layer the discovery of the Local Bubble on top of the great escape, the narrative that our position in the cosmos is a statistical anomaly yet fully explained by chance begins to buckle. We did not merely escape to a quieter neighborhood; it appears the neighborhood was specifically cleared for our arrival. This new address provided a secondary gift that we touched on a moment ago: visibility. Far from the blinding dust and light of the stellar nurseries, we possess a pristine, unobstructed view of the cosmos. We are currently located in one of the only spots where we can survive long enough to build telescopes, while the sky is also clear enough to use them. By chance, this is highly improbable; by design, it is essential.

And then we must consider another aspect of the timing of our arrival in this local bubble. This protective void we find ourselves within formed only 14 million years ago—syncing perfectly with the emergence of hominids and the rise of the modern human mind, and the advent of intelligent observers. A fortunate confluence of eventsfor every human currently on the planet. Projects such as the James Webb, and the preceding Hubble telescope may have never been proposed had we found ourselves in a patch of sky where our ability to see great distances was severely compromised.

The Formation of the Protoplanetary Disc & Ignition of Our Sun

As the nascent solar system traveled to its current location, the gas within our soon-to-be protoplanetary cloud condensed and ignited our sun. It is estimated that 99.86% of the protoplanetary disc which formed our solar system became our Sun[18]. The rest of our solar system, including asteroid belts and their contents, dwarf planets, terrestrial planets, gas giants and even a theorized 9th planet that was sacrificed and lost to space, is made up of the 0.14% of what remained in the protoplanetary disc.

The collapse of the protoplanetary disc begins angular momentum causes it to rotate faster and faster, like an ice skater spinning while pulling their arms closer to their body. This angular momentum causes the entire cloud to flatten out in a disc like the rings of Saturn. The increased momentum, collapse, and flattening causes the pressure and friction of the innermost contents in the disc to attain the critical pressure and heat necessary to cause the hydrogen atoms to fuse together and ignite. The blast from this star creation event, -the creation of our sun- pushed the lighter gaseous elements to the outer ring of our protoplanetary disc while the heavier elements were able to withstand this blast and, due to gravitational effects, moved *inward* and closer to the newly created star, providing the needed materials for planet building for our inner solar system planets.

The Formation of the Gas Giants & Terrestrial Planets

The gas giants formed first, and due to how protoplanetary discs evolve, and they had to do this *very* quickly by geological time—they had approximately 5 million years to grow large enough to capture sufficient material to become a gas giant. Jupiter was first, then Saturn, Neptune, and Uranus.

To form a gas giant, the first task is to accumulate a planetary core at least 10 times the mass of Earth to provide sufficient gravitational strength to sweep up and accumulate hydrogen within the solar system before it is blasted outward to the far reaches of the planetary circumference. Not 1, *but 4 gas planets* needed to grow to this minimum size, in addition to what is believed to be a 5th gas giant that, according to computer models, was created, played a key role in the location and evolution of the Earth, and having completed its work, was sacrificed by ejection beyond the reach of the sun's gravitational influence, doomed to travel the milky way as a rogue planet.

Research at first indicated that for the 4 remaining gas giants in our solar system, it was not possible for them to grow large enough to capture sufficient gas. But then scientists found a breakthrough method these planets likely used to facilitate this process, *-chondrules-*. Miniscule grains of heavier elements, the size of wheat germs, that have been flash heated by the sun into droplets of molten rock nanometers to a centimeter in size. It is believed that there was an immense number of these *chondrules* in the early solar system that rained down upon the small rocky proto planets, pulled in by their gravity, and seemingly designed for the purpose of rapid and efficient planet accumulation. Because these objects were relatively very slow moving, they could be efficiently swept up by protoplanets. Computer simulations have confirmed that chondrules allowed planetesimals–the germ of planets ranging from 1 km in size– to grow 1,000 times faster than they could without chondrules. The hypothesis of chondrules, if verified, solves the mystery of how—Jupiter, Saturn, Neptune, and Uranus—grew *large* enough *fast* enough to become *gas giants*.

Once our solar system had the 5 gas giants—the above 4 and ejected 5th, the next task for our solar system was to account for the terrestrial planets. In order of appearance, moving in order outward from the Sun, there is Mercury, Venus, Earth and Mars (Pluto is no longer considered a planet). Terrestrial planets formed in this region instead of gas planets because they were closer to the Sun where there were heavier elements that could withstand being blasted to the outer part of the planetary disc when the sun ignited. Similar to the regions where the gas giants were created, it is believed a richness of *chondrules* were present to aid in the formation of the terrestrial planets.

> "Pebbles in a dusty disk orbiting the star continuously collided and sometimes stuck together. After tens of millions of years, these agglomerations had built up into planetesimals — the beginnings of the planets. Other rocky fragments remained, crashing into their larger kin and leaving deep craters. Over time, the Solar System thinned out, leaving something like the configuration we see today[19]."

This process of planet-building helped to create the life-inhabited Earth, but this process was also helped by a series of improbable, unimaginably complex dropping movements of the gas giant planets within our solar system. The idea that our solar system initially formed and looked more or less like it does now when at the beginning, with perhaps minor movements is incompatible with what computer simulations have shown was likely a complex, interactive, dynamic and counter-intuitive series of movements and configurations required for our solar system to be in its current state. All of these movements are also vital to life on Earth. The movements of the gas giants are explained in two complex and interrelated hypotheses of solar system formation that have supplanted over a century of speculation. The models are called, *The Grand Tack* and *The Nice Model.*

Solar System Formation: From Kant-Laplace to the Grand Tack and Nice Models

Earlier, we mentioned the Kant–Laplace Nebular Hypothesis (KLNH), independently developed by Immanuel Kant and Pierre Laplace. Refined over the years with the addition of new information, the KLNH began to break down in the 1990s as questions arose that it could not adequately answer. The most significant of these questions came from three main sources:

-Analysis of information from observations of exoplanets by the Keppler telescope

-Information derived from Computer Simulations

-Further research and information from the observation and study of our solar system

With the vast increase in computing power over the last 30 years, scientists have been able to develop simulations that test competing hypotheses of planetary formation. The challenge to the KLNH, as revealed by these three sources, is that our solar system is far too complex to be explained by the model's simplistic assumptions. The major discovery that unsettled the KLNH was the realization that our solar system could not have formed simply in place, with planets maintaining the positions in which they initially accumulated. Simulations reveal that, for our solar system to appear as it does today, significant movement, reordering of planetary positions, and several highly improbable events must have occurred.

The implications of this shift are far-reaching. For two centuries, the KLNH served as the primary mechanistic explanation for our solar system, reinforcing the notion that nothing unusual had taken place in its history. When Napoleon asked Laplace why he did not mention God in his explanation of the cosmos, Laplace famously declared that he had "no need of the God hypothesis" because in his mind celestial mechanics and orbital stability was a simple, inevitable, clockwork process[20]. If you have a cloud of dust and gravity, you inevitably get a solar system like ours.

Modern simulations have shattered this clockwork certainty. We now know that the inevitable outcome of natural planet formation is not an orderly, life-supporting system like ours, but at least in the case of our solar system, a chaotic process of improbable movements, and fortuitous conditions clearing the way for a life-bearing planet. By revealing that our solar system is not the norm of physics but an exception to it, science has simultaneously weakened the inevitability argument and the statistical probability argument for our good fortune. We are not the standard output of the machine. Nor are we merely the product of reasonable chance, but rather the result of converging anomalies that exerted unusual forces and ultimately allowed for the rise of humanity.

The Grand Tack (Jupiter migrating inward and reversing course, thus reshaping the inner solar system) and the Nice Model (gas giant gravitational instability that shifted their orbits and scattered icy bodies outward) revealed that our solar system did not come to exist in isolation. It was necessarily embedded within a larger structure whose properties and evolutions directly shaped its possibilities. Therefore, the scale of our inquiry must widen as we shift our focus from local mechanics to all of the cosmic conditions required to result in the architecture of our habitable solar system.

The Grand Tack Model & The Nice Model

Modern models of solar system formation suggest that the giant planets did not form where we find them today. Instead, their migrations reshaped the architecture of the entire system. The Grand Tack and Nice models describe these dynamical rearrangements. We begin with a brief overview of each before examining their implications in detail.

The Grand Tack model (3 papers)[21] describes early solar system dynamics that took place between approximately 1 to 10 million years after the formation of the solar system.

Jupiter, followed by Saturn, migrated *inward* toward the inner solar system, became locked in orbital resonance, and then "tacked," like sailboats, reversing course and migrating outward to their present positions in the outer solar system. This hypothesis accounts for several distinctive features of our planetary architecture. If correct, the Grand Tack set the initial dynamical conditions for everything that followed in the solar system's evolution. Before this model was proposed, the arrangement of the planets appeared intuitively orderly to early astronomers. However, as our understanding of planetary formation has advanced — particularly through observations of exoplanetary systems — it has become clear that our solar system is highly unusual. The counterintuitive elegance of the Grand Tack is best appreciated when viewed; several high-quality visualizations are available online.

The Grand Tack in Detail

There are a number of unusual characteristics about the arrangement of our solar system, and the Grand Tack is one of those unusual set of actions. In the early solar system, Jupiter was pulled inward by the currents of protoplanetary gas, dust, rocks, and planetesimals. This movement caused Jupiter to fall into a near Mars-distance orbit at a time when Mars did not yet exist. Saturn then began its own movement into the inner solar system following the gravitational interactions that induced Jupiter's migration. The movements of Jupiter and Saturn are where the name "Grand Tack" originated, as these planets "tacked" like ships moving inward toward the inner solar system. Once within the inner solar system, Jupiter and Saturn fell into resonance, where for every three orbits of Jupiter, Saturn made two orbits. This interaction—orbital resonance—is a highly stable and difficult-to-disrupt relationship once established, although later in the Grand Tack evolution, it is broken by large-scale changes in the surrounding protoplanetary environment we will discuss shortly. At the time this resonance served an important role allowing Saturn to gravitationally maintain its relationship with Jupiter, keeping Jupiter from getting much closer to the Sun than the current orbit of Mars.

To the modern observer, this scenario seems immensely counter-intuitive, because today, Jupiter orbits safely in the outer solar system, hundreds of millions of miles away from Mars. But the Grand Tack suggests that in the chaotic early days of the protoplanetary disc, Jupiter did not stay in its original orbital plane, and because it didn't, several hazards, including asteroids and planetesimals were scattered outward to the benefit of Earth.

Driven by drag from thick nebular gas, the young Jupiter spiraled inward, crossing the asteroid belt and arriving at 1.5 AU—the approximate distance where Mars sits today. It effectively moved deep into the inner solar system. It was only the formation of Saturn, and the subsequent gravitational tug-of-war between the two giants, that pulled Jupiter back outward toward its current location. This inward-then-outward journey is the "Tack" of the Grand Tack.

Grand Tack Structural Effects on the Inner Solar System

Stage 1: Jupiter forms roughly 3–5 AU (AU stands for Astronomical Unit and it is equal to the distance between the Earth and Sun) from the Sun and rapidly accretes gas, becomes massive and begins to migrate to the inner solar system due to interaction with the protoplanetary disk.

Stage 2: Jupiter migrates inward to ~1.5 AU, near Mars' current orbit scattering planetesimals limiting the material available to form Mars (The small Mars problem is one of the most difficult to solve problems in the formation of our solar system)

Stage 3: Saturn forms and being less massive its migration inward is faster, allowing it to catch up with Jupiter.

Stage 4: Jupiter and Saturn become locked in a 2:3 orbital resonance. For every 2 Jupiter orbits, Saturn completes 3. This gravitational interaction reverses their inward migration and their migration changes directions.

Stage 5: The Jupiter and Saturn resonance causes them to migrate to their present location within our solar system, redistributing planetesimals into the asteroid belt

Stage 6: Migration halts, Jupiter and Saturn stabilize at wider orbital distances, and later dynamical evolution takes place (the Nice Model)[22].

Grand Tack Outcomes:

-Mars was diminished in size

-The inner asteroids and planetesimals which could have caused many extinction level events on Earth were ejected.

-Earth found its current location squarely within the solar system habitable zone, where water is a liquid, solid and gas, which is an extremely narrow range of temperatures in comparison with the temperature range in our universe and in our solar system.

-The Earth received plentiful water to populate its oceans, lubricate the tectonic plates, form the carbon cycle, precipitate a weather cycle, provide a cradle for species, and, as hypothesized, although it appears quite impossible under any plausible scenario, Earth became the location where life began.

-Liquid water is *only* possible in the solar system's habitable zone but at the same time, as Mars can attest, *it is of little* value in the habitable zone if one finds themselves on a planet incapable of holding onto water or an atmosphere because there is no way to keep it from being stripped off by the Sun.

-The Grand Tack model explains the asteroid belt between Mars and Jupiter. The gas giants disrupted the formation of the asteroid belt, which otherwise may have formed into another planet-, and the Grand

Tack also explains why Mars is *much smaller* than models anticipated it should be. This is what is known as, "the small Mars problem[23]." With the Grand Tack model, the location of Jupiter, for a time in the orbit that Mars now occupies, would have scattered most of this material that could have been incorporated within Mars. Even without the Grand Tack it is estimated that Mars should be an estimated 1.1 to 1.5 times *more* massive than the Earth, instead it is .16 times the mass of the Earth[24].

-Under the Grand Tack model, Earth and Venus were the beneficiaries of this scattering as they were able to grow within the richest terrestrial planet making material in the solar system. And clearly, Earth was able to obtain *the richest material* for itself thus making it a life supporting, life sustaining, resource rich planet[25].

In hindsight, we can view the Grand Tack as a massive terraforming mission, on the largest scales imaginable, with a lifesaving mission to benefit Earth in several ways, calling in the services of the 4 largest planets in our solar system, and possibly a 5th, acting in concert, *as if* they were guided with the intention, purpose, and agency, to transfigure the solar system for the benefit of life on Earth. If our universe was indeed a meditation as proposed, the movements of these gas giants to create a survivable inner solar system was surely a key part of it.

The Grand Tack is not the only model accepted as a solution to planetary formation. It represents one of several competing models including: (1) Low-Mass/Narrow Annulus Model (no large Jupiter migration), (2) Early Instability Models (Modified Nice Framework), (3) Pebble Accretion + Disk Evolution Models. Although there are different models there is consensus on several points, such as planetary migration, gas disk interactions, giant planets influence terrestrial planet formation, and the early solar system was dynamic and violent. There is no consensus that Jupiter migrated to 1.5 AU, or that the Grand Tack is the only or best explanation. The bottom line is at this stage it is agreed that the formation of our solar system is a chaotic, dynamic process and small differences in initial conditions can produce multiple pathways, and multiple pathways can produce our present-day solar system[26].

The Nice Model

The Nice Model picks up roughly 500 million years after The Grand Tack solar system formation process. The model proposes that the giant planets did not form in their current, widely spaced orbits. Instead, they likely formed in a much more compact configuration closer to the Sun[27].

The Nice model is far less successful than the Grand Tack in gaining acceptance. This is partly due to the fact that one of the features it attempts to explain does not currently hold consensus as having occurred at all. The Nice model has been undermined by questions surrounding the evidence from the time period where the *late heavy bombardment* (LHB) is alleged to have taken place.

According to the model, this compact configuration was inherently unstable. Approximately 500 million years after solar system formation, gravitational interactions — likely involving Jupiter and Saturn — destabilized the system. This triggered a chaotic scattering event that flung Uranus and Neptune outward into a disk of icy debris. In some simulation studies, an additional fifth giant-planet–mass body is sometimes included in early configurations, which may later be scattered or ejected during these violent dynamical interactions. This is not a required feature of the model, but it can improve statistical matches to observed planetary orbits in some simulations.

This process helped clear debris from the inner solar system and is proposed as one possible contributor to the Late Heavy Bombardment (LHB), allowing the planets to eventually settle into their current stable orbital positions.

The Late Heavy Bombardment (LHB) is hypothesized as a period of intense asteroid and comet impacts across the inner solar system. During this time, millions of tons of rocky and icy material are believed to have struck the terrestrial planets. Some researchers suggest that this bombardment may have delivered significant amounts of Earth's present water and organic precursor materials. These impacts would have been extremely energetic, potentially reshaping planetary crusts and influencing early surface chemistry. The exact intensity, duration, and even the existence of the LHB remain subjects of intense scientific debate, as direct geological evidence on Earth is limited due to later tectonic and erosional processes making earlier evidence, if it indeed exists, difficult to discern. Critics point out that if the bombardment had happened as vigorously as proposed, over such a short period of allotted time, there would be more evidence on Earth from this period. The proposed period of the LHB would have also placed a restriction on when life could have arrived, as the impact of a LHB upon an Earth believed to have been full of life would have likely been a sterilizing extinction level event, or, at least set the progress of life nearly back to the beginning[28].

The evidence for the LHB is largely missing or erased. It also appears to conflict with a time when life on Earth had already established itself. If it did happen, then there is a lot to explain about how Earth avoided extinction. If it did not happen then equally there is no obvious answer to how Earth received a copious amount of water, is substantially rocky and suffused not only with a number of elements required for a stable planet, but also for a rich magnetosphere, crust, and for technological life as well.

The Nice Model also helps explain several features of the current solar system architecture, including the formation of the Oort cloud, the Kuiper belt, and the capture of Trojan asteroids associated with Jupiter and Neptune. Trojans are small bodies that share a planet's orbit but do not orbit the planet itself. The Nice Model works backward from current planetary positions to reconstruct earlier solar system configurations.

Even with the general acceptance of major aspects of the Nice Model, we do not yet have a model that explains every fine detail of the current arrangement of planets, moons, meteors, and other solar system structures. The advent of advanced computer simulations has been unable to resolve many mysteries. In some respects, the complexity of the problem has deepened, as simulations have revealed previously unknown problems and a vast increase in possible solutions and more chaotic sensitivity. The nice model itself deals with chaotic gravitational dynamics, where small changes in starting conditions produce huge differences in outcomes. With greater computing power, the model reveals its "sensitive dependence on initial conditions".

The implication of this mystery is important to our hypothesis. If the emergence of a life-supporting planetary system were a robust and inevitable outcome of physics, our simulations should reproduce such systems with relative ease, we should be able to plug in initial conditions and have the simulation create our universe. But this is not the case.

This is similar to an earlier problem we witnessed with simulations concerning population III star formation, where the simulations couldn't produce the initial round of stars without metals, even with significant tweaks and fixes programmed in. What these models suggest is the formation of a planetary system capable of supporting a planet like Earth appears to require a highly constrained sequence of dynamical and physical conditions. The fact that our best models rely on rare or highly specific initial configurations suggests that our solar system may be the product of a tightly constrained process that remain beyond our understanding, rather than a statistical or random outcome of physical laws.

But if our solar system, within our galaxy, held within the larger universe is a teleological meditation, then like all of the other fortuitous conditions we have investigated, under the banner of variables and special problems, we have yet another instance where its very possible that only an instantiation can rescue. But if we are saying the universe is a meditation, and that is what we are saying, then that needed instantiation is already built-in along with all of the other fortuitous conditions within the divine meditation. In this case we will never find a dynamical model that can result in our life-inhabited planet.

The Nice Model nonetheless successfully explains several large-scale mysteries of solar system structure, including the distribution of the giant planets, the formation of distant reservoirs of small bodies such as the Oort cloud and Kuiper belt, and the capture of Trojan asteroids associated with Jupiter and Neptune. A Trojan asteroid is a small rocky body that shares a stable gravitationally balanced orbit with a planet, either ahead of or behind the planet as they orbit the Sun.

Nice Model Recap

-The Nice Model explains later outward planetary migration.
-Giant planets likely formed in a compact configuration closer to the Sun.
-Gravitational interactions between Jupiter and Saturn destabilized the system.
-Uranus and Neptune were scattered outward into their current positions.
-Some simulations include a temporary fifth giant-planet–mass body that may have been ejected.
-This dynamical instability may have contributed to the Late Heavy Bombardment.

As the large-scale architecture of the solar system settled into its present configuration through the combined dynamical effects of planetary migration and orbital instability, attention shifts from the motion of planets to the materials they delivered and redistributed within the protoplanetary disk. The same gravitational interactions that shaped the final orbits of the giant planets also influenced the population of smaller bodies—planetesimals, asteroids, and meteoritic material—that would later play a crucial role in the chemical and thermal evolution of the inner solar system. Among the most significant contributors to Earth's internal heat and long-term geological activity is the radioactive isotope potassium-40, a naturally occurring heat source embedded within Earth's mantle and crust. Along with asteroids and icy bodies, the gradual accumulation of radioactive elements helped shape Earth's internal core convection, and long-term magnetic and tectonic activity. In this way, the story of planetary motion transitions into the story of planetary composition—how cosmic dynamics ultimately became planetary geology.

Rocks, Metals and Radioactivity

Over the next 30 million years Earth grew close to its present size as the intense bombardment of material continued. At this point it is believed that the entire Earth melted from the inside out. Exactly how this happened is a puzzle, but there is substantial evidence that it took place due to the accumulation and incorporation of hot radioactive elements within Earth's core. One of the much-needed heavier elements creating the Earth melting effect is a critical amount of potassium-40, which plays an essential role in maintaining heat in the Earth's core which is crucial to life and will be discussed in greater detail later within this variable. The half-life of isotope potassium-40 means that as galaxies and the universe age, the metallicity of galaxies and our universe also *rises*. This means that in the future there will be less of these short-lived isotopes available. This factor plays a role in the potential life-sustaining properties of all stars and planetary systems to form in the future, and it is also a critical factor in the inevitable heat death of our universe.

This finding highlights a precarious balance in cosmic timing. Because Potassium-40 has a specific half-life, its abundance in the universe creates a constrained window for habitability. If Earth had formed later in the galaxy's history, the universe would have been too 'cold' (depleted of these isotopes) to melt the core, depriving Earth of the substantial benefits, many of them life enabling for our planet, of Potassium-40.

The melting of the Earth's accumulated layers of metals and radioactive elements acquired in layers of meteors over millions of years, allowed nickel, iron and other metals to become vital ingredients to maintaining Earth's molten core. Once within the Earth's core, these elements further separated into layers. First an iron/nickel core in the center, surrounded by a thick 1,800-mile mantle composed of silicate rock composing 67% of the Earth's mass, and 84% of its volume[29]. Soon after Earth became largely molten—driven by the lack of a substantial atmosphere and the extreme temperature contrast between the several-thousand-degree interior and the near -270°C temperatures of surrounding space—a thin crust began to form and gradually thickened over time.

Near Circular Orbits

The orbits of Earth and Venus have very low eccentricities (near circular orbits), significantly lower than those of Mercury, Mars, and most asteroids. The eccentricities of planetary orbits within our solar system are relatively low when compared to distributions commonly produced in early planetary formation simulations. N-body simulations—a dynamical system modeling gravitational interactions among multiple bodies—

conducted by J.E. Chambers and G.W. Wetherill modeling the final stage of terrestrial planet formation produced Earth-like planets with higher orbital eccentricities (e ~ 0.15)[30]. This is notable when compared with Earth's measured orbital eccentricity of approximately 0.0167.

> "More recent simulations using a larger number of planetary embryos have failed to overcome this problem, despite producing systems of terrestrial planets that are like our own in other respects. Here, I suggest that large eccentricities are the norm, while almost-circular orbits like those of Earth and Venus are the exception[31]."

The conclusions from this result are not insignificant, if not perhaps over-stated within the above paper:

> "With this information, it seems very unlikely that stable planetary systems, in which a small Earth-like planet resides in the habitable zone, exist in any other galaxy in our universe[32]."

Earth's orbital configuration can therefore safely be described as far lower than many early formation simulations would predict and highly ordered relative to many known planetary systems. The most important effect of this finding is how our orbital configuration contributes to Earth's stable seasons, climate patterns, and relative long-term environmental regularity. While this does not necessarily imply that such orbits are statistically anomalies, it does suggest that Earth occupies a particularly and perhaps unusually stable region of orbital parameter space.

This circularity is not merely aesthetic. Because our orbit does not swing in and out, and because our large Moon stabilizes our axial tilt at a steady 23.5 degrees, Earth partakes of the benefits of this tilt while also avoiding the irregularity of planets such as Mars.

Such stability affords life on Earth the gift of time. Complex life requires eons of relatively stable temperatures to evolve. If Earth's tilt were allowed to vary wildly—as is the relative norm in the universe–our climate would be far more likely to swing into inhospitable temperatures above and below what is conducive to life, perhaps tipping the vital climatic norms Earth has enjoyed, to life ending catastrophes during the extreme climate events of the past.

The Earth-Moon system acts as a type of stabilizing gyroscope, maintaining the steady, generous, but non-threatening climate band necessary for civilizations to rise. What we find in our circular orbit, tilt and resonance with the moon, is Earth has once again found itself in a condition of cosmically 'engineered' stability. This contrasts with many of the environments we find in other solar systems and even on other planets within our own solar system. The 'default' setting of the universe is chaos, planets incapable of life, or in locations outright hostile to life, while our specific setting is within what appears to be a curated stability[33].

A Thought Experiment on Moons

Imagine reading a science fiction novel about an advanced civilization orbiting a star much like our own. Their world is dominated by a vast moon. It is immense, luminous, and impossibly close. It governs climate, tides, long-term stability, and shields the planet from cosmic hazards. What it provides is astonishing and beyond simple fortune.

At the heart of this novel lies a deep mystery central to this fictional culture. Despite centuries of scientific progress, vast computational power, and mastery of physics, no one within this culture, not even its scientific elite, can determine where their Moon came from.

As a reader of science fiction, you naturally expect that at some point the veil of this mystery will be removed. You expect a revelation of sorts. And you would not be wrong to surmise that this revelation would fundamentally change this culture's understanding of how they came to be, and what their purpose is within the cosmos.

Now here is the twist. This story is not science fiction. It is a scientific fact that accurately describes our current level of understanding with our Moon. The origin of the moon remains a mystery. We do not possess a fully resolved or historically certain account of its origin. All manner of models have been proposed, and none satisfy all of the known constraints. And as we have learned more through successive ages, the paradox has sharpened rather than dissolving.

What we do know is factual and more astonishing. Without the Moon, Earth would be profoundly different. Its axial stability, tides, climate, and evolutionary trajectory would all drastically shift. Remove the Moon, and life, let alone complex life, likely never would arisen.

Our Moon

Earth's long-term habitability is the result of several interlocking physical and dynamical factors with its Moon, many of which have been previously mentioned. The Earth–Moon system represents a highly specialized configuration that contributes directly to Earth's long-term rotational and climatic stability[34]. But there is a perplexity in this relationship. The origin of the Moon has remained a conundrum and none of the current hypotheses have been able to determine the origin of the Moon.

Historically, there have been three main hypotheses proposed for the existence of the moon. They are ***The Fission Theory, The Capture Theory,*** and ***The Giant Impact Theory*** which we will examine later in the chapter. In later years hybrid hypotheses have been added, ***The Co-formation Theory*** and recent ***Synestia Theory*** are prime examples.

In the modern age, armed with new technologies, we have been able to narrow down the conditions necessary to explain our Moon. As is typical in our narrative, additional information has imposed additional constraints upon the hypotheses while narrowing the field of possibilities for the origin of the moon.

At present, each solution has either crippling problems or hypothesizes events that face astronomical improbability.

Moon Rocks Do Not Solve the Mystery

It was believed that lunar rocks retrieved from the Moon missions of the 1970s would resolve the issue of the Moon's origin, but instead, what scientists were able to determine is that the Moon appears to be neither possible as a part of the Earth, nor possible as having nothing to do with the Earth. The isotopic ratios are similar, but crucially off. The moon rocks we retrieved revealed a contradiction with isotopic anomalies. The Earth moon system is chemically unusual. It isn't perfectly identical, nor is it totally different. It is what is referred to as "selectively similar" and puzzlingly so, requiring us to come up with exotic models with numerous add-on refinements to fall in line with the evidence we have from the isotopes. But even these models raise serious questions. Astronomer Irwin Shapiro famously quipped, "The best explanation for the Moon is observational error—the Moon doesn't exist."

The 1984 Kona Conference

Irwin I. Shapiro (1929–2013) was not a peripheral voice in planetary science. A Harvard astrophysicist and long-time director of the Harvard-Smithsonian Center for Astrophysics, Shapiro was a member of the National Academy of Sciences and internationally known for discovering the "Shapiro time delay," one of the classical experimental confirmations of Einstein's General Relativity. His career placed him at the center of gravitational physics and precision astronomical measurement. Perhaps the most famous Dr. Shapiro quote is difficult to source, although it is found frequently when discussions of the Moon's oddity come up. He allegedly quipped, "The best explanation for the Moon is observational error—the Moon doesn't exist," he was expressing the frustration of an entire field that after Apollo, and the retrieval of Moon rocks, the mystery or the origin of the Moon had become harder, not easier, even impossibly so.

That tension came to a head at the 1984 Kona Conference on the Origin of the Moon[35] a major international meeting that brought together proponents of all major formation models: fission (the Moon spun from a rapidly rotating Earth), capture (the Moon formed elsewhere and was gravitationally trapped), co-accretion (Earth and Moon formed side-by-side), and the then-emerging giant impact hypothesis. Apollo samples had introduced serious discrepancies. Oxygen isotopes in lunar rocks were nearly identical to Earth's—unexpected under capture scenarios. The Moon's low iron core and volatile depletion challenged co-accretion. Angular momentum constraints strained classical fission models. Every theory failed somewhere.

By the end of the conference, the giant impact hypothesis stood tallest—not because it resolved every contradiction, but because it accommodated more of the data than its competitors. It explained the Moon's low iron content and high angular momentum better than the alternatives. But even then, it required refinements. Simulations suggested the Moon should be composed largely of impactor material, which conflicted with the isotopic similarity to Earth. Subsequent decades have introduced high-energy impact variants, post-impact equilibration models, and more recently Synestia scenarios—each designed to reconcile lingering inconsistencies[36].

In that sense, the field did move forward after Kona, but not by eliminating paradox. Rather, the dominant model evolved through successive adjustments to account for isotopes, volatiles, angular momentum and heat (thermal history). The field of Moon origin would argue that the Moon's origin is not unknown—but it is not fully settled, but the stark reality is the Kona conference crystallized a reality: the most familiar body in our sky had forced planetary science into a humbling position. The data were real. The models were imperfect. And work remained and remains to this day to reconcile the evidence.

Since the Kona conference, our instruments have improved, simulations have grown more powerful, and isotope measurements have become extraordinarily precise. But the core tensions remain: Earth–Moon system must still be accounted for, and the paradox has not disappeared.

The paradox has instead deepened to the point of being practicably unreachable by first principle physics alone.

Technical Details Complicating Resolution of the Earth Moon System

To understand why the mystery of the moon has been so difficult to solve in the past and even more impossible to solve in the present, we need to conduct an assessment of the scale of the problem from the main categories that have complicated the mystery.

In the chaotic and violent environment of the early solar system, each planet formed within a specific accretion zone, developing a unique chemical and isotopic signature. Every planetary body—planets, meteors, asteroids, and comets—possesses its own chemical fingerprint. For the Moon and Earth to possess nearly identical, but technically divergent isotopic ratios is statistically improbable. The data suggests that there is something unusual about the relationship between the Earth and Moon that defies simple probabilistic expectation intuited by isotopic similarity. And despite the confident assurances of researchers working on this issue that a solution is all but completed, more than forty years after the Kona Conference, resolving the origins of the Moon continues to challenge our best technology , our best minds and the most robust computer simulations.

There are many physical parameters that make it exceedingly difficult to formulate a single coherent evolutionary pathway for the Moon. This makes it very hard for origin scenarios to speak to all of the known constraints.

1. **Isotopic Similarity**: The isotopic similarity between the Earth and Moon is improbable because it requires either an impacting body very nearly chemically identical to Earth, or extremely efficient post-impact mixing within a vaporized planetary disk. This is statistically narrow because planetary embryos typically develop distinct isotopic histories depending on their location within the solar nebula. The probability of two colliding bodies originating from nearly identical nebular regions and resulting in a vaporized disk that reassembles with near, but not identical chemical histories is extremely low.
2. **Angular Momentum Balance**: The angular momentum balance is improbable because the collision between Earth and the impacting body (named Theia in the Earth Moon research community) would need to occur within a narrow range of impact angles, mass ratios, and early Earth rotational states. Collision simulations produce either excess or insufficient angular momentum. Achieving the precise angular momentum and strike geometry necessary to produce the present Earth–Moon orbital relationship is dynamically difficult, bordering on near-impossible.
3. **Mass Ratio Problem**: The Moon is unusually large relative to Earth compared to most planet-satellite systems. Modern simulations struggle to produce enough orbiting debris to assemble the Moon without overproducing angular momentum or vaporization

4. Volatile Depletion: Volatile depletion between the Earth and Moon is improbable because it requires extremely high-energy, high-temperature vapor disk formation while simultaneously preserving isotopic similarity. This combination is physically possible, but each required parameter works against the other, which results in the net result occupying a very small region of feasible parameter space.

5. Theia Composition Uncertainty: The composition of Theia, the name for the hypothesized impactor of Earth is uncertain and improbable because it requires either that Theia was unusually Earth-like chemically, or that post-impact mixing erased most compositional differences between Earth and proto-lunar material. Neither scenario is impossible, but both are statistically restrictive[37].

6. Iron Core Size: The Moon's iron core is exceedingly small, possibly as little as 1 or 2%, of lunar mass. By contrast the Earth has an iron core of approximately 30% of its mass. Earth-Moon formation models suggest that a giant impact would mix material violently enough to disrupt the layering from the inner core to the surface level of both the Earth and Moon. but this is not what we observe. This combination, in light of an assumed violent collision, is difficult to reconcile in Moon creation models.

7. The Crust Thickness: The crust thickness of the near and far side of the Moon differ significantly. The far side has much thicker crust at approximately 50 to 70 kilometers, whereas the near side of the Moon has an average thickness of 30 to 40 kilometers. There are several plausible scenarios for this difference, but we simply do not understand enough about initial conditions to confirm that our hypothesis is the definitive explanation[38].

Probability Summary

The probability of all The 7 above constraints being satisfied simultaneously requires an extremely narrow range of initial conditions and physical processes acting together in highly improbable ways. And the addition of each constraint removes several possibilities from the field of play. By the time we've gotten to number seven the bucket of possibilities is completely empty. And this is only part of the problem We not only need to find a solution for our Moon from this over-constrained set of parameters but we must also fulfil an obvious yet unspoken 8th constraint, and that is the problem of the present-day function of the Earth/Moon system. It does no good to solve the problem of the Moon's origin, if by the time we emerge from this list of constraints the Moon we are left with cannot do all of the things our Moon does for us. This is the final blow. Whatever model is offered must explain not only how the Moon was created, but the long term and present-day mechanics of the Earth Moon system, which includes a number of highly unusual and improbable conditions. Some of these conditions are Earth's magnetic field protection, tidal stabilization, ocean dynamics, axial tilt stabilization, long-term climate stability, geological activity support, core dynamo and magnetic shield reinforcement, eclipse geometry, and biological and evolutionary timing effects (lunar cycle influence on biological rhythm) among others.

The Moon As Enigmatic Teacher

We are left with a glaring mystery: the most visible object in our night sky is the one thing we can provide no explanation for. There is no apparent solution that can satisfy all of the needed conditions.

To be an intelligent observer is to be a student of the cosmos. If the universe is a school, the Moon has been one of our primary instructors. It has the rare combination of being both the object of mythology and spiritual belief, while also being an undisputed foundation of order, regularity and scientifically demonstrated benefits.

Long before humanity understood the physics of stars, we understood the rhythm of the lunar cycle. Its phases taught us the concept of the month, seasons, cycles and the year; its movement taught us navigation; its predictable face taught us that the heavens were not capricious and chaotic, but ruled by a deep, objective, observable order.

The Moon's role in the modern age extends far beyond what it meant to ancient cultures. It has served as the essential laboratory for our greatest leaps in physics.

Gravity: It was by calculating the orbit of the Moon that Newton validated his law of universal gravitation.

General Relativity: It was the Moon's ability to block the Sun during the 1919 eclipse that allowed astronomers to see starlight bending around the Sun, proving Einstein's theory.

Spectroscopy: It was during solar eclipses—made possible by the Moon's perfect size and distance—that scientist first isolated the element Helium in the solar spectrum.

The Moon has been a vital catalyst that transformed Earth from a planet that could have life, to a planet with life, to a planet populated by intelligent observers of the universe. The importance of the Moon for human beings cannot be overstated and sobering.

We began our inquiry into the Moon with Dr. Irwin Shapiro's 1984 quote, "The best explanation for the Moon is observational error—the

Moon doesn't exist." We end with a quote summarizing the as of 2024, in the abstract to the paper *A Systematic Survey of Moon-forming Giant Impacts. II. Rotating Bodies*, "In the leading theory of lunar formation, known as the giant impact hypothesis… while several distinct Moon-forming impact scenarios have been proposed… none of these scenarios have been successful at explaining the full set of known constraints, especially without invoking one or more controversial post-impact processes[39]."

There are many competing hypotheses for the Moon. We will not cover them all, nor will we go into depth, but we will spend time highlighting the main claims and the main problems with the most prominent theories. It is important to note that each piece of each competing narrative is physically possible (not impossible), but when each model is evaluated as a whole, the end result cannot represent all known constraints of our Earth Moon system. The Moon origin problem is a multi-constraint problem with no current closed-form solution.

The Fission Theory

First proposed in 1879 by George Darwin, the astronomer son of Charles Darwin, the fission theory held that the Earth and moon initially started off as one body, and that the moon was thrown off as a separate body due to the rapid rotation of the Earth. This hypothesis requires an astonishingly rapid orbital period for the Earth resulting in ~2.5-hour long days. George Darwin explained:

According to Darwin, the rapid rotation of the Earth created an interior oscillation multiplied by the oscillation of the solar tides resulting in the eventual splitting off of the moon, after which a braking effect would take place to slow down the Earth's rotation.

The flaws in this theory are several, and it was not long before the fission theory was demonstrated to be incompatible with the facts.

If the Earth had indeed been spinning fast enough to create a bulge that would eventually form a separate object, analysis indicates the bulge would result in a wobble in the Earth's spin which would in turn cause the Earth to *slow down* rather than continue to break apart. Fission would never take place.

A second problem is shared with nearly all other moon-from-Earth system hypotheses—if the moon had indeed been flung off from the Earth, the moon would be in orbit around the Earth's equator, unless some additional event is proposed to demonstrate that some event radically altered the Earth's tilt after the creation of the Moon. Our moon is not in this relationship with our Earth. Presently the moon's orbit is tilted approximately 5 degrees compared to the plane of the Earth's orbit around the Sun, which means that it's tilted about 18.5% from the 23.5-degree tilt of the Earth in relation to its orbit around the Sun. The physics of fission and the orbit of the moon are sufficient to *rule out* the fission theory of the moon's creation.

The Capture Theory

This theory proposes that the Earth *captured* the moon as it entered its orbit from regions unknown. This theory has several difficult problems. The biggest problem is that our moon is incredibly large in comparison to the size of Earth, so large that it's the largest moon of any planet in our solar system in relation to its host planet. The moon is approximately 1% of the mass of the Earth and although this doesn't necessarily rule *out* any theory, it requires conditions that are unlike the creation of any other moon in our solar system. The size of the Moon also places difficult constraints on all creation hypotheses. For the moon to have been captured by the Earth's gravitational field, it would have needed to be going *extremely slowly*, at *a perfect angle*, and in most models passed through some sort of medium such as a dense cloud of gas or dust in order to slow it even further so that it would have any chance of being "captured".

If all of these conditions were satisfied, the moon capture theory would require a host of additional finely tuned and perfectly aligned factors. But given all of these constraints, the most difficult problem for the Moon capture theory is the issue of isotopes. As we mentioned, the Earth and the Moon have many identical isotopes, and yet, others which differ, this perplexity has been the bane of all moon creation hypotheses and in this case it makes it very unlikely that the moon just happened to be floating by, under perfect conditions and was captured by the Earth.

The Co-formation or Condensation Theory

This theory argues that the moon was forged alongside Earth, in the protoplanetary cloud of the primitive solar nebula. Had such an event occurred, then this answers why the Earth and Moon would be composed of similar materials and isotopes in addition to being in a similar location; however, the moon and Earth are not similar in density which would *not* be the case under the co-formation theory, as they would have similar cores had they accreted together within the protoplanetary disc. The Earth and moon are also, puzzlingly, not composed of the *exact* same materials. The Earth has a significant iron core which is absent from the moon. The absence of compositional similarity, despite isotopic similarity, is another problematic constraint for all moon formation hypotheses. For the co-formation or condensation theory the devil is in the details, and the details are where the co-formation hypothesis succumbs to a compositional challenge it cannot answer.

The Giant Impact Theory

This theory proposes that a giant object crashed into the Earth, causing the Earth and the object to destroy one another, mix their ingredients, and reform into an Earth-Moon system. This hypothesis is what the Kona conference settled on, famously because at the time, it was closer to explaining the constraints than the other hypotheses, even though it did not satisfy all constraints, rendering all other proposals (at the time) *less likely*. The giant impact hypothesis explains how an Earth/moon type system could form, but like all other proposals it cannot adequately address the constraints that have been imposed for all of the conditions we find.

For the giant impact theory to be responsible for our moon, a giant impact would have needed to be *at least* the size of Mars. That theoretical object has been named Theia. The impact is theorized to have occurred 4.5 *billion* years ago—within the first 100 million years after the solar system was formed[40]. It requires that the Earth is in a condition where it has the *perfect* rate of spin, angle, speed and location for the precise, high velocity impact of an impact that is very nearly head-on, but not precisely head-on. Mathematical analysis and modeling has found that only a precise impact conforming to the above conditions could result in our Earth-Moon system. But this is not the biggest problem. Analysis has shown that if the moon was created in this way, the moon would be composed mostly of *the impactor* (Theia) and not what the Earth is currently composed of. Once again the isotopic similarity between the Earth and moon rears its ugly head and rules this hypothesis out.

Analysis has shown that there is *less than* a 0.1% to 1% chance (1 out of 1,000 to 1 out of 100) that any proposed *Theia* narrative could have had the same isotopic profile as Earth[41].

Synestia Theory

Synestia Theory is a radical Moon origin hypothesis co-developed by Sarah T. Stewart, a MacArthur genius grant awardee and Harvard's Earth and Planetary Science graduate student Simon Lock[42]. The term, "Synestia" received half of its name from *Hestia*, the goddess of the hearth, and the other half of its name from the word, "*syn*" (together).

The Synestia hypothesis proposes that our Earth, rather than looking as it does presently, was once in a distorted flattened oblate state with flattened north and south pole and a wide bulging equator. This is due to the Earth being brought into a rotational period ("day") of ~2-3 hours due to angular momentum from accretion and a number of well-placed and accumulated collisions. This in itself would be the product of an improbable number of precise and finely tuned collisions. Then, while in this condition the Earth would have been impacted by Theia, the hypothesized large impact that created the Earth Moon system. This impact would have been unimaginably violent, more of an obliteration, so extreme that a significant portion of Earth would have been vaporized while a large fraction of Earth's silicate mantle would have melted with the rest being liquefied. The result was not a planet, but a Synestia—a swirling, superheated donut shaped vapor of rock and molten rain[43].

Inside this chaotic cloud, the material for the Earth and the Moon was thoroughly mixed, erasing the boundaries between the two worlds.

Synestia Details

So what we really have is an incredibly exotic theory that doesn't explain the full history of Earth or the impactor, doesn't result in the isotopic components of the Moon, and doesn't explain how the donut shaped Earth came to be in such an exotic condition. Finally, like many of our current materialistic theories (inflationary, string theory, many worlds interpretation, evolution, etc.) it erases all evidence that it has taken place. Additionally, it takes place on timescales so short that no one can reasonably expect to ever see a synestia event, ever, anywhere. Then we have the highly improbable nuts and bolts of the concept. Synestia events require trajectories, speeds and impact angles that are all highly, highly improbable. If this theory sounds familiar it's because it is a quintessential materialistic theory.

It advances the materialists narrative and is almost completely unfalsifiable and can never be observed. Yet somehow, from this impact and mixing we can find the emergence of the Moon and Earth. Amazingly synestia describes their isotopic similarity—but only to a degree. There are additional inputs to this long shot hypothesis that are required to emerge with not only a Moon and a host planet, but our Moon and Earth[44]. All of the other dynamics required to make Synestia theory play out come from giant impact models, N-body orbital simulations and thermal evolution models.

Moon Origin Summary

We have endeavored to explore why the origin of the Moon is so hard to work out. We have investigated the competing constraints, and the different models and each theory fails for different reasons. No imaginable scenario can explain our moon, not even the highly improbable synestia theory. In the end, despite it's highly improbable set up and result, synestia theory does not result in our Earth or Moon without additional elements. 40 years after the original Kona conference (1984) and over 50 years after Apollo astronauts first brought back Moon rocks (1969) the mystery has only deepened. The only part of this mystery that is understood, is how the Earth benefits from our Moon. That is what we will investigate next[45].

An Unusually Large Moon

By mass, our moon is the largest in our solar system in relation to its host planet. The Moon is also the fifth largest of the ~290 moons in our solar system, and the most unusual in its overall quality and characteristics of all moons in our solar system. Three of the largest moons—Io, Callisto and Ganymede—orbit Jupiter which has a diameter of 88,846 miles. Rounding out the top 5 largest moons is Titan which orbits Saturn with a diameter of 74,898 miles. Notably, Uranus at 31,763 miles diameter, and Neptune at 30,775 miles diameter, do not have a moon as large as Earth at 7,926 miles diameter.

Our unusually large moon stabilizes the Earth and exerts a gravitational influence that provides many life critical benefits. One of these benefits is in the moon's ability to raise the ocean tides. Tides mix nutrients from the land with the oceans, creating the fecund *intertidal zone*. This is where a good portion of the landmass of the Earth is periodically immersed in seawater. The moon also exerts one-third of its tidal energy along rugged areas of the deep ocean floor, and it is believed that this may be a main driver of ocean currents… circulating enormous amounts of heat[46] and providing a highway of sorts for fish, animals, plankton, weather systems, and seafaring.

The Earth's Molten Core & Earth's Magnetic Shield

Yet another one of the Moon's most enduring contributions to the Earth's habitability is less visible than tides or axial stability: The Moon helps sustain the geodynamo that generates our protective magnetic field

Earth's magnetosphere is one of the most important large-scale protective factors contributing to the planet's long-term habitability. The magnetosphere is generated by the convection of the electrically conducive outer liquid iron core outside of our solid inner core. Iron is highly fit for this role due to both its stability as a metal and magnetic properties on the entire core: roughly 30 per cent of Earth's total mass is made up of both liquid and solid sections driving the strength of the magnetosphere. And our iron core meets all three requirements for Earth's magnetic field under *Dynamo Theory*[47].

(1) A sufficient volume of electrically conductive fluid. The molten iron core of Earth fulfills this requirement.

(2) Sufficient energy must exist within the system to drive convection. For Earth, the heat escape from the inner core to the mangle is the source of convection energy.

(3) Rotation; as the Earth spins, the Coriolis effect twists the rising metal into spiral coils which organize the liquid into the geodynamo that generates the magnetic field.

The magnetosphere protects our atmosphere from erosion caused by the charged particle solar wind, particle radiation, coronal mass ejections from the Sun and cosmic rays from deep space.

It is believed that erosion caused by solar winds is the likely cause of Mars being stripped of its atmosphere. Neither Mars, Venus, nor Mercury have magnetospheres and if Mars had an atmosphere at one time as believed, it would have certainly lost it once it lost its magnetosphere[48]. The magnetosphere also keeps at bay unwanted energy from all cosmic sources that are unable to penetrate the twin donut shaped inner, and outer zones of the Van Allen radiation belts. These belts were discovered in 1958 by James Alfred Van Allen, an early space scientist at the University of Iowa[49].

The Moon plays a secondary role in the Earth's magnetosphere through its gravitational and tidal influence. This interaction, from the beginning, has transferred rotational and orbital energy estimated at trillions of watts to Earth's interior and potentially the outer liquid core.

"(The Earth's) mantle deforms elastically due to tidal effects caused by the Moon. The researchers show that this effect could continuously stimulate the motion of the liquid iron alloy making up the outer core, and in return generate Earth's magnetic field. The Earth continuously receives 3,700 billion watts of power through the transfer of the gravitational and rotational energy of the Earth-Moon-Sun system, and over 1,000 billion watts is thought to be available to bring about this type of motion in the outer core. This energy is enough to generate the Earth's magnetic field[50]."

There is an existing paradox for the Earth's geodynamo—molten iron core on the outside of our solid iron core—to be at its current 3,800 degrees C, the entire Earth would have needed to have been molten 4 billion years ago and the core at this time would be required to be at 6,800 degrees C, a temperature and condition that cannot be explained. Under the tidal moon effect, the Moon provides a significant amount of the needed vital energy to the Earth's core, drastically reducing the required initial temperature of the Earth's core from the theorized 6,800 degrees C to 4,100 degrees C, thus cooling only 300 degrees C in the intervening time period. The constant infusion of billions of watts of energy to the Earth's geodynamo, maintains its magnetosphere, thus protecting Earth' s atmosphere and oceans from being stripped away, thus safeguarding all of the life on Earth. But this is not enough to resolve the paradox. There are several paradoxes currently at play that we will address momentarily.

Through an active magnetic core, a tectonically active crust builds mountains, subducts old sea floor, and recycles the carbon dioxide in the atmosphere, all of which make Earth more habitable. This action is facilitated by the continuous presence of liquid water on Earth's surface. The chemical reaction of water with minerals weaken the crust, providing lubrication that allows the crust to bend without breaking[51]. Plate tectonics is vital to the carbon cycle. This cycle involves the travel of carbon from the biosphere, hydrosphere, atmosphere and geosphere back into the Earth, and eventually returning to these sources.

A new paper has proposed that despite the Moon's assistance in maintaining the Earth's molten core, the molten core itself should not exist. In a paper titled, *Earth's Inner Core Nucleation Paradox*, scientists found that for Earth to have a molten core, the inner solid core would require "supercooling". They have found that the barrier for supercooling is uncrossable. Further, the study found that other explanations are not plausible

> "The conventional view of Earth's inner core is that it began to crystallize at Earth's center when the temperature dropped below the melting point of the iron alloy and has grown steadily since that time as the core continued to cool. However, this model neglects the energy barrier to the formation of the first stable crystal nucleus, which is commonly represented in terms of the critical supercooling required to overcome the barrier. Using constraints from experiments, simulations, and theory, we show that spontaneous crystallization in a homogeneous liquid iron alloy at Earth's core pressures requires a critical supercooling of order 1000 K, which is too large to be a plausible mechanism for the origin of Earth's inner core[52]."

At its core, the paradox is a nucleation problem. A common physical analogy is supercooled liquids, which can remain in a liquid state well below their freezing temperature if no seed crystal exists to initiate phase transition. According to the above 2018 study, the Earth's core appears to face the same problem except at a planetary scale. For Earth's inner core to have formed through spontaneous crystallization of molten iron, the core would have had to cool roughly 1000 Kelvin below its melting temperature to overcome the nucleation energy barrier. Current geophysical models do not support such extreme supercooling.

In practical terms the paradox entails the Earth being too warm to have formed its solid inner core through spontaneous nucleation alone. Without an initial solid seed to trigger crystallization, the core would be expected to remain entirely molten without abatement. Yet, this is not what we observe. The solid inner core — the geodynamical engine supporting Earth's magnetic field — is undeniably present, and also undeniably unexplained.

What binds these paradoxes together is a singular, staggering commonality: they should not exist.

Under any standard model of probability or first-principle physics, the Earth's interior should have collapsed into one of two "Natural" states:

(1) a perpetual liquid ball with no magnetic shield (Dead), or

(2) a frozen solid lump with no tectonic or volcanic recycling (Dead).

Instead, we observe a third, "Forbidden" state: a stable, multi-billion-year coexistence of solid and liquid that defies the 1,000-degree temperature gap and the energy barriers of nucleation.

These are mechanical requirements—highly specified, non-random interventions in the structure of the planet. There seems to be only one reason for them both to exist, when they should not, and that is to promulgate the thread of teleology throughout the Algorithm—the creation of a life-habitable and life-inhabited planet.

The formation of the inner core requires a transition that the system should not be able to reach, while its continued existence requires a balance that is unstable and should not exist. The system must both originate and persist under constraints that prevent both. The uncomfortable reality is these paradoxes sit at the foundation of the geodynamo.

Near the End Check-In The Divine Meditation & The Anthropic Principle

Thus far in this chapter, we have examined a series of mysteries — all conducive to life, with some only partially understood through conventional scientific models and others, still at the level of mysteries. A clear pattern is beginning to emerge as we have gathered evidence from fields including inner solar system dynamics, planetary science, and Earth system science.

To support life as we know it, we appear to have acquired, or benefited from, the following conditions from Earth's formation through the present:

A Universal habitable zone — the region around a star where liquid water can persist.

A Galactic habitable zone — a region of the galaxy with lower exposure to catastrophic radiation and stellar hazards.

Birth within a Spiral Galaxy — which has more stable star formation environments than other types of galaxies.

Emerged from a metal-rich protoplanetary disk — Our Earth was enriched by multiple prior supernova events, providing heavy elements necessary for planetary differentiation, technological potential, and long-term radiogenic heating (including isotopes such as potassium-40 for internal planetary heat maintenance).

Migrated away from our Star Nursery and major cosmic hazards — including nearby stars, transient sterilization events, and destabilizing gravitational perturbations from other stellar bodies (sometimes described in orbital migration literature as the "great escape").

Became safe for life due to a series of unusual gas giant planet movements — described in models such as the Grand Tack and Nice models, which suggest Jupiter and Saturn migrated significantly during early solar system formation and performed needed work to make our inner solar system habitable and stable for hundreds of millions of years.

Acquired atmospheric observability — including an improbably clear atmosphere and favorable placement within a spiral arm region that minimizes excessive cosmic radiation and allows us a near unobstructed view of the entire cosmos.

Obtained an unreasonably circular and stable planetary orbit — reducing extreme seasonal and climatic instability that would otherwise threaten life.

Fallen into resonance with an unusually large natural satellite — the Moon provides tidal regulation, climate stabilization, and ecological rhythm that cycles carbon, creates new environments and provides long-term stability for life.

Obtained a life-facilitating axial tilt and rotation rate — producing day-night cycles conducive to biological and climatic stability.

Generated a robust functioning magnetosphere — protecting the atmosphere and oceans from solar wind stripping and high-energy particle radiation.

The world of materialistic science would interpret this convergence of improbable constraints as nothing more than a statistical certainty somewhere within our universe given the size of the universe. But what we have demonstrated is that life cannot be adequately explained as the product of blind material processes operating under first-principles. At every stage within our divine meditation, what is required for life either exceeds the probabilistic capacity of the universe, or it is functionally and probabilistically impossible.

But this brings up the central paradox. Our universe has the conditions necessary for life, i.e. a habitable universe, stars that produce elements, chemistry, fitness for biology, the ability to produce life habitable galaxies, solar systems and planetary systems– but, what is required to take these perfect life conditions and turn them into actual life is in a sense encrypted beyond possibility. What is required is an actor, with the key for encryption that can take these possibilities for life and articulate them across impossible probabilistic thresholds.

In this sense, just when it comes to the Earth Moon system life requires what can reasonably be described as over half a dozen conditions that exceed the probabilities capacity of the universe, and another over half dozen conditions that are functionally impossible (miracles) that the fundamental conditions required for them to exist, cannot be derived from first principle physics at all. These events are not simply rare, they represent a complex, interactive, collaborative, cumulative, layered constraint topology, where multiple requirements must be satisfied simultaneously.

We are fortunate that we live in an age where we have hard scientific information to show demonstrate the constraints required for our making and the impossible nature of our existence.

Introduction to the Final 1/3rd of our Meditation

If the life conductive conditions discussed so far are the 'hardware' required to keep us alive, the next section can be interpreted as the 'software' designed to help us see our place within the heavens to a new level, a level that requires the first robust found of scientific instruments. We will see in our upcoming subjects that both the solar system and the earth were not merely made to allow and maintain life, but also to teach it and show life how to become intelligent, and to reveal the inner machinations of the cosmic engine.

Moon Eclipses & Spectroscopy

The position of the moon in relation to the Earth and Sun provides the Earth with "perfect eclipses" where the moon covers the entire sphere of the sun while not obstructing our view of the Sun's chromosphere and corona. The chromosphere is the second outermost layer of the Sun's atmosphere extending from 250-1,300 miles above its surface. This view is afforded to us because of an improbable alignment between the Sun and

Moon. Thus, both layers of the suns outer portion are made available for us to view.

The Chromosphere: The reddish layer extending 250–1,300 miles above the surface.

The Corona: The pearl-white outer atmosphere extending millions of miles into space.

The sun is roughly 400 times larger than the Moon, but it is also 400 times further away from Earth, making the Moon the ideal size to entirely cover the sun entirely during a solar eclipse. Aiding in this effect is the fact that the Sun and moon are *the two most perfectly round objects* in the entire solar system leaving the visible chromosphere perfectly visible for scientific information.

Of our solar system's 8 planets and 176 confirmed moons, the Earth/Moon combination is the *only one* with perfect eclipses. The speed of the moon's orbit, combined with the speed of the Earth's rotation also renders our perfect eclipses suitably long to make important measurements. The Earth will continue to experience perfect or close to perfect eclipses for the next 250 million years, -a period of time that also matches the time when the emergence of intelligent observers on Earth (*now*). This is a time period comprising just 5.5% of the current age of the Earth, and it is a condition that will only exist for 2.5% of the total lifespan of our sun (10 billion years).

Perfect eclipses allowed scientists to minutely examine absorption and emission lines of electromagnetic radiation from the sun which allowed us to understand the chemical composition of the Sun. By using this same method, humanity was empowered to deduce the composition of the photons of *any light emitting object* in our universe with spectroscopy, a science that was formalized in 1881.

Aiding this effect is the shape of the bodies themselves. The Sun and the Moon are the two most perfectly spherical objects in the solar system. If either were significantly oblate (squashed), the 'fit' would be imperfect, compromising the observation. This alignment is unique in our neighborhood. Of the solar system's 8 planets and 176 confirmed moons, the Earth/Moon combination is the only one that experiences perfect eclipses.

However, this phenomenon is not permanent; it is a fleeting window in cosmic history. We are currently in a 250-million-year epoch where this alignment is possible. This represents just 5.5% of the Earth's current age and only 2.5% of the Sun's total 10-billion-year lifespan.

The implication is profound: Intelligent observers (humans) have emerged on Earth at the exact moment when the viewing conditions for the universe are perfect. We have arrived just in time for the show.

The Electromagnetic Spectrum, Vision, Light and Characteristics of The Earth

> "The electromagnetic spectrum (EMS) consists of many different types of radiation. The two most familiar forms of light and heat such as our sun primarily emits. (Heat is infrared or IR radiation, which we experience as warmth as our sun shines on the skin.) But there are many other kinds such as radio waves, microwaves, ultraviolet, X-rays and gamma rays. All the various different types of electromagnetic radiation including light and heat or infrared flow through space in the form of energy waves analogous to ripples on the surface of a pond[53]."

The above quote points out something so obvious but that we would never know if our science had not the ability to view a wide range of suns. Our sun provides with the exact radiation that heats our Earth perfectly, while also providing us with information that tallows us to discern the chemical composition of not only our sun, but any distant object we can observer.

Aiding in this, our Earth has an *exceptionally clear* atmosphere for cosmic observation, allowing all of the *helpful electromagnetic radiation* needed for photosynthesis, visual light, and the transmission/reception of radio waves to fall directly on the surface of the Earth. While at the same time, our atmosphere is amazingly adept at blocking *harmful electromagnetic energy* that would have made first life on Earth difficult and constituted an ongoing hazard to life and speciation. This condition constitutes an unusually beneficial trifecta of advantages for a race destined to be technologically advanced.

To understand the miracle of filtering provided by our atmosphere, we must first understand the chaotic ocean of energy in which we swim. The universe is saturated with radiation, flowing through space in the form of energy waves known as the Electromagnetic Spectrum (EMS).

This spectrum is incomprehensibly vast. On one end, we have Gamma Rays, with wavelengths as small as $\mathbf{10^{-17}}$ meters—a fraction of the diameter of an atomic nucleus. On the other end, we have Low-Frequency Radio Waves, which can stretch for a hundred thousand kilometers from crest to crest.

Between these two extremes lies a chaotic jumble of X-rays, microwaves, and ultraviolet radiation. Most of this energy is either useless for biological life or actively destructive to it. Yet, right in the middle, there is a tiny, narrow slice of frequencies that we call 'Visible Light' (380 to 700 nanometers).

Visibility

The fact that we can see at all is due to the specific architecture of our planet's atmosphere. It functions not merely as a blanket, but as a highly intelligent filter. If our atmosphere were too thick (like Venus), it would block all useful energy. If it were too thin (like Mars), it would fry us with radiation. Instead, Earth possesses an 'Atmospheric Window'—a precise opacity setting that achieves a beneficial trifecta:

The Shield: It blocks the deadly gamma, X-ray, and far-UV radiation.

The Window: It is transparent to the specific narrow band of visible light required for vision and photosynthesis.

The Signal: It is transparent to radio waves, allowing for the transmission of information and the study of the cosmos.

This transparency is a precondition for a technological civilization. We are a race of plant-eating, light-dependent observers who can view the heavens only because our atmosphere clears a path for the stars to shine through.

Vision

The composition of the human eye leverages the qualities of visual light. The complex engineering of optical vision matches with the fitness of biological structures and the elements that compose these visual structures to allow us to see. The range of the electromagnetic spectrum is vast. Wavelengths in the visual electromagnetic radiation (the visual spectrum) range from 380 to 700 nanometers. Gamma rays may be as little as $\mathbf{10^{-17}}$ "only a fraction of the diameter of an atomic nucleus[54]," whereas "Some extremely low frequency radio waves maybe a hundred thousand kilometers from crest to crest[55]."Just outside of the visual light spectrum it is theoretically possible to "see" microwave, x-ray and there are a number of animals that can see within the ultraviolet (butterflies, reindeer, some birds and bees) and infrared (some snakes, fish, mosquitos, beetles, etc).

The final piece of this puzzle is the human eye itself. It is not enough to have light. The complex engineering of the human eye is perfectly matched to the window of light our atmosphere filters in. While some creatures have evolved to see into the ultraviolet spectrum (butterflies, reindeer) or the infrared (snakes, mosquitoes), the human optical system is tuned to the peak output of our Sun and the maximum transparency of our air. It is a lock-and-key relationship: the biological instrument (the eye) is calibrated exactly to the physical conditions (the light) allowed by the planetary filter (the atmosphere).

Amazingly, the light we need for vision, plants and plankton need to photosynthesize, and what is required to warm our Earth, melt ice from ice ages and facilitate chemical reactions in living cells is afforded by our atmosphere.

Technological Observation

Our technology benefits from our atmosphere as well, as if according to plan. Radio signals easily pass in and out of our atmosphere, allowing our technology to "see" the universe in vastly different wavelengths unavailable to our eyes. This balance is critical, surgical even. If our planet were significantly *larger*, our gravity would be such that our atmosphere would contain gasses which would alter our ability to absorb light, and our ability to see out past our atmosphere[56]. On the other end, if our planet were significantly *smaller*, our gravitational effect would not be sufficient to hold onto our current atmosphere, and we would be unable to filter out the various forms of harmful light.

> "Only rocky planets of approximately the size of the Earth, with the same gravitational pull, can hold an atmosphere consisting of heavy gasses like oxygen and nitrogen and support advanced, active carbon-based life forms like ourselves. If they (planets) are bigger, they retain the primeval gasses like Jupiter. If they are smaller, they lose their atmosphere like the Moon… The same gravitational force which retains the vital oxygen for life also imposes an absolute limit on the size of terrestrial organisms[57]."

The size of our planet is perfect, the composition of our atmosphere is perfect, and the properties of light for chemistry (photosynthesis) and human eyesight are also perfect. All of these factors—physics, chemistry, the electromagnetic spectrum and biology— have to be in perfect alignment for there to be life, complex life and intelligent observers. There is no way around it. The anthropic principle is mute on this subject. It can say that life can look back at the conditions that allow intelligent life, but it has nothing to say about how perfectly in alignment these factors must be, nor does it have anything to say about how limitless our abilities extend, the opportunities of Earth and the cosmos, or the perfection of biological innovation must be. There is no reason why any of these coincidences must all align so exquisitely. They are all independent factors.

The Earth's Resources - A Comprehensive Overview

When we survey the physical composition of our planet, we find a catalog of resources that goes far beyond the basic requirements of survival. Earth appears to be stocked with a suspiciously utilitarian inventory—a 'starter kit' for high-tech development that includes everything from the most basic fuels to the most exotic conductors.

If Earth were merely a biological preserve, we would expect water, simple carbon chains, and sunlight. Instead, we find a layered inventory that sustains not just life, but industry:

Energy: A diversity of dense fuels (coal, oil, gas) and kinetic forces (wind, solar, hydro).

Infrastructure: Natural logistics systems like rivers, jet streams, and ocean currents that facilitate trade and movement.

Restoration: Planetary-scale recycling systems, from the water cycle to plate tectonics, which refresh the surface crust.

Technology: An accessible crust rich in the specific elements required for the Information Age—silicon for chips, lithium for batteries, and carbon allotropes for advanced materials.

There is literally nothing required for a spacefaring, digital civilization that is missing from this inventory. Crucially, Earth did not create this bounty; it inherited it. There is literally nothing that is required for life and hi-tech life that is missing from our inventory. The sun is responsible for all of this bounty, except for the hydrogen and helium which served as the initial fuel for stars to create this endless bounty then combined in limitless molecules that represent all that we encounter. The production list of our stars? It is the periodic chart and all that we can see, hear, feel, touch, and experience. It is who and what we are.

To understand how unusual Earth is, we must look at the ingredients list of the universe. In the wake of the creation of the universe, the cosmos was composed almost entirely of two simple elements: Hydrogen (~75%) and Helium (~25%).

Vital Elements - Earth Edition

Hydrogen is the most abundant element in our universe, produced in the earliest epoch of our universe. In the early universe roughly 3/4th of the universe was hydrogen and the rest largely helium. Hydrogen however only makes up only 0.14% of Earth's crust by weight, and less than 1% of the total weight of the Earth. Hydrogen has an atomic weight of 1.01, which makes it extremely light with one proton and one electron. So, although it is plentiful, it has very little mass. This means although the Earth is covered in oceans and has enormous ice caps at the poles, in addition to lakes, rivers, streams and clouds, water only accounts for 4.8% of the weight of the Earth. And since oxygen has an atomic weight 15.99 compared to hydrogen's 1.01, with 2 hydrogen atoms to each oxygen atom of water, each molecule of water is roughly 87% oxygen. This means that 99.86% of the weight of the Earth was produced by stars (hydrogen is plentiful, but it's so light it contributes very little to the weight of the planet).

Suns are therefore responsible for nearly 100% of Earth's internal and external energy sources, and nearly 100% of all chemical energy on Earth (wood, coal, oil, gas), and virtually 100% of our technology by mass.

Earth, therefore, is a rare concentration of stellar debris. We are living on a planetary-scale accumulation of 'processed' material, gathered from the ashes of catastrophically obliterated stars. Without this specific concentration of heavy elements, what we now recognize as Earth would instead be a gas cloud made of the two lightest elements. No magnetic core to shield us, no carbon to scaffold biology, and no silicon to build our technology.

Vital Resources

Humanity at first used natural raw products such as stone, wood and animal products to forge its technologies, weapons and for food, clothing, shelter, and safety. All of the metals needed for the copper; bronze and iron age were present and near the surface. Iron and steel for the industrial revolution. Coal, wood, and oil, at first found within plants and from whales, domesticated and game animals, and then within the Earth derived from vast reserves of deep ground petroleum.

These vital resources literally lubricated the rapid rise of societies. The sun created the Earth, providing the raw materials for Earth. The sun provides the energy for Earth to power its currents, rivers, atmosphere, and molten core/magnetosphere. The sun provides the energy for photosynthesis for the original atmosphere terraforming cyanobacteria. The sun provides energy for the plankton that serve as the first link in the food chain in all of our oceans. The sun provides the energy of wind, rain, plate tectonics and ocean currents that ground down the original rock to make organics

Let us now make full note of what has been provided to Earth; a well-stocked inventory is only a part of the vital equation required for our well-resourced home planet. The sun provided this inventory, while also continuing in its role as the engine to power the life of the Earth itself. While the Earth's internal heat (radioactive decay) drives plate tectonics, it is the Sun that powers almost every dynamic system on the surface:

The Atmosphere: Solar heat drives the wind, rain, and ocean currents that erode rock into fertile soil.

The Biosphere: Solar energy powers the photosynthesis of plants and plankton—the base of every food chain.

The Fuel: Even our 'underground' energy is actually derived from solar. Coal, oil, and natural gas are simply ancient sunlight, captured by prehistoric plants and compressed by geology into dense chemical batteries.

The sun and the rays of the sun, directly radiated to Earth, are our cosmic umbilical cord. Never severed, and one that can never be severed if there is to be life upon our surface.

Organic Fuels: Oil, Gas & Coal

The leading hypothesis for the origin of Oil and Gas and Coal falls under the *organic origin* theory. Petroleum and gas are byproducts of the early plankton that died and accumulated on the sea floor approximately 300 million years ago, and were buried by layers of sediments, and cooked by heat and decomposition. With the increased pressure of geological processes, this organic layer was transformed into petroleum and gas that either migrates to the surface or is held in underground reservoir traps underground.

The formation of coal took a slightly different pathway. It formed under the same general time scales (300 million years) from the biomass of the carboniferous period when vast swamp forests covered large regions of the planet along with towering ferns, primitive trees, reeds, mosses, and dense vegetation accumulated in thick organic layers[58]. Once again, over millions of years layers upon layers formed due to the pressure, heat and geological processes many of these processes have been previously touched upon in this chapter. This is what led to chemical changes within the wood where oxygen was forced out leaving rich carbon deposits of coal. The trapping of carbon dioxide from the air due to photosynthesis and containment underground had a drastic effect upon the atmosphere as billions of tons of carbon was pulled out of the atmosphere and the ratio of oxygen steadily increased, giving rise to increased tree growth, carbon trapping, coal and new life forms.

The energy produced by coal is a highly efficient fuel, some 4 times greater than the equivalent amount of wood. In essence, Oil, Gas and Petroleum represent stored chemical energy, like gigantic well-placed batteries, created by the sun and photosynthesis for modern industrial applications. What would humanity have done without these vast stores? Vehicles, manufacturing, ocean going ships, trains, and heat have all been indispensable to the technological rise of complex societies.

One of the more remarkable and little-known aspects of coal formation involves a biological gap in Earth's early ecosystems. During the Carboniferous period, fungi and bacteria capable of efficiently decomposing lignin and cellulose had not yet been developed by biology (evolved)[59] . This has been referred to as the fungal lignin-decomposition gap hypothesis. Fungi perform approximately 3/4th of the work and bacteria most of the remaining 1/4th. This temporary gap in the microbiome allowed massive amounts of plant biomass to accumulate without fully decaying.

In essence, fossil fuels function as portable reservoirs of concentrated solar energy, capable of powering machines that reshape landscapes, construct megacities, terraform the planet and transport vast quantities of material across oceans and continents. Nearly every sector of modern civilization depends upon the extraordinary energy density contained within hydrocarbons.

And there is more. The application of petrochemicals is not restricted to fuel sources.

Beyond fuel and physical goods, petroleum is fundamentally a chemical feedstock built from complex hydrocarbon molecules that are refined into thousands of industrial and medical materials resulting in pharmaceuticals, synthetic rubber, cosmetics, fertilizers, detergents, paints, lubricants, and components in electronics manufacturing. In modern industry, the chemical petroleum foundation provides the raw building block for everything from medical packaging and surgical tools to barriers of all types. Without this added utility of petrochemicals, many medical advancements would have been stunted.

Based upon 2015 data, all fuel reserves are estimated to last for 200 years from when humans *first began* using them. Specifically, Oil reserves are predicted to last *30-51 more years, coal, *70-114 years, and natural gas, *40-53 years. These estimates are on the lower end, and assume no new reserves are located—these estimates are independent of current resources.

Seen from this perspective, these resources function almost as a civilizational bridge—providing the concentrated energy required to ignite industry, mechanize agriculture, and build global infrastructure while humanity develops the knowledge and technology necessary to transition toward more advanced energy systems such as hydroelectric, nuclear, solar, wind, and future sources not yet fully realized[60].

Terraforming the Earth

When first life began, Earth was hostile, rocky, and far from the rich and lush planet we know today. The land was filled with volcanoes that spewed water vapor, methane, ammonia, carbon dioxide and hydrogen sulfide. This era of Earth is referred to as *the Hadean era,* named after Hades, the Greek God of the dead and the underworld. In this age, our Sun was

71% as strong as it is now, but due to geothermal activity, and a significant contribution of greenhouse gases, the temperatures of Earth would have ranged from a high of 700 to 2000 degrees Celsius and in some environments, possessed temperatures low enough to enable thermophiles and hyperthermophiles to thrive in temperatures of approximately 80 degrees Celsius (176 degrees Fahrenheit).

The atmosphere of the early Earth was chemically hostile to life as we know it. While ultraviolet light split small amounts of water vapor to create a trace oxygen level of roughly 1%, this was far below the threshold required for aerobic respiration. The transition from a suffocating world to a breathable one was not a geological accident, but a biological project engineered by cyanobacteria.

These microscopic organisms, forming dense mats known as stromatolites, spent hundreds of millions of years photosynthesizing—literally exhaling the atmosphere we breathe today. To grasp the sheer scale of this terraforming effort, compare it to the Industrial Revolution. Since 1760, human activity has raised atmospheric carbon dioxide by a fraction of a percentage point, yet the climatic impact is profound. In contrast, these bacteria raised atmospheric oxygen from near-zero to heights of 35%. They did not merely tweak the climate; they fundamentally replaced the chemical composition of the sky.

However, this renovation came at a steep price. For the first billion years, the oxygen produced was absorbed by iron dissolved in the oceans, effectively "rusting" the world. Once these chemical sinks were full, oxygen flooded the atmosphere. For the dominant lifeforms of the time—anaerobic bacteria—oxygen was a deadly poison. This triggered the Great Oxygenation Event, a mass extinction that wiped out the vast majority of life on Earth and banished the survivors to the deep oceans. Simultaneously, the rising oxygen reacted with atmospheric methane (a potent greenhouse gas), stripping away Earth's thermal blanket and plunging the planet into a global ice age. Yet, this "catastrophe" was the essential prerequisite for us. By clearing the stage of simple anaerobes and filling the lungs of the planet, the event paved the way for the rise of eukaryotic life—complex, energy-hungry organisms that would eventually diversify, (in ways other than Darwinian means) into the animal kingdom.

Photosynthesizing, carbon dioxide producing algae forms the first/initial link in the *ocean food chain*. This chain provides energy and food for every life form directly or indirectly, and algae has performed this vital link/role in the chain of life from the very beginning. Algae also serves a vital back-end role as well by processing the wastes produced by other life forms. A curious and helpful outcome was precipitated by the surplus of oxygen. Some of this oxygen went into the upper atmosphere, where it was radiated by UV light and created the ozone layer. The ozone layer blocks UV radiation, and this process also facilitates the colonization of the lands of the Earth by life.

For a staggering 1.5 billion years after eukaryotic life first appeared, the Earth remained a quiet, microbial world. It wasn't until the Ediacaran period (635–541 million years ago) that the first emergence of a new type of creature began to populate the seafloor; complex, soft-bodied lifeforms.

These creatures were truly alien. Unlike modern animals, they had no heads, tails, or limbs. They did not walk, crawl, or swim; they were largely stationary, quilted mattresses of tissue that absorbed nutrients directly from the water. They resembled modern algae, fungi, or lichens more than animals—a 'beta test' for multicellular life that existed in a peaceful, predator-free garden before the explosion of complexity that was to come.

The Meditation Ends – Human vs. Cosmic Meditation

We have now completed our reflection on the nested layers of the cosmos — from the universe, galaxies, and solar system to the Earth–Moon system, and finally to life itself. Each layer depends on precise physical conditions and rare “non-negotiables” that make life possible.

In human contemplative traditions that utilize meditation are largely imaginative, perceptual and mindful. The human mind uses imagery, symbolism, and focused attention to quiet internal noise and perceive reality more clearly. The 1000 Buddhas standing upon a tower, each juggling spheres, represents the disciplined balancing of thought, awareness, and perception within the mind.

But cosmic meditation is different. The cosmic mind does not meditate upon imagination. It meditates upon reality itself — and what it contemplates becomes physical existence. The spheres being juggled in cosmic meditation are not symbolic. They are actual processes maintained by gravitation, electromagnetism and thermodynamics. The balance of galaxies, stars, and planetary systems is not imagined into being — it is continuously enacted through physical law.

Zooming down to Earth, the meditation becomes increasingly intricate. Every human contains over 30 trillion human cells and roughly an equal number of microbial companions, each performing exquisitely timed chemical, genetic, and metabolic processes governed by homeostasis. Across all life on Earth, these trillions of cells form vast interdependent networks — a living contemplation of matter organizing itself into energy and information.

Within each cell, molecular systems behave like tiny meditators of their own. DNA is replicated with astonishing precision, proteins fold into functional structures, and energy is stored and released through biochemical molecules like the citric acid cycle and photosynthesis.

The planetary meditation continues at larger scales. The Earth–Moon system stabilizes climate and tides through gravitational interaction with the Moon, while Earth's molten iron core generates the protective magnetic field that shields life from cosmic radiation. The result is a cosmic meditation of staggering complexity — a universe and biosphere that do not imagine reality but continuously generate and sustain it through nested layers of physical law, biological organization, and energetic balance.

From the orbits of galaxies to the molecular choreography within a single cell, the cosmos appears as a vast, self-sustaining contemplation of matter, energy, and information unfolding across time.

Chapter Summary Discussion

Having reached the end of our chapter, *the Divine Meditation*, we can now answer the central question of our investigation. Are the extraordinary conditions that allow Earth to host humanity simply a rare statistical outcome within a vast universe? Our investigation shows they are not.

The conditions that make Earth habitable—the structure of our solar system, the presence of our Moon, the resources of our planet, and the delicate relationships between Earth, the Sun, and the surrounding planets—cannot reasonably be explained as the product of chance. Taken together, these conditions exceed the probabilistic capacity of our universe to produce them through unguided processes alone.

Our good fortune is therefore more than fortunate. Statistically speaking, it is a miracle. The evidence points toward some form of intervention or guidance operating alongside the physical processes of the cosmos.

This conclusion follows the same pattern we discovered in every previous variable and special problem we examined. Again and again, we encountered conditions that are extraordinarily rare, highly specific, and deeply interdependent.

These findings also challenge several principles that modern science often treats as starting assumptions. Ideas such as the Anthropic Principle, the Copernican Principle, and the Principle of Mediocrity encourage the expectation that our location, our history, and the conditions that allow life should be typical rather than exceptional.

Our investigation points in the opposite direction. The conditions required for life are not ordinary outcomes of physics and probability. They appear extraordinarily rare, and when considered together they exceed the probabilistic capacity of the universe to produce them by chance.

But the pattern we uncovered goes even further. For Earth and humanity to exist requires a sequence of layered, interdependent events—each improbable and each required for the next. These rare events become prerequisites for one another. Each unlikely condition enables the next.

What we observe is therefore not a single anomaly, but something more remarkable: a chain of structured improbability. These are nested dependencies in which no single condition can be evaluated in isolation. Only the cumulative picture reveals the full magnitude of the improbability involved.

The Value of *the Divine Meditation*

Up until this point we have evaluated each variable and special problem in isolation. Treating all prior conditions as givens, so that we can evaluate the probability of only the condition at hand.

In this chapter, however, we are unable to do this. We have considered a convergence of factors that draw upon discoveries and conditions in all previous chapters: our universe, our galaxy, our star, Earth, the planets, chemical elements, life, Earth's biodome, etc. Because of this, we must depart from our usual method of evaluation.

We already have the value of all previous chapters in hand, and the verdict is clear; either the probabilities involved exceed the capacity of the universe, or meaningful odds cannot be calculated at all under first-order physics.

Given this cumulative pattern, the value of the Divine Meditation can only be expressed in one way.

Value of the Divine Meditation = *M (miracle-impossible)*

---

[1] (W. Baade, F. Zwicky, On Super-Novae, Mount Wilson Observatory, Carnegie Institution of Washington and California Institute of Technology, Pasadena, March 19, 1934)12.

[2] G. Gonzalez, J.W. Richards, The Privileged Planet, Pg 152, 2020 Regnery)

[3] (G. Gonzalez, J.W. Richards, The Privileged Planet, Pg 152, 2020 Regnery)).

[4] (G. Gonzalez, J.W. Richards, The Privileged Planet, Pg 152, 2020 Regnery)14

[5] (The Habitability of Large Elliptical Galaxies, Daniel P. Whitmire, MNRAS 3/30/20)15.

[6] (P. Dayal, C. Cockell, K. Rice, An. Mazumdar, The Quest For Cradles Of Life: Using The Fundamental Metallicity Relation To Hunt For The Most Habitable Type Of Galaxy, 8/21/15, The American Astronomical Society, The Astrophysical Journal Letters, Vol810, Num.1)16.

---

[7] (H. Ross, Improbable Planet, Baker Books, 2016)17.
[8] (G. Gonzalez, J.W. Richards, The Privileged Planet, Pg. 165 Regnery Gateway, 2004)
[9] G. Gonzalez, J.W. Richards, The Privileged Planet, Pg. 165 Regnery Gateway, 2004)19
[10] (G. Gonzalez, J.W. Richards, The Privileged Planet, Pg. 165 Regnery Gateway, 2004)
[11] (Mass Extinction And The Structure Of The Milky Way, M.D. Filipovic, J. Horner, E.J. Crawford, N.F.H. Tothill, 9/19/13, Serbian Astronomical Journal, arXiv:1309.4838)21.
[12] -Nuclear Astrophysicist and Researcher Dr. Maria Lugaro of the Konkoly Observatory)
[13] (C.J. Lada, E.A. Lada, Embedded Clusters in Molecular Clouds. 2003, Annual Review of Astronomy and Astrophysics, 41, pgs 57-115, doi: 10.1146)
[14] (D. Raghavan, H.A. McAlister, T.J. Henry, D/R. Gies, R.J. White, D.W. Latham, G.W. Maracy, B.D. Mason, A Survey of Stellar Families: Multiplicity of Solar-Type Stars. 2010, Astrophysical Journal Supplement Series, 190(1), Pgs 1-42).
[15] (Hyperphysics, Georgia State University, stellar neighbors of the sun, the closest star, hyperphysics(dot)phy-astr(dot)gsu(dot)edu/)
[16] (M. Starr, The Solar System Exists Inside a Giant, Mysterious Void, And we Finally Know Why, spacealert.com, January 12, 2022)36
[17] (Ibid, quoting Joao Alves, paper author in M. Star, The Solar System…) Original Research quoted: (C. Zucker, A.A. Goodman, J. Alves, S. Bialy, M. Foley, J.S. Speagle, J. Grobschedl, D.P. Finkbeiner, A. Burkert, D. Khimey, C. Swiggum, Star Formation near the Sun is Driven by Expansion of the Local Bubble, Nature, January, 2022; 601(7893):334-337.PMID: 35022612)37
[18] (M. Woolfson, The Origin and Evolution of the Solar System, Astronomy & Geophysics. 41(1):12. 11 July 2020, https://onlinelibrary. wiley.com/doi/full/10.1046/j.1468-4004.2000.00012.x)22
[19] (A. Mann, Bashing Holes in the Tale of Earth's Troubled Youth, News Feature, Jan 24, 2018, Nature 553, 393-395 (2018), doi: https:// doi.org/ 10.1038/d41586-018-01074-6)25
[20] C.C. Gillespie, The Edge of Objectivity, 1960, Princeton University Press)
[21] [1] F. Masset, M. Snellgrove, "Reversing type II migration: resonance trapping of a lighter giant protoplanet", Monthly Notices of the Royal Astronomical Society, Volume 320, Issue 4, February 2001, Pages L55–L59, [2] A. Pierens, S.N. Raymond, "Two Phase, Inward-Then-Outward Migration of Jupiter and Saturn in the Gaseous Solar Nebula", Astronomy & Astrophysics, Vol: 533, Sept 2011, A131 [3] G. D'Angelo & F. Marzari, "Outward Migration of Jupiter and Saturn", The Astrophysical Journal, Vol: 757, No. 1, Sept 2012)
[22] (K.J. Walsh, A. Morbidelli, S.N. Raymond, D.P. O'Brien, A.M. Mandell, A Low Mass for Mars from Jupiter's Earth Gas-Driven Migration, Nature 475, No7355: Pgs 206-209), and (A. Morbidelli, K.J. Walsh, The Grand Tack Model: A Critical Overview, Annual Review of Earth and Planetary Sciences 44: 275-301)
[23] (Wetherill 1991; Chambers 2001; Raymond et al. 2009).
[24] Wetherill 1991; Chambers 2001; Raymond et al. 2009)
[25] (Walsh, K., Morbidelli, A., Raymond, S. et al. A low mass for Mars from Jupiter's early gas-driven migration. Nature 475, 206–209 (2011).)
[26] (K.J. Walsh, A. Morbidelli, S.N. Raymond, D.P. O'Brien, A.M. Mandell, A Low Mass for Mars from Jupiter's Early Gas-Driven Migration, Nature 475, No7355: Pgs 206-209), (A. Morbidelli, K.J. Walsh, The Grand Tack Model: A Critical Overview, Annual Review of Earth and Planetary Sciences 44: 275-301)
[27] (3 papers, [1] (R. Gomes; H. F. Levison; K. Tsiganis; A. Morbidelli (2005). "Origin of the Cataclysmic late Heavy Bombardment Period of the Terrestrial Planets"

Nature. 435 (7041): 466–9), [2] (Tsiganis, K.; Gomes, R.; Morbidelli, A.; F. Levison, H. (2005). "Origin of the Orbital Architecture of the Giant Planets of the Solar System" Nature. 435 (7041): 459–461), [3] Morbidelli, A.; Levison, H.F.; Tsiganis, K.; Gomes, R. (2005). "Chaotic Capture of Jupiter's Trojan Asteroids in the Early Solar System" Nature. 435 (7041): 462–465.)

[28] (A. Mann, Bashing Holes in the Tale of Earth's Troubled Youth, News Feature, Jan 24, 2018, Nature 553, 393-395 (2018), doi: https:// doi.org/ 10.1038/d41586-018-01074-6)

[29] (K. Lodders, The Planetary Scientist's Companion, 1998, B. Fegley, New York, Oxford University Press)41.

[30] Chambers, J.E. and G.W. Wetherill, G.W. (1998) Making the Terrestrial Planets: N-Body Integrations of Planetary Embryos in Three Dimensions, Icarus, in press.).

[31] (How Special is Earth's Orbit? J.E. Chambers, American Astronomical Society, DPS meeting #30, id.21.07 (Bulletin of the AAS) Vol.30, p.1051)43.

[32] (Ibid, p.1051)44.

[33] (G. Gonzalez, J.W. Richards, The Privileged Planet: How our Place in the Cosmos is Designed for Discovery, 2004 Regnery)

[34] (J. Laskar, Jacques, et al. "Stabilization of the Earth's Obliquity by the Moon." 1993. Nature 361: 615–617).

[35] (W.K. Hartmann, William K., R.J. Phillips, G.J. Taylor, eds. Origin of the Moon. Houston: Lunar and Planetary Institute, 1986. Conference held in Kona, Hawaii, 1984.),

[36] (R.M. Canup, Dynamics of Lunar Formation, in Treatise on Geophysics, 2007,42:441-475. updated review)

[37] (D.J. Stevenson, A.N. Halliday, The Origin of the Moon. Philos Trans A Math Phys Eng Sci 13 September 2014; 372 (2024): 20140289)

[38] (D.J.Stevenson A.N. Halliday, The Origin of the Moon. Preface. Philos Trans A Math Phys Eng Sci. 2014 Sep 13;372(2024):20140289, PMID: 25114319), (R.M. Canup, K.Righter, K. Dauphas, K. Pahlevan, M. Ćuk, S.J. Lock, S.J. Stewart, J. Salmon, R. Rufu, M. Nakajima, T. Magna, Origin of the Moon, 2023, Reviews in Mineralogy and Geochemistry, 89(1): pgs 53-102)

[39] (T. Meier, C. Reinhardt, M.Timpe, J. Stadel, b. Moore, A Systematic Survey of Moon-forming Giant Impacts. II. Rotating Bodies, 2024, The Astrophysical Journal, Vol 978, No. 1)

[40] (N. Angier, Revisiting the Moon, The New York Times, September 7, 2014)48.

[41] (K. Pahlevan, D. Stevenson, Equilibration in the Aftermath of the Lunar-forming Giant Impact, Earth and Planetary Science Letters 262 (3-4): 438-449. Bibcode: 2007E&PSL.262..438P)49

[42] (S.J. Lock, S.T. Stewart, M.I. Petaev, Z. Leinhardt, M.T. Mace, S.B. Jacobsen, M. Cuk, (2018), The Origin of the Moon with a Terrestrial Synestia. Journal of Geographical Research: Planets, 123, 910-951. https://doi.org/ 10.1002/2017JE005333)50.

[43] (S.J. Lock, S.T. Stewart, The Structure of Terrestrial Bodes: Impact Heating, Corotation limits and Synestias, J. Geophys. Res. Planets, 122, 950–982, doi:10.1002/2016JE005239)

[44] (S.T. Stewart, Where Did the Moon Come From? A New theory, Sarah T. Stewart YouTube, Ted Talk YouTube Channel, March 26, 2019)

[45] (S.T. Stewart, Where Did the Moon Come From? A New theory, Sarah T. Stewart YouTube, Ted Talk YouTube Channel, March 26, 2019), (C. Ready, Was The Moon Created from a 'Donut Earth'? July 22, 2019, Launch Pad Academy, YouTube Channel), (S.J. Lock, S.T. Stewart, M.I. Petaev, Z.M. Leinhardt, M.T. Mace, B. Stein, M. Cuk, The Origin of the Moon Within a Terrestrial Synestia, Journal of Geophysical Research. 123(4): 910 910. S2CID 1191845200)

[46] (G. Gonzalez, J.W. Richards, The Privileged Planet, Regnery, Page)

[47] (G.A. Glatzmaier, P.H. Roberts, P. H. A Three-Dimensional Self-Consistent Computer Simulation of a Geomagnetic Field Reversal, 1995, Nature, 377(6546), 203–209)
[48] (G. Gonzalez, J.W. Richards, The Privileged Planet, 2004, Regnery, Pg. 57).
[49] (climate.nasa.gov Feature, August 3, 2021, Earth's Magnetosphere: Protecting Our Planet From Harmful Space Energy).
[50] (D. Andrault, J. Monteaux, M. Le Bars, H. Samuel, The Deep Earth May Not Be Cooling Down, Earth and Planetary Science Letters, Vol. 443, June 1, 2016, Pgs 195-203)
[51] (G. Gonzalez J.W. Richards, The Privileged Planet, 2004, Regnery, Pg. 55)
[52] (Earth's Inner Core Nucleation Paradox, Ludovic Huguet, James A. Van Orman, Steven A. Hauck, II & Matthew A. Willard, Earth and Planetary Science Letters, Volume 487, April 1, 2018, Pgs 9-20, article may be located online under the title: "Researchers: Standard Explanation For How Earth's Core Formed Is Not Possible")
[53] (M. Denton, Children of light, Pg. 21, The Privileged Species Series, Discovery Institute Press, 2008, Seattle)
[54] (M. Denton, Children of Light, Discovery Institute Press, Pg. 22),
[55] (Ibid, Denton,Pg. 22)).
[56] (M. Denton, Children of Light, Discovery Institute Press.
[57] (M. Denton, Children of Light, Discovery Institute Press, Pg. 105)63
[58] (The Privileged Planet, G. Gonzalez, J.W. Richards, Regnery, 2004).
[59] (D.J. Hibbett et al., The Paleozoic Origin of Enzymatic Lignin Decomposition Reconstructed from 31 Fungal Genomes. Science, 2012, 336(6089), pgs. 1715–1719).
[60] (R.M. Hazen, The Story of Earth, 2012)

Prelude to Our Final Two Chapters

Now we turn our attention to the final stage of our investigation, where a simple fact emerges: the universe did not simply reveal the information we have been using within our investigation. These facts have been culled from the collective effort of humanity within all scientific fields. We have this knowledge because humanity had to wrest this vital information from existence by scientific inquiry.

This simple fact illustrates our point. Humanity is not just a race that endeavors to investigate reality. Humanity is a race of intelligent observers driven to investigate reality.

Reality can only be experienced and known by intelligent observers within reality. And to understand reality requires a level of technological ability and scientific acuity. But as we learned in the divine meditation, to be an intelligent observer requires a transparent atmosphere, a planet safe from hazards, the right planetary conditions, a rich, diverse, interdependent biosphere, and the resources to manufacture the technologies needed for exploration—metals, silicon, fuel, rubber, synthetics, engines, etc.

But to create these things requires a prerequisite level of technological ability, and to practice science requires a prerequisite level of intelligence and agency.

All of these factors culminate in the category of consciousness. It is the particular qualities of human consciousness along with a myriad of additional factors that has driven humanity to create complex cultures, to develop spoken, written, and digital languages, and to engage in trade, academics, experimentation and the dissemination of knowledge.

All of these qualities and outputs are impossible without the specific capacities of human consciousness.

Therefore, consciousness is the next investigation we are obligated to engage in. Consciousness is where the path we have been following leads, beginning with the creation of the universe, moving through the rise of humanity, to our latest chapter on *the divine meditation*, the common thread running through all of our discoveries is the persistence, intelligence and strategies of human consciousness.

Special Problem No. 4: Consciousness and Youness

Part I: SCIENTIFIC PERSPECTIVES

Throughout this book, we have traversed the architecture of the universe—from the sub-microscopic mechanics within the cell to the foundational quantum nature of the atom. We have also widened our aperture to capture the most transformative epochs of cosmic history: the birth of the universe, the CMB, and the ignition of the first stars. Yet, beyond these observable domains lies a frontier invisible to telescope, microscope, or detector. It is a realm accessible only through the technology we carry within: the dual phenomena of consciousness and Youness.

In consciousness, the observer and the observed converge. It is the crucible where our subtlest imaginings collide with reality, transmuting them into lived experience. Because we possess this duality, the universe is transformed; it becomes knowable, discernible, and self-aware. In this vital exchange, the cosmos encounters its own reflection, and another miracle takes place. For the first time, the observer becomes knowable to itself by interacting with the universe.

This is the ultimate duality of our existence: the reality we inhabit, and the consciousness that perceives it.

Our investigation into the Divine Algorithm has consistently shown that physical laws and probability cannot fully account for the human experience. Like the "variables" and "special problems" we encountered in our study of the cosmos, consciousness refuses to be reduced to a purely material construct. This is not an irony; it is a design. We are the culmination of universal intent: beings capable of grasping that we were instantiated, and that in our own nature, we mirror the architecture of the cosmos. Our search for the Divine Algorithm is, ultimately, a search for the truth of the human animal, how it was made, what it is composed of, and the ultimate revelation—that we ourselves are made up of something that is exquisitely impossible.

Our conclusion for this chapter is something we already arrived at long ago. Contrary to most western and all scientific materialism understandings of consciousness it is not reducible to physical matter and probability. Consciousness cannot be emergent, any more so than our universe, or the impossible creation of the first life form, or the equally exquisite operation of DNA, protein synthesis, replication and the mysteries of the mitochondria. The difference is, in examining consciousness and Youness, we are now looking at our problem of the impossibility of form from the inside, and not from without.

That is the stunning realization. Our very nature—every last bit of it, and what we think of as being ourselves, is the foundation of the entire mystery we have been chasing throughout our investigation into the Algorithm

Defining Consciousness and Youness

In our understanding **consciousness** is the integrated, multidimensional "show" of human existence—a comprehensive, real-time cinematic interface that synthesizes sensory data, cognitive faculties, memory, and the temporal flow of past, present, and future into a unified narrative. While it involves our senses and thoughts, we err if we attempt to define it by reducing it solely to these characteristics[1]. Consciousness is also the foundational gateway through which all experience enters and ideas are formed. It is so subtle that we often "melt" into it, losing our self-consciousness as we engage deeply with life, yet it is also the vessel for the full spectrum of human reality—from the mundane to the transcendent states of awe, epiphany, and revelation.

If consciousness is the screen upon which the experience of reality is projected, Youness is the eternal witness seated before it.

Current ontological frameworks suffer from a persistent category error: they routinely collapse the "show" into the "witness," bundling the content of experience together with the entity having the experience. While mainstream theories—even those grappling with the "Binding Problem"—hint at a unified point of perception, they fail to explicitly identify the Witness as a functional, distinct requirement of reality. They describe the mechanisms of the screen but ignore the one seated in the chair.

To correct this, we must define Youness as a discrete phenomenon. **Youness** is the ineffable center of the individual—the deeper evaluative field that observes not only the outward flow of reality but the unfolding of the self, including its own internal reactions and cognitive breakdowns. While consciousness provides the raw, fragmented data of being, Youness provides the depth of presence, the unchanging witness, and the critical interface between the experience of the world and the essence of the person living it.

Without this distinction, the materialist remains trapped in a circular loop: they describe the screen but cannot account for the viewer. By naming Youness as a distinct function, we resolve the incoherence of current models, finally separating the phenomenon of being from the agent who witnesses it.

Definitions Are Claims

These definitions are more than mere terminology; they are ontological claims positioned against the "Hard Problem" of consciousness. By identifying that neither consciousness nor Youness can be located or explained by material instruments, we move past the limitations of purely physicalist models. We acknowledge that consciousness is not merely a byproduct of biological machinery, but an anomaly that refuses to collapse into a materialist construction. Because we cannot find it in matter, we must recognize it as a foundational requirement of reality itself—a technology that allows us to inspect, participate in, and understand the universe we inhabit.

This realization reveals a strange and profound symmetry at the center of our investigation: consciousness makes the universe, the universe, through multiple, impossible steps, evolves beings that possess consciousness, and through that consciousness, we finally grasp the nature of the reality that created us. This circularity is the narrative engine of the Divine Algorithm, suggesting that we are meant to discover the universe's own mechanism for understanding its design. We reach this state of understanding not through abstract theorizing alone, but through direct engagement. Just as one might experience the distinction between the "show" of thought and the "witness" of the self through the stillness of Zazen meditation, our entire inquiry serves as a bridge. It invites the reader to step back from the sensory screen of consciousness and come into direct contact with the settled, observational reality of Youness, where the true nature of our existence becomes discernible.

Fundamentally and undeniably, you are one of the things that our reality was meant to produce. Precisely, exactly you. This is the impossible mystery of consciousness and Youness we will endeavour to examine to its very limit.

Groping for Meaning

For centuries, thinkers have wrestled with what consciousness means. Is consciousness merely the product of the human body, i.e., biological machinery? —neurons exchanging electrochemical signals within the astonishing complexity of the brain? Or is the inner experience of consciousness indicative of something deeper, something far more profound, and subtle, and elusive in its construction?

This question is as technical as it is fundamental. It forms the essential paradox of consciousness that is debated among competing ontologies. This debate finds its best articulation in the question, *"could purely objective physical processes give rise to the intimate, enduring, subjective feeling of life itself?"*

This question has become one of the most persistent problems in modern thought. And although it is a settled question within our understanding, there is a furious debate taking place as to its nature. How this debate is understood and how these ideas are communicated, is as consequential as an understanding can be. Our understanding of ourselves changes our perspective, our understanding and even the nature of our lives.

How consciousness is conceived informs how individuals think of themselves, and ultimately what the nature of life and reality is ultimately about.

To say consciousness is an illusion, or that we are biological machines, or to say that we have souls and an immaterial divine nature, all of these ideas hold vast power over how individuals view themselves and the nature of their lives.

The Gap

Despite the lack of a universal definition, all inquiry eventually strikes the "explanatory gap"—the fundamental disconnect between physical brain processes and the qualitative feeling of being alive. This can be a difficult concept to grasp, because it would not be apparent that there would be a problem created by how humans natively experience reality[2]. Materialism fails because it cannot account for why our experience feels like anything at all. To see why biology alone is insufficient to explain our experience as conscious beings, we need only consider a simple thought experiment: if we were to construct a robot that perfectly replicated every physical input, output, and reaction of the human brain, it would still lack the internal witness. It would possess the data, but it would have no one home to experience it.

Robots Have No Subjectivity

To grasp the fundamental error of the materialist perspective, we must hold it against the mirror of a robot thought experiment. Imagine a machine constructed from the same atomic building blocks as a human, equipped with every sensor necessary to process the world. Such a machine would navigate reality with perfect efficiency, yet it would do so in total darkness. It would process electrical inputs as raw data, but it would possess no internal screen, no integrated "show" of experience, and no capacity to feel the weight of its own existence. It would be a processor of information, but it would not be a participant in life however closely it mocked human interaction.

Materialists argue there is no functional difference between this machine and a human. By their logic, if we are merely complex biological robots, our experience should also occur in the dark—a mechanical feedback loop devoid of any subjective "what it is like" to be. The fact that we do not exist in such a state—that we are instead immersed in a seamlessly integrated, conscious, and emotional reality—is not a fluke of biology. It is the evidence that the materialist framework is fundamentally incomplete. And yet materialist dissemble furiously to refute this inescapable trap. We will examine many facets of their façade in this regard.

Throughout our investigation into the Divine Algorithm, we have repeatedly ruled out chance and mechanical law as sufficient causes for life's diversification and the architecture of the cosmos. To suggest that human consciousness is merely an emergent property of a biological robot is to ignore the very "Special Problem" we have been tracking: the presence of the subject. A robot may act, but only a human possesses a Youness—a core witness that evaluates, experiences, and stands apart from the data stream (and even when nothing happens at all, it still witnesses). This is not just a disagreement over definitions; it is an acknowledgment that our reality is inherently incompatible with a purely materialist construction. We do not inhabit a world of mechanical shadows; we inhabit a world of meaning, observed by an entity that is fundamentally more than the sum of its atoms.

Introducing *Youness* - Augustine of Hippo (354 - 430)

We are not the first to realize that the key to reality lies within. St. Augustine of Hippo, positioned as a bridge between the ancient and modern worlds, recognized this centuries ago. In his Confessions, he urged: "Do not go outside yourself; return within yourself. Truth dwells in the inner man[3]." Augustine understood that the structure through which the world becomes known is the inner witness—the very observation we have identified as Youness.

Augustine famously noted that people travel to wonder at mountains, seas, and stars, yet pass themselves by without a second thought. This remains our primary obstacle: the "universal show" of consciousness is so overwhelming, so dense with sensory and cognitive data, that we lose the subtlety of our own subjectivity. We are so busy observing the world that we forget to observe the observer, a kind of blindness we have to our own 'double-consciousness.'

By keeping Augustine's inward focus in mind, we can conduct the necessary experiment: peeling back the overwhelming, ever-present flood of consciousness to make direct contact with the inner witness. This illustrates precisely why our bifurcation of consciousness into the dual, wedded phenomena of "consciousness" (the show) and "Youness" (the observer) is vital. Without this distinction, we remain lost in the movie of our lives, forever missing the screen—and the witness—that observes and interacts with the show, and makes the experience possible.

An Encounter with Youness

You can test the distinction between consciousness and Youness in the quiet of a darkened room. Hold your thumb and index finger together and ask, "What is aware of this?" Two things emerge: the sensation itself, which is consciousness, and the presence that recognizes it, which is Youness. Notice that this sensation does not feel located inside your head; it is experienced exactly where it occurs, at the fingertips. This spatial, sensory reality contradicts the default brain-centered model. If we were merely mechanistic, brain-based observers, our experience would feel like an informational report—a secondary "output" of neural processing telling us 'out there, two of your fingers are being held together, their exact coordinates are…'—rather than an immediate, direct engagement with thc world.

As philosopher Edmund Husserl emphasized, consciousness is always intentional; it is intrinsically "of" something, pointing toward the world rather than observing internal representations from within the skull[4]. This outward relation is a fundamental structure of our existence[5]. Consciousness brings us this information, but Youness acts as the witness and the evaluative field that makes it relevant, gaining wisdom and shaping the self through the interplay of dcsire and necessity.

This leads to the Integration Problem—the mystery of how the brain combines massive, disparate sensory and cognitive processes into a single, unified experience[6]. While vision, memory, language, and emotion occur in distinct neural regions, our experience is never fragmented. We perceive a coherent scene, not a disjointed flood of data. Science cannot identify any structure that performs this unification.

This is closely related to the Binding Problem, where disparate inputs like color, shape, and motion are fused into a singular reality without any identifiable biological mechanism acting as the integrator[7]. Youness does what the integration problem and binding problem can hardly articulate.

It is only a problem when at attempt is made to reconcile the phenomenon within a materialist's frame.

Beyond mere observation, Youness exerts an inexplicable agency. We can direct our focus—picking one bird from a flock against the clouds—with an exactitude that defies materialist explanation. If the self were merely an illusion, what commands these actions, and why do they feel so perfectly synchronized with our desire? Consciousness provides the "show," but it is an indescribable, indelible aspect of the individual that directs the focus. The fact that we navigate this with seamless fidelity, without ever feeling a violation of our intent, remains a sublime mystery—one that is irreconcilable with any model that views the human animal as a biological robot.

The Integration Problem

Our earlier experiment—simply touching two fingers together—reveals a fundamental truth: we experience the world as integrated beings, not as brains processing information. Our intuition tells us we exist in the world, yet mainstream models defy this by treating experience as a series of secondary representations confined to the skull. This focus on the brain as the "center" of experience is the source of modern science's primary struggle with consciousness. It leads to the Hard Problem of Consciousness, an explanatory gap where neurobiology fails to account for the qualitative "feel" of subjective experience[8].

This failure manifests as the Generative Paradox: how do purely local neural interactions generate a unified, intrinsic, first-person experience?[9] The fallacy lies in the shared premise that the brain holds the key to resolution. Attempts to solve this by positing a "homunculus" within a "Cartesian theater"—a small observer inside the brain commanding the show (more will be said on this later)—are widely recognized as fallacious, as they collapse into infinite regress or fail to explain the immediate unity of the experience[10]. There is no identifiable brain region or network that performs this integration; the paradox persists because consciousness directs, binds, and generates with a precision that cannot be traced to material structures. It is simply what our immaterial understanding of consciousness does, much like how RNA polymerase in mere seconds can find a specific gene on a 3 billion long string of letters making up DNA without a map, memory or any sensory surfaces whatsoever. It is simply what it does.

The scientists asks, 'if the brain is not the integrator, what is?' The answer must be a system capable of constructing, interpreting, and manipulating information with self-intimacy: consciousness itself.

Consciousness does not merely reflect reality; it serves as the category that resolves these paradoxes. The "self" is a total concept, occupying the entirety of the biological form, which explains why we experience the world through our whole body rather than as a localized informational report within the brain.

This leads us to the subtle, enduring nature of Youness. Even when we strip away the layers of personality, history, and ego—a state sometimes described in Eastern traditions as "no self" or "emptiness"—the observer remains. Whether in deep meditation or states of ego dissolution, the witness to that disappearance is always present. How else would we know such events took place? If we peel away the "clothing" of accumulated worldliness, we are not truly dissolved; we are returned to the core. Youness is the persistent, unattached witness that precedes the self, demonstrating that while our attachments and memories may shift, the observer remains fundamentally intact.

Augustine of Hippo (354 - 430) Redux

Youness is exactly what Augustine of Hippo discovered when he undertook an experiment akin to our own. In his Confessions, he reflected inward, systematically stripping away external realities and sensory content to focus on the self as the ultimate observer[11]. In doing so, he discovered the persistent "I"—the first-person witness that remains when all else is removed. Augustine identified this watcher as the core of his identity: it is "I" plus observation.

This approach is strikingly similar to the methodological doubt René Descartes would employ over 1,200 years later. Like Augustine, Descartes eliminated everything that could be doubted, arriving at the witnessing cogito—"I think, therefore I am." Both thinkers confirm that Youness is an unrelenting presence; it is always there, even in silence, always at the ready.

This is the vital contact between the individual and the universe that characterizes the true human experience. We do not describe our lives as mechanical, informational data-processing taking place inside our skulls. We describe life as an intimate, full-contact meeting between the individual and reality itself. Whether through Augustine's inward reflection or our own modern inquiries, those who endeavor to identify what lies within bear witness to the same fact: the observer is the fundamental ground upon which all experience is known.

Direct Experience vs. Interpretation

By design, consciousness functions to provide an individual observer with extreme fidelity to the physical world. Our senses, proprioception, and cognition are aligned to place us directly within reality, providing a "show" that feels unmitigated and untranslated. This leads to a profound question: What if our sensory apparatus is not a mere interpretive device, but a grand instrument designed to facilitate direct engagement with reality itself and not just mere representation, or a sensory surface to nerve to neuron processing state of awareness?

Mainstream neuroscience often counters this, citing sensory injury or disability as proof that our perception is merely a "transmission" susceptible to breaking. But within our framework, injury does not prove the senses are interpretive; it simply shows that the system—the biological bridge facilitating our direct contact—has been damaged. The consequence is a loss of clarity, not a loss of truth. If we assume the brain merely interprets reality, we fall back into the same "Hard Problem" that materialism cannot solve: how do neurons firing in the dark generate a first-person, world-anchored point of view?

We must reject the notion that humanity experiences a mere interpretation of reality. Consider the scale of the Divine Algorithm: the universe has undergone a stunning series of instantiations and crossed countless probabilistic thresholds to produce an intelligent observer. Why would the universe go to such immense lengths to create a sentient being, only to leave its contact with reality to the chance of a "fragile, secondary interpretation"? or worse yet, to come this far, only to be unable to directly experience reality itself, having to settle for mere interpretation or secondary representation?

That would be a structural failure, not a sublime design.

The "miracle" of our existence is not just that we are here; it is that we are here as direct witnesses. Consciousness is the culmination of the universe's effort to know itself. To suggest that we are locked behind a barrier of interpretation is to misunderstand the intent of the algorithm. We are not processing symbols of reality; we are experiencing the reality itself, and our capacity to do so is the final, crowning miracle of the entire cosmic architecture. We must refute this at the outset of our investigation if we are to make our way to the truth.

Modern Views of Consciousness - Edmund Husserl (1859–1938)

Edmund Husserl's work provides a striking phenomenological validation of our dual-layered model. By employing the concept of epoché—a methodological suspension of judgment regarding the existence of the external world—Husserl stripped away the "psychological self," including personality, history, and memory[12]. What remained after this reduction was what he termed the "transcendental ego." This pure subject of experience is exactly what we have identified as Youness: the persistent watcher that remains when all else is negated.

Husserl found what we have just described, that this observer is defined by intentionality; it does not merely receive data, but directs consciousness toward the objects of experience, whether those objects are real, imagined, or abstracted from history[13]. Crucially, Husserl observed that "all consciousness is consciousness of something," noting that this directedness is an inherent structure of the mind. Even in the state of "nothingness," the witness remains present, silent, and ever-ready to focus.

This confirms our bifurcation: consciousness provides the "show," while Youness provides the intentionality. Unlike a machine, which processes inputs in the dark, the human observer is synonymous with intention. Even when our focus is subtle—thinking without words or observing the silence—the inner self is actively engaged in the act of pointing toward experience. This inseparability of observation and intention is the final evidence that the "self" is not a byproduct of material processing, but the foundational agent of reality itself.

Pierre Teilhard de Chardin (1881–1955)

Finally, we find a powerful confirmation of our hypothesis in the work of Pierre Teilhard de Chardin, who viewed the evolution of the universe through the lens of both paleontology and spiritual necessity. He famously observed: "We are not human beings having a spiritual experience; we are spiritual beings having a human experience[14]."In this statement, Teilhard de Chardin identified the very tension that drives the "Hard Problem"—the intuitive sense that the totality of human experience defies the limitations of pure, mechanistic biology.

Teilhard de Chardin argued that material and spiritual energy are not separate, but complementary—that in the final analysis, they must be facets of a single energy operating within the universe. He proposed that the soul is not an emergent byproduct of the brain, but a "focal point of transformation" through which the universe acts[15].

This perspective aligns perfectly with our investigation. The "problem" of consciousness arises precisely because we are looking for its origin in biological matter, when it actually flows from an immaterial essence. We have moved from the "show" of consciousness to the "witness" of Youness, and finally to the recognition of the soul as the active, focal point of the universal Algorithm. We are not merely observers of a mechanical system; we are the biological culmination of the universe's effort to turn inward, transform its own energy, and finally know itself.

Rene Descartes' (1596 - 1650) *Discours De La Méthode* (1637)

To understand the persistent nature of the observer, we must revisit the foundation laid by René Descartes. In his Discourse on the Method (1637) and Meditations on First Philosophy (1641), Descartes sought a bedrock of truth that could withstand any possible doubt. He initiated this by engaging in a radical thought experiment: he systematically "convinced" himself that there was absolutely nothing in the world—no sky, no earth, no minds, and no bodies. (R. Descartes, Discourse on the Method of Correctly Conducting One's Reason and Seeking Truth in the Sciences, 1637)

In this state of total negation, Descartes arrived at his first principle of knowledge: cogito ergo sum, or "I think, therefore I am." (R. Descartes, Meditations on First Philosophy, 1641) In this moment of reduction, he stripped away the "universal show" of consciousness to find the irreducible truth of the witness.

Critics often mistake the distinction between this witness and the physical form for a form of "dualism." This is a category error. A violin is not a dualist entity; the wood (matter) and the music (the expression) are distinct, yet they are inseparable components of a single reality. The human being is not two things held together by a mystery; we are a unified manifestation of what the Divine Algorithm has illustrated, which is all that we think of as being material (inanimate), is made real and sustained, just as what is alive (biology) is made real and sustained by an instantiating consciousness. In our reality, the physical form is the necessary vessel for the expression of the soul. At the point of transition—death—the expression ceases and the vessel remains, but to confuse the instrument with the music is the fundamental mistake of the materialist.

By isolating the witness from the contents of consciousness, Descartes confirmed what we have seen across our investigation: the inner subject is the silent, ever-present foundation of existence. This cogito is not merely an intellectual assertion; it is the realization that the observer is the necessary precondition for any knowledge at all. Youness remains at the ready, even when the world itself is stripped away.

To find this truth Descartes induced a state of meditative mindfulness and placed himself in the condition where he found, "...there is absolutely nothing in the world, no sky, no earth, no minds, no bodies[16]." in a sense approximating the ignorance of a newborn and the uncertainty of experience without anchor. Then, while in this space, he found that something then appeared.

> "I have convinced myself that there is absolutely nothing in the world, no sky, no earth, no minds, no bodies. Does it now follow that I too do not exist? No: if I convinced myself of something then I certainly existed. But there is a deceiver of supreme power and cunning who is deliberately and constantly deceiving me. In that case I too undoubtedly exist, if he is deceiving me; and let him deceive me as much as he can, he will never bring it about that I am nothing so long as I think that I am something. So, after considering everything very thoroughly, I must finally conclude that this proposition, I am, I exist, is necessarily true whenever it is put forward by me or conceived in my mind[17]."

Descartes found that even in the midst of immense skepticism and doubt, and even if within the possibility of complete deception, there was always something that could not be denied, and that was the self. Whether real, imaginary or otherwise, the self was the self. And it was more foundational than any possible label that could be applied to it (real, illusion, epiphenomenon, etc.).

> "While we thus reject all of which we can entertain the smallest doubt, and even imagine that it is false, we easily indeed suppose that there is neither God, nor sky, nor bodies, and that we ourselves even have neither hands nor feet, nor, finally, a body; but we cannot in the same way suppose that we are not while we doubt of the truth of these things; for there is a repugnance in conceiving that what thinks does not exist at the very time when it thinks. Accordingly, the knowledge, I think, therefore I am, is the first and most certain that occurs to one who philosophizes orderly[18]."

What Descartes could not deny was that thought, whether it was in affirmation of the self, negation, or deception, there is an *I am*, and therefore an underlying self.

If we examine Descartes' argument, we can excavate that he is not saying that he exists simply *because* he thinks. Clearly, thought is his proof, but what he is reaching for is far more nuanced. Thought is dependent, and its identification exposes a pre-existing something *below* thought that allows thought and manifests as thought. What this is, is what erupts as thought, and thus substantiates Descartes ontology.

In this construction we can see a direct reference to our identification of youness, at its root, as a dispassionate observer. An un-erasable ever-present substantiating ephemeral condition, underlying all cognition and that which observes consciousness and all that plays out in the arena of personal experience. This is what forms the basis that allows cognition, awareness, expression and its various additional products to manifest.

Descartes' claim thus goes much deeper than the common understanding derived from "I think therefore I am". Descartes pays homage to this idea by changing his original phrase "*cogito ergo sum*" to a more expansive and inclusive, "*Cogito ergo sum, dubito, ergo sum - vel quod idem est - cogito, ergo sum.*" which translates into, "I think therefore I am as, I doubt, therefore I am - or what is the same - I think, therefore I am." This expands his original declaration into a number of self-affirming states within the inner subjective space that affirms the irrefutable self, yet signifies that both, and neither is ultimately responsible for his realization:

> "By the word thought, I understand all that which so takes place in us that we of ourselves are immediately conscious of it; and, accordingly, not only to understand (INTELLIGERE, ENTENDRE), to will (VELLE), to imagine (IMAGINARI), but even to perceive (SENTIRE, SENTIR), are here the same as to think (COGITARE, PENSER). For if I say, I see, or, I walk, therefore I am; and if I understand by vision or walking the act of my eyes or of my limbs, which is the work of the body, the conclusion is not absolutely certain, because, as is often the case in dreams, I may think that I see or walk, although I do not open my eyes or move from my place, and even, perhaps, although I have no body: but, if I mean the sensation itself, or consciousness of seeing or walking, the knowledge is manifestly certain, because it is then referred to the mind, which alone perceives or is conscious that it sees or walks[19]."

Here Descartes goes very deep, and as it turns out, deeper than what he is given credit for in the contemporary western understanding of consciousness. In the above statement, Descartes retrenches and articulates that what he means by thought is "all that which so takes place in us that we of ourselves are immediately conscious of it[20]," in other words, consciousness and the thing that exists beneath consciousness and is the recipient of consciousness and that in turn manifests actions and expressions through consciousness, is none other than the I/Self. The I/Self is not consciousness per se. It works *with* consciousness as is our claim.

In this construction we see the defeat of the materialistic idea of consciousness as illusion or epiphenomenon. Taken literally, we find that Descartes explicitly demolishes this idea, as thought itself, whether it be dream, imagination, and by extension, illusion, delusion or even psychosis, at its root, the act of sensing, "or consciousness" ("the sensation itself") validates itself. At this point whatever label is applied to it is already too late. The self has been validated.

Descartes extends this thought by saying that understanding (*intelligere*), will (*velle*), imagination (*imaginari*), and perception (*sentire*) equally impacts our consciousness and makes us conscious of the I/Self. This finding by Descartes is not new in the history of man. This understanding accords with eastern spiritual perspectives that distinguish between experience (consciousness and the one who has the experience, known as the I/Self (youness).

Descartes' Principles of Philosophy & The Buddha

In *Principles of Philosophy* Descartes clarifies this point again in article 9. "What is meant by 'thought' " he writes, "I take the word 'thought' to cover everything that we are aware of as happening within us, and it counts as 'thought' because we are aware of it[21]." But these "thoughts" as Descartes terms them, are not discerning 'things', they are 'informational things. The discerning thing is the I/Self that receives these experiences.

The Buddha illustrates the similarity of this identification within his own understanding hundreds of years prior:

> "It is not really the ears or eyes that hear or see. Was it so, then, as the Buddha asks, why do not the dead see and hear just as much as the living? What hears and sees is not the sense-organ but Self the unattainable. The sense organs are instruments the Self uses for Itself. And when it hears, its hearing reaches the end of the universe which has no ears corresponding to ours. So, with the rest of the senses. It is not the particular sense alone that hears or sees. When It hears I hear, you hear, everybody, every being hears. It is for this reason when I attain enlightenment the whole universe attains it. The Unattainable is attained as unattainable – this is the experience not of the psychological or logical self, but the Unattainable Self[22]."

Descartes' investigation culminated in a realization that unified Eastern and Western insights: the I/Self—what we call Youness—is the foundational anchor of existence. By establishing the self, he was able to validate the existence of the universe itself; the world exists because there is a witness to experience it.

Yet, Descartes recognized that he did not create himself, nor the objective Truth he encountered. He understood that existence, awareness, and what we have discovered within the "Divine Algorithm" of physical laws could not arise from the void or emerge from chance. His ontological proof of a Creator is often lazily reduced to the idea of a "perfect being," but his actual argument was rooted in the undeniable reality of his own existence and the architecture of the world. To Descartes, the seamless integration of mind and reality—the fact that we can apprehend Truth at all—points to an Instantiator.

This is why Descartes remained untroubled by what later generations labeled the "mind-body problem." As philosopher Jonathan Westphal notes:

> "Though we do find in the "Meditations" itself the distinction between mind and body, drawn very sharply by Descartes, in fact he makes no mention of our mind-body problem. Descartes is untroubled by the fact that, as he has described them, mind and matter are very different: One is spatial and the other not, and therefore one cannot act upon the other. Descartes himself writes in his Reply to one of the Objections:[23]"

Descartes was not struggling with the interaction of two substances; he was indicating the dead end of strict materialism. He knew that the bridge between the immaterial soul and the material body was not a biological mechanism, but the existence of a higher power that maintains the universe and imparts life. This mirrors the ancient Hindu concept of Atman (the true self) being of the same essence as Brahman (the absolute reality)—a union sustained by creation itself.

For Descartes, the "problem" of interaction only exists for those who refuse to acknowledge the immaterial nature of the individual and the necessary power that instantiates reality. Ever since, his work has served as the foundation for Western exploration, yet contemporary thinkers often miss his most important point: the witness and the world are not two separate things fighting to interact—they are two aspects of a singular, divinely authored expression.

Charting the Path Ahead

Having bifurcated our subject into the "show" of consciousness and the "witness" of Youness, we have anchored our investigation in the deep insights of our predecessors—from the inward introspection of Augustine and the radical phenomenological reduction of Husserl, to the foundational cogito of Descartes and the evolutionary vision of Teilhard de Chardin. These thinkers provide the lens through which we will now map the path ahead.

In the chapters that follow, we will bridge the gap between this ancient intuition and contemporary inquiry through a three-stage investigation:

The Materialist Critique: We will begin by examining the views of contemporary materialist interpreters of the mind. By exposing the conceptual incoherence of the brain-as-machine model, we will demonstrate that the materialist framework remains not just unresolved, but inherently unresolvable.

The Philosophical Landscape: We will then engage with the modern dialectic, focusing on the work of David Chalmers, Daniel Dennett, and Thomas Nagel. By probing core concepts—from "philosophical zombies" to the "black box conjecture"—we will strip away common blind spots and clear the terrain for our own hypothesis: that consciousness is not a product of the brain, but the fundamental stage upon which reality is authored.

The Contemplative Integration: Finally, we will widen our perspective to include the profound insights of Eastern contemplative traditions. Drawing from Vedic philosophy and the masters of the Far East—including Osho, Bankei Yotaku, and the classic lineage of Bodhidharma—we will explore dimensions of awareness that remain largely invisible to Western science.

Taken together, these perspectives allow us to approach the central mystery from every angle. We will clarify the role that consciousness and Youness play in the structure of reality itself. By the end of this exploration, we will see why this phenomenon is not merely an evolutionary accident, but the culmination of the universe's unfolding intent—as exhibited by the Divine Algorithm—and the true sanctuary where the human soul originates.

Contemporary Views on Consciousness

David Chalmers famously opened his 1995 inquiry into the nature of the mind with a sobering assessment:

> "Consciousness poses the most baffling problems in the science of the mind. There is nothing that we know more intimately than conscious experience, but there is nothing that is harder to explain... Many have tried to explain it, but the explanations always seem to fall short of the target[24]."

This shortfall is not a failure of intelligence; it is a structural limitation of the materialist framework. Cognitive scientists Harry Haroutioun and Carlos Montemayor underscore this current paralysis:

> "Despite the advancement of research on consciousness, researchers tend to accept the view that the qualitative character of conscious contents (and the first-person perspective) are irreducible to scientific methodology. Thus, many researchers don't touch this topic[25]."

Because physicalists are constrained by a methodology that acknowledges only physical laws and matter, they are fundamentally unable to derive an underlying state that is immaterial in nature. This is the heart of our claim. When confronted with the reality of consciousness, the materialist is forced to retreat into one of two intellectually impoverished corners: either they insist that consciousness is a mysterious "emergent property" of the brain (or the universe itself, via panpsychism), or they argue that it is a systemic misperception—an illusion that, in some sense, does not truly exist.

The following snapshots, curated for public consumption, illustrate how these two failing strategies define the current landscape of the debate.

Steven Pinker, Cognitive Psychologist

"The philosophical problem of sentience or qualia or sometimes called the hard problem of consciousness. I think (it) might ultimately be a quirk of our own way of analyzing the world—that is the mind reflecting on itself is naturally going to be puzzled by some aspects of itself. We know from neuroscience that there is no aspect of consciousness that does not have some physical correlate…There's no ESP, there's no life after death, there's no mysterious action at a distance. It's all information processing in neurons. Why it [consciousness] should 'feel like something' to me, -to be that network of neurons-, I don't think we have a satisfying answer to, and it may not be a scientific puzzle at all[26]…"

Sabine Hossenfelder, Theoretical Physicist

"Personally, I don't think that to describe our observations, our experience of us thinking requires anything more than particle physics. I'm happy to leave the understanding of consciousness to neurobiologists or whatever those fields are called. But I also don't think that we need to add anything to the fundamental laws of nature that we have collected in physics. I think it's sufficient." (YouTube: Big Think, Do humans have souls? | Physicist Sabine Hossenfelder, Jan 15, 2023)

Roger Penrose, Mathematician, Physicist

"...I'm a physicalist so I believe what's going on in our heads is obeying the same laws of physics which is going on around us. There's nothing, special injected into our heads which is, -gives us- a quality which is beyond the laws of physics. I'm, -I'm a physicalist. I believe it's the same physics in here [pointing towards head] as out there, and everywhere. Yeah. Maybe different parts… are emphasized in different ways…

…whatever consciousness depends upon… …how general relativity relates to Quantum Mechanics." (YouTube, The Institute of Art and Ideas, Roger Penrose on Quantum Mechanics and Consciousness | Full Interview, Mar 9, 2024)

Sean Carroll, Theoretical Cosmologist

"What about consciousness? What about my thinking? Surely you don't believe that that's just a bunch of atoms obeying the core theory of physics? So, yes I do! We can't prove it, right? This is something we believe is true, but as a conjecture that we think is reasonable all I can do is give you some clues, some signposts along the way[27]."

Brian Greene, Theoretical Physicist, Mathematician

"...how can mindless thoughtless particles come together in a configuration that somehow yields the inner sensation of thought, feeling, emotion? how can they possibly generate that if they themselves don't have any intrinsic version of those conscious qualities from the get-go?... we will recognize, as we do with life, that consciousness is nothing but particles coursing through a gloppy gray in our particular structure head and within that motion those particles generate the conscious sensations that we all experience. To my mind, when you describe consciousness in the manner that we have just now, I don't think that in any way diminishes consciousness[28]."

Richard Dawkins, Evolutionary Biologist

"Well I always confess myself I'm baffled by it [consciousness]. I mean I do see it as a deep and profoundly difficult problem. I am committed to the view that there is nothing there other than physics. There's nothing there other than as you say atoms and electrons… we start with physics and then we get chemistry and we get a process of Darwinian natural selection which gradually builds up nervous systems. Step-by-step they get more and more complicated… I can't see any other way[29]."

Sam Harris, Philosopher, Neuroscientist

"I'm not arguing that consciousness is a reality beyond science or beyond the brain or that it floats free of the brain at death. I'm not making any spooky claims about its metaphysics. What I am saying, however, is that the self is an illusion. The sense of being an ego, an 'I', a thinker of thoughts in addition to the thoughts. An experiencer in addition to the experience… that sense of being a subject, a locus of consciousness inside the head is an illusion. It makes no neuro-anatomical sense. There's no place in the brain for your ego to be hiding. We know that everything you experience – your conscious emotions and thoughts and moods and the impulses that initiate behavior – all of these things are delivered by a myriad of different processes in the brain that are spread out over the whole of the brain[30]."

Nicholas Humphrey, Neuropsychologist

"Sometimes I say that consciousness is an illusion and the reason for saying that is it seems to us that it exists outside of time and space and matter as defined by physics. It can't be so. I'm a materialist. This is being produced by my brain or your brain and so I believe that in the end the explanation must come down to what's going on inside our heads, but consciousness doesn't seem to have those qualities because it makes us think of ourselves as being something outside of matter, of having metaphysical significance perhaps, which sets us above and outside the mundane world... I hope it's an illusion I should say because as a scientist I want to believe that we can explain it in terms of physics and chemistry[31]."

What emerges from these perspectives is not a solution, but a shared strategy of evasion. Whether by dismissing the hard problem as a mere 'quirk' of our cognitive architecture (Pinker), insisting on the sufficiency of current particle physics despite acknowledging the lack of a bridge (Hossenfelder, Penrose, Carroll), or reducing the lived experience of the 'I' to a neuro-anatomical illusion (Harris, Humphrey), these thinkers are united by a common commitment: they prioritize the materialist methodology over the reality of the phenomenon itself. By framing consciousness as a problem that *must* be solved within the constraints of atoms and neurons—even when they admit that atoms and neurons contain no intrinsic qualities of thought or feeling (Greene, Dawkins)—they reveal the inherent circularity of their position. They do not explain consciousness; they explain it away, effectively conceding that if the 'I' cannot be found in the wetware of the brain, the 'I' must not exist.

David Chalmers & Panpsychism

In 1994, philosopher David Chalmers modernized the Cartesian mind-body problem by formalizing "The Hard Problem of Consciousness"—the challenge of explaining how physical processes in the brain yield subjective experience[32]. By contrasting this "Hard Problem" with the "easy problems" of sensory discrimination and cognitive reportage[33], Chalmers highlighted the fundamental deadlock of modern science.

To escape this deadlock, Chalmers has increasingly turned to panpsychism: the hypothesis that material reality possesses an intrinsic level of "proto-consciousness." The claim is that when matter reaches a certain threshold of complexity—whether in synapses or microtubules—this proto-consciousness "leaps" into the coherent experience we recognize as human consciousness.

> "Panpsychism, taken literally, is the doctrine that everything has a mind... we can understand panpsychism as the thesis that some fundamental physical entities have mental states. For example, if quarks or photons have mental states, that suffices for panpsychism to be true, even if rocks and numbers do not have mental states. Perhaps it would not suffice for just one photon to have mental states. The line here is blurry, but we can read the definition as requiring that all members of some fundamental physical type (all photons, for example) have mental states... I will understand panpsychism as the thesis that some fundamental physical entities are conscious: that is, that there is something it is like to be a quark or a photon or a member of some other fundamental physical type[34]."

Chalmers proposes this as a hybrid solution—a way to satisfy the physicalist's demand for causality while acknowledging the immaterial nature of the mind. However, this is less a scientific hypothesis than a strategic retreat. It attempts to circumvent the core premise of materialism: that a property cannot spontaneously appear without an underlying foundation. If consciousness is a feature of physical reality, it must exist at every level of organization; if it is not, it cannot emerge.

Panpsychism fails because it merely engages in a "stacking of uncertainties." Because our most exhaustive measurements of particles and fields have revealed no such "consciousness signal," the theory relies on unobservable micro-properties to explain the mystery. It diffuses the problem into an untold number of quantum-scale bits, hoping that quantity will somehow replace quality. But the biggest failure is structural: Panpsychism explains nothing about the "I." It provides no account of how a cohesive, subjective self is constructed, nor how this self settles the dilemmas that define the Hard Problem.

It is a red herring—a speculative expansion of a problem that it does nothing to solve.

Finally, panpsychism traps its proponents in a fundamental contradiction. If they insist that consciousness is an inherent, real feature of the universe, they cannot simultaneously dismiss it as an "illusion" or an "epiphenomenal artifact"—a defence used by materialists like Steven Pinker to minimize the significance of the mind. A phenomenon cannot be both a fundamental property of matter and a deceptive trick of the brain. By attempting to have it both ways, the panpsychist reveals the intellectual exhaustion of the materialist project: when faced with the undeniable reality of the witness, they would rather invent a universe of "conscious photons" than admit that the soul is not made of matter.

Materialists Summary

Materialism is not merely incomplete; it is structurally paralyzed. It remains trapped in a binary that it cannot navigate: either consciousness is a fundamental reality—in which case it demands an explanation that matter, by definition, cannot provide—or it is an illusion, which is a fundamentally incoherent proposition. One cannot dismiss the very mechanism (experience) that provides the evidence for the material world itself. As Descartes demonstrated, the "cat" of experience has already escaped the "bag." The existence of the I/Self (Youness) is the primary data point of reality, and it is validated by the sheer fact of its operation.

The problem of consciousness is not a failure of current scientific instruments; it is the death knell of the materialist paradigm. Because a coherent material solution is fundamentally impossible, the "third way" is no longer an optional hypothesis—it is the only rational avenue of inquiry. The immaterial architecture of the soul must be explored, as this is the only framework that accounts for the observer without relying on the impossible alchemy of turning dead matter into living witness. We are no longer waiting for science to catch up to the soul; we are following the evidence where it inevitably leads.

Daniel Dennett & The Materialists

Daniel Dennett the Oxford-trained, American philosopher rose to prominence in the late 1960s with his analytical work *Content and Consciousness.* Known for his strict adherence to materialist, naturalistic and Darwinian principles, these perspectives are clearly represented in his views on consciousness. He argues that our seamless, "movie-like" experience of reality is a fundamental misinterpretation, which he derisively labels the "Cartesian Theater." To Dennett, there is no central observer, no inner sanctum, and no "you" watching the show. Instead, he offers the "Multiple Drafts Model"—a chaotic, distributed process of asynchronous neural events that somehow, miraculously, results in the human experience.

Dennett's stance is not a deduction born of direct evidence; it is a forced rationalization required by his materialist dogma. He acknowledges the richness of experience only to attack it as a user-interface error. In his own words:

> "Visual stimuli evoke trains of events in the cortex that gradually yield discriminations of greater and greater specificity. At different times

> and different places, various "decisions" or "judgments" are made; more literally, parts of the brain are caused to go into states that discriminate different features, e.g., first mere onset of stimulus, then location, then shape, later color (in a different pathway), later still (apparent) motion, and eventually object recognition. These localized discriminative states transmit effects to other places, contributing to further discriminations, and so forth[35]."

In an interview for *Philosophy Overdose* on YouTube, (May 22, 2023, The Illusion of the Cartesian Theater), Daniel Dennett explains,

> "If we are going to have a successful theory of consciousness it was going to have to be counter-intuitive. Hard to believe at first because our intuitions tell us so strongly, facts, -facts about our mind that just aren't facts. And I think the most potent and seductive is the idea that inside our heads, somewhere, there is a sort of a show going on. Colors and shapes, sounds and itches and tingles are happening in a special private subjective arena and we are the audience. Each of us is the audience for that set of apparitions, appearances, representations. I call that place the Cartesian Theater. It's the mind as Descartes conceived of it in the 17th century. There's no such place. It seems to stand to reason or has to. 'Look I can close my eyes and there I am. I'm in that place.' Well you think you are, that's what you think is going on. Here's what's really going on.... You're leaving something important out, and it's this hunch that there's something more beyond the spike trains and the coordinating activity of the nervous system. There's got to be this extra wonderful layer that makes life worth living... and they stopped the process right there. And they don't ask the next question. What I call the hard question. 'And then what happens?.. ...then they, -they start cataloging whatever it is that's in consciousness. 'And I've got this thought, that feeling, and a little itch of my toe, and and.'... well they should start taking seriously all the stuff that happens next, or that can happen next... once you've got that catalog and you've really done a good job of getting that catalog ask yourself whether and isn't that what being conscious is, -being responding in all these ways why do you suppose there's this extra material mysterious thing that's at the nexus that is the cause of this? It's a very powerful feeling but it's, -I argue it's just a mistake. The cause of all your reactions your emotional reactions your memories are just more spike train combinations in your nervous system that, you're adding one cause too many by putting in the qualia, the internal subjective mysterious Whatchamacallit, isn't actually doing any work[36]."

Notice the thievery in this logic. Dennett claims he has a good theory simply by dividing by zero—by defining the witness out of existence because it doesn't fit his neural spike-train model. He admits the feeling of being a self is powerful, yet he demands we reject it because his ontology has no room for a soul. He does not identify a flaw in the experience of consciousness; he identifies a flaw in the existence of his paradigm and chooses to sacrifice the former to save the latter.

Dennett's argument is an exercise in intellectual bankruptcy. He commits the ultimate materialist theft: he retains the functional reality of consciousness—the persistent self that makes decisions, writes books, and debates philosophy—while simultaneously denying its conceptual existence. He treats human consciousness as if it were a faulty computer program, yet he relies on the full, integrated weight of that faulty consciousness to construct his entire argument.

Dennett is in the position of a man standing inside a house, insisting that the house does not exist, all while enjoying the shelter of its roof. He claims the "I" is an illusion, but who is the "I" making the claim? If the Cartesian Theater is a mistake, then Dennett's own book is a product of a non-existent author. By calling consciousness an illusion, he is one step too late; the cat of experience has already escaped the bag. An illusion is, by definition, an experience. To categorize consciousness as an illusion is to validate it as a reality that is being experienced.

Dennett does not have a theory of consciousness; he has a manifesto of denial. He does not solve the hard problem; he attempts to ignore the observer out of existence, hoping that if he labels the most fundamental truth of our lives a mistake, he can continue to force the round peg of the human soul into the square hole of his neural spike trains. It is a philosophy that survives only because its author refuses to acknowledge the he is there, witnessing the show.

The Subtle Reality of Youness

Materialists are not alone in their struggle to define the elusive nature of Youness, but they are fundamentally mistaken in their conclusion. Their mistake arises from conflating "illusion" with "non-existence." While Eastern traditions like Buddhism's Anatta (not-self) and Hinduism's Maya (illusion) describe the self as impermanent, they never suggest that the subjective reality is empty or a "mistake."

For the materialist, "illusion" implies that experience is a lie to be debunked. For the Eastern traditions, "illusion" describes the transitory, misinterpreted nature of our surface-level attachments. They do not assert that the reality of experience is a falsehood; rather, they argue that our apprehension of it is incomplete. Where Dennett seeks to collapse the witness into nothingness, these traditions seek to strip away the "surface reality" to reveal the underlying essence. In this sense, the contemplative traditions aim to enrich and elevate the witness, while materialism seeks only to discard it.

The difficulty in encountering Youness lies in its inherent subtlety. It is not a "thing" to be captured, but the invisible stage upon which everything else occurs.

The Upanishads capture this subtlety vividly:

> "...It cannot be perceived by the senses; a finer spiritual sight is required… It is subtler than the subtle, because it is the invisible essence of everything; and It is greater than the great because It is the boundless, sustaining power of the whole universe; that upon which all existence rests." -The Upanishads[37]

This is why history is filled with the accounts of seekers who encountered only silence and were unable to find themselves within it. The difficulty of the "Hard Problem" is not merely a failure of neurobiology; it is a testament to the fact that the observer is the precondition for all knowledge. You cannot look at the thing that does the looking. The problem is refined and subtle and often the inner aspect of Youness must be encountered several times to be recognized.

The good news, however, is that this silence is not an absolute barrier. It is a threshold. Through rigorous contemplative commitment—and facilitated by methods ranging from meditation to the strategic use of traditional catalysts—the barrier between the seeker and the witness can be dissolved. This encounter is the historical cornerstone of human wisdom. In the next section of our inquiry, we will move from the theoretical deadlock of Western philosophy to the practical, time-tested methods used across traditions to encounter Youness directly. It is a reality that need not remain a mystery to those willing to engage in the search.

More Bats, Zombies & Illusions

In 1974, Thomas Nagel famously asked, "What is it like to be a bat?[38]" His point was simple: subjectivity is irreducible. We can map every synapse in a bat's brain, yet we remain forever locked out of the qualia of its experience. This is the "problem of other minds," a direct consequence of the Cartesian divide. We can observe bodies, but we can never verify the witness inside them. Nagel expounds in "What Is It Like to Be a Bat?":

> "Our own experience provides the basic material for our imagination, whose range is therefore limited. It will not help to try to imagine that

> one has webbing on one's arms, which enables one to fly around at dusk and dawn catching insects in one's mouth; that one has very poor vision, and perceives the surrounding world by a system of reflected high-frequency sound signals; and that one spends the day hanging upside down by one's feet in an attic. In so far as I can imagine this (which is not very far), it tells me only what it would be like for me to behave as a bat behaves. But that is not the question. I want to know what it is like for a bat to be a bat. Yet if I try to imagine this, I am restricted to the resources of my own mind, and those resources are inadequate to the task… The best evidence would come from the experiences of bats, if we only knew what they were like. So, if extrapolation from our own case is involved in the idea of what it is like to be a bat, the extrapolation must be incompletable. We cannot form more than a schematic conception of what it is like." -Thomas Nagel, *What Is It Like To Be A Bat?*[39]

David Chalmers later inverted this realization into the thought experiment of the "philosophical zombie"—a being physically and behaviorally identical to a human in every way yet possessing no inner subjective life. The zombie is the ultimate stress test for materialism: if you can conceive of a being that functions perfectly without consciousness, then consciousness cannot be a mere byproduct of physical function.

Materialists are left with an impossible choice: they must either admit that consciousness is a fundamental reality that their physics cannot explain, or they must argue that consciousness itself is an illusion.

This is where the materialist framework descends into incoherence.

When a materialist labels consciousness an "illusion," they are attempting a logical sleight of hand that fails upon inspection. To label something an "illusion" is to admit that an experience is taking place; you cannot be deceived by a non-existent experience. If you are experiencing an illusion of consciousness, then you are, by definition, conscious. The distinction between appearance and reality collapses because, in the domain of the subject, the feeling is the fact. If you feel you are a self, you are a self. To categorize that experience as an "illusion" is a category error of the highest order—it is an attempt to explain away the primary data point of our existence because it refuses to conform to a pre-existing physicalist bias.

This difficulty infects all materialist accounts. Whether they lean on panpsychism, emergence, or eliminativism, they all share a singular failure: they can map the machine, but they remain silent on what it feels like to be that machine. They treat the observer as a variable to be divided out of the equation, ignoring the fact that without the observer, the equation itself could never be written.

The Hard Problem persists, not because our science is incomplete, but because our paradigm is fundamentally misaligned. As long as the "witness" is treated as a byproduct of matter rather than the precondition for it, the materialist will continue to find themselves explaining away the very thing they are trying to explain. They are not solving the mystery; they are simply refusing to look at the witness that is doing the asking.

The Black Box Conjecture

The materialist end-game, when stripped of its academic veneer, results in a vision of the human being as an isolated, ghostly passenger trapped in a simulation.

If we take the materialist model seriously—as Pinker, Harris, and Feldman Barrett do—we are forced into what we call the Black Box Conjecture. This model posits that consciousness is entirely confined to the skull. We do not interact with the world; rather, we interact with a neural simulation of it. All sensory input is reduced to a dark, silent box of neural activity, where the brain collates signals into a user-interface experience—a VR headset of our own construction. Within this experience, the human being sits, entirely closed off from reality, experiencing a second order representation of actual reality.

The consequences of this model are not just scientific; they are existential:

The Death of Objective Truth: As cognitive scientist Donald Hoffman argues in his Interface Theory of Perception, if evolution shaped our senses strictly for survival, the probability that we perceive reality as it actually is, remains zero[40]. (We are not interacting with the world; we are interacting with an evolutionary "avatar" that obscures reality rather than revealing it.

The Epiphenomenal Ghost: Under this interpretation, our conscious experience is purely epiphenomenal—a secondary byproduct of neural processing. When you "decide" to raise your hand, that "want" is not a cause; it is a late-arriving echo of a physical process that occurred microseconds before you were aware of it. We are passengers in a vehicle set to permanent autopilot, clutching a steering wheel that isn't connected to the axles.

The Exile of the Subject: As Dr. Lisa Feldman Barrett confirms, the brain is "trapped in its own dark silent box." It receives outcomes, but it has no knowledge of the objective truth that caused them.

"The question of, 'does your experience match reality or are you really experiencing reality in its dark objective truth?' and the answer is simply: 'Not exactly.' Reality for us is what we can sense with our sensory surfaces, and what we can make sense of with the signals in our brain. The signals in our brain are as necessary for our experience of reality as the sensory signals that come from the world… your brain, my brain, everybody's brain is trapped in its own dark silent box called your skull. And the brain has no knowledge of what is going on around it in the world or in the body, right? Because it is in this skull and it is receiving sensory signals from the sensory surfaces of your body. These signals are the outcomes of some changes in the world or in the body – but the brain doesn't know what the changes are, it's receiving the outcomes[41]…"

The conclusion is absolute: under the Black Box Conjecture, humanity as a locus of causal agency does not exist. We are not pilots of our biological vehicles; we are spectators of a pre-recorded neural broadcast, trapped within a skull, fundamentally exiled from the reality we think we inhabit.

Yet, this horrifying specter collapses upon inspection. The Black Box Conjecture relies entirely on the premise that consciousness can be reduced to neurons and chemical signals. But this has never been validated. It is a house of cards built on a foundation that does not exist. Brain scientists have been unable to demonstrate how matter generates the subjective witness, which is precisely why the Hard Problem remains an immovable obstacle.

The Black Box Conjecture is the *reductio ad absurdum* (reduction to absurdity). of materialism: it asks us to believe that the human being is a total illusion, created by a brain that is itself a total illusion, perceiving a world that is also an illusion. It is a closed system of mirrors.

We have seen what has gone into the making of our reality, from its instantiation at the CMB, through all of the steps and levels of emergence across the entire cosmos to create our world and Earth, and life upon it as a meditation. Why would this creative source go to so much trouble to create such a sublimely perfected and beautiful world, only to have the final product of their Algorithm, the thinking, feeling, passionate and delightfully engaged human animal locked within an inescapable black box, never truly engaging with the world it has so incandescently ordained? What would be the advantage of having this being living within this world and biological form, feeling the self and the universe, and, more importantly, observing it immediately, completely, transcendently at a distance so tantalizingly close yet unable to make the final contact? Are we to believe that the creative source that has achieved miracle after miracle finally found a truth state that they were impotent to pull off as the finality upon the complete conclusion of their reality? The answer must be no. To find this condition at the end of a perfected chain would be more inexplicable than all of the miracles we have yet witnessed.

As horrifying as the specter of the Black Box Conjecture is for the human race, it collapses to irrelevance upon inspection. What it assumes—that consciousness can be reduced to neurons and chemicals in the brain—cannot be done. Brain scientists have been unable to validate this hypothesis, and this is the reason why the hard problem of consciousness exists to begin with. The data we have been able to gather from the brain increasingly indicates something other than emergence and illusion explains consciousness.

All Bets Are Off – Science Can't Find Consciousness

The mystery of consciousness is not a temporary oversight of modern science; it is a structural inevitability of the materialist paradigm. In every sector of inquiry—from the origin of life to the nature of the mind—we encounter the same pattern: when science approaches the "core" of the phenomenon, the materialist explanation dissolves.

We see this most clearly in the failure of the "promise of instruments." For decades, the materialist wager was that if we simply built a bigger microscope or a faster computer, we would eventually catch the spark of consciousness in the act. They were wrong. The more granular our measurements become, the further the target retreats.

The clearest indictment of this failure is the Chalmers-Koch 25-year bet. In 1998, neuroscientist Christof Koch bet philosopher David Chalmers that neuroscience would explain consciousness within a quarter-century. During that period, brain imaging exploded, connectomics advanced, and we mapped the brain with unprecedented precision. Yet, in 2023, Koch conceded the bet. Despite an avalanche of data, the hard problem remained exactly where it started[42].

The same pattern of failure emerged in recent adversarial testing of leading consciousness theories, such as Integrated Information Theory (IIT) and Global Neuronal Workspace Theory (GNWT). Rather than locating a center of decision-making, large-scale studies found that neural signals are distributed and diffused, baffling the very models meant to explain them[43].

The result is an inescapable paradox: the better our tools, the more the materialist hypothesis is falsified.

This is not a scientific milestone; it is a systemic collapse. When a model predicts that X causes Y, and we develop the technology to watch X and find no trace of Y, the correct scientific response is to abandon the model. Instead, the materialist establishment has engaged in a form of intellectual defiance. They double down on the promise of future technology, re-label the missing consciousness as an illusion, or dismiss the observer as a quirk of neural processing.

This is no longer science; it is a paradigm commitment. It is a refusal to acknowledge that the "gap" in their research is not a temporary technical hurdle—it is a categorical boundary. They cannot solve the problem, so they attempt to redefine the problem out of existence.

But seen another way this dead end is actually a gateway. The fact that the materialist approach has been stymied for decades is not a cause for despair; it is a signal.

The evidence is now sufficient to invalidate a purely mechanistic view of the brain. When we stop trying to force the witness into the machine, the "blocking" of the reductionist path becomes a liberation. The way forward is wide open, provided we are willing to follow the evidence—rather than the dogma—to its logical conclusion: that consciousness is not an emergent byproduct of matter, but a primary reality that matter itself cannot contain.

The human legacy has known this for millennia. Before we were blinded by the promise of the "dark, silent box," cultures the world over possessed a profound, direct understanding of the nature of the witness. We have simply been looking in the wrong place.

We Have Exhausted the Materialist's Toolkit

In generations past, the mystery of life was never reduced to a gross material equation. The ancients recognized that the reality of selfhood—a spark erupting from nowhere, housed in living flesh, and tethered to an expiration date—demanded an immediate, ever-present solution. Therefore, the answer cannot be hidden in a distant particle or a future-dated technological breakthrough. It must be accessible now, reachable by will, intention, and direct observation.

The individual is the nexus where the universe meets awareness. As we analyzed in our inquiry into abiogenesis, the gap is not merely chemical; it is a profound disparity in category. It is animation, intention, and information—a reality that proceeds by constraint rather than blind chance.

While modern science became enthralled with the technology of machines, ancient traditions remained masters of the technology of consciousness. They understood that contemplation was not a passive activity, but a rigorous, iterative discipline used to refine one's identity. Today's obsession with scanning gray matter to view consciousness as patterns and pulses is a fantasy—a category error that mistakes the shadow for the object.

The Youness we have defined is the true technology of the human being. It is the window through which we see the world, the mirror reflecting who and what we are, and the repository of reason and constraint. It is where the pupil of the cosmos asks the deepest questions and receives the most profound answers:

What is this world? It is our home and our teacher.

What is the nature of this experience? It is our life and an opportunity.

From where do these questions arise? From everywhere and from nowhere at all.

This is the domain of consciousness stripped of dogma, paradigm, and stipulation. The perplexity we feel arises not because consciousness is a mystery, but because we are the very thing we are attempting to know. Coming to know ourselves cannot be achieved by the application of outward instruments. It must be encountered with the original technology of mind seeking mind. As the Chandogya Upanishad declares:

"Tat Tvam Asi" — "Thou art That."

We are the answer we seek. Youness is fixed in both directions: equally adept at looking outward at the world and inward at the self. It is the sacred portal through which all things arise and all things are settled.

In considering this, we must clarify one final question:

Is consciousness a miracle?

We have encountered many miracles in our inquiry, but consciousness is not one of them. It is certainly miraculous, and it cannot be explained by physics or probability, but consciousness cannot be a miracle if our lives are to have meaning.

Everything that consciousness facilitates—what we witness, evaluate, and learn—all of these can be miracles. But consciousness itself is the ground floor. It is the foundation from which all else proceeds. It is the seed from which the flower blooms, and it is the witness that observes the bloom. It cannot be a miracle because it is the condition for the existence of miracles. So maybe we are playing word games, but the point is insoluble. Consciousness is what our entire reality is predicated upon, therefore it must pre-exist all that we have previously encountered within The Divine Algorithm as "miracles."

The oldest traditions in humanity traced the shape of this condition. They bore witness to the twin mysteries: the universe without and the individual experience within. This is the thread of continuity we have been searching for—a lineage of profundity and spiritual sustenance that contrasts so sharply with the reductionist dead ends we have dismantled: the "hard problem," the "Cartesian theater," and the "black box conjecture."

The reductionist road has reached its terminus. The path forward—the path of the witness—is wide open.

Part II: ALTERNATIVE PERSPECTIVES ON CONSCIOUSNESS

The history of human thought—from Imhotep and Aristotle to The Buddha and Descartes—is not a collection of disparate opinions, but a singular, unfolding testimony to the reality of the Witness. While these thinkers differed in their conclusions, they all operated from the same "ground floor": the immediate, unassailable fact of consciousness.

We (humans) share an objective intersubjectivity. While our lives, cultures, and eras differ, the fact of our experience is identical. We inhabit the same physical reality, governed by the same laws, viewed through the same fundamental aperture of awareness.

This shared stage is the only reliable data we possess. Yet, Western inquiry has suffered a fatal divergence: it has mistaken the map for the territory.

The critical error of the modern West is the attempt to reduce consciousness to description. We have sought to contain the infinite breadth of experience within the finite container of language, forgetting that language is merely a proxy—a shadow cast by the thing itself, like Plato's allegory of the cave. By prioritizing the description over the encounter, we have effectively shuttered the window of direct perception.

Logical Positivism

This catastrophe was codified by Logical Positivism, a movement that sought to purge the human experience of anything it could not measure. Its central dogma, the Verification Principle, declared that any statement not testable by empirical observation or formal logic was not merely false, but "nonsensical[44]."

The Verification Principle is not just an academic error—it is a scalpel that removes the very things that make life worth living.

This wasn't merely an abstract exercise in philosophy; it was an act of intellectual impoverishment. Under this principle, the most profound aspects of human life—love, beauty, moral devotion, and the weight of grief—were relegated to the status of life static. Because you cannot measure the intensity of a mother's love with a spectrometer or calculate the truth value of a vow of loyalty with formal logic, the Positivist dismisses them as cognitively meaningless.

To accept their framework, one must be prepared to argue that a sonnet, a prayer, or the feeling of justice is no more meaningful than the random static of a radio out in the rain. They did not just sharpen the tools of science; they narrowed the scope of human experience to the size of a laboratory dish, effectively declaring that if an experience cannot be quantified, it does not exist.

Such movements are incredibly disingenuous. While the architects of such schemes enjoy all the things mentioned above; love, devotion, beauty, the appreciation of art—they dismiss them outright in their drawing room debates. It's much the same with materialist's conceptions of consciousness. Adherents of such beliefs live as passionate, feeling beings yet they argue to the public, and their students they the human animal is nothing more than biological robot, made by accident and destined for a void at the end of life. These individuals don't feel the pain of the horror they've created, precisely because the ineffable beauty of the human animal they attempt to excise is hardwired into their souls.

Logical Positivism was not a scientific advancement; it was an act of intellectual censorship. Its application to consciousness created a dual-level fallacy:

1. **The Purge of Qualia**: By restricting the realm of "knowledge" to raw, external data, the Positivists banished qualia—the very subjective experience that constitutes the "Hard Problem"—from the realm of relevant inquiry. They defined the most essential aspect of human existence as "nonsense."
2. **The Linguistic Fallacy**: They attempted to reduce the "thing-in-itself" to a linguistic representation, operating under the delusion that words could achieve a 1:1 correspondence with reality. They act like words are a net that can trap reality, as if one could write a perfect enough sentence, it could 'up the truth. Why drink a glass of water when we have written the entire concept down so precisely it serves as a substitute for the 'thing-in-itself'? Words are tools we invented that are mapped onto reality, they are not reality itself. In their attempt to explain the mystery of consciousness using only definitions and logic, they haven't actually figured anything out. They've just built a mind palace in place of the reality they don't have the language to reconcile.

In their zeal for a sterile "scientism," the Positivists succeeded in banishing the "feeling of life" from the domain of knowledge. They did not solve the mystery of consciousness; they legislated it out of existence. By declaring that which cannot be "pointed to" as irrelevant, they built a house of knowledge with no foundation. They traded the direct, lived experience of Youness for a dictionary of empty labels.

The result is those of the western academic mind have become experts at describing the cage, all while denying the existence of the bird.

The Reductionist Trap

The treatment of qualia within materialist frameworks is presented as a scientific inquiry; it is actually an act of erasure. Solving the hard problem of subjective experience by stripping away the defining qualities of subjective experience until only the physical event remains is not an advance, it is a wholesale retreat. And this is the core of the reductionist trap: to demand that qualia be explained as a byproduct of gray matter, and in the effort to locate the physical origin of qualia, completely fail into a precise category error.

They don't actually explain what it feels like to taste a strawberry, feel love, or see a sunset. Instead, they just strip those experiences down until they are nothing but cold, hard data—the anatomy of the eye, the mri image of a brain looking at a sunset, the physiology of the mouth, and the spectrographic representation of a strawberry. To declare, 'this chemical was released by the brain,' the underlying assumption is to describe the physical machinery, the experience has been articulated. But they've actually just thrown the experience away and replaced the feeling of living with a list of technical bullet points.

They transmute a subjective experience into a material object, a process that is logically impossible. This heist is executed through two consecutive errors of misattribution:

1. **The Materialist Transmutation**: They attempt to relegate the ineffable—the feeling of the thing—to a mechanical construct. They have not refined qualia; they have discarded it through redefinition.
2. **The Linguistic Fallacy**: They rely on language as a 1:1 map of experience, operating under the delusion that a description can embody the "thing-in-itself."

This is not a scientific "refinement"; it is a systemic act of denial. The logical positivists and their materialist heirs believe they have made consciousness "scientific," but they have simply blinded themselves to the data. By mandating that only the measurable exists, they ensure that the "sight" of the witness is permanently forbidden.

The result is a grotesque substitution: they treat the map as if it were the territory. A map of New York may detail its location, dimensions, and infrastructure, but it contains none of the things that make the city "New York"—the commerce, the architecture, the pulse of human life, and the vibrant sensory explosion that defines the experience.

People love New York; they do not love its coordinate on a topographical chart. Yet, the materialist insists that because they have described the map, they have successfully captured the city.

This is a vicious circularity.

The Hard Problem cannot be resolved by brute force, redefinition, or the semantic gymnastics of logical positivism. These are not scientific solutions; they are defensive maneuvers designed to protect a paradigm that cannot account for the witness. When we prioritize the description over the experience, we lose the only thing that actually exists. We are left with a universe of perfectly described, perfectly empty maps, while the city itself—the living, breathing reality of Youness—is left entirely unvisited.

Moving From Physicalism

We are now departing from the Western obsession with descriptive mechanism to return to the only primary data we have: Direct Experience.

Consciousness is not an object to be measured; it is the field in which all objects appear. It is not "information processing"; it is the living stage upon which the mixing bowl of the individual encounters the world. This is where consciousness makes itself so undeniable. When it becomes a vital eruption of Youness in the midst of qualia—the precise, impossible moment of contact where the Witness meets the field of reality.

To understand this is to understand that language is a secondary technology. It arrives long after the fact of reality. It is an abstract system of symbols that is utterly hollow until a living, conscious being breathes life into the very words to capture and communicate direct experience.

Words are not truth; they are pointers. As D.T. Suzuki famously highlights from the understanding of Zen, language is like a finger pointing at the moon; the moon (the experience of seeing the moon) is not the finger (the words about the moon), but without the pointing, we may never look up and see the moon[45]. This is the proper context, because if we mistake the finger for the lunar surface, we learn nothing.

The materialist trap is exposed by their own definition of biology. They insist that the brain is a machine—a biological computer. If this were true, truth would be a matter of data transfer. One could simply "download" the experience of enlightenment or the qualia of existence through a series of correct propositions. Yet, we know this is impossible. We do not need, nor is it possible to download the experience of first love or the weight of a pregnant silence; we must live it.

AI, The Ultimate Robot, The Fallacy of The Cage

This reveals the final collapse of the AI/Materialist analogy: If biology were merely a "machine," a robot could hear the words of a spiritual master and possess the experience of enlightenment and see the world anew. But the robot remains a void. It lacks the Witness. It lacks Youness. It processes symbols, but it possesses no referent for them. It can't. This is the machine that is forever locked within the black box conjecture, never to escape. Never to experience reality directly, because there is no witness, and there is no consciousness.

The Western approach has spent its energy measuring the cage, believing that if it describes the bars precisely enough, it will eventually understand the bird. It is time to look away from the bars and toward the sky. We are not the code; we are the consciousness that perceives the code. We are the answer that we seek (Chandogya Upanishad).

Vedanta – Âtman & Brahman in our Ontology

The Eastern traditions—Buddhism, Vedanta, and the perennial wisdom of the world—have known that a description is not reality. That the finger pointing at the moon is not the moon. That qualia is the experience itself, and that the greatest mysteries of life, can only be solved by the living, with consciousness and the witness of Youness. They do not seek to imprison experience in a cage of nomenclature. They use language as a vehicle to transport the seeker toward the summit, but they acknowledge that at the peak, words fall away.

We must now turn to a framework that does not treat consciousness as an emergent property of matter, but as the ground of all reality.

In the Vedic tradition, this is found in the non-dual relationship between Brahman (the infinite field) and Âtman (the localized soul).

These are not two distinct things. They are the same reality viewed through different apertures. As the ocean is the same substance whether measured as a vast, unified body or as a single drop at the crest of a wave, Brahman and Âtman are identical in essence. This is the Vedic solution to the Hard Problem of consciousness: the observer and the observed are not separate entities interacting in space; they are one, arising from the same field of consciousness.

The universe is not composed of "stuff" that happens to produce awareness—the biological form with consciousness; the universe is composed of the same fundamental substance as the observer—chemistry. As the Upanishads articulate, reality is an infinite field of potentiality that flickers into form, and substance and change, and humans arise from this very same essence:

"That which cannot be spoken, perceived, grasped, or thought—that alone is Real."

"That which is the essence of the self of all beings."

"Sat-Chit-Ananda—being, consciousness, bliss."

Âtman is the irreducible fact of self—aware, present, and primary. As the text declares, Ayam Ātmā Brahma—"This Self is the Absolute[46]."

In this view, the "Black Box" of the materialist is exposed as a misunderstanding of **Maya—the appearance of the world, the illusion**.

Maya is not "unreal," but it is not independent. It is the field of form, change, and multiplicity that arises within consciousness. The ancient verses of the Ṛigveda confirm this: "Indra, through the power of Maya, assumes different forms[47]."

The materialist insists that we generate consciousness within our skulls through gross anatomy. Vedanta reverses this: **we appear within the field of consciousness as physical beings**. Nothing is known outside this field. No object, law, or structure is encountered apart from it.

To clarify this doctrine, we look to P.D. Shastri:

"In other words, there is only one Reality, call it Brahman or Atman — what you will, and the world around us which appears so real is not so. This is the central thought which has been so admirably expanded and developed in various ways in the Upanishads, and what we call the doctrine of Maya is nothing more than an attempt to explain this fact in detail, to show how it is impossible for the world to be anything more than an 'appearance' as distinguished from 'Reality,' Which strictly speaking is only Brahman[48]."

The physical body and the individual within it are veils draped over the screen of consciousness. When these veils are transcended, they collapse into the qualia of existence—the raw, unmediated experience of reality without form.

This is the Third Way, It does not require us to (1) deny our senses, nor does it (2) require us to adopt the nihilism of the "Black Box." (3) It requires only that we recognize our true location: we are not the contents of the skull; we are the field that contains the universe. The materialists are staring at the shadows on the wall, convinced they are the screen itself. Vedanta reminds us that we are the light.

The Conclusion Is Immediate.

The conclusion is immediate and inescapable: Consciousness is not a byproduct of the universe; it is the precondition for the universe to appear.

The actors are the thing. Not the stage. The true foundation is the universal field of consciousness—what the Vedic tradition identifies as **Maya**.

If Maya is the illusory stage, Lila is the reason the stage exists. **Lila is the divine play**—existence as a spontaneous, overflowing, and joyful manifestation. It is the overflow of abundance. It is Brahman choosing to manifest as the "Many" simply because it can.

We must now bridge this metaphysics with the biology we have previously surveyed.

Consider the mechanisms we examined earlier: the precise architecture of DNA, the high-fidelity machinery of transcription and translation, the rhythmic cycling of the citric acid cycle, and the spinning molecular motor of ATP synthase in the mitochondrial membrane.

These are not merely mechanical accidents occurring in a vacuum. **Brahman is the animating** power behind the trillions of discrete, synchronized actions occurring in every cell of your body. It is the intelligence that sustains the life cycle of every organism.

It expresses itself as the localized Âtman in every spark of Youness—in every moment you witness the world, you are witnessing Brahman observing itself through the aperture of your consciousness.

This is the "divine meditation" we previously explored. It is the inexhaustible, simultaneous awareness that animates the development of the embryo and the rotation of the galaxy alike. It is not a force that "pushes" matter from the outside; it is the field in which matter arises, sustains itself, and eventually dissolves.

Every identifiable thing—from the subatomic flicker of a particle to the vast sweep of a planetary orbit—is located within this meditation. It is the single, unified source of all activity. We have spent chapters documenting the complexity of the machine, but we were looking at the mechanics of the Lila. We now recognize the hand of the player.

The materialist stares at the machinery and asks, "How did this happen?" The witness observes the meditation and understands, "It happens because it is sustained." The play is not a mystery to be solved by dissection; it is a reality to be lived.

Brahman & Atman – The Final Click

If the identity between yourself (Âtman) and the Absolute (Brahman) has yet to fully register, it is likely because the implications are too profound to immediately grasp. We must overcome the illusion of separation. This truth is best encapsulated in the doctrine of Bhedabheda: the simultaneous reality that you are both a distinct, localized being and yet, in essence, the entirety of the Whole. We appear separate, but we are identical in substance—localized flickers within the divine meditation.

When you recognize yourself as Âtman, you understand that your existence is not a statistical accident or a biological byproduct. It is a deliberate expression—a chosen participant in the Lila of Brahman. This does not make the self tenuous or illusory; it renders the self eternal. Because your Youness is a direct manifestation of the Absolute, you are more real, and more foundational, than any materialist model could ever permit.

This realization dissolves the existential dread that haunts the modern materialist. The question of "how you came to be"—and the haunting fear that you might never have existed at all—simply evaporates. Within the infinite, eternal field of Brahman, the individual does not "start" in the sense of being assembled from parts. They arise.

You did not begin as an inert object that suddenly acquired the "miracle" of consciousness. You were highlighted as a distinct light within the play. If you exist now, you have always existed, held by the only power that grants existence—Brahman. You were conceived as an integral part of the play and then revealed to yourself as Âtman by the only Source that exists.

This is the non-dual reality: Brahman and Âtman are the same, viewed from different directions. The observer is the Absolute looking at itself through the aperture of your own life.

This is the "Good News" that has served as the bedrock of wisdom in the East for centuries. It is the final answer to the "Hard Problem" of consciousness. The problem was never that the brain could not create the mind; the problem was the mistaken belief that the mind was trapped inside the brain. Once the illusion of the "Black Box" is shattered, the Witness remains. You are not the passenger in the vehicle, and you are not the machine. You are the Light that makes the entire play visible.

Vedanta Redux

Vedanta does not merely describe a religion or spirituality. In its philosphy explains our reality. This is not to 'choose sides' among religions. In our use of Vedanta, we are seeking a truth condition for the relation between the human, consciousness and the inner observational quality of

Youness. For any philosophy or tract of writing to capture the truth of this condition as Vedanta has so well and in such a multifaceted and perfected puzzle, is a boon to us all.

Vedanta is philosophy as much as it is religion and spirituality and a way of life, and a truth condition. It functions equally well on all three levels. As with any great ontology when lived to its logical end it provides more than a roadmap. It provides the constraints of the topography that the map is drawn against.

Its articulation captures the dynamic of consciousness that has eluded the western cultural zeitgeist from the very beginning to this day. (there are indeed great western masters, but they prove to be the exception and not the rule)

We have included many ideas from many parts of the world within our investigation. We have borrowed from many spiritual conceptions as well as religions. Our point that we must arrive at is the collective human experience as it relates to consciousness. And to articulate this experience, and its understanding, we must draw from all threads that go into the fabric of our condition, selecting those that are the brightest, most robust, and more importantly, most likely to endure under the task assigned.

A Particular Kind of Blindness

Modern western science is structurally blind to the hard problem. And it performs a category error by attempting to locate a mythical structure within the brain responsible for consciousness, and the inner observer we call ourselves.

Western science is beyond fantastic at nearly all things. It has more than earned its due: In the hard sciences it has described the rules (physics/chemistry) in the finest possible detail. But when it comes to consciousness, it is effectively silent or wrong, or stumped on what is making the music or who wrote the score of consciousness, or why the arrangement makes us 'feel' the way we do. Science has measured every dimension of the instrument and transcribed the score and run it through complex algorithms attempting to identify the player.

Vedanta found the essence, the player, their instrument, the dancehall centuries ago and has been commenting ever since.

We know that as humans we do not either consciously, nor unconsciously do these things, nor do we have the executive power to control and coordinate these molecular/cellular activities. So how do they happen? Thus, in the inexplicable moment to moment activities of life, we can see another example of how Âtman and Brahman are one and the same. There can be no talk of, or possibility of separation, nor of true death, merely transfiguration from dwelling within a living form, and dwelling as a purely spiritual form.

The Universality of Consciousness As Represented By The Vedas

The origination of these concepts—*Âtman, Brahman,* and *Maya* as found in *The Upanishads*—are told in the legend of the original seers—*Dhrishtas*—it was they who came upon the truths, while in a state of deep meditative communion with God, which is called, *Shruti*, or, "that which was heard."

As the story is told, what was heard was represented word for exacting word and transcribed into four complete texts by sages. This is what is known as *The Vedas*.

Just as **Youness** arises, and thoughts arise, the *Vedas* arose within the *Dhrishtas*.

> "No date for the origin of the Upanishads can be fixed, because written text does not limit their antiquity. The word *Sruti* makes that clear to us. The teaching probably existed ages before it was set down in any written form[49]."

> "The name Upanishad has been variously interpreted. Many claim that it is a compound Sanskrit word U-pa-ni-shad, signifying 'sitting at the feet or in the presence of a teacher'; while according to other authorities it means 'to shatter' or 'to destroy' the fetters of ignorance. Whatever may have been the technical reason for selecting this name, it was chosen undoubtedly to give a picture of aspiring seekers 'approaching' some wise Seer in the seclusion of an Himâlayan forest, in order to learn of him the profoundest truths regarding the cosmic universe and God[50]."

Âtman and *Brahman* are Vedic/Hindu words that represent the universality of an immaterial self—**Youness**—as the recipient of conscious experience.

As *Âtman* is described in the Vedas, it is as if Descartes lifted passages from *The Upanishads* to use in his *Discourse* and *Meditations*.

> "That by which a mortal perceives, both in dream and in waking, by knowing that great all-pervading Âtman the wise man grieves no more. In these verses the teacher tries to make plain that all knowledge, as well as all sense perceptions, in every state of consciousness–sleeping, dreaming or waking– is possible only because the Self exists. There can be no knowledge or perception independent of the Self. Wise men, aware of this, identify themselves with their Higher Self and thus transcend the realm of grief[51]."

Having described reality's layers through the Vedas, we can now describe how reality is correctly layered, and illustrate how this layering is not a mere idea; rather, it is forced and surfaced by all that we have learned in our investigation thus far.

The System

This system—the way in which reality actually works—differs from Western conceptions which have it exactly backwards, attempting to place the power as the jewel in the crown, rather in its proper place as the crystal through which the light of the divine reflects all things, including crowns and jewels. This backwards arrangement is the reason why the Western conception ends in contradiction, and why the ultimate reality—consciousness—is relegated to being an illusion: not an illusion like *Maya*, but an actual illusion, meaning it does not exist, has no power, and is mistaken as being of a real nature, when in fact, it is nothing.

Western thought attempts to place consciousness as the illusory jewel in the crown of biology. It has the arrangement exactly backwards. Consciousness is the crystal through which the light of the divine, which is the same as this crystal, reflects all things, including the jewels, crowns, and the kings who wear them.

The Inversion

We now arrive at the decisive turning point.

The dominant framework within Western science proceeds in a particular order: reality is understood as fundamentally physical, composed of matter and energy governed by laws. From this, increasingly complex structures emerge—chemistry from physics, biology from chemistry, neural systems from biology—and finally, at the far end of this chain, consciousness appears. In this model, consciousness is treated as a product, an outcome, something generated by processes that are themselves non-conscious.

From within the physicalist's paradigm, this appears coherent, unavoidable. Upon closer inspection, a critical problem arises. All access to these proposed foundational layers—physics, chemistry, biology—is mediated through observation, measurement, and interpretation; however, the product takes place as experiences. Data, the creation and use of instruments, the imagination to make models, the foundation of understanding to derive equations, all appear within consciousness. So, there is no access to a world outside of experience.

What there is—the only thing there is—is experience itself.

And this experience takes place in a very specific place. And that place is where the world is presented, interpreted, and known… through consciousness as it collides with **Youness**. This introduces a structural asymmetry: the materialist model claims that consciousness emerges from matter. Yet all knowledge of matter is already given within consciousness. (Consciousness is the foundation, not the other way around.)

This insight, that consciousness is not the final product of a material universe, but the foundation from which that universe proceeds, is not merely a modern philosophical conclusion. It was articulated with extraordinary precision nearly two thousand years ago by the Apostle Paul:

> "For from Him and through Him and to Him are all things. To Him be the glory forever. Amen"
>
> -Romans 11:36

In a single sentence, Paul names the full structure of what the Algorithm has uncovered.

(1) From Him – consciousness as the origin, the source from which all reality proceeds.
(2) Through Him – consciousness as the sustaining medium, the field within which all material reality exists and operates.
(3) To Him – consciousness as the destination toward which the whole enterprise of Creation is directed: intelligent observers capable of recognizing the pattern that authored them.

Origin, sustenance, and purpose – the entire architecture that we have been unravelling in these pages, was compressed into twelve words by a man writing two millennia before the first telescope was pointed at the sky. What the Vedantic tradition calls Brahman, what this book calls the Divine Meditation, and the one Paul called Lord – all name the same transcendent reality from radically different starting points. The convergence itself is part of the evidence. No single tradition invented this insight; each arrived at it through its own methods, in its own language, across its own history. That they agree is not a coincidence. It is the Algorithm making itself known through every available channel.

The Fruitless Search

Searching for the source of consciousness within matter is a fundamental error in logic. It is circular: this is to ask the contents of awareness to explain the awareness itself. It is like a character in a book trying to find the author by looking between the printed letters.

If, however, we reverse the direction, the problem dissolves.

Rather than beginning with matter and attempting to derive consciousness, we begin with consciousness as the undeniable ground of all experience. From this standpoint, consciousness is not an output; it is the field within which all outputs appear—the *Maya* from which the energy of *Brahman* is transformed, and impermanent forms appear, change, and take on form. In this orientation, matter, biology, and neural processes are not generating consciousness; they are phenomena appearing within it.

The world is only ever known as experience within consciousness.

This inversion is not speculative—it is structurally compelled. One cannot step outside of consciousness to verify a world that exists independently of it, because any such verification would itself occur within consciousness. Reality is "trapped" within *Brahman*. Nothing can arise outside of it, independent of it, or as an independent view upon it. From here, our layered ontology becomes clear:

Consciousness is foundational.

The biological emerges from it. First, from the consciousness of instantiation that created our world and a series of productions (hydrogen, stars, elements, chemistry and biology). Second, consciousness is foundational as the biological form has been constructed as a seat for human consciousness and sentient consciousness to observe and live within reality.

It is therefore the case that neither consciousness, nor **Youness** can possibly 'emerge' from the biological form. They arrive within it and lord over it.

This inversion also settles a question long ignored by materialist frameworks: the ultimate source of structure, order, and the laws governing reality. Physics, the standard model, chemistry, biology, and life itself all presuppose constraints, patterns, and lawful regularities that cannot arise by blind emergence alone. Each instance, each cosmic expansion, from the formation of stars and galaxies, the emergence of life and consciousness, requires not just order, but the intervention of a guiding intelligence embedded in the fabric of reality—consciousness itself.

Consciousness supplies the supra-ordinal principle that transforms possibility, and at times the apparently impossible, into actuality. The given framework for this principle:

-Constraints give rise to structured appearance.

-These structures manifest as what we call physical reality.

-Biology emerges as one such structure within this appearance.

-Neural systems arise within biology.
-Thought and language arise downstream as interpretive and symbolic tools.

At no point in this chain does consciousness suddenly "appear." It was never absent. It is present at every stage. This is why our experience of the world as individuals "feels" like we are experiencing the world directly, physically, intimately, fully, and not as some kind of convergence of bound-up sensory information within gray matter. Because we do experience the world directly, physically, intimately as *Âtman* in the play of *Lila*.

The error, then, is in assuming that correlation implies generation—that the lower layer of biology and neural activity produces the higher, when in fact the entire structure is appearing within the higher.

And this leads to an unavoidable conclusion: what we have taken to be unexplained as emergent features of reality—the universe appearing within the CMB, its expansion rate, the first generation of stars and galaxies, the appearance of dark energy, the formation of solar systems and galaxies, the emergence of life, and finally consciousness—are not successive problems arising from a blind material substrate. They are expressions within consciousness itself. This has been demonstrated within every chapter as the one synthesis that brings clarity. It is the one synthesis that resolves the paradox of probability, possibility, and the seeming violation of physical law, under one fully explanatory ontology.

The error is directional, and complete, not merely technical. We have assumed that the level below—biology—is responsible for consciousness. It would be an equal error to assume that consciousness has risen from the bottom to be expressed through biology. Rather, biology exists as a vehicle, a specialized interface, a vibration through the foundational medium, through which consciousness is experienced within a structured, physical reality.

And words as the final carrier of this signal? Words cannot be final. They are only possible because consciousness and its constraints created the architecture for them to exist, be relevant, and carry meaning. Words are the shadow, the foundational signal that allows for objective truth to be propagated among observers.

This is where our inquiry, the direction of our entire hypothesis of *The Divine Algorithm*, culminates. The preceding analysis has not only shown why this inversion is coherent, but why articulations that are alternatives to the foundation necessarily must fail. Any attempt to ground consciousness in something outside of itself must rely on tools—perception, thought, language—that already presuppose consciousness.

In this sense, no competing framework can pass the full constraint of rational and structural scrutiny without collapsing back into the very ground it seeks to explain.

In our model, reality unfolds in layers—consciousness first, then constraints, then physical reality—and the laws of physics as expressions of these constraints. At every stage, across every Variable and Special Problem, we observe cross-domain statistical anomalies exhibiting highly specified outcomes. The dominant framework has no explanation. The evidence speaks to the application of consciousness combined with an instantiating power that can find highly specified outcomes operating within, yet transcending, the probability space of physical laws. This hypothesis provides a single unified answer: a shared principle with the power to address all anomalies, with Goal-Directed Selection—to maintain and establish a life-allowing universe and to bring intelligent observers into that universe.

This is the crucial thread—consciousness manifesting conscious beings who are then capable of recognizing the very pattern that authored them.

This is where the algorithm culminates: the universe, life, and consciousness… that can reflect upon itself, observe the algorithm, and see itself as the cause.

Consciousness as the Foundation - Bhagwan Shree Rajneesh (Osho)

Nearly forty years ago, after a spiritually transformative experience, I was conducting my own search into the nature of the human soul (consciousness). While at my student job shelving books, I found a volume at The Suzzallo Library at the University of Washington with the intriguing title, The Grass Grows by Itself (1976). It was written by an individual who at the time went by the name of Bhagwan Shree Rajneesh, now known as Osho. Within this book were a series of transcribed discourses from a retreat he was operating out of Poona, India.

The title intrigued me, because grass figures prominently in tales of spiritual awakening. For many seekers, grass becomes divinely beautiful in the context of a transcendent experience. This happened to me as well, in the midst of a soul-shattering transcendent experience. While all was blissful and perfected, and all inquiries answered in the affirmation of truth, I looked on the ground, and in the moonlight, I could hear the voice of each and every dew-covered blade of grass rejoicing in God's glory.

Returning to my regular life was incredibly painful. There was little support in the 1980s. How could anyone understand without the benefit of having had the transcendent experience? So, when I found this little book, it was manna. It filled my soul and let me know I was not alone. The world I experienced was real, vibrant, and sat as the foundation to the one we experienced in the day-to-day.

Individuals in an elevated spiritual state are commonly arrested by the energy and beauty of grass. It may sound counter-intuitive for the simplest plant form to be so robust as a spiritual beacon, but it is a theme that recurs frequently in elevated states—near-death experiences or those induced by LSD, psilocybin, DMT, or deep meditation. Individuals in these states report unusual colors, divine light clinging to dew; they hear a divine sound rising up from the grass, a chorus of praise crying out to the glory of The Creator.

Bhagwan was a professor of philosophy at Raipur Sanskrit College and Jabalpur University before he began teaching his unorthodox methods of meditation. What follows is a selection from The Grass Grows by Itself that captured my attention and signaled that my experience was no fantasy:

> "Physicists have been working for three hundred years to find the base, the substance of matter, and the deeper they reached, the more they were puzzled. Because the deeper they groped, the less and less substantial matter was; the less and less material matter was. And when they really stumbled upon the source of matter, they simply couldn't believe it, because it was against all their conceptions. It was not matter at all: it was simply energy. Energy is non-substantial. It has no weight. You cannot see it. You can only see the effects of it; you can never see it directly. Eddington, in 1930, said that we were in search of matter, but now all new insight into matter shows that there is no matter, it looks more and more like a thought and less and less like a thing. Suddenly the insight of Buddha became very, very significant again, because Buddha did the same with human matter, the human stuff. Physicists were trying to penetrate matter in an objective way to find out what was there inside it, and they found nothing. Total emptiness. And the same was discovered by Buddha in his inner journey. He was trying to find out who was there inside - the substance of human consciousness - but the more he penetrated, the more he became aware that it becomes more and more empty. And when he suddenly reached to the very core, there was nothing. All had disappeared. The house was empty. And around this emptiness everything exists. Emptiness is your soul. So, Buddha had to coin a new word which had never existed before. With a new discovery you have to change your language. New words have to be coined, because you have revealed new truths and old words cannot contain them.

> Buddha has to create a new word. In India people had always believed in the reality of the soul, ATMA, but Buddha discovered that there was no soul, no ATMA. He had to coin a new word - ANATTA. ANATTA means no-self. The deepest hidden in you is emptiness - a state of no-self. You are not; you only appear to be[52]."

When I began to read, I found that this thing we have named Youness was being discussed. In its depths, it was nothing; yet, I was still there to witness it. Or, paradoxically, I was not there and still witnessed it. Even in nothingness, we are there to witness it.

The concept of anatta is unambiguous. The self is nothingness at the core, yet a witness remains. This paradox is a quirk of life on the physical plane, where we have no access to our spiritual content prior to our instantiation as living beings. This is why when we darken the earthly lights of our personal history, thought, and feeling, we find… nothing. The veil prevents the seepage of the deeper spiritual self, thus anatta—no-self—yet the witness remains.

This physical world, the world of the veil, inherently holds a number of paradoxes: wave-particle duality, the vacuum of space from which matter arises, or the mirror—which reflects all objects but has no image of its own. We live in a world where the deepest truths appear as paradoxes. They are unavoidable when reality is observed from a perspective that penetrates the immaterial nature of things.

Bhagwan's words revealed what I grappled with. But to the world at large, what he stated was a concept, an idea—not a reality the world lives within. Even when the physicists of the Golden Age were bewildered by quantum mechanics, they saw its truth, but it remained irreconcilable. They called it a duality. They confronted it as an unresolved paradox, never seeing it as a harbinger for an underlying truth.

Where the physicist sees a barrier, the Sadhu sees an invitation. For them, the Hard Problem is not a wall to ram their head against, but an entry to the Absolute where the viewer and the illusion merge, and the Western artifice disappears into the fantasy it arose from.

The words of Bhagwan validated my experience: the shattering of my soul, the emptiness I confronted, the bliss that issued forth, and the profound confusion upon my return. It was so rare to hear Youness (anatta, Âtman) being discussed in any context. Everything about The Grass Grows by Itself carried the fragrance of a deeper truth. And so I returned to it many times.

Sunyata

Lama Surya Das points out in *Awakening the Buddha Within*[53] that what we think is not who we are. Thoughts are like the weather: unpredictable, impermanent, and subject to change. Even in an undisciplined mind, where thoughts and feelings fluctuate, something remains constant.

Lama Surya Das asks: what is the individual when thoughts cease? In between thoughts, do we cease to exist? No. There is always something silent, observing in the absence of input, thought, or sensory experience.

This observer is the one aspect of self that is not impermanent. As he writes: "This inner light is unbound, untrammeled, and immaculate. It can be relied upon; it can be depended upon. It is perfect, inherently wise and warm, free and complete from the beginningless beginning. Actualizing that luminous, formless, and intangible core is what awakening is all about."

This principle is *Sunyata*, a central concept in *Mahayana Buddhism*[54] Sunyata enables a complete deconstruction of all that consciousness senses and allows sole focus on the inner observer—emptiness or voidness—that is not limited to the self but encompasses everything. "Everything" in this context means there is nothing behind appearances; all phenomena arise as if from emptiness and return to the unborn nature of emptiness.

"It Cannot Be Perceived By The Senses"

> "Although this Âtman dwells in the heart of every living being, yet it is not received by ordinary mortals because of its subtlety. It cannot be perceived by the senses; a finer spiritual sight is required… the thought must be indrawn from all external objects; the mind and body must be under control; when the whole being thus becomes calm and serene, then it is possible to perceive the effulgent Âtman. It is subtler than the subtle, because it is the invisible essence of everything; and It is greater than the great because It is the boundless, sustaining power of the whole universe; that upon which all existence rests."
>
> —The Upanishads[55]

> "Knowing that the senses are distinct (from the Âtman) and their rising and setting separate (from the Âtman), a wise man grieves no more. A wise man never confounds the Âtman, which is birthless and deathless, with that which has a beginning and end. Therefore, when he sees his senses and his physical organism waxing and waning, he knows that his real Self within can never be affected by these outer changes, so he remains unmoved."
>
> —The Upanishads[56]

Chao Chou and Nan Chuan

In the opening of, *Zen: The Special Transmission*[57], *Osho* begins with the parable of *Chao Chou* and *Nan Chuan.* In this story he speaks of awakening to "the authentic self" and "unborn mind." The unborn mind, coined and made famous by Bankei, is known by many names. In this parable, it is called *Tao.*

Chao Chou asked Nan Chuan, "What is the Tao?"

Nan Chuan Answered, "The ordinary mind is Tao"

Chao Chou then asked, "How can one approach it?"

Nan Chuan replied, "If you want to approach it, you will certainly miss it."

Chao Chou: "If you do not approach it, how do you know it is the Tao?"

Nan Chuan: "The Tao is not a matter of knowing, nor a matter of not knowing. To know is a delusory way of thinking, and not to know is a matter of insensibility. If one can realize the Tao unmistakably, his mind will be like the great space —vast, void, and clear. How then can one regard this as right and that as wrong?"

Upon hearing this remark, Chao Chou was immediately awakened[58]."

Nan Chuan told Chao Chou that the "ordinary mind is Tao." When Chao Chou asked how to approach it, he was told: "If you want to approach it, you will certainly miss it."

That is the dilemma. We are surrounded, suffused, and bathed in this ever-present awareness, yet we are perplexed how to "wake up" to what we are already experiencing.

Whether spiritual seeker or materialist, we choose one of the following: live in the world of appearance—neurons, biology, synapses—or accept the invitation to the deeper reality that creates them.

Upon hearing the concise words of Nan Chuan words, Chao Chou was immediately awakened.

Yet, we remain unawakened even though we have just read the same words.

We are left behind because we the words are not the container of truth rather than the pointer. In the capable hands of Nan Chuan, he chose the words wisely, like a diamond cutter, knowing what to say and when, to bring realization to Chao Chou.

If we attempt to approach the Tao (or any transcendent truth) with the intellect, failing to realize that the intellect is the very thing that obscures it (the message must penetrate beyond to the inner witness of Youness)

If words were the final carrier of truth, Gutenberg's press would have rendered every monastery, meditation hall, and act of contemplation obsolete centuries ago.

Zen master Bankei realized this revolutionary truth by stripping away all ritual to proclaim the single teaching of the "Unborn mind." He found that to attain this perspective requires *no effort*—not the absence of trying, but a withdrawal of the effort to *do* something; a receding by subtraction to the thing-in-itself. Bankei nearly died attempting to "approach" it, only finding the Unborn Buddha-Mind when he finally relinquished the struggle to find it.

Bankei

17th century Zen Master Bankei Yotaku revolutionized Zen by stripping away what could be seen as the rituals and superstitions of Zen at the time to proclaim the single teaching of the "Unborn mind" to the common man, and all who wished to hear. In his search he nearly died by asceticism and endless meditation, refusing to desist until he had become enlightened. Near death he found the unborn mind, the emptiness within that the Buddha called *Anatta*, already functioning within him.

Bankei's realization—that to approach the Absolute is to miss it—is the final nail in the materialist coffin. The materialist paradigm attempts to "approach" consciousness by looking at the machine, but as Bankei discovered, the perspective required for truth is not found in the doing, but in the receding to the thing-in-itself.

Bankei affirms the key to awakening:

> "What I called the 'Unborn' is the Buddha-mind. This Buddha-mind is unborn with a marvelous virtue of illuminative wisdom. In the unborn, all things fall right into place and remain in perfect harmony. When everything you do is done according to the Unborn, the eye that sees others as they are opens in you, and you know in your own mind that everyone you see is a living Buddha." –Bankei[59]

Summary, Special Problem #4

Consciousness and Youness are complete enigmas to the materialist paradigm.

Consciousness cannot be explained, nor can it be understood how it arises—or how it can exist.

Ancient understandings have recognized these phenomena within each individual as the culmination of universal intent: *Brahman* and *Atman*

(Hinduism), *Anatta* and *Sunyata* (Buddhism), No Mind or the Unborn Mind (Zen and Daoist insight), and the *Vishvarupa*, the universal form in which consciousness encompasses all.

This is what David Chalmers calls, "the hard problem." and why Thomas Nagel points to the irreducibility of "what it is like."

It all comes down to three realities:

The knower: the human being

The known: the universe and all it produces

KNOWING: the pivot, the golden eruption.

The entire effort of humanity emerges in relation to *knowing*. And as Descartes says, we may not know anything, but we know *something*, and because we know the thought of something, we know that *we are* (I think, therefore I am). From this our entire experience as living beings within a world at large can be validated, by ourselves, by this knowing.

The chance of immaterial consciousness and Youness existing within our universe as a physical emergence, based on probability and first-order physics is *zero*. It cannot be explained.

Value of Special Problem #4: Consciousness and Youness:

M (miracle-impossible)

Our investigation continues with another closely related matter as we drill down into the granular nuances of consciousness.

The consciousness of biology itself operates no less powerfully and with no less perfection than human consciousness. It is a level of awareness far more capacious than the relatively simple contemplations of an individual human. In its minuteness, and its ability to carry out the most precise, exacting and necessary functions in life, with rapidity and in unimaginable prolificity, we can rightfully frame cellular consciousness as being far more pervasive and unresolvable than human consciousness.

Cellular consciousness, which had to have been present from the first living organism, as the very thing that defines life itself, is an open secret among those that study molecular biology, yet it is rarely discussed openly in scientific papers. This dimension of life up until now has been allowed to be subsumed within the mechanistic actions, pathways, and processes endemic to life itself. But this is untenable.

Our investigation continues by drilling down into this granularity. We must understand both biological function and the necessary intelligence—the animating force—behind the consciousness of the cell.

[1] (D. Papineau, H. Selina, Introducing Consciousness, Pg. 6, 2010, Totem Books, Ltd)
[2] (D. Papineau, H. Selina, Introducing Consciousness, Pg. 6, 2010, Totem Books Ltd)
[3] (Augustine, Confessions, Book X)
[4] (E. Husserl, General Introduction to Pure Phenomenology, 2012, Routledge)
[5] (E. Husserl, General Introduction to Pure Phenomenology, 2012, Routledge)
[6] (S. Dehaene, Consciousness and the Brain, 2014, Viking Press), (D.J. Chalmers, The Conscious Mind: In Search of a Fundamental Theory, 1996, Oxford University Press)
[7] (C. O'Callaghan, Binding and the Unity of Consciousness, 2014, Oxford University Press)
[8] (D.J. Chalmers, The Conscious Mind: In Search of a Fundamental Theory, 1996, Oxford University Press)
[9] (G. Tononi, Integrated Information Theory (IIT), 2004, BMC Neuroscience, 5:42. 2004), (D.J. Chalmers, The Conscious Mind: In Search of a Fundamental Theory, 1996, Oxford University Press)
[10] (D. Dennett, Consciousness Explained, 1991, Little, Brown and Co.)
[11] (H. Chadwick, trans., Augustine of Hippo, Confessions, Book X, 1991, Oxford University Press)
[12] (E. Husserl, General Introduction to Pure Phenomenology, 2012, Routledge)
[13] (E. Husserl, General Introduction to Pure Phenomenology, 2012, Routledge)
[14] (P. Teilhard de Chardin, The Phenomenon of Man, 1959, Harper Perennial Modern Thought)
[15] (P. Teilhard de Chardin, The Phenomenon of Man, 1959, p. 63)
[16] (R. Descartes, Discourse on the Method of Rightly Conducting One's Reason and of Seeking Truth in the Sciences The Meditations Part II, 2nd Meditation),
[17] (R. Descartes, Discourse on the Method of Rightly Conducting One's Reason and of Seeking Truth in the Sciences The Meditations Part II, 2nd Meditation)
[18] (Rene Descartes, 1596-1650, Principles of Human Philosophy, Part 1, Article 7 Translated by John Veitch, LLD.D, The Classical Library, 2002)
[19] (Rene Descartes, 1596-1650, Principles of Philosophy, Part 1, Article 9, Translated by John Veitch, LL. D, The Classical Library, 2002)4
[20] (ibid, Principles of Philosophy),
[21] (Ibid, Descartes, Principle IX)
[22] (D.T. Suzuki, Ed R.M. Jaffe, Selected Words of D.T. Suzuki, Volume I: Zen, Pg. 193, 2015, University of California Press, Oakland, CA)
[23] (J. Westphal, Descartes and the Discovery of the Mind-Body Problem, The MIT Press Reader, adapted from J. Westphal's The Mind-Body Problem, 2016, The MIT Press)
[24] (D. Chalmers, Facing up to Consciousness, 1995, p. 1)
[25] (H. Haroutioun, C. Montemayor, Attention: Finding Perspective on the First-Person Perspective, 2021)
[26] (YouTube, Mercatus Center: Steven Pinker on the Mind and Consciousness | Conversations with Tyler, Nov 3, 2016)
[27] (Youtube: The Institute of Art and Ideas, What is Consciousness, Sean Carroll.Dec 2, 2018)
[28] (YouTube, Pioneer Works, Brian Greene on Consciousness, Apr 23, 2020)
[29] (Youtube, Pangburn, The Physics of Consciousness, Richard Dawkins and Brian Greene, May 31, 2023)
[30] (Youtube, Big Think, Sam Harris: The Self is an Illusion, September 16, 2014)
[31] (YouTube: Closer to the Truth Is consciousness an illusion? Episode 1002, Jul 15, 2020)

[32] (D. Chalmers, Facing Up to the Problem of Consciousness, 1995)
[33] (Ibid, Facing up to the Problem of Consciousness),
[34] (D. Chalmers, Panpsychism and Panprotopsychism, 2013)
[35] (D. Dennett, Consciousness Explained, Pg. 8, (google books) 2017, Little, Brown and Company)
[36] (Philosophy Overdose on YouTube, May 22, 2023, The Illusion of the Cartesian Theater, Daniel Dennett)
[37] (Ibid, The Upanishads, Pg. 60)
[38] T. Nagel, What Is It Like to Be a Bat? The Philosophical Review, Vol. 83, No. 4 (Oct. 1974), pp. 439, Duke University Press on behalf of Philosophical Review)
[39] (T. Nagel, What Is It Like to Be a Bat? The Philosophical Review, Vol. 83, No. 4 (Oct. 1974), pp. 439, Duke University Press on behalf of Philosophical Review)
[40] D. Hoffman, The Interface Theory of Perception, 2015, Bulletin & Review, Vol. 22, No. 6, pp.1480-1506), (D. Hoffman, The Case Against Reality: Why Evolution Hid the Truth from Our Eyes, 2019, W.W. Norton & Company).
[41] (L.F. Barrett, Big Think, The Well, How Your Brain Creates Reality, Explained By A Neuroscientist: Your "Social Reality" isn't An Absolute Reality. A Leading Neuroscientist Explains Why. bigthink.com/ the-well/ what-is-reality/)
[42] (M. Lenharo, Decades-Long Bet on Consciousness Ends—And It's Philosopher 1, Neuroscientist 0, Nature, 2023).
[43] (O. Ferrante, U. Gorska-Klimowska, S. Henin, R. Hirschhorn, A. Khalaf, A. Lepauvre, L. Liu, et al. (2025). Adversarial Testing of Global Neuronal Workspace and Integrated Information Theories of Consciousness. Nature, 642(8066). doi:10.1038/s41586-025-08888-1), (L. Melloni, L. Murdrik, M. Pitts, C. Koch, et al., Marking the Hard Problem of Consciousness Easier, Science, May 28, 2021, Vol. 372, Issue 6545, pp. 911–912)
[44] (A.J. Ayer, Language, Truth and Logic, 1936, Victor Gollancz, Ltd).
[45] (D.T. Suzuki, Essays in Zen Buddhism, 1949, Rider and Company). (D.T. Suzuki, Essays in Zen Buddhism, 1949, Rider and Company).
[46] (P. Olivelle, trans, The Early Upanishads: Annotated Text and Translation, 1998, Oxford University Press).
[47] (S.W. Jamison, J.P. Brereton, trans, The Ṛigveda (6.47.18), 2014, New York: Oxford University Press).
[48] (P.D. Shastri, The Doctrine of Maya in the Philosophy of the Vedānta, Pgs 47-49, 1911, Luzac and Co. London)
[49] (Ibid, The Upanishads, Pg. 15)
[50] (Ibid, The Upanishads, Pg. 14)
[51] (Ibid, The Upanishads, Pg. 74)
[52] (Osho, The Grass Grows by Itself, Pg. 22, Zen Talks given from 2/21/75 to 2/28/75, English Discourse Series, 8 Chapters, 1976)
[53] (L. Surya Das, Awakening the Buddha Within, 1997, Pg. 82, Broadway Books)
[54] (L. Surya Das, Awakening the Buddha Within, 1997, Pg. 121, Broadway Books).

[55] (Ibid, The Upanishads, Pg. 60)

[56] (W. Paramânanda, The Upanishads, Volume 1, Third Edition, 1919, Pg. 88, from the original Sanskrit text, translated and commented upon by W. Paramânanda, The Vedanta Centre, Boston, Mass,)
[57] (Osho, Zen: The Special Transmission, Talks given from 01/07/80 am to 10/07/80 am English Discourse series, Rajneesh Foundation Intl, 1984)47
[58] (Osho, Ibid, Pg. 2)48
[59] (Norman Waddell, Unborn: The Life and Teachings of Zen Master Bankei, 1622-1693) (N Waddell, Unborn: The Life and Teachings of Zen Master Bankei, 1622-1693, Pg. 59, 2001, North Point Press)

Variable 15: Cell Consciousness

The foundation of life is the living cell.

At some point in Earth's history, the first cell had to emerge.

If we inhabited a world governed strictly by first principles of physics and the laws of chemistry, this transition would be a matter of cause, effect, and sheer necessity. But this is the problem. The requirements of this transition are so great it would guarantee that no life whatsoever would arise (see chapter on 'abiogenesis).

But we do not live in a world dictated solely by physical law. We live in a reality authored by Consciousness, which established physical law and matter as the precursor to life. Under this framework, it is not necessary for life to stumble blindly, tumbling through coincidence and probability to achieve the literal impossible. Life was programmed from the start. This reality is why material biology cannot reconcile any of the things that cellular life can achieve.

Life is far more than the structural functions of biology. It also represents an inexplicable animating agency having crossed the Rubicon from the non-living to the living. And a host of informational content that is not information alone, but information in action, including DNA, RNA, macromolecules, metabolic pathways, body plans, growth regulatory networks, the sugar code, and sexual reproduction among several dozen highly regulated networks that require information, action and consciousness to execute.

At present, cellular consciousness manages all of the above aspects. Yet, to initially instantiate them, some form of *creation nexus* is required—a concentrated event that integrated these elements within the microscopic confines of a cell. Like the other 'Special Problems' in our Algorithm, the method remains unknown, we only know that it was done and that cellular consciousness orders the operation and continuation of all life.

Under The Microscope & Through The Telescope

Under the microscope, the cell reveals itself not as a passive collection of chemicals, but as a system of intricate coordination. DNA is transcribed into RNA, proteins are assembled and folded into precise functional structures, energy is harvested through mitochondria, and signaling networks regulate the internal state. These processes occur continuously and simultaneously—an integrated system where countless molecular events must unfold in exact sequence, location, and proportion.

To describe this level of organization, we introduce the concept of *cellular consciousness.*

This does not imply reflective thought as humans enact with their consciousness and Youness; rather, it refers to a fundamental biological awareness and the capacity of living systems to coordinate and preserve the integrity of the organism. This requires more than mere *bioenergetics* (energy to drive cellular operations); it requires an animating force that is intelligent, intentional, and purposeful in its action. A *category* of movement that speaks to an untold number of levels in advance of mere 'movement' itself.

Within this framework, cellular consciousness is the organizing illumination through which life maintains its structure, ensuring molecular operations proceed in harmony rather than dissolving into chemical disorder.

The work of cellular consciousness is such that it can hardly be attributed to macromolecules that are in some cases several atoms in width and length. What is required is a kind of broadcast network that animates the entire network, and all living networks. Once again, how this takes place is unknown, but, we do know the informational capacity of a macromolecules is not enough to be able to navigate the cell, and the actions required independently.

Seen in this light, the cell is the primary arena where consciousness first manifests as 'life' within the physical world. From the first single-celled organism, increasingly complex forms of awareness have emerged, culminating in the reflective self-consciousness of human beings. Yet, the move from cellular to human consciousness is not a mere transition—it is a new *instantiation*. While related, they represent entirely different dimensions of consciousness-reality.

This truth reveals the deepest mystery known to the molecular investigator: the cosmos we have been studying from the outside is also encountered, directly, from within. The invisible principles we call Dark Matter and Dark Energy—the vast, unobserved scaffolding of the universe—are not merely structural remnants of an uncertain Big Bang. Instead, they are the macro-scale expressions of the same consciousness that orders the cell. In this view, what physics identifies as "dark" forces, our ontology recognizes as the foundational constraints first instantiated during the era of the CMB—the very ordering principles that allow a universe to exist and, eventually, to observe itself. One way or another, these dark influences are the product of the pre-existing order instantiated into existence, in the first instance, by consciousness.

The Hit

Cells are possessed of *consciousness*; a fundamental force historically called *the life force* that imbues cells with the cellular coordinating activity that creates and maintains the life of the cell.

This is not speculation, or a belief, this conclusion is forced by the evidence, which is substantial and has been hypothesized by many luminaries in the field of molecular biology as an observational necessity.

Cellular consciousness facilitates the expression of the information encoded within the DNA, RNA, and the epigenome, by executing the instructions they contain. By every conceivable measure, DNA is entirely inert; it possesses no inherent power to animate or execute. A fundamental paradox arises: the very macromolecules DNA encodes are the same ones it relies upon to act. If DNA is the library, and the cell is its representation, the macromolecules serve as the librarians, construction crews, and operators—yet there is no discernible material structure within these molecules that explains their ability to perform such complex, coordinated roles.

Life therefore depends on three irreducible enigmas:

Cellular Consciousness: The coordinator of cellular operations.

Individual Consciousness: (As discussed in Special Problem #4, which extends to all living things in some form, not just humans).

Informational Content: The complex digital symbolism of the nucleic acids.

None of these three can be derived, deduced, or 'naturally' organized by the physics and chemistry attributed to our universe. This omission is not a matter of missing data; it is an ontological boundary. As previously documented, this explains why humanity still lacks a functional, agreed-upon definition of life. If cellular consciousness and the source of genetic information had been resolved through materialism, would we not have a robust definition by now? The answer is a definitive 'Yes.' This glaring omission in modern biology serves as negative confirmation: these three pillars remain unresolved precisely because they are orphaned from a materialist origin. The answer is they do not arise from the mechanical emergence of an unknowing universe; they are the principles that instantiate it.

The mechanical investigator's search for a physical cause is a category error. We do not need to explain the 'how' of cellular coordination any more than we need to explain the 'how' of galactic rotation; both are the automated operations of an unknown thinker. And the necessity of both is not speculation, or conjecture, *we witness them.*

In comparison to the vast, teeming architecture of the cosmos, where galaxies outnumber the grains of sand on our shores, the management of a single cell's metabolic pathways is miniscule. We are afforded the luxury of this total automation so that our consciousness may remain unburdened. We do not ask how a character in a book manages their physical health; we recognize that the Author has automated the character's biology so the story can focus on the character's journey.

This entire structure—what is called *Maya*—is the stable stage designed for the education and journey of consciousness and Youness. We are the characters granted the freedom to live, specifically because we are the beneficiaries of Supreme grace in having the impossible task of sustaining the life that contains us maintained from above. And its silent agent is cellular consciousness.

No Definition

In the college textbook, *Campbell's Biology*, there are 37 entries under *life*, but there is no definition for life nor is there an entry for *life* in the glossary. The closest this text comes to providing a definition for life is on page 2 of the first Chapter, where it states, "What is life? Even a child realizes that a dog or a plant is alive, while a rock or a car is not. Yet the phenomenon we call life defies a simple, one-sentence definition[1]."

While this statement is provisionally true, it does not accurately relate the magnitude of the current dilemma. At present the definition of life requires understanding, and until this is achieved there can be no definition.

Campbell's biology, a key academic text, is not alone in being devoid of a definition for life. There is currently no definition for life that is scientifically supported, agreed upon, and represented in text in the world of science.

Another collegiate textbook, *Molecular Biology of the Cell*, another primary text, also has no entries for *life*, but in Chapter 1, the book begins, "The surface of our planet is populated by living things–curious, intricately organized chemical factories… Yet our ancestors, knowing nothing of cells or DNA, saw that all these things have something in common. They call that something 'life', marveled at it, struggled to define it, and despaired of explaining what it was or how it worked in terms that relate to nonliving matter[2]."

There is nothing that could be called a definition of life offered in this book.

The *Stanford Encyclopedia of Philosophy online* has an entry for *Life* that comes to the following point; "It is difficult to non-controversially define life… the nature of life continues to be a hotly debated topic[3]."

Erwin Schrödinger, who we discussed in an earlier section, is a Nobel laureate for his founding of the *Schrödinger wave function*, he is also the author of the 1944 classic, *What Is Life?*[4] In this book Schrödinger is unable to answer the definition of life, instead he states, "Life seems to be orderly and lawful behaviour of matter, not based exclusively on its tendency to go over from order to disorder, but based partly on existing order that is kept up[5]."

What Schrödinger means is that unlike all other systems and arrangements known to man, *the second law of thermodynamics* appears to have no relevance when it comes to this thing we call *life*. The *second law*, which is the inexorable tendency towards entropy, i.e., disorder, is notably absent, or somehow removed from the equation of life. Life, in its robust nature, is the exact opposite of betrothed to entropy. It is self-healing, reparative, resilient, and engages in multiple systems to fend off threats and enact countermeasures and proactive mechanisms to maintain itself.

There are known instances where systems exist that are able to maintain small pockets of high order within larger systems where the second law of thermodynamics holds. The primary reason given for this achievement is that the energy of the sun is able to provide the requisite power to reverse this law within small special pockets of expected entropy, which means that for the overall system in wider context, entropy is not violated. But to attribute this known phenomenon to the cellular consciousness present in life would represent what is called a categorical fallacy of misattribution.

The power of our sun, and of the stars, has been known to our universe from the beginning, and has had no power to reverse this law, except under one observed condition, and that is where its energy serves as the first input in the chain of life. From there, it is no longer relevant as the cause of life's resistance to entropy. In other words, the only possible explanation is circular. It must invoke the very thing it seeks to explain as the cause. So it is not "the sun" that reverses entropy, it is *life* in the observation of the biology of Earth that alone has 'figured out', in a teleological sense, how to take the energy of the sun and reverse entropy. In terms of life, the sun is just the power outlet, and life is the machine that came with the knowledge of how to plug itself in. The sun gets no credit for creating life any more than the power company can claim credit for the intellectual innovation of silicon valley.

Schrödinger Redux

Schrödinger, after making his initial statement on life, asks, "What is the characteristic feature of life?" he then answers, "When it (Life) goes on 'doing something', moving, exchanging material with its environment, and so forth, and that for a much longer period than we would expect an inanimate piece of matter to 'keep going' under similar circumstances[6]." This is an observation, not a definition. And to make note that matter 'keeps going' longer than expected is a profound reductionism. It mistakes an ultra-marathon for a defiant slow-motion leakage. We are not talking about minutes, or hours, but in some cases several thousands of years. Notably, what Schrödinger left out is the show—the mover evident between the lines:

Life is not just animation; it is information-based, highly specified, coordinated, collaborative, and intentional. It is the observation of a specified agency—a billion nearly error-free actions undertaken for the single unified goal of the maintenance of what we observe to be life."

Life is the observation of the necessary *movements* by *consciousness*—meaning, not just animation, but information based, highly specified, coordinated, collaborative, timely, intentional, and nearly error free actions, processes and "movements", all undertaken for the single unified goal of the continuance, and maintenance of what we observe to be life, devoid of any discernible possible material attribution.

That is the working definition when the facts are laid on the table, and the gaps are analyzed. After all the bets and bluffing, when all the cards have been drawn and the biologists are forced to reveal their hands, we finally get to evaluate what we see when we turn our textbooks to the definition of life. The world of modern science holds no winning cards. Instead, they hold cards with no suit, and no rank.

In the entire history of humankind, there is only one thing that meets the definition for what *life* is observed to do, and that is consciousness, embedded within living organisms. This consciousness has previously only been known to be represented in human consciousness, but the functional requirements of the cell demand its presence as the primary substrate of biology. It is the internal coordinator that physics cannot account for but observed and undeniable. We see it in action serving as the very definition of life.

Only consciousness has the capacity to guide, direct and coordinate movement. It is not physical law (physics, chemistry), nor random motion causing molecules to collide together; nor can it be "DNA" that guides these movements. And although the word *consciousness* is left unspoken by Erwin Schrödinger, consciousness is evident, necessary, and irreducibly required for life. "He that will remain unnamed" is known by all just the same.

The Brownian Ratchet

If we rely solely on physical laws and the inert library of DNA to explain cellular activity, the mathematical foundation of biology collapses. Without a timely, coordinating consciousness, cellular movements would be governed strictly by Brownian motion—the random, disordered, and entropic bombardment of molecules[7].

The dominant framework attempts to escape this chaos by invoking a top order add-on to Brownian Motion known as the 'Brownian Ratchet[8].' This is the remarkable claim that random thermal noise can be "rectified" into highly specified, coordinated navigation simply by "the asymmetrical shape of a protein". They invoke molecular motors, such as kinesin, 'walking' along a cytoskeleton as a mechanical inevitability of biased diffusion[9]. Yet, when they see how these proteins move, they encounter a contradiction: the cell is a chaotic, hurricane of random collisions, yet, defiantly, and inexplicably, the proteins within it move with the precise, rhythmic march of a soldier under command. So, what did they really describe? Disordered shaking and where the shaking ends, but what is missing is the very thing that brings coherence to the "ratchet". They have no identified driver to the destination. We have nothing. Nothing has been explained. Just more stuff, more observation but once again, the power is missed[10].

Target Acquisition

To demonstrate the necessity of cellular consciousness, we need only look at a single, fundamental challenge: the *Target Acquisition Problem.* Consider the task of a ribonucleoprotein we investigated in our chapter on DNA. It must locate a specific gene sequence—a needle in a haystack—hidden within a two-meter-long strand of tightly coiled, packaged DNA. It must find this specific gene exactly when it is needed, edit it, splice it, and ferry it to a manufacturing site where thousands of mRNAs and tRNAs converge. It must then oversee the rapid assembly of proteins and enzymes, folding them into complex tertiary shapes, and delivering them to precise locations within the cell. There is no purely physical explanation for this choreography. No chemical affinity, vibration, or natural law has the inherent capacity to decide which gene is needed, locate it among billions of base pairs, and coordinate its production with thousands of other independent molecules. The physical links explain the mechanics, but they do not explain the management.

To rely on materialism in this role would be akin to imagining a swarm of autonomous drones with no sensors, no software, and no communication links—drones that cannot store or process instructions—yet they collectively navigate a complex cityscape, picking up and delivering thousands of packages, to precise addresses, while avoiding collisions, and coordinating with one another with perfect timing. These drones deliver up to 3,000 packages per minute with an error rate of less than one wrong delivery for every 100 million packages delivered, yet here is the twist: no addresses are provided, and no instructions are given about what is needed, what is in the packages or when the packages are needed. The result? each drone picks up and delivers the exact package needed, when, where and to whom it is required flawlessly. This is the impossible challenge materialism attempts to explain. It is a fool's errand. The paradigm contains no puzzle solving tool to stand up to the challenge.

This is the reality of DNA replication. It operates with a fidelity that defies the randomness of chemistry, boasting error rates as low as one per *100 million letters*[11]. To attribute this speed, precision, and intent to "blind physics" is not science; it is a leap of faith greater than any religion, yet this leap is taken by all who throw their lot in the possibility of a solution from such insufficiency. This is their defense against cellular consciousness. This is their alternative to what they cannot identify.

This example is key. Because it cuts to the heart of why no definition of life is available. To have a definition, we must understand what is happening physically. And the only way this is possible is if researchers are able to attribute a specific part of each macromolecule—whether an amino acid side chain, an enzyme domain, a "structural motif"—with the capacity to hold a mapping of large portions of the cell, or to receive and transmit signals that precisely localize it within the cell. But this is not all. What must also be explained in our above example: how these macromolecules are guided to identify specific gene sequences, edit and splice them, transport them, catalyze their manufacture, fold the resulting proteins, and deliver products exactly where and when they are needed.

Experiments reveal that DNA-binding proteins, such as transcription factors, locate specific sequences among billions of base pairs at astonishing speed and with remarkable fidelity[12], with no apparent explanation whatsoever. These are observed facts: the proteins do it, the rates and accuracy are measured, and the molecules perform the functions required. Beyond this, there is no attribution whatsoever: no physical explanation exists for how these molecules could possibly encode or act on the informational maps to guide their movement is another absolute unknown[13].

An important point of clarification. Within scientific literature on the target search problem, you will find statements that give the impression that the issue has been partially solved or that meaningful progress has been made. In A. B. Kolomeisky's, *Physics of Protein–DNA Interactions:*

Mechanisms of Facilitated Target Search (2011), the problem is described as being, "not fully resolved[14]." In reality, what has been documented within the literature are local chemical interactions—binding and catalysis—which amounts to known chemistry, which is a category removed from how molecules are directed, coordinated, or timed across the cell as a unified system.

For another example, in the context of DNA replication fidelity, we find statements such as: "Although significant progress in understanding protein search and recognition of targets on DNA has been achieved, detailed mechanisms of these processes are still strongly debated[15]." implying that meaningful progress has been achieved towards the salient issues at hand, and that several contending hypotheses for the target search problem are being debated; when in actuality nothing salient regarding the specific problem of target search has been established, besides the standard insufficient chemistry resulting from the beginning and end of the process (not the search itself).

And more recently, in S. Park et al's 2021 paper, *A Mini-Review of the Diffusion Dynamics of DNA-Binding Proteins,* we find the statement: "… the quantitative understanding of the whole target-search process remains elusive[16]." In this context, the use of the word, "whole" as a stretch, is perhaps technically correct, but that it implies that almost "whole" or even partial understanding has been achieved, is definitively not the case. Nothing of the salient issue has been solved.

Existing work fails to explain:

Mechanisms for global coordination across the cell

How processes are selected, prioritized, and timed in real time

Structures mapping or directing system-wide activity

How thousands of molecular events integrate as a coherent operation

Erwin Schrödinger, without the aid of our modern high-powered instruments, reached the same foundational conclusion with regards to the target search problem, although the term did not exist 80 years ago. Given the limitations of biology within the materialist paradigm—physics and chemistry alone—and what was known about the required operations of the cell in the early 1940s, Schrödinger knew there was no first-principles, physics-based explanation for what was observed. Schrödinger grappled with the fact that life appeared within biology, throughout biology, and as the activity of biology, yet was never identifiable within the matter of biology itself. He wrote, "We must therefore not be discouraged by the difficulty of interpreting life by the ordinary laws of physics… We must be prepared to find a new type of physical law prevailing in it. Or are we to term it a non-physical, not to say a super-physical, law?[17]" By elimination, no cause grounded in the physical structure of the cell remained a viable candidate, leading him to posit the possibility of a "super-physical law." This missing principle is what we identify as cellular consciousness.

Before the structure of DNA was known it was imagined that there was something about the genome that could speak to the animation of the cell. With the eventual understanding that DNA was inert, no part of the cell could be attributed as the source of this animating power.

Life is evident throughout biology yet cannot be located within matter itself—leading Schrodinger to posit a principle beyond conventional physics even without knowing what was in the nucleus, and without qualifying his remarks that the eventual solution would be found in Brownian motion or any other non-specific means. Schrödinger's proposed missing principle aligns with what we identify as cellular consciousness.

Do Cell Signaling Networks Explain 'Cellular Consciousness?'

Are cell signaling networks responsible for what appears to be cell consciousness?

The answer is no. Cell signaling networks are required in biology for single cell life to become multi-celled complex life (sexual reproducing life). Cell signaling networks come into play because complex life must somehow be called upon to perform together with a singleness of purpose and coordination that is not possible in the level of life below. Cell signaling networks govern processes where the cell interacts with itself, other cells, and the environment in specific ways These networks are largely made of three components: signals, receptors, and effectors[18].

Cell signaling networks don't have anything to do with how proteins and enzymes interact with DNA, nor do they control the step-by-step logic of the sequence hypothesis—the precise 'typing' of the genetic code into functional proteins—nor, as a category, the discrete intracellular functions within cells that make up complex life. These actions—DNA transcription, translation, protein synthesis, and protein folding—are not signaled. They can't be. They tell the cell what to do, not how. In detail, proteins and enzymes have no independent signal receptors with which to be coordinated from the soup of genome targeting, to the nuts of protein/enzyme delivery. Amino acids, even if they could be signaled to begin the process of folding, ultimately produce a protein that is folded entirely and uniquely based on its specific amino acid sequence. No external signal can substitute for this internal, sequence-driven logic[19]. Signaling, the kind we are talking about, relies on the 'shaking' of diffusion, a process far too slow to coordinate DNA replication, which happens at a staggering 1,000 nucleotides per second. Finally, signaling networks are themselves composed of complex proteins that require a pre-existing navigator to build them. It is a circular category error to use a signaling network to explain the origin of the very protein-based machinery required to transmit and receive those signals[20]. The signal is permission; the Intelligence to do what is required would have to be within the doer, not the messenger.

DNA's Ghost in the Machine & Base Pairing Enigma

DNA is an enigma on multiple fronts with three of the primary enigmas being: (1) the organization of its information into a four-letter digital, complementary code, (2) the level of coordination of its transcription and translation machinery, and (3) the reliance upon a high-speed, nearly flawless, precision-guided system of specialized macromolecules.

As is true for all of our detailed descriptions, what is important to our point will be summarized at the end.

To begin with, the DNA double helix is not simply a pair of identical copies. It is composed of two *complementary* strands of digital information. The four-letters of the code are *A,C,G,* and *T*. Pairing rules dictate that *A* only pairs with *T*, and *C* only pairs with *G*, held together with hydrogen bonds. This results in no letter ever pairing with itself, and each letter pairing with only 1 other letter. This is how the code on one strand is complementary to the code on the opposite strand.

DNA is arranged in an antiparallel orientation. This means the two strands run in opposite directions: the 5′ head of one strand is aligned with the 3′ tail of the other. The physical layout is flipped, such that the two strands are directional like two lanes of a highway where the traffic flows in opposite directions. This allows the 'digital' rule for how the information is read to remain absolute: both strands must be read and built from the 5′ end to the 3′ end.

Another detail is that DNA is not joined on the ends, it is joined in the middle, once again, letter by letter, according to the pairing rules. This means when it is read, or copied, the two strands must be 'unzipped' in the middle.

Another complication to how the two strands are read is because the strands are side-by-side but facing opposite ways, as the DNA "zipper" opens on the *Leading strand*, the machinery moves move forward into the opening, reading the strand as it goes, while on the Lagging strand, the machinery is oriented in reverse. To follow the 5' to 3' rule, it has to wait for the zipper to open, jump backward, and build in small, discrete segments—

called *Okazaki fragments*, section by section, seamlessly by the efficiency of the machinery.

The genes that are copied are called *Sense* genes. The *Sense* gene is the functional "sentence" or blueprint. It is the specific sequence of letters that, when read in the correct 5′ to 3′ direction, provides the instructions to build a protein. One of the counter-intuitive parts about the DNA double-helix is that although both strands carry the same complementary information, only one of the two strands holds the *Sense* sequence for a specific gene; the other is the *Antisense*. This is where another counter-intuitive wizards' trick takes place. Because the *Sense* gene is in the correct letter order, the RNA polymerase that reads the gene can't read from the *Sense* gene, it has to read from the *Antisense* gene. By reading the *Antisense* 'mirror', the pairing rules ensure the resulting RNA is a perfect digital copy of the original *Sense* blueprint. There is only one small biochemical substitution quirk left to discuss, because the *Antisense* gene is used to make the *Sense* gene, but it is being translated into RNA, RNA uses a different letter from DNA. It substitutes *U* (Uracil) for the *T* (Thymine) in DNA's *A,C,G,* and *T* code.

How we've just described DNA, in granular detail, is not how DNA is usually explained. It is usually said to be a double helix composed of two copies of the same genetic information. This gives the false impression that each strand is an exact copy of the other. Even when the pairing rules are revealed, it often remains unstated that the strands are not identical. But then how the strands are joined, by the pairing rules, defies intuition. How could two exact copies be joined to each other, in opposite directions, and yet no letter in the 3 billion letters of the code would ever pair with itself, and in addition the letter *A* only pairs with *T* and the letter *C* only pairs with *G*? As we have revealed, the answer to this seeming impossibility is that the strands are not identical replicas. They are complementary, mirror-image reflections. But in classrooms, and general texts, the only students that would ever find this out would be the ones that asked the teacher after class because they were left perplexed with the counterintuitive nature. The structure, and therefore much of the majesty of the double helix is condensed to emphasize the point of "two copies". But by truncating the message of life itself, an opportunity is missed to provide the germ of realization that something is amiss and that all this apparent strangeness may hold a message that can resolve confusion into revelation.

Now we've arrived at the point in our explanation where we can get to the ghost in the machine that cannot be reconciled by any material explanation, the *Target Search Problem* redux.

The RNA polymerase, and its search party of *Transcription Factors*[21] are able to find and copy genes by somehow knowing exactly what gene it needs, and how to get to it—a virtual needle in a haystack—with knowledge of its 4 letter Antisense code, the orientation it is in on its given strand, and how the entire search party can travel to it. And since the gene is only one strand of the two, the search theoretically involves 6 billion individual bases, not the 3 billion it would be if DNA was two exact copies as often advertised. The needle in the haystack the RNA polymerase is searching for would require 4,000 bales of hay, and a search that should take years or perhaps a lifetime is somehow inexplicably reduced to mere seconds.

How this all happens cannot be represented. Scientists can identify the genes and document their location, what they do, their protein make up, and their role in the cell. They keep extensive databases on all that takes place as well as the parts and players. But to do this requires immense resources in technology to sequence, research, observe, experiment upon and understand genes.

The recent collapse of the *Junk DNA* myth reveals that the wizard behind protein synthesis isn't just reading the gene it needs, it is conscious of its entirety, and navigating a vast, functional landscape where every letter plays a role in the choreography. Not that *Junk DNA* answered to anything, because to any observer, and surely to a mindless macromolecule, the entire double helix is a mass of the same 4-letter code in a seemingly disordered arrangement. But somehow these macromolecules, in whatever life form they find themselves in, just know the genome of their host in toto. But this is not all. The transcription factors—some 30 to 50 different ones[22]—coordinate with each other, and RNA polymerase each doing its task, in order, seamlessly, on site, at the speed of a 747 turbofan—some 50 to 100 nucleotides per second—without minds, appendages, and no sensory surfaces to see the code, feel, sense, or synchronize with it yet they recognize it with a precision that defies the biophysics of their constitution and chaos of their molecular environment.

The standard biological narrative relies on *Facilitated Diffusion*[23], a constellation of actions such as shaking, 'hopping' 'jumping', and Brownian motion to explain how the RNA polymerase finds its target. They argue that the protein doesn't just 'float' as in a 3-dimensional search; it 'slides' along the DNA in a 1-dimensional manner, within this hypothesis, making the search far less complex, like a train on a track vs. a plane in the sky. However, this creates a *mechanical paradox*: for a protein to slide fast enough to be useful, its binding to the DNA must be weak, but for it to stay on the track, its binding must be strong[24].

To claim random motion can find the correct gene in the molecular storm the library is situated within is a mathematical absurdity. In a 3-

billion-unit genome built of only four repeating letters, a random search would be hopelessly lost. The target gene can effectively be anywhere, meaning on either strand, forward, backward, and in any location. To imagine the effort required, if we were to blow up the RNA polymerase to the size of a dime, then the 2.1-meter-long genome would be 1,484 miles long—the distance from New York City to Dallas, Texas. Trying to find one target gene would be the equivalent to finding a three-foot section of curb somewhere along this distance, while being hit and cajoled about the cell by thousands of molecules every minute, and no clear "vision" to the target.

But the search is far more difficult than driving down the New York to Dallas highway. The DNA 'road' from New York to Dallas is not empty; it is cluttered with millions of other proteins acting as roadblocks. A sliding polymerase, as proposed, would hit a stop sign, known as a 'roadblock' every few nanometers, forcing it back into the 'hopeless' 3D search.

And there is more. The road, instead of being flat, is coiled six times over into chromatin, hiding gene inside a compressed labyrinth of proteins. This coiling compresses the equivalent 1,500 mile road into a 7-mile high and wide gordian knot. To find the target hopelessly entangled coils must be navigated.

For the human genome *Pioneer Transcription Factors*[25]are the ones that identify the gene site in advance, but even so, in all of this, the primary thing that is required is not accounted for. Something is driving the system, allowing it to find its target in seconds, as if pulled by a biological magnet of sorts, except, no such thing exists[26].

The Genetic Code is Entirely Digital and Arbitrary

One of the reasons DNA can have the rules it has for how each strand is composed, how the four-letters are paired, and in how genes are read, is that the code itself is truly digital. The four-letters of *A*,*C*, *G*, and *T* aren't actually letters, they are bases acting as digits, and they could have been called numbers. Where modern computers use 0's and 1's the genetic codes use numbers 1-4 (with a different molecule for "4" in RNA). This means the code itself is arbitrary up until the assignment of amino acids by the ribosome. But even in this assignment of amino acids to their specific 3-letter combination of letters, there is no chemical reason for this assignment. Chemistry only comes into play once an amino acid chain is folded, that's when its shape and chemical properties become functionally relevant.

In the flow of the Sequence Hypothesis the assignment of amino acids is the final step of Translation.

Up until that point the letters could have been in any order, and any set of 3 letters. What is important is the assignment of amino acids. Such that knowing what amino acid is needed, someone with adequate resources could determine the needed sequence on DNA for all of the proteins and enzymes that have been identified.

Bottom line: the assignment of a specific 3-letter DNA code to a specific amino acid is chemically arbitrary. We reviewed this issue earlier under the topic, 'The Frozen Accident.' The frozen accident refers to the fact that there is no physical or chemical reason why the codon "G-A-G" must equal "Glutamate" any more than any set of three letters must call for any of the amino acids. DNA is the coding convention used throughout the living world, with several notable derivations that use a few different codons (i.e. candida yeast, plant mitochondria, mycoplasma, acetogenic bacteria, etc.) There is no living thing on Earth that does not use this basic code. This similarity of code is one of the main premises cited by evolutionary theory as proof of common descent of all life from the LUCA (last universal common ancestor). But this is in no way proof of a pre-existing chemical affinity for amino acids that was lost to history. An alternate explanation for the use of this foundational code is the use of a singular, optimized linguistic standard for genetic information and life.

The arbitrary nature of the code is an accepted fact in biology:

> "The code by which the sequence of bases in nucleic acids determines the sequence of amino acids in proteins is, essentially, arbitrary ..." — F.H.C. Crick, *The Genetic Code,* 1968[27]
>
> "The genetic code is digital and discrete." —B. Alberts, et al, Molecular *Biology of the Cell* (4th ed.), 2002[28]
>
> "There is no chemical/physical necessity for the current code; it is a convention enforced by the translation apparatus." —F.H.C. Crick, *The Genetic Code. The Mind of a Chemist,* 1968[29]
>
> "There is no intrinsic chemical affinity that determines which amino acid is assigned to which codon." —C.R. Woese, *On the Evolution of the Genetic Code,* 1965[30] (
>
> "The code could have been otherwise in principle." And, "Several alternative genetic codes exist ... for example, in vertebrate mitochondrial genomes... UGA codes for tryptophan instead of stop." —M. Nirenberg, *Deciphering the Genetic Code: Breaking the Code,* 2004[31]

Because the code is arbitrary yet functional, its origin cannot be explained as an inevitable chemical outcome. As we mentioned, there is no chemical necessity requiring any particular amino acid be assigned to any particular 3-letter combination of letters.

Another important note is that contrary to evolutionary theory, the DNA, RNA to amino acid to protein coding convention cannot arise piecemeal or gradually. It requires a fully integrated configuration to function at all.

If we approach the coding convention of first life, or more specifically the first instance of self-replicating life, we find the same conclusion. Whether the origination of this convention is framed as arriving as an "RNA-world" based system, or original DNA "world" based system, the coding convention could only have instantiated as a high-fidelity, fully operational system from the outset. There is no coherent pathway for a gradual emergence of a coding convention such as this, nor for a transition between fundamentally different coding frameworks, as in the proposal that there was an "RNA-world" that somehow transitioned to a "DNA-world." That is a conjecture.

How an arbitrary code became utilized as a genomic system—one that, through transcription and translation, maps to amino acids and yields functional proteins (themselves exceedingly rare among possible sequences)—is likely unexplainable in purely physical terms, because there is no underlying physical necessity driving it.

But there are additional problems. As we reviewed in an earlier section the chance of happening upon a functional protein sequence of amino acids by chance would exhaust the current age of the universe.

Then there is the protein folding problem. To fold a short protein of 100 amino acids would take $\mathbf{10^{27}}$ years, whereas the universe is only 1.38 to $\mathbf{10^{10}}$ years. This is because each of the 100 amino acids has 3 possible folding positions for a total of $\mathbf{3^{100}}$. If the search went through $\mathbf{10^{13}}$ different configurations per second, that's how we arrive at $\mathbf{10^{27}}$ years. But once again, the same thing we find in DNA Target Search, this process defies physics and odds and takes place in 1-100 microseconds (1 million microseconds in a second). The protein folding problem with 100 amino acids, is what Cyrus Levinthal used for his model protein which became known as *Levinthal's* Paradox that we previously investigated[32].

Yet the arbitrariness of the code does help account for DNA's unusual structural features, including its highly specific base-pairing rules, and the existence of a 'negative' copy of complementary code, as well as the freedom of codes to be on either strand, written front to back or back to front. Because the code is not chemically constrained, it permits the implementation of highly non-intuitive organizational principles, so long as the resulting amino acid sequences are correct at the end, and this can be achieved by assignment on both ends—code to amino acid and amino acid to code. Something made this assignment.

It is as if the entire genetic system was established with translation fidelity, high speed, structural elegance, and quality control as primary constraints, and only afterward were specific coding assignments fixed. The speed is not just a show, it is necessary. The demands of the cell are such that if maintenance and life maintaining operations were not at the speed at which they were observed, life would not be possible. The cell produces 2,000 to 10,000 proteins per minute[33]. And it has to be able to respond to environmental changes with protective genes almost immediately, and all of these thousands of proteins need to be coordinated with one another and in correct order[34].

There is a prodigious level of fidelity, precision, and multi-layered symmetry in protein synthesis that is extraordinarily difficult to reconcile under any naturalistic framework. But this difficulty is not shared under a consciousness-based system. Consciousness is a dimension where all of the needed requirements: intelligence, awareness, agency, planning, coordination, are not left unexplained.

DNA Contains No Second-Order Assembly Instructions.

DNA codes for proteins and enzymes, not for the assembly of the cell itself.

Although DNA provides the instructions for the parts required to assemble a cell, it contains no instructions for how to assemble them into an actual cell structure. To use an analogy: if DNA were a stadium blueprint, it would include the specifications for all the parts: girders, cement pours, chairs, electrical wiring, doors, kitchens, bathrooms, phones, scoreboard, etc. But it would not include the instructions for how all these elements need to be assembled into a functioning stadium. This means that the primary organizing structure of the physical body is somehow being constructed with no instructions.

How is this happening? This is another explanatory gap for evolution, where some other system provides the missing information and action.

A New Problem for Evolution & DNA

The materialist model for how cells build themselves is that the instructions are inherent in the parts themselves (self-assembly). DNA doesn't supply the plans because no plans are needed. All the parts possess specific chemical affinities that essentially dictate their own organization.

This is a claim of *emergent complexity*, The idea that a windstorm can pass through a junkyard and a 747 will emerge because the bolts fit the nuts and 'fitness' dictates that it makes more sense to be an airplane than a car, or a dozen washing machines. But this hand waving doesn't work. In the cell we see work being done, we see parts being assembled. ATP is used. Specialized motors like kinesin and dynein 'walk' parts on microtubules to their needed location. These parts aren't randomly bumping and sliding into place.

The more reasonable hypothesis is that the entire process works off the template of an existing cell, guided by the same ghostly intelligence that has done everything else. Membranes and organelles are grown from existing ones. They don't 'appear,' they are duplicated and assembled. This is observed when cells divide. The process harkens back to the idea of epigenetics, embryology and the sugar code where complex information is provided in concert with DNA. This reveals the inexplicable cooperation 'gap' of DNA and several distinct forms of additional cellular information. Something is coordinating these affairs. It is so curious that materialists, who insist upon attribution, say nothing is amiss, while those who are allegedly invoking a 'God of the gaps' are demanding the receipts.

Biochemist Dr. Franklin Harold argues, DNA is a list of parts, the bricks but not the map, pointing out that something has to tell the bricks where to go[35]. He allied this the Coordinate Paradox in his 2005 paper, *Molecules into Cells: Specifying Spatial Architecture*. He found that there is a "hierarchy of physiological processes that bridge the gap" between molecules and the cell. This organization is not in DNA; the cell itself provides the context for expression[36]. Geneticist Dr. Michael Levin has similarly argued for the existence of a *bioelectric navigator*[37], where electrical gradients tell cells where to go and what to build, independent of their genetic instructions. These bioelectric scaffolds serve as a navigational source, telling cells where to go and what to build independent of the parts provided by DNA. His research in *Bioelectricity in Development, Regeneration, and Cancer* illustrates a top-down information system[38].

Materialism largely ignores non-DNA derived assembly information. They invoke emergent feedback loops. Effectively stating that it just happens, because their model works on evolution at the level of DNA. If changes in DNA can't solely address speciation, and all other things, and multiple other forms of discrete forms of information are demanded, the prospect of naturalism in the evolutionary process is fundamentally broken. But still we demand, what coordinates this? We are offered a GPS (Bioelectric Fields) and a highway (Microtubules), but no Driver. Compasses and shipping lanes don't sail ships. There must be a captain. We have identified the voltage switch and chemical tag but have failed to even recognize the transduction gap of how mindless proteins—possessing no signal reception devices translate silent electrical gradients into precision guided journeys. This missing device is no different from the host of machinery that zeros in on a specific gene on the 1,500-mile gordian knot of coiled DNA. This is the agency of the system that turns the various forms of information into action[39].

Does DNA Possess Consciousness?

There are those that hypothesize that DNA *is* conscious or that it possesses consciousness[40]. They cite the *interaction* between DNA and the macromolecules that service it as evidence of consciousness. They also deduce that DNA must be conscious in some way because DNA contains the genes that encode for the structure of the human brain, and the human brain, in their understanding, gives rise to human consciousness. This is a hypothesis we reject. Although we concede that the consciousness of the cell extends to DNA, it does not emanate from it.

To preserve the materialist worldview, scientists have scrambled to find a physical explanation for this agency. If cells are communicating and coordinating, there must be a physical signal—a wire, a frequency, or a vibration—that explains it without invoking "mind."

A prime example of this effort is a 2013 paper by Dr. Laurieanne Dent of Pepperdine University, titled *A New Model of Intracellular Communication Based on Coherent, High-Frequency Vibrations in Biomolecules*. The paper's premise was a promising lifeline for materialism: it posited that the mysterious "life force" or coordination within the cell was actually caused by high-frequency vibrations. If true, this would reduce consciousness to simple mechanics.

However, the investigation hit a wall. Despite the attractive logic of the hypothesis, the study could not validate it. The conclusion of the paper—a document intended to support a physical mechanism—ultimately left the mystery unsolved. This 2013 failure is notable: science went looking for the physical strings for the puppet of cellular animism and found only empty space. If the mechanism of life were purely material, presently, there would be no possible hiding place for it.

> "So complicated are subcellular processes, that in recent years computer animators have been employed to help us visualize these sophisticated molecular machines and processes at work. When watching an animation of, say, gene expression, we see biomolecules 'flying around', apparently guided to their targets; protein transcription factors glide to their specific DNA-binding sequences, mRNA transcripts seem to be directed through nuclear pores toward ribosomes for translation, tRNAs land on a ribosome and perfectly align with mRNA to make the anticodon-codon hybrid. What we cannot appreciate from these animations are the blinding speeds at which these processes occur. DNA replication occurs at a rate of 50

nucleotides per second in humans... How, then, do these biomolecules locate each other with such accuracy and rapidity?[41]"

And from the abstract:

"Biochemistry alone is insufficient to explain how widely separated biomolecules locate and move toward one another with accuracy and speed[42]."

Make note: *"we see biomolecules 'flying around', apparently guided to their targets"* How does this occur? The posited hypothesis that somehow "high-frequency vibrations" shake and cajole these biomolecules into such high-speed perfected collisions, at the inexplicable rate of "50 nucleotides per second" when at best unguided vibrations would create accidental collisions, with precious few being the desired kind, is absurd on the face.

If the animating life force and apparent consciousness of cells was reducible to biology, then whatever it could be reduced to would be the definition of life. This same force would also form the core of our understanding of abiogenesis, speciation and human consciousness.

In the paper we quoted above[43], Dr. Laurienne Dent states, "Classically, Brownian motion has been invoked as the mechanism" that guides molecules to "walk" to their needed locations[44] For the knowledgeable, this is a huge red flag. Brownian motion is *random* and caused by random fluctuations. It is not purposeful, nor directed, and therefore has no power to cause molecules that need to be together, to get together, by chance collisions, at the correct three-dimensional orientation, at the precise time they are needed, in the correct order, the millions of times these collisions are required.

The speed of this action is noted in this same paper, thus nullifying the proposal of Brownian motion.

"Forty years ago, it was shown that the targeting of the *lac* repressor (a DNA-binding protein in bacteria) to its DNA-binding site occurred up to 1,000 times faster than the predictions of diffusion and random collision[45]." Instead of providing evidence that Brownian motion is a viable cause, the paper concludes with a confession, "While this study provided no evidence for our hypothesis that living, nucleated cells have a vibration that may originate in the nucleus and cause coherent cytoplasmic motion... I strongly believe the attraction-at-a-distance mechanism is based on a resonance principle[46]."

Evidence for the Consciousness of Cellular Systems

Barbara McClintock, cytogeneticist, stated in her 1983 Nobel acceptance speech (Physiology/Medicine):

> "The responses of genomes to unanticipated challenges are not so precisely programmed. Nevertheless, these are sensed, and the genome responds in a discernible but initially unforeseen manner[47]."

Barbara McClintock also recounts that a cell has the ability to detect *broken ends* of DNA, direct them *towards* one another and unite them so that they are correctly oriented.

> "They (cells) make wise decisions and act upon them… In addition to modifying gene action, these elements can restructure the genome at various levels, from small changes… to gross modifications… such as duplications, deficiencies, inversions, and other more complex reorganizations[48]."

Take note. In a Nobel acceptance speech we hear the words, 'responses,' 'sensed,' 'responds,' 'wise decisions,' and 'act' with regards to the cell. How can proteins do all of this if as Dr. Laurienne Dent hypothesizes, random motion or a resonance principle was the cause? Dr. McClintock is talking about an entirely different class of reactions. If purely physical caused motions were the cause, we would not be hearing about cells doing unforeseen things. These actions would instead be *foreseen* and *anticipated,* and Dr. Laurienne Dent's findings would be entirely different. The problem isn't evidence; it's the fact that the only hypothesis that can speak to our observations defies naturalism. Although Dr. McClintock's lecture does not refute a reductionist hypothesis, it does render it implausible.

There are many other observations of consciousness culled from the world of science, often stated far more directly than the remarks of Dr. Barbara McClintock.

Perhaps the most striking comes from Dr. Lynn Margulis, a National Medal of Science recipient and the revolutionary founder of *Endosymbiotic Theory*. Margulis did not view the cell as a mere biological machine, but as something exhibiting cognitive-like properties. In her seminal 1967 article, *On The Origin of Mitosing Cells,* she writes: "The evolutionary antecedent of the nervous system is 'microbial consciousness'." and that "the components fused via symbiogenesis (of eukaryotic and bacterial cells) are already 'conscious' entities... Even remnants of 'microbial mind' can be inferred from behaviors of thriving microorganisms[49]." For Margulis, consciousness was not a late evolutionary addition, but something rooted at the very beginning of life, consciously and with something best described as *mind* however primitively, at the cellular level, as a foundational spark.

Michael Denton, in *The Miracle of the Cell* discusses several aspects of intelligent cellular behaviors, with dimensions of awareness, purpose, selection, and goal-orientation, particularly in ciliate cells.

Denton points out that ciliates display courtship rituals, pre-conjugal mating 'dances', reciprocal learning, and even 'deceit' and 'cheating' in communicating their reproductive fitness[50]. He also describes the observed intelligence of leukocytes—white blood cells—which are able to sense and 'chase down' bacteria within a blood smear to destroy them[51].

The implications of Margulis's "microbial mind" and Denton's "cellular courtship" expose an entirely new level of organization in biology that resists collapse into bottom-up, mechanistic description. These accounts broach a hardened boundary not unlike the hard problem of consciousness, yet at a far more foundational level—one that takes place prior to brains, neurons or any centralized nervous system.

For over a century, we have been told that consciousness is a late-stage byproduct of neural complexity—a phenomenon that only emerges when the brain becomes sufficiently complex. But when we examine observations from cytology and microbiology, what emerges is not a simple machine-like picture of life, but one in which cells exhibit persistent, adaptive, and highly coordinated behaviors. Whether described as agency, responsiveness, or proto cognition, these patterns suggest the organizing principle of life cannot be reduced to mechanism alone but instead is a precondition.

Across disciplines as diverse as immunology, microbiology, neuroscience, and information theory, researchers have encountered cell behaviors that defy traditional mechanistic explanations. Here we find cells that sense, respond, adapt, anticipate, remember, and even make decisions based upon changing conditions.

Below is a selection of observations drawn directly from the scientific literature. Individually, they are provocative, taken together, they form a pattern that falls in line with the concept of consciousness being foundational.

> Cells can decide when to divide, what to become, where to go, and what to eat.
>
> —Jane Kondrev, Leaving It Up to Chance: How Cells Make Decisions, (2014)[52]

> Science speaks of intelligent molecules that react directly to external stimuli and change their shape.
>
> —Michael Nash, A Single Molecule in Sight, (2013)[53]

All levels of studies on cells imply that individual cells are capable of forming memories despite the lack of neural networks.

> "Clearly brains and nerves are not the last word on memory."
>
> —Jenniver Frazer, Can A Cell Remember? (2021)[54]

> The fact that T-Cells have memory (can remember and respond specifically to pathogens) and the fact that memory is formed, stored and actualized within the immune response is what allows humans to resist and become immune to infection.
>
> —Nathan D. Pennock, et al, T Cell Responses, (2013)[55]

> Cells make decisions such as whether to commit to duplication and even change their "mind". If yeast cells have enough nutrients to replicate, they will, but if the nutrients become depleted, they can reverse course, stop and the DNA will return to the nucleus. Thus, regular feedback of information is being processed by a decision-making capacity.
>
> —Deniz Irvali, When Yeast Cells Change Their Mind, (2021)[56]

Any theory of consciousness must explain its presence in the simplest life forms. As Baluska and Reber note, consciousness has its "original instantiation" in prokaryotes. These single-celled organisms display anticipatory behavior—predicting environmental changes, such as preparing enzymes for future food sources—and social intelligence (using Quorum Sensing to count their population and coordinate group attacks).

> Since all life is based on cells, any theory as to how consciousness emerged must be able to explain how it emerged from the simplest cells. Consciousness evolves from the 'original instantiation' in prokaryotes.
>
> —Frantisek Baluska et al, Sentience and Consciousness in Cells, (2019)[57]

> Single celled organisms display learning, risk management, memory, anticipation. Learning which means that the sentience of cells must be accepted.
>
> —Bhakti Niskama Shanta, Life and Consciousness (2015)[58]

> Predator microbes can lie in wait, anticipate, and attack in ways that mirror the existence of higher animal forms. The roots of human intelligence lie in individual cells, for they are sentient, ingenious, and autonomous.
>
> —Brian J. Ford, The Secrets of Intelligence Lie Within A Single Cell, (2010)[59]

The conscious experience resides in individual cells, not groups. Jonathan Edwards argues that single neurons have a form of sentience because they solve the "binding problem"—physically integrating thousands of distinct signals (sight, sound, thought) into a single, unified electrical state:

> The individual cell is the observer; the brain is merely the colony.The conscious experience resides in cells, not groups of cells, Neurons have some form of sentience.
>
> —Jonathan C.W. Edwards, Is Consciousness Only A Property Of Individual Cells? (2005)[60]
>
> Microorganism consciousness as observed in microtubules have the ability to quantum compute and this form of consciousness may have arisen from this mechanism.
>
> —J. Shashi Kiran Reddy, Understanding The Emergence Of Microbial Consciousness, (2017)[61]
>
> Proteins are able to form networks that are incredibly complex and monitor chemical reactions which allow them to communicate, "think" and convey upon the cell something which closely approximates the cognitive ability of the brain, including comprehension.
>
> —Robin P. Araujo et al, The Topological Requirements For Robust Perfect Adaptation In Networks Of Any Size, (2018)[62]
>
> Every strand of DNA in every cell is damaged between 50-100 thousand times a day. Cells detect this and send out repair proteins to quickly repair the damaged section.
>
> —Kristiane Sjogren, How Our Cells Repair Their Damaged DNA, 92015)[63]

This mechanism was discovered by three scientists who received a Nobel Prize in Chemistry in 2015.

Electrical engineer, Dr. Perry Marshall, author of Evolution 2.0 had an epiphany when he viewed the similarities between DNA activity and his own work on the ethernet. He states:

> "The similarities were almost scary. Encoding, decoding, error detection, error correction, checksums, layers. On and on." And "Organisms cut, splice and rearrange their DNA, performing astonishing engineering feats in real time… feats that put even the world's smartest software coders to shame."
>
> —Perry Marshall, Are Cells Intelligent, (2017)[64]
>
> In Marshall's view, there is an intelligence behind the cell that is not the intelligence of the cell itself. He uses another computer analogy:
>
> "Even if a computer is automatically generating emails, they always originated from a conscious source."
>
> —Perry Marshall, Why Do Many Scientists See Cells as Intelligent?, (2020)[65]

> White blood cells are believed to be sentient, capable of feeling and possess a rudimentary form of consciousness although they have no central nervous system or brain.
>
> —David Turell, ID Explained: Are Cells Intelligent, (2021)[66]

At this point, the conclusion is no longer tentative. What stands before us is not a collection of chemical events, but a system that behaves as if it knows—of itself, of its environment, and of what must be done. And whatever else this may ultimately be called, it cannot be accounted for by mechanism alone.

Across every level of observation, the cell exhibits behaviors that cannot be coherently reduced to passive chemistry. It senses, evaluates, decides, remembers, anticipates, and acts—consistently, reliably, and with a precision that holds across layers of extraordinary complexity. These are not the properties of a system merely undergoing reactions; they are the signatures of agency.

Biological Information, Consciousness & Life

In the spring of 2011, a symposium at Cornell University convened to confront what they called "the enigma of biological information": the observation that the information underlying life cannot be reconciled with purely naturalistic explanations.

The symposium brought leading experts together from across disciplines: physics, information theory, computer science, molecular biology, and mathematics. The findings resulted in a 500-page body of work with contributions from nearly thirty researchers who all shared the same recognition: the origin and organization of biological information remain fundamentally unresolved within existing frameworks.
The themes that emerged from the conference:

> Information is foundational to life, not a byproduct of it.
>
> Information cannot be reduced to its material system.
>
> No existing framework explains its origin [the origin of information] or function[67].

Information is indispensable to biology. When joined with the two additional major enigmas, consciousness and the life force, what they all share in common is their inability to be materially attributed.

The General Introduction to the proceedings captures the full scope of this challenge. It moves the definition of life away from static chemistry and toward dynamic communication, revealing that the "information" of life permeates every interaction within the cell:

> "If we define information more broadly as 'all that which is communicated,' the information within a living cell is much greater than its DNA sequence. All the components of the cell, including all the RNA and protein molecules, are continuously communicating with each other. It is recognized that there are hundreds of thousands of different types of interactions within the cell's 'interactome,' and most of these interactions in one way or another involve communication... it is very important we do not forget that most biological information in the cell is above and beyond the genome. On a level entirely above and beyond all this communicated information within the cell, information is also being communicated between cells, and between organisms. On a still higher level, we have the little-understood biological information that underlies the human mind, our own intelligence, and human consciousness. All of this is biological information![68]"

In this sense, we can say that all of material reality is, at its core, information. Biology sits atop a vast, precarious stack of physical conditions.

The Foundation: the universal low-entropy state, cosmological constants and dark energy

The Framework: the Standard Model of physics and the fundamental forces governing quantum and classical mechanics

The Encoding Medium: hydrogen, helium, heavy elements, and the chemical affinities that enable carbon-based life.

Each of these variables is informational, a value selected from an infinite set of possibilities—all required for the four great macromolecules of life to emerge, including the nucleic acids of DNA and RNA.

DNA depends on the alphabet of carbon, carbon depends on the grammar of physics, and physics depends on the initial conditions of the universe. In this sense, the physical universe is hardware, life is software, and consciousness is the user working within the system. We live in a system where everything—all matter, all physical forms, "know" what to do. Biologically, cellular consciousness proceeds largely without conscious intention. This is what allows the universe to culminate in the human mind, unencumbered and free to serve as the universal observer, with little regard for the complex operations of cellular consciousness and cellular operations.

This is another great enigma. That human consciousness sits atop a vast landscape of trillions of cells, each of which is completing tens of thousands of actions per second, complex, dynamic, coordinated, quantum even, and life sustaining. That the individual consciousness of the human animal can remain completely unaware with no consequence. Sleeping, unconscious, or engaged in activities that divert all of our attention, and yet this action, these cells, continue to do what must be done for the organism's survival.

And all of the evidence shows us that if our universe is information, then information can only come about by the application of consciousness. Materialism invokes "emergence" but this comes far too late and ignores a critical point: whatever emerges must already exist as potential. The material ontology is thus inverted. To account for a universe like ours, consciousness must precede material form, anticipating the creation of life, technology, planets and galaxies. The very constraints of existence must be imposed prior to the instantiation of the medium we call our universe. *The Divine Algorithm*—that is, the series of intractable problems etched across the landscape of our reality—requires intelligence from the very outset. Consequently, the notion of 'emergence' is insufficient—a failed explanatory process. What is required is a form of studied navigation, knowledgeable of the possible universal configurations, bound up with the causal primacy to bring a life allowing medium into existence. This is what allowed intelligent life to exist with perspicuous specificity and exactness. Now that we have reached our destination, and traversed the impossible landscape, we can say that all of this was assured from the outset.

The conclusion is clear. The narrative of emergence provides no useful information. It is science turning an uninterested eye towards conditions and constraints that are all but demanded. And then there is the inversion. Consciousness must precede material form, if that material form is information, there is no escape.

The answer to the question of when the material of our physical reality becomes information, is "at the beginning". Logic holds that only consciousness—or a hypernym far more encompassing—can intelligently navigate the macromolecules of life to complete their biological tasks.

Consciousness, in its various forms, is evidenced by the instantiations (universe, expansion, stars, life, etc.) within our universe, and necessitated by the data of our physical reality. But consciousness is what modern science has been forced to leave unnamed, an omission which leaves gaping explanatory voids in the scientific endeavor.

The Fitness of Cells & The Fitness of The Environment Redux

We have established that the essence of life—the genetic code and the consciousness that directs it—requires an Author. But what about the environment we reside in? If our Algorithm is truly all-encompassing, we should expect to find its signature not just in the complexity of biology, but in the fundamental properties of matter itself. We should find that reality is not merely compatible with life, but in anticipation of it, preconfigured in advance for capacities that would emerge billions of years later. This observation—that the universe seems "pre-rigged" for biology—is a chemical fact. As Michael Denton notes *in The Miracle of the Cell* (2020), "On any consideration, the evidence conveys the irresistible impression that the properties of the atoms have been crafted with incredible precision to perform highly specific functions upon which the life of the cell depends[69]." Recall Doctor Lawrence Henderson's 1913 classic, *The Fitness of the Environment*, his findings on the "fitness" of the living cell was such that the perfect facility and myriad requirements of the environment in meeting this fitness could no longer be considered a remarkable coincidence. He found this evidence of a large-scale conspiracy in the convergence of needed conditions that accident was an insufficient explanation, for this "coincidence" extended all the way down to the bottom. Henderson's key insight being biology didn't randomly solve how to wrest life from existence, rather the universe solved the problem of biological life in advance. This is why in choosing the title, *The Fitness of the Environment*, Henderson was drawing a direct reference to the Darwinian concept of, "survival of the fittest." The mechanism of selection presupposes a prior condition: that the environment must already possess the capacity to generate viable forms. Without that prior fitness, there is nothing to select, nothing to refine, and no path to life. There can be no talk of "the fittest" if an environment that has already been programmed with the capacity to create the fittest.

> "The fitness of the environment results from characteristics which constitute a series of maxima — unique or nearly unique properties of water, carbonic acid, the compounds of carbon, hydrogen, and oxygen and the ocean — so numerous, so varied, so nearly complete among all things which are concerned in the problem that together they form certainly the greatest possible fitness. No other environment consisting of primary constituents made up of other known elements, or lacking water and carbonic acid, could possess a like number of fit characteristics or such highly fit characteristics, or in any manner such great fitness to promote complexity, durability, and active metabolism in the organic mechanism which we call life[70]."

We can see that a world made to serve the needs of life was not the final target, the final target was creating the physical structure required for the life force of consciousness to inhabit biology. Consciousness, like the fix of matter, atoms, and biology, had to have been applied in the beginning. Life and consciousness cannot arise all-of-the-sudden, 'at the top' the plan of fitness had to have been laid at the moment of creation if it is ever to arise at all.

The material fitness of our universe provides a complete shopping list of chemical properties, available three-dimensional structures and conformations needed for every imaginable biological system that has or will come into existence. Its most pointed form is within the 4 elements known as CHON, carbon, hydrogen, oxygen and nitrogen, and with the addition of phosphorus and sulfur to make up CHNOPS. These are not merely the ingredients of life—they define the narrow and highly constrained chemical window within which life is even possible.

> "The unique fitness of many of the atoms in the first quarter of the periodic table for the design of the cell is stunning. On any consideration, the evidence conveys the irresistible impression that the properties of the atoms have been crafted with incredible precision to perform highly specific functions upon which the life of the cell depends. Together, the collective fitness of the properties of these atoms for the cell make up what I term a 'primal blueprint' for the design of the carbon-based cell, a blueprint laid down in the order of things from the beginning of the universe[71]."

Fitness is not just compatibility. It is the optimization of an imposed environmental constraint as prearranged by consciousness to serve as the seat of consciousness in all its myriad forms, within the biology of the living human species.

Chapter Summary

Materialist's theory requisitions only unguided, blind, processes to assemble biology. But this is not what works. Guided, conscious, intelligent and highly precise macromolecules are the only ones up to the task of assembling, maintaining and monitoring biology. We now know the chances of life, with all of the assistance of the fitness of the environment, is not possible given the potential age of our universe and the resources required by chance to contrive it. Biologically relevant molecules will never arrive. Not because they are rare, but because the pathways required to produce them are not accessible under unguided material conditions. Neither life, nor speciation is possible, even with all of the fitness and foreshadowing of the capacities of matter if not for the addition of a series of consciousness-like instantiations, as we have found at every juncture. This is the engine of our universe.

This missing vital ingredient, the life force itself, cannot be assembled. Matter can be arranged, but it does not awaken. The transition from non-life to life is not a construction—it is an event. All material preconditions can be satisfied—and life still does not occur. The universal engine of instantiation must act again to cross the Rubicon. Henderson asked the question, "To what extent do the characteristics of matter and energy and the cosmic processes favor the existence of mechanisms which must be complex, highly regulated, and provided with suitable matter and energy as food? If it appears that the fitness of the environment to fulfill these demands of life is great, we may then ask whether it is so great that we cannot reasonably assume it to be accidental, and finally we may inquire what manner of law is capable of explaining such fitness of the very nature of things[72]."

Henderson recognized the question—but not its full implication: that such fitness does not merely invite explanation, it demands one. A perfectly prepared universe that never produces life is indistinguishable from one that was never intended to, and also one that doesn't exist at all. What isn't observed is nowhere to be found. Without consciousness, matter remains inert, information remains unexpressed, and the universe remains silent.

Value for Variable 15: Cell Consciousness

The modern biological narrative has reached a profound structural limit. A limit that has been with the sciences since prior to biology being a formal discipline. Darwin struggled to divine it; his predecessors were baffled by it. But the burden fell on Darwin's generation because the reality of an intelligence far beyond the biological was becoming impossible to ignore, and religious explanations all too eager to answer the question. The moment to strike had come. Even though they knew not the mechanism, nor had they any hope of understanding cellular biology or genetics, evolutionary theory was co-introduced with Alfred Russel Wallace and rammed through by the generation after Darwin in a state of objective ignorance, while still locked outside of the black box of biological secrets.

What if Thomas Henry 'Bulldog' Huxley and Charles Darwin had been sat down with a microscope and shown the inner workings of the genome, the precision of cell division, and the high-speed turbine of ATP synthase? Had they known that DNA was a digital, front-to-back mirror image, complementary and yet distinct, would they have remained satisfied with randomness as their God of the gaps? We can catalog the proteins and map the structure of the organelles within the cell, but the most essential variable remains strategically unnamed. The 3.8-billion-year-old masterpiece of agency we call the cell, has been recast in their minds as nothing more than a chemical factory.

In every millisecond of its existence, the cell solves tens of thousands of 'hard problems' of vitality through precise, inexplicable navigation. To look at a cell and see only chemistry is like looking at a symphony and seeing only the ink of the score—missing entirely 'what it is like' to experience Mozart's *Symphony No. 40 in G minor*. This is the fallacy.

It's time to enter reality and acknowledge the navigator at the helm—cellular consciousness.

Current biological paradigms fail to provide a reductive materialist explanation for the observed teleological behavior—the 'aboutness' of macromolecular action. While modern synthesis accounts for proteins and mechanics, it remains insufficient in addressing the high-fidelity, goal-directed coordination integral to cellular efficiency. This vitality represents the life force itself: the very definition of life. It is the characteristic that mandates: the consumption of ATP, electrons to move by quantum jumps through the electron transport chain, cellular division to take place, and RNA polymerase and its convoy of 30+ assistive macromolecules to target a single gene, attach, and begin transcribing DNA. On the highest level of complex intelligent life, it provides the living foundation that consciousness and Youness reside within. Cellular consciousness is a fundamental "hard problem" that bridges the enigma of biological information with biological intelligence and biological action.

The Value of *Cell Consciousness*: *M (miracle-impossible)*

Meta-Discussion

Consciousness makes its presence known in our universe through several ordered expressions. These appear to us as distinct levels and successive layers of complexity, each brought into being through the process of instantiation.

In truth, the underlying consciousness and its expressions are the operation of **a single entity**, and the appearance of layers and complexity are manifestations of nonduality, this entity in action across all levels of our reality.

Constraint: The deepest and most foundational expression of consciousness is the existence of constraint itself. This expression is prior to matter, energy, and the laws of physics; it renders our reality in a structured and restricted form. Of all possible states, only a narrow and highly specific subset is permitted. Some refer to the imposition of selective constraint as *fine-tuning*.

Where a fundamental condition possesses a specific value, it is evidence that a selection has taken place. A fundamental constraint has been imposed with specificity and exactness. A principle capable of performing that imposition and settling upon that constraint made a choice. At the universal scale, this level of selection is non-arbitrary. It is forward-looking, informational, ordered, and goal directed. As new levels of instantiation proceed forward from the setting of constraint, the forward-looking nature of the constraint becomes evident, and the need for each layer becomes evident. The actions of consciousness at this level results in describable conditions and represents a capacity to precisely determine what is allowed and what is not.

Constraint is the foundation, the first expression of consciousness in our reality.

Universal: The Universal consciousness is the structured expression of constraints, and the ordering of reality into a coherent and stable system. This is the level at which the laws of physics, such as the cosmological constant, the expansion rate, the baryon asymmetry, the horizon and what we perceive as the influence of dark energy and dark matter operate. These are positive features of existence—its realized framework set by prior constraint–and the required needs of a reality such as ours.

The universe exists within a precisely defined and enforced order to produce a precisely defined and enforced outcome—universal observers.

Material: This is the base layer of material reality—a coherent system of relations that represents our physical reality. We recognize the base layer of material reality in the fields and particles of the *Standard Model*, and the governing influences of gravity, dark energy, dark matter, quantum mechanics and classical laws.

Just as the mapping of symbol and sound to our real world creates language, constraint provides the mapping for the formal 'grammar' of the atomic realm. The first instance is what we identify as the primordial hydrogen and helium—the first two letter of the full chemical alphabet.

Constraint creates the universe space, which determines the primordial material, differentiated into a range of manifestations used to create the physical structures of the universe. These are guided by mostly unseen forces (constraint) —gravity, dark energy, dark matter—and imbued with the layers of fitness and molecular potency and chemical versatility for molecules, chemistry, and biology.

Human Consciousness: The localized expression of perception and awareness. The aperture through which we experience the self and the physical world.

Individual – "Youness": The most perplexing, subtle, profound, and capacious expression of consciousness. This is the subjective experience of being—the domain of identity, personality, intention, free will, and choice. It is what makes the individual a distinct center of experience: a conscious agent, uniquely and individually made. Alone and independent within the cosmos. One against the heavens. A form of consciousness engaged in a human experience. Not an emergence, not an illusion, not an epiphenomenon.

What appears here as personal awareness is the same selecting principle seen at the highest level (consciousness), now localized and experienced from within as the paradox among all paradoxes: the individual is made, both physically by constraint—as the culmination of constraint—and immaterially as consciousness and Youness itself, given to the individual as the *diamond sutra* of existence. All of the cosmos was made for the experience of this singular creation.

How something can be both made and given is the mystery of all mysteries. That is what each and every independent life form represents—a made and given, impossible gift.

Instantiations

From the beginning, the pattern is consistent:

The instantiation of the universe

The formation of Population III stars

The formation of galaxies and cosmic structure

The sudden appearance of life and cellular consciousness

Speciation & Diversification, each requiring a separate instantiation

The instantiation of consciousness in complex life

At each stage, what appears is a crossing, a transformation, and a transition into a new domain of organization that cannot be reduced to, nor arise from the prior state by 'law' or 'probability' alone.

These instantiations are events: the realization of constrained possibilities at successive levels of complexity—the pattern of *The Divine Algorithm.*

Unification

It must be recognized that all we have listed above—constraint, universal, material, consciousness and the individual—are not independent forms, but manifestations of a single underlying consciousness. Unity, nonduality, One in form, content and even in appearance. Plurality is better understood as expression across scale, from the level of constraint to the level of individual experience.

The consciousness that orders the universe is not separate from the consciousness that operates within the cell or the consciousness through which the individual experiences the world. Fractals.

This continuity is reflected in the unresolved features of physics—the measurement problem, non-locality, and the presence of hidden information—which point to a deeper unity beneath what appears to be separation and perplexity.

What we encounter as personal awareness is not an isolated phenomenon, but a localized participation in the same selecting principle that governs reality at every level.

In the final analysis, the unity of consciousness is most directly evidenced in experience itself. All else is pretense, confinement, constraint, and potential as precursors to the universal show—Maya, and the play of life—Lila, to Brahman—the one and only sustaining power, and Atman—the instantiation of Brahman as the individual, which is one and the same as Brahman. This immediacy would not be possible if human consciousness were not continuous with the same underlying reality that gives rise to all other expressions of consciousness. In our world, the ability to relate to something, in whatever way that takes place, is proof of unity. This is how it operates, so that the individual experiences reality, in a vital contact of consciousness meeting consciousness through the appearance of the universal show. That's what our lives are.

And what of the material? Is it not real? Does this mean it does not exist?

Material analysis depends on separate, clearly defined categories and the relationships between them. If the chain of explanation is followed without interruption and without the imposition of artificial boundaries, the distinctions between categories dissolves. What appears as distinct domains, such as physics, atoms, chemistry, biology and consciousness, are continuous expressions of a single underlying reality. The apparent division was never absolute, only provisional, as it must be in a realm solely constructed for purposes of consciousness and conscious experience.

Transitioning From Constraint To Objective Truth

Having validated *the Divine Algorithm* we move from the mechanism of creation to the medium that allows it to be known at all. Our validations are born of Constraint. By first principles physics and probability, we distinguish the possible from the impossible, not by belief, but by the constraints that define existence itself.

Materialism functions as a selective ideology, acknowledging only the constraints that conform to physicalism while discarding the full set of constraints and their implications, namely the need for immateriality in every meaningful level of reality.

To draw conclusions from anything less than the complete set of universal constraints is to fall into the trap of Plato's allegory, mistaking the shadow of physical form for the cause of what casts the images.

Constraints do more than resolve the *Algorithm*; they provide consciousness with access to objectivity. This field is an anomaly to the materialist, being that it is immaterial, yet accessible to consciousness. But it is a logical necessity if there is to be objective truth. Without an objective substrate certain knowledge is a logical impossibility.

To understand how knowledge is distilled from probability, we must now define the bridge. Something we rightfully call *The Evaluative Field* presented as our closing piece.

---

[1] (L.A. Urry, M.L. Cain, S.A. Wasserman, P.V. Minorsky, J.B. Reece, N.A. Campbell, Campbell Biology, Pg 2, Series, Pearson, 11th edition, October 19, 2016).
[2] (B. Alberts, A. Johnson, J. Lewis, M. Raff, K. Roberts, P. Walter, Molecular Biology of the Cell, Pg.3, Fourth Edition, Garland Science, March 1, 2002)
[3] (C. Mariscal, "Life", The Stanford Encyclopedia of Philosophy (Winter 2021 Edition), E.N. Zalta (ed.), URL plato(dot) stanford(dot) edu/ archives/ win2021/ entries/ life/)
[4] (E. Schrödinger, What Is Life?, 1944, Cambridge University Press)
[5] (E. Schrödinger, What Is Life? Chapter 6, Pg. n23, 1944, Cambridge University Press)
[6] (E. Schrödinger, What Is Life?, Chapter 6, Pg. n23, 1944, Cambridge University Press)
[7] (A. Einstein, Investigations on the Theory of the Brownian Movement, 1905, Annalen der Physik, 322(8), 549–560).
[8] (C.S. Peskin, G.M. Odell, G.F. Oster, Cellular motions and thermal fluctuations: The Brownian ratchet, 1993, Biophysical Journal, 65(1), 316–3240.
[9] (J. Howard, A.J. Hudspeth, R.D. Vale, Movement of microtubules by single kinesin molecules,1989, Nature, 342(6246), 154–158. (The first experimental visualization of a single molecular motor "walking" in a directed, processive manner).
[10] (W. Hwang, M. Karplus, Structural basis for power stroke vs. Brownian ratchet mechanisms of motor proteins, 2019, Proceedings of the National Academy of Sciences (PNAS), 116(44), 22152–22161).
[11] (R.E. Johnson, R. E., et al. Fidelity of Human DNA Polymerase η. 2000, Journal of Biological Chemistry 275, 7447–7450).
[12] (R.E. Johnson, R. E., et al. 2000),
[13] (R.E. Johnson, R. E., et al. 2000)
[14] (A.B. Kolomeisky, Physics of Protein–DNA Interactions: Mechanisms of Facilitated Target Search, 2011, Physical Chemistry Chemical Physics, 13(6), 2088-2095).
[15] (A.B. Kolomeisky, Physics of Protein–DNA Interactions: Mechanisms of Facilitated Target Search, 2011, Physical Chemistry Chemical Physics, 13(6), 2088-2095.),

[16] (S. Park, O.C. Lee, J.H. Jeon, J.-H. A Mini-Review of the Diffusion Dynamics of DNA-Binding Proteins: Experiments and Models, 2021, Journal of the Korean Physical Society, 78, 408–426.)
[17] (E. Schrödinger, What Is Life? Chapter 7, Pg. n27, 1944, Cambridge University Press)
[18] (L.A. Urry, M.L. Cain, S.A. Wasserman, P.V. Minorsky, J.B. Reece, N.A. Campbell, Campbell Biology, 2016, 11th ed., pp. 212-239 Pearson).
[19] (L.A. Urry, M.L. Cain, S.A. Wasserman, P.V. Minorsky, J.B. Reece, N.A. Campbell, Campbell Biology, 2016, 11th ed., pp. 212-239 Pearson).
[20] (L.A. Urry, M.L. Cain, S.A. Wasserman, P.V. Minorsky, J.B. Reece, N.A. Campbell, Campbell Biology, 2016, 11th ed., pp. 212-239 Pearson).
[21] (B.J. Greber, E. Nogales, The Structure of Eukaryotic Transcription Pre-Initiation Complexes and Their Functional Implications, Subcell Biochem. 2019;93:143-192. PMID: 31939151)
[22] (B.J. Greber, E. Nogales, The Structure of Eukaryotic Transcription Pre-Initiation Complexes and Their Functional Implications, Subcell Biochem. 2019;93:143-192. PMID: 31939151)
[23] (M. Slutsky, L.A. Mirny, Kinetics of Protein-DNA Interaction: Facilitated Target Location in Sequence-Dependent Potential, Biophys J. 2001 Dec;87(6):4021-35, PMID: 15465864)
[24] (M. Slutsky, L.A. Mirny, Kinetics of Protein-DNA Interaction: Facilitated Target Location in Sequence-Dependent Potential, Biophys J. 2001 Dec;87(6):4021-35, PMID: 15465864)
[25] (K.S. Zaret, Pioneer Transcription Factors Initiating Gene Network Changes. Annu Rev Genet. 2020 Nov 23;54:367-385. PMID: 32886547)
[26] (B. Alberts, A. Johnson, J. Lewis, M. Raff, K. Roberts, P. Walter, Molecular Biology of the Cell (4th ed.), pg. 175-177, 301-306, 2002, Garland Science), (L.A. Urry, M.L. Cain, S.A. Wasserman, P.V. Minorsky, J.B. Reece, N.A. Campbell, Campbell Biology, 2016, 11th ed., pp. 212-239 Pearson).
[27] F.H.C. Crick, The Genetic Code, 1968, Nature, 1968 reprinted in The Mind of a Chemist).
[28] (B. Alberts, A. Johnson, J. Lewis, M. Raff, K. Roberts, P. Walter, Molecular Biology of the Cell (4th ed.), 2002, Garland Science)
[29] F.H.C. Crick, The Genetic Code. The Mind of a Chemist, 1968, University of Washington Press)
[30] C.R. Woese, On the Evolution of the Genetic Code, 1965, Proceedings of the National Academy of Sciences, 54(6), 1546–1552)
[31] (M. Nirenberg, Deciphering the Genetic Code: Breaking the Code, 2004, Science, 306(5698), 1498–1502)
[32] (C. Levinthal, How To Fold Graciously, 1969, Mossbauer Spectroscopy in Biological Systems: Allerton House, Illinois, editor, P. Debrunner, J.C.M. Tsibris, E. Munck, 22-24, University of Illinois)
[33] (R. Milo, R. Phillips, 2015, Cell Biology by the Numbers, Garland Science).
[34] (M. Slutsky, L.A. Mirny, Kinetics of Protein-DNA Interaction: Biophys J. 20004 DEc; 87(6): 4021-35. PMID: 15465864)
[35] F.M. Harold, The Way of the Cell: Molecules, Organisms and the Order of Life, 2001, Oxford Univ. Press)
[36] (F.M. Harold, Molecules into Cells: Specifying Spatial Architecture. 2005, Microbiol Mol Biol Rev 69:).
[37] (M. Levin, Bioelectric signaling: Reprogrammable circuits underlying embryogenesis, regeneration, and cancer, Cell, 2021, Cell, Vol 184, Issue 8, Pp. 1971-1989, Elsevier)

[38] (G. Zhang, M. Levin, Bioelectricity in Development, Regeneration, and Cancers, Cell Bio: Joint Meeting Ameri. Soc. Cell Biol, & Euro MOlec. Bio. Org, Bioelectricity, 6:1)
[39] (Y. Schiffmann, Self-Organization in and on Biological Spheres, 2007, Progress in Biophysics and Molecular Biology, Volume 95, Issues 1–3), (B. Alberts, A. Johnson, J. Lewis, M. Raff, K. Roberts, P. Walter, Molecular Biology of the Cell, 4th Edition, Garland Science, 2002)
[40] (J.K. Grandy, The Three Dynamic Levels of DNA Consciousness, International Journal of Arts & Sciences, ISSN: 1944-6934, (6(3): 313-327, 2013 UniversityPublications(dot) net)
[41] (Ibid, L. Dent, A New Model of Intracellular Communication…)
[42] (Ibid, L. Dent, A New Model of Intracellular Communication…)
[43] (Ibid, L. Dent, A New Model of Intracellular Communication…),
[44] (Ibid, L. Dent, A New Model of Intracellular Communication…)
[45] (Ibid, L. Dent, A New Model of Intracellular Communication…).
[46] (Ibid, L. Dent, A New Model of Intracellular Communication…).
[47] (Barbara McClintock's 1984 Nobel Prize Lecture, nobelprize(dot) org/ uploads/ 2018/ 06/ mcclintock-lecture(dot) pdf)
[48] (Barbara McClintock's 1984 Nobel Prize Lecture, nobelprize(dot) org/ uploads/ 2018/ 06/ mcclintock-lecture(dot) pdf)
[49] (The Conscious Cell, Lynn Margulis, Annals of the New York Academy of Science,, April 2001, 929:55-70, pubmed(dot) ncbi(dot) nlm(dot) nih(dot) gov/ 11349430/)
[50] (The Miracle of the Cell, P. 18-19, Michael Denton).
[51] (The Miracle of the Cell, P. 18-19, Michael Denton)
[52] (Leaving It Up to Chance: How Cells Make Decisions, Contemporary Science Issues and Innovations March 18, 2014, Belmont Media Center, Jane Kondev, Ph.D. Brandeis University..)
[53] (A Single Molecule in Sight: Intelligent Molecules, January 17, 2013, Dr. Michael Nash, Nanosystems Initiative Munich)
[54] (Can A Cell Remember? Jennifer Frazer, Scientific American, Opinion, May 28, 2021, scientificamerican .com/ article/ can-a-cell-remember/)
[55] T Cell Responses: Naive To Memory And Everything In Between, Advances in Physiology Education: 2013, Dec: 37(4): 273-83, Nathan D. Pennock, Jason T. White, Eric W. Cross, Elizabeth E. Cheney, Beth A Tamburini, Ross M. Kedl, 10.1152/advan.00066.2013)
[56] (When Yeast Cells Change Their Mind: Cell Cycle "Start" Is Reversible Under Starvation, Deniz Irvali, Fabian P. Schlottmann, Prathibha Muralidhara, Iliya Nadelson, N. Ezgi Wood, Andreas Doncic, Jennifer C. Ewald, Nov 2, 2021)
[57] (Sentience and Consciousness in Cells: How the First Minds Emerged in Unicellular Species, Frantisek Baluska, Arthur R, February 4, 2019)
[58] (Life and Consciousness - The Vedantic View. Journal of Communicative and Integrative Biology, 2015 Sep-Oct; 8(5): e1085138, online: Oct 9, 2015, Bhakti Niskama Shanta)
[59] (The Secrets of Intelligence Lie Within A Single Cell, Brain J. Ford, New Scientists, April 21, 2010, newscientist(dot) com/)
[60] (Is Consciousness Only A Property Of Individual Cells?, Jonathan CW Edwards, Journal of Consciousness Studies, Volume 12, April 2005, No. 4-5, pp60-76)
[61] (Understanding The Emergence Of Microbial Consciousness: From A Perspective Of Subject-Object Model -SOB: J. Shashi Kiran Reddy, Contzen Pereira, Journal of Integrative Neuroscience 2017; 16 (s1): S27-S36. Doi: 10.3233/JIN-170064)
[62] (The Topological Requirements For Robust Perfect Adaptation In Networks Of Any Size, Robyn P. Araujo, Lance A. Liotta, Nature Communications, 9, Article Number: 1757 (2018) May 1, 2018.)

[63] (How Our Cells Repair Their Damaged DNA: Researchers Find Important Mechanism That Allows Cells To Repair Damaged DNA. (This) Could Answer Questions About Cancer Development, Kristian Sjogren, November 11, 2015. Sciencenordic. com/ biology-denmark/ how-our-cells-repair-damaged-dna/ 1425278)
[64] ("Are Cells Intelligent?" Dr. Perry Marshall, At Evolution 2.0, March 2, 2017)
[65] (Perry Marshall, in, Why Do Many Scientists See Cells as Intelligent?: October 7, 2020, mindmatters .ai/ 2020/ 10/ why-do-many-scientists-see-cells-as-intelligent/)
[66] (ID Explained: Are Cells Intelligent (Introduction) March 7, 2021, David Turell, https:// mindmatters. ai/ 2020/ 10/ why-do-scientistts-see-cells-as-intelligent/)
[67] (R.J. Marks II, M.J. Behe, W.A. Dembski, B.L. Gordon, J.C. Sanford, General Introduction, Biological Information, New Perspectives, Proceedings of the Symposium, Cornell University, USA 31 May - 3 June, 2011, WorldScientific.com)
[68] (R.J. Marks II, M.J. Behe, W.A. Dembski, B.L. Gordon, J.C. Sanford, General Introduction, Biological Information, New Perspectives, Proceedings of the Symposium, Cornell University, USA 31 May - 3 June, 2011, WorldScientific.com)
[69] (M. Denton, The Miracle of the Cell, p.11, Discovery Institute Press, 2020)
[70] (L. Henderson, The Fitness of the Environment, Pg. 272, 1913, [originally] The MacMillan)
[71] (M. Denton The Miracle of the Cell, The Privileged Species Series, 2020, Introduction, Discovery Institute Press)28
[72] (L. Henderson, The Fitness of the Environment, Pg. 39 [last page of chapter 1], 1913, [originally] The MacMillan)

Interlude No. 3: The Evaluative Field

Abstract

The existence of an immaterial, objective constraint field is necessary for intelligent observers to know and evaluate objectively. The only way this field can work is if it exists not as a thing, but as a field that is contacted, and interfaced and provides the user with the ability to evaluate propositions from an objective matrix. This is what allows humanity to arrive at individual and collective truth.

Truth-recognition cannot be produced by language, cognition, or physical structure. Language can propose. Cognition can model. Physical systems can cause and compute. But none of these can generate the objective distinction between correct and incorrect representation.

If error is real, then correctness must be real. And if correctness is real, then evaluation cannot be a late-arriving biological invention (language, symbol, sound) or a social convention. It must be prior to any act of judgment, and independent of the observer.

The paper introduces the *Evaluative Field*: a preexisting constraint-space that makes truth, falsity, and knowledge intelligible. It argues that attempts to reduce evaluation to physical structure inevitably collapse into circularity or smuggle in the very standard they claim does not exist. Knowledge acquisition is therefore not the production of truth but contact with a standard that already existed.

We Begin with a Question

Think of how interesting it is that we can come to know what we previously didn't know. This question appears innocent, but it contains one of the deepest fractures in philosophy. If learning is real, the human mind must be capable of acquiring truths it did not previously possess. How does this happen? Any attempt to explain knowledge acquisition necessarily falls into one of three categories, none of which can be the source of truth:

Language creates truth
The human mind generates truth
Truth comes from reality itself

Language Cannot Create Truth

Language cannot certify truth. It maps onto reality but exists primarily as a medium for communication and self-expression: ideas, beliefs, fiction, delusion, psychology, and emotion. Language can propose, signal, describe, or transmit—but it cannot, by itself, *validate* correctness. It can justify or plead for certification, but the act of certification requires something beyond words.

If truth could arise from language alone, coherence would equate to correctness. But coherence is not truth—it is internal consistency. False statements can be grammatically flawless, internally consistent, compelling, and even universally accepted. To certify truth reality must first be engaged, then mapped into language, which can then be checked for validity. Language comes in too late to be evaluative.

Cognition Alone Cannot Generate Truth

If truth resided solely within the mind, it would collapse into psychological states: conviction, coherence, confidence, or preference. Cognition would either need to pre-contain all truths or generate them from nothing. Either path undermines objectivity. If the mind merely houses latent truths, recognition becomes contingent on chance triggers. In either scenario, error becomes unintelligible: there is no "wrong," only "different."

The problem is clear: cognition alone cannot produce truth. It is mapped onto a reality that is external, and it relies on that reality for evaluation. Language and cognition are interdependent, but neither can certify correctness without collapsing into circularity.

Reality is not Evaluation

Reality is not a statement. Reality contains no untruths: a rock is never wrong; a tree is never false. Error resides not in reality itself but in how reality is represented (language) or comprehended (cognition). Mere existence of a referent does not explain why one statement succeeds while another fails. Reality is the target, not the act of evaluation. And yet error occurs. Correction occurs. Learning occurs. This necessitates an objective mediating structure—a space in which multiple mappings can be evaluated against a common referent, where correctness becomes detectable.

From our discussion, two claims emerge:

An evaluative structure must exist for truth-recognition to be possible. If error and correction occur, there must be an *objective* distinction between correct and incorrect mappings.

This structure cannot be language or cognition. Language proposes; cognition maps. Neither can certify correctness without circular self-validation. The evaluative field is the precondition that makes truth- and error-recognition intelligible.

The evaluative field is the logically necessary third layer. Without it, distinctions between "I believe this" and "this is correct" collapse. Once the field is engaged, understanding shifts: problem-solving is not merely a recombination of language or rearrangement of mental states. It is the effort to constrain ideas and test them against the field. The characteristic "click" of coherence and relevance emerges—a marker of alignment with the objective structure of correctness—but unlike the false click of narrative satisfaction or self-preservation (of one's beliefs, paradigm, dogma, etc.), this click is only treated as evidential when it survives constraint-testing against the field. Alignment can then be iteratively cross-checked: premises against reality, known facts against known untruths.

Understanding is not manufactured by symbols. Understanding is not invented by, nor does it reside in cognition. Understanding must first be encountered to be represented. Words, mapped onto the world, interface a mind, also mapped onto the world, conjoined with the evaluative field—an entity that resides outside language and cognition, that enables truth recognition. Analysis of prior knowledge is insufficient to generate new knowledge. That cannot be the origin account, because it presupposes what it claims to explain. If we return to the beginning of the individual—the origin of understanding—we find that "existing knowledge" is not yet even representable. Language is unmapped. Cognition is immature. Symbolic representation has not yet stabilized into anything like a one-to-one correspondence with the world.

To claim that existing knowledge bridges the gap of truth certification is therefore to enter a loop that fails at the point of origination. In the state of abject ignorance into which all humans are born, cognition alone cannot originate knowledge without representation, and representation cannot originate without cognition.

And neither can certify knowledge without the evaluative field. Only the field can rupture this circularity and without the field, any attempt to create knowledge is a conjectural journey that may or may not converge on truth.

The Asymmetry of Truth: The One-Way Street

Systems of thought can be infinitely refined toward truth, but when they are forced toward incorrectness, they inevitably fracture. This is because truth is the state of minimal representational friction—a ground state where an idea aligns so perfectly with the objective field that it requires no external energy to persist. Falsity, by contrast, is an excited state of information; it is a disturbance of equilibrium that demands constant cognitive input to survive the friction of misalignment.

The Evaluative Field functions as a ground state of objective resistance to falsity. This creates a fundamental asymmetry in how information interacts with the field through the process of iteration.

Because the Field is the invariant objective standard of correctness, incorrect representations can be systematically iterated toward truth through the removal of 'noise' or the addition of coherence. Iteration is an extractive process that strips away distortion introduced by the observer. As an idea converges toward the substrate, it becomes increasingly self-sustaining. Truth, therefore, is the state in which an idea no longer requires intervention to remain stable—it aligns with the field.

The reverse is structurally impossible. Starting from correctness, one cannot iterate *constructively* towards falsity. This is not iteration, rather, degradation. The signal becomes progressively degraded. Truth, when aligned with the substrate, occupies a state of minimal representational friction. Movement away from alignment introduces friction, which equals error. It is already at the lowest possible state of 'representational friction.' Any movement away requires error.'

Every coordinate of the field is tethered to and constrained by the whole. Consequently, the concept of a 'perfected lie'—one indistinguishable from truth—is a structural impossibility. To introduce a single falsehood without creating global friction would require rewriting the entire substrate of the field to accommodate the local error. The substrate cannot be rewritten, it is pre-symbolic, immaterial, and non-local. Errors do not exist in the field, they exist in representation, within the world of ideas. Consequently 'local', 'material' and 'symbolic' untruths generate an unavoidable contradiction that exposes the mismatch. What is true, was true from the outset, and the outset lies prior to ideas, prior to language, prior to life, and prior to existence. Untruths of any sort become evident, when they are incorporated within conjecture. They will limit the ability of the conjecture to find footing in the real world. This often requires workarounds to smooth the tension they introduce.

The Logic Trap: Truth is a state of perfect efficiency. Truth represented in words or symbols is near-perfect, but being representation, it remains one step removed from the evaluative substrate. It requires no external support to persist. Falsity, by contrast, is an unstable, excited state of information—a disturbance of equilibrium. Representational error introduces a 'friction' that must be overcome by one holding the incorrect idea. This work manifests as added complexity, such as 'promissory notes' for future evidence, and cognitive workarounds, consensus or dogma construction. Error is inefficiency: it requires ongoing input to maintain coherence, yet no amount of energy can integrate error in truth. It demands to persist alongside objective truth. Without this input, false representation collapses into belief, or dogma. The Field does not accommodate error–it cannot. The burden remains on the holder of the false idea to simulate the appearance of local coherence; however, the false idea will remain under pressure from field.

Subtle Points

Coherence, by itself, is not sufficient to establish a truth condition. Purely cognitive theories collapse into subjectivity when forced to account for language as truth. Truth requires a mediating structure—neither symbol nor mental state, but an objective constraint-space imposed by reality, above both.

Some may object: "Isn't your evaluative field just the brain computing and updating itself?" Brains compute. Environments provide feedback. But computation rearranges symbols; feedback triggers adaptation. Neither produces the *normative* distinction between correct and incorrect. Optimized outputs are not thereby true. "Error," "learning," and "correction" presuppose a space in which "better" and "worse" mappings exist as real constraints.

Materialist orthodoxy resists what is plain because it restricts, by definition, what counts as real—thereby excluding the evaluative field, which is necessarily immaterial. With this imposed limit, the mind self-polices: avoiding truths deemed "not allowed." Then, when confronted with the evaluative field, the reflex is to relocate it—into language, the brain, social agreement, categories, optimization—anywhere it can fit within the materialist frame. This is not refutation; it is philosophical retreat.

Nearly the entire western academic world operates under materialist assumptions that prohibit immaterial cause. Though necessary for truth-recognition, the field is structurally prohibited from explanation by the dominant paradigm because it is an example of something that has immaterial grounding. Materialist interpretations are uniquely disruptive in the derivation of the evaluative field but by using constraint-first logic we can peel away the demand for materialist layers. It is simply necessary, irrespective of dogma.

The Key Move in Identifying the Evaluative Field

Identifying the evaluative field requires freeing ourselves from the assumption that it must arise as a physical law or necessary material construct. Our goal is not paradigm alignment or belief verification; it is deriving constraints, interpreting objective data, and originating the foundations of knowledge. With this orientation, we are freed from false constraints.

By progressively identifying what the field cannot and must be, we narrow its contours and can triangulate its location. Necessarily, it occupies an immaterial space. Because it is available to all intelligent observers at any time for any evaluative application—the evaluative field must therefore be ever-present, inexhaustible, and non-local. Its content must preexist any possible inquiry; it cannot be generated by the act of evaluation itself. It is, in effect, eternal—encompassing all "knowables" across the universe's multi-billion-year existence.

Accessing the Field

We can now ask how the field can be accessed. Access must also be immaterial and immediate, as previously identified. These constraints tell us about the field—but when we reverse the perspective, they also tell us something about human consciousness. For consciousness to access immaterial evaluation, consciousness must have immaterial aspects. If we did not already know this (as shown in the chapter on consciousness and Youness), we would know it now based on the constraints surfaced by the evaluative field.

It can thus be seen that by using a constraint-application method, and building upon what we have already established (the immaterial nature of consciousness and Youness) we can witness the generative power of the field in real time: we access it to apply constraints, and in doing so, we simultaneously gain knowledge about the field itself—building a traversable scaffolding, constraint by constraint.

Think of the evaluative field like a structure that runs parallel to and similar to reality itself: reality exists independently of any observer, yet we each experience it subjectively. Reality presents as material, yet it is not exhausted by material description. The evaluative field is slightly different. It is immaterial and mediates truth, knowledge, and correctness among individuals—without a material mechanism. In both cases, the tension between objective constraints (reality and the field) and individual experience becomes an intuitive, natural experience, not puzzling.

The evaluative field thus can be said to erupt at the same time the constraints are laid for our universe, which is in the beginning. To have a reality with intelligent observers, if those observers are to acquire knowledge and evaluate truth, then the evaluative field must be instantiated.

The Evaluative Field Is Strictly Necessary

We can now use our conclusion—the existence of the field—to impose additional constraints, such as:

If the evaluative field exists, any account of knowledge acquisition must presuppose it. And any account of knowledge acquisition that omits the field cannot explain how truth is recognized.

These constraints derive from earlier insights: any account relying on language, cognition, perception, consensus, computation, experience, optimization, or feedback as the source of truth is inadequate and incoherent as a generator of truth.

The evaluative field allows truth-recognition, and this capacity enables us to apply an objective standard of correctness.

If an account claims to explain knowledge acquisition without the field, it implicitly relies on something functionally equivalent. To be coherent, it must smuggle in the field under another name, locating it in a structure—language, brain, perception, or consensus—things that cannot generate a universal normative standard of truth. Such an account is necessarily false.

Prevailing alternative accounts of knowledge acquisition fail, and fall into several categories, each of which we have already shown to be inadequate as sources of evaluation. These accounts—relying on language, cognition, computation, sensory experience, or consensus—fail to meet the necessary constraints inherent in the evaluative field.

Historical philosophers and experts on knowledge acquisition have failed to identify the true constraints. This is compounded by the fact that modern academic hypotheses are prohibited from conceiving of the process of knowledge acquisition as being immaterial in nature which it most certainly is.

The popular justified-true-belief theory fails (Gettier problems), and contemporary epistemologists admit that the basic questions remain unresolved. Some thinkers treat knowledge as primitive, effectively conceding that their frameworks cannot separate truth-recognition from the assumed structure. The evaluative field resolves these structural problems and explains why they fail insofar as they attempt to ground truth without an evaluative constraint structure. The field must be named; handwaving is insufficient. Constraints revealed by the field itself predict that any account ignoring, denying, or failing to conceive of it as it actually operates will inevitably fail.

Platonic Thought Revisited

Plato saw that truth could not be grounded in sensory flux—Forms, or eternal structures of correctness, were necessary. But he could not demonstrate the mechanism by which finite minds access these truths. Without such access, Forms remain metaphysical postulates. The evaluative field provides the mechanism: error and correction reveal the necessity of evaluation. Words propose, minds map, the field evaluates, and recognition occurs. The field introduces a non-representational constraint layer that cannot be reduced to symbols, beliefs, or physical processes.

Gödel's Incompleteness Theorem

Gödel's theorem is not merely a statement about the limits of mathematics; it is a formal proof of the Evaluative Field's necessity. Gödel demonstrated that any sufficiently powerful consistent formal system contains true statements that cannot be proven from within the system's own rules. This "incompleteness" arises precisely because symbolic representation—whether in the form of mathematical notation, language, or neural firing patterns—is fundamentally a closed loop, and in all respects a representation of what is pre-language, and pre-representational within the Evaluative Field. The Field itself holds the objective evaluation—the completeness that Gödel indicated cannot be completed solely by a system (removed from the field).

A system of symbols can never reach out and grab the truth of its own axioms. It requires a "meta-system" to recognize what is true but unprovable. The Evaluative Field is that meta-system.

What Gödel was actually indicating is that truth is not a product of the arrangement of symbols; it is a quality recognized by the mind through its connection to the Field. Any attempt to ground truth in symbolic representation or material computation will necessarily fail because these systems lack the external, non-local infrastructure required for objective evaluation.

The Field completes Gödel's theorem by providing the missing "outside." It explains why we can see a truth that a formal system cannot prove. We are not simply calculating; we are accessing a non-representational constraint layer that dictates correctness from beyond the symbolic. Without the Evaluative Field, Gödel's unprovable truths would remain forever invisible; with it, they become the very evidence that the material and symbolic worlds are nested within a higher, immaterial order of evaluation.

The Paradox of Representation and the Evaluative Field

Paradox is not a feature of reality, but a symptom of the friction between symbolic representation and the Evaluative Field. The attempt to compress the multi-dimensional, non-local nature of the evaluative field into the linear, binary constraints of language or logic, causes a self-imposed representational system glitch. This glitch is what the symbols provide as paradox.

Paradoxes like Schrödinger's Cat or Russell's Paradox only exist because our representational systems require a “this” or “that" state. In contrast, reality, as it operates within the evaluative field, is non-local and fully integrated. It has no paradox. It doesn't "decide" between “this” “that” “A” and “Not-A” until it is forced into the bottleneck of a material or symbolic frame.

For Example:

1. **Quantum Superposition**: Is the natural state of the field before it is collapsed into a single, linear data point.
2. **The Liar's Paradox**: "This statement is false" does not break reality, rather, it breaks symbolic representation, and whatever brain attempts to reconcile it as a “this” or “that” statement. In truth, the Liar's Paradox is a proof, of the fallibility of language in the face of the field. Gödel showed us that the map (words) is not the territory; the evaluative field is the territory, and reality has no falseness, nor paradox. It is the perfect representation of constraint (arising from the Evaluative Field).

Paradox takes place when a system has reached its limit and it reaches (in representation) to represent un-representable—immaterial constraint/evaluation/Truth itself. With this understanding, we can say: “Gödel says paradoxes are the unprovable truths that emerge when a system tries to look at itself. The evaluative field provides the objective vantage point that resolves the loop. We can also say: “Plato says The Forms" are the non-local archetypes within the evaluative field. The Shadows on the wall are the linear (two-dimensional), paradoxical representations we struggle to reconcile.

Retrieval From The Field

Plato was unable to provide an account for how evaluation occurs, or how his Forms were known. He posited that cognition somehow accesses objective truth. He came up with the idea of Anamnesis, (Plato, The Dialogues of Meno and Phaedo, 1997, editor, J.M. Cooper, Hackett Publishing) a ‘reminiscence’ suggesting a prior existence within the forms, with incarnation in the next life having the ability to recover what was once known. In the Evaluative Field, contrary to Plato's ideas, there are no separate entities such as Truth or Justice in residence. The field is necessarily prior to the world of knowledge, ideas, minds, symbols or language. The field is what makes all of these possible. The field is the scaffolding from which beings are and their thoughts, are allowed to exist, evaluate, and to access objectivity.

The field is accessed through a process of approximation, error, correction, internal dialectic and iteration. The inquirer enters the field first through thought, then, under imaginative problem solving, knowledge and objectivity is extracted. Contact is immediate but the extraction is a struggle, and an evolution of understanding takes place while on a separate track a struggle of articulation and representation ensues. Because progress is iterative, the field yields specificity in direct proportion to the request. In a sense, the question becomes the answer. Famous examples from the world's most transformative thinkers provide insight into this contact.

Albert Einstein described his insights as arising from "visual and muscular" images that he wrestled with for years before he was able to translate them into famous language and then into his famous papers. In a letter to Jacques Hadamard, published in *The Psychology of Invention in the Mathematical Field*[1], Einstein noted, "The words or the language, as they are written or spoken, do not seem to play any role in my mechanism of thought. The psychical entities which seem to serve as elements in thought are certain signs and more or less clear images which can be 'voluntarily' reproduced and combined... This combinatory play seems to be the essential feature in productive thought—before there is any connection with logical construction in words or other kinds of signs which can be communicated to others[2]."

Henri Poincaré described a similar process: after weeks of concentration and singleness of thought, the solution in the form of an epiphany hit him all-of-the-sudden while stepping onto a bus. He recalled that his *subliminal self* had been evaluating against a "standard of aesthetic beauty" before insight granted him favor. In *Science and Method*[3] specifically the chapter "Mathematical Creation." Poincaré describes the creative process in four stages: Preparation: which is conscious intense effort where one is immersed in the problem, gathering facts and attempting solutions. Incubation: a period of rest or work on other subjects, while the subconscious mind continues the primary effort. Illumination: the "a-ha!" moment where a potential solution flashes into conscious awareness. Verification: where the insight is tested and verification is attempted. Poincaré emphasized that the subconscious was capable of testing countless combinations and selecting the most elegant and useful ones[4].

Nikola Tesla claimed he could create and test his inventions entirely within his mind. His imagination conjured the constraints so completely that when he actually built his inventions the physical representation was literally an "after thought". Tesla wrote, "My method is different. I do not rush into actual work. When I get an idea I start at once building it up in my imagination. I change the construction, make improvements and operate the device in my mind... Invariably my device works as I conceived that it should, and my experiment comes out exactly as I planned it. In twenty years, there has not been a single exception[5]."

Johannes Kepler discovered the laws of planetary motion which he described as a "sacred frenzy" of "thinking God's thoughts after Him" wherein he recognized the truth of elliptical orbits. When he had the thought, he checked it against his work, and the 'click' of alignment verified his intuition[6].

When accessed as a formal request for structure and constraint, the field can, in very rare instances, yield fully coherent, articulated thoughts. Yet, without the effort and work on behalf of the individual to master their craft, and place themselves in the position of cutting-edge inquiry, and translate these extractions into coherent form, what is accessed may never be realized by the world at large. Without these demonstrations and commitment, learning and knowledge would collapse.

Constraints and the Accumulation of Knowledge

Each new discovery is not invented but forced by the power of narrowing constraints, producing a chain of necessary inferences anchored to what is already true. This is why the method of knowledge acquisition must begin by examining evidence and enumerating what must be true. Everything else—paradigm-alignment, false restrictions, consensus, mythology, authority, tradition—is irrelevant.

But notice, trial, error, and accumulation (of knowledge) presuppose the evaluative field. Failure in this sense, is only meaningful, and definable because correctness exists independently.

Why the Field is Evaluative, not Corrective nor Coercive

This point is crucial for clarity, the problem of, "if I can be wrong, how can I ever be sure I'm right?" The perceived difficulty of explaining how knowledge acquisition coexists with false belief is likely the one place where critics of knowledge acquisition theories imagine they might be forced to compromise. They may imagine that either the field is perfect, and we never make mistakes, or we make mistakes, which means the field doesn't exist. Although the individual may stumble, the field — and the potential for knowledge — remain steadfast. Because a high precision ruler is used incorrectly and is thought to yield an incorrect result, this does not call into question the precision of the ruler. In this way, the field silences the threat of unidentifiable misrepresentation. Because the standard is steadfast, error correction is not merely a possibility among many, it is a given, as continued iterative correction converges toward truth. Truth is guaranteed in principle wherever incorrectness conspicuously stands in its place. The field is not a crystal ball, but a forensic filter, providing the gravity of truth that is experienced as an inexorable pull toward continued refinement. Through these iterations, the truth will eventually be found out. Think Keppler, Newton, Einstein, Schröedinger, Bohr, Hubble, Dirac, and Feynman.

The history of inquiry is a graveyard of "incorrectness standing in place of truth." The Field will continue its offerings, and correctness, eventually moving the following ideas from being placeholders of truth to footnotes in history many of which we have already espoused within our narrative.

Knowledge over ignorance, and successive iterations of understanding will always remain possible, despite inaccurate, misapprehended, and false constructions.

What The Field Isn't

The field exists, yet individuals still fall into error. Access is not privileged to intelligence; presence does not guarantee clarity or insight. Evaluation can be redirected, constrained, or hijacked by survival instincts, social pressures, identity, and emotion. Other obstacles are built into language, framing, and perspective, creating many potential traps that can constrain reasoning.

These "pitfalls" are necessary for the individual, but not for the field. If the evaluative field were coercive or guarded against error, then error would be impossible, yet error is both common and stable, and in many respects productive, forming the early iterations of knowledge.

Reality itself, novel environments, distinct interiority, free will, and individuation require that the field be accessible, flexible, interrogative, and malleable without being mandatory. The field enables truth-recognition. It contains no will to enforce it; will is a characteristic of the individual, and that suffices. The field makes evaluation possible but does not replace agency.

Even brilliant minds can defend falsehood with precision while the field remains active, optimizing for criteria other than truth if accessed at indirect angles. Its existence is a condition of intelligibility, not a universal enforcement engine. It is the scaffolding upon which error, insight, and understanding coexist, revealing not only the structure of truth but also the freedom inherent in conscious evaluation.

The evaluative field can be real, objective, infinite in utilization, and universally available, yet still compatible with delusion, ignorance, self-deception, and persistent error. It is a capacity, a faculty, a navigable structure. Intelligence amplifies skill, but not correctness: even the most capable minds can fail if evaluation is captured, redirected, or constrained by identity, language, reward or emotional biases.

Why would reality be built this way? Because consciousness and Youness—the presence of intelligent observers—serve priorities beyond correctness. Experience, agency, and desire necessarily coexist with truth. The instant perception of all truth, in contrast to desire, will, and agency, would compete with—and potentially partially or wholly collapse—beliefs, override choice, and flatten the richness of engagement. The universe appears designed to be intelligible, but only through the tension between correctness and the dimensions of conscious life.

The Evaluative Field & The August Wheel

Earlier we introduced the August Wheel. We can now map it onto the evaluative field.

Even the most intelligent mind can wander into delusion, confusion, or persistent ignorance. The evaluative field exists, but its presence does not guarantee correct orientation. Intelligence amplifies reasoning and coherence, but opposing mental forces—irrationality, dogmatism, rationalization, intellectualization, and dissembling—can still misdirect judgment.

The field is not designed to prevent the selection of wrong paths. It remains active, but navigation can be captured, redirected, or falsely constrained.

This raises the question: If the evaluative field exists, how do we know where we are relative to truth or error? The August Wheel answers this. It is both a compass and a diagnostic tool, mapping human cognition relative to the evaluative field. We may not always know our approximation to truth, or whether we have found actual knowledge versus self-preservation, but there are signs.

Ignoring red flags in scientific evidence, assuming paradigms over the products of science, dismissing the testability or observability of a concept, or violating the constraints of physical law and probability all indicate that our approach to truth is compromised. The same applies to categorical limits in chemistry, atomic theory, quantum mechanics, and consciousness studies. The more one ignores objective evidence, the more their path diverges from knowledge.

These dynamics become even more pernicious when mislabeled: when paradigms dominate and what is false or impossible is treated as possible, probable, or objective fact, or constraint, or untouchable dogma, none of this impinges on the wheel itself. These are socio-cultural or psychological distortions—obstacles in human reasoning, not failures of the field.

On the wheel, the Golden Path represents the disciplined ascent from *petite* truths to *meta* truths, guiding the journey from ignorance to transcendence. By following it, one can navigate the interplay between error, insight, and understanding, recognizing the structure of truth while honoring the freedom inherent in conscious evaluation.

The vertical axis runs from Ignorance to Transcendence. All humans are born into a state of complete ignorance. Transcendence is not mystical—it represents alignment of belief and reality with the evaluative field: this is coherent, grounded, and robust understanding.

The horizontal axis captures misalignment. Leftward lies Delusion (belief intersecting fantasy); rightward lies Confusion (not knowing or being unable to distinguish what is true). Delusion and confusion represent "falls": too far left, one is captured by delusion; too far right, one is overwhelmed by confusion. Intelligent minds are especially vulnerable because internal coherence can make these states feel stable, even elegant. Social reinforcement, consensus, reward, or dogma can further mislabel these states as truth.

The characteristic "click" of perceived alignment is not always a marker of truth. It can reflect genuine understanding, but it can also signal alignment with one's own delusion or confusion. In such cases, the "click" may be accurate, and aligned within that internal frame, but this coherence must not be mistaken for alignment with the field.

The Wheel does not claim rigid geometric necessity. Truth and belief are not mathematically fixed positions; they are orientations represented on the wheel—a tool for understanding both hindsight and, for insightful, real-time cognition.

The Evaluative Field & The Origin of Physical Laws

The evaluative field is not merely a convenient concept—it is unavoidable and necessary. Physical existence itself relies on constraint. Reality cannot instantiate without correct restraint: physical laws require consistency, and matter cannot persist or interact meaningfully without boundaries separating "this" from "not-this," or "cause" from "effect." Constraint implies a standard of correctness. For particles, fields, and forces to obey stable, repeatable laws, a framework in which those laws can hold must already exist. A universe cannot "settle into law" without preexisting lawfulness. Without embedded "rightness," there is no stable identity, no persistence, no repeatability, and no intelligibility.

The evaluative field is the medium in which lawfulness is possible. It underpins both the recognition of truth by observers and the existence of coherent reality. Physical laws do not emerge from matter; they are expressions of correctness instantiated by the field. The lawful behavior of matter and energy depends on this pre-existing framework.

Even the act of arguing against the field, or denying its existence, occurs within it. Claims, syllogisms, constraints, and standards of coherence, relevance, and correctness all presuppose the field. The evaluative field—and the preexistence of constraints—forms the inescapable foundation of intelligibility.

Truth & The Evaluative Field

The evaluative field is not a "truth beam." It does not continuously correct the mind in the way gravity continuously inhabits a falling object. The field is ever-present, but access to it is conditional. Put simply: the field is always available, inexhaustible, and indefinitely corrective—but it is not coercive, pre-emptive, or independent of individual will, and free choice any more than the existence of a ruler makes inaccurate measurement impossible. A standard can only be correct when it is applied.

If evaluation were compulsory, error would be structurally impossible. But error—belief, dogma, paradigm, myth—is not only possible; it is ubiquitous, stable, and often socially reinforced. Therefore, the field cannot operate by forced correction. It must operate as a constraint-space that is *available* rather than imposed: a map that can be consulted, but not one that automatically pilots the observer.

This also explains why elegance can deceive. Palatial coherence, filigreed obfuscation, encrypted dissembling, and labyrinthine cleverness can generate a closed internal satisfaction–a cathedral within the mind–that substitutes for field-contact. Where the model is internally complete, the field's pressure becomes faint; where gaps cannot be bridged, it remains mute. And this is the key tell. The mind's tendency to mistake linguistic and mathematical elegance for field-contact is eminently understandable—especially when buttressed by the endorsement of the greatest minds within a given discipline. But no elegance, flourish, or majesty of speech can traverse the untraversable: the substitution of beauty for Truth. The field cannot correct this. No additional cleverness, refinement, or coherence can perform this transformation. The missing traversal is not a technical gap; it is categorical. Consequently, the field remains silent. The prose continues, carrying gaping wounds—passages remain suspended at irresolvable impasses.

Science & The Field

Even if the evaluative field cannot be continuously corrective, it still provides knowledge that can be applied by supplying constraints. It is a map. The usefulness of the field is greatly enhanced by using additional tools, provided by the field, to discern truth from error. Science is such a tool. It supplies principles and external constraints:

Observability
Experimentation
Verification
Confirmation
Evidence
Reproducibility
Prediction
Falsifiability

Science doesn't "ask" whether one agrees — it forces convergence by public constraint.

Science is a tool born of the field—a technology for interacting with it. Its purpose is to make reality discernible and actionable for humans. Science tells us how to use the field to render our universe intelligible. It translates immaterial constraints into observable, testable, and repeatable phenomena. Without the field, science could not exist. The field provides the underlying constraints, and science formalizes them, creating structured methods.

The field requires no diagrams when the words are precise. Why? Because the field is the map. Any secondary representation is a step removed; precision in words is how the diagram is surfaced. Higher-resolution (precision) contact means finer-grained alignment with the evaluative field. The more precisely language conveys constraints, the closer the observer comes to direct contact with the field. It can be approached but never obtained.

Elegance, however, can be a solvent. Elegance dissolves the reader's contact with the field. Instead of direct constraint-contact with the field, the mind is interacting with the model's internal harmony—which feels like truth but is not. Coherence masquerades as correctness. Each representational layer inserted between mind and field reduces precision, increasing susceptibility to plausible falsehoods.

Darwinism, the Big Bang, evolution—are elegantly constructed. Their internal coherence explains away implausibility, softens red flags, and converts "we don't know" into "we basically know," and "we will eventually know fully", all while failing the harder tests of observability, representation, experimentation, and validation. This illustrates the deception of elegance. The more internally coherent a false model becomes, the harder it is to detect by intuition alone.

Throughout history, alignment and misalignment with the field has left its fingerprints on human insight. Dirac, Einstein, Poincaré, and Rutherford exemplify alignment, while Einstein (in quantum mechanics and the Big Bang), Darwin, Guth, Dennett, and Everett illustrate misalignment. These are non-coercive bullseyes and misfires—cases where minds either phase-locked precisely with immaterial evaluation or, conversely, misattributed partial truth, error, and dissonance (allegiance, dogma, paradigm) as contact with the field.

These examples illuminate a profound truth: consciousness does not generate truth—it accesses it. Success and error are measurable in terms of alignment or misalignment with the field's generative structure. Every precise insight is a bullseye in phase-lock; every misfire reveals friction introduced by representational layers, elegance, distraction, or incomplete precision. The evaluative field remains the constant detector, guiding human understanding when the mind correctly interacts with its constraints. Even in error, it is redemptive: nothing eliminates the field. It remains ever-present, ever active, and inexhaustibly capable of offering corrective iteration.

Unusual and Extraordinary Applications of The Field

Unusual and extraordinary human cases suggest a comprehensive and robust applicability of the field. Waking, dreaming, hallucination, meditation, severe brain injury, or cognitive disability do not eliminate the field. Psychedelic experiences, near-death events, or symbolic breakdown during such experiences may change how the field is accessed, but do not destroy it, not limit access. Whether it exists or is available is not a question. It is foundational

Savants display access to structures most minds do not easily reach. Mediumship claims, telepathy reports, and remote viewing claims whether one accepts them or not—indicate a far-reaching application of the field. The calculation of numbers, calendar dates, days, years, and activities, language acquisition, the recitation of Pi to arbitrary lengths (tens of thousands), artistic recall, memory and recall of whatever is read, heard, or seen point towards a conceptual convergence: information and correct relation are not confined to ordinary sensory channels. Being immaterial, access may indeed be unrestricted.

Even if these cases are disputed, the philosophical point remains: if any such access occurs, it would not create the field, nor replicate it, nor eliminate its necessity. It would represent another mode of participation in it.

At this point something sobering becomes clear. Even if one rejects all extraordinary claims, the argument stands. Every critique, every objection, every denial of the Evaluative Field already necessarily presupposes it. Argument relies on logic, meaning, truth and falsity, representation, or correctness. Even the claim "there is no objective evaluation" must use evaluation to convey meaning. There is no attack outside the field, because the act of attacking already stands inside the very structure being denied. This means the field is not merely prior, universal, or invariant. It is inescapable. To deny it is to use it. To challenge it is to inhabit it.

---

[1] (J. Hadamard, The Psychology of Invention in the Mathematical Field, 1954, Dover Publications),
[2] (J. Hadamard, The Psychology of Invention in the Mathematical Field, 1954, Dover Publications)
[3] (H. Poincaré, Science and Method, 1952, Dover Publications),
[4] (H. Poincaré, Science and Method, 1952, Dover Publications)
[5] (N. Tesla, My Inventions: The Autobiography of Nikola Tesla, 1982, Inner Light Pub.)
[6] (J. Kepler, Harmonices Mundi (The Harmony of the World), 1619, translated by E.J. Aiton, A.M Duncan, J.V. field, Philadelphia, American Philosophical Society, 1997)

Section 1: Values & Validations of The Algorithm

Now, at the end of our systematic review we have one final task to perform, and that is to evaluate the *Variables* and *Special Problems* to find a way to represent the chance that the existence of our universe aligns with the hypothesis and paradigm of materialism or is irrefutable evidence of an instantiating power within reality.

There is no precise number we can arrive at, partially because for several *Variables* the exact values cannot be known, while for several other *Variables* and *Special Problems* things are required that cannot be reconciled with physical law, and thus, have no probability that can be measured.

Practically speaking, the very first event in our *Algorithm*, the instantiation of the universe at the CMB, being inexplicable in any material way, proves the point at the outset, seeing as all other things flow from this event.

Our method was however to disregard all that happened prior to the level in which the focus problem presented itself, and evaluate the probability anew. By this method, we still arrive at more impossibilities than not.

***Variables & Special Problems* Values:**

Possible Values are:

> ***Miracles are denoted with the label*** **<u>M-Miracle</u>**. No process or event within the material paradigm (first principle physics) can bring about the condition.
>
> ***Functional Miracles are denoted with the label*** **<u>fM-Functional Miracle</u>**. fM-Functional Miracles are events that have odds that exceed the probabilistic capacity of the universe (PCOU) determined at less than 1 chance in $\mathbf{10^{140}}$.
>
> ***The value of "<u>Inevitable</u>" is noted and will be explained.***

Special Problem #1: Beginning of the Universe: **<u>M-Miracle</u>**

Variable #1: Baryon Asymmetry: **<u>M-Miracle</u>**

Variable #2: Initial Entropy State of Universe, 1 in $\mathbf{10^{10^{123}}}$. This exceeds the PCOU: **<u>fM-Functional Miracle</u>**

Variable #3: The Horizon Problem: The Penrose number for the low entropy state, is derived from the thermalization of the CMB and is given as $\mathbf{10^{10^{123}}}$. This exceeds the PCOU: ***<u>fM Functional Miracle</u>***

Variable #4: Cosmological Constant 1 in $\mathbf{10^{120}}$ finely tuned with The Expansion Rate of The Universe (x 1 in $\mathbf{10^{120}}$) equals 1 in $\mathbf{10^{240}}$. This exceeds the PCOU: **fM-Functional Miracle**

Variable #5: The Fine Tuning of Gravity: 1 part in $\mathbf{10^{168}}$. This exceeds the PCOU: **fM-Functional Miracle**

Variable #6: Fine Tuning and Anomalies of The Standard Model of Physics: **M-Miracle**

Variable #7: Fine Tuning of Carbon Resonance, 1 in $\mathbf{10^{143}}$. This exceeds the PCOU: **fM-Functional Miracle**

Variable #8: Stellar Nucleosynthesis (Population III Stars Without metals), the chance of stars forming is effectively zero. Meaning, out of any arbitrary number of simulations, say $\mathbf{10^{140}}$ or greater ($\mathbf{10^{10^{123}}}$) none would result in stars: **fM-Functional Miracle**

Variable #9: The CMB As The Instantiation of The Universe: **M-Miracle**

Variable #10: CHNOPS: no odds can be calculated. According to Atomic Theory, the fitness of the CHNOPS elements cannot be otherwise. however improbable the fitness: *1:1,* **Inevitable**

Special Problem #2: Abiogenesis:1 chance in $\mathbf{10^{100,000,000,000}}$. This exceeds the PCOU: **fM-Functional Miracle**

Variable #11: DNA, Sequence Hypothesis 1 chance $\mathbf{6^{3,000,000,000,000}}$ x 20,007 x $\mathbf{10^{74}}$: **fM-Functional Miracle**

Variable #12: Mitochondria, Cellular Respiration, CAC, ETC & ATP Synthase, DNA—nuclear & mitochondrial + separate coding conventions, Quantum jumps, proton pumping, etc. This exceeds the PCOU: **fM-Functional Miracle**

Variable #13: Chirality & Organic Molecules: 1 chance in $\mathbf{10^{100,000,000,000}}$. This exceeds the PCOU: **fM-Functional Miracle**

Special Problem #3: The Evolution of the Species, ~ 5 to 50 billion separate speciation events: **fM-Functional Miracle**

Variable #14: The Divine Meditation: **M-Miracle**

Special Problem #4: Consciousness: **M-Miracle**

Variable #15: Cell Consciousness: **M-Miracle**

Final Tally: *1 Inevitable*, 7 M-Miracles, & 11 fM-Functional Miracles

The chance of our universe being the produce of the materialist's hypothesis is Zero, (0).

Summary

In a probabilistic, bottom-up universe, the path to complexity is a chain that is only as strong as its weakest link. For the materialist paradigm to hold, every variable and special problem listed above must be bridged by unguided processes. Our investigation has revealed insufficient—and missing—links to provide any bridge for unguided processes to serve as the 'glue' that holds our reality together.

Mathematically, any number multiplied by zero is zero. In the context of universal history, a single event with zero probability halts the materialist design permanently. This means a single chain, as in The Divine Algorithm, with a single multiple of Zero, invalidates the entire proposition. The materialist proposition for our existence stands at zero.

This fact also holds relevance for the 'inevitability' rating we arrived at for our Variable of CHNOPS. CHNOPS is only 'inevitable' because we are examining it in a vacuum, with no consideration of the factors that have gone into its possibility and thus, 'inevitability.' In truth, CHNOPS rests upon a foundation of four M-Miracles and six fM-Functional Miracles. This means in the final meta-analysis, CHNOPS is also an M-Miracle, as are all variables and special problems following the special problem of the beginning of our universe.

Overall, we have identified eight M-Miracles and eleven fM-Functional Miracles. From first principles, the materialist hypothesis suffers from a lack of causal capacity in explaining any part of our universe. The universe we inhabit is not a product of law, nor chance. Something else is required to instantiate, maintain, and join the chain of miracles within our universe from its inception to its destined product: biological humans with human consciousness and 'Youness.'

Next, we move on to the Validations of The Divine Algorithm.

We have identified multiple internal self-validations of the hypothesis whereby the nature of our universe, by constraints, inherently validates *The Divine Algorithm*.

Validations of The Divine Algorithm

Validation 1 Intelligent observers must reach a pre-requisite level of technological ability to identify and recognize *The Algorithm.*

Result: The Algorithm is *inevitable* for all groups of intelligent observers throughout the universe that have reached the pre-requisite level of technological ability to recognize and decipher *The Algorithm*. Prior to this stage of technological development, *The Algorithm* will remain hidden.

Validation 2 *The Algorithm* is made up of a pattern of *Miracles* to a materialistic view, and as instantiations that defy organizational reason to consciousness aligned ontologies. The instantiations extend across every level of organization and every discernible part of physical reality. As a group they are not localized or isolated, but present throughout the entire structure of the universe. This pattern contains the detailed technological and scientific information required for the rise of intelligent observers within the universe.

Validation 3 *The Algorithm* is unlike any message, signal or pattern discovered, is not written, nor is it a signal, nor is it represented in any symbolic representation. It is not contained or constrained by any physical space, nor held in any particular location. It is not "propagated" and it does not come *from* anywhere; however, when it is recognized and understood, it avails itself to being translated into the language or mathematical representation of any group of intelligent observers that has discovered *The Algorithm.*

Result: The Algorithm is an ingenious unambiguous non-symbolic, non-representational message unlike any ever discovered.

Validation 4 The identifying characteristics of *The Divine Algorithm* are consistently the same. They are events etched into the fabric of the universe that cannot be accounted for by first principles physics or probability under any paradigm. Even for cultures receptive to consciousness in the universe, they are unmistakable for the levels of reality they give rise to.

Validation 5 *The Algorithm* acts through highly specified, non-random events beyond the reach of chance. They are precisely the kind of intervention required, at precisely the points in our universal history required to (1) maintain a life-permitting universe, (2) prevent the breakdown of a life-permitting universe, (3) enable the emergence of life, and (4) produce intelligent observers within our universe.

Result: The Variables and *Special Problems* are evidence of the *exact* series of miracles/events required to set a new universal condition that allows intelligent observers to manifest, from the inherent fitness of the universe.

Validation 6 Each of the 19 parts of *The Algorithm* are collaborative, cooperative, interdependent, sequential and synergistic with respect to one another. As each part makes itself known within our universe, ever increasing functional capacities and functional hierarchies are created within our universe, and each of the 19 instantiations in *The Algorithm* set the stage for the next instantiation in *The Algorithm*. The final revealed hierarchy is the realization of the Algorithm, and all that it implies.

Validation 7 Because *the Algorithm* exceeds the maximum constructive capacity of any lifeform within our universe; meaning it cannot be the cause of any life form within our reality. The pattern can only be attributed to a designing consciousness that set the constraints for the universe and brought it into existence.

Validation 8 The properties required to produce the structure of *The Algorithm* necessarily reflect the properties and capacities of the source of *The Algorithm*. These characteristics function as an evidentiary dossier and genetic imprint for the force behind *The Algorithm*. The power behind *The Algorithm* reveals itself through what *The Algorithm* achieves.

Result: This force must be eternal with regards to our universe having set the constraints for the universe and having created the universe and being active within the universe in the present.

This force must be omnipotent with regards to our universe, having created the universe and having the power to enforce conditions that set constraints that span the entire observable universe.

This force must be omniscient with regards to the universe, having the understanding to contrive the universe and manipulate it over a series of instantiations, that synergize capacities set within reality by constraint at the very beginning, to result in our present-day universe.

This force must be immaterial and nonlocal with regards to our universe, seeing as the universe has no possible material cause or material limitation, and that the creative power behind our universe possesses the power of immaterial instantiation upon the entire universe in the form of several forms of immaterial consciousness: constraint, universal, material, instantiation, individual, cellular, etc.

Validation 9 Specific to humanity, as long as the universe is viewed through the paradigm of scientific materialism *The Algorithm* will remain invisible and thought of as a perplexing set of disconnected anomalies or intractable problems. The belief in counter-narratives to *The Algorithm* will serve to limit human progress, and our ability to accept the invitation of *The Algorithm* to know the primary cause and foundation of our universe. The falsification of scientific materialism is a necessary step for humanity to discern *The Algorithm*. In this respect, materialism functions as a limiting, self-sealed framework that prevents recognition of *The Algorithm*.

Validation 10 *The Algorithm* is *only* meaningful to intelligent observers. As a thought experiment, if the universe were exactly as it is, but it had no intelligent observers, there would be nothing within the universe that could be "aware" of, or identify *The Algorithm*, yet *The Algorithm* would still exist (within the thought experiment). For this reason, *The Algorithm* can only be interpreted as an overt message for intelligent observers. Only intelligent observers can observe it, identify it, discern it and comprehend its meaning. It is therefore not only meaningful to intelligent observers, but it also exists (along with the universe it produced) only in service to intelligent observers. In our thought experiment (which can only be conducted by beings that think) because *The Algorithm* exists without intelligent observers, the paradigm of scientific materialism is falsified within the thought experiment universe.

> *Result*: The falsification of materialism is not a quirk of human understanding. It is a scientific fact.

Validation 11 *The Divine Algorithm* is far too complex for humans to have possibly anticipated, and far too comprehensive to be a product of human imagination. *The Algorithm* is too unambiguous in its makeup to be misperceived or misinterpreted. This means *The Algorithm* is neither anthropomorphized, a system of belief, or figment of the mind's ability to create patterns of meaning, or a quirk of human cognition. It is a real, existing, relevant and meaningful pattern within our universe.

With its multiple forms of validation *The Algorithm* is the *kind* of message that humanity, and all intelligent observers in our universe will be able to recognize, by reason, constraint, and characteristic, as something that could have only been conceived by an omniscient, omnipotent, eternal, immaterial consciousness and organizing power behind the universe.

Validation 12 *The Divine Algorithm,* in what it represents, what it achieves, and in how it conveys its message is in perfect alignment with the oft-named qualities of an all-powerful, all-knowing creator of the universe as described in the historical record. This is not a coincidence. Independent cultural and experiential traditions converge on descriptions consistent with the characteristics derived from *The Algorithm*. Even in cultures that have non-theistic spiritual practices, there exists a prolific record of individuals within that culture having had *spiritually transformative events*. In these events, participants report encountering a creative consciousness which is one for one with what we have identified as the creative capacities and creative source of *The Algorithm*—an eternal, immaterial, omnipotent, omniscient creative consciousness.

These are not separate ideas. It represents a convergence and a conclusion reached from independent directions.

Validation 13 Once surfaced within a culture, the individual within that culture will have the concept of *The Algorithm* available for them to consider on their own terms outside of the social view. It will then be directly accessible to the individual, independent of institutions, language, culture, belief, or prior knowledge. Each individual can then freely choose to understand and accept *The Algorithm* or reject it. No other "approval" is required. *The Algorithm*, once known, is a direct apparel to the individual.

> *Result:* Once discovered *The Algorithm* is a direct appeal to each individual, circumventing all systems of universal explanation, or dogma or control. A prisoner in a dungeon, or an ascetic in a mountain cave, knowing no language, written or spoken, and cut off from any contact with the world, can have a spontaneous experience where they receive the message of a creator, the same message contained within *The Algorithm*. The message of *The Algorithm* therefore is functionally irrepressible and expressed in multiple forms and relevant particularly to individuals.

Validation 14 Any group of Intelligent Observers that discover and decipher *The Algorithm* would have the information needed to recognize that it is universally applicable to all life within our universe. This implies that all spiritualities, religions or scientific conjectures that posit a single immaterial creative source for the universe, converge upon the power behind *The Algorithm* as being that single creative source.

This conclusion is an outgrowth of (1) The nature of the instantiations that make up *The Algorithm*, (2) *The Algorithm's* translatability into any system of representation and, (3) the necessary characteristics and capacities required to be responsible for the instantiations that make up *The Algorithm*.

The Algorithm represents validation of all spiritual ontologies that align with its principles. Individual cultural differences are to be expected. Once the core idea has been obtained and it is used to form the central spine of spiritual ontologies, it joins together all ontologies that posit a single creative source.

Validation 15 The Divine Algorithm proposes a single coherent cause that replaces multiple disconnected, ad hoc explanations for the intractable anomalies within our existence. In The Algorithm, we have a single explanation vs 19 conjectures for scientific materialism.

Validation 16 The Algorithm refines the Anthropic Principle and the Copernican Principle (the Principle of Mediocrity) and the Cosmological Argument. It validates that our universe was intended for life and that the position of humanity and Earth is privileged, as are any form of life, or intelligent beings throughout the universe. This realization is available for any race that identifies *The Algorithm*.

Life on Earth is not an accidental happening on a lonely insignificant rock. Intelligent life is not incidental, and the universe is not merely compatible with life, it is structured for it and calls it forth. Life on Earth is the culmination and final product of *The Algorithm*. Our place and time in the universe is not arbitrary. We exist at the exact point in time, and age of humanity where *The Algorithm* can be read.

The Anthropic principle:

(*The definition that has been invalidated*): Our observations of the universe are limited to and selected by the fact that they are only possible in a universe where we are possible, and where we exist. We should not be surprised to observe these facts.

> *Result*: This is a tautology of *The Algorithm* twisted to manipulate a meaning that is the exact opposite of what is true, while reaching the same conclusion as the truth. *The Anthropic Principle* is sophistry, a subversion of reality. It uses the inevitable outcome achieved by instantiation to deny the necessity of the Instantiator.

Principle of Mediocrity (aka The Copernican Principle):

(*The definition that has been invalidated*): What is to be expected, and what is observed is mediocrity vs. that which is superlative or special. Nothing within our universe can be considered exceptional, privileged or intended. There is no goal or purpose in our universe. Our location and existence as living beings on earth is insignificant.

The Cosmological Argument Redux:

The existence of a creative consciousness (instantiator) is a necessary inference derived from the functional output of what has been instantiated.

Validation 17 One of the strangest aspects of *The Algorithm* is it is constructed and hidden in a way no human would imagine designing. It is not concealed by simplicity, or complexity alone. It is concealed in a highly unusual location, within impossibility itself. Each problem within it appears unsolvable. Complicating *The Algorithm*'s difficulty each of its 19 parts are broken up and spread throughout space, time and category, conveying the illusion that there is no connection between its member *Variables*. But this is where the genius of *The Algorithm* is revealed. Its counterintuitive structure is singular and purposeful. We have only to recognize the commonality of its member parts and recognition will be forced by necessity. And the moment that recognition happens, what at first appears as intractability will be recognized as design: then the solution cannot be unseen, as the entire Algorithm falls into place in chronological order, from the beginning of the universe to the present.

This Concludes Section 1 of our Investigation.

We have followed probability to its limits and reached the boundary of material explanation. What remains is constrained to an unavoidable set of facts regarding our universe.

Having recognized the pattern of *The Algorithm*, this is what we have endeavored to provide.

The Algorithm is not a hypothesis among many—it is the only framework that accounts for the full set of constraints, transformations, and outcomes observed within our universe.

To draw conclusions from anything less than the complete set of universal constraints is to fall into the trap of Plato's allegory, mistaking the shadow of physical form for the cause of what casts the images.

Constraints do more than resolve the *Algorithm*; they provide consciousness with access to objectivity. This field is an anomaly to the materialist, being that it is immaterial, yet accessible to consciousness. But it is a logical necessity if there is to be objective truth. Without an objective substrate certain knowledge is a logical impossibility.

To understand how knowledge is distilled from probability, we must now define the bridge. Something we rightfully call *The Evaluative Field.*

The Divine Algorithm

Book Three: *Transcendence*

The Divine Algorithm

TABLE OF CONTENTS

Book Three: *Transcendence*

Page	Chapter
587	Introduction
588	1. Mystical Experience & Spiritually Transformative Experience
597	2. Maya & The Communication of God Consciousness
606	3. The Prophets Experience
626	4. The Physical/Mentally Triggered Mystical Experience
659	5. The Biologically Triggered Mystical Experience
682	6. The Near Death Experience & Out of Body Experience
700	7. Transcribed Near Death Experiences, Atheists & Skeptics
737	8. Miracles
748	9. Is the Creator in Book 3, the Instantiator in Books 1 & 2?
751	10. For The Future
756	Acknowledgments

Introduction

"For since the creation of the world God's invisible qualities–his eternal power and divine nature–have been clearly seen, being understood from what has been made, so that people are without excuse." -Romans 1:20

"Jesus said, 'Understand what is right in front of your face, and all the secrets will be revealed to you. There is no secret that will not be revealed." -Gospel of Thomas 5

Review

Section 1 established The Algorithm as our new paradigm: a reality where an immaterial consciousness is the universe's primary cause, and the operational mode of the universe is instantiation

In Section 2, we apply this framework to the human record. By investigating written and oral traditions—from prophets, oracles, shamans and 'regular folk'—we will demonstrate that the creative force behind the physical "miracles" of the cosmos is the same force that inspired all human spirituality.

Through this investigation, we will arrive at a unified conclusion:

Common Origin: The universe, life, and the human soul were created by the same process and force.

Universal Access: Multiple pathways have been made available for humanity to commune with this intelligent force.

Immaterial Validation: We will provide sufficient evidence to demonstrate the human soul is a non-local, immaterial entity.

1 The Mystical Experience & The Spiritually Transformative Experience

> "There are many types and many names for experiences that can share common features and be catalysts for spiritual transformation: near-death experiences (NDEs), near-death-like experiences (NDLEs), out-of-body experiences (OBEs), visions, spiritual emergencies, awakenings, kundalini, enlightenment, exceptional human experiences (EHEs), pre-birth memories, past-life experiences, nearing death awareness (NDAs), after-death communications (ADCs), empathic or shared near-death experiences, peak experiences, etc. STEs include or may be called numinous, noetic, transcendent, transpersonal, mystical, anomalous, religious, paranormal, parapsychological or ecstatic experiences[1]."

The Mystical Experience is synonymous with The Spiritually Transformative Experience. They differ only in that *the mystical experience* is a reference to the experience itself, while the *spiritually transformative experience* is in reference to the transformational effects in consequence of the experience for the individual.

An example is provided below.

> "This state, when the veil has vanished and yet the body remains for some time, is what the Vedantists call the Jivan Mukti, the living freedom. If a man is deluded by a mirage for some time, but one day the mirage disappears, –if it comes back again the next day or at some future time, he will not be deluded. Before the mirage first broke, the man could not distinguish between the reality and the deception. But when it has once broken, as long as he has organs and eyes to work with, he will have to see the image, but will no more be deluded. That fine distinction between the actual world and the mirage he has caught, and the latter cannot delude him any more."
>
> -Swami Vivekananda, The Vendanta Philosophy, address (1896)[2]

The spiritual transformation that accompanies the mystical experience represents a rebirth and a new life of understanding for the individual.

> "Oneness of mind and will with the divine mind and will is not the future hope and aim of religion, but its very beginning and birth in the soul. To enter on the religious life is to terminate the struggle. In that act which constitutes the beginning of the religious life—call it faith, or trust, or self-surrender, or by whatever name you will—

there is involved the identification of the finite with a life which is eternally realized. It is true indeed that the religious life is progressive; but understood in the light of the foregoing idea, religious progress is not progress towards, but within the sphere of the Infinite. It is not the vain attempt by endless finite additions or increments to become possessed of infinite wealth, but it is the endeavor, by the constant exercise of spiritual activity, to appropriate that infinite inheritance of which we are already in possession. The whole future of the religious life is given in its beginning, but it is given implicitly... Though he is not exempt from temptation and conflict, [yet] in that inner sphere in which his true life lies, the struggle is over, the victory already achieved. It is not a finite but an infinite life which the spirit lives. Every pulse-beat of its [existence] is the expression and realization of the life of God."

-John Caird, An Introduction to the Philosophy of Religion (1958)[3]

When Caird speaks of "terminating the struggle" he is referencing the dismantling of the system of Energy Expenditure that maintains the false, materialist view of the self as a mechanism, and the "I" as an impotent, entrapped witness caught within an arbitrary world, where meaning is stillborn. And by "victory" Caird points directly to the system's collapse into its most efficient state: the Zero-Resistance Substrate of Truth. What we have previously identified as the *evaluative field* of Truth.

The struggle is the friction of incoherence. Victory resonates with the Substrate of Truth.

William James - Varieties of Religious Experience

Medical doctor and philosopher William James is recorded as a founding father of the modern practice of psychology. He was a prolific lecturer in multiple disciplines, expounding on the subjects of psychology, physiology, philosophy, anatomy, biology, and medicine over the course of three decades at Harvard University. James was also the founder and organizer of the schools of pragmatism and functional psychology. Among his many endeavors, James studied the mystical spiritual experience in his classic text, Varieties of Religious Experience (1958). In James's view, the mystical experience was the core experience of the religious experience. James viewed these experiences as being conveyed by the precondition of an altered state of consciousness and/or a receptive condition for a mystical experience.

"One may say truly, I think, that personal religious experience has its root and centre in mystical states of consciousness; so for us, who in these lectures are treating personal experience as the exclusive subject of our study, such states of consciousness ought to form the vital chapter from which the other chapters get their light."

-William James, *The Varieties of Religious Experience* (1958)[4]

In his study of religious experience, James saw the need to delineate the mystical nature of the personal religious experience from the common day use of the term, "mystic." James therefore restricted the use of the term, "mystical" to events that met the definition for the "4 main characteristics" of a mystical experience.

Four Main Characteristics of the Mystical Spiritual Experience (William James)

The Western Understanding of The Mystical Experience in Two Parts: William James & W.T. Stace

1. Ineffability: The subject of (the mystical experience) defies words and expression, and that no adequate report of its contents can be given in words. It follows from this that its quality must be directly experienced to be known. The experience cannot be imparted or transferred to others.

Ineffable: *Incapable of being expressed, cannot be uttered, cannot be conveyed in words. Indescribable. Incapable of being expressed; indescribable or unutterable.*

2. Noetic quality: Although very similar to states of feeling, mystical states also have the quality of being a state of knowledge. Mystical states are states of insight into truth that are normally off limits to discursive intellect. These feelings/experiences are illuminations and revelations that are full of significance and importance, and as a rule they carry with them a curious sense of authority, -that they are coming from a foundation of understanding-. These two characteristics, -revelation and authority- if they are inclusive of any experience, according to James, entitle that experience to be called mystical.

Noetic: *of, or based on the intellect or an intellectual quality*

Two other qualities are less sharply marked by James, but are usually found in mystical experiences, these are:

3. Transiency: Mystical states cannot be sustained for long. Except in rare instances, half an hour, or at most an hour or two, seems to be the limit beyond which they fade into the light of the common day to day experience. Often, when faded, their quality can but imperfectly be reproduced in memory.

4. Passivity: Although the oncoming of mystical states may be facilitated by preliminary voluntary operations, such as fixing one's attention, or going through certain bodily performances (physical rituals or practices), or in other ways which manuals of mysticism prescribe, when the characteristic consciousness of the mystical experience has set in, those who experience the event feel as if their own will were in abeyance. They may sometimes feel as if they were grasped and held in consciousness and attention, by a superior power while the direct communion and information download is in process.

> "Mystical states, strictly so called, are never merely interruptive. Some memory of their content always remains, and a profound sense of their importance. They modify the inner life of the subject between the times of their recurrence... These four characteristics (ineffability, noetic quality, transiency, and passivity) are sufficient to mark out a group of states of consciousness peculiar enough to deserve a special name and to call for careful study. Let it then be called the mystical group."
>
> –William James, The Varieties of Religious Experience (1958)[5]

Philosopher, Walter Terence (W.T.) Stace, author of *Philosophy and Mysticism*, (1960)

Walter Terence Stace also defined the characteristics of the mystical experience in a similar way to James and described the mystical experience as achieving "Godhead", or -partaking of the nature of *God-mind*, or *God-presence*. This concept is synonymous with James' notion of "noetic quality" and "passivity" where information is transmitted while the individual feels held at attention by a superior power. These ideas also conform with the Hindu concept of "Atman with Brahman", where the individual (Atman) is joined in oneness with Brahman (God, God nature, or, the unnamable, ineffable source of all creation)[6].

Qualities of the Mystical Spiritual Experience (W.T. Stace)

The qualities of the mystical experience as described by Stace are categorized by the following 7 qualities

(1) Unitary consciousness: the experience of a unifying vision where the individual is felt to be one with the creative source and all things. There is only *one thing, one experience and one reality*, and the illusion of separation gives way to a higher truth of collective consciousness.

(2) Nonspatial and nontemporal awareness: the feeling that there is only one moment and one space. This is a logical extension of the product of unitary consciousness.

(3) Sense of reality and objectivity: what is experienced feels *real*, and is described as lucid, and even *meta-real* or more real than ordinary reality. It is experienced as an objective, yet mystical.

(4) Blessedness: the experience feels blessed, and the individual feels joy, satisfaction, fullness and completeness. There is a deep feeling of gratitude, and thankfulness for being allowed to have this special experience, sacred knowledge and direct communion with the whole. This quality of blessedness is also transferred to all things, living and non-living in our existence. The world and everything in it is experienced as a sacred gift.

(5) Sacredness: the experience is apprehended as holy, sacred and divine. This is attributed to the experience of *God* or *God* Consciousness within the mystical experience. The sacredness is the reason the mystical experience is considered a *religious* experience, and/or, the validation of prophetic vision and the foundation for the world's organized religions.

(6) Paradoxicality: in reference to the mystical experience paradoxicality can refer to the introduction of a paradox or paradoxes, *"a statement* (about a concept, experience or quality) *that is seemingly contradictory or opposed to common sense and yet is perhaps true"*[7], or paradoxicality can also mean, *"not being the normal or usual kind*[8]*."* Both are true at the same time in context with the mystical experience. Paradoxes are experienced within a state of consciousness, within the mystical experience, that is not "normal" and even, singular/ highly unusual[9].

(7) Ineffability: what is experienced, is an experience that is incapable of being expressed fully in words. Words are the "wrapper" that refer to the ineffable experience.

> "The many degrees of the mystical ecstasy culminate in what Hindu theology has described as sat-chit-ananda, being-awareness-bliss; here for an eternal moment all contradictions are reconciled, all questions answered, all wants irrelevant or satisfied, all existence encompassed by an experience that is felt to define ultimate reality: boundless, timeless and ineffable."
>
> -Lester Grinspoon, J.F. Bakalar *Psychedelic Drugs Reconsidered,* (1979)[10]

In paradoxicality within the mystical experience, emotions, perceptions, thoughts, awareness, are super-natural, hyper-lucid and all of existence is unified, as in:

Unitary consciousness and yet being aware of being an individual.

Time dilation, eternal moment, and yet, a progression of events and ordered happenings (linearity) and awareness of one's personal (linear) history.

Awareness of the material realm and the realm of soul/consciousness

Awareness of the collapse of all dualities

The mystical experience may be described through words, as we have detailed above, but the experience, its impact, its consilience, its profundity, cannot possibly be given either in words. The experience is beyond normal experience and speaks of a different form of consciousness and a newfound "override" capacity within consciousness to commune directly with Godhead. The experience is not in existence; it is within consciousness.

"The realization of enlightenment, or Buddhahood, is difficult. And it is also not difficult. It is difficult if you start looking for it. It is not difficult if you simply sit down, settling within yourself in calmness, quietness, being just purely aware. Then you are the Buddha, then you are the enlightenment. It is not that you become enlightened, it is not your becoming. It is your very being, it is you in your simplest, spontaneous nature. Enlightenment is your self-nature. Once somebody attains to enlightenment, the greatest difficulty is to convey it to those who are living in darkness and who have never seen any light. It is almost like talking about light to blind people. One enlightened master has been reported to have said, "My whole effort of conveying my experience is just like selling glasses to blind people." Hence, many of those who have attained have remained silent, and those very few who have spoken, know that their words cannot carry their enlightenment, its beauty, its joy, its fragrance; that the moment the experience is translated into words, something essential dies. Only a corpse reaches to the other person. But out of compassion, hoping against hope, a few enlightened people down the ages have made every effort to convey to you that life is not all that you think it is. It is much more, infinitely much more. But no enlightened man has ever written a single word, for the simple reason that the spoken word has a certain warmth, and the written word is absolutely cold."

–Osho, Bodhidharma: The Greatest Zen Master (1987)[11]

The mystical experience does not belong to the realm of physical sensations, abstract thought, or logical reasoning. It is an experience with the *ALL* from which *all* things are given birth and *all* things are made sensible.

The sage or mystic on the mountaintop, or in the monastery, can give the seeker an indication of what to look for and clues and instructions on how to take additional steps forward, however, the nontemporal, immaterial, paradoxical and ineffable nature of the mystical experience requires a *personally* initiated experience and the employment of unorthodox, and at times absurd methodologies. Absurd or unorthodox relative to how mathematical, academic, and time/space/distance problems are solved.

The mystic indicates the location of the spiritual treasure, revealing that it has occupied the center of our consciousness all along. While we have been searching outwardly for value, the true richness is found through the inner witness to experience. This 'treasure' is where the contact is made with the substrate of existence—the source itself. In the words of Meister Eckhart, the 13th-century Christian mystic:

> "The eye through which I see God is the same eye through which God sees me; my eye and God's eye are one eye, one seeing, one knowing, one love." -Meister Eckhart

This is the interesting thing. Where Osho talks of *"selling glasses to blind people"* and where Meister Eckhart talks of *"...my eye and God's eye are one eye"*, we find that what comes into resolution may appear to be outward—that which is seen, but in truth what comes into resolution is ourselves. Right? Because the objective world remains unchanged, invariant, in every measurable way. What has shifted is the internal resolution of the witness—the aperture through which the All beholds itself, not the 'seeing eye' rather, the resolution of the witness within. This is where the contact is made. In our recognition, in our identification, and in our knowing.

> "...a Sufi mystic, Dhun-nun had a disciple... One day, Dhun-nun gave him a stone and told him to go to the market, to the vegetable market, and to try to sell it. The stone was very big, it looked beautiful. But the master said: 'Don't sell it, just try to sell it: observe, go to many people and just report to me how much we can get for it from the vegetable market.' The man went. Many people looked at it and they thought: It can be a good show-piece, our children can play with it, or we can use it as measures for our vegetables. So they offered, but just a few small coins, like ten paisa. The man came back. He said: 'At the most, we can only get ten paisa for it - and responses were different, from two paisa to ten.' The master said: 'Now you go

to the gold market and ask people there. But don't sell it, just enquire how much.' From the gold market the disciple came, very happy, and he said: 'These people are wonderful. They are ready to give a thousand rupees for this. Responses were different, from five hundred to a thousand rupees.' The master said: 'Now you go to the jewelers, but don't sell it.' He went to the jewelers. He couldn't believe it. They were ready to offer fifty thousand rupees. And when he wouldn't sell, they went on increasing the offers - they reached one hundred thousand rupees. But the man said: I am not going to sell it. They said: 'We offer two hundred thousand rupees, three hundred thousand rupees, or whatsoever you say. But sell it!' The man said: 'I cannot sell. I am just to enquire.' He couldn't believe it - these people were mad. He himself thought that the price that was offered in the vegetable market was enough. He came back. The master took the stone and said: 'We are not going to sell it, but now you know that it depends on you, on if you have the touchstone, the understanding. You go on asking questions and you live in the vegetable market. You live in the vegetable market and you have the understanding of that market. Then you ask for valuable secrets: you ask for diamonds. First become a jeweler, and then come to me. Then I will teach you.'

-Osho, The Grass Grows by Itself, (1976)[12]

---

[1] (American Center for the Integration of Spiritually Transformative Experiences, [aciste.org], What is a Spiritually Transformative Experience (STE) 2022 ACISTE)
[2] (Swami Vivekananda, The Vedanta Philosophy, an address before the graduate philosophical society of Harvard University, 4th edition, Pg 15, (1896) 1901)
[3] (J. Caird, An Introduction to the Philosophy of Religion, London and New York, 1880, pp. 243-250, and 291-299, much abridged': Within, W. James, Varieties of the Religious Experience, (1958), Lecture XVIII. Philosophy. Pgs. 452-53)4
[4] (W. James, Varieties of Religious Experience (1958), opening to Lectures XVI And XVII. Mysticism.)
[5] (W. James, The Varieties of Religious Experience, 1958, Pg. 370-71)6
[6] (Concepts and core definitions from William T.Stace, Mysticism and Philosophy, Lippincott, 1960):
[7] (Merriam-Webster Online: Paradoxical)8
[8] (Merriam-Webster Online: Paradoxical)9.
[9] (W. T.Stace, Mysticism and Philosophy, Lippincott, 1960)
[10] (L. Grinspoon, J.F. Bakalar, Psychedelic Drugs Reconsidered, Pg. 13, Basic Books, 1979])10.
[11] (Osho, Bodhidharma: The Greatest Zen Master, 7/5/87 to 7/14/87, English Discourse series, 20 Chapters, 1987, Chapter #1, Chapter Title: To Seek Nothing Is Bliss, 7/5/87)12

[12] (Osho, The Grass Grows by Itself: Talks on Zen, given 2/21/75 to 2/28/75, English Discourse Series, 8 Chapters, 1976)13

2 Maya & The Communication of God Consciousness

> Bhagavad Gita, Chapter 7, Verse 13: "Deluded by the three modes of Maya (ignorance, passion and goodness), people in this world are unable to know Me, the imperishable and eternal."

Section 1 of *The Divine Algorithm* could be rightfully described as a discussion concerning the illusory aspect of existence as embodied within 19 Variables and 4 Special Problems. The source of this illusory aspect is the source that has instantiated our reality. This is the one that manifests physical form and the biological life that plays host to consciousness and the spiritual entity within that holds our individuality.

Section 2 of *The Divine Algorithm* is where we investigate the power behind the instantiating power that created the multitudinal levels of existence. This power is what we are attempting to describe as it has been described in oral and written history. This power is the primary cause of all *manifestations;* a point illustrated stepwise in Section 1.

Maya is an illusion, but the opposite of a misperception. It is "real", but a manifestation of a deeper energy that enlivens material form. *Maya* is an essential paradox wherein each level of existence is its own inscrutable and inexplicable mystery nested within yet another *inexplicable, highly constrained, rule bounded, and perfected* mystery.

In the mystical experience, we find confirmation that Maya—the world of appearance—is one of energy bound by energy. In the view of science, physical reality (Maya), at its most fundamental scales, can be measured, but cannot be found as a physical 'thing'—a virtual impossibility in material terms.

Materialism, which only acknowledges that which is 'physical', is therefore confronted with the non-physical foundational reality underlying everything it calls 'material', because what it describes arises without identifiable origin and is ordered by rules that are observed but not located as objects. In this sense, the underlying structure is parameter-based and free in form.

Materialism has no account of the origin of this energy, nor of the source of the rules that govern it. We find here a confirmation, 5,000 years after the Bhagavad Gita, that Maya is a reality, and that materialism is a limited veneer.

The Key

The Divine Algorithm is our message, key and fulcrum into knowledge of *Maya* from a scientific perspective. It is our wedge between the appearance of a definitive materialistic reality, and the veneer that impresses a material reality upon our sense-state and the immaterial projection of reality that lies beneath it.

The mystical experience is the point at which the instantiating power removes its mask, revealing waves of transcendent consciousness. These waves—manifested as color, biological life, geological structures, stars, and atoms—wash away our ignorance and direct us to the basis of physical reality, which is fundamentally constructed out of pure constraint, observed by pure consciousness as what appears as physical form.

> "Maya is fundamentally inscrutable: we don't know why it exists and we don't know when it began. What we do know is that, like any form of ignorance, Maya ceases to exist at the dawn of knowledge, the knowledge of our own divine nature." -Vedanta philosophy

The vision of reality beyond form, and the special transmission from God Consciousness to Human Consciousness is the basis of The *Spiritually Transformative Experience.*

> "Maya is the veil that covers our real nature... Brahman is the real truth of our existence: in Brahman we live, move, and have our being. 'All this is indeed Brahman,' the Upanishads—the scriptures that form Vedanta philosophy— declare. The changing world that we see around us can be compared to the moving images on a movie screen: without the unchanging screen in the background, there can be no movie. Similarly, it is the unchanging Brahman—the substratum of existence—in the background of this changing world that gives the world its reality. Yet for us this reality is conditioned, like a warped mirror, by time, space, and causality— the law of cause and effect. Our vision of reality is further obscured by wrong identification: we identify ourselves with the body, mind, and ego rather than the Atman, the divine Self. This original misperception creates more ignorance and pain in a domino effect: identifying ourselves with the body and mind, we fear disease, old age and death; identifying ourselves with the ego, we suffer from anger, hatred, and a hundred other miseries. Yet none of this affects our real nature, the Atman. Maya can be compared to clouds which cover the sun: the sun remains in the sky but a dense cloud cover prevents us

from seeing it. When the clouds disperse, we become aware that the sun has been there all the time."

-The Vedanta Society[1]

And

"There are two views among Hindu scholars relating to the concept of Maya. In some philosophical systems, Maya refers to the mysterious power or the cosmic energy of the Supreme Being with which He projects the universe from Himself. In other philosophical systems, Maya is the thought of a cosmic illusion or ignorance, Avidya, that deludes the individual soul (Atman) into forgetting its own divine nature. This forgetfulness of its true nature further causes the soul to mistakenly identify itself with the body and mind, assume individuality, and thus subject itself to pleasure and pain in the phenomenal world." -Dr. Venugopal Menon, All My Thoughts, Spiritual and Religious (2020)[2]

What have we learned in our journey thus far?

We learned that in every form of material reality, there lies an inscrutable and impossible *Maya*-illusion at its foundation. This is not typical deception; it is an illusion that is our reality. It manifests as the atom, the molecule, and the biological cell, yet functions as a transitional phase-state of the omnipotent consciousness—the instantiating power behind our universe.

We see, and experience, and live within the multiple forms and projections of this energy, within the *Maya*-illusion of three-dimensional space, and the *Maya*-illusion of time.

Section 1 traces *Maya* in its multitudinal forms, each variable and special problem exposes a projection —a definitive indicator of an unseen constraint field that intervened, as Primary Cause to structure the mystery.

A universe *from nothing.*

Matter from energy that should have *canceled itself out.*

A perfectly thermalized horizon, that *is not possible.*

An initial entropy state that is beyond *the limit of chance.*

A cosmological constant that is *120 orders of magnitude off* from prediction.

A finely tuned set of fundamental interactions, that is precisely *what is required* for our reality, and also, absolutely improbable.

An impossible to complete, *unresolvable, and paradoxical* standard model of physics, that remains incomplete and incapable of being completed.

Gravity and time that rule our existence as two exacting interacting forces, yet *cannot be traced to anything, nor attributed to anything nor quantized (irreducible, non-material, non-reductive).*

Hydrogen gas that collapses into dense clouds of *the exact size, proportion and balance of interactions* to ignite, burn, and create our periodic chart of elements, because of a *highly improbable* carbon resonance that *coincidentally,* provides the material for later generation stars to form our *life-allowing solar system* and *life itself.*

The *unique fitness* of Carbon for biology and life.

The *impossibility* of abiogenesis.

The *impossibility* of chirality.

The *impossibility* of DNA.

The *impossibility* of the mitochondria, the electron transport chain, and the citric acid cycle.

The *impossibility* of speciation.

The *impossibility* of our Moon, Earth / Moon system and Solar system.

The *impossibility* of *consciousness.*

The *impossibility* of *Youness.*

Each of these is a point where the materialist ruleset breaks down, revealing the invisible hand of the Instantiator enforcing the specific resonance required for youness to exist.

The tell-tale seed of Maya is present in every stage of our reality, from initial instantiation to the final product of youness. Every level of our existence—every 'thing'—is untraceable and unattributable in its provenance to anything else within our perceived realm. Causation, specifically Primary Cause, is ruptured at the foundation.

Our existence is a house of cards that does not fall simply because Brahman holds them. This makes the architecture sturdier than any robust form of physics or engineering; Brahman did not make Maya to fall, but to stand. Make no mistake: Brahman is the foundation—not physics, not mass, nor the non-existent force of gravity.

How did the authors of the Vedas know this? They knew because they had seen it. Once experienced, the bell of truth cannot be 'un-rung.' This is the logical one-way valve: Materialism to Mysticism is a structural upgrade in bandwidth. Mysticism back to Materialism is a structural impossibility. You cannot 'un-know' the source of the projection once you've seen the projector. It would be like asserting the map is the territory after having sown and reaped from the land itself.

Scientifically, our existence cannot be said to diverge in any meaningful way from the characteristics of Maya. In this reality, we are magical and mystical by definition—realizing that these concepts are elevated in rigor far above any materialist construction. Objectively and inarguably: we have never witnessed anything other than magical mysticism.

> This is the source of the Swami Vivekananda quote:
> "If a man is deluded by a mirage for some time, but one day the mirage disappears, –if it comes back again the next day or at some future time, he will not be deluded. Before the mirage first broke, the man could not distinguish between the reality and the deception. But when it has once broken, as long as he has organs and eyes to work with, he will have to see the image but will no more be deluded." -Swami Vivekananda, The Vedanta Philosophy, speech (1896)[3] (

Every level of our existence and every *'thing'*, is untraceable—unattributable in its provenance to anything else within our perceived existence. Causation, specifically Primary Cause, is ruptured at the foundation. Our existence is a very particular 'house of cards'—a form of *Maya t*hat does not fall simply because *Brahman* keeps them upright. This makes the 'cards' sturdier than the most robust material firmament. *Brahman* did not intend for *Maya* to disintegrate our reality, but to facilitate the mysticism of physical existence. Make no mistake: Brahman is the foundation—not physics, not physical forces, not mass, nor the non-existent force of gravity. Our existence, if we attempt to view it as a complete materialist ontology, is the very definition of *impossible*. It is *Brahman* that makes our reality beyond the possible, the probable, or the conceptual. It is existing, functional, representational, and profoundly meaningful. Logically, and by constraint (the evaluative field) nothing exists without the instantiating power of *Brahman*.

Scientifically, our existence cannot be said to diverge in any meaningful or fundamental way, from every quality and characteristic described by the Hindu concept of *Maya*. In our reality, and as living beings, we are *Magical* and *Mystical* by definition, while also realizing that these concepts are elevated in rigor and justification far above any possible materialist construction. Objectively and inarguably: we have never witnessed anything other than magical mysticism.

There is however one thing within our physical world that is impossible, and that is for any human to view, or know, or experience, or think, or live or love anything that is not miraculous and intimate in nature to Brahman.

> "Physical science, then, cannot formulate laws outside of Maya, the very texture and structure of creation. Nature herself is Maya; natural science must perforce deal with her ineluctable quiddity. In her own domain, she is eternal and inexhaustible; future scientists can do no more than probe one aspect after another of her varied infinitude. Science thus remains in a perpetual flux, unable to reach finality; fit indeed to formulate the laws of an already existing and functioning cosmos, but powerless to detect the Law Framer and Sole Operator. The majestic manifestations of gravitation and electricity have become known, but what gravitation and electricity are, no mortal knoweth."
>
> -Paramahansa Yogananda, *Autobiography of a Yogi*, (1946)[4]

Romans 1:20 reads:

> "For since the creation of the world God's invisible qualities—his eternal power and divine nature—have been clearly seen, being understood from what has been made, so that people are without excuse."

We are without excuse. The data dictates that "God's invisible qualities" are the mechanical reason for "what has been made."

> "Maya is the veil that covers our real nature and the real nature of the world around us." -Swami Vivekananda[5]
> "Maya is Nature herself—the phenomenal worlds, ever in transitional flux as antithesis to Divine Immutability." -Paramahansa Yogananda[6]
>
> "Maya is the thought of a cosmic illusion or ignorance, Avidya, that deludes the individual soul (atman) into forgetting its own divine nature." -Dr. Venugopal Menon[7]
> "(Maya connotes) 'a magic show, an illusion where things appear to be present but are not what they seem' " -James G. Lochtefeld, Ph.D.[8]
>
> "Maya is the veil of transitoriness in Nature, the ceaseless becoming of creation; the veil that each man must lift in order to see behind it the Creator, the changeless Immutable, eternal Reality." -Paramahansa Yogananda[9]

"Maya refers to the mysterious power or the cosmic energy of the Supreme Being with which He projects the universe from Himself." - Swami Vivekananda[10]

"Maya is the principle of relativity, inversion, contrast, duality, oppositional states" -Paramahansa Yogananda[11]

No aspect of our reality is possible as a material construct; conversely, every level of our reality is a physical impossibility. Yet, these levels exhibit a recursive dependency—each impossible layer providing the specific scaffolding required for the next level to exist.

And then there is this:

If all of the levels of reality were not impossible in the precise, calibrated ways, existence, as we know it, would not be possible. The stability of our realm is predicted on the precision of what materialism labels as its physical failures—the very thing it attempts to serially patch over with ad hoc conjectures. Instead, what they represent are Maya, as she projects levels upon levels upon levels upon our reality in the form of laws, forces, interactions, and dimensionless quantities within our existence.

"(Maya) The delusionary power inherent in the structure of creation, by which the One appears as many. Maya is the principle of relativity, inversion, contrast, duality, oppositional states... The Sanskrit word Maya means 'the measurer'; it is the magical power in creation by which limitations and divisions are apparently present in the immeasurable and inseparable."

-Swami Paramahansa Yogananda, Autobiography of a Yogi (1946)

At the subatomic scale, this embodiment vanishes into a vacuum so absolute it defies material description. We lack the linguistic or mathematical tools to quantify a 'thing' that possesses no inherent substance.

The mystical experience is the truth behind the illusion. Our existence is not a trick; it is a curriculum. We live within a 'cathedral of teaching' where divine energy animates both the witness and the medium simultaneously. The clues to the mystery are embedded in the fabric of the realm, ensuring we recognize the Truth and recognize we are living within the very body of the instantiator, witnessing as the instantiator's divine energy and life force animates our world and all sentient beings.

Life is many things, but deception is not one of these things. It is not a deception. If it was meant as a deception then we would know nothing of God and we would not be given the mystical experience, or spiritual knowledge, nor any of the entryways and portals of communion with Truth. And without this access we would not be able to build societies, cultures, and institutions of learning and knowledge.

If this realm were a true deception, the materialist fantasy would be seamless. The missing forces, particles, and logical discontinuities of Section 1 would not exist. The 'Instantiator' would have closed the loops, conveying the sense of a complete, self-contained, and law-bound atheistic material existence.

But this is not the case. The closure the materialist seeks does not exist. The Algorithm therefore represents Grace. It prevents us from mistaking the objective fantasy of the materialist proposition for the truth of the divine substrate. Our world, and the embedded Algorithm we find within it, rescue us from being lost on The August Wheel—that is, unknowing what is true (illusion) and believing in objective fantasy (delusion). It instead points the way towards the Golden Path where belief merges with Truth and the entire duality collapses into transcendence.

The Language of God and the communication of God Consciousness

If our existence is consciousness emanating from an infinitely creative and intelligent being, then everything we experience in our reality is a form of communication from this entity. When we test this idea against our reality, we find it to be True. Everything in our existence is a carrier of information—encoded data that can be read, and decoded. When decoded, it functions as a training program for the witness of inner consciousness. In this context, the world as information that teaches us what it is, and what we are is an objective, constraint-based fact.

This means that what we label as science, is actually the study of divine embodiment, and of divine communication. The overlapping rules and forms of arrangements are not mere layers of structure; they are distinct 'languages'---logical structures of information rich archives expressed through forces, energies, fields, chemistries and biological forms.

Our world functions as a dynamic library of the Instantiator's communication—an ongoing, unceasing, and ever-revealing transmission of proximal, and then ultimate reality.

---

[1] (unknown, Vedanta Society of Southern California, The Concept of Maya, vedanta.org/ what is vedanta - the concept of maya)14

[2] (Dr. Venugopal Menon, All My Thoughts, Spiritual and Religious, The Concept of 'MAYA' - the Cosmic Illusion, posted December 21, 2020, menonlifetimethoughts.com the concept of maya the cosmic illusion)
[3] Swami Vivekananda, The Vedanta Philosophy, an address before the graduate philosophical society of Harvard University, 4th edition, Pg 15, (1896) 1901)

[4] (P. Yogananda, Autobiography of a Yogi, Chapter 30, 1946, The Philosophical Library)

[5] (Ibid, Swami Vivekananda, 1896)
[6] (Ibid, Paramahansa Yogananda, 1946)
[7] (Ibid, Dr. Venugopal Menon, 2020)
[8] James G. Lochtefeld, The illustrated Encyclopedia of Hinduism: A-M., 2002 The Rosen Group. Pg 433)
[9] (Ibid, Paramahansa Yogananda, 1946)
[10] (Ibid, Swami Vivekananda, 1896)
[11] (Ibid, Paramahansa Yogananda, 1946)

3 The Prophets Experience

Overview

In our inquiry into the power behind our universe, we examine the prophetic experience. Throughout the annals of human history, prophets have reported communion with a force described as the primary mover of our reality. These accounts are vital to our investigation; they provide a source of insight that matches the characteristics of the power we seek to decode. In what follows, we will examine the experiences of acknowledged spiritual masters and their descriptions of this encounter. Later, we will analyze the substance of their experience and the specific information conveyed through it. This overview illustrates the 'first layer' of contact and the biographical markers that preceded their universal engagement.

The Buddha

It is said that Siddhartha Gautama, the Buddha had his own spiritually transformative experience early in life, and that the profound insight gathered from this experience placed him upon a spiritual path that culminated in his inevitable enlightenment many years later.

The Prophecy

At his birth Siddhartha was prophesied to be a great teacher. He was born the son of a king and a queen named Maya, also referred to as Mahamaya, who died seven days after his birth. The reason given for her death has several interpretations: the completion of her Earthly destiny; a necessary condition given that her womb could no longer be used for anything else; or, upon giving birth to the enlightened one, her immediate rebirth in Tushita Heaven as a male to watch over Siddhartha.

It is said that when Siddhartha was born, he immediately took 7 steps to the north, and with each step lotuses blossomed under his feet. After the seventh step, he pointed with one hand to the sky and one towards the earth and declared, "I am the chief of the world. This is my last birth."

This statement served as confirmation to the King of a prophecy given prior to Siddhartha's birth. The hermit-sage Asita examined Siddhartha and found the 32 marks of a great man, and 64 Brahmin scholars—who had interpreted Mahamaya's dream of a white elephant with six tusks entering her side—predicted a two-path destiny: he would either become a *Mahapurisa*, a Great Being ruling by power, or a *Chakravartin*, a Universal Monarch who rules the world through spiritual insight.

In a calculated attempt to override the prophecy, Siddhartha's father quarantined him within the palace grounds, attempting to shield him from the nature of the human condition and the understanding that would trigger his ascendency into one of two prophesied destinies.

This effort failed when the Buddha, a well-known master of all he studied, grew curious and ventured outside of the enclosure of the palace he had been forbidden to leave. There, he encountered: a sick man, an old man and a man's corpse. After this he saw a man who was content—a wandering monk. These observances have forever been known as the four sights when the Buddha's destiny was engaged, and he soon set off to solve the mystery of the sufferings by disease, age and death.

It was not long after this that the Buddha, in earnestness of enlightenment and understanding, subjected himself to six years of learning, wandering and fasting to precipitate an inward spiritual revelation to solve the riddle of the four sights.

After these trials, having not found resolution, Siddhartha found himself in a grove, in a weakened and frail state. It was then he realized that self-torture was not the path. He accepted a nourishing offering of milk-rice from a woman named Sujata, who had originally come to the grove to make an offering to the tree spirits. This simple meal provided the strength he needed for the final struggle.

Enlightenment

The Buddha then bathed in the nearby river—this was the last food and bath he would have for seven weeks. Refreshed, he retreated to the comfort of a Pippala tree—forever after known as the Bodhi tree, or *Tree of Awakening*—and vowed not to rise until his search was complete.

That night many evil thoughts entered his mind. He resisted these spirits and finally saw the cause of evil and suffering, and the secret to being released from all suffering.

The Three Insights:

(1) seeing thousands of his own births and deaths.

(2) the "divine eye," of how *all* beings are reborn according to their karma.

And

(3) The Four Noble Truths (*Dukkha, Samudaya, Nirodha, Magga*)

With this understanding the Buddha, at age 35, emerged victorious and solved the riddle of sorrow, unhappiness, suffering, old age and death.

From the Pali Canon

> "When my concentrated mind was thus purified, bright, unblemished, rid of imperfection, malleable, wieldy, steady, and attained to imperturbability, I directed it to knowledge of the destruction of the taints. I directly knew as it actually is: 'This is suffering'; I directly knew as it actually is: 'This is the origin of suffering'; I directly knew as it actually is: 'This is the cessation of suffering'; I directly knew as it actually is: 'This is the way leading to the cessation of suffering.' When I knew and saw thus, my mind was liberated from the taint of sensual desire, from the taint of being, and from the taint of ignorance. When it was liberated, there came the knowledge: 'It is liberated.' I directly knew: 'Birth is destroyed, the holy life has been lived, what had to be done has been done, there is no more coming to any state of being."
>
> -Majjhima Nikaya i 247–249 (MN 36) (1995)[1]

Enlightenment was his.

After his victory, Siddhartha meditated under a Banyan tree for seven days to enjoy the bliss of his newfound liberation.

The Lightning Strike

We have presented the lightning strike of insight—the spiritually transformative and mystical experience of the Buddha in its raw form. The concretizing of his realization into the "Three Marks of Existence," the "Five Aggregates," "Dependent Origination," the "Three Jewels," the "Four Dharma Seals," the "Four Noble Truths," and the "Eightfold Path"—this was all ashes compared to the moment when the man, starving and weak, a biological form with an inner witness of *youness*, was fighting, groping, and striving to make contact with the Evaluative Field, and then succeeding. That is it. That is the moment. That is what is available to every being.

We speak of the lightning strike, not the bottle that attempts to capture it. The vision, the inner witness, the inner transformation, the contact with the Field of Constraint, the dissolution of Maya, and the collapse of duality. This experience is endemic and available to all humanity. It is the experience of ultimate importance. This is the prophet's experience. It is not "spirituality," ontology, religion, or ritual, but the pure form of contact and communion between the divine and the human biological organism—with the consciousness that originated the **Constraint Field.**

This is the ontological truth our entire narrative has been driving toward: instead of a mere instantiation across the entirety of our physical reality, it is a transmission from Creator to creation.

"Religion" is the concretized record of the contact; it is not the contact itself. The Prophet is the biological witness. By framing the Buddha and all prophets as biological forms with an inner essence and an inner witness, we remove them from the pedestal of the supernatural and place them into the natural—as one of the Instantiator's instantiations.

The importance of this is not to diminish the prophets, but to uplift humanity. The experience is repeatable rather than merely admirable. For those who have experienced direct contact—spirit to spirit—they may not be prophets, but in their own quiet way, that contact supplants everything they have learned, read, or heard, serving as the ultimate touchstone of their spiritual understanding.

Analysis

After Siddhartha renounced his princely life, and prior to his enlightenment, he subjected himself to severe asceticism that pushed his body to the brink of collapse. In this weakened state, he recalled a spiritually transformative experience from his childhood. While sitting beneath a rose-apple tree as a boy, he had spontaneously entered a state of profound peace and clarity.

It was this key insight, recalled at his most desperate hour, that allowed him to find the path to the cessation of suffering. He realized that enlightenment was not a product of physical torture, but a return to that primary, undistorted state of consciousness.

> "(The Buddha's) father had taken him to watch the ceremonial plowing of the fields before the planting of the next year's crop... he looked at the field that was being plowed, he noticed that the young grass had been torn up and that insects and the eggs they had laid in these new shoots had been destroyed. The little boy gazed at the carnage and felt a strange sorrow, as though it were his own relatives that had been killed. But it was a beautiful day, and a feeling of pure joy rose up unbidden in his heart.... It is essentially a moment of ecstasy, a rapture which takes us outside the body and beyond the prism of our own egotism... This surge of selfless empathy had brought him a moment of spiritual release... Instinctively, the boy composed himself and sat in the asana position, with a straight back and crossed legs. A natural yogin, he entered into the first jhana, a trance in which the meditator feels a calm happiness but is still able to think and reflect... Years later, just after he had cried, with mingled

> optimism and despair, 'surely there must be another way to enlightenment! 'Gotama recalled this childhood experience... 'Could this,' he asked himself, 'possibly be the way to enlightenment?'..."
>
> –Karen Armstrong, *Buddha* (2001) (K. Armstrong, *Buddha*, 2001, Penguin Books Ltd.)19

This early experience and taste of bliss became a template within his mind, and it served as a signpost indicating that a different path was available and a new vision of the world existed to be enjoyed. A new doorway to the spiritually transformative experience, created by a lightning moment of insight, triggered by gazing at the carnage of a plowed-up field, was displayed within the inner show of his consciousness, and joined with and transcended by his *youness*.

This experience had not been taught, spoken of, or ritualized prior to the moment of its occurrence. He found it within, and having found it, it became revelation.

The spontaneous joy unleashed by this realization provided *the Middle Way*—a path between the extremes of the palace and the forest. When later actualized beneath the Bodhi tree, the method, revealed by his reverie from the field, would have world-changing effects, causing waves of spiritual manna to resound to this very day.

Later teachers of Buddhism suggest this early experience opened the Buddha's spirit so that, after six years of searching—armed with the memory of that golden moment, he was able to find victory through a new application of effort. Won after severe trials and finalized through a simple meal, a bath, and repose—was recognized by the Buddha as the correct way: *the Middle Way* toward enlightenment.

> "Instead of torturing our reluctant selves into the final release, we might be able to achieve it effortlessly and spontaneously. Could nibbana be built into the structure of our humanity?"
>
> -Karen Armstrong, *Buddha* (2001)[2]

Effortlessly and spontaneously—just as the Buddha had experienced at the plowed field.

Nibbana (Nirvana) is the highest spiritual state in Buddhism. The word comes from *nir* (stop) and *và* (to blow); thus, Nibbana is the "blowing out" of the fires of the mind—the cessation of suffering fueled by greed, hatred, and ignorance[3].

The teachings of the Buddha, preserved in the Pali Canon, represent the methodization of his life from birth to enlightenment, as recalled through his teachings from age 36 to 80, over the subsequent decades as a complete system formalized into the *Four Dharma Seals, the Four Noble Truths, and the Noble Eightfold Path.*

The Prophet Mohammad *ṣallā -llāhu ʿalayhī wa-sallam*

The birth of Islam came about by similar means as the birth of Buddhism, Judaism and Christianity. Mohammad—*may the blessings and peace of Allah be upon him*—during the month of *Ramadan*, would fast and pray in a cave. During Ramadan 610AD, while in the *Cave of Hira* doing prayers and reflection, this particular retreat was a *Tahannuth*: a deliberate act of shedding the "profane" self to become a vessel for the "Divine."

It certainly was effective. He was visited by the archangel *Gabriel (Jibril)* who embraced the Prophet and commanded him *"Iqra!'* (Read / Recite). *The Prophet* replied that he could not—"I am not a reader"—The command was then repeated. This is when the first five verses of the Surah Al-Alaq (The Clot) were revealed.

> "Recite in the name of your Lord who created— Created man from a clinging substance. Recite, and your Lord is the most Generous— Who taught by the pen— Taught man that which he knew not." – Jibril (Quran 96:1–5)

By commanding the use of 'the pen' a significant event was indicated. 7th Century Arabia was largely an oral culture. But this revelation and command was a direct edict to compile the words, and wisdom, logic and constraint field of revelation. This living dialectic of knowledge was meant to be recorded, studied, analyzed and its wisdom continually extracted from its divine architecture.

This was called *The Clot* (Al-Alaq) by later scholars to indicate that special covenant when the seed of Islam had been fertilized and planted and would grow into the living word. In Islamic cosmology, the *Supreme Pen* is the first creation –the instrument through which the possibilities of existence are marked upon the "Preserved Tablet".

The Pen is therefore the divine bridge, the device of translation from direct communion and the recorded word.

The Process

Throughout history, the mystical experience has served as the essential spark for the birth of the world's religions. When the methods of the prophets are exposed, a clear pattern emerges: retreat, isolation, fasting, and a systematic withdrawal from the profane to facilitate an encounter with the divine.

These are not merely religious rituals. They are the physical mechanics of clearing the outer and inner show of consciousness to allow for a direct transmission.

"We will show them Our signs in the horizons and within themselves until it becomes clear to them that it is the truth." (Quran 41:53)

Muhammad—like Jesus, the Buddha, and Moses—followed this trajectory. Each of these figures accessed the one true well of divinity, imparting essential truths that are manifested not as concepts, but as living realities.

"Indeed, We will cast upon you a Heavy Word (Qawlan Thaqila)." (Quran 73:5)

The revelation is terrifying because it is an act of biological re-alignment. Muhammad did not expect his commission; he was deeply troubled, fearing he was losing his mind—a reaction common to those who have been forcibly reordered by the Truth of what they encountered. When Waraqa ibn Nawfal identified the Namus that visited Moses, the mission was confirmed. The "good news" was not just a new teaching; it was the proclamation that there is a source, a power, and an order to all that machinates on the physical plain.

The path of the Prophet differs from that of the individual seeker because it carries a commissioned task. The Prophet is the biological bridge. By embracing the "Heavy Word," they are reordered; their thoughts are restructured by the Truth of their encounter. In embodying this task, they become more than human teachers—they become the biological witnesses who hold the touchstone for the rest of humanity.

Guru Baba Nanak (1564-1606)

The prophet of Sikhism, *Guru Baba Nanak,* became enlightened when, according to differing legends, he either plunged into a river before daybreak for his daily bath and prayer, or, he left his clothes at the riverbank and retreated into the forest. Both legends are mythical in the representation of retreat into the forest (the unknown, to encounter ultimate self and ultimate truth. The allegory of entering water (being born, made anew, refreshed). In both legends for Guru Baba Nanak, the river and nearby area was searched, and he was not to be found until he emerged 4 days later wholly changed from an encounter with God.

Nanak said he had been in a divine trance where God provided him with a cup of nectar and stated:

"I am with you. I have made you happy and those who shall take thy name. Go and repeat My Name and teach others to do the same. Remain uncontaminated by the world. Practice the repetition of My name, charity, adulations, worship and meditation. I have given you this cup of Nectar, a pledge of My regard."

-Eleanor Nesbit, *Guru Nanak,* (1999)[4]

Guru Nanak, filled with love, and serenaded by the spontaneous music of the divine, sang *The Mool Mantra,* the first several lines of the *5,894* verses in the *Guru Granth Sahib*—the central holy religious scripture of Sikhism:

One universal creator God. The name is Truth. Creative and Personified. No fear. No hatred.

Image of the undying, beyond birth, self-existent.
By guru's grace chant and meditate

The Guru Granth Sahib then continues:

True in the primal beginning. True throughout the ages. True here and now. O Nanak, forever and ever true.

By thinking, he cannot be reduced to thought, even by thinking hundreds of thousands of times. By remaining silent, inner silence is not obtained, even by remaining lovingly absorbed deep within. The hunger of the hungry is not appeased, even by piling up loads of worldly goods. Hundreds of thousands of clever tricks, but not even one of them will go along with you in the end. So how can you become truthful? And how can the veil of illusion be torn away? O Nanak, it is written that you shall obey the hukam of his command and walk in the way of his will.

By his command, bodies are created; his command cannot be described. By his command, souls come into being; by his command, glory and greatness are obtained. By his command, some are high and some are low; by his written command, pain and pleasure are obtained. Some, by his command, are blessed and forgiven; others, by his command, wander aimlessly forever. Everyone is subject to his command; no one is beyond his command. O Nanak, one who understands his command, does not speak in ego.

-Siri Guru Granth Sahib Ji, Translation by Dr. Sant Singh Khalsa (2013)[5]

The Christ

In the Gospel of John, Jesus speaks of a realization that transcends his physical birth. His awakening wasn't to a new truth within the world; rather, it was a condition he was born into from beyond the veil.

He was born under the star of Bethlehem, fled to Egypt, and then at twelve—in the Temple Incident—astonished his teachers with his understanding. He replied to his parents, 'Did you not know that I must be about my Father's business?'—thus declaring his greater purpose.

He next appears at thirty, after eighteen years of silence as a carpenter's son. Then, in rapid succession, he is baptized by John, retreats to the desert for forty days, performs the miracle at Cana (water into wine), calls his disciples, and begins his teaching.

In the second year, his popularity grows: he heals the sick, feeds five thousand, walks on water, and delivers the Sermon on the Mount.

Matthew 17: 4 - 5

> "Peter said to Jesus, 'Lord, it is good for us to be here. If you wish, I will put up three shelters—one for you, one for Moses and one for Elijah." While he was still speaking, a bright cloud covered them, and a voice from the cloud said, 'This is my Son, whom I love; with him I am well pleased. Listen to him!'"

And then we are suddenly at Passion Week, with his triumphant Entry into Jerusalem, the Last Supper, betrayal, crucifixion, and resurrection.

In terms of the bifurcation of Atman and Brahman—which are one and the same—that we have used to understand the relationship between *existence* (the all) and *the individual*—Jesus speaks from a lifelong knowledge that his bridge was never broken. He is the Atman that never struggled to realize its identity as Brahman. He rejected, or more precisely failed to see from the outset, the notion of separation.

His path is not the hero's journey—a circuitous search for self-understanding and the culmination of an anointment. Rather, Jesus represents birth as culmination. His story begins with illumination and proceeds as a journey where he illuminates others.

John 8:58

> "Very truly I tell you... before Abraham was born, I am!"

John 10:30, 38

> "I and the Father are one... believe the works, that you may know and understand that the Father is in me, and I in the Father."

John 17:5, 8

> "And now, Father, glorify me in your presence with the glory I had with you before the world began... For I gave them the words you gave

me and they accepted them. They knew with certainty that I came from you."

John 14:9–10

"Anyone who has seen me has seen the Father... Don't you believe that I am in the Father, and that the Father is in me? The words I say to you I do not speak on my own authority. Rather, it is the Father, living in me, who is doing his work."

Prophecy of the Prophets – Supernatural Indications

Interestingly, the prophets Guru Nanak, Christ, The Buddha and The Prophet Mohammad all had a similar incident occur in their early lives of what would normally be considered of a supernatural nature.

As a youth, when The Buddha was resting under a tree and had his first experience where he viewed God's creatures in the churned-up Earth, it was said that although the day was hot and the sun was strong, the shade of the tree he was under *did not move* with the sun. It stayed in a position to protect the young Buddha from its rays.

"I recalled seated in the cool shade of a rose-apple tree, while my father the Sakyan was working, I had entered and dwelt in the first jhāna... And while the sun moved on, the shadow of that rose-apple tree stood still, not leaving the body of the Bodhisatta."

–Bala Janamsakhi, Sakhi of the Tree's Shadow

Guru Nanak is said to have slept under a tree and similarly, his tree was the only one where the shadow *did not* change positions.

"Rai Bular looked and saw that the shadows of all the other trees had shifted as the sun moved across the sky, but the shadow of the tree under which Nanak lay remained stationary. It stood fixed, like a canopy, shielding the boy's face from the scorching rays."

–Bala Janamsakhi, Sakhi of the Tree's Shadow

It was also said that The Prophet Mohammad was observed in a caravan crossing the desert, and that he was protected by the shade of one small cloud.

"He saw the Messenger of God in the caravan when they were approaching, and a cloud was overshadowing him among the people. Then they came and halted in the shade of a tree near the monk. He looked at the cloud when it overshadowed the tree, and the branches of the tree leaned over the Messenger of God until he was in their shade."

—Ibn Ishaq, The Life of Muhammad

Christ was not covered by a cloud and shielded by the Sun, rather, for several days prior to his birth, the star of Bethlehem appeared *above his place of birth* and served as a guide to the three wise men, who miraculously were led to a single individual, despite the star's position in the celestial heights.

> Saying, Where is he that is born King of the Jews? for we have seen his star in the east, and are come to worship him…
>
> -Matthew 2:2
>
> When they had heard the king, they departed; and, lo, the star, which they saw in the east, went before them, till it came and stood over where the young child was.
>
> -Matthew 2:9

Throughout human history, celestial events have served as harbingers, delivering messages of world-changing significance. Yet, our modern inquiry reveals that the most 'miraculous' signs are not found in the momentary suspension of shadows, but in the very architecture of the universe itself. When compared to the formation of galaxies or the precise ordering of the heavens, the stationary shadow pales in scale—and yet, both share the same signature: a conspicuous favor and a guiding grace that suggests a power operating far beyond the limits of human understanding.

We see these privileged conditions in the formation of Earth within a star nursery and in our solar system's improbable journey through innumerable hazards into a vast, protective void. We see it in the inexplicable Grand Tack—the "manicuring" of our position by the gas giants—and in our faithful attendant, the Moon. This mechanically indispensable satellite, which by all standard models should not exist, contributes in every possible way to the habitability and rhythm of our lives.

The Power experienced and described by the prophets inhabits these same capacities. This instantiating force, through its edict, embeds attributes of its own nature within the fabric of existence. Whether it is the stabilization of a planet's axis or the protection of a prophet beneath a tree, we are witnessing the same coherent movement: a power that does not merely create but actively maintains the privileged conditions necessary for life and for the Truth to be realized.

A Different Understanding of the Ultimate

Buddhism, as a spiritual practice, differs from Old Testament-based beliefs and the Vedic Hindu tradition from which it emerged. Most interpretations do not ascribe the operation of the universe to an immortal soul or a creator God; yet the Pali Canon frequently references rebirth, celestial realms, and various spirit entities. The Buddha himself is recorded as having been visited by these incarnations throughout his life and prior to his enlightenment under the Bodhi tree.

While the theistic "God" is absent from the Buddhist framework, treating this absence as the defining characteristic overlooks the point at the heart of Buddhism—and, by extension, Christianity, Islam, and other God-fearing traditions.

The Buddha avoided theistic labels to prevent the mind from clinging to a "Self" (*Atman*), yet his descriptions of the Unconditioned (*Asankhata*) align with the "Power" identified in other traditions—a Truth that is, as D.T. Suzuki notes, is *akalika* (beyond time). In this view, everything in the "profane" world is conditioned; it possesses a cause, a beginning, and an end. The "Power" is the singular element that is unconditioned. Whether referred to as "The Father" or "The Deathless," these terms serve as different pointers toward the same Constant in the universal equation.

Paradox naturally arises when rational understanding attempts to penetrate these depths. This power—the Truth—experienced as personal revelation, simply "is." and our description of it, often falls along anthropomorphic articulations. Ultimately, our understanding must necessarily transcend labels because the foundation of existence is constraint and Truth based. It is pre-language and pre-representational; it exists beyond what can be captured as a formal "concept."

The link between these traditions is the Buddhist concept of enlightenment, or Nirvana:

> "...an actual insight into truth... as immediately presented transcendental consciousness as the Dharma directly perceived (*sanditthika*), beyond time (*akalika*), personally experienced (*ehipassika*), altogether persuasive (opanayika), and understood each for themselves (*pacattam veditabbo viññuhi*)... enlightenment is the ultimate truth realized in oneself through one's own efforts." - D.T. Suzuki, *Essays in Zen Buddhism* (1949)[6]

What is received by discursive understanding in any spiritual practice is effectively a different pathway to a single destination. These are varying methods of acknowledging—and experiencing—a higher truth.

Across major traditions, prophets are recorded as having the same fundamental spiritually transformative revelation. While followers utilize different rituals, scriptures, and prayers to commune with this power, the practices are linked at the point of the Spiritually Transformative Experience.

"Enlightenment"

Not all practitioners of religions or spiritual systems undergo the "enlightenment" experience; in fact, it is regarded in nearly all contexts as exceedingly rare. Nonetheless, this event forms the lynchpin—the Heart Sutra, as it were—of all major traditions. Its most prominent characteristic is that the transformation does not, and cannot, occur in the "outer" realm of words, texts, or remonstrations. Rather, it finds its home in the most sacred inner sanctum of the individual. It is an ineffable "what it is like" event—a contact that resonates with the entirety of one's being.

Whether a belief system centers on a Creator or excludes the possibility of a God, the experience of enlightenment or Nirvana is uniformly described as being directly perceived, beyond time, and altogether persuasive[7]. It represents a collision with an underlying field from which all things emanate, where all truths converge, and where time collapses into an immediate "now."

It is as if the conditions required for our existence—the ruleset, the inherent perfection, and the Truth that allows the One and the all (duality) to exist—collapse into its most relevant state—that of a human experience. In this state, the architecture of reality is not merely amenable to cognition; it is experienced as the superlative and perfected food of contemplation.

"Enlightenment" is endemic to the human animal—a potentiality of the species that is not beholden to, or exclusive to, any single religion, ideal, scripture, or practice. It is the antithesis of a container, ritual or incantation. It is a transformation of consciousness catalyzed by an encounter with the ultimate within the human form, a state that reveals the underlying mystery of life and the perfected nature sufficient to represent it in all of its known forms (the biological, chemical and atomic with ineffable immaterial consciousness as the command module)

In this context, to categorize one system of belief as "the experience of God" and another as "Godless" is perhaps technically accurate, but it misses the fundamental point emphasized across all spiritual paths. Whether Godless or God-fearing, the unitary experience of spiritual transformation transpires entirely within the individual consciousness. Once this experience takes hold, the specific triggers—whether rituals, beliefs, or cultural frameworks—are transcended. They dissolve into a unitary oneness, for it is a timeless experience: immediate, whole, and complete unto itself.

We often mistake the finger pointing at the moon for the moon itself.

We categorize, we debate dogmas, and we build elaborate linguistic fortresses to protect our specific "version" of the truth. But these are merely the scaffolding of the human ego. In the rare moment of the spiritually transformative experience, the scaffolding is not just removed; it is annihilated. The "human animal" remains, but the "human narrative" vanishes along with what we pray to, worship, its representations, scripts, myths, etchings, and precious 'sacred' items.

What is left is a consciousness stripped of its filters. In our everyday existence, we perceive the world through the narrow slit of time and space—a survival-based "user interface" that keeps us focused on the immediate and the profane. But the enlightened state represents a total system failure. The "rule set" of our existence—the physical laws, the temporal flow, the very separation between self and other—collapses into a singular, self-evident point of origin.

This is where the universe erupts. This is where the edict that binds matter, dictates fitness, and has been helped by additional miracles to become our modern-day world—all of this, as a personal experience of contact from our soul and all souls, all things, the All.

It is the moment the Atman realizes it isn't just *connected* to the Brahman; it is the moment the Atman realizes that the Brahman is the only thing that ever truly existed—and we are it. The bridge isn't crossed—the bridge disappears, because there was never a gap to begin with.

> "In Christianity we seem to be too conscious of God, though we say that in Him we live and move and have our being. Zen wants to have even this last trace obliterated. That is why Zen followers advise us not to linger even where the Buddha is and to pass quickly away where he is not. All the training of the monk in the Zendo in theory as well as in practice, is based on the notion of 'meritless deed'.
>
> Poetically this idea is expressed as follows.
>
> 'The bamboo shadows are sweeping the stairs,
> But no dust is stirred:
> The moonlight penetrates deep in the bottom of the pool,
> But no trace is left in the water.' "
> -D.T. Suzuki, Essays in Zen Buddhism (1949)[8]

Zen Buddhism wishes to *lose*, rather than to *gain*. *"Non-attainment, therefore, in Zen is positive conception."* and D.T. Suzuki explains that the acquisition and accumulation of knowledge is the opposite of the Zen state of non-attainment where enlightenment may be found. Lao-tzu says:

"Scholars gain everyday while the Taoists lose everyday[9]."

The process of transmuting the ineffable, mystical experience into "rules" of teaching, and a religious understanding and practices for continual contact with the divine through daily living, is the difference between the mystical experience of spirituality and the creation of *a spiritual practice* and *canonized religion.*

In this we can also make sense of the reluctance to discuss such empty ritualistic "distractions" .

> "The Buddha was quite to the point where he thus derided all those philosophers in vain talkers of his day, who merely dealt in abstractions, empty hearsays, and fruitless indications. Zen therefore wants us to build a staircase right at the front of the very palace into whose upper story we are meant to mount up. When we can say, 'This is the very personality, this is the very house,' we have the satori interviewed face-to-face and realize by one's self. (Ditthe va dhamme sayam abhiññā sacchikatvā.)" -D.T. Suzuki, Essays in Zen Buddhism (1949)[10]

The Common Tongue

The Buddha was raised as a North Indian/Nepalese Hindu prince. His father hid him behind palace walls, but he saw the old, the infirm, the sick, the dead and the poor as well as their sufferings. This vision/experience transformed the Buddha and eventually lead to his victory.

Being Hindu, there was the small matter of the caste system. Only those at the top of the caste were given the knowledge of God and information of a religious nature. This knowledge required the understanding of *Sanskrit*, whereas the common tongue was *Pali*. Hinduism had also, in the intervening millennia become largely concerned with esoteric involvements such as animal sacrifices, ceremonies, formalized rituals and ritualistic prayers.

The Buddha's journey of understanding took him to several teachers, but he found no real answer to suffering. He then resolved himself to find this answer through meditation, having recalled the moment in his youth, when in repose he found a state whereupon he viewed the churned up Earth and the lifeforms of insects that appeared to him as sacred beings filled with sacred life. The Buddha attained his enlightenment. He then immediately began to teach and gather devotees.

Coming from a religion that was kept to the highest class, and bogged down by esoteric pursuits, he spoke and taught *in the common tongue—Pali.* He exhorted his followers to forget rituals. Be a good person, practice good works, right speech, right thought, and follow the eightfold path *and liberation can be yours*, personally, unmediated by any clergy.

Many have opined that the reason the Buddha never discussed God is that having witnessed the corruption of his own native religion in the hands of religious pundits, he saw God and the circus that attends the worship of God as unhelpful to the message he wished to deliver.

There are many who attempt to make sense of Buddhism in their own understanding and from the perspective of their own deistic system of belief, because it is far too confusing to them to imagine that the Buddha holds no regard for, nor 'believes' in an all-powerful creative consciousness or more precisely., a personality that attends this consciousness. And it is this distinction that perhaps is the source of the confusion.

> "...with Zen followers all literature was like a finger pointing at the moon, and there was not much in itself that will actually lead one to the seeing of one's own inner nature; for this seeing was a realization which must be attained by one's own personal efforts apart from the mere understanding of letters. All Buddhist sutras including the Lankāvatāra, Vajrasamādhi, and Vajracchedikā could not be of much help to the real earnest seeker of the truth so long as his idea is to grasp the naked facts with his own ungloved hand. This was possible only when his own inner consciousness opened by itself, from within, through his whole souled efforts. Literature is helpful only when it indicates the way, it is not the thing itself.
>
> -D.T. Suzuki, Essays in Zen Buddhism (1949)[11]

A clear example is presented in the event of the Buddha's enlightenment. Here, we find the hallmark signatures of a deep communion—a totalizing oneness within the mystical experience. This signals to all seekers that there is a power, a purpose, and a unifying truth beneath the human condition. It reveals a foundation to creation where liberation is not merely a philosophical goal, but a functional reality. This was the Buddha's gospel: that suffering is a byproduct of the filtered interface, and liberation is the act of returning to the Unconditioned Source.

Unity of the Prophets Experience

> "Since time immemorial, the attention of mankind has converged on the idea of the spiritual basis of existence. There had never been a single society recorded in history which had not a kind of religious institution. It is the very essence of the nature of man not to be satisfied with the purely material aspect of life. The whole annals of human thought and evolution itself are undeniable evidence of the attempt of man to transcend the world of matter and reach the region of the spirit... The purpose of religion is to awaken in man the longing to transcend matter. The desire to transcend the world of matter and explore the region of the spirit, is something which is innate in man, lying deep in the core of his being... The aim of religion is to awaken this longing in man so that he may attain a particular kind of consciousness which will make him arrange his life according to certain patterns of behavior and thinking."
>
> -Professor Dr. Syed Hussein, *Progressive Islam. Vol 1* (1954)[12]

One God or Many Gods?

Each prophet emerges from an encounter and declares: there is one God, or There is one truth. This claim appears again and again across traditions. Yet history presents us with many religions, many practices, and many names for The Divine. One cannot help but to become aware of a great contradiction as there seems to be a series of Gods, and each is proclaimed to be *the one true God*. At the surface, this appears as contradiction—multiple Gods, each asserted as the one true source.

That conclusion, however, assumes we are comparing interpretations. We are not.

We propose that these individual spiritual practices, rather than being viewed as an essential contradiction when viewed together as differing spiritual understandings, should instead be taken individually, 'literally', and at full face value in the content of their individual message. When examined this way, a consistent structure begins to appear—not in doctrine, but in the mechanics of the experience.

This is not a claim that all reports are identical. The spiritual world isn't a classification system. It is an underlying consciousness, an immaterial, non-local field of constraint. What is received is dependent upon what is asked. And what is received is understood in subtle ways, necessarily, based upon ones understanding, and expectation. All descriptions and experiences need not be strictly the same. For example, we can look at the Greyson NDE Scale: not every feature must be present, but when a threshold of core markers is met, the experience is recognized as belonging to the transcendent encounter.

The encounter also has key, classic aftereffects that attend the profound re-ordering of one's understanding and contact with the evaluative field itself. To bring rigor to this observation, further along, we will be examining the work of researchers who have attempted to categorize these "system failures" of the ego and the consistency of its reported dimensions. They include:

A Singularity: Each proclaims one true God (one experience of truth).

First Cause: Each encounter is known as the power behind all creation.

Ineffable Quality: The encounter possesses the quality of incomprehensible, ineffable truth—beyond the bandwidth of language—which is always a step removed.

Filtered Identity: In the presence of this entity, the individual knows themselves as a temporary experiential filter, a brief aperture through which the infinite views the physical realm.

Temporal Collapse: Every person who encounters this entity describes a realization that collapses time into an eternal "Now"—no past, no future, only the presence.

Universal Architect: This singular consciousness is known as the one that instantiates the physical laws of the universe, the cosmos, and life itself.

Consciousness & Youness: This truth cannot be found in the outer world. It is known within, as a direct internal encounter of the Source with the individual consciousness.

Ego Annihilation: In the face of this all-powerful, all-knowing entity, the ego disintegrates. Every artifice is burned away, and the deepest truth of the individual is exposed.

These are not contradictions. They represent consilience and specificity. We see a matching of attributes, characteristics and description among the Prophets, and the many individuals who have become enlightened, the many who have had mystical experiences, and spiritually transformative experiences, or those who have encountered what they identify as *God*.

There is no contradiction except for those looking for one. And all that can really be found is the differing cultural lenses of spiritual expression and differing individual understandings. For example, If a group of individuals each stand on different hills to watch the rising sun, when they gather, do they ask: *"Which of us has seen the real sun? Which hill is the True vision? Which of us is mistaken?"* Although the vantage points differed, the sun was identical. They all felt the same warmth; they all saw a glare upon which they could not directly gaze. They each had different thoughts and may have had different realizations.

We are talking about 'reviews' not the show itself.

In the declaration of *one true God*, each instance is the "One." God is not an objective experience. God can only be a subjective one by an individual human. A spiritual encounter that only the filter of consciousness and youness is designed to experience and make contact with directly.

Many voices, many times. One call, one caller.

There is only *One*.

Sikhism
There is but one God whose name is true. He is the creator, immortal, unborn, self-existent.
Islam
Allah – there is no deity except Him, the Ever-Living, the Sustainer of [all] existence.
Sufism
All this is God. God is all that is.
Baha'i
He, in truth, hath throughout eternity been one in His Essence, one in His attributes, one in His works.
Judaism
The Lord is God in heaven above and on the Earth beneath; there is no other.
Hinduism
He is the one God hidden in all beings, all-pervading, the Self within all beings, watching over all worlds, dwelling in all beings, the witness, the perceiver.
Christianity
There is one God and Father of all, who is above all, and through all, and in you all.
Buddhism is a special case:

> "Whether you say 'The kingdom of God is within' as Jesus did (in Luke 17:21) or that we all have innate Buddha nature as Tibetans do, in the end doesn't it come down to the same thing: We are all lit up from within as if from a sacred source."
>
> -Lama Surya Das, *Awakening The Buddha Within* (1997)[13]

There is a parable taught in some form by nearly all eastern and middle eastern cultures that illustrates the essential unity of all religions while also accounting for the appearance of their being in conflict and essential difference. This parable is of the blind men and the elephant. In it, there are several blind men in a room with an elephant. Each of the men grabs a hold of a different part of the elephant. One holds a tusk. One holds the trunk, one holds the leg, and the other the tail. If the version of the story has more blind men, then the ear and body of the elephant is grasped as well. The men then all begin to communicate to figure out what they have got a hold of. Each describes a different experience altogether.

In some of the versions of the story, the men almost come to blows due to their disagreement. In truth, as we know, they are all holding different parts of the same elephant. The only difference between this parable and our understanding of the world's spiritual understandings, is that the prophets of the world rather than grabbing a hold of a different part of the elephant, God has grasped the consciousness of each of the prophets and related a personal and intimate message to the limits of each individual's understanding. The prophets do not describe different parts of the same form. They describe an encounter with the totality of the one Truth.

---

[1] (Bhikkhu Bodhi (Trans.), The Middle Length Discourses of the Buddha: A Translation of the Majjhima Nikaya, (Wisdom Publications, 1995), pp. 332–343)
[2] (K. Armstrong, Buddha, 2001, Penguin Books Ltd.)
[3] (From Dhammawiki.com: Nibbana (Pali))
[4] (E.M. Nesbitt, G. Kaur, Guru Nanak, Bayeux Arts Incorporated, 1999)
[5] (The Siri Guru Granth Sahib, Kulbir Singh Thind, MD & Singh Sahib Sant Singh Khalsa, MD Hand Made Books, 2013 Tucson Az, USA)
[6] (D.T. Suzuki, Essays in Zen Buddhism, First series, Pg. 61-62, 1949, Grove Weidenfeld)
[7] (D.T. Suzuki, Essays in Zen Buddhism, First series, Pg. 61-62, 1949, Grove Weidenfeld) .
[8] (D.T. Suzuki, Essays in Zen Buddhism, First Series, Pg. 352, Grove Weidenfelt, 1949)26
[9] (D.T. Suzuki, Ibid Pg. 351)27
[10] (D.T. Suzuki, Essays in Zen Buddhism, First Series, Pg. 264, Grove Weidenfelt, 1949)28
[11] (D.T. Suzuki, Essays in Zen Buddhism, First Series, Pg. 188, Grove Weidenfelt, 1949)
[12] (Prof Dr Syed Hussein Alatas, Progressive Islam Vol 1 No 4-5, 1954, Amsterdam.)29
[13] (L.S Das, Awakening The Buddha Within, 1997, Broadway Books)

4 The Physically/Mentally Triggered Mystical Experience

The mystical experience may be entered through many avenues. What matters here is not the diversity of paths, but the nature of the precipitating event. And to be clear, not all mystical experiences are identifiably 'triggered.' Sometimes, they just happen, even if the experiencer can describe what was happening prior, there is no identifiable cause.

In this section we are focusing on endogenous (intrinsic) vs exogenous (extrinsic) triggers.

Endogenous triggers arise from within the organism itself. They are initiated through natural physiological and psychological processes—breath, posture, exhaustion, sensory deprivation, repetition, surrender. No foreign agent is introduced. The body-mind system alters its own state using its native mechanisms.

Exogenous triggers are imposed from outside the organism. Substances such as psychedelics (e.g., LSD, DMT, psilocybin, ayahuasca) chemically intervene in neural signaling, producing altered states through pharmacological means rather than intrinsic regulation.

The practices we will examine—whether the isolation of a cave, the rotational exhaustion of Sufi whirling, a conscious breakdown in the face of a new oral or written articulation, or the ordeal of the Sun Dance—function as endogenous biological keys. They push the organism toward a threshold where ordinary cognitive structuring fails. In contemporary terms, this often corresponds to a disruption of the brain's default mode network— (the ego's headquarters) to go offline, while the "mind" encounters the base operational layer of reality.

When this structure destabilizes, what remains is a direct encounter with the underlying field of awareness prior to conceptual division, abstraction or secondary (or tertiary) representation, such as writing, speaking, analysis, etc.

In this sense, the "mystical" is not something added to experience, but what becomes apparent when the ordinary filters of experience are defeated and fall away—through processes native to the organism itself or discovered by generations of cultural practice.

To begin, we start with Matsuo Basho's Old Pond, or Ancient Pond Haiku, one of the most famous Zen Haikus of all time, as translated by D.T. Suzuki:

Into the ancient pond

A frog jumps

Waters sound

This Haiku is not an actual record of awakening. It is part of a composition session where the frog was the subject. By historical record, Basho composed the poem in a hermitage. The *haiku* occurs nowhere near water and out of earshot of any frog splash. Yet, as an example, it is classic, where the beauty of witnessing a natural event at an opportune moment, pregnant with natural metaphor and allegory, causes a crashing of ego and awakening to emptiness.

Next is an actual record of awakening in the classic Zen tradition:

Zōbō pursued only literary studies before someone warned him that that was not the way to ultimate truth. Then he went to a Zen master and learned to contemplate emptiness. It took a long time for Zōbō to attain realization. Eventually, he got to the point where he was so absorbed in concentration that he forgot to eat and sleep. One night as he was sitting quietly, unawares Zōbō fell asleep from exhaustion. When his Zen teacher struck him to wake him up, all at once he realized enlightenment. Zōbō was scarcely over 60 years old when he passed away in 1840. On the brink of death, he wrote a final verse:

Zōbō at 60! Here's my real state:
Where eight clouds are standing, I piss at the sky. It's a wonder, and a pity too,
I didn't kill all the imitation Zen in the world.

After Zōbō's death, the emperor of Japan titled him Zen master, spiritual mirror shining alone.

-Thomas Cleary, Zen Antics (1993)[1]

Zōbō, after long contemplation to the point of disappearing from the attendance to his own needs, collapsed in exhaustion. A sudden strike of the staff brought realization. Thus we see the long preparation and study of Zen under instruction and the preparatory altered state of consciousness, brought about by meditation, contemplation and the fragile state of exhaustion, combined with the well-timed fateful staff strike by the Zen master. This is what released the essence of Zōbō's "ripe fruit" thus elevating *him onto the* spiritual plane. This event is a classic example of a mentally triggered enlightenment.

It is typical in Zen stories of enlightenment, where a well-timed strike, shout, slap, simple word, or, perhaps the sound of a frog splashing into the water, releases one into the transcendent realm.

Nature, animals, art, and music, are conducive to subtle, yet profound altered states of consciousness. These can be used by an experienced master as opportunities for insight.

Zen Is

Zen is interesting in that the road to enlightenment is very straightforward, explicitly non-mystical. To get on the path and stay on the path until the end of the road is reached is the business of the monk. How long does it take to reach the 'end'? More often than not the 'end' is never reached. The road is notorious for being endless.

Zen presents awakening as simple but unforgiving: no ritual escalation, no trance induction, only unassisted awareness in ordinary consciousness. Nothing is added to carry the mind across; whatever fails does so in full sobriety. Enlightenment appears to take place within events that have no identifiable 'trigger'. There are only occasions, never mechanisms. The well-timed strike of the master's staff is not a 'trigger' otherwise, a could simply beat a room of new monks into realization. The causative event is almost invariably an everyday thing, not rare, but well-timed and entirely common, even mundane incidents where awareness remains without interpretation.

For this reason, even across long monastic lineages, full realization is rare: if in a single monastery 3 or more are recognized as 'attaining' over decades of teaching the monastery will be 'famous'. Not because the path is complex, but because nothing assists in it—and crossing over is undetermined, so to have several realize, is an indication that between the master, and the practice, the flame burns bright enough to light the way.

It can be said that Zen constructs a transmission field where different descriptive modes all converge on the same experiential claim: realization is unassisted, structurally non-triggered, and rare precisely because it cannot be mechanically produced.

Additional examples of statements of profound insight obtained by a transformative experience caused by mentally / physically based methods of contemplation are provided in the following examples:

The Tao Te Ching by Lao Tzu

The Bhagavad Gita

The 99 Names of Allah from the Quran

Matthew 17:1-13 from the New Testament

and from Sri Ramakrishna

The Great Master, 4th Edition, by S. Saradananda

Chapter One from *The Tao Te Ching* by Lau Tzu

Tao (The Way) that can be spoken of is not the constant Tao...
The name that can be named is not a constant name.
Nameless, is the origin of Heaven and Earth;
The named is the mother of all things.
Thus, the constant void enables one to observe the true essence. The constant being enables one to see the outward manifestations. These two come paired from the same origin.
But when the essence is manifested,
It has a different name.
This same origin is called "The Profound Mystery."
As profound the mystery as it can be,
It is the gate to the essence of all life

The Way of Taoism, founded by Lau Tzu, reaffirms the ineffable nature of the exalted state. And indicates that the 'nameless" is a field underlying the physical, and words, and things, where "*the constant void enables one to observe the true essence.*"

From *The Bhagavad Gita* (*The Song of God*) 10:12-15

Arjuna said: You are the Supreme Brahman, the ultimate, the supreme abode and purifier,
the Absolute Truth and eternal divine person. You are the primal God, transcendental and original, and You are the unborn and all-pervading beauty. All the great sages such as Narada, Asita, Devala and Vyasa proclaim this of You, and now You Yourself are declaring this to me.
O Krishna, I totally accept as truth all that You have told me. Neither the Gods nor demons,
O Lord, know Thy personality. Indeed, You alone know Yourself by Your own potencies,
O origin of all, Lord of all beings, God of gods, O Supreme Person, Lord of the universe!

This is a song to Brahman. Here we see the beauty and difficulty involved in forming a description of the ineffable. A problem the Buddhist and Taoist avoid. They describe the flavor, the essence, the subtlety of the path, leaving blank the destination. Often, the destination is described in negatives—what it is not—using words to point at what is beyond words. In this respect the reference to, "*the unborn*" is fully descriptive. It is potential, power, consciousness, from which 'all' emerges. 'Un–born,' once again, pre-language, pre-representation, and pre-abstraction.

The 99 names of Allah from the Quran (the first 20)

The Beneficent (He who wills goodness and mercy for all His creatures)

The Merciful (He who acts with extreme kindness)

The Eternal Lord (The sovereign lord, the One with the complete dominion, the One whose dominion is clear from imperfection)

The Most Sacred (The One who is pure, so far removed of any imperfection.)

The Embodiment of Peace (The One who is free from every imperfection)

The Infuser of Faith (The One who witnessed for Himself that no one is God but Him)

The Preserver of Safety (The Protector. The One who witnesses the saying and deeds of his creatures)

The Mighty One (the strong, the defeater who is not defeated)

The Omnipotent One (The One that nothing happens in His dominion except that which He willed)

The Dominant One (the One who is clear from the attributes of the creatures and from resembling them)

The Creator (the One who brings everything from non-existence to existence)

The Evolver (The maker the creator who has the power to turn the entities)

The Flawless Shaper (The One who forms His creatures in different pictures)

The Great Forgiver (The forgiver, the One who forgives the sins of his slaves time and time again)

The All-Prevailing One (The dominant, the One who has the perfect power and is not unable over anything)

The Supreme Bestower (the One who is generous in giving plenty without any return. He is everything that benefits whether halal [permissible] or haram [forbidden])

The Total Provider (the sustainer, the provider)

The Supreme Solver (the opener, the reliever, the judge, the One who opens for His slaves the closed worldly and religious matters)

The All-Knowing One (The knowledgeable; the One nothing is absent from His knowledge)

The Restricting One (the constrictor, the witholder, the One who constricts the sustenance by His wisdom and expands and widens it with His generosity and mercy)

The "ineffable "in "99 names", leaving no stone unturned. Often the challenge of naming what Allah is, is so daunting, and ultimately impossible, "God" is often described only by what God *is not*. But in the 99 names, those who have been enveloped in the ecstasy of union will certainly recognize many defining characteristics.

Matthew 17:1-13, The Transfiguration of Jesus

1 After six days Jesus took with Him Peter, James and John the brother of James, and led them up a high mountain by themselves.
2 There He was transfigured before them. His face shone like the sun, and His clothes became white as the light.
3 Just then there appeared before them Moses and Elijah, talking with Jesus.
4 Peter said to Jesus, "Lord, it is good for us to be here. If you wish, I will put up three shelters–one for You, one for Moses and Elijah."
5 While he was still speaking, a bright cloud enveloped them, and a voice from the cloud said,
"This is My Son, whom I love; with Him I am well pleased. Listen to Him!"
6 When the disciples heard this, they fell face down to the ground, terrified.
7 But Jesus came and touched them. He said, "Get up. Don't be afraid."
8 When they looked up they saw no one except Jesus.
9 As they were coming down the mountain, Jesus instructed them, "Do not tell anyone what you have seen, until the Son of Man has been raised from the dead."
10 The disciples asked Him, "Why then do the teachers of the law say that Elijah must come first?"
11 Jesus replied, "To be sure, Elijah comes and will restore all things.
12 But I tell you, Elijah has already come, and they did not recognize him, but have done to him whatever they wished. In the same way the Son of Man is going to suffer at their hands."
13 Then the disciples understood that He was talking to them about John the Baptist.

A direct communication from the creator to the Prophet. There are many who *hear* the voice of the creator. Rarer still is the one who is communicated with directly, where others see and hear, and they are further instructed, guided and gifted with methods, abilities and talents that give proof of their divine link. But it is the rare individual, who beyond enlightenment, who beyond communion, is given instruction and Good News to give to others.

From: ***Sri Ramakrishna The Great Master*****, 4th Edition, by S. Saradananda, The Spiritual Aspirant, (6) Spiritual Eagerness and the First Vision; 13. The description of the Master's attainment of the first vision of the universal Mother. The Master's yearning at the time:**

> He used to say, "There was then an intolerable anguish in my heart because I could not have Her vision. Just as a man wrings a towel forcibly to squeeze out all the water from it, I felt as if somebody caught hold of my heart and mind and was doing so with them. Greatly afflicted with the thought that I might never have Mother's vision, I was dying of despair Being in an agony I thought that there was then no use in living this life. My eyes suddenly fell upon the sword that was there in the Mother's temple. I made up my mind to put an end to my life with it, that very moment. Like one mad, I ran and caught hold of it, when suddenly I had the wonderful vision of the Mother, and fell down unconscious. I did not know what happened then in the external world–how that day and the next slipped away. But, in my heart of hearts, there was a flowing, a current of intense bliss, never experienced before, and I had the immediate knowledge of the Light that was Mother."
>
> On another occasion the Master described to us in detail his wonderful vision spoken of before. He said, "it was as if houses, doors, temples and all other things vanished altogether; as if there was nothing anywhere! And what I saw was a boundless infinite conscious sea of light! However far and in whatever direction I looked, I found a continuous succession of effulgent waves coming forward, raging and storming from all sides with a great speed. Very soon they fell on me and made me sink to the unknown bottom. I panted, struggled, and fell unconscious." The Master told us that at the time of his first vision he saw a conscious sea of light. But what about the divine Mother's form consisting of pure consciousness only–the form of Hers with hands that give boons and freedom from fear? Did the Master then have the vision of that form also in that sea of light? It appears that he had, for as soon as he had the slightest consciousness at the time of his first vision, he, we are told, uttered repeatedly the word, "Mother" in a plaintive voice.

-Sri Ramakrishna, The Great Master, 4th Edition, (1952)

The great sage Ramakrishna, continuing in his fits of passion and refusal to be denied his ecstasy with "Mother". There may have never been a sage more fitful, committed and magnetic (and dramatic) than Ramakrishna.

We will end with a final Zen lesson from *Master Kakushin* who learned his Zen in China and then returned to Japan with the formless lotus blossom of Truth. The blossom that emerges from the surface, opens, follows the sun and submerges to repeat the process.

> "The (emperor) Go-Uta (invited) Kackushin to a special imperial villa to teach Zen. The master said, "The Buddhas understand mind; ordinary people misunderstand mind. The source of all Buddhas is one; the realm of misunderstanding and understanding divide. Without depending on one on another power, you can know by inherent capacity. If you want to arrive at Buddhahood, you must look into your own mind" -Thomas Cleary, *Zen Antics* (1991)[2]

The secret of all existence is contained within the master Kackushin's decree: the source of all Buddhas is one, but the realm of understanding and misunderstanding divides. Our consciousness is the only non-illusion, the only aperture through which Truth is entered. This is the core of the Physically/Mentally Triggered Mystical Experience.

From a materialist perspective, our entire lives are lived within an illusion of time and physical law. This is the lie. Physics does not have the final say. Life does. More precisely consciousness does.

Consider the sunset—a daily event. When we watch a sunset from the beach we are seeing an illusion. The Earth's atmosphere acts like a massive lens, bending the sun's light around the curvature of the planet. By the time the bottom of the sun appears to touch the horizon it has already dipped entirely below our field of view two minutes to four minutes ago.

Science also tells us that any sound we hear, happened in the past, equal to the time it took the speed of sound to get to us whether it is an inch away (less than a hundred microseconds) or a thousand miles. Even our sense of touch, taste, and gazing into the eyes of a loved one are all, if "science" is to be taken literally, in the past, if even by milliseconds. Yes. the lover that touches you, kisses you, holds you, that is all, if even imperceptibly, in the 'past'. But is that 'true'? These conditions aren't just for humans, *but* living things are the only things in the universe that can 'read' these signals, experience them, and know 'what they are like', and because of this, we are also experiencing the universe, however distant, however our perception operates, exactly and precisely as intended, or at least, in the only way they can be experienced, and, we are the only things that can experience it—subject to the physics (underlying laws) of the universe, and Einstein's relativity.

So in going over Zen, Buddhism, Christianity, Islam, and other interpretations of the breakdown of the mental sieve to penetrate into the evaluative layer of the *unborn mind*, we find that different culturally distinct contemplative technologies (meditation, ritual, prayer, austerity, paradox training) produce similar breakdowns of discursive cognition, resulting in a shared experiential mode where awareness is no longer mediated by language or conceptual interpretation. We find ourselves instead at the 'base layer'.

Our thoughts, at all times, but particularly in the operating system of the base layer are exactingly immediate and *transcendent* in the *now*.

This is what makes the scientific explanation of perception, time and consciousness a *lie*.

> The sun sets when we see it set.
>
> The sound happens when we hear it.
>
> The touch of a mother upon her child, the taste of a delicious fruit in our mouths, and gaze of our lover deep into our soul all happen when we experience them happening. Not in the past, nor in the future for 'other observers.'
>
> The past is the illusion, the incomplete half of a duality that terminates in consciousness. The past is *the dream* and these explanations, while "technically correct", are tricks and games. They are the description of waves. *Consciousness* is the shore. *Youness* is the firmament. Existence does not exist, until we experience it, until it joins with a living source, and our experience makes everything real, from the mists of *Maya*.
>
> Whatever it is that thoughts are, and whatever it is that we are, and whatever it is that consciousness is, is the only immediate, fully present, happening, "now" experience we have ever had. Maya exploits this fact, to create the heavens and Earth, for none of it has to exist "out there", it only has to exist, "in here", within us, when the information of the heavens makes contact with our inner being. This validates that the great mystery and the great answer must always lie within. Even if it were 'without' it would mean nothing, change nothing, and fail to be understood until it made contact within.

This is our key. Looking inward, clearing our mind of the *monkey mind* that thinks and chatters (another obstructive phantom). When silenced, we plunge into the clarity of our unadorned immediate self. Here is the seat of all experience and the mirror where the signal is received.

This is the effort behind the physically and mentally triggered mystical experience. If the vital connection can be made between what is and what we are, like a diamond cutter's precise blow, the mind is defeated. The walls crash. The divine revealed. Insight achieved. Truth released. Dualities collapse. Paradoxes resolved. Ignorance obliterated.

> *THIS was a thought that I wanted*
> *THIS was the thought that I needed*
> *THIS was the thought that seeded my mind, and deleted my mind, and repeated within my mind and thereby completed my mind,*
>
> *This is when everything blossomed.*
>
> *Everything I ever imagined, pondered upon, every answer I could have hoped for was within this thought. Tomorrow will be the first day of my life.* - Damon Eaves, *EVZ-YOYO Esplanade!* (1989)[3]

The methods employed and conditions used to obtain the divine vision, may be *anything* originating in this world, any event, any thought, or any action. These are our tools. They are provided by the training program of life. If we remember the essential nature of *Maya*, then we know / may realize that the sword of *Maya* also cuts both ways. All things are equally illusory and divine. Therefore, for our consciousness, in the effort to defeat the essential illusion, and to expose enlightenment and *samadhi*, all objects may be turned *inside out*. All objects are equally potent in their ability to serve as keys to realization when they are in resonance with our consciousness. This accounts for the many "unintended" mystical experiences, similar to the ones recounted above, that have occurred by unexpected means, as in the following story of Ramakrishna in his youth around age 6 or 7 years old when he saw white cranes against a black cloud.

> "I was following a narrow path between the rice fields. I raised my eyes to the sky as I munched my rice. I saw a great black cloud spreading rapidly until it covered the heavens. Suddenly at the edge of the cloud a flight of snow-white cranes passed over my head. The contrast was so beautiful that my spirit wandered far away. I lost consciousness and fell to the ground. The puffed rice was scattered. Somebody picked me up and carried me home in his arms. An access of joy and emotion overcame me. . . . This was the first time that I was seized with ecstasy."
>
> -Romain Rolland, *The Life of Ramakrishna,* (1930)[4]

Later, Ramakrisna says of this experience,

> "Since that day, I've been an altogether different man. I began to see another person within me."

This experience is known by his followers as the first ecstasy of Ramakrishna.

There are innumerable stories of individuals being seized by ecstasy suddenly and without expectation, with no meditation, prayer nor even previous religious inclination. The divine resonance has somehow been struck, the divine key of all keys has flown into the lock and a new realm has been entered.

This type of example is contrasted with the innumerable stories from China and Japan which recount those who have studied for years, or decades and then left their teachers and retreated into the woods, or a cave in an unpopulated location, and have committed themselves to breaking through to the mystical experience at whatever personal costs were required, and even in acceptance of death. Then, just at the moment when they are about to collapse, or are at the height of sickness or fever, they had an experience of the highest order, a communion with the divine known as *satori* in Japanese.

Satori

> "Without the attainment of satori no one can enter into the mystery of Zen. It is the sudden flashing of a new truth hitherto altogether undreamed of.. the whole edifice has now come to the ground, when behold a new heaven is opened to your full survey... Religiously this is a new birth, and, morally, the revaluation of one's relationship to the world. The ladder now appears to be dressed in a different garment which covers up all the ugliness of dualism, which is called in Buddhist phraseology delusion (māyā) born of reasoning (tarka) and error (vikalpa)."
>
> -D.T. Suzuki, *Essays in Zen Buddhism,* (1949)[5]

The Satori of Dōken

> "Under Daiye (Tai-hui, 1089-1163), the great Zen teacher of the Sung dynasty, there was a monk named Dōken (Tao-ch'ien) who had spent many years in the study of Zen, but who had not yet delved into its secrets, if there were any. He was discouraged when he was sent on an errand to a distant city. A trip requiring half a year to finish would surely be a hindrance rather than a help to his study. Sogen (Tsung-yüan), one of his fellow monks, took pity on him and said; 'I will accompany you on this trip and do all that I

can for you. There's no reason why you cannot go on with your meditation even while traveling.' They started together."

One evening Dōken despairingly implored his friend to assist him in the solution of the mystery of life. The friend said: 'I am willing to help you in every way, but there are five things in which I cannot be of any help to you. These you must look after yourself.' Dōken expressed the desire to know what they were. 'For instance,' said the friend, 'when you are hungry or thirsty, my eating of food or drinking does not fill your stomach. You must eat and drink yourself. When you want to respond to the calls of nature, you must take care of them yourself, for I cannot be of any use to you. And then it will be nobody else but yourself that will carry this corpse of yours [i.e. the body] along this highway.' This remark at once opened the mind of the truth-seeking monk, who, transported with his discovery, did not know how to express his joy. Sogen now told him that his work was done and that his further companionship would have no meaning after this. So they parted company... What was it, one may ask, that flashed through Dōken's mind when his friend gave him such matter-of-fact advice?

-D.T. Suzuki, *Essays in Zen Buddhism,* (1949)[6]

As alluded to by D.T. Suzuki, Zen Buddhism is full of stories of those who have struggled their entire lives and have not attained, and then a simple utterance, a jerk of the nose, or a seemingly innocuous statement unlocks the most coveted of all experiences.

There are also stories of great masters who are not strictly of the Zen tradition, who attained only after years of effort and fruitless attempts at "effortless" effort.

"The notion of sudden insight plays a special role in Zen Buddhism. The ultimate goal of Zen is 'satori," which Japanese scholar D.T. Suzuki explained as 'acquiring a new viewpoint for looking into the essence of things.' A practitioner may pursue Zen assiduously for many years before achieving satori, but if satori is achieved, it will likely be triggered by some seemingly irrelevant event. According to Suzuki, 'An inarticulate sound, an unintelligent remark, a blooming flower, or a trivial incident such as a stumbling, is the condition or occasion that will open his mind to satori.' He states that this 'new viewpoint' will be acquired in an instant: 'Satori is the sudden flashing into consciousness of a new truth hitherto undreamed of. It is a sort of mental catastrophe taking place all at once, after much piling up of matters intellectual and demonstrative. The piling has reached a limit of stability and the whole edifice has come tumbling to the ground, when, behold, a new heaven is open to full survey.'"

-John Kounios, *The Eureka Factor,* (2015)[7]

In the struggle to attain satori, individuals have used culturally revealed practices, such as: meditation, readings, mortifications, rituals, asceticism, trials, endurance, sleep deprivation, retiring to huts and caves and practicing near 24 hour a day meditation. All that can be imagined has been pursued to unlock the mystical experience.

The Special Case of *Thatanka Iyotake*

The famed Unkpapa Lakota Holy man, *Thatanka Iyotake*, more commonly known as *Sitting Bull,* was the son of *Returns Again* and *Her Holy Door*. In his culture, he communed with *Wakan Tanka*, the great spirit and sacredness of/within all things. Within his tradition there were many ways in which to invoke and live within the protection and guidance of spirit entities. The most well-known of these rituals was the *sun dance,* as described by Thatanka Iyotake's Great-grandson, Ernie LaPointe. The stories he relates had been told to him, in turn, in the oral tradition of his people[8].

Thatanka Iyotake's entire life was lived in the protection and guidance of spirit entities. He was *chosen* from his earliest days and guided and communicated with as he navigated his life plan in aid of his survival and the survival of his people.

By all accounts he was one of the most powerful holy men of his time, and also the greatest warrior and strategist of his age of any culture. His most well-known spirit visions were after he had performed the consciousness altering *sun dance,* also known as, *Wiwang Wacipi* (gazing at the sun as you dance). This was taught to him through example by his mentor and 2nd father, *Four Horns*. Prior to the dancing part of the ritual, the dancer undergoes purification in sweat lodge and fasts and bathes themselves in sweet grass smoke, sage and/or cedar and is then painted head to toe with traditional symbols and designs, while wearing eagle feathers, bells, rattles or other noise making objects to ward off evil spirits. Offerings of meat, tobacco and other items are made to the sacred tree set in the middle of the dance.

When the dance begins, the dancer stares at the sun and moves in a circular motion while tied to the tree with leather straps in the center of a circle that has been specially selected for the ritual. The other end of the leather straps are connected to sharp bones or sticks that have pierced the flesh and have been pushed through the upper body. Dozens of pieces of flesh have also been cut from the dancer's arms in order to soak a blanket in blood (a red blanket) in honor of the creator spirit. As the dancer moves progressively through this dance he both stares at the sun and forcibly pulls and rips the bones that are piercing his body until they are free and no longer connected. To increase the altered state of the dancer, sharp edged gravel is traditionally added to the inner ground to multiply the suffering. Sweat and blood is spilled throughout the ceremony[9].

Sometimes a buffalo head is substituted for the sacred tree in the center, or, dragged behind by the dancer.

When *Thatanka Iyotake* had performed this dance for some time, he then agreed to be pierced on his back and chest and suspended by skewers above the ground. When he did this he remained suspended through the night. He was found in the morning and cut down by his friend. He whispered that he had had a vision and then he recounted it.

In his vision, he was on the ground and told to lie still by a woodpecker as a bear inspected his face. He was then transformed into a young man who encountered a wolf with two arrows in his side. He removed the arrows and was promised to be known by all nations for his deed. He was next an adult seated next to a *Wicasa Wakan* (medicine man), and told he had great honor, had been blessed by the *Wakan Tanka* (the great spirit and sacredness of/within all things) and that his responsibilities would be greater than his blessings.

This vision, of greatness and renown was fulfilled as today, and in his lifetime, the entire world knows the name *Sitting Bull,* and his deeds.

Sitting Bull shocked America and the entire western world by obliterating the U.S. Cavalry at *the battle of the greasy grass,* a location known by the U.S. Cavalry as *Little BigHorn* where not one of General George Custer's 350 cavalry men was left alive[10].

The sun dance is a particularly difficult ritual. The pain, fatigue and sun induce an altered trance state after several hours to several days according to the stamina of the practitioner. At some point, the individual collapses from exhaustion, dehydration, hunger, lack of sleep and other deprivations, only then, does the vital communion, and desired vision play out.

The vision given to Thatanka Iyotake wasn't the only vision he received in his lifetime. He also had a vision prior to the battle of the greasy grass approximately 10 days before the battle.

He once again *performed the sun dance. 50 pieces of flesh were taken from each arm, and his feet were thoroughly cut by the sharp gravel. He danced until the sun went down and then rose the next morning. *Thatanka Iyotake*'s prayers were answered and he had his vision. In it many long knives (cavalry) had fallen into the camp. They were upside down with their feet above their heads and they had no ears. He was explicitly told that none of his people could take any belongings from the dead cavalrymen.

This vision of a great victory was fulfilled, but sadly, the people did not heed his instruction from the vision because they took the belongings from the dead and the dying long knives. The night of the victory there was a great celebration, but Thatanka Iyotake did not partake because he was sad for what his people had done in defying his vision. The darkness soon enveloped his people as the reservation and near genocide of his people was not far off.

There are many stories in the legend of *Thatanka Iyotake*, and many have to do with his bravery, compassion, the soundness of his plans and strategy but also, his uncanny and inexplicable capacity of being protected from harm on repeated occasions despite his habit of placing himself in the direct line of fire of his enemies[11].

It is said that before the battle of the greasy grass, Sitting Bull rode his horse back and forth across the front lines of the U.S. Cavalry well within range of the soldiers, but despite dirt flying and shots ringing the air, neither Sitting Bull nor his horse were struck.

Like all prophets, shamans and holy men, he had been anointed and performed miracles equal to the power of his birth.

In the example of *Thatanka Iyotake*, we once again are reminded of the extreme asceticism and "Suffering & Sacrifice" as described earlier in this chapter in several Zen stories of enlightenment, and, in how the Buddha's six-year search and starvation and privations and endurance in meditating created an awakening.

Suffering, or, in the hands of a skilled practitioner, spiritual mortification, as a precondition, in practice, in ritual, and in entering an alternative state of being, is oftentimes the causative 'fire' that frees the individual from the material, biological world, and transmutes them into a receptacle for spiritual insight. Such visions are well known and repeated themes in the record of awakening taught to seekers across cultures.

The Sweat Lodge & The Evaluative Field

The Native American sweat lodge ceremony is a technology of the soul. While the Western world dismisses ritual as superstition, the healer has seen first-hand how an extreme condition is an effective way to bypass the analytical mind and force contact with the divine.

As we mentioned earlier, prior to Sitting Bull's sun dance, he underwent a purification and other rites, including a sweat lodge ceremony.

I had the opportunity to take part in a first peoples sweat lodge led by a holy man, his assistant and a small group of about 7 participants.

The people who had gathered for this sweat had a number of reasons for attending. They each hoped to emerge with help and insight into what they were grappling with. In most cases the desired resolution was very specific. For instance, one woman was going through a very difficult period of time with her young adult daughter. Each person was on their own journey and on the day of the ceremony, it was unavoidable that we all had become travelers along a collective journey as well.

Invariably, the effective part of the sweat lodge was the pressure of suffering—the literal burning of flesh and lungs by superheated steam. The fear of irreparable harm was the primary obstacle. In modern contexts, ceremonies are often sanitized to 'ensure safety,' a strange quality to strive for in a precipitative ritual where faith in the healer is the only safety net. One only has to look at the Sun Dance—fifty strips of flesh cut from the arms, bodies suspended from posts by skewers in the chest, sharpened rocks to make the 'dance' a near unendurable affair.

There is a mechanical distinction between a conceptual expansion and a somatic demolition. The sweat lodge is a high-bandwidth confrontation with the survival drive that forces the analytical mind to abdicate. A seventy-five percent experience—one that does not break the ego or silence its dialectic—is a survey of the map; it is not the territory. To reveal the underlying architecture of the 'Unborn,' the system requires more than an intellectual assent—it requires a total starvation of the ego monologue and dialectic. Many of the great prophets had to achieve in this same manner. They had their revelation, but only after deprivation and defeat of the internal dialectic. It is a 'democratic' process in many ways.

We were in the hands of an extremely skilled healer—a leader capable of gauging the individuals involved and stripping us down to our essential core over multiple rounds of what can only be called suffering, while also keeping us entirely safe. Anyone could leave at any time.

During the application of water to superheated river stones pulled from a roaring fire, there was singing and chanting led by the holy man. With each round, we were broken down by the steam and unrelenting heat that shrank time to a series of small, linked moments where the only objective was survival. It grew more intense the longer we progressed. The escalation in heat and duration from round 1 to 4 and beyond was notable. At first, I was bracing myself. By the fourth round, my thoughts became about surviving the moment. At the beginning of the 5th round the holy man advised me to wear a sheet (bedsheet) for some form of protection. I declined (neither he nor his assistant were wearing one) and then he suggested it again. I took the 'hint' and wrapped myself in a sheet along with the one other remaining participant and we were happy for it. As advised, we turned towards the outside of the lodge and as instructed by the leader, breathed the cooler air coming in from the bottom. We were instructed to think about our issues (prayer) through the rounds. I could slowly feel all the things I thought about myself, and that I thought I wanted out of this experience melting away. I simply could not think of anything while trying to 'survive.' This essential change, having one's 'issue' snatched from one's hands in the midst of an overwhelming stimulus was a key factor. Strangely, when I thought of my life, in that heat, and in that pain, and my increasing fear that when I walked out my skin would be hanging off of my burnt limbs, things got very simple. Just survive. Be thankful for each breath. At some point this has to end. No. Be thankful for the cooler air, that you are not alone, that this will end.

Because everything else I thought of as being a problem had been burned off, I saw my 'issue clearly.' It was just standing there. I was solely responsible for it. I was the one creating and sustaining the 'fire and steam' that sustained it. I was the cause of my own 'suffering' in my life. I had the power to change what was 'wrong.' I could go around, over, under, or even through the issue. In the light of the purifying fires, it was nothing more than a figment of how I conceived of my life and my obstacles.

But there were still two more rounds to go. The holy man looked at me and said, "these last two rounds are going to be hot." as if the others weren't, which to my ears meant, "you can leave if you want."

The two rounds it was only me, the holy man, and the 'bringer' of superheated rocks. It was so hot and such a disaster of steam, I *knew* my skin was cooked and gone and would likely fall off when I emerged. I was going to need to get to the hospital as soon as this thing was over, and I contemplated what I'd tell people from my hospital bed when the asked why I didn't leave.

But none of that happened. I was intact. We were all laid out on the damp ground, fully drained of power when it was over. And I can say I've rarely been as exultant and happy to be alive.

There was a clarity to this realization of my 'problem' in the context of the sweat lodge. Before, I didn't know I was blocking my way and creating my own problems. The sweat lodge, through the haze, smoke, steam, and suffering brought an immediacy that made the dynamics plain and clear. The ceremony showed me the most direct path forward. If a 'problem' represents a 'blockage'. There had to be a thousand solutions. It was clear. This burning had brought me to the door of truth in a way I hadn't imagined possible. I thought of the woman with the problem with her daughter. She can't be in conflict with her daughter if she refuses to be in conflict. She can live in a way where the boundaries are clear as are the consequences. No fights, conflicts, arguments, just 'cause and effect.' Love, care, kindness, respect, can all persist in the midst of whatever is going on.

Later in life I noticed this same clarity could be reached in running. On occasions when I was in a good rhythm, I could feel my mind elevate to a space where I could think of any issue and suddenly all of the blockages were removed. The clutter that made the problem appear like a mountain, or complex, or 'multi-step' was no longer apparent. There was only a simple solution in its place.

Meditation could bring this state. Many individuals have discovered a space where their access to the field is greatly facilitated, easier, more immediate.

So, what is happening here?

Although the sweat lodge, running, meditating etc., are not mystical experiences, what is being contacted is the same: the evaluative field. From spiritual figures to scientific insight—whether the Buddha or Einstein—this same contact appears. Anytime there must be evaluation and Truth, there must be a pre-existing objective standard. Immaterial, non-local, and inexhaustible. Otherwise, truth cannot be accessed, propositions cannot be evaluated, and understanding cannot be immediate, inexhaustible and specific. Not that all penetrations into the realm of thought make contact with the field. If one adheres to a dogma, then they are accessing a local, constructed architecture. The August Wheel tells us, there are two 'falls.' When one is unknowing as to what is true the fall is into illusion. When what one believes in is objective fantasy the fall is into delusion. It is only when what we believe aligns with what is true, is transcendence touched and the field accessed and the belief confirmed, or, the corollary, an open proposition and appeal is made, and the field, being fed the rudiments of the issue, answers with a clear path and truth—perhaps new knowledge—as in a Buddha, or an Einstein.

When we engage in these activities—and these culturally discovered and outlined methods—we tap into the well of wisdom. And this 'tapping' illustrates an interesting quirk of Truth. In gaining access to a deeper truth, *we come to embody the wisdom we seek.* And when we do this the world can be seen in an alternative substrate. Beyond *Maya* and *Lila*, in the context of the unborn, the pre-linguistic, pre-symbolic, field of constraint. The same immaterial boundaries where the laws of physics, constants, and the field of instantiation finds its power. Our world is *Maya*, but it is a specific kind as we have noted. Made up of interlocking parts of the intelligent observer 'training program'. The Fibonacci code of the rose, photosynthesis, fruiting bodies, pollination, fragrance, beauty, light. The entirety of life can be seen as a 'mirror', a 'consciousness substrate', navigated by access with the evaluative field, that reveals as much about what is being observed as what is doing the observation. Science, when done correctly, and unchained from fallacious dogma, illusion and delusion, is the key that unlocks the secrets of 'the evaluative field.'

And if The Divine Algorithm is correct, and the culmination of the universe from the first moments was to produce intelligent observers, it follows that we have the ability to contact the field in a proliferation of ways to retrieve all possible knowledge and truth. The mystical and transformative and spiritual experience are special forms of access. The 'final boss' event of the field.

The field is not a quirk of the human condition. It is the completion of the circle initiated at the origin of the universe. The system produces the observer, and under the right conditions, the observer comes into contact with the system of constraint that serves as the basis of their being. In that moment, what is revealed is not an idea, not a model, not a ruleset, but the underlying nature of reality as experience—the moment where the 'youness' of the individual collides with the consciousness of the field and the instantiating power behind the universe. This contact and the experience it provinces is the "what it is like" of ultimate reality. Clearly, it cannot be reduced to matter or mechanism, for these things are several steps away information from the field of constraint. The field is *the cause* of particles, structures, and processes, but it is not 'caught' within them, reliant upon them, or beholden to them. Quite the opposite. Cause always proceeds effect.

The Revealed Self

The revealed self, in contact with the field is something that Rinzai Zen master Bankei (1622-1693) discussed as our previously introduced concept of "unborn mind".

> "Although you arrived in the world with nothing but the unborn Buddha-mind, you fell into your present deluded ways as you were

growing up, by watching and listening to other people in their delusion. You picked all this up gradually, over a long period of time, habituating your mind to it, until now your deluded mind has taken over completely and works its delusion unchecked. But none of your deluded thoughts was inborn. They weren't there from the start. They cease to exist in a mind that's affirming the Unborn." – Norman Waddell, The Unborn: The Life and Teachings of Zen Master Bankei (1982)[12]

The fire and steam perform a mechanical stripping of this habituation. By starving the deluded mind of its narrative fuel, the 'Unborn' isn't created—it is revealed as the only thing that was there from the start. The sweat lodge, and its efficiency in breaking down the ego, pride, stubbornness, pettiness, and in separating the unborn mind from the physical body shocked me.

Is it a coincidence that this ceremony uses the four sacred elements—earth, air, fire and water—with life, consciousness and youness to achieve the magical transfiguration of the soul? No.

At the end of the sweat lodge I was wiped out and felt I would surely remain in this condition for several days, but less than a half hour later, after a short cold shower, I felt energized and invigorated and as alive and awakened as I'd ever felt. This was another unexpected secret of the technology.

Is it a coincidence that this ceremony uses the four sacred elements—earth, air, fire, and water—combined with the vital contact of consciousness and "youness" to achieve the transfiguration of the soul within the evaluative field? No.

It is the ultimate low-tech, high-fidelity mirror of reality. Made up of the things every group of being accesses and uses every day. The genius of the first peoples was to use these essential elements to construct a pressure cooker to precipitate the breakdown of the ego in a ritual that forces a direct collision between the universe and the observer.

What the sweat lodge revealed through experience, can be described in mechanical terms.

Integrating The Signal - The Second-Order Disjunction

There is a clarity to the recognition of one's "problem" in states of heightened resolution. But the return to ordinary life can be a second-order disjunction depending on the depth of the resolution. Perhaps no disjunction takes place when one realizes that with regards to a particular problem, they are the 'issue'. But when the realization of a greater magnitude, such as, 'we are all one', 'life is like a dream compared to contact with the evaluative field', or 'we are all in the midst of a play of Lila in an illusion of Maya', the 'return' to our previous lives can certainly be described as a 'second-order disjunction'. For the monk or devotee, the return is stabilized by an already-formed container; the monastery is a technical environment designed to sustain the flame. For the layperson, the common world is an environment in which few flames are seen, and the momentum of society is to blow those flames out.

People call this state ineffable, but that is a misnomer. Ineffability is merely the result of attempting to describe a non-shared experience. To someone who has never seen an animal, playing with a kitten is ineffable. To someone without the constitution for emotion, love is ineffable. "Ineffable" simply means a lack of common reference. Yet the architecture of the transcendent surely exists; between two individuals who have accessed the same operational layer, it is not ineffable at all.

It is the recognition of an alternative substrate of reality—encountered in an extended present where sequential thought collapses and the system is apprehended as a total field rather than a chain of events.

It Has Been Foretold That God's Nature is Within Us

Socrates (469-399 bce) believed that God's nature was within us and available to us through a life of *virtue*. Jesus referred to himself and his methods as *the way*. Mohammad, (may the blessings and peace of Allah be upon him) believed himself to be the appointed messenger of *the one true God of Abraham*, commanded to teach the people by *the pen* so that they would know the divine truth that resided within. The Buddha felt that he had been given *the middle path* to propagate to the people so that they would have the key to release themselves from the wheel of suffering.

The prophet's mandate—in all of these examples—is the dissemination of a 'key'. They do not merely offer comfort; they provide the technical means—the 'Way,' the 'Middle Path,' the 'Pen'—required for the individual to initiate their own contact with the Field.

Perhaps one of the best examples of the mystical experience of Prophets, as it differs from all others who experience mystical insight and enlightenment, can be found in *the Vision of Universal Form* from *the Bhagavad Gita*.

In *The Bhagavad Gita, the song of God*, the great text of Hinduism, all possible *"God(s)"* are manifestations and forms of one divine all-encompassing source of *Brahman*. In *the Vision of Universal Form, where Krishna* is revealed to *Arjuna* in full measure:
Verse

1 Arjuna said: Having heard the supremely confidential spiritual knowledge, which You have revealed out of compassion for me, my illusion is now dispelled.

2 I have heard from You in detail about the appearance and disappearance of all living beings, O Lotus-eyed One, and also about Your eternal magnificence.

3 O Supreme Lord, You are precisely what You declare yourself to be. Now I desire to see Your divine cosmic form, O Greatest of persons.

4 O Lord of all mystic powers, if You think I am strong enough to behold It, then kindly reveal that imperishable cosmic form to me.

5 The Supreme Lord said: Behold, O Parth, My hundreds and thousands of wonderful forms of various shapes, sizes, and colors.

6 Behold in Me, O scion of the Bharatas, the (twelve) sons of Aditi, the (eight) vasus, the (eleven) rudras, the (twin) Ashwini Kumars, as well as the (forty-nine) maruts and many more marvels never revealed before.

7 Behold now, Arjuna, the entire universe, with everything moving and non-moving, assembled together in My universal form. Whatever else you wish to see, observe it all within this universal form.

8 But you cannot see My cosmic form with these physical eyes of yours. Therefore, I grant you divine vision. Behold My majestic opulence!

9 Sanjay said: O King, having spoken thus, the Supreme Lord of Yogi, Shree Krishna, displayed his divine and opulent form to Arjuna.

10-11 In that cosmic form, Arjuna saw unlimited faces and eyes, decorated with many celestial ornaments and wielding many kinds of divine weapons. He wore many garlands on His body and was anointed with many sweet-smelling heavenly fragrances. He revealed Himself as the wonderful and infinite Lord whose face is everywhere.

12 If a thousand suns were to blaze forth together in the sky, they would not match the splendor of that great form.

13 There Arjuna could see the totality of the entire universe established in one place, in that body of the God of gods.

14 Then, Arjuna, full of wonder and with hair standing on end, bowed his head before the Lord and addressed Him, with folded hands.

15 Arjuna said: O Shree Krishna, I behold within Your body all the gods and hosts of different beings. I see Brahma seated on the lotus flower; I see Shiva, all the sages, and the celestial serpents.

16 I see Your infinite form in every direction, with countless arms, stomachs, faces, and eyes. O Lord of the universe, whose form is the universe itself, I do not see in You any beginning, middle, or end.

17 I see Your form, adorned with a crown, and armed with the club and disc, shining everywhere as the abode of splendor. It is hard to look upon You in the blazing fire of Your effulgence, which is radiating like the sun in all directions.

18 I recognize You as the supreme imperishable being, the Ultimate Truth to be known by the scriptures. You are the support of all creation; You are the eternal protector of Sanātan Dharma (the Eternal Religion); and You are the everlasting Supreme Divine Personality.

19 You are without beginning, middle, or end; Your power has no limits. Your arms are infinite; the sun and the moon are like Your eyes, and fire is like Your mouth. I see you warming the entire creation by Your radiance.

20 The space between heaven and Earth and all the directions is pervaded by You alone. Seeing Your wondrous and terrible form, I see the three worlds trembling in fear, O Greatest of all beings.

21 All the celestial gods are taking Your shelter by entering into You. In awe, some are praising You with folded hands. The great sages and perfected beings are extolling You with auspicious hymns and profuse prayers.

22 The Rudras, Adityas, Aasus, Sadhyas, Vishwa Devas, Ashwini Kumars, Maruts, Ancestors, Gandharvas, Yakshas, Asuras, and Siddhas are all beholding You in wonder.

"The beauty about divine knowledge is that the more it is churned the more nectar it produces, and hence the world has undoubtedly been enriched by these insightful commentaries."

-Swami Mukundananda, *The Bhagavad Gita, The Song of God, Chapter 11* (2019)[13]

In *The Vision of Universal Form* we are presented with the vision of God as creation itself.

The Divine Algorithm speaks of repeated "scientific" versions of *The Vision of Universal Form*. Repeatedly in our journey we are confronted with situations where, *"...you* [we] *cannot see My cosmic form with these physical eyes of yours."* because we cannot see the universal creator's intervention—the instantiations that make up the algorithm. We do not know how it was achieved, by what means, what transpired, or how the impossible, miraculous transformation took place. We instead are confronted with what has been achieved. We find only the final wisps of the magician's smoke and the new thing taking form. This is what leaves us bewildered as to how it could have been achieved.

"Physical eyes" in this sense means more than just our eyes, and includes everything we use, including technologies, that assist us in peering into the substance of the universe. *"You cannot see..."* means the power of creative consciousness, and not the substance itself.

Are we forbidden to know this substance? No. We are invited to understand it. The Divine Algorithm is the invitation.

We are each the cosmic form. We not only get to see it, but we are also it. We are living, breathing, thinking beings made of the same magic that wrought the heavens.

The prophet's mandate is the dissemination of the key to this understanding. They do not merely offer comfort; they provide the technical parameters—once again, the 'Way,' the 'Middle Path,' the 'Pen'—required for the individual to initiate their own contact with the Field.

Our existence, as described in *the Bhagavad Gita* is a personally constructed diorama and monument to the creative powers of *Brahman*. And as *a universal form*, the universe itself is the model we study to understand the consciousness of the creator. When we look out upon the heavens, we are viewing what is in effect, an MRI of sorts for the body and consciousness of the creator. It is their handiwork, and by viewing Their patterns, we can know their quality and also ourselves.

> Sri Krishna is about to reveal to Arjuna the most staggering vision: the vision of the One in the many, the many in the One, the vision that shows that all are the One. This vision alone explains and justifies the apparent contradictions of the relative world and reconciles and antinomies of justice and mercy, fate and free will, suffering and divine love. It reveals all that was, is, and shall be. It relates and unifies the diversity of the world. Seeing God in all and all in God, man overcomes all doubt and perplexity and submits to the divine will. Thus even the most terrible thing in the world loses its terror and even the most hateful thing becomes a mode of the Lord's manifestation. The man blessed with this vision accepts the world with an all-embracing joy and great courage and goes forward with sure steps to discharge his appointed task. He sees all

things in one view, and not in a divided, parietal, and bewildered fashion.

-Swami Nikhilananda, *The Bhagavad Gita* (1944)[14]

Insight Redux

This brings us to the nature of insight. An insight is an experience of sudden clarity or recognition. Its authority emerges from the convergence between internal realization and external verification: when brought into contact with reality, the clarity of understanding holds in the ways the insight itself anticipates. This is how many scientific insights enter the historical record. The experience often arrives as intuitive truth, accompanied by a sense of resolution or completeness. But only when that recognition works in reality does it take its place as established knowledge.

But there is a telling asymmetry to the nature of insight: the feeling of correctness precedes external proof. In this sense, validation occurs twice—first internally, as a convergence within the mind, and then externally, as confirmation in the world. The fact that this internal convergence can occur in advance of demonstration suggests that the mind is not generating *truth* arbitrarily but encountered the confirmatory space of the evaluative field.

We can recall the classic story of Archimedes and Hiero II of Syracuse as a canonical example of this pattern, where insight precedes and correctly anticipates its verification.

King Hiero II of Syracuse summoned Archimedes to solve a problem. The King had commissioned a new crown from a goldsmith but was uncertain whether the crown had been made of pure gold. Archimedes' task was to confirm if the crown was made of pure gold, but he was told he could not damage or harm the crown in any way.

Archimedes was puzzled with the problem for some time. Then one day while stepping into a tub filled with water, he observed the water rise as his body displaced the water. Archimedes instantly had an insight on how to measure the amount of gold, jumped up, and ran out of his house naked while yelling, "Eureka!", the Greek word for, *"I have found it."*

The insight of Archimedes was revealed by displacement; since gold is heavier than silver, if the volume could be known, then determining if the crown was pure gold would simply require taking the given volume, translating it into gold, and weighing it against the crown. If the two measures matched, then it would have to be solid gold. If the volume was greater than foreign material had been included within the crown.

But how did Archimedes suddenly arrive upon the answer? conscious work on the problem and hours spent puzzling over the problem set the stage for his realization. Even when he was not conscious of the mental process of problem solving, his mind was still occupied and looking for a solution. By the time he stepped into the tub, he *recognized* the solution and the insight caused his exclamation.

Humanity has come upon a number of cultural practices and methods to precipitate insight. In order to generate a problem-solving *insight* a certain process is generally followed. First, one must fully understand, or have a near expert level of understanding of the foundations of the subject that they are seeking an insight in. If one is looking for insight into their own lives, then there is a high level of 'self-experience' that serves as the basis. If they are seeking knowledge that has not been previously obtained by the discipline, their 'credentials' either in a formal setting, mentorship, autodidactic, apprenticeship or unorthodox course must be in order. They must 'assign' themselves a crucial piece of work. There must be assiduous and passionate dedication to solving the issue, most often a singleness-of-purpose. Last, each individual will have their own particular method for attacking their subject that leans into their individual strengths. History is rife with scientific luminaries who have made astounding insights: Max Planck (energy quanta), James Maxwell (Maxwell's equation), Albert Einstein (relativity), and Sir Isaac Newton (The Principia). Their methods, walks, staying up for days, hot baths, retreat to country estates, are known only to them and their intimates. Once the stage is set, and the foundational work has been done, and one's effort is fully and obsessively engaged, if insight is to happen, it often comes upon the individual suddenly. There is no given timetable. As in the desire for enlightenment, some of these pursuits are lifelong. Either slowly chipping away with a series of small insights or working out the connective tissue between a 'large' insight and how it manifests within the larger scheme.

The process is different for insights facilitated by rituals and practices that are designed for spiritual communion insight, but I would argue that the structure is largely the same irrespective of subject, meaning that in a sense, the luminaries we recalled above, are more like shamans and wizards than they are like their contemporaries. They all had a particularly perceptive talent, inexplicable gifts and also an affinity for deep understanding within their chosen fields. For the shaman and the world changing scientist alike, insight takes place at the peak moment of their experience, when they are caught between two worlds, at the boundary of understanding and reaching into the abyss, and instead of retreating, they instead "let go" and gave themselves over to the finality of understanding.

For those who have partaken of psychedelics, such as LSD, mescaline, DMT or psilocybin, the moment of 'let go' is similar in nature with 'peaking', when the experience culminates and there is nothing to be done but to give over to it and accept one's fate. At the moment when one feels control slipping away, it is often, if not always the best course to let go, and fall into deep acceptance that another power has taken over. For those that use insight in the way psychonauts use peak experiences, the 'let go' function of insight is key.

But of equal, if not greater value, is set and setting, also known as the environment.

> "It's well known to the experienced that medium to high doses of psilocybin mushrooms, given advantageous internal and external conditions–often called set and setting–can provide experiences of stunning insight, visions of great beauty, an abundance of life, contact with spirit entities, and authentic mystical experiences completely beyond the boundaries of the separate ego."
>
> – Dr. James Fadiman, *The Psychedelic Explorer's Guide* (2011)[15]

The factors of "set and setting" include the space, expectations, mood, psychological state, and one's guide or company, and other seemingly peripheral factors in the psychedelic journey. The sanctity of set and setting can be seen in the 'mad' scientists and obsessed inventors that fly off the handle if their work area, work environment, or workstation is disturbed in any way. When psychedelic experimentation is the order of the day, if an individual has a 'bad trip' they always know why. The central theme of the bad trip presents the reason. Dr. James Fadiman, author of *The Psychedelic Explorer's Guide*, and a contemporary of Ram Dass and Timothy Leary, recalled that his favorite explanation for a bad trip was, "everything was ok until the car burst into flames."

Western Approaches to Insight

For a more scientific approach towards insight, we can look to the example of Henri Poincaré, the French philosopher, physicist and mathematician who we explored in an earlier section. In his book, *The Foundations of Science*[16], He discusses the perplexing process of insight in the section titled, *Science and Method, Chapter III, Mathematical Creation*. Poincaré talks about the striking appearance of sudden illumination upon a subject after "long, unconscious prior work[17].") where one works hard, takes breaks and continually returns to puzzle solving. For Poincaré it is the conscious and direct work *combined* with the unknown work of the subconscious mind that allows sudden insight to take hold. The combined process becomes an ongoing meditation and contemplation. Poincaré states, *"These sudden inspirations... never happen except after some days of voluntary efforts which (have) appeared absolutely fruitless and (hence) nothing good seems to have come, where the way taken seems totally astray*[18]*."* Poincaré also makes an important point of how the insight that is experienced is an emotional experience of ecstasy. Poincaré is speaking specifically about mathematical problems, but the concept applies to insight, and not math, and as such is inclusive of any puzzle solving engagement that is similar to the process he describes. In insight the individual is seized in a profound and all-encompassing way often after searching over years, decades or in the case of many Zen monks, a lifetime. The pursuit, whether long or short, is dogged and determined.

> "...Among all the stimuli for our senses, for example, only the most intense fix our attentions, unless it has been drawn to them by other causes. More generally the privileged unconscious phenomena, those susceptible of becoming conscious, are those which, directly or indirectly, affect most profoundly our emotional sensibility. It may be surprising to see emotional sensibility invoked a propos of mathematical demonstrations which, it would seem, can interest only the intellect. This would be to forget the feeling of mathematical beauty, of the harmony of numbers and forms, of geometric elegance. This is a true aesthetic feeling that all real mathematicians know, and surely it belongs to emotional sensibility."-Henri Poincare, *The Foundations of Science* (1904)[19]

It is not always the case that a great deal of work has taken place in order to have an insight. And it's also not always the case that a great emotional feeling must accompany an insight. A penetrating insight, with no apparent previous work, that was accompanied by a great burst of emotion was previously discussed in the example of *Sri Ramakrishna* at his vision of white cranes upon a cloud blackened sky.

> "...a flight of snow-white cranes passed over my head... An access of joy and emotion overcame me. . . This was the first time that I was seized with ecstasy[20]."

Penetrating insights can occur at any time and for any reason, but they always involve a profound truth that is highly relevant for the individual, and some triggering event that releases the coveted insight that in rare cases ascends to a transcendent experience, or even enlightenment.

We could construct a set of 'Truth stairs' to represent the progression up to and past insight. At the bottom are *solutions* and the feeling of *completion*, next we have *epiphany* and *insight*. Next, the *transcendent experience*, and then the *spiritually transformative experience*, next we have *full enlightenment* and finally at the end we have *ascended masters* and *prophets* of which only a handful have ever existed, who are in communion with *Godhead* and also have been born to serve as living, breathing spirit guides for the earthbound.

In the teachings of prophets and masters, their stories tell of paradoxical speech, often in the vein of the Socratic method, 'drawing out' the inherent understanding of those that seek their guidance. Questions from seekers are often met with more prescient questions from the master. These questions urge seekers to turn inward to consult the resources of their inner selves. And the questions the seekers receive in return often serve to reorient the inquiry into its proper, or more productive form, pointing a more direct path from seeking to finding. Often the deepest questions need only to be turned inside-out. The seeker is literally standing upon the answer, but instead of seeing the truth, they question it. The master's question forces a 180-degree turn, revealing that the seeker is already standing on the foundation they claim to be looking for. These are 'complex' or 'loaded' questions where the master has no choice but to turn the question around rather than to confirm the faulty assumption.

This is a very important point that cannot be emphasized enough. When one has become conditioned to life on the profane level, when one is confronted with transcendent truth, strangely, one's reaction can often be to doubt. The insight appears true, but it is far too disruptive and comprehensive to digest, so there is a hesitancy. I had such a moment myself. I was standing on top of the greatest truth I had ever encountered, and within, I could feel myself saying, 'No.' not because it wasn't true, but because it was simply too large and too complete. The answer was something I chased but never expected to catch. In the moment of my doubt, the affirmative emotion of the insight slowly began to shrink away and fade into oblivion. Ironically, the fact that the insight began to fade made me realize it was real and true. I immediately accepted it, and it came upon me in full, and from that day I have been forever changed. When a truth is "too complete," it threatens the very structure of the one perceiving it. The hesitancy isn't a lack of truth or a true doubt of the truth, rather it's a reaction to the incomprehensible nature of its scale. This is a profound observation—that the truth can and will recede from doubt shows us several things about the *Evaluative Field.* It will answer. And it will at times answer in measure to the inquiry, but it can also answer back in a manner *far greater than the level of inquiry. But, in all cases, the individual retains free will.

And this brings up the final aspect of profound insight. The bell cannot be unrung. There are times when great insights are made, and when this insight is made, the individual can be completely sober, or perhaps psychoactive or ritualistically assisted, but the revelation is such that the individual feels 'high'. This is an interesting phenomenon, because sometimes this feeling of being 'high' does not recede. Over the next several hours, or days, the individual does not 'come down'. After several more days or weeks or months, the individual never 'comes down'. What has happened is this new state of understanding has become the new reality—the new baseline.

Life can be lived as a series of escalating realizations, where each successive level of insight is not merely experienced, but permanently integrated into the structure of one's spiritual progression.

Great masters are poised to strike at the heart of the seeker's inquiry. It may appear as something of a game to us, but it is born of the necessity that the master's answer is an experience unto itself, and not mere words. The prophets are master jewelers of the soul who know how to cut and polish the jewel of individual existence so that it is clear enough to serve as a mirror for the training program of reality.

Socrates referred to the process of teaching as *midwifery* and compared it to the growth of a child within their mother's womb.

Einstein's Insight Generator

To return to a subject we visited earlier, it is documented that Einstein used at least 12 well-known thought experiments to solve key mysteries within existence. What is remarkable about many of Einstein's thought experiments is their simplicity. He imagined things that were so simple, yet he knew they held the key germ to solving mysteries that had never been solved before.

This appearance of 'simplicity' is entirely deceptive because the issues Einstein contemplated were among the most perplexing of his age. If his experiments were 'simple' it is because they carried with them the key attribute of being reduced to their fundamental logical and epistemological constraints where the truth has been exposed as residing within three closely related, interacting layers, as appearance, as method, and as a truth condition.

<u>Special Relativity</u> *(Pursuing a beam of light)*:

> "If I pursue a beam of light with the velocity c (velocity of light in a vacuum), I should observe such a beam of light as an electromagnetic field at rest though spatially oscillating. There seems to be no such thing, however, neither on the basis of experience nor according to Maxwell's equations. From the very beginning it appeared to me intuitively clear that, judged from the standpoint of such an observer, everything would have to happen according to the same laws as for an observer who, relative to the earth, was at rest. For how should the first observer know or be able to determine that he is in a state of fast uniform motion? One sees in this paradox the germ of the special relativity theory is already contained."
> -Albert Einstein, *Autobiographical Notes* (1952)[21]

So, what is happening here? What is Einstein doing?

Einstein's special relativity does not begin as an equation-driven derivation, but as a reduction of reality into a pure constraint scenario: an observer chasing a beam of light at velocity *c*. In doing so, he removes everything non-essential until only the relational structure remains—observer, motion, and light. At this level, the problem is no longer mathematical, it becomes directly evaluable. It is a situation that can be held in consciousness and tested for coherence *against itself*. Within that evaluative space, a contradiction immediately appears: the observer cannot meaningfully detect uniform motion in the way classical intuition assumes.

It is from this internal collision of constraints—not from formal calculation—that the germ of special relativity emerges. *Only after this insight is stabilized* does Einstein translate it back into the secondary language of physics: equations, transformations, and formal structure. But the decisive step is not mathematics—it is the successful compression of reality into a form that can be directly interrogated within the evaluative field.

This is not an isolated case, but a recurring structure in Einstein's method. To see the applicability of the pattern we can point to two additional examples:

Relativity of Simultaneity ***(Lightning bolts, observers, and embankment and a train):***

Einstein used a thought experiment involving two observers, one on an embankment and one on a train observing the same lightning strikes from an equidistant location. The motion of the train leads one observer to see one lightning strike as occurring before the other, whereas objectively, they were the same. This illustrates relativity of simultaneity[22].

<u>Equivalence Principle</u> *(Painters and Elevators)*:

Einstein describes his breakthrough of imagination in his painter's thought experiment as "the happiest thought of my life". A pure *ecstasy*, if he ever experienced such a thing, "*...Because for an observer in free-fall from the roof of a house there is during the fall*—at least in his immediate vicinity—*no gravitational field.* Namely, if the observer lets go of any bodies, they remain relative to him, in a state of rest or uniform motion, independent of their special chemical or physical nature. The observer, therefore, is justified in interpreting his state as being 'at rest[23].'"

---

[1] (T. Cleary, Zen Antics 100 Stories of Enlightenment, Shambala, Boston & London, 1993)
[2] (T. Cleary, Zen Antics 100 Stories of Enlightenment, Shambala, Boston & London, 1993)
[3] (D. Eaves, evz-yoyo Esplanade!, 1989)
[4] (R. Rolland, The Life of Ramakrishna, Vol 1, Pg. 26, 1930 Pub. M. Mukherjee, Calcutta)33
[5] (D.T. Suzuki, Essays in Zen Buddhism, First Series, Pg. 261, Grove Weidenfelt, 1949)34
[6] (D.T. Suzuki, Essays in Zen Buddhism, First Series, Pg. 241-2, Grove Weidenfelt, 1949)35
[7] (J. Kounios, M. Beeman, The Eureka Factor: Aha Moments, Creative Insight, And The Brain., Pgs 25-26, Random House LLC, 2015)36
[8] (E. LaPointe, great-grandson of Sitting Bull, Sitting Bull His Life and Legacy, Digital Edition, Gibbs Smith, 2011) (N. Philbrick, The Last Stand, Custer Sitting Bull and the Battle of the Little Bighorn, Viking, 2010)37
[9] (W.C. Sturtevant, Gen. Editor, Handbook of North American Indians, Vol 12, 13 part 1 & 2, Smithsonian Institution, Washington, 2001) & (M. Crummett, Sun Dance, The 50th Anniversary Crow Indian Sun Dance, Falcon Press, 1993) & (J.G. Jorgensen, The Sun Dance Religion, Power for the Powerless, 1972, University of Chicago Press)38.
[10] (E. LaPointe, great-grandson of Sitting Bull, Sitting Bull His Life and Legacy, Digital Edition, Gibbs Smith, 2011)
[11] (W.A. Graham, The Story of the Little BigHorn, Custer's Last Fight, The Century Company, 1926)
[12] (N. Waddell -translated- THE UNBORN: The Life and Teachings of Zen Master Banket 1622-1693 Revised Edition, Pg. 136, "The Dialogues" North Point Press, 1982)41
[13] (S. Mukundananda, The Bhagavad Gita, The Song of God, Chapter 11: Yog Through Beholding The Cosmic Form of God, (The Vision of Universal Form) Translation/Commentary by Swami Mukundananda, Opensource, CCO 1.0 Universal, archive.org, 2019)42
[14] (S. Nikhilananda, The Bhagavad Gita - With Notes, Comments and Introduction by Swami Nikhilananda, 1944, Pg. 255)43

[15] (J. Fadiman, The Psychedelic Explorer's Guide: Safe, Therapeutic, and Sacred Journeys, Chap 15, Pg. 281, Park Street Press, 2011)44
[16] (1921, The Science Press),
[17] (Ibid Poincare, The Foundations of Science, Pgs 389-94
[18] (Ibid Poincare, The Foundations of Science, Pgs 389-94).
[19] (Ibid Poincare, The Foundations of Science, Pgs 389-94)48
[20] (R. Rolland, The Life of Ramakrishna, Vol 1, Pg. 26, 1930 Pub. M. Mukherjee, Calcutta)49
[21] (A. Einstein, in, Autobiographical Notes, from P.A. Schilpp, Albert Einstein-Philosopher Scientist (2nd ed), New York, Tudor Publishing, 1952)50.
[22] (A. Einstein, Relativity: The Special and the General Theory (15th ed), New York, Crown Publishers, Inc, 1961)51.
[23] (A. Einstein, Document 31: Ideas and Methods. II. The Theory of General Relativity, (1920) in M. Janssen, R. Schulmann, I. Robert, L. Jozsef, C. Lehner, D.K. Buchwald (eds.), The Collected Papers of Albert Einstein. Volume 7: The Berlin Years: Writings 1918-1921 (English translation supplement) Calif. Institute of Tech., 2018)

5 The Biologically Triggered Mystical Experience

Across cultures human beings have discovered that the doorway to mystical states can be opened not only through discipline, ritual, persistence and deprivation, but through biology itself.

Certain substances encountered in nature serve as intentionally laid portals to higher states of consciousness. They collapse the distance to the state itself, synergizing with the seeker's intent to bridge the profane and the transcendent.

At effective doses, these compounds cause a rapid reconfiguration of perception, identity, and meaning that places the individual into direct contact with a deeper order of experience. Combined with pre-established rituals, rhythms, ceremonies, and shamanic intervention, a functional highway to the divine has been established on Earth.

Each substance carries a distinct signature—a way of seeing, revealing, and structuring the encounter. Across cultures and generations, individuals report consistent visual and emotional patterns, along with the sense of encountering something not of their own construction, a 'voice' or 'perspective' from the biological substance itself, the strength of which is dependent on substance, dose, and intention. These experiences may include presences, characteristic perspectives, and recurring forms of guidance that instruct, correct, challenge, or reorganize perception. These effects extend across what has been mapped as cognition, ego structure, sensory integration, and the core "problems" of consciousness: for example: the 'binding problem,' qualia, the 'explanatory gap,' 'neural correlates,' the 'unity problem,' and the mystery of the self. These layers are not merely altered but reorganized and reinterpreted within the experience.

Within this framework, these substances are not random triggers, 'glitches', nor accidental, or incidental artifacts of biology. They appear in our environment as distributed interfaces, embedded within the natural world, each tuned to reveal a different aspect of the same underlying reality, precisely calibrated to the environments and cultures that host them. The rituals, cosmologies, and practices that arise around them are not incidental; they are the cultural stabilizations of repeated contact.

In this sense, the *biologically triggered mystical experiences* keyed by these substances are not merely entryways, but they serve as key parts of the message itself—biology, as it were, speaking to itself, through the substrate of consciousness and constrained interpretation.

When we look at the distribution of bioavailable psychoactive and psychedelic substances, we find there are very few cultures that do not have access to these substances, and their availability, in multiple forms, is prolific.

What is striking is not merely their presence, but their repeated use as instruments of insight. Across vastly different environments and traditions, they are used to arrive at experiences that share a recognizable structure—suggesting that what is being accessed is not random but patterned.

This pattern is not confined to ancient or indigenous contexts. It appears just as clearly in the modern world, including among those trained in the most rigorous academic disciplines.

Two such individuals—Harvard-trained psychologists Dr. Richard Alpert and Dr. Timothy Leary —found that a particular kind of chemical trigger did not merely alter perception, but catastrophically disrupted their existing conceptual frameworks, allowing them to encounter a deeper layer of reality directly.

Prior to these experiences, each had achieved significant academic distinction. Alpert, a Stanford-educated psychologist (Ph.D. from Stanford University), held multiple appointments at Harvard and was a prominent figure in personality and social psychology. Leary, a Ph.D. clinical psychologist from UC Berkeley, served as Director of Research at the Kaiser Foundation Hospital in Oakland, was a faculty researcher at Harvard, and authored the widely used text *The Interpersonal Diagnosis of Personality*[1].

The following excerpt, drawn from Ram Dass' *Be Here Now*[2], captures an insight facilitated through their early psychedelic work with psilocybin in Mexico, specifically in the Oaxaca highlands in the Huautla de Jiménez region. This area, associated with Mazatec ceremonial use of psilocybin mushrooms. This experiential period formed part of the early groundwork of what became known as the *Harvard Psilocybin Project,* initiated by Timothy Leary and Richard Alpert.

> "One night while drinking together, we plotted a trip across North and South America, and when I said I flew a plane, he [Leary] said, 'Great, we'll fly in your plane.' and I said, 'wonderful', and neglected to tell him that I had only a student license... I got my license and an airplane on the same day and flew to Mexico the next day in a death defying leap. When I got there I found that Timothy had done some other type of flying, just about the week before. Frank Baron, who was a psychologist at Cal, and old friend of Tim's, had introduced

him to an anthropologist in Mexico and they had come to know about the Teonanacatl, the flesh of the Gods, the magic mushrooms of Mexico, which one obtained from Crazy Juanna, a woman up in the mountains who ate the mushrooms all the time. Contact was made with her and the mushrooms were obtained. Tim had eaten nine of these mushrooms—so many male and so many female mushrooms—with a group of others around a swimming pool and had had a profound experience. He said, 'I learned more in the six or seven hours of this experience than I had learned in all my years as a psychologist.' That is a strong statement!"

-Richard Alpert (Ram Dass), *Be Here Now* (1974)[3]

Humans possess a lock released by the biological key of psilocybin and other psychedelics. This was the key insight by Leary and Alpert. Their academic careers provided the necessary precondition for illuminating this fact. They then used their substantial understanding as a fund of knowledge to interrogate against the *Evaluative Field*. Their prior learning was not replaced, but force multiplied and illuminated from an entirely new perspective and reflected back to them as articulable insight.

Reports of mystical-type experiences date back many centuries (e.g. in the case of Rumi or St Teresa of Avila), if not millennia (i.e. in the case of Plotinus)[4] . Mystical experiences have occurred in the course of structured spiritual or religious practices as well as in cases in which there was no direct intention to have such an experience. Mystical experiences are uniquely interesting and important to study because they are sometimes associated with abrupt, substantial, and sustained changes in behavior and perception[5]. Furthermore, the authoritative sense of interconnectedness and sacredness that defines such experiences suggest that mystical experiences may be foundational to the world's ethical and moral systems[6]. While there are countless reports of profound spiritual and mystical experiences that have occurred in the absence of psychoactive substances, historical evidence abounds for the role of psychoactive substances in ceremonial approaches to facilitating such experiences[7]. Further, descriptions of naturally occurring mystical experiences are strikingly similar to profound spiritual experiences occasioned by hallucinogenic substances[8]

Biological compounds with unique consciousness-altering effects preceded the advent of humanity. They were 'keys' that pre-existed the creation of the 'locks' they would free within the human mind. Their psychedelic nature is an objective, inherent property—unrealized until the capacity to articulate and label the experience emerged. The ubiquity of these specified neural receptors across the animal kingdom reveals a pre-existing biological architecture. This is not an evolutionary accident—within this ontology, there is no 'evolution' nor 'accident.' All things are teleological from the outset; what appears instead is structural readiness. All things are teleological from the outset—but a structural readiness. When triggered, the resulting cascade of consciousness-inducing data streams confirms the mechanical function of the system. And the information conveyed by the experience is both validation and proof.

The primary "lock" for this chemical key is the 5-HT2A receptor (Serotonin 2A receptor), located on the "apical dendrites of pyramidal neurons in Layer 5 of the neocortex[9]." These neurons serve as the brain's information hubs, responsible for integrating high-level cognitive data and sensory feedback. When a compound like psilocin binds to this receptor, it induces a conformational shift in the receptor that triggers a unique intracellular cascade. Recent evidence suggests that this binding occurs within intracellular compartments of the neuron, not just on the surface. In other words, the key psychedelic substance doesn't just insert a key on the lock at the door of the neuron; it enters the cell and works directly on the internal machinery. It's a biological courier that has the credentials to bypass the cell's external security and has the credentials to bypass the cell's external security and the molecular form required to remodel the neuron from within. Current theory suggests this is why the psychedelic experience is "neuroplastic" meaning, it can result in permanent shifts in perspective. To this extent the psychedelic molecule can be seen as a 'blue collar' neural electrician that has rewired the neuron[10].

Think of a bee collecting pollen at a cherry blossom. When this pollen makes contact with the stigma of another cherry tree, it does not function as a mere signal or "message" to begin producing fruit. The pollen grain is targeted information; it germinates the cherry tree, performing work by delivering male gametes directly to the ovary. This genetic data is then incorporated into a new cellular blueprint. These genetic instructions are then incorporated into a new cellular blueprint, enabling the production of fruit that would otherwise lack the required information. Similarly, the psychedelic molecule is not a 'tag you're it' for the brain. The mushroom acts as the "bee," the active psychedelic molecule is in the role of the "gamete," and the absorption and subsequent molecular cascade function of psilocybin is the "sexual reproduction" and "fruit" of the experience. In both cases, we observe biological 'information missiles' performing molecular work upon targets designed to receive them. Without this level of complexity, and work, and specificity, the neuroplasticity required to produce an experience with lasting change would not be possible.

Humanity was designed to have these experiences. Psilocybin was designed to deliver them.

Ram Dass & Timothy Leary

For scientists and researchers like Ram Dass and Timothy Leary, the "cacophony of qualia"—the overwhelming flood of sensory and cognitive data as described by recreational psilocybin users—was not a mere side effect, or a 'light show' or a 'hallucination.' It was a forensic bypass of the brain's native censorship. Unlike a recreational user, distracted and overwhelmed by shifting colors and distorted sounds, the trained intellect of Dass and Leary recognized this state for what it was: the stripping away of profane defense mechanisms to reveal a vital, objective data set and underlying substrate of reality.

This is where the "network collapse" described by researcher Robin Carhart–Harris[11] becomes critical. By mechanically forcing the brain's default mode network to drop its narrative filters, the brain is momentarily freed from its own biases. Carhart-Harris explicitly identifies this transition as a shift toward a more primitive, unconstrained style of cognition, noting:

> "The results of the present study support the view that the subjective effects of psilocybin are caused by a decrease in the activity and integrity of the brain's most highly connected hub-nodes... This may be described as a 'disintegration' of the brain's highest-level control centers, allowing for a state of unconstrained cognition."
>
> -Robin Carhart-Harris, *Neural Correlates of the Psychedelic State* (2012)[12]

For Dass and Leary, this collapse allowed the vast fund of psychological knowledge they had accumulated over the course of a lifetime of brilliance to be interrogated directly against the raw reality of the experience. Then, the "blue-collar" neuroplasticity identified by researcher Lindsay M.V. Vargas in her paper on *Psychedelics Neuroplasticity*[13],allowed them to concretize their insight, perception, experience and cognitive integration into a permanently rewiring of their brains and the creation of a new psychological ontology. Their insights weren't just seen or experienced as simple qualia; they were retained and built upon as new constraints, new truths, and new knowledge, pulled from the underlying substrate of our reality, transforming their academic theories into witnessed realizations. Timothy Leary recorded one of his journey's:

> I opened my eyes. I was in heaven. Illumination. Every object in the room was a radiant structure of atomic-god-particles. Radiating. Matter did not exist. There was just this million matrix-matrix lattice-web of energies… Everything hooked up in a cosmic dance. Fragile. Indestructible… And the incredible shattering discovery. Consciousness controlled it all. Or (to say it more accurately), all was consciousness. I was staggered by the implication. All creation lay in front of me. I could live every life that had ever been lived, think every thought that had ever been thought. An endless variety of ecstatic experience spiraled out around me. I had taken the God-step. -Timothy Leary, *High Priest* (1968)[14]

Ram Dass and Timothy Leary were willing to take substantial academic and reputational risks in the face of what they had discovered. They had both reached the pinnacle of academic achievement but were seen by many as pushing the boundaries too hard, and 'stretching out' the name of Harvard in unsanctioned ways. In 1962 Timothy Leary's lecturing duties were suspended, In 1963 Richard Alpert was dismissed and Leary's contract was not renewed.
Perhaps this Ram Dass quote best summarizes the entire episode:

> "The game is not about becoming somebody; it's about becoming nobody." -From the Harvard Crimson, July 9, 1963, by Efrem Siegel[15]

> "President Pusey announced on May 27 [1963] that the Corporation had terminated the appointment of Richard Alpert as assistant professor of Clinical Psychology and of Education for giving hallucinogenic drugs to an undergraduate in violation of an agreement with the University. The dismissal, the first in Pusey's ten years at Harvard, climaxed more than a year of off-and-on controversy over the work of Alpert and an associate, Timothy Leary, Lecturer on Clinical Psychology, with psilocybin." -Efrem Sigel, (1963)[16]

Tim Leary, Richard Alpert and Ralph Metzner authored a book, The Psychedelic Experience: A Manual Based on The Tibetan Book of The Dead (T. Leary, R. Metzner, R. Alpert, The Tibetan Book of The Dead, Citadel Press, 1964). The book outlines five levels of the psychedelic experience. It is the last 2 levels—the five levels may mix, recede or come in waves, etc.—where the mystical aspect of the experience has the opportunity to be enacted within one's consciousness.

> **Level 1**: Light synesthesia and heightened awareness, state change but feelings of being "in the same world".

Level 2: Visual perception of objects breathing, warping and other visual effects, strong emotional responses, closed eye visuals.

Level 3: visualization of fractals, tracers, and strong hallucinations, deep thoughts of the self and the universe, thought loops, high sensitivity to physical stimuli, auditory disorder, uncontrollable laughter.

Level 4: Powerful, otherworldly visuals, visions, dissolution of reality, out-of-body experiences, profound realizations.

Level 5: Ego "death", communication with otherworldly beings, total eradication of reality, audiences with god-like beings, lucid/sober mental state[17]

By their own model, Leary and Alpert emphasize that the primary psychedelic opportunity resides in the final two stages. At these final levels, the experience shifts from a mere alteration of the world—where reality is buckling and shuddering, but still recognizable—to a total reinterpretation of consciousness. Here, the mind is freed by the biological key to come into direct contact with the Evaluative Field. As the ego dissolves, this higher state takes over, projecting its lucidity back onto the profane world and transforming our previously held conceptions into witnessed realization of the world within our world.

Psychedelic Compounds

The main psychedelic compounds are also known as hallucinogenic drugs. While hundreds of compounds are broadly classified as "hallucinogens" they are ordered in five main classes—cannabinoids, tropanes, piperidines, and empathogens and our target class, psychedelics.

The *classes of psychoactive compounds:

Cannabinoids: Sedative, analgesic, and perceptual-dampening effects (e.g., THC, CBD).

Tropanes: Deliriants that block acetylcholine, a neurotransmitter for memory and focus (e.g., mandrake, nightshade, datura, cocaine).

Piperidines: Dissociative and stimulant-linked compounds (e.g., PCP, Ritalin, Demerol).

Empathogens: Compounds that trigger large-scale serotonin release, with downstream effects on oxytocin and dopamine systems (e.g., MDMA)

Psychedelics (Tryptamines/Phenethylamines): Act primarily via 5-HT2A receptor agonism, producing a high-resolution reorganization of

perception and cognition (e.g., psilocybin, DMT, ayahuasca (DMT + MAOIs), LSD, mescaline).
(Drugs are grouped mainly by their molecular structure. The effects listed are general effects. Cocaine, Ritalin, and mescaline share structural features with their respective groups but do not necessarily produce the same primary experiences).

Although the placement of substances like Ritalin, Demerol, and cocaine in this classification may seem "off"—grouping cocaine with deliriants like nightshade and datura (the Tropanes), or placing Ritalin (a stimulant) and Demerol (an opioid) into the same class as PCP (a synthetic dissociative)—this categorization is based strictly on their molecular architecture rather than their sensory effects.

Often drugs are grouped intuitively on what they do to our mood or our body, however, in our investigation of consciousness and the nature of the mystical experience, we are prioritizing the structural classification. There are several unique molecular features, and actions that are specific to psychedelics that create their effect—mechanisms so ingenious they effectively override the "biological noise" of other substances on our list.

Psychedelics—specifically Tryptamines and Phenethylamines—are drugs that possess a physical architecture that serves as a high-resolution access filter to the underlying substrate of reality. Psychedelics are more than mere drugs, they are literal, biologically engineered keys primed to unlock the 5-HT2A (serotonin) receptor in specific targeted ways. Experience and research have shown us that this lock is the very gateway to the divine, and that psychedelic compounds were designed and distributed throughout our environment specifically to perform this task.

Living Doorways In The Living World

Biological *Life* is divided into two *Empires* also referred to as *Superkingdoms*. These Empires are separated into *Prokaryote* (simple, single-celled organisms like bacteria), and *Eukaryotic (complex, multi-celled) life*, including complex organisms like mammals, fungi, and vascular plants—from the ancient, cone-bearing pines and redwoods, to like oaks and roses. There are several key differences between the two *Superkingdoms* of life.

Eukaryotic life holds its DNA in a nuclear envelope. Prokaryotic life *does not*. Its DNA is held in a single circular chromosome coiled within the cytoplasm. Furthermore, the organelles of the prokaryote are not enclosed in membranes, whereas eukaryotic organelles are shielded by their own individual membranes. Prokaryotes have a single chromosome. Eukaryotes have chromosomes in *pairs*. Humans, who are Eukaryotic, have 23 pairs of chromosomes. Prokaryotes utilize the process of binary fission (a copy is made and the cell splits in half) for cell division. Eukaryotes perform cell division by *mitosis,* a multi-stage high-resolution operation where the nuclear envelope is dissolved, and the DNA is packaged and split by individual spindles.

The characteristics that divide these two forms of life are clear, yet they contain "hidden features" that disrupt the pattern, such as mitochondria within Eukaryotic cells, they undergo division by fission much in the same way as a bacterium. In our review of these aspects—DNA, mitochondria, paradoxes among both—in Section 1, we have learned that whoever created our existence, and created biological life, likes to perform unexpectedly, with hidden features that confound any ability to categorize life as solely a material, mechanical, physical, chemically guided process.

The word Prokaryote is from ancient Greek *pró* (before) and *káruon,* which means nut, kernel, or in this usage, nucleus. Eukaryote is also Greek and comes from the word *eu* (well, good) and the same *káruon* used in prokaryote.

In these two *Empires,* -Prokaryote and Eukaryote-, there are either six, or seven *Kingdoms* depending on the classification system utilized. Bacteria and Archaea are in the Prokaryotes empire. *Protozoa, Chromista, Plantae, Fungi* and *Animalia* are in the Eukaryote empire. In terms of the Availability of Biological (Psychedelic) Doorways to the mystical experience, the most important Kingdoms for our discussion are the last three in the Eukaryote *Empire, -Plantae, Fungi* and *Animalia-,* because within these three, all of the naturally occurring psychedelic substances may be found.

A Psychoactive substance is:

> "...a chemical substance that changes functions of the nervous system and results in alterations in perception, mood, consciousness, cognition or behavior. These substances may be used medically, recreationally, or spiritually; to purposefully improve performance or alter one's consciousness; as entheogens for ritual, spiritual, or shamanic purposes, or for research[18]."

A Psychedelic substance is synonymous with a *hallucinogenic* substance, and they are both considered to be

> "a class of psychoactive substances that produce changes in perception, mood and cognitive processes. Psychedelics affect all the senses, altering a person's thinking, sense of time and emotions[19]."

The list below is a list of the most well-known and widely available sources of psychedelic compounds.

Kingdom Animalia: Frogs, Toads, Fish & Sponges

2 types of psychoactive toads: the cane toad (*Rhinella marina*), and the Sonoran Desert toad (*Incilius alvarius*). Both produce *bufotenine*, a DMT derivative.

Frogs of the *phyllomedusa* genus: The bicolor tree frog also known as the giant monkey frog.

Up to 15 different known types of psychoactive/ hallucinogenic fish. One hypothesis for the hallucinogenic properties of the fish is that the active ingredients are derived by the fish from the accumulation of toxins digested from macroalgae.

At least 6 species of sea sponges—notably those in the genus *Smenospongia*—have been found to produce DMT and 5-MeO-DMT.

Kingdom Fungi: Fungus

Rye ergot fungi is a parasite that latches onto rye, wheat, barley, oats and wild grass. Famously, the synthetic form of ergot amides is what LSD is based upon.

There are over 200 species of psilocybin-containing mushrooms distributed globally, primarily within the genus *Psilocybe*.

Kingdom Plantae: Plants

5 main types of hallucinogenic cactus (mescaline): Peyote, San Pedro, Peruvian Torch, Bolivian Torch, and *Echinopsis Zamnesiana*.

2 kinds of Iboga (root bark) contain *ibogaine,* a psychedelic: the Iboga root bark and *Tabernanthe iboga* are an extract of this rainforest shrub.

5 types of tropane alkaloid-producing nightshades. (*solanaceae, scopolamine,* and *hyoscyamine*): *datura, atropa belladonna, hyoscyamus niger, mandragora officinarum* (including *scopolamine*, and *hyoscyamine*). Note: while tobacco (*Nicotiana*) belongs to the Solanaceae family, it lacks the hallucinogenic tropanes.

Khat, a shrub (*cathine* and *cathinone*) can cause delusions and hallucinations

4 primary sources of DMT producing plants: *Chacruna* (ayahuasca), *Jurema* (bush), Yopo tree (seeds) and *Chaliponga/chagropanga* (vine) (Lysergic acid amide, or LSA) Morning Glory & Hawaiian baby woodrose

Cannabis[20]

When we zoom out and look at this biological inventory, we are confronted by a statistical anomaly. While conservative estimates often cite a billion cataloged species of life presently on Earth, more recent estimates suggest the number may be closer to one trillion[21].A vanishingly small fraction—approximately 200 species of fungi and a handful of specific plants and animals—carry psychedelic keys. But, in another quirk, rather than these lifeforms being nearly impossible to find, their deployment in the environment closely follows human development. Simply put they thrive where humans live. *Psilocybe cubensis*, the common magic mushroom, grows directly from the waste of domesticated cattle[22]. Not coincidentally, cattle were domesticated at the same time human civilization accelerated. Ergot—the raw material for LSD—latches onto the rye and wheat that built human civilization. In the modern world, the most potent forms of magic mushroom prefer the wood chip beds of city parks and suburban gardens to the wilderness. Psychedelics thrive in the debris of cleared land, farming, and timber processing.

And producing the psychedelic molecule is costly[23] Complex molecules like Psilocybin, Mescaline, or DMT require a multi-step, high-energy biological process that does nothing to further the survival of the plant. These compounds do not reliably deter insects[24] nor attract pollinators, and do not protect the organism from being eaten. In fact, many insects and animals consume them without any negative effect on their fitness, and the waste products of those that eat these plants do little to propagate the lifeform, except in the case of domestic cows. There is only one real beneficiary of the costly production of psychedelic molecules and that is the serotonin receptors built into the human brain. These plants are literal portals to the underlying substrate of our reality. That is their purpose.

The production of the psychedelic molecule requires a specialized, multi-step, energy (ATP) and resource-hungry (cofactors like S-adenosylmethionine) assembly line[25]: 4-steps in Fungi, 6-8 steps in Peyote, 2-3 steps in plants, and in Ergot, a 14-gene operation. The organism pays a high metabolic tax to create a molecule that it cannot use, but that fits perfectly into the human serotonin receptor. And there is another paradox. In the case of psilocybin, scientists discovered that the cluster of four genes required to make the molecule is found in two entirely different, and unrelated species of fungi. To explain this identical complexity in unrelated fungi, Neo-Darwinists resort to the concept of *Horizontal Gene Transfer*[26]—the idea that a massive, 13,000-base-pair genetic 'package deal' jumped from one species to another by accident.— There is no evidence this 'jump' ever took place. The only 'justification' is the usual Darwinists circular justification— "we see it, Darwinism is true, therefore it must have happened in a Darwinian manner—*Horizontal Gene Transfer.*"

Darwinism is once again left at the altar. Their concept is that individual genes slowly evolve one digital letter at a time according to fitness, eventually finding novel proteins from the void. This process has no place whatsoever in understanding two psychedelic mushrooms with the exact same 13,000-base-pair suite of instructions, so they make up *Horizontal Gene Transfer*. No such thing took place. The reason why the same suite of genes is found in two different species of mushrooms is because that's how they were instantiated, as predetermined by the ontology of The Divine Algorithm. Once again, we see 'the universe' placing an Easter Egg within our reality to confound any and all contending ideas that stray from The Truth. This feature provides additional support for our ontology. Species do not evolve, they are instantiated. And in the case of psychedelic mushrooms, they are instantiated with the instructions to create psychedelic compounds that can attach to the 5-HT2A serotonin receptor, so that humans can have a biologically triggered mystical experience.

But the evidence for intentional production and instantiation has an even better example in the existence of *Claviceps purpurea*—Ergot fungus.

The existence of the Ergot fungus effectively collapses any materialist explanation for the existence of psychedelic gateways. From an evolutionary perspective, the manufacturing process defies every law of biological thrift—directly contravening the law of *Selective Pressure* to show that Ergot is an enormous energy drain, yet the organism spends energy it doesn't have for a result it doesn't use and that doesn't benefit it in any way. It violates *Genetic Drift*; since it's not useful for survival, it should have fallen into disuse, yet it existed for approximately a hundred million years prior to the arrival of modern man. *Cost of Complexity* is violated—why does Ergot have a 68,500 base pair gene cluster dedicated to the production of psychedelic molecules that do nothing for its survival? Finally, there's Dollo's Law of Irreversibility—which states that once a complex pathway is lost it cannot likely be re-created, yet we find the exact same structure in two distinct, unrelated species. This not only violates Dollo's Law, but the naturalistic principle that inspired it—that there is no and can be no outside blueprint for the production of psychedelic molecules.

While it can be understood that a mushroom in a forest might use psilocybin as a minor deterrent, the Ergot fungus possesses zero autonomy; it is a parasite. It can't grow in the soil or survive in the wild. It must latch onto grains that humans used to build their civilizations. By placing the literal portal to the underlying constraint field responsible for our reality onto the most essential human food supply, the instantiating power behind our existence and intelligent observers ensured that the psychedelic key would be harvested, consumed and perform its appointed duty.

Work by T. Haarmann, et al. (2007), details the production of LSD by Ergot in a paper titled, *The Ergot Alkaloid Gene Cluster of Claviceps Purpurea,* published in *Applied and Environmental Microbiology*[27]. What follows is from the abstract:

> "Here we report the identification and characterization of the entire ergot alkaloid (eas) gene cluster from Claviceps purpurea. The cluster consists of 14 genes spanning a 68.5-kb region. Gene expression analysis and targeted gene disruption of several genes in the cluster, including dmaW, easC, easD, easE, and easF, confirmed their role in the biosynthesis of the lysergic acid [LSD] backbone. The high degree of conservation of this complex gene cluster across different lineages suggests a highly specialized metabolic pathway that has been maintained despite the high energetic cost to the organism."
>
> -T Haarmann et al, *The Ergot Alkaloid Gene Cluster in Claviceps Purpurea: Extension of the Cluster Sequence and Intra Species Evolution* (2005)[28]

Indigenous Ayahuasca Ceremony

> "The experience resulting from the use of psychedelic drugs are often described in religious terms. They are therefore of interest to those like myself who, in the tradition of William James, are concerned with the psychology of religion. For more than thirty years I have been studying the causes, the consequences, and the condition of those peculiar states of consciousness in which the individual discovers himself to be one continuous process with God, with the universe, with the ground of being, or whatever name he may use by the cultural conditioning or personal preference for the ultimate and external reality. We have no satisfactory and definitive name for experiences of this kind. The terms 'religious experience', 'mystical experience' and 'cosmic consciousness' are all too vague and comprehensive to denote that specific mode of consciousness which, to those who have known it, is as real and overwhelming as falling in love."
>
> -Alan Watts, *Psychedelics and Religious Experience* (1968)[29]

In an interview in Newsweek titled, *Ayahuasca Can Help You Communicate with Plant Spirits, and Heal Your Body and Mind*[30] researcher of plant-based medicines and botanist Chris Kilham was asked about his views on the interface between the hallucinogenic ayahuasca plant and human consciousness. Below are parts of the interview:

Douglas Main: And why did you first decide to take ayahuasca?

Chris Kilham: "[Dennis McKenna] told me about a shaman in Peru…. I showed up to this place with three nights to spend…. So, we had a ceremony and I drank some of the brew… but I was still pretty unprepared for the staggering intensity of ayahuasca. I started having lavish visions, seeing the wheels of the universe turning. And then all of the sudden I was talking to my mother… After conversing for a couple minutes, the grief was gone, and it hasn't come back."

Douglas Main: "Is it understood how ayahuasca works?"

Chris Kilham: "No it's Not… If you want to dive into the deep end of the Olympic ayahuasca pool, though, there's the spirit explanation. Shamans who work with spirits of the plants insist, as I do, the plants are assisting us…"

Douglas Main: "I do think that a lot of people, perhaps scientists especially, wouldn't think that it's crazy to talk of 'plant spirits.'"

Chris Kilham: "Well, all I can tell you is that in my work I'm called by plants all the time. And when I encounter scientists or herbalists or doctors who work with plants, especially the medicinal value of plants, I always ask them, 'Are you working for the plants, or are the plants working for you?' And they pretty much always say they are working for the plants. There's a sense of being called, being enlisted by the plants[31]…"

The indigenous knowledge of a remote Amazonian tribe represents a bewilderingly complex understanding of plant biology, particularly with hallucinogens. This evidence was provided by Jeremy Narby in his book, *The Cosmic Serpent: DNA and the Origins of Knowledge*[32].

"The main enigma I encountered during my research on Ashaninka ecology was that these extremely practical and frank people, living almost autonomously in the Amazonian forest, insisted that their extensive botanical knowledge came from plant-induced hallucinations. How could this be true?" J

-Jeremy Narby, The Cosmic Serpent (1999)[33]

The rest of the article explains how this enigma is deepened by the sheer complexity of the ayahuasca brew. For millennia, indigenous shamans have combined two specific plants—a hallucinogen containing dimethyltryptamine and a secondary plant containing a monoamine oxidase inhibitor—that, when boiled together, bypass the human digestive system's natural defenses to allow the hallucinogen to reach the brain.

> "So here are people without electron microscopes who choose, among some 80,000 Amazonian plant species, the leaves of a bush containing a hallucinogenic brain hormone, which they combine with a vine containing substances that inactivate an enzyme of the digestive tract, which would otherwise block the hallucinogenic effect[34]."

The mathematical improbability of such a discovery, given the vast number of Amazonian plant species, has long confounded mainstream science. Even the most distinguished ethnobotanists have questioned whether such a sophisticated biochemical solution could have been reached through simple trial and error, or if it represents an access to knowledge that defies the standard materialist model of discovery.

Psychedelic Problem Solving

If the Amazonian tribes used these substances to solve the biological puzzle of the forest, modern researchers began to wonder if the same interface could be applied to the technical puzzles of the silicon age.

Dr. James Fadiman—psychologist, psychedelic researcher, and contemporary of Ram Dass and Ken Kesey—adds a critical dimension to this intrigue of humans being "taught" or shown knowledge by the psychedelic substances they consume. Among Fadiman's findings, during a research project in the Menlo Park area near Stanford University, Fadiman recruited highly motivated professionals to test whether psychedelics could resolve high-level technical challenges requiring breakthrough creative solutions. Participants accepted into the study were highly motivated.

The subjects were administered a 'cognitive dose'—not a transcendent or 'heroic' quantity, but just enough to catalyze an altered state. The results represented a statistical outlier in the history of human problem-solving: out of 44 attempted problems, only four participants were unable to find a solution. The study concluded that the key was the precise alignment of set, setting, intention, and the psychedelic agent.

Sam Harris interviewed researchers James Fadiman for his *Making Sense Podcast where they discussed set and setting*[35]:

> **Sam Harris**: "How do you think about the concept of set and setting?"
>
> **James Fadiman**: "Well what we know is that psychedelics are incredibly sensitive to set or your mental attitude and setting–the

literally physical situation you're in, and that includes the other people who are around. And the main reason why I wrote that book, 'The Psychedelic Explorers Guide' is for the first two chapters which detail how to become a good journeyer, how to set yourself up the best way possible, and also how a guide should behave, and what knowledge they should have, because it's not a simple question, but fundamentally the question is always, 'Do you feel safe? Is it truly safe? Is it private so you won't be disturbed? And do you have some idea of what the doses are like? And do you have some intentions?' And perhaps that's what set is about–which is, what do you intend and it is likely that you will have what you intend, but [rather] you'll have something else that goes with it. So, all of those are important and I think the way I've said it is, it's set, setting and situation–which is, what's your life situation? If your life situation is very difficult, even if you're in a lovely setting, your life situation is going to impinge on that, and you're going to be in trouble..." -Dr. James Fadiman, Sam Harris Making Sense Podcast (2021)[36]

For those seeking a naturally derived gateway for consciousness-to-consciousness communion with the instantiating power behind our reality, the psychedelic substance functions as a dual-action mechanism: a simultaneous subtraction and addition. This is the vital combination. By blocking the habitual filters—the cognitive boundaries that define and limit our sensory reality—the substance acts as a chemical solvent for our rigid definitions of self. Simultaneously, it adds a previously unavailable perspective, inducing a synesthetic permeability where ideas intertwine and entirely new contexts emerge.

This process of inhibiting a "preventive filter" while providing a biochemical catalyst forms the architecture of the psychedelic interface. The active molecules enter the mind like a specialized program: one half deactivates the brain's security system, while the other installs an infusion of otherworldly connectivity. They move with unerring precision, expressly engineered for this specific biological architecture.

The universe operates through this same multi-purposed, interacting design. If all life is built with intent, then the psychedelic experience is the Signal. It is a consciousness signal embedded in the first moments of the universe, surviving to this day—manifesting through the layers of universal intent into a biological form carrying the blueprints and the command to manufacture the 5-HT2A Key. This is the allure and the promise that captured the spirit of an entire generation.

How Psychedelics Work

Psychedelics as a class are defined by their specific interaction with the serotonergic system. They all hold the key to hijack the 5-HT2 serotonin receptor master lock: 5-HT is shorthand for serotonin, 2A is the specific receptor subtype that serves as the gatekeeper for sensory integration[37] . To be clear, there is no dedicated 'psychedelic receptor', rather the specific serotonin receptor (5-HT2A) is what psychedelics brute force attack, and all of the major psychedelics hold the molecule specifically shaped to hijack the 5-HT2A serotonin receptor.

By binding to the 5-HT2A protein, the psychedelic molecule doesn't just 'influence' the brain; it effectively bypasses the usual filters of the ego, forcing the effective reality prediction centers to stay open, flooding the mind with data that normally would not be able to get in, and due to an unusual 'lid' feature of the molecule itself, the psychedelic molecule locks into place within the receptor for hours after the drug clears the physical body. The "Lid" is the 'extracellular loop 2 (EL2) that folks over, trapping it in the receptor, causing it to fire continuously for hours[38].

Normally, the brain "guesses" what is in its immediate environment to preemptively define the space for consciousness and take on the role of interpreting reality. When the 5-HT2A key is activated, this system recedes or 'breaks'. The brain is forced to *look* at the world without guessing, and contextualization, or its usual shortcuts (familiarity, etc). This allows the experience to be new, novel, fresh, in the moment and with a characteristic lucidity of an 'extra' level of reality[39].

A New Language Is Required

These characteristic perceptual shifts, so characteristic of the psychedelic experience, was something psychiatrist and psychedelic researcher Humphry Osmond and writer Aldous Huxley, author of *A Brave New World* and *The Doors of Perception*, realized required a new language. The two met after Osmond struck up a correspondence with Huxley who requested a dose of psychedelics. Ormond obliged by driving to Los Angeles from Saskatchewan to hand deliver Huxley's first psychedelic dose.

> "Huxley concocted 'phanerothyme' by combining the Greek words for 'to show' and 'spirit' as well as the tagline: 'To make this mundane world sublime, Take half a gram of phanerothyme.' Osmond, however, came up with his own phrase: psychedelic. He combined the Greek words for psyche (for mind or soul) and deloun (for show), along with his own rhyme: 'To fathom Hell or soar angelic/Just take a pinch of psychedelic.' Osmond said the term meant 'mind manifesting' and that it was 'clear, euphonious and

uncontaminated by other associations.' Much like a user drops acid, Osmond dropped the term 'psychedelic' during a meeting at The New York Academy of Sciences in 1957."

-Nick Fetty, *The Origin of the Term 'Psychedelic'* -Nick Fetty, *The Origin of the Term 'Psychedelic'* (2024)[40]

The Meta Point

Out of the vast sea of psychoactive possibilities, there is only one specific class of molecule that serves as a high-resolution filter to the underlying substrate of consciousness. This is a substrate where our understanding of reality runs in parallel with the scientific facts established by our investigation into *The Divine Algorithm.* As we established in our summary, there is a pre-existing, underlying substrate to our reality where the underlying ruleset, objective constraint field, and instantiating consciousness behind the Algorithm resides. The 'unborn' plan of our existence holds its intent and its teleological plan. In the existence of the psychedelic molecule, we find a series of direct portals, allowing direct contact by humans with the pre-existing consciousness substrate.

In both mystical experience and biologically triggered states, humans do not merely "imagine" a new world; they experience the intuitive, lived reality of the layer that holds and sustains the reality we inhabit, and from which our physical reality erupts.

This is the same field where the great masters have found their gospel, their truth and their enlightenment. The Bhagavad Gita, The Old Testament, the Quran, the New Testament, and the Pali Canon are all doctrines, captured by the ascended masters from their experiences within this substrate.

Other Considerations

While the pharmacological effects of a given substance can be measured and dosages dialed in for the general population, the nature and meaning of the experience itself remains a flexible variable. This variability is often a reflection of the complex, layered architecture of human identity. The bifurcation of the *Minimal Self* and the *Narrative Self* developed by philosopher Shaun Gallagher in his paper, *Philosophical conceptions of the self: Implication for Cognitive Science*[41], is helpful in understanding the impact of different levels of ego-attachment.

The Narrative Self is the self that is more likely to readily disintegrate in the face of a profound psychedelic contact. The 'narrative' is the 'story' we make up about ourselves, our social status, professional titles, skills, labels, significant events, and what society and individuals mirror back to ourselves and that we learn to lean upon in presenting ourselves. the social status, labels, and mirrored expectations we lean upon. In the face of 5-HT2A activation, this narrative is often the first to disintegrate.

Conversely, the minimal self is what becomes more prominent. This is our immediate selves, in the now, that we carry and cannot shed. Often this is unspoken, pre-language, 'felt', and experienced as the 'what it is like' of individual selfhood. It is self-existing and self-justified. It is a 'knowing' of the self that has been persistent across our lifetime. It simply 'is' 'us' without assertion. As articulated in the REBUS model[42], the psychedelic 'master key' relaxes the high-level priors of our narrative identity. For those with high Psychological Flexibility, this is not a destruction, but a forensic validation of the core 'Youness' that exists beneath the simulation. They simply slip out of the narrative 'glove' to reveal the persistent substrate that has been there all along.

The minimal self can be quite surprising, fresh, and transformative when the 5-HT2A site has been activated and the environment itself has become newly causative, and the underlying substrate of reality has been revealed. Depending on one's familiarity with their minimal self this activation, or detachment of the self-ego-narrative can be experienced as a violent disintegration, or, conversely, one can slide into the state almost unknowingly, like a child (appear childlike) and become almost unrecognizable to others in the minimal state. The self that has been built up, is peripheral to the individual, and more akin to a kind of narrative glove they easily slip out of for there was no real attachment to the accumulated narrative. It simply 'was' as the minimal self 'is'. Both extremes of attachment and non-attachment can be observed in a group of psychedelic seekers. Shedding, or non-attachment with the narrative self can also be learned, and the attendant psychological flexibility can be cultivated and enhanced by meditative or mindfulness practices, and, by psychoactive and psychedelic experiences where the shedding of ego, narrative, attachment, beliefs, concreteness, labels, desires, etc. has become an all too familiar state of affairs.

Other factors that are impactful for the nature of psychedelic experience include personality, preparation, expectation, intention, spiritual depth and famously, set, and setting.

The individual biological response and sensitivity to the drug itself is a significant factor. There are individuals who, for any number of reasons, are highly reactive, and are easily induced into altered states of consciousness on psychedelics, or by suggestion, regardless of sobriety or chemical influence[43].

Psychedelic Research

In the year 2000, 2 Johns Hopkins researchers, Roland Griffiths and William Richards, conducted a double-blind study where 36 volunteers participated in 2 sessions. In one of the sessions, they received psilocybin and in the next a placebo. 61% of those receiving psilocybin reported a "complete mystical experience" that satisfied all of the criteria as documented by W.T. Stace in *Mysticism and Philosophy*[44] These criteria—detailed earlier in Part 1 (The Mystical Experience & The Spiritually Transformative Experience)—include: sacredness, noetic quality, deeply felt positive mood, ineffability, paradoxicality and transcendence of time and space[45].

Twenty-five years earlier, in 1974, Walter Clark conducted an experiment where LSD was given to eight subjects, 9 to 11 months later he asked them to rate the intensity of their experience on a scale of "1" (least) to "5" (most) in several categories. The most common rating was "5," which was equated with, "beyond anything ever experienced or even imagined" for elements such as: timelessness, spacelessness, paradoxicality, presence of God, ultimate reality, blessedness and peace, mystery, and rebirth[46].

Clinical research has demonstrated the ability to provide measurable, biologically facilitated mystical experiences to "regular" individuals in controlled trials who by and large have no expectation, preparation, script, or ongoing spiritual practice. This experience appears to be available virtually *on demand*, thus facilitating insights and perceptions which would not otherwise be available in the course of a 'normal' lifetime.

These key insights have recently been leveraged to trigger the elusive and much coveted therapeutic insight and scientifically verified improvement for a number of intractable mental health conditions including depression, anxiety, post-traumatic stress disorder (including war trauma), and trauma. These insights are of the kind that is conducive to evidence based and measurable mental health improvement far beyond established evidence-based best practices[47].

There are many large research studies currently being conducted on the efficacy of these compounds in various medicinal and therapeutic applications. Two of the organizations that have made keeping the public informed as to their efforts, are The Johns Hopkins Center for Psychedelic & Consciousness Research and MAPS (Multidisciplinary Association for Psychedelic Studies).
Johns Hopkins psychedelic and consciousness research site states:

> "The Center for Psychedelic and Consciousness Research at Johns Hopkins works to advance the scientific understanding of

psychedelics and their potential for treating mental health disorders, enhancing well-being, and expanding our understanding of consciousness...Our research is investigating therapeutic effects in people who suffer a range of psychiatric, neurological, and other conditions including substance use disorder (smoking, alcohol, other misused drugs), existential distress caused by life-threatening disease, major depressive disorder, anorexia nervosa, Post-Treatment Lyme Disease, depression associated with Alzheimer's Disease, Obsessive-Compulsive Disorder, chronic pain, PTSD, and more. " -Johns Hopkins Center for Psychedelic and Consciousness Research, (2026)[48]

The center has produced over 150 peer-reviewed articles and studies in scientific journals (as of April 2026). Conditions they have studied include addiction, existential duress caused by life-threatening disease, treatment-resistant depression, opioid addiction, Alzheimer's disease, Post-traumatic stress disorder, post-treatment Lyme disease syndrome, and anorexia nervosa.

MAPS (Multidisciplinary Association for Psychedelic Studies, www.maps(dot)org) states, "Founded in 1986, the Multidisciplinary Association for Psychedelic Studies (MAPS)... develops medical, legal, and cultural contexts for people to benefit from the careful uses of psychedelics and marijuana." They further note: "We envision a world where psychedelics and marijuana are safely and legally available for beneficial use, and where research is governed by rigorous scientific evaluation of their risks and benefits." (as of June 2025)

---

1 (T. Leary, Interpersonal Diagnosis of Personality; A Functional Theory and Methodology for Personality Evaluation, 1957, Ronald Press.).
2 (R. Dass, Be Here Now, 1971, Lama Foundation),
3 (R. Alpert, Be Here Now, Pg. Crown Publishing, 1974)
4 (Stace, 1960a).
5 (Miller & C'de Baca, 2001).
6 (Huxley 1947)...
7 (Schultes, Hoffmann, & Ratsch, 2001).
8 (Roberts, 2001). (F.S. Barrett, R.R. Griffiths, Classic Hallucinogens and Mystical Experiences: Phenomenology and Neural Correlates, [Introduction], Current Topics In Behavioral Neuroscience. 2018; 36: 393-430.).

[9] (M.V. Vargas et al, Vargas, M. V., et al., Psychedelics Promote Neuroplasticity Through the Activation of Intracellular 5-HT2A Receptors, 2023, Science, 379(6633), 700-706).
[10] (M.V. Vargas et al, Vargas, M. V., et al., Psychedelics Promote Neuroplasticity Through the Activation of Intracellular 5-HT2A Receptors, 2023, Science, 379(6633), 700-706).
[11] (R. L. Carhart-Harris, et al. Neural Correlates of the Psychedelic State as Determined by fMRI Studies with Psilocybin, 2012, Proceedings of the National Academy of Sciences, 109(6), 2138-2143)
[12] (R. L. Carhart-Harris, et al. Neural Correlates of the Psychedelic State as Determined by fMRI Studies with Psilocybin, 2012, Proceedings of the National Academy of Sciences, 109(6), 2138-2143)
[13] (M.V. Vargas et al, Vargas, M. V., et al., Psychedelics Promote Neuroplasticity Through the Activation of Intracellular 5-HT2A Receptors, 2023, Science, 379(6633), 700-706)
[14] (T. Leary, High Priest 1968a pp 324-328 [edition dependent], World Publishing Company)
[15] (E. Siegel, Harvard Year [1962-1963], The Harvard Crimson, 1962-1963, thecrimson(dot)com)
[16] (E. Siegel, Harvard Year [1962-1963], The Harvard Crimson, 1962-1963, thecrimson(dot)com)
[17] (T. Leary, R. Metzner, R. Apert, The Psychedelic Experience: A Manual Based on The Tibetan Book of The Dead, Citadel Press, 1964)
[18] ebi.ac.uk, ChEBI, checmial entities of biological interest, dictionary of molecular entities.)60
[19] (adf.org.au, ADF, Alcohol and Drug Foundation)61
[20] (C. Ratsch, A. Hoffman, The Encyclopedia of Psychoactive Plants: Ethnopharmacology and Its Applications, 2005, Park Street Press), (C.S. Grob, J. Grigsby, (editors) Handbook of Medical Hallucinogens, 2022, Guilford Publications) (G. Samorini, Animals and Psychedelics: The Natural World and the Instinct to Alter Consciousness, 2002, Park Street Press)
[21] (K.J. Locey & J.T. Lennon, Scaling Laws Predict Global Microbial Diversity, 2016, Proceedings of the National Academy of Sciences (PNAS), 113(21), 5970-5975).
[22] (H.T. Reynolds, et al., Horizontal Gene Cluster Transfer Increased Hallucinogenic Mushroom Diversity, 2018, Evolution Letters, 2(2), 88-101).
[23] (J. Fricke, F. Blei, & D. Hoffmeister, The Biosynthetic Pathway of Psilocybin, 2017, Angewandte Chemie International Edition, 56(40), 12352-12355).
[24] (M. Meyer, J. Slot, The Evolution and Ecology of Psilocybin in Nature, Fungal Genet Biol., 2023, Jun:167:103812)
[25] (J. Fricke, F. Blei, & D. Hoffmeister, The Biosynthetic Pathway of Psilocybin, 2017, Angewandte Chemie International Edition, 56(40), 12352-12355)
[26] (H.T. Reynolds, et al., Horizontal Gene Cluster Transfer Increased Hallucinogenic Mushroom Diversity, 2018, Evolution Letters, 2(2), 88-101)
[27] (T. Haarmann, C. Machado, Y. Lübbe, T. Correia, C.L. Schardl, D.G. Panaccione, P. Tudzynski, The Ergot Alkaloid Gene Cluster in Claviceps Purpurea: Extension of the Cluster Sequence and Intra Species Evolution, 2005, Phytochemistry. Jun;66(11):1312-20).
[28] (T. Haarmann, C. Machado, Y. Lübbe, T. Correia, C.L. Schardl, D.G. Panaccione, P. Tudzynski, The Ergot Alkaloid Gene Cluster in Claviceps Purpurea: Extension of the Cluster Sequence and Intra Species Evolution, 2005, Phytochemistry. Jun;66(11):1312-20)
[29] (A. Watts, Psychedelics and Religious Experience, California Law Review Vol. 56, No. 1 (Jan. 1968), pp. 74-85, Calif. JSTOR)62.

---

[30] (Douglas Main, Ayahuasca Can Help You Communicate with Plant Spirits, and Heal Your Body and Mind, Newsweek, January 1, 2015,, newsweek(dot)com),

[31] (Douglas Main, Ayahuasca Can Help You Communicate with Plant Spirits, and Heal Your Body and Mind, Newsweek, January 1, 2015, newsweek(dot)com)

[32] (J. Narby, The Cosmic Serpent: DNA And The Origins of Knowledge, 1999, J.P. Tarcher/Perigree)

[33] (J. Narby, The Cosmic Serpent: DNA and the Origins of Knowledge, 1999, J.P. Tarcher/Perigree)

[34] (J. Narby, The Cosmic Serpent: DNA and the Origins of Knowledge, 1999, J.P. Tarcher/Perigree)

[35] (Sam Harris, Making Sense Podcast: March 23, 2021, Psychedelics and the Self: A Conversation with James Fadiman (Episode #242):

[36] (Sam Harris, Making Sense Podcast: March 23, 2021, Psychedelics and the Self: A Conversation with James Fadiman (Episode #242)

[37] (S. Ali, X. Tian, K.A. Cunningham, J. Zhou, Targeting the 5-HT2A Receptor for Developing Psychedelic Drugs and Beyond, J. Med. Chem. 2025, 68, 10, 9831–9835).

[38] (S. Ali, X. Tian, K.A. Cunningham, J. Zhou, Targeting the 5-HT2A Receptor for Developing Psychedelic Drugs and Beyond, J. Med. Chem. 2025, 68, 10, 9831–9835).

[39] (R.L. Carhart-Harris, and K.J. Friston, K. J., REBUS and the Anarchic Brain: Toward a Unified Model of the Brain Action of Psychedelics. 2019, Pharmacological Reviews, 71(3), 316-344.)

[40] (N. Fetty, The Origin of the Term 'Psychedelic' April 4, 2024, The New York Academy of Sciences, Blog Article, nyas(dot)org)

[41] (S. Gallagher, Philosophical Conceptions of the Self: Implications for Cognitive Science, 2000, Trends in Cognitive Science, 4(1):14-21.)

[42] (R.L. Carhart-Harris, K.J. Friston, REBUS and the Anarchic Brain. Pharmacological Reviews, 2019. 71(3), 316-344.),

[43] (H.M Dourron, H. Copes, D.H. Grossman, M. Bradley, C.R. Nicholas, J.A. Turner, G. Allen-Clark, M. Gossage, P.S. Hendricks, How Do Psychedelics Impact People with a History of Non–Affective Psychosis? A Qualitative Study. Front Psychiatry, 2025 Dec 9;16:1716545, PCID:12722819).

[44] (W. T.Stace, Mysticism and Philosophy, Lippincott, 1960).

[45] R. Griffiths, W. Richards, U. McCann, R. Jesse, Psilocybin can occasion mystical-type experiences having substantial and sustained personal meaning and spiritual significance. [2006] Psychopharmacology [Berl], 187 [3], 268–283; discussion 284–292).

[46] (W. Clark,Hallucinogen Drugs Controversy, 1974,. In S. Radoubo-Thomas, A. Villeneuve, And C. radouco-Thomas, Eds. Pharmacology, Toxicology, and Abuse of Psychotomimetics (hallucinogens). Quebec: Les presses de l'Université Laval. Pg. 411-418. 1974 [Thanks to L. Grinspoon, J.F. Bakalar, Psychedelic Drugs Reconsidered, Basic Books, 1979]).

[47] (E. Vermetten, E. Krediet, T. Bostoen, J.J. Breeksema, R.A.Schoevers, W. van den Brink, Psychedelics in the treatment of PTSD. Tijdschr Psychiatr. 2020;62(8):640-649. Dutch. PMID: 32816292.), (N. Schimmel, J.J. Breeksema, S.Y. Smith-Apeldoorn, J. Veraart, W. van den Brink, R.A. Schoevers RA. Psychedelics for the treatment of depression, anxiety, and existential distress in patients with a terminal illness: a systematic review. Psychopharmacology (Berl). 2022 Jan;239(1):15-33. PMID: 34812901.) (J.J. Breeksema, A.R. Niemeijer, E. Krediet, E. Vermetten, R.A. Schoevers, Psychedelic Treatments for Psychiatric Disorders: A Systematic Review and Thematic Synthesis of Patient Experiences in Qualitative Studies. CNS Drugs. 2020 Sep;34(9):925-946. PMID: 32803732.)58

[48] (hopkinspsychedelic(dot)org/who-we-are/ about April 2026)

6 The Near Death Experience & Out-of-Body Experience

According to The International Association for Near-Death Studies, Inc (IANDS) "The term 'near-death experience' (NDE) was coined in 1975 in the book *Life After Life* by Dr. Raymond Moody, M.D., Ph.D."

> "An NDE is a distinct subjective experience that a minority of people report after a near-death episode. In a near-death episode, a person is either clinically dead, near death, or in a situation where death is likely or expected... Many near-death experiencers (NDErs) have said the term "near-death" is not correct; they are sure that they were in death, not just near-death… the term "near-death" can be understood to mean 'actually in the first moments of death though not permanently dead.' NDErs have reported two types of experiences. Most NDErs have reported pleasurable NDEs. These experiences involve mostly feelings of love, joy, peace, and/or bliss. A small number of NDErs have reported distressing NDEs. These experiences involve mostly feelings of terror, horror, anger, isolation, and/or guilt. Both types of NDErs usually report that the experience was hyper-real—even more real than earthly life…"
>
> -IANDS, *What Is A NDE?* (2022)[1]

The Near-Death Experience Research Foundation (NDERF) website provides thousands of pages of information about NDEs and their definition of the near death experience is simple:

> *"A lucid experience associated with perceived consciousness apart from the body occurring at the time of actual or threatened imminent death."*

The near-death experience (NDE) represents a critical anomaly that fractures the boundary between materialist dogma and verifiable data. It is a unique intersection of four conditions that fall far outside the purview of the materialist paradigm:

(1) the out-of-body experience
(2) the mystical experience
(3) the spiritually transformative experience, and
(4) documented clinical death.

While many mystical states are purely internal, the NDE is unique for producing veridical data. A veridical NDE involves the acquisition of specific, accurate information regarding the physical environment that the individual could not have perceived through their biological senses, particularly because their body is in a state of clinical death or deep unconsciousness. This form of knowledge—veridical—functions as a forensic record of awareness operating without an identifiable biological substrate.

When this information is subsequently cross-referenced and confirmed as accurate by third parties (close family members, medical personnel, witnesses, etc.) or objective data (such as EKG, audio transcription, or off-site vision, video and photographic evidence), the NDE ceases to be a subjective anecdote. It becomes a verifiable, scientifically amenable bridge between an independent individual consciousness and our physical reality. The nature of veridical NDEs provides that there exists a reality signal, for both the individual, and the perception of our physical reality that remains active even when the biological 'hardware' is offline.

Critically, these experiences are not beyond the reach of the scientific method as alleged by materialistic skeptics; they are data points that require curiosity and a non-dogmatic approach to the nature of consciousness. This purview of investigation logically extends to the "impossible" (by materialistic standards) knowledge states seen in savant syndrome—whether acquired through trauma or present from birth. In these cases, the brain does not create genius; rather, a structural or functional shift in the neural architecture allows the individual to natively access the evaluative field in an unusually coherent manner. The savant is not calculating primes or memorizing symphonies in the traditional sense; they are accessing the pre-existing, objective field of information that the normal brain specifically evolved to filter out.

This filtering concept is further validated by the phenomenon of non-local perception, or "telepathy", often witnessed in non-speaking autistic "spellers" and their family members who report consistent, veridical telepathic coherence between one another. In this example we can see that when the standard biological "hardware" for language is bypassed or unavailable, the consciousness often defaults to a more fundamental mode of transmission. This isn't merely a subjective feeling of connection, but a documented transfer of specific, external information that defies any sensory apparatus, or proximity. It once again validates the existence of the evaluative field. For certain individuals—and those bonded to them—the walls of the individual mind are uniquely permeable, allowing for a direct, mind-to-mind resonance, by way of this field, that materialism cannot account for. This style of communication also indicates that not only is there objective evaluation, universal constraints and Truth in the field, but as creations (instantiations) of reality, individual human entities are also to be found, in their fullness, within this field.

Perhaps the most startling fracture in the materialistic paradigm is the capacity for *seeing without eyes*, a form of extra-retinal perception demonstrated by individuals who can perceive and navigate their environment with total accuracy despite being blindfolded or biologically sightless. Whether it is the legendary feats of *Kuda Bux* or those who have trained with and been taught the method of *seeing without eyes* by groups, and modern veridical accounts from NDEs and *intuitives*, the implication is profound: the image of reality in the way it exists, is not a product of the eye and optic nerve, but a direct data stream that can be perceived by the consciousness itself.

Each of these layers—the savant's omniscience, the telepath's resonance, and the "eyeless" vision—represents a forensic record of objective information overriding biological limitations, proving that the signal is independent of the station.

Critically, none of these areas are outside of the purview of scientific exploration and validation. Science is the systematic pursuit of the truth behind the data—a process of incremental discovery that follows the evidence wherever it leads; therefore, there is no phenomenon, material substance, observation, experience, or immaterial reality in our universe that lies outside of its purview. We simply must release science from the shackles of the materialist paradigm and allow it to return to its original mandate: the fearless investigation of reality in all its complexity.

Science has no requirement for knowing "how" these events take place to say that they did indeed occur, nor is there any scientific obligation to attribute these experiences to a specific physical substrate before they can be acknowledged, by science, as facts. The contrary is true. Such events are where the scientific method excels: because it proceeds by incremental knowledge—documenting the "what" of phenomenon while leaving the "how" for later synthesis—there is no obligation within the practice of science to resolve all possible theoretical issues, particularly those that are a threat to the paradigm of materialism, at the moment of discovery. We have long accepted the reality of gravity, quantum entanglement, and the efficacy of medicines—such as aspirin, which remained a mystery for 74 years before its mechanism of action was correctly identified, or even the placebo effect—without having a total, biologically underpinned, undisputed understanding.

The primary obstacle to the acceptance of NDE data is not a lack of empirical evidence, rather, it is the fact that the existence of these events fundamentally falsifies the materialist construct. Consequently, NDEs are relegated to a prohibited class of inquiry. One will repeatedly hear the common refrains of, "we don't understand the phenomenon.", or "more research is needed.", or "low oxygen, hypoxic brain." These are disingenuous defense mechanisms in action. By the materialist's own definition, a phenomenon that suggests consciousness is independent of brain function *cannot* be verified, not because the tools of science are insufficient, but because verification would require the total dissolution of the current materialist's scientific paradigm. Science is not the problem. The dogma of materialism is.

Near death experiences include spiritual events, and reports of an otherworldly existence at the time of clinical death. If the other categories of human experience we have investigated of mystical experiences, physically induced experiences, psychedelically facilitated experiences, and God consciousness as experienced by the Prophets are *pathways* towards communion with the divine, the Near Death Experience represents the one-way teleportation technology to the seat of universal consciousness.

In the NDE, communication with a number of spiritual entities have been reported, from deceased relatives, miscarried children, lost animal companions, spiritual guides, historical prophets, ascended masters, saints, angels, higher dimensional beings, all the way up to the intelligent all powerful creative, instantiating source referred to as the creator of all things—what humans refer to as *God*.

While the content of these communications is often described in the language of the sacred, often in unspoken, pre-representational forms, the mechanical reality of the experience remains grounded in a verifiable phenomenon.

To understand the legitimacy of these claims, we must return to the forensic evidence that anchors the NDE to our physical reality.

> "Veridical near-death experiences are NDEs in which people reportedly out-of-body have observed events or gathered information that was verified by others upon the experiencer's return to a conscious state."
>
> -IANDS, *Key Facts About NDEs* (2020)[2]

The onset of a NDE is often sudden and occurs either immediately prior to a traumatic death, at the moment of clinical death or in close proximity to an event that can cause death or grievous bodily harm. For many, the transition from biological life to clinical death is so instantaneous that the awareness of the transition itself is bypassed. There is no 'departure' sequence; there is only the sudden realization of a new vantage point.

It is not uncommon for NDErs, upon the commencing of their experience, to become aware of their consciousness in a perspective where they are viewing their own near-death incident, without at first appreciating that fact—nor the reality that what they are viewing is their own earthly body. This is another distinction between NDEs and OBEs (out-of-body experiences) in that OBE subjects are nearly always aware of the transitional state from being in their body—to out of their body. While a standard OBE is often characterized by the shock of displacement, the NDE is most often marked by a profound, inexplicable calm. The NDEr does not feel 'displaced' from their body; they are simply 'released' from it, while often unaware that they have made a transition—the primacy is on the seamless experience of one's consciousness, not the association of that consciousness with the physical form. As we have asserted from our investigation into Consciousness and Youness, consciousness is the native state, not the state of biological form with consciousness.

Many NDEers report the experience of viewing themselves from above or from a distance from their physical bodies, feeling safe and complete and intact in their consciousness, feeling no pain, or discomfort, as the scene around their death is viewed. Rather than to be shocked or surprised, they often find the scene merely 'interesting,' viewing their own physical form with the detached indifference one might feel toward a discarded snakeskin or a heap of used clothing.

This early observational phase—the 'local' portion of the experience—functions as a forensic waystation. It is the final anchor to our physical reality before the unfolding of the consistently documented 'classic' aspects of the NDE.

The Out-of-Body Experience (OBE)

The out-of-body experiences (OBEs) which frequently accompany NDEs are also sudden, but unlike NDE's they can take place when there is no threat of death whatsoever. Indeed, they are more perplexing, as there often can be no obvious identifiable trigger, nor need for mental 'preparation,' nor the desire for an OBE, or even knowledge that OBEs are possible. This is counter-factual to the claim that individuals are manufacturing these events or confabulating them into existence. One cannot manufacture a defense mechanism they didn't know existed against an event that isn't a threat. The clinical record is clear: OBEs are not the product of psychiatric pathology—by definition—a confabulated OBE is, objectively, not an OBE. And for those that have had an OBE, there is no clear, consistent history of dissociation, hallucination, or delusional thought among experiencers. This fact does not keep these experiencers from being pathologized by the medical establishment and dismissed by the scientific establishment. Nor does it prevent the experiencers from being accused of making up their issues. This social gaslighting doesn't just mislabel the objective reality of the event; it effectively severs the individual from professional support and personal support at the precise moment they are attempting to process a reality-shattering experience.

Although they can happen 'anytime', there are times more prone to their occurrence. They have been noted to occur during altered states of consciousness, such as rituals, particularly during deep meditative states, yoga, chanting, deep breathwork, trance-type dance, deep focus or deep relaxation, or even just before or after waking. Although OBEs are frequently an element within a NDE, they manifest during 'peak' physiological thresholds—moments where the biological system is pushed to its absolute limit. Whether through the intensity of childbirth, the trauma of assault, or the deliberate sensory overrides of mortification rituals, the signal from the underlying substrate of reality creates a breach in one's biological firewall in situations where death is not an imminent hazard.

OBEs Are A Different Phenomenon Than NDEs.

> "I distinguish out-of-body experience from being in the heavenly realms or visionary realms. The OBE in my definition happens right in the human realm typically in the room or nearby... And then when you go beyond that into another realm, that then becomes a visionary heavenly realm.... So. I hope that's a good introduction."
>
> -Alex Ferrari, *Next Level Soul Podcast with William J. Peters* (2023)[3]

OBEs are largely devoid of mystical elements and the characteristic insight, blessedness, noetic and paradoxical qualities typical of mystical experiences, and near death experiences. However, the implications of the OBE are nearly every bit as profound because there is no possible way to explain the experience as being in concern with day-to-day life on the physical plane.

In an OBE the individual is outside of their body, they are entirely lucid and the sense of their consciousness is indistinguishable from that of the normal waking state of consciousness. Often the vision of the world is reported as being blurry or unfocused. Frequent OBEers report having to use conscious intent to command one's 'spirit eyes' to "focus!" in order to achieve a clear vision.

As mentioned previously it is not uncommon to experience mild or significant panic either during the OBE, or particularly when one finds that there is difficulty in returning "inside" of one's physical body. Such experiences are generally short-lived, lasting seconds to several minutes and rarely lasting longer.

Occasionally the OBEer will report seeing other entities in outline or unfocused in the periphery. Unlike NDEs it is rare that OBErs experience anything like spirit communication, visions, information downloads, feelings of bliss or euphoria. If the OBE contains these elements, then it will be considered to be a *mystical* experience *with* an OBE, or vice-versa, and not referred to as simply an OBE. It is estimated that approximately 5% (~5%) of the population will experience at least one OBE in their lifetime, although some sources estimate that a much larger percentage of the population has had, or will have an OBE[4].

Because NDEs and OBEs cannot be accommodated by any physicalist construction, a cottage industry of materialist 'researchers' has emerged. This industry is characterized by denying the perception of the experiencer and mischaracterizing their reports as being objectively mistaken, impossible or the result of low oxygen 'hypoxia-induced hallucinations.' This strategy gives the general public the impression that these reports have no basis, no veridical evidence, and no elements that cannot be made amenable to competing, physicalist interpretations. With all supportive evidence stripped from these accounts, the reports sound like fantasy and the reporters sound confused, or perhaps motivated by attention.

These dismissals inevitably fail, seeing as the phenomenon is exactly as reported; the substituted explanations objectively cannot account for the forensic data. There are a number of competing physicalist hypotheses, all different, none complete, and none that address the phenomenon. What they are good at is muddying the field. The exchange and discussion of these ideas confuse the public and while their attention is taken up by debating the merits of the meritless, the content of the reports and transformational quality of the experience upon the reporters is left behind.

What these alternative models have in common is the attempt to trap the reality of immaterial consciousness within the hardware of the brain–something that demonstrably cannot be done. Among the incoherent offerings are:

(1) Daniel Dennett's Multiple Drafts hypothesis: (which argues consciousness is an illusion created by the brain 'editing' competing streams of data).

(2) Microtubule Consciousness: (known as Orch-OR) co-developed by Nobel prize awarded physicist Roger Penrose.
(3) Mathematical Information Theory (IIT): which suggests that if enough neurons are wired together, consciousness will eventually 'emerge.'
(4) Epiphenomenalism: which treats the silent immaterial witness at the heart of identity as an abstract 'user interface' created by a brain.

These are red herrings that deflect attention away from the authentic phenomenon of OBEs and NDEs.

Near Death Experience Investigation & Research

Physician, psychiatrist and New York Time best-selling author Dr. Raymond Moody, came from obscurity into the public with the release of his book, *Life After Life: The Investigation of a Phenomenon - Survival of Bodily Death* (1975). This book is uniformly identified by NDE researchers as signaling a new age in consciousness studies. Hundreds and perhaps thousands of NDErs have identified Moody's book as life-changing because it confirmed that their experiences were real, and that hundreds of others who had them as well. They found out in unison that they were not alone.

With this book, an experience that defied the limits of science, yet had deep roots in history, broke upon the collective zeitgeist and for the first time demanded to be taken seriously and understood as a real phenomenon.

Prior to Dr. Moody's book, many of those who had experienced NDEs felt shunned from telling their story and relating their experience. For many even their friends and families were largely unable to come to terms with the reality of the experience even when these experiences resulted in recorded death within medical facilities.

> "...there was almost no research in the area of near-death experiences. In fact, there was almost no professional interest in the subject... Most doctors ignored these occurrences if they were told them by their patients. Sometimes they even thought their patients were 'crazy' and recommended psychiatric or hospital care..."
>
> -Dr. Raymond Moody, *Life After Life* (1988)[5]

Many NDErs had great difficulty integrating the profound life experience of the NDE into their day to day lives. This was compounded by disbelief—often from the very people they turned to for support, including religious clergy and even mental health professionals. The combination of overwhelming, often ineffable information and the perplexed or dismissive reactions of others left many in a beleaguered state. An NDE, however profound, impactful, or revelatory—however complete in its ability to render even the most vile and heinous acts on Earth into a cohesive sense of "ok-ness"—remains difficult to reconcile once the individual is plunged back into the very conditions, they emerged from. This is further complicated by another phenomenon that often accompanies the NDE which is a persistent internal pressure to realign one's life: relationships, occupation, desires, habits, diet, and daily structure, so that it aligns with the truths revealed within their encounter.

The American Center for the Integration of Spiritually Transformative Experiences (ACISTE) has documented that this friction stems from a profound internal paradox: while the encounter represents a total world-view completion, it simultaneously triggers a complete ego fracture. In most cases, this necessitates a radical reordering of one's fundamental orientation to existence and introduces a previously unknown factor into their personal and professional relationships. ACISTE identifies the resulting impact through the following common challenges[6]:

-Processing a radical shift in reality
-Accepting the return (to life) and "homesickness" for "home" (the spiritual realm of the NDE)
-Issues related to sharing the experience (with others)
-Integrating new spiritual values with Earthly expectations
-Problems dealing with newfound psychic abilities and psychic sensitivities
-Increased sensitivity to electricity, chemicals, smells, sounds, etc.
-A yearning to find and live one's purpose

Additional NDE challenges/aftereffects were identified in the study, *Pleasurable Western Adult Near-Death Experiences*[7]. The psychological, spiritual and psychological aftereffects of NDE include:

Psychological:
Loss of fear of death
Profound shifts in fundamental values
Increased self-worth and meaning and purpose in life
Physical:
Changed bodily functions
Malfunction of electrical devices in NDEers vicinity
Social:
Relationship stress from NDEr's psycho-spiritual shifts
Shift towards service-oriented careers[8].

Just as in the world of hard science—particularly physics and cosmology—there has been, within the medical establishment, a strong preference for reductionist explanations. In many cases, nearly any ad hoc or improbable hypothesis is entertained before granting credibility to first-hand accounts and what the overwhelming evidence has revealed.

Although this attitude has shifted somewhat—with the rise of legitimate academic research, dedicated organizations, support groups, conferences, books, and the proliferation of NDE accounts through podcasts and online media—the broader establishment has shown limited curiosity. From the release of *Life After Life* in 1974 to the present, aside from a handful of committed researchers and institutions, there remains a reluctance and a taboo against the serious investigating of the implications of phenomenon, knowledge, abilities, history, meta-physics, spirit entities as well as insight regarding the meaning, purpose, and the origin of the physical plane provided by NDE accounts. These accounts are often framed by experiencers as pointing toward something foundational—and if they are to be believed, even in small measure, this information set must be recognized as a kind of "holy grail" and key to our physical existence.

Yet the very nature of NDE experiences works against them in a sense. Their coherence, intensity, and perceived perfection trigger a well-worn heuristic: if it sounds too good to be true, it probably is. To this extent skepticism, from the surface level is readily understood. What is harder to explain—perhaps inexplicable within the mandate of science—is the extent to which these accounts are dismissed outright by those who most strongly assert themselves as scientists and as arbiters of truth in the world we inhabit.

Dr. Moody's book allowed NDErs to speak about their experience openly. In doing this, many NDErs believe they are fulfilling a directive they received within their NDE to "tell others" and share their experience so that humanity will *know* that there is an afterlife, and that the human soul exists, and that there is a profound and deep purpose to existence and an all knowing intelligent consciousness lies behind existence, and that consciousness made us, and loves each of us dearly. But perhaps more importantly, every aspect of an individual NDE also falls under this directive. This means that every part of the NDE experience is to be disseminated. Not to be lost in this information is that the instantiating power behind our reality, has made itself fully known, as an unimaginably complete and powerful entity within the NDE experience. And this power has not only created our realm, but untold numbers of realms. And if NDE experiences are to be believed, this effectively scales the immaterial realm of existence and instantiating power behind it beyond any possible human-centric model. Our universe is just one instantiation among what can be understood as an infinite set of additional instantiations.

This is where The Divine Algorithm finds its final resolution. All that we have discovered in our investigation thus far, falls in line with the information, experience and downloaded knowledge collectively harvested by all forms of spiritual communication, but particularly those that take place within the NDE experience.

In *Life After Life* Dr. Moody names several pioneering NDE researchers: Dr. Melvin Morse, specializing in child NDE; Dr. Michael Sabom, who started out as an NDE skeptic; Michael Grosso who explored the NDE experience to find links with 'philosophical truths'; Dr. Kenneth Ring, who Dr. Moody says, "legitimized" his work; and Robert Sullivan, who in his prior life was in "the plastics business" but was so curious as to the NDE phenomena, he interviewed 40 combat veterans who reported NDEs.

For all of the above pioneers we owe a deep debt of gratitude for their curiosity, openness and persistence by following the evidence wherever it leads, in the best traditions of the practice of science, even at the peril of their professional reputations. They faced the constant threat of ostracism and ridicule from the established scientific and academic cottage industries.

Their fearless work has created a foundation and a platform that has allowed present day NDE research to flourish.

Dr. Moody's findings have since been reproduced and validated by subsequent NDE studies conducted by professional, impartial researchers that do not have a prior agenda to "debunk" NDEs or have entered the field in order to 'debunk' NDE claims.

Dr. Moody's initial investigation categorized the NDE into "9 elements of the NDE experience":

A Strange Sound*:* A buzzing, or ringing noise, while having a sense of being dead

Peace and Painlessness*:* Pain vanishes and they experience peace when they leave the body.

Out-of-Body Experience*:* The Sensation of rising up and floating above their own body/feeling of being in a spiritual body.

The Tunnel Experience*:* Being drawn into darkness through a tunnel/reaching a realm of radiant golden-white light.

Rising Rapidly into the Heavens: Instead of a Tunnel, some people report rising suddenly into the heavens and seeing the Earth.

People of Light: The dying meet people who glow with an inner light. Often they find friends and relatives who greet them.

The Being of Light: The dying often meet a powerful spiritual being whom some have identified as God, Jesus, or some religious figure.

The Life Review: The being of light presents a panoramic review of everything they have ever done. They come away feeling that love is the most important thing in life.

Reluctance to Return: The being of light sometimes tells the dying that they must return to life. Other times, they are given a choice[9].

The University of Virginia Division of Perceptual Studies provides further evidence that the core features of the NDE experience are consistent among subjects:

Univ. of VA features of the NDE:

Feeling very comfortable and free of pain
A sensation of leaving the body, sometimes being able to see the physical body while floating above it
The mind functioning more clearly and more rapidly than usual
A sensation of being drawn into a tunnel or darkness
A brilliant light, sometimes at the end of the tunnel
A sense of overwhelming peace, well-being, or absolute, unconditional love
A sense of having access to unlimited knowledge
A "life review," or recall of important events in the past
A *preview* of future events yet to come
Encounters with deceased loved ones, or with other beings that may be identified as religious figures[10]

The NDE characteristics uncovered by Dr. Moody and The University of Virginia echoes highly correlated aspects of the NDE researched by NDERF (Near Death Experience Research Foundation) researcher, Dr. Jeffrey Long, MD's book, *Evidence of an Afterlife*[11].

***Dr. Long has* identified 12 elements of the NDE experience in his scientific survey/study of 613 NDErs** (percent of the 613 who affirmed each element in parentheses):

Intense and generally positive emotions or feelings (76.2%)
Out-of-Body Experience (OBE): Separation of consciousness from the physical body (75.4%)
Heightened senses (74.4%)

Encountering a mystical or brilliant light (64.4%)
A sense of alteration of time or space (60.5%)
A return to the body, either voluntary or involuntary (58.5%)
Encountering other beings, either mystical beings or deceased relatives or friends (57.3%)
Encountering or learning special knowledge (56%)
Encountering otherworldly ("heavenly") realms (40.6%)
Passing into or through a tunnel (33.8%)
Encountering a boundary or barrier (31%)
Life review (22.2%)[12]

In Dr. Long's study, widely discussed elements of the NDE such as "passing into or through a tunnel" (33.8%) and "encountering other beings, either mystical beings or deceased relatives or friends" (57.3%) are relatively low in comparison to the popular conception that all NDEs are characterized by "going into the light" and seeing deceased relatives. Typically, each NDE will have several aspects of the NDE as reported above, a few elements that are unique to a subset of NDErs, and only on very rare occasions, *all* NDE elements are reported. There are core patterns, themes and core events, but the individual narratives appear to be specific to the experiencer.

The final pillar among the major U.S.-based NDE research organizations is IANDS (International Association for Near-Death Studies, Inc.). Their documented nine characteristics of the NDE closely mirror the findings established by Dr. Moody, UVA, and Dr. Long:

Intense emotions: commonly of profound peace, well-being, love; other NDEs marked by fear, horror, loss.

A perception of seeing one's body from above (called an out-of-body experience, or OBE), sometimes watching medical resuscitation efforts or moving instantaneously to other places.

Rapid movement through darkness, often toward an indescribable light.

A sense of being "somewhere else," in a landscape that may seem like a spiritual realm or world.

Incredibly rapid, sharp thinking and observations.

Encounter with deceased loved ones, possibly sacred figures (the Judges, Jesus, a saint) or unrecognized beings, with whom communication is mind-to-mind; these figures may seem consoling, loving, or terrifying.

A life review, reliving actions and feeling their emotional impact on others.

In some cases, a flood of knowledge about life and the nature of the universe.

Sometimes a decision to return to the body[13].

Across the four scales presented, there is a striking alignment in the reported and documented characteristics of the NDE.

As research has advanced, the NDE has emerged as a common and widely reported phenomena. When extant data is extrapolated, "*Surveys taken in the US*[14] *and Germany* [http(colon)//www(dot)nderf(dot)org/number_nde_usa(dot)htm]" suggest that anywhere between 4% and 15% of the general population has experienced an NDE.

Across the world, and spanning all cultures, NDE elements are shared with no deviation regarding age, education, intelligence, language, race, religion, or mental and medical condition. This uniformity—reported by disconnected individuals who have often kept their experiences secret for decades—is the opposite of what one would expect to find if the NDE were a manufactured phenomenon. It suggests a universal experience rather than an imaginative event bound to cultural artifacts.

To ensure methodological integrity and mitigate the inclusion of fabricated or idiosyncratic accounts, the primary investigators cited in this review employ standardized psychometric instruments for validation. The definitive tool for this purpose is the Near-Death Experience Content (NDE-C) Scale, globally recognized (and utilized) under the more informal name, the *Greyson Scale*[15].

The Standard NDE Content (NDE-C) Scale (The Greyson Scale)

Since its inception in 1983, The Greyson Scale has been the standard tool to validate NDEs for research purposes, distinguishing them from non-NDEs or accounts that lack sufficient clinical data. The primary objective of the scale is to ensure all researchers are addressing the same core phenomenon. The scale is based on NDE narratives classified into four groups: paranormal, psychological, physiological, and environmental.

Each of the sixteen items has three possible ratings, with 0, 1, or 2 points available. While the maximum score is 32, any NDE with a score of 7 or higher is considered validated for research purposes. As the data shows: "Among a criterion group of NDErs, the mean score on this scale was 15 with a standard deviation of 7.84; we therefore use a score of 7 or greater (1 standard deviation below the mean) as the cut-off point for identifying an experience as an NDE[16]." The scale has been translated into over 20 languages "and used in hundreds of studies worldwide. It remains the primary method for identifying and quantifying the depth of a near-death experience[17]." Two major NDE organizations in the United States, NDERF and IANDS, both utilize the standardized NDE-C (Greyson Scale) to ensure rigorous consistency across their research.

The NDE Content Questionnaire

(1) Did time seem to speed up or slow down?
(2) Were your thoughts speeded up?
(3) Did scenes from your past come back to you?
(4) Did you suddenly seem to understand everything?
(5) Did you have a feeling of peace or pleasantness?
(6) Did you have a feeling of joy?
(7) Did you feel a sense of harmony or unity with the universe?
(8) Did you see, or feel surrounded by, a brilliant light?
(9) Were your senses more vivid than usual?
(10) Did you seem to be aware of things going on elsewhere, as if by extrasensory perception (ESP)?
(11) Did scenes from the future come to you?
(12) Did you feel separated from your body?
(13) Did you seem to enter some other, unearthly world?
(14) Did you seem to encounter a mystical being or presence, or hear an unidentifiable voice?
(15) Did you see deceased or religious spirits?
(16) Did you come to a border or point of no return?[18]

The Greyson Scale has been subjected to Rasch Analysis—also known as Modern Test Theory—by statistical researchers Rense Lange and James Houran. Rasch Analysis is "a mathematical approach for quantifying latent traits based on the fundamental assumption that a subject's response to an item is a function of the difference between 1) his/her abilities and 2) the characteristics of the item... and has been adopted by the fields of clinical and public health research as a tool for investigating diverse health outcomes[19]..." The rigorous statistical analysis performed by Lange and Houran yielded a definitive conclusion: the Greyson Scale is valid. This confirms that the scale isn't just capturing 'subjective scatter' or cultural artifacts; it is successfully measuring a consistent, underlying reality. "For those with true near-death experiences (NDEs), Greyson's (1983, 1990) NDE Scale satisfactorily fits the Rasch rating scale model, thus yielding a unidimensional measure with interval-level scaling properties[20]."

Dr. Bruce Greyson has also investigated the NDE experience in relation to mental illness to determine if any correlation exists. Generally, people who report NDEs exhibit the same prevalence of mental illness as the general population. Over the course of a year, Dr. Greyson applied his NDE scale to every patient admitted to his psychiatric clinic who reported being near death. He found that the percentage of validated NDEs in this psychiatric group (20%) matched the percentage found in the general population. Furthermore, Greyson looked at mental illnesses that could theoretically be associated with NDEs—such as PTSD and Dissociative Disorder—and found the rates were equal to the general population. These psychiatric diagnoses did not apply to people with near-death experiences more than anyone else[21].

In addition to the three primary NDE research hubs cited in this section—Dr. Moody, MD, Ph.D., The University of Virginia, IANDS and NDERF—there are dozens of additional researchers, authors, and investigators who have pioneered NDE research and instilled much-needed scientific rigor within the discipline:

P.M.H. Atwater, prolific NDE author of over 15 books

Dr. Peter Fenwick, MD, King's College

Dr. Bruce Greyson, MD, UVA Division of Perceptual Studies

Dr. Jeffrey Long, MD, NDERF

Jody Long, Esq., NDERF

Robert & Suzanne (Carol) Mays, IANDS

Dr. Gary E. Schwartz, Ph.D., Eternea

Dr. Ian Stevenson, MD, UVA Dept. of Psychology; and many others, including three highly influential members of the NDE community and original founders of IANDS:

Dr. Kenneth Ring, Ph.D. Lecturer, and Author of, *Life at Death* (1980) and the ground-breaking study on blind NDErs called, *Mindsight* (1999)

John Audette

Dr. Michael Sabon, M.D.

---

[1] (IANDS.org, What is a Near-Death Experience? December 8, 2022)
[2] (IANDS.ORG, Page: Key Facts About Near-Death Experiences - Veridical NDES, © 1996-2020 IANDS)

[3] (Next Level Soul with Alex Ferrari: A mind, Body & Soul Podcast, Jan 23, 2023 NDE Stories: Nobody Wants to Talk About My Experience on the Other Side with William J. Peters.)65

[4] (O. Blanke, N. Faivre, S. Dieguez, Leaving Body and Life Behind: out of Body and Near-Death Experience, in S. Laureys, O. Gosseries, G. Tononi, eds. The Neurology of Consciousness: Cognitive Neuroscience and Neuropathology. 2nd Ed. Elsevier; 2016: 323-347.) & (O. Blanke, out-of-body experiences and their Neural Basis. BMJ. 2004; 329(7480):1414-1415)

[5] (R. Moody, Life After Life: Investigation of a Phenomenon - Survival of Bodily Death, HarperOne, 1975 & R. Moody, The Light Beyond: New Explorations by the Author of Life After Life, Bantam Books, 1988)67.

[6] Footnote: Each subheading below is dealt with in depth at the eACISTE website located at: https(colon) / /aciste(dot) org/ about-stes/ common-challenges-following-an-ste/

[7] (Noyes, Fenwick, et all 2009)128

[8] (R. Noyes, P. Fenwick P, et al, After effects of pleasurable Western adult near-death experiences, The Handbook of Near-Death Experiences: Thirty Years of Investigation, pp 41–62. Edited by Holden JS, Greyson B, James D Santa Barbara, CA, Praeger/ABC-CLIO, 2009.)129

[9] (R. Moody, Life After Life: Investigation of a Phenomenon - Survival of Bodily Death, HarperOne, 1975 & R. Moody, The Light Beyond: New Explorations by the Author of Life After Life, Bantam Books, 1988)69

[10] University of Virginia, Division of Perceptual Studies, Near-Death Experiences (NDEs) https://med.virginia.edu/perceptual-studies/our-research/near-death-experiences-ndes/)

[11] (J. Long, P. Perry, Evidence of the Afterlife: The Science of Near-Death Experiences, HarperOne, 2010)71

[12] (Dr. Jeffrey Long, MD with Paul Perry, Evidence of the Afterlife: The Science of Near-Death
Experiences, HarperOne, 2010)72

[13] (IANDS, Characteristics of a Near Death Expereince https://www.iands.org/ndes/about-ndes/characteristics.html)74

[14] (Perera, M. et.al.(2005). Prevalence of Near-Death Experiences in Australia. Journal of Near-Death Studies, 24, 109)75, Australia (Knoblauch, H. et.al. (2001). Different Kinds of Near-Death Experience: A Report on a Survey of Near-Death Experiences in Germany. Journal of Near-Death Studies, 20, 15-29.)

[15] (C. Martial, J. Simon, N. Puttaert, O. Gosseries, V. Charland-Verville, A.S. Nyssen, B. Greyson, S. Laureys, H. Cassol, The Near Death Experience Content (NDE-C) Scale: Development and Psychometric Validation, Conscious Cogn. 2020 Nov;86103049. PMID 33227590)77.

[16] (IANDS, Important Research Articles, Greyson NDE Scale, Quantifying The Phenomenon, Sat, June 11, 2022, https: // www. Iands .org/ research/ nde-research/ important-research-articles/ 698-greyson-nde-scale. html)78.

[17] (Dr. Bruce Greyson, Near-Death Experience Pioneer #494, Podcast, Skeptiko: Science At The Tipping Point, March 31, 2021).

[18] (B. Greyson, The Near-Death Experience Scale: Construction, reliability, and validity. Journal of Nervous and Mental Disease, (1983), 171(6), 369–375. https:// doi(dot)org/ 1097/ 00005053-198306000-00007)80

[19] (G. Rasch, Probabilistic models for some intelligence and attainment tests (Expanded ed.). 1980Chicago: University of Chicago Press.)81

[20] (R. Lange, B. Greyson, J. Houran, A Rasch Scaling Validation Of A 'Core' Near-Death Experience. Br J Psychol. 2004 May;95(Pt 2):161-77. PMID: 15142300.)82 (Dr. Bruce Greyson, Near-Death Experience Pioneer #494, Podcast, Skeptiko: Science At The Tipping Point, March 31, 2021)

---

[21] (Dr. Bruce Greyson, Near-Death Experience Pioneer #494, Podcast, Skeptiko: Science At The Tipping Point, March 31, 2021)

7 Transcribed Near Death Experiences, Atheists and Skeptics

Entering the Dimension of the NDE - Introduction

The verified, transcribed NDE accounts in this section are drawn from a diverse range of individuals. These accounts document spiritually transformative events, mystical experiences, encounters with an all-powerful instantiating intelligence, and interactions with deceased loved ones, spirit entities, miracles and other inexplicable happenings (as viewed from the lens of a physicalist paradigm).

We will then highlight the NDEs of two prominent atheists. The second account is meticulously documented and involves the renowned philosopher of logical positivism and atheist A.J. Ayer. His experience—and the subsequent shift in his perspective as reported by close associates—is critical to our discussion.

At the conclusion of this section, we will provide a technical analysis of a direct effort on behalf of the academic establishment to debunk NDEs, framing the process through a direct correspondence between a team of atheistic academics consisting of UC Philosophy Professor John Martin Fischer and his co-author Benjamin Mitchell-Yellin on one side, and the NDE research team of Robert and Suzanne Mays.

NDE Accounts

> Fair Use Notice (17 U.S.C. § 107) - Notice of Transformative Purpose: The NDE accounts that follow are utilized strictly as data to support scholarly commentary and a transformative argument for a new ontology and paradigm that challenges the materialist framework. This commentary is non-substitutive of the original media referenced. Readers are encouraged to reference the sources, as the full context of the accounts will not be within our analysis.

Martin was 14 years old and experimenting with dangerous drugs. He overdosed. Martin describes his heart beating erratically before it stopped. He lost consciousness and had this NDE:

> *I had been floating or flying, but now I was standing on the floor of a hall looking towards the far end of it. God was there. Or more accurately, a great Fire of Light that I understood represented or stood for God. But it was also really God. I can't explain this paradox; It was just something I had to accept. But there wasn't a problem because there was no intellectualizing, just the raw naked experience of the presence of God*[1]*.*

I can't describe how I felt. I was awed, struck dumb, amazed, overwhelmed. But these words don't describe what a profound experience this was. God was totally beyond anything I had ever imagined, –beyond language and thought, utterly beyond, but at the same time right there in front of me, mysteriously near and accessible. This can seem so lame now; total love, total goodness, total knowing.

And God spoke to me! Not with any words but directly to my mind. And what he said was so profound; I can't repeat it in words, but the gist of it was: "You don't really know yourself, do you?" God's question seems simple but also had other meanings, like: "You don't know anything, do you?" And "Nobody knows anything, do they?" Meaning that we don't know how to love, how to exist, with the full and complete awareness and acceptance of the fact that God IS. What could I say? I had to admit that no, I didn't know anything. After all, I was standing there right in front of God, who knows everything! So I had to go back. God knew it, and I knew it. End of discussion. I was disappointed. I thought I would go into the Light, but dutifully and sadly turned around to go back[2].

An NDEr who died from a cardiac arrest at age 4 with memories verified from one of the earliest (in terms of age of subject) life reviews in the NDERF (NDE research foundation) archives:

I felt like I was part of everything and everyone. I just floated up and could see other rooms. I later detailed conversations that there is no way I could have known. I lifted out of the hospital and continued ascending. I began browsing through time. I saw events from my short life. I later detailed things that occurred before I could even talk. These were events I should not be able to remember but I do and in great detail. I kept ascending, I felt at peace. There were no questions or unknowns. Time wrapped in on itself. There was no past, present, or future as we see it here. Everything happened now and all at one time. I felt no fears or worries. I began drifting towards a beautiful light and I wanted to touch it[3].

An NDE by a member of the Hindu faith who finds that Love is the "cosmic composition":

I realized that the real goal of the mind-created life game was to remove the veil of delusion and ignorance and reach the ultimate Truth of one's reality in life itself. This Truth, this self I was bathing in and which had encompassed my being, was nothing but love. The truth was love; pure unconditional, universal love, that is almost non-achievable in human terms. It was the fabric of this consciousness of which I was a part. Yes, because love is God and God is love. It was very clear that the cosmic composition was made up of nothing but love! Love is the God-glue that is holding all together. I realized I was not just inside the love but rather, one with it, this universal body of love, of God[4].

Consider the following NDE excerpt, which details a subject's ability to observe specific events occurring far from their physical body—events later confirmed by the very individuals under observation. Accounts of this class are categorized as 'Veridical NDEs.' These are not merely subjective memories; they are instances where experiencers acquire verifiable information that they could not have obtained by any normal, biological means. They represent a forensic breach of the materialist perimeter.

I see my body in the bed and it is very still. My friends Megan and David are now in another room, a waiting area. I am omnipresent both with myself in my hospital room and in the waiting room. David and Megan are discussing that if I survive this, I need to have a person that can make my medical decisions if I cannot. This conversation and the location of it have been confirmed. My attention is shifted back in the room. The conversation in the waiting room between my friend and my son and what is going on in my room and the ICU are happening all at once, but I am able to shift my attention between the two[5].

Sir Alexander Ogston, the eminent Scottish military surgeon and discoverer of the *Staphylococcus* bacteria, provides a definitive veridical account that defies materialist reduction. As a Fellow of the German Surgical Society, Ogston's career was defined by the highest order of clinical rigor. In 1900, while battling typhoid fever during the Boer War, Ogston experienced repeated out-of-body occurrences where his consciousness moved independently through the facility. He observed specific events in distant wards and on the far side of the hospital grounds—data points he later confirmed with accuracy upon his recovery.

"Mind and body seemed to be dual, and to some extent separate. I was conscious of the body as an inert tumbled mass near a door; it belonged to me, but it was not I. I was conscious that my mental self used regularly to leave the body... until something produced a consciousness that the chilly mass, which I then recalled was my body, was being stirred as it lay by the door. I was then drawn rapidly back into it, joined it with disgust, and it became I, and was fed, spoken to, and cared for." And then, "In my wanderings there was a strange consciousness that I could see through the walls of the building, though I was aware that they [the walls] were there, and that everything was transparent to my senses I saw plainly, for instance, a poor RAMC [Royal Army Medical Corps] surgeon, of whose existence I had not known, and who

was in quite another part of the hospital, grow very ill and scream and die; I saw them cover his corpse and carry him softly out on shoeless feet, quietly and surreptitiously, lest we should know that he died... Afterwards, when I told these happenings to sisters (nuns), they informed me that all this had happened just as I had fancied[6]."

from Dr. Moody's book *The Light Beyond: New Explorations*, we are provided with another rare occurrence where a veridical NDE was shared between two deceased individuals:

I was terribly ill and near death with heart problems at the same time that my sister was near death in another part of the same hospital with a diabetic coma. I left my body and went into the corner of the room where I watched them work on me down below. Suddenly, I found myself in conversation with my sister, who was up there with me. I was very attached to her, and we were having a great conversation about what was going on down there when she began to move away from me. I tried to go with her but she kept telling me to stay where I was. "It's not your time," she said. "You can't go with me because it's not your time." Then she just began to recede off into the distance through a tunnel while I was left here alone. When I awoke, I told the doctors that my sister had died. He denied it, but at my insistence, he had a nurse check on it. She had in fact died, just as I knew she did[7].

One of the most robust veridical NDE cases on record was documented by Kimberly Clark, MSW, a clinical assistant professor at the University of Washington. Clark's report is unique in its evidentiary weight, as it involves a subject who, while clinically dead, acquired specific, external information regarding events occurring far beyond the physical range of their unconscious body—data that was later physically verified with accuracy.

(Maria)... was a migrant worker who, while visiting friends in Seattle, had a severe heart attack. She was rushed to Harborview Hospital and placed in the coronary care unit. A few days later she had a cardiac arrest and an unusual out-of-body experience. At one point in this experience, she found herself outside the hospital and spotted a single tennis shoe sitting on a ledge of the north side of the third floor of the building. Maria not only was able to indicate the whereabouts of this oddly situated object but was able to provide precise details concerning its appearance, such as that its little toe was worn and one of its laces was stuck underneath its heel. Upon hearing Maria's story, Clark... went to the location described to see whether any such shoe could be found. Indeed it was, just where and precisely as Maria had described it, except that from the window through which Clark was able to see it, the details of its appearance that Maria had specified could not be discerned. Clark

> *concluded, "The only way she could have had such a perspective was if she had been floating right outside and at very close range to the tennis shoe*[8]*."*

Many NDE accounts include the extraction of precise, veridical intelligence that serves a critical functional purpose upon the subject's return: it acts as an ontological anchor, tethering the returning consciousness back to the material reality it must once again inhabit.

> *"During my NDE following a cardiac arrest, I saw both my dead grandmother and a man who looked at me lovingly but whom I didn't know. Over ten years later my mother confided on her deathbed that I'd been born from an extramarital affair; my biological father was a Jewish man who'd been deported and killed in World War II. My mother showed me a photograph. The unfamiliar man I'd seen more than ten years earlier during my NDE turned out to be my biological father*[9]*."*

A significant veridical case is documented in *Near-Death Experiences: The Rest of the Story* by P.M.H. Atwater. This account details a medical doctor's experience within a hospital setting—a confirmed out-of-body occurrence where the subject acquired specialized, veridical knowledge that was physically inaccessible to his unconscious form.

> *While clinically dead, George Rodonaia, a physician himself, was drawn to the newborn section of the hospital adjacent to the one where his body was. A friend's wife had just given birth to a daughter. The baby cried incessantly. As if possessed of X-ray vision, he scanned the baby's body and noticed that her hip had been broken after birth. Upon reviving, and after his tongue returned to normal size so he could speak... George told the doctors about the baby's fractured hip, and added, "The hip is broken because the nurse dropped her." Sure enough, X-rays showed the break, exactly as he had described. When the nurse was confronted, she admitted to dropping the child and was fired*[10]*.*

This next NDE is world-famous, involving Anita Moorjani. It involves an NDE experience where special abilities or information (veridical) were conferred by the NDE event to the deceased (*"Inspired"* YouTube channel, *Famous Case of Anita Moorjani, author of, 'Dying to Live':*

> *February the 2nd 2006 I went into a coma and the doctors basically said that these were my final hours. The tumors I had some of them were the size of lemons. I had cancer for four years and it had metastasized, so these tumors were all over my lymphatic system. From the base of my skull all around my neck under my arms, my breasts, all the way down to my abdomen... On February the 2nd (2006), I fell into a coma and the doctors said that these were my final hours.... But as I was in the coma, unbeknownst to all of them, my eyes were closed and I was completely in this coma, but I was aware of*

everything that was going on around me. I could hear, and see, and feel everything that was happening... Not only could I hear what they were saying, it was like I could feel what they were feeling. I was completely empathic to what they were feeling. I could see and hear things that were beyond the room that my physical body was in. I realized that I could even see my physical body. It was like I had become separate from my body, and it felt incredible. It felt just absolutely amazing... As I went deeper into the experience it reached a point where the choice of whether to come back or not. I encountered my deceased father. My father had died 10 years prior and I felt that he was there to greet me. He wanted me to know that it wasn't my time but even though it wasn't my time, I still felt I had the choice... I actually understood how every choice and every decision that I had made in my life, in every moment in time, had led me to that point and I could see it and in that awakening... It's like everything was happening simultaneously... I started to open my eyes and my family was thrilled and then they called the doctor and I greeted the doctor by name. I said, "Good afternoon Dr. Chan." And he was surprised I knew his name because he only started tending to me after I went into the coma. So he said, "How did you know my name?" And I was surprised he asked that question, because to me it just kind of seemed normal. I just said, "Aren't you the doctor that came in this morning?" And I told him the procedure of exactly what he'd done... And he said, "But you were in a coma when I did that. How do you even know I did that?" and that's when it started to transpire with everyone that something had happened[11]*...*

In this testimony, a young boy dies from a Judo throw accident and receives a "download" of information and is then introduced to God. He is later told by his priest that his NDE was probably the work of the Devil: A&E, YouTube, *I Survived Beyond and Back*, "Ray":

In front of me were these giant titanic golden columns rising up like Jack and the Beanstalk... it felt like it was a gate or an entrance to something. I was greeted by a being or a spirit. I don't know what this thing was. He said, "I want to show you some things." and there was what I can only call a download. It was like billions of questions were answered in just a moment. All of creation opened up and shimmered and I understood heavy stuff... after the big download this being said to me, "I want to show you who you really are." and he introduced me to God face-to-face. What God looked like was when you look at a photograph of a galaxy. Each one of those little pinpoints of light was God and each one of those little points of light was worshiping saying, "We love you. I love you." It was beautiful, blissful, unconditional, pure love... I was told by God that it was not my time and I had to go back. And I just, "not, no no no..."... I was sent back to my body... (the parish priest) told me that since I didn't meet Jesus, I couldn't say Jesus was there, it has probably been a trick of Satan. I was crushed. I remember thinking, "if the Devil is that good, I don't have a chance[12]."

Jayne Smith's NDE resulted in two massive Information Downloads 'imported' into her consciousness. *(Jayne Smith's Profound NDE,* YouTube "NDE Video"*):*

Suddenly what I think of as a block of knowledge came into my being. And I began to know things I didn't know before. It was being fed into me as if I were a receiving station. I had no thoughts of my own at all while this was going on. And what was being transmitted to me was that I was an immortal being. That I had always existed. I always would exist. I was eternal. I was indestructible. Um- There's nothing that never really could happen to me to wipe me off the face of the Earth or any other place. and I knew that that was true. I didn't doubt anything that the light was giving me, because somewhere inside of me I knew the light was, -I didn't think, "Oh, the light was gone." I didn't think that, -I just felt that the light was from God. I have feelings, not thoughts. And I knew anything the light was giving me was the absolute truth, and I just kind of floated with this knowledge of my indestructibility.. And then suddenly a second block of knowledge descended upon me and what I knew within that block of knowledge was that the world is perfect. And when I say world, I mean, the totality. The allness, that everything... and I lived with, -with this feeling for a while knowing that all the things that we think are cruel, and horrible and and all the wars that are going on, knowing that somehow this is part of the perfect –of the perfect plan, that it's part of what God is working out and that I didn't personally have to worry about it[13]...

The next case documented by *Shaman Oaks* provides a *life review* where the deceased was allowed to view his *life plan*—the intended trajectory for his existence. He was able to see that he had not achieved what he was supposed to and had to return to make things right. *(Man Shocked After Taking His Own Life,* YouTube, "Shaman Oaks"*):*

There was this blueprint of my life and there was challenges put in place for me to overcome. And it was obvious that I didn't pass any. It was all reset. Nothing I did actually accomplished anything and all of these challenges were still there. They were, you know, unfinished. For me it was just like it's not a disappointment but for me it was like, "Oh man I couldn't even get one thing right on this list."... my younger brother died a year before he came up to me and his bubbly self was like, "Wow that was quite the experience, quite the journey." in a very jokingly manner. And then he's like, "I want you to come over here. I want you to follow me." Then he brought me into this lightroom and this is where there were three beings. There were two women and there was one male guide. So the two women, they had like Victorian clothing. And then the guy had tribal clothing. So I do believe that these people were part of the soul family or guides. They did have a glow to them kind of light energy, but everything out there was light energy. It's just degrees of how bright and how not bright... These beings told me that you have a job to do. You're not done. You have to complete things. You have a better life that could be waiting for you. You don't have to go back. We're not forcing you to go back. But we highly, highly recommend you go back. From that level, -the soul level- there was no choice it was obvious that I'm going back. Because I just saw how badly I screwed everything up and I'm gonna, I want to go fix it... After that there was this amazing kind of holographic image that completely surrounded me in this room and it was my future, and it was like basically the future of Earth. It was very different. It was changed so much that it didn't make sense. Say I was about 50 years old and you can tell that things were not what they are now. Something big changes the Earth... for me what really helped me was finding something, finding a purpose, finding something more to live for. Finding that meaning. You have to give meaning to life. It's not up to life to give your life meaning, it's up to you to give life meaning[14].

Following a horrific car accident, Michaela, achieved found herself in a state of extra-physical observation while unconscious in the Emergency Room. While in this state she was able to report back high fidelity data from the hospital cafeteria: *(A&E, Youtube, I survived Beyond And Back, "Michaela")*:

Attending ER Trauma Surgeon:

There was a call through the 911 center and directly to my pager... We found Machaela, who was in the backseat of a vehicle and had suffered severe trauma... she was having very slow respirations, less than eight per minute... Michaela's injuries, that you could see visibly, which were very gruesome... In fact I wondered if she was near death at this time.

Michaela:

... The next thing I know I was up in the corner of a hospital room looking at my body lay there. When you're having an out-of-body experience it's like watching your life unfold in front of you like a movie. I remember seeing my mom and dad in the cafeteria. It was like a bench style seating and my grandma and my gram were sitting across from my parents. And my dad is a smoker. Anyway, he said he was gonna go have a cigarette, because he just wanted to get some breathing room and get out of there. And it was funny because my grandmother, my mom's mom who would never, has never, and

would never have a cigarette in her life, was like, "Oh, I need one too. I'm Gonna have one too." and then my other grandma was like, "Yeah me too." And neither my gram or my grandmother would have smoked a cigarette.

Michaela's Mother:

Mikayla said to me, "Mom, you and dad and grandma and gram were in the cafeteria. You were talking about me, and dad got up to go have a cigarette, and grandma says, "I'm coming to have a cigarette with you." There was no way that she could have known that that even happened unless she was there.

Michaela:

To have been through something like that it really makes me think differently about time here on Earth... When I think about death I think more about, it's a celebration. It's almost like graduating to another level I guess is how I think of it. It changed everything about my life[15]*...*

Following a lethal exposure event during an ice climbing expedition, Peter Panagore transitioned from the material plane into the realm of the Divine where he reported a direct encounter with 'creative consciousness' and observed the *Veil* that separates the physical plane from the afterlife.*(Peter Panagore, YouTube, Guy Lawrence 8/16/2020):*

...but I haven't gone through the door yet. "Do I have to go?" And God says, "It's time for you to come home. Come home now. Go through the door." And I say, "But I've got to, you know, can we talk about this?" I, my parents are suffering, they lost my sister and I can't leave them to suffer. And in an instant, I'm swept across heaven to the edge of our universe, where I can see with God all around me and inside me all of the Earth at once, like a hologram. And I can see every single 7 billion persons in it doing their 7 billion things at the moment that I'm looking and I could see every single one of them, and I can see every single one of them is covered by a veil. I can see every single one of them is beloved. I can hear the voice say to me "In the way that I love you now, I have always loved you, and all is well and will be well because of my love." which is 10 trillion times more than I can possibly explain. It's so much, so it healed all of me, I had no suffering. I was, I was, I couldn't understand suffering, except for when God showed me my parents' faces and showed me their suffering. That they would endure in life until they die[16]*d...*

Atheist NDEs

Self-avowed atheists represent a critical "control group" in NDE research. Following an NDE, nearly all renounce their former materialism to profess three core ontological shifts: (1) the existence of a human soul unbound to the body; (2) the reality of an all-powerful, all-knowing spiritual consciousness; and (3) the understanding that life extends beyond the confines of terrestrial time and space.

If NDEs were merely self-fulfilling neuro-chemical delusions, the atheist data set should yield a high percentage of "non-events" (experiences failing to meet clinical criteria) or provide a forensic debunking of the phenomenon. Instead, we find the opposite. To date, there is no known record of an individual who has met the objective, externally validated criteria of the Greyson Scale who has subsequently disavowed the reality of the event.

This is a vital distinction: the Greyson Scale does not merely filter for belief; it authenticates the presence of a specific, non-local phenomenon using standardized, objective measures. Despite the immense pressure of the materialist paradigm, not a single authenticated experiencer has come forward to attribute their NDE to drug intoxication, delirium, or hypoxia. The readout remains consistent: the atheist experience is identical to the believer's, resulting in a total collapse of their former framework.

No atheist has returned to explain the "mechanics of the delusion"—because the experience is not a delusion.

The Next Two NDEs Are Atheist Experiences

The case of Howard Storm documents the total collapse of a lifelong materialist worldview during a state of clinical death. Upon death Storm experienced a descent into a hellish realm of profound isolation and predatory entities. Somehow remembering a piece of a prayer, his appeal was answered by a higher power, and a 'rescue':. *(Howard Storm's Testimony, YouTube, JuJu Fischer of Men.):*

> *My faith was materialism. What I believed in, all of my phd friends, and all of my friends believed that if you can't measure it, see it, weigh it, count it, it simply doesn't exist. So when I was dying people asked me if I prayed, and it's like, I'm trying to explain to you, the last thing in the world I would have done is pray... I asked for morphine. I asked for a doctor, because of the pain in my duodenum, which was leaking hydrochloric acid and other delightful digestive juices. I was dissolving myself, I was digesting myself on the inside... I'd seen two doctors in emergency and they were very nice but they didn't do anything; they just sent me over to surgery. They told me I had an hour to live... I went unconscious, awoke and I was standing there. I felt absolutely physically more real, more alive, and completely healed than I'd ever known... So like in our vision we see 180 degrees if you have two eyes. Well I was seeing way more than 180 degrees. I'm going, "Oh that's so*

weird."... And then I noticed in the bed, mostly covered by a sheet... was a person. And I looked at the person who was facing my wife away from me, and to my horror it bore a remarkable resemblance to me. Now I knew rationally that that wasn't me, because I was standing there... I mean that's crazy stuff. That's schizophrenia to say that you were standing over yourself looking at yourself... So then I'm thinking, "How come it looks so much like me? Because could it be a coincidence? No, that's ridiculous." So I'm trying to think of a scenario. So what I came up with, which I realized was absolutely ludicrous, was that the French hospital personnel had made a wax replica of me. I realized, they didn't have the time to do that, and that takes a lot of skill, because it was a very good replica. And what would be their motive? You know? I mean, why would they go to all that trouble and expense[17]..."

Howard then began to hear voices in the hallway, calling him by name. Expecting the hospital staff for his surgery, he entered a gray, desolated realm—a state of separation from the divine defined by the malice of those inhabiting it. He was subjected to a brutal, predatory torment that stripped him of his humanity until he was left as nothing more than "roadkill." In the final extremity of his despair, a command to pray surfaced, and his desperate invocation of the divine shattered the darkness. He was met by a light of impossible intensity, from which hands reached out to heal him instantly, replacing his agony with a love so profound it defied language. As he was held, he realized he was being brought toward the very source he had spent his life denying. Upon arriving at the threshold of that light, his telepathic dialogue with the Creator dismantled his shame; he learned that his most private thoughts were known, and that this transparency was not a condemnation, but an invitation into his rightful place.

During a holographic life review, he did not just see his past, but felt the impact of his actions on others, learning that the "curriculum" of existence is the command to love. Even when he argued to stay in that state of grace, the Creator sent him back to the physical world with a final, diagnostic truth: that the world is a mirror, and one will find only what they carry within their own heart[18].

While Howard Storm provides the raw, experiential collapse of the materialist framework, (Howards full experience is far longer and involves many different aspects, only a small portion of which can be captured for our purposes) our next account comes from the architect of 20th-century *Logical Positivism*. Alfred Jules (A.J.) Ayer, perhaps the most influential atheist philosopher of his era, produced two articles detailing an experience that fundamentally challenged the 'verifiability principle' he spent his life defending. This review includes his primary accounts, notes from his treating physician, and a forensic ontological analysis.

***Atheist philosopher Sir A.J. Ayer, (1910-1989)**

A.J. Ayer was a Fellow and Dean of Wadham College at Oxford, Fellow of the British Academy, gained acclaim in his mid-20's with publication of his acclaimed book, *Language, Truth and Logic* (1936), which expounded on the school of thought known as, "logical positivism" a quote is included below to give us a flavor of his philosophy and thoughts on God:

> *"To test whether a sentence expresses a genuine empirical hypothesis. I adopt what may be called a modified verification principle. For I require of an empirical hypothesis, not indeed that it should be conclusively verifiable, but that some possible sense-experience should be relevant to the determination of its truth or falsehood. If a putative proposition fails to satisfy this principle, and is not a tautology, then I hold that it is metaphysical, and that, being metaphysical, it is neither true nor false but literally senseless. It will be found that much of what ordinarily passes for philosophy is metaphysical according to this criterion, and, in particular, that it can not be significantly asserted... that men have immortal souls, or that there is a transcendent God[19]."*

One year prior to his death, the preeminent philosopher A.J. Ayer underwent a critical NDE, documenting his findings in a column titled 'What I Saw When I Was Dead' (published in *The Sunday Telegraph* and *The Spectator*). A central component of his account is the testimony of a young, Oxford-educated physician—an expert witness whose clinical expertise provides substantial authority to the event. The following excerpt begins after Ayer, having initially discharged himself following a pneumonia-related hospitalization, was forced to seek re-hospitalization:

> *"Fortunately for me, the young doctor who was primarily responsible for me had been an undergraduate at New College, Oxford, while I was a Fellow. This made him extremely anxious to see that I recovered; almost too much so, in fact, for he was so much in awe of me that he forbade me to be disturbed at night, even when the experienced sister and nurse believed it to be necessary[20]."*

Ayer was closely monitored, but by his own admission, his attempt to consume 'smuggled' food led to a critical failure; he choked on a piece of fish he had 'carelessly tossed' into his mouth, triggering a total cardiac and respiratory arrest. Clinical observation by the ward sister and the attending physician confirmed that Ayer was deceased for a duration of four minutes. Upon his resuscitation, Ayer provided the following data regarding his experience while offline:

> *"I was confronted by a red light, exceedingly bright, and also very painful even when I turned away from it. I was aware that this light was responsible for the government of the universe. Among its ministers were two creatures who had been put in charge of space. These ministers periodically inspected space and had recently carried out such an inspection. They had, however, failed to do their work properly, with the result that space, like a badly fitting jigsaw puzzle, was slightly out of joint. A further consequence was that the laws of nature had ceased to function as they should. I felt that it was up to me to put things right. I also had the motive of finding a way to extinguish the painful light. I assumed that it was signaling that space was awry and that it would switch itself off when order was restored. Unfortunately, I had no idea where the guardians of space had gone and feared that even if I found them I should not be able to communicate with them. It then occurred to me that whereas, until the present century, physicists accepted the Newtonian severance of space and time, it had become customary, since the vindication of Einstein's general theory of relativity, to treat space-time as a single whole. Accordingly, I thought that I could cure space by operating upon time. I was vaguely aware that the ministers who had been given charge of time were in my neighborhood and I proceeded to hail them. I was again frustrated. Either they did not hear me, or they chose to ignore me, or they did not understand me[21]."*

It is a profound irony that Ayer—the man who spent his life arguing that the nature of reality could only be accessed through empirical verification—was given a diagnostic task he was fundamentally incapable of performing.

When he attempted to 'fix' the jigsaw puzzle of space-time, he was not merely hallucinating; he was undergoing a clinical demonstration of the limitations of the materialist framework. He was attempting to apply a Newtonian wrench to a quantum, conscious architecture. The reality he encountered was not built on cause-and-effect mechanics, but on the *Evaluative Field*—an entity that does not respond to logic, but to consciousness itself. In that 'broken space,' Ayer's tools were not just insufficient; they were nonsensical.

Ayer says that this experience may have been "delusive" but he believes it to be real because he was supplied with information from his "French friend" that their mother had suffered a cardiac arrest and that she recounted that she "must stay close to the light." Ayer concludes that he believes the experience to be truthful, saying there is, "rather strong evidence that death does not put an end to consciousness" A remarkable statement for an avowed materialists and atheist. Ayer concludes:

> *"So there it is. My recent experiences have slightly weakened my conviction that my genuine death, which is due fairly soon, will be the end of me, although I continue to hope that it will be. They have not weakened my conviction that there is no God. I trust that my remaining an atheist will allay the anxieties of my fellow supporters of the British Humanist Association, the Rationalist Press Association and the South Place Ethical Society[22]."*

A.J. Ayer's published account in *The Telegraph* is not the end of the matter. His experience, as recounted in the newspaper article, is unusual for NDEs. He reports spiritual beings having "failed to do their work properly" and that he tried to "fix" space because they had rendered it like "a badly fitting jigsaw puzzle". If this was all that Ayer had reported, it would not be enough to be verified as an NDE on the Greyson Scale. But this was not all of the information from his experience. It was rumored that there was more information Ayer did not disclose concerning his experience while "dead". Ayer, meanwhile, perhaps due to negative feedback to his article, felt moved to either correct, clarify or in some way retract his account of August 28, 1988.

Mr. Ayer's second column *"Postscript To A Postmortem"* was published in *The Spectator*, October 15, 1988:

> *"My purpose in writing a postscript to the article about my 'death', which I contributed to the 28 August issue of the Sunday Telegraph, is not primarily to retract anything that I wrote or to express my regret that my Shakespearian title for the article, 'That undiscovered country', was not retained, but to correct a misunderstanding to which the article appears to have given rise."*

> *"My motive for writing the original article was twofold. I thought that my experiences had been sufficiently remarkable to be worth recording, and I wished to rebut the incoherent statement, which had been attributed to me, that I had discovered nothing 'on the other side'. Evidently, my having discovered something on the other side was a precondition of my having completed the journey. It follows that if I had discovered nothing, I had not been there; I had no right to imply that there was a 'there' to go to. Conversely, if there was evidence that I had had some strange experiences, nothing followed about there being 'another side'. In particular, it did not follow either that I had visited such a place, or that I had not[23]."*

Ayers then all but retracts his original declaration, *"I have since been told, rightly or wrongly, that it* (my brain) *would not have functioned on its own for any longer period* (4 minutes) *without being damaged. I thought it so obvious that the persistence of my brain was the most probable explanation that I did not bother to stress it. I stress it now. No other hypothesis comes anywhere near to superseding it..."*

This 'clarification' remains a curious retraction—if it can be categorized as such—given that Ayer concludes his secondary statement with the following observations:

> *But our concepts are not sacrosanct. They can be modified if they cease to be well adapted to our experience. In the present instance, the change which would supply us with a motive for altering our concept of a person in such a way as to admit the possibility of reincarnation would not be very great. All that would be required is that there be good evidence that many persons are able to furnish information about previous lives of such a character and such an abundance that it would seem they could not possess the information unless they themselves had lived the lives in question*[24].

The thought is curious because in introducing reincarnation, he directly contradicts the foundational logic of his book, *Language, Truth and Logic*, Pgs. 2 and 72-76[25] where he rejects the spirit realm, the afterlife and God, so one could conclude that reincarnation, in any form, is an equally incompatible concept, and impossible with respect to his views. In this contradiction we could draw the conclusion that since this statement was made after his NDE, Ayer had undergone somewhat of a conversion, if not in spirit, then at least in drawing the loop of inclusion in his philosophy.

This artifact is not the end of the curious case of A.J. Ayer. There is more.

Neither in his original article, nor his subsequent retraction did Ayer fully report his near-death experience as he had recounted it to others. It is Ayers' raw, unredacted experience that lies at the heart of the matter, and Ayers' actual experience altered his view of life, death, and God. These are not trivial matters to a philosopher—they represent a total ontological collapse.

Dr. Jeremy George provided corroborative information of Ayer's NDE. Dr. George was 'the young doctor' referred to in Ayer's original column, 'What I Saw When I Was Dead[26].'...The young doctor who was primarily responsible for me had been an undergraduate at New College, Oxford, while I was a Fellow...'. According to Dr. George, who was attending Ayer on the night he died for 4 minutes, he questioned Mr. Ayer immediately after his experience and he was given a far more forensic account than was divulged in Ayer's article from August 28th,

1988—one that revealed a profound departure from his published skepticism:

> *"... By a strange coincidence, Dr. George had been a student at New College, Oxford in the 1970s when Ayer was at the college as Wykeham Professor of Logic... Dr. George gave Ayer emergency oxygen and put him immediately in the intensive care unit, where his condition improved. 'He would not have survived the day.... The hospital notes state: 'cardiac arrest with bradycardia, and asystole, but was resuscitated'... Very discreetly, I asked him, as a philosopher, what was it like to have had a near-death experience? He suddenly looked rather sheepish. Then he said, 'I saw a Divine Being. I'm afraid I'm going to have to revise all my various books and opinions[27].'"*

Make of that statement as you will. A fuller exposition of the article by William Cash fills in more blanks on Ayers' experience as related by Dr. Jeremy George. We will resume that thread shortly.

Ayers' NDE is significant due to *who* he is, the fierceness of his previous atheistic beliefs, and his change in tenor after his experience. As we have seen in previous examples of atheists that have had NDEs, the experience itself appears to be the one solution to the blockages of mind caused by the conclusion that there is no God, nor spiritual plane. There is certainly a spiritual plane, but we should not diminish the role of *mind* in constructing impasses within the mind based on belief, and perhaps an unconscious allegiance to what may appear to be the final word on such matters, but in truth is *confusion*, or a *delusion* or simply *ignorance*, as indicated by *The August Wheel*.

To give some background, Dr. Jeremy George was a senior consultant in the Department of Thoracic Medicine at London University's Middlesex Hospital. When he was interviewed by Mr. William Cash, he had the official hospital notes of A.J. Ayer. The interviewer, writing for the National Post on April 28, 2009, adds that at the time of the incident on June, 1988, when the eminent 77-year-old British philosopher, arguably the most influential 20th century rationalist "after Bertrand Russell" "died", Mr. Cash recalls Ayers' experience, "bears a very curious resemblance" to reports from a study at Southampton General Hospital, of near death experiences of 63 survivors of cardiac arrest, published in the monthly international/interdisciplinary medical journal of the European Resuscitation Council. This evidence is, according to Dr. Peter Fenwick, the leading consultant to the Institute of Psychiatry in London, *"...the first medical evidence that proves the mind can continue to exist after the body is clinically dead, and that a form of afterlife is now scientifically explainable."* Dr. Fenwick noted those who survived and had experiences had reported being *"changed... Those who were religious found their faith renewed. Those who had no faith often acquired at least a belief in some form of afterlife[28]."*

William Cash's article continues after Ayer made his comment to Dr. George that the would have to, *"revise all my various books and opinions." because he had seen a "Divine Being"*. The narrative continues:

> *'He clearly said 'Divine Being,' ' said Dr. George. 'He was confiding in me, and I think he was slightly embarrassed because it was unsettling for him as an atheist. He spoke in a very confidential manner. I think he felt he had come face to face with God, or his maker, or what one might say was God.'... Certainly it does seem very odd that Ayer, in either of his two detailed articles, did not so much as mention his conversation with Dr. George about having to rewrite all his books and works... According to Freddie* [Ayer], *and his newspaper piece, the first conversation he remembered having was with his ex-lover Beatrice Tourot, who was sitting on his bed. They spoke in French, with Ayer saying: 'Did you know that I was dead ? It was most extraordinary, my thoughts became persons[29].'*

After his experience, according to Cash, Ayer, like the 63 survivors of cardiac arrest, underwent a noticeable change in personality. According to Dee Wells, his twice married wife, he became much nicer, less boastful, and took an interest in other people and began to notice the beauty of natural scenery. Cash continues:

> *"What is also undeniably true -- and has never been reported on -- is that at the end of his life, Freddie spent more and more time with his former BBC debating opponent, the Jesuit priest and philosopher Frederick Copleston, who was at Freddie's funeral at Golders Green crematorium. 'They got closer and closer and, in the end, he was Freddie's closest friend,' said Dee. 'It was quite extraordinary. As he got older, Freddie realized more and more that philosophy was just chasing its own tail[30].'*

Materialist and atheist NDEs possess unique evidentiary weight because they dismantle the materialist paradigm from within. When the subject is a lifelong denier—someone who has actively 'refuted' the phenomenon as a delusion—their subsequent renunciation serves as a total collapse of the materialist defense. There is no clearer validation of the data than the conversion of the primary skeptic.

In the final analysis, truth is validated by the Experience of Youness—the individual's direct encounter with the Evaluative Field. As the case of A.J. Ayer demonstrates, when the presentation of non-local consciousness becomes undeniable, the previous ledger of facts, figures, and materialist authorities is rendered obsolete. The arena of direct experience does not merely suggest a new reality; it instantiates it.

The Effect of the NDE

For those that have had an NDE, research has shown that it is a *meaningful and transformative* event. Nderf.org has administered a 100-question survey to 834 NDErs to gather information on aftereffects of the NDE:

"meaning and significance of our Earthly lives" BEFORE the NDE experience:

26.9%, "are meaningful and significant"
19.9%, "are probably meaningful and significant"
24.6%, "possibly meaningful and significant"
13.1%, "are not meaningful and significant"

"meaning and significance of our Earthly lives"AFTER the NDE experience:

78.1%, "are meaningful and significant," +51.2%
7.7%, "are probably meaningful and significant," + 12.2%
8.2%, "possibly meaningful and significant," +16.4%
4%, "are not meaningful and significant," - 9.1%[31]

This information forms an unusual pattern for *a single experience, ranging from minutes to at best an hour or two, out of decades of experiences in life.

Perhaps an even more significant consideration is how these results could be expected to compare to alternative hypotheses for NDEs. If NDEs were the result of dreams, sudden onset delusions, hypoxia, medical complications, mental defects or self-fulfilling prophecies, they would not be expected to have a significant positive effect on one's viewpoint on the meaning and significance of "our Earthly lives[32]."

Dissembling Attempts to Explain NDEs and Disavow The Human Soul

The death knell of efforts to debunk NDEs lies in the existence of veridical accounts. These reports provide the most potent—and insurmountable—evidence for the existence of an immaterial human soul, and its inherent ability to travel, observe, and record consciousness. This is wholly inexplicable within the materialist framework. The validation of an immaterial soul follows the signature pattern of the other mysteries we have thus far investigated: a discontinuity or inexplicable instantiation is 'solved' by the existence of a purposefully immaterial level of existence. Unlike our previous discoveries, however, veridical accounts of the human soul necessarily usher in a complex, unimaginably large, and nearly incomprehensible afterlife realm—the embodiment of the elusive 'Theory of Everything' that serves as the missing key to our reality.

We call consciousness and Youness the final two products of *The Divine Algorithm*, but the human soul is something different. Youness must be constructed and awakened in the course of life, making it a universal 'product' of this reality. But our souls are not a product of this existence, nor of *The Divine Algorithm*. Our souls are an import from another realm of immaterial sentience and perfection, crafted by the very same creative consciousness and instantiating power that underlies our reality.

Science In Name Only: The Materialists Denial of Falsifying Information

We can classify the systematic attempts to debunk the NDE phenomenon as a direct implementation of the "Project Blue Book" playbook used for UFOs.

Project Blue Book was the codename for the study of UFOs by the US Air Force from 1952-1969. Project Blue Book operated under two primary directives: 1) to assess if the phenomena posed a threat to institutional security, and 2) to process and categorize the incoming data. After 17 years, 12,618 reports, and hundreds of thousands of man-hours, the project arrived at a consolidated conclusion: that the vast majority of reports were attributable to conventional, natural causes.
The findings were reported as:

(1) No UFO reported, investigated, and evaluated by the Air Force ever gave an indication of being a threat to our national security

(2) There was no evidence submitted to, or discovered by the Air Force that sightings categorized as "unidentified" represented technological developments or principles beyond the range of modern scientific knowledge;

And

(3) There was no evidence indicating that sightings categorized as "unidentified" were extraterrestrial vehicles[33].

While the third conclusion is founded, the first two were of a predetermined nature. Project Bluebook was, by design, beset by incompetence, shoddy work, and from its inception held a mandate to debunk all reported incidents if possible and, in any event, determine that the rest are inconclusive. *The Robertson Panel*, assembled to advise Project Bluebook issued this explicit directive. J. Allen Hynek called the Project

Bluebook, "The Society for the Explanation of the Uninvestigated." But Project Bluebook was not the final word on UFOs by the U.S. Government.

On June 25th, 2021, U.S. Pentagon press secretary John Kirby issued an immediate press release on UFOs, now referred to as "unidentified aerial phenomenon" (UAP). This release marked a formal acknowledgment of a persistent physical anomaly, stating:

> *"Today the Director of National Intelligence delivered to Congress a preliminary assessment on unidentified aerial phenomena (UAP) and the progress that the Intelligence Community and the Department of Defense UAP Task Force has made in understanding this threat... Incursions into our training ranges and designated airspace pose safety of flight and operations security concerns, and may pose national security challenges. DOD takes reports of incursions – by any aerial object, identified or unidentified – very seriously, and investigates each one... The report submitted today highlights the challenges associated with assessing UAP occurring on or near DOD training ranges and installations. The report also identified the need to make improvements in processes, policies, technologies, and training to improve our ability to understand UAP[34]."*

All along, from prior to Project Bluebook, the U.S. government knew a significant percentage of reported UFO phenomena were real and could not be explained by any known technology or naturally occurring anomaly. The about face of June 2021 was an admission that UFOs are real, they have always been real, and Project Bluebook was indeed a coordinated misinformation psyops campaign. Contrary to the belief that UFOs were not possible, and the assertion that scientific evidence had found reasonable explanations for nearly all accounts, the U.S. government had in its possession, irrefutable scientific evidence for the existence of UFOs, and, reportedly, multiple craft and / or craft and physical debris, crash retrievals, and remains in its possession. This evidence included military and civilian pilot accounts; on the ground observations; photos; videos; radar and other scanning technologies.

The release confirmed that the United States in fact had in their possession voluminous firsthand understanding of the capabilities of UFOs.

It was not *science* that required or dictated that Project Bluebook become a high-level, disinformation psyops campaign. And it was not the scientific method that failed to provide the gathered evidence as confirmatory findings to the public. The opposite is true. It was science that had proven the existence and capabilities of UFOs, that confirmed the data, scrutinized and made sense of the reports, cross-referenced information sources, sifted through the data, applied rigorous reasoning and scientific understanding to rule out alternative explanations, and allowed the U.S. government to draw solid conclusions.

The science we have developed, can facilitate the acquisition of knowledge in all investigated fields, by an ever-widening diversity of technologies and strategies.

Project Bluebook in many ways is the template for how the NDE phenomenon has been treated up to now. And in the about-face of the U.S. government in acknowledging UFOs, we have a preview for how NDEs will be treated in the future.

Just as the U.S. government held a 'voluminous first-hand understanding' of UAPs while publicly disavowing them, the materialist establishment currently sits on a mountain of verified NDE data. The 2021 UAP admission is the blueprint for the coming NDE disclosure: the moment when the 'uninvestigated' becomes the 'undeniable.' Scientism is not the protector of the scientific method; it is the administrative wall built to delay the acknowledgement of the *Evaluative Field*.

An Example of a Coordinated Attempt to Falsely Debunk NDE Phenomena

John Martin Fischer is an American philosopher and Distinguished Professor at the University of California, Riverside. He is also one of a handful of highly esteemed University Professors of the University of California System. There are only 38 individuals who have held this distinction in the University of California System since 1960.

Dr. Fischer is the recipient of a $5.2 million grant to study issues related to immortality. His statements on the subject are interesting and illustrate the institutionalized cognitive dissonance that seizes an individual who believes that consciousness is a brain-based emergent phenomenon, who must then consider the irrefutable evidence of the existence of a non-material consciousness and human soul.

Dr. Fischer believes that NDEs occur, but he's less certain as to their etiology. He is thus burdened by materialist constraints to propose a material solution to NDEs while also attempting to falsify the existence of a human soul, all while attempting to use the practice of good science as his guiding light.

Below is an excerpt from a Q&A involving Dr. Fischer.

> *"There are some near death experiences... where we can independently verify. There are famous cases where people are under general anesthesia or perhaps*

partial anesthesia... And when they regain wakeful consciousness after the anesthesia wore off the report conversations of the doctors or the nurses... and they couldn't have known this really without having some sort of out-of-body experience, or at least it seems that way. So those are the really fascinating ones. There's a famous case with someone was having an operation and when he regained wakeful consciousness the nurses were wondering where his dentures were, and he said, "Oh you put them over there," (the dentures were in the drawer indicated) and that happened when he was not awake... ...there are many reports where we can verify the contents as being true. The problem is we can't be sure how the individual got the information. Did the information come in some understandable, naturalistic way?, or did it come supernaturally? It is frustrating and challenging... My interest first of all, I would like to study, if I can, much more in depth the reports of near-death experiences where they are accurate, and they are reporting things that actually took place. I'd like to study and try and figure out how the information got through to the individual... my main point is to try and figure out the meaning of near-death experiences and what they can teach us about how we should live our lives. ...only about 8 to 10% of the people in near-death experiences of contexts, like they've had a cardiac arrest or they've been in a car accident or they've fallen, only about 10% report a near-death experience. Why should that be if the soul was different from the body?[35]"

Dr. Fischer is clearly conflicted, wanting to debunk NDEs while simultaneously acknowledging they contain information valuable to explore. In his various writings and presentations, his actual position remains ontologically elusive.

Below is a comprehensive formal challenge from the *Journal of Near-Death Studies* titled, *Understanding Near-Death Experiences: A Rejoinder to Mitchell-Yellin's Response*[36]. In this paper, Robert and Suzanne Mays—renowned NDE researchers for the last 40 years—reply to Benjamin Mitchell-Yellin (a frequent collaborator with Dr. Fischer). This exchange follows a prolonged dialectic initiated by the Mays'original critique of the Mitchell-Yellin/Fischer book, *Near-Death Experiences: Understanding Visions of the Afterlife*[37].

In the article Robert and Suzanne Mays respond directly to assertions and hypotheses concerning NDE phenomenon made by Mitchell-Yellin and Dr. Fischer that call into question the authenticity of NDEs. The Mays' also address several suspect strategies Mitchell-Yellin & Fischer use in their book,

In the article, the Mays' respond directly to assertions that call into question the authenticity of NDEs. They also expose the laxity regarding serious questions that demand serious inquiry that Mitchell-Yellin & Fischer utilize in their book. *Near-Death Experiences: Understanding Visions of the Afterlife*[38]

(1) Reliance upon generalized *Ad Hoc* explanations for a broad range of NDE elements and for specific components of individual experiences:

> *"Our main criticism of Fischer and Mitchell- Yellin's book (2016) is that it explains NDEs in terms of ad hoc hypotheses... (a number of separate distinct and incident specific explanations for elements within NDE accounts) ... The more serious problem comes when even a few additional NDE cases are examined (than the 4 discussed by Mitchell-Yellin & Fischer) .... Considering that there are over 100 cases of paranormal phenomena from NDEs documented in the book, 'The Self Does Not Die*[39]*', our contention is that there would be an explosion of ad hoc hypotheses to cover even a small portion of these cases... a multi-factor physicalist explanation becomes untenable and improbable, because it must rely on so many additional ad hoc hypotheses*[40]*."*

2. The intellectual obligation incurred to provide a comprehensive overall explanation for NDEs:

> *"Mitchell- Yellin (2017) stated that 'there is no good reason to insist that all aspects of all NDEs be explained by a single factor, nor is there good reason to insist that simpler explanations are more likely true than complex ones' (p. 102). We suggest that the requirement for a 'single explanation' for NDEs is not a preference but is, rather, a requirement of the empirical data. In our analysis of NDEs, in our critique, and in this rejoinder, it is clear that NDEs occur in widely various and complex circumstances, yet the experience itself is quite consistent, with highly consistent elements and intensity of each element. In an earlier paper (Mays & Mays, 2015), we argued that this evidence from NDEs shows that a 'single explanation' of NDEs is actually supported by the data... the physicalist explanation of NDEs really needs to account for the consistency of NDEs*[41]*..."*

3. The categorical mislabeling of NDE subjects as experiencing: delusions, illusions, hallucinations, and dream states—objectively erroneous classifications that ignore the structural consistency of the data.

Mays and Mays also call out Fischer and Mitchell-Yellin for using, "an *apophasis— to *imply,* not to state directly, but nevertheless to indicate, that elements of NDEs are inaccurate and therefore are no different from hallucinations, illusions, dreams, or delusions[42]." Mitchell-Yellin & Fischer state:

"We shall use the term 'accurate' throughout the remainder of the book to mean 'truthful' or 'corresponding to external reality.' An accurate experience depicts reality as it really is. Not all real experiences are accurate. Hallucinations, illusions, dreams, delusions— these may all be real experiences [apophasis]. *People really do have them. But they are not accurate experiences. They do not correspond to reality*[43]*."*

4. The strategic suspension of inquiry—delaying any acknowledgment of an inexplicable phenomenon until or unless a materialist cause can be manufactured, a perpetual deferment designed to protect the paradigm from the evidence.

Mays and Mays discuss one of the standard materialist approaches to information it cannot explain. The strategy is to delay rendering any conclusion until a materialistic explanation is found, which may never be forthcoming. This is the equivalent to a chess player indefinitely delaying a move prior to being check-mated in order to explore "solutions", and refusing to continue the game until such a solution is found. At present, there are an excess of *on-hold* areas of scientific exploration in the field of NDEs and many which we have previously explored, that have been effectively placed in a perpetual holding pattern, waiting for a reductionist answer that is not, and cannot be made. Mays and Mays point out:

"Throughout their book, Mitchell- Yellin & Fischer (2016) appealed to the ultimate success of scientific progress to provide physical explanations that, in the future, will fully explain NDEs.... In their response to this charge, Mitchell-Yellin clarified (their position)[44]*"*

Mitchell-Yellin and Fischer state:

"We were trying to show that there is no good argument in favor of supernaturalism on the basis of NDEs. Part of this involves trying to show— and this gets at the appeals to the progress of science— that there are good reasons to think that there may, in the future, be good physical explanations of the relevant phenomena. But again... this point is not to claim that supernaturalism is false. It is to claim that NDEs do not prove, or even provide good evidence for the claim, that supernaturalism is true[45]*."*

The Mays' conclude their rejoinder by exposing the fundamental insolvency at the heart of the Mitchell-Yellin & Fischer position. They leave the authors with what amounts to an ultimatum regarding their refusal to engage with the data:

"Scientific inquiry, — at least inquiry that is unimpeded by physicalist bias and dogma— can yield a conclusion regarding which theory of NDEs is most valid, whether that theory is purely physical or is nonphysical... Our criticism

> *is that the authors have failed to provide adequate physical explanations of a wider selection of NDEs— and appeals to promissory materialism are unconvincing[46]..."*

Counter-Proposals & Skeptical Rejoinders for the NDE Phenomenon In 4 Parts

> *"Out of respect for people's ability to generally understand reality, if skeptics want to represent that NDEs are not real, they should present strong evidence that NDEs are not real. Over 20 different 'explanations' of near-death experience have been suggested by skeptics over the years. If there were one or even several 'explanations' of NDE that were widely accepted as plausible by the skeptics, there would not be so many 'explanations'. The existence of so many 'explanations' suggests that skeptics have not presented an explanation of NDEs that is accepted by consensus of the skeptics themselves. Skeptic arguments neither addressed the consistency of the accounts nor provide hard evidence to disprove their validity[47]."*

While the integration of an NDE is a profound personal struggle, it is further complicated by a standard set of institutional rebuttals. These skeptical tropes aim to impeach the veracity of the experience by reducing it to a biological or psychological error.

Part 1: NDEs Are Not Possible - Consciousness As A Material Construct, & Illusion

Part 2: Low Brain Oxygen Levels/Cerebral Hypoxia & Other Medical Conditions

Part 3: Veridical NDEs

Part 4: Cultural/Religious Expectation

Part 1: NDEs Are Not Possible - Consciousness As A Material Construct, & Illusion

From the reductionist perspective, when the mind is working fully and environmental data is processed through the senses, the resulting gestalt *is* our consciousness. There is no need for a super-ordinal 'master.' The idea of a commanding identity existing above this momentary collation is, in this view, fundamentally untrue—a persistent illusion. This perspective was thoroughly reviewed in *Special Problem #4: Consciousness.*

By defining consciousness as an emergent byproduct, and with NDEs and OBEs defined as illusions, or brain glitches, materialism is effectively absolved from solving something it claims cannot and does not exist. Those who claim to have experienced them are simply mistaken.

Materialism assumes that once the brain is mechanistically mapped, the etiology of these 'mistaken' reports will be revealed.

This hand-waved response does not address the persistent failures of the materialists' model to account for the many issues surrounding consciousness—qualia, the binding problem, intentionality, and the explanatory gap.

We find validation for the insistence on the brain-based etiology of all human experience within the field itself.

Daniel Dennett, provides the quintessential defense of this reductionist closure in his paper, *The User-Illusion of Consciousness*[48]. In his view, once the physical mechanics are described, the inquiry is over:

> *"Functionalism is, after all, a species of behaviorism, the conservative default position of all the physical sciences: once you've explained all the external and internal behaviour of something (volcanoes, the weather, the rotation of galaxies, the ribosome's ability to read DNA, the folding of proteins), you get to declare victory, since there is nothing left to explain*[49]*."*

By equating human consciousness with the mechanical behavior of a volcano or a protein fold, Dennett effectively closes the door on the observer. He further dismisses the "Hard Problem" by arguing that "phenomenality" is a trick of the hardware: In Dennett's view, once you've explained *all of the parts*, and *all of the interactions*, you've explained *all of the things*. Citing Mark Solms, Dennett attempts to dismantle the 'epistemic intimacy' we have with our own experiences, but by using the allegory of sugar not being sweet, rather, being a 'lie' of the tongue, he reveals the germ that infects his thought. He is essentially arguing that the *experience* of the thing is less real than its *description*.

> "Solms tackles the traditional but dubious assumption of epistemic intimacy… 'So, Chalmers thinks the dual aspects of 'the phenomenal features of the world' and 'the physical' are in information itself — at its source — not in the equipment of the participant observer. This is like thinking that sweetness is intrinsic to the molecular structure of glucose[50]."

His footnote following this quote cites the analogy with glucose drawn from Hurley, Dennett and Adams (2011), but he doesn't quite follow through all the way to the user-illusion. Solms nicely picks out 'the equipment of the participant observer'. It is there that the illusion of 'phenomenality' is to be found, in just the same way the sweetness of sugar is not to be found in the structure of glucose but in the dispositional structures activated by the detection of glucose. Do not try to sprinkle subjective honey on those informational structures, and do not try to attribute 'phenomenal' visual properties to the informational structures of the visual cortex ('mental paint' or figment as I have called it)[51]."

For Dennett, the question David Chalmers proposes as being the hard problem misses the point by introducing a false level of "phenomenality" to something that has none. Because they deny the reality of the hard problem of consciousness, there can be nothing phenomenal about bits of information that are processed and apprehended by the brain.

This is the tragedy of the materialist perspective: to maintain the integrity of the model, they must devalue the very consciousness that allows their model to exist. The observer is in effect forced to participate in their own erasure. By defining "what it is like" as a mistaken byproduct they have done far more than offer a complete and closed system for the understanding of the magnificence of the human condition, tragically, and clearly without awareness, they have effectively exiled the observer from their own reality.

For more on this topic return to *Special Problem #4*.

Firsthand Experience & The "One Way Effect"

The nature of NDEs as primary sensory, experiential, subjective perceptions means that they can only truly be known as a personal experience, not unlike the experience of enlightenment. And like enlightenment, the idea that an outside consciousness can arbitrate the reality of the experience is nonsensical. Because of the profound nature of such spiritually transformative consciousness experiences, there is a well-documented *"one-way" effect* we have previously discussed that occurs with those that have OBEs and NDEs. Once the bell has been rung, it cannot be unrung. It cannot be unheard.

When one experiences an OBE or NDE, whether one's prior ideas on these events was one of a skeptic, physicalist, reductionist, atheist, scientists, or an enlightened spiritual master, once an individual has had an OBE or NDE experience, whatever disbelief or doubt in the reality of the occurrence vanishes in the first moments of the experience. The converse holds true as well. There are *no known cases* of an individual who previously believed OBEs and NDEs were real who had one, and then decided that their experience was not real, or fake or as Dennett suggests, a misperception.

If the assertions of disbelieving individuals held any validity, then once these individuals had these experiences, they would then have evidence-in-hand to deconstruct the arguments and firsthand understanding to articulate why individuals were mistaken, and where the critical misalignment, misperception, untruth, or incompatibility resides.

This does not happen. When deniers have NDEs, they are converted with fidelity and totality. Ideologically driven denials, especially those based on deeply held beliefs, dogmas, paradigms, melt. The fidelity of NDEs, OBEs and STEs are impervious to nearly all forms of contravening 'evidence'. They are category errors, attempting to explain non-local, immaterial substance, with biological and chemical operations.

First-hand experience is the reason why I know OBEs are real, and knowing that they are real, they provide the pathway to the reality of the NDE experience.

A Personal Out of Body Experience

I have had an out of body experience. It was neither pleasant, nor a part of a mystical experience. This event occurred when I was about 17 years old. I was laying down on my bed and became aware that my right arm had suddenly gone numb. Thinking that my arm had fallen asleep, I attempted to move it. Strangely, I could feel myself "moving" my fingers and flexing my arm, but looking at my arm, it did not move or flex at all. Having never had this experience, I thought it was an unusual case of my arm falling asleep. I then attempted to lift my arm, and once again, I felt my arm lifting up but looking at my arm, it did not move. At this point I knew that somehow; the sensation of my arm had become disconnected from my physical arm.

I began trying to "insert" the sensation of a hand and arm "into" my arm, as if my numb arm were a long sleeve I could slide into like a glove. Despite several attempts, it just didn't work. I could slide my arm in, but every time I lifted it, the sensation went up while my arm remained in place. I couldn't re-establish the connection. I decided that since my other arm wasn't asleep, if I pushed myself up quickly that would snap my numb arm back into place. But when I pushed up forcefully with my left arm, instead of my body rising up off the floor, my entire consciousness rose up *out* of my body, and I found myself hovering several feet above myself looking down upon my body from behind and above. An instant sense of surprise and panic set in. My vision was also unusual. It had a slight milky white foggy-ish haze as I recall. My body sense felt exactly as it did in my body, but obviously without the burdens of gravity or biology. I was panicked. It was not pleasant. I wasn't suffused with love feelings, or knowledge any different than my normal life. knowledge. The common trope of 'flying around' was unavailable to me; I was gripped by the singular, practical fear that possibly, this was how people died. They drifted out of their bodies and couldn't get back in. Clearly, I knew nothing of the phenomenon to guide me. This was in the early 1980's.

After a few moments I realized only the upper part of my body, from my knees up, had lifted out. The rest of my body was in place. I then tried to pull myself down using my legs as an anchor using the 'muscles' of my midsection. Strange as it seemed, as I bore down, feeling the midsection effort, I slowly started easing down. Once I was fully back, I tried to get up with both hands again and everything was connected. It was over. I ran downstairs to tell my mother. She didn't really say much.

At the time I didn't know there was a debate on the nature of consciousness or the reality of out of body experiences. My sense is that there wasn't a debate that had crystalized to the extent it has in the present day. Later, when I became aware of the debate on consciousness, I didn't think of my experience as evidence that could refute it. I didn't know what had happened. Perhaps my brain 'projected' my image outside of my body. But now that I know more, that's not possible. The very notion invalidates materialists' ontology, and has no possible structural, biological means.

Consciousness, for the scientists that investigate it, has nothing to do with the soul. In truth, as we have discovered, consciousness involves the mystery of how everything we experience in life is represented by physical systems. It is the animating force that defines life. It is what guides macromolecules and ribosomes and kinesis and the ATP synthase and electron transport chain and citric acid cycle. It is how we operate our bodies, laugh, love, live, eat, and wake. Consciousness is the substrate that makes the physical accessible, experienceable and manipulable. It is the indispensable feature that allows a universe to be not only coherent but known and observed.

Such events as OBEs and NDEs happen regularly in human experience. Earlier we cited the figure ~5% of the population is estimated to have had an OBE[52]. The statistical dissonance between reported OBEs and NDEs is a revealing anomaly. Since an NDE fundamentally requires an out-of-body state, the reported figures for standalone OBEs should logically be much higher. This discrepancy is likely a function of the nascent, fragmented state of the research, but it also points to a significant reporting bias.

For the average individual, a standalone OBE is not only perplexing—forcing the question *"What happened?"*—but it lacks the narrative scaffolding of an NDE. Without a medical crisis, there are no attending personnel to record the event. Without a revelatory or otherworldly encounter, there is no "story" to tell. What remains is simply a personally difficult-to-digest experience with no official, widely known agency to record it.

No one is asking the question, so the data remains uncaptured. We are likely looking at a vast, silent population of experiencers who have seen the seams of the biological interface come apart but have no 'official' bucket to hold the content of their experience.

Part 2: Low Brain Oxygen Levels/Cerebral Hypoxia & Other Medical Conditions

One of the popular counter-explanations for the NDE phenomena is cerebral hypoxia, or lack of oxygen to the brain. When we turn to the medical literature to cross check the signs of low brain oxygen against *The Standard NDE Content (NDE-C) Scale*, there is no correlation.
The symptoms for low brain oxygen, in order from mild hypoxia to severe hypoxia, are:

Difficulties with complex learning tasks

Reductions in short-term memory

Reduced ability to move/decreased motor control

Difficulty in paying attention

Difficulty in making sound decisions

The Standard NDE Content (NDE-C) Scale (Greyson Scale) combines 16 measures:

(1) Did time seem to speed up or slow down?
(2) Were your thoughts speeded up?
(3) Did scenes from your past come back to you?
(4) Did you suddenly seem to understand everything?
(5) Did you have a feeling of peace or pleasantness?
(6) Did you have a feeling of joy?
(7) Did you feel a sense of harmony or unity with the universe?
(8) Did you see, or feel surrounded by, a brilliant light?
(9) Were your senses more vivid than usual?
(10) Did you seem to be aware of things going on elsewhere, as if by extrasensory perception (ESP)?
(11) Did scenes from the future come to you?
(12) Did you feel separated from your body?
(13) Did you seem to enter some other, unearthly world?

(14) Did you seem to encounter a mystical being or presence, or hear an unidentifiable voice?
(15) Did you see deceased or religious spirits?
(16) Did you come to a border or point of no return?[53]

From the above comparison it can be seen that the symptomatic profile of hypoxia represents a fundamental degradation of the consciousness signal, while the NDE represents a sharpening of it. Attempting to link the two is equivalent to the claim that cutting a fiber-optic cable would reliably result in a tenfold increase in bandwidth.

To evaluate the validity of the near-death experience, one must first confront a staggering observational paradox: why does the human mind report its most expansive, high-fidelity clarity at the exact moment the human brain enters its most profound state of clinical decay? This is due to a clinical conflict between two distinct biological states of oxygen deprivation, and the NDE event itself, which is categorically removed from either.

The Key is the distinction of states:

Voluntary Hypoxia: Managed oxygen deprivation where the "metabolic cleaning crew" remains active.
Acute Circulatory Failure: A catastrophic hard system crash where the system turns toxic.
The NDE: A state of hyper-lucid consciousness that persists after the biological engine has seized. It is categorically removed from voluntary hypoxia and acute circulatory failure.

The NDE operates in another dimension of consciousness altogether because it ignores the mechanics of the "metabolic cleaning crew." A breath-holder utilizes a functioning heart to maintain the heart/blood/brain cleaning loop, scrubbing CO2 and neutralizing lactic acid. In contrast, cardiac arrest terminates this relationship instantly. Without hydrostatic pressure (a beating heart) to circulate the system, the brain's environment turns toxic in mere seconds. This toxicity triggers a total electrical shutdown—a desperate attempt by the brain to preserve its cellular integrity at the total cost of its conscious function resulting in the above-mentioned difficulties: "reductions in short-term memory" "reduced ability to move/decreased motor control" and "difficulties paying attention" and "making sound decisions."

This is why CPR is taught as a time sensitive race to artificially circulate the vascular system to mitigate a terminal buildup of metabolic waste that begins the moment the heart fails[54].

The timeline is unforgiving. In humans, the EEG begins to change within eight seconds of cardiac arrest and reaches a "flatline" within an average of 15 to 20 seconds[55]. Within this 20-second window, the brain's oxygen reservoir is exhausted. The lack of blood pressure, combined with the surge of CO_2 and lactic acid, forces a state of metabolic paralysis and an emergency "circuit breaker" trip in the neurons[56].

This 20-to-30-second window is the "Hard Wall" of materialism.

If consciousness were merely a byproduct of the biological engine, the NDE experience would immediately begin to degrade mere seconds after the heart stopped and completely terminate the moment the EEG went flat and the blackout began. Clinical physiology is clear: complete circulatory arrest causes total inactivation of the brain's control centers within minutes, but the functional "shutdown" of conscious electrical activity is almost instantaneous[57].

Yet, it is precisely within this window of toxic shutdown and physiological collapse that the most detailed, veridical NDE data is recorded. The subject is retrieving high-fidelity information from a state that, by every law of medical physics, is no longer equipped to process it. One is a system losing its power and its cleaning crew; the other is a consciousness operating at peak capacity. They are moving in opposite directions.

> "Recent studies in cardiac arrest survivors have indicated that although the majority of cardiac arrest survivors have no memory recall from the event, nevertheless approximately 10% develop memories that are consistent with typical near death experiences. These include an ability to 'see' and recall specific detailed descriptions of the resuscitation, as verified by resuscitation staff (veridical NDE). Many studies in humans and animals have indicated that brain function ceases during cardiac arrest, thus raising the question of how such lucid, well-structured thought processes with reasoning and memory formation can occur at such a time[58]."

When we scan the medical literature for other conditions associated with the creation of fantastic, hyper-lucid alternative states of consciousness and a particular class of deaths, there is no condition where these two events are correlated. NDE's are only correlated with one factor and that's death, whether it is actual or imminent death.

Cerebral hypoxia is not a medical finding derived from evidence; it is not a plausible reason for NDE accounts. When individuals succumb to low brain oxygen levels, their consciousness experience is reduced and degraded, not increased and marked by lucidity. It is objectively absurd to assign a condition defined by cognitive collapse as the precipitatory cause of hyper-lucid consciousness.

Part 3: Veridical NDEs

We have covered veridical accounts severally, so this will be brief. The 'Veridical' NDE is the smoking gun of immaterial consciousness: it is the retrieval of objective information from a state of total physiological collapse.

"Veridical perception, in which the near-death experiencer reports seeing or hearing events during their NDEs that, given the condition and/or position of their physical bodies, should have been impossible to perceive but are nevertheless corroborated as accurate[59]."

In veridical NDEs the information that is reported could have only been obtained if the following were true:

(1) The observer is detached from the biological anchor while maintaining high-fidelity awareness, operating independently of the physical location of the body.

(2) Their consciousness was not a function of physiology/biology/physics

(3) The NDEer has access to a fund of intelligence, knowledge and a level of omniscience

Some examples of these types of information are:

-*Information* about future events *that have come to pass*

-*Information* about past events that could not have been known

-*Information* about individuals they have never met and/or had access to their memories, emotions, thoughts, and perceptions

-*Information* from another worldly realm, where past extra-physical and/or extra sensory information is known

-*Information* which could have only been known if the individual's consciousness existed in a non-material medium capable of sustaining and maintaining their individual integrity and ego functions: memory, sense of time, context, language, both verbal and written as well as thoughts, self-integrity, personality, knowledge, etc.

This last point proves that the core components of identity—memory, language, and self-integrity—are not biological 'byproducts' that dissolve upon death; they are fundamental properties of the observer that persist even when the physical transmitter is offline.

Part 4: Cultural & Religious Expectations

The charge that NDEs are merely self-fulfilling prophecies or religious fantasies is contradicted by the global data set. Validated NDEs have been recorded across every ethnicity, culture, and language on Earth. While a practitioner of a specific faith may interpret an encountered being of light through their own theological lens, the core elements of the encounter—the non-local perspective, the hyper-lucidity, and the veridical accuracy—remain identical. Cultural themes are merely the outermost layer; the NDE itself is a universal architecture operating independently of the subject's expectations.

If the NDE were merely a psychological "self-fulfilling prophecy," it would follow a social template—a cultural script formulated to fit the individual's internalized stories and religious upbringing. However, historical evidence provides a sharp rebuttal: this structure existed long before the public was aware of it. Prior to the release of Dr. Raymond Moody's Life After Life, there was no popular knowledge of the NDE. There was no universal "expectation" to meet, or a singular prophecy to be fulfilled. It was only after Moody's book that individuals understood their private experiences aligned with a global, standardized phenomenon.

This is further confirmed by the "one-way effect." Atheists and agnostics frequently report these experiences, often resulting in a profound spiritual conversion. Conversely, there is a notable lack of cases where religiously affiliated individuals abandon their faith after an NDE. Instead, their concept of the divine becomes expansive. They come to understand that this power is not confined to their specific tradition; it is the high-fidelity, generative engine of all possible instantiations—the primary configuration from which all spiritual beliefs and the physical realm itself emerge.

---

[1] (J. Long, P. Perry)85
[2] (J. Long, P. Perry, God and the Afterlife, Pg. 81-2, HarperOne, 2016)86
[3] (NDERF #4538)87
[4] (NDERF #3899)88
[5] (NDERF #7469)89
[6] (B. Greyson, A Doctor Explores What Near-Death Experiences Reveal about Life and Beyond. St. Martins, 2021 Pgs 70-71)90
[7] (Ibid Dr. Moody)91
[8] (D. Linn, S. Fabricant Linn, M. Linn, The Gifts of Near-Death Experiences, Hampton Roads, 2016, Pg. 21)92
[9] . (P. van Lommel, Consciousness Beyond Life, HarperOne 8/9/2011, pgs 32-33)93

[10] (P.M.H. Atwater, Near-Death Experiences: The Rest of the Story, HRPUB, 2011, pg. 97-98)94
[11] (Near Death Experience Showed Me Something I Could've Never Imagined | Anita Moorjani (NDE), 8/3/2021, YouTube Channel, "Inspired")95
[12] (A&E, I Survived Beyond and Back, "Ray" 11/212011 YouTube Channel: "Brian Garrett")96
[13] (Jayne Smith's Profound NDE - Near Death Experience. YouTube Channel: "NDE Video" 7/22/2021)96
[14] (Man Shocked After Taking His Own Life & Being Shown His Future | NDE, YouTube Channel: "Shaman Oaks")97
[15] (A&E, I Survived Beyond and Back, "Michaela" 3/11/2011 YouTube Channel: "Brian Garrett")
[16] (The Ultimate Adventure; An Ice Climbers Near Death Experience (NDE) | Peter Panagore, YouTube Channel: Guy Lawrence 8/16/2020)100
[17] (Howard Storm's Testimony | You Think hell doesn't exist? This will change your mind.. JuJu Fisher of Men, YouTube, 2/11/2022)101.
[18] (Howard Storm's Testimony | You Think hell doesn't exist? This will change your mind.. JuJu Fisher of Men, YouTube, 2/11/2022)101.

[19] (A.J. Ayer, Language, Truth And Logic, Pg. 2, 1936, publisher unknown, pg numbers refer to this edition: https://antilogicalism.com/wp-content/uploads/2017/07/language-truth-and-logic.pdf , or search: A.J. Ayer, Antlogicalism, Language, Truth And Logic, pdf)102
[20] [Ayer, A.J. (28 August 1988). "What I Saw When I Was Dead." The Sunday Telegraph. Reprinted as "The Undiscovered Country" in The Meaning of Life (1990) and The Philosophy of A.J. Ayer (1992)]103
[21] [Ayer, A.J. (28 August 1988). "What I Saw When I Was Dead." The Sunday Telegraph. Reprinted as "The Undiscovered Country" in The Meaning of Life (1990) and The Philosophy of A.J. Ayer (1992)]104
[22] [Ayer, A.J. (28 August 1988). "What I Saw When I Was Dead." The Sunday Telegraph. Reprinted as "The Undiscovered Country" in The Meaning of Life (1990) and The Philosophy of A.J. Ayer (1992)]105
[23] (A.J. Ayer "Postscript To A Postmortem" was published in The Spectator, October 15, 1988)106
[24] (A.J. Ayer "Postscript To A Postmortem" was published in The Spectator, October 15, 1988)107
[25] (A.J. Ayer, Language Truth and Logic, Ibid, Pg. 2, 72-76)
[26] (A.J. Ayer, Ibid),,
[27] (William Cash, The National Post March 3, 2001, What Happened When The World's Most Famous Atheist Had A Near-Death Experience)110
[28] [Cited Study: S. Parnia, D.G. Waller, R. Yeates, P. Fenwick, A Qualitative and Quantitative Study of the Incidence, Features and Aetiology of Near Death Experiences in Cardiac Arrest Survivors, Resuscitation, Pages 149-156 Volume 48, Issue 2, February 2001)]111.
[29] W. Cash, The National Post March 3, 2001, What Happened When The World's Most Famous Atheist Had A Near-Death Experience) 112
[30] W. Cash, The National Post March 3, 2001, What Happened When The World's Most Famous Atheist Had A Near-Death Experience)113
[31] (From NDERF.org: Near-Death Experience After Effects: This page contains 834 near-death experiences (NDEs) and their responses to a series of NDERF survey questions "The following questions are about changes in your life that might have occurred after your experience." For each concept, near-death experiencers

(NDErs) were asked to select the best response expressing their beliefs before their NDEs, and at the time they shared their NDEs with NDERF.)114

[32] *footnote: it is not uncommon that NDErs have had previous or multiple NDEs, which appears to make them much more likely to have an NDE in future situations where they face a life threatening situation or near death event.

[33] (E.U. Condon, D.S. Gilmore (ed), Final Report of the Scientific Study of Unidentified Flying Objects, Bantam Books, 1968, https://files.ncas.org/condon/)115

[34] (U.S. Department of Defense, Press Release, Immediate Release, Statement by Pentagon Press Secretary John Kirby on Unidentified Aerial Phenomena Assessment, June 25, 2021, defense.gov)116

[35] (Professor Dr. John Martin Fischer, Univ of Calif-Riverside YouTube video: Near-Death Experiences: What we know about death, immortality, and near-death experiences, -Univ of Calif-Riverside YouTube Channel, Pub. June 12, 2019)117

[36] (Journal of Near-Death Studies, December 2017/Winter 2017, 36(2): 66-99).

[37] (Oxford Press, 6/1/2016).

[38] (Ibid Oxford Press, 6/1/2016):

[39] (T Rivas, A. Dirven, R.Smit, The Self Does Not Die 2016)120

[40] (Journal of Near-Death Studies, December 2017/Winter 2017, 36(2): 66-99.)121

[41] (Journal of Near-Death Studies, December 2017/Winter 2017, 36(2): 66-99.)122.

[42] (Journal of Near-Death Studies, December 2017/Winter 2017, 36(2): 66-99.)123.

[43] (Near-Death Experiences: Understanding Visions of the Afterlife Fischer & Mitchell- Yellin, 2016, p. 35)124 *Footnote: Apophasis is a rhetorical device wherein the speaker or writer brings up a subject by either denying it or denying that it should be brought up. Accordingly, it can be seen as a rhetorical relative of irony.

[44] (Journal of Near-Death Studies, December 2017/Winter 2017, 36(2): 66-99.)125:

[45] (Near-Death Experiences: Understanding Visions of the Afterlife, Fischer & Mitchell- Yellin, 2016, p. 106)126.

[46] (Journal of Near-Death Studies, December 2017/Winter 2017, 36(2): 66-99.)127:

[47] (J. Long, P. Perry, God and the Afterlife, Pg. 194, HarperOne 2016)

[48] (D. Dennett, The User-Illusion of Consciousness, Journal of Consciousness Studies, 28, No. 11–12, 2021, pp. 167–77)

[49] (D. Dennett, Ibid, pp. 168)131

[50] (M. Solms, The Hidden Spring: A Journey to the Source of Consciousness, 2021, p. 260, New York: Norton)'

[51] (M. Solms, (2021) The Hidden Spring: A Journey to the Source of Consciousness,.p. 260, New York: Norton) (D. Dennett, Ibid, pp. 174)133

[52] (O. Blanke, N. Faivre, S. Dieguez, Leaving Body and Life Behind: out of Body and Near-Death Experience, in S. Laureys, O. Gosseries, G. Tononi, eds. The Neurology of Consciousness: Cognitive Neuroscience and Neuropathology. 2nd Ed. Elsevier; 2016: 323-347.) & (O. Blanke, out-of-body experiences and their Neural Basis. BMJ. 2004; 329(7480):1414-1415)134.

[53] (B. Greyson, The Near-Death Experience Scale: Construction, reliability, and validity. Journal of Nervous and Mental Disease, (1983), 171(6), 369–375. https://doi(dot)org/ 1097/ 00005053-198306000-00007)135

[54] (A.C. Guyton, J.E. Hall, Textbook of Medical Physiology, 2020, 14th Edition, Elsevier).

[55] (H.L. Clute, W.J. Levy, Electroencephalographic Changes During Brief Cardiac Arrest in Humans. Anesthesiology, 1990, 73(5), 821–825.) .

[56] (A.C. Guyton, J.E. Hall, Textbook of Medical Physiology, 2020, 14th Edition, Elsevier).

[57] (A.C. Guyton, J.E. Hall, Textbook of Medical Physiology, 2020, 14th Edition, Elsevier).

[58] (S. Parnia, P. Fenwick P. Near Death Experiences in Cardiac Arrest: Visions of a Dying Brain or Visions of a New Science of Consciousness. Resuscitation. 2002;52(1):5-11.)136

[59] (J.M. Holden, Veridical Perception in NDEs, in J.M. Holden, B. Greyson, D. James (Eds), The Handbook of Near-Death Experiences: Thirty Years of Investigation, 2009, (pp.185-211). Praeger / ABC-CLIO)137.

8 Miracles

In our previous chapter we highlighted several NDEs where an even more miraculous event than surviving death, being outside of the physical body, and encountering the creative force behind reality has taken place. The historical record, and NDE research has also uncovered NDE accounts where *miracles* have taken place.

Definition: By miracles we mean the combination of two things:

(1) A suspension of local physical constants that defies any materialist reduction, and;

(2) A non-linear (the result does not follow a predictable, logical, or causal path), beneficial outcome that re-orders the individual's trajectory within the Field and also, by association, reorders our entire reality.

When a miracle takes place It's not simply the case that something impossible has happened. Every NDE is an 'impossible' event for materialism. We are highlighting the fact that in some NDE events, reality itself has been reconfigured in such a way to benefit the individual, and this benefit has no possible material explanation.

The event is not just probabilistically highly unlikely (impossible), but a real-time instantiation beneficial to the individual has taken place within reality itself. A literal miracle (impossible benefit) within a miracle (an NDE).

We now turn our attention to *miracles* as a class of phenomena that validates the existence of an immaterial realm, and a power beyond our reality that can transform, change, influence and alter the fate of all those within our reality, *at will*.

A miracle is a systemic re-calibration; by altering the fate of even one individual, the Algorithm effectively re-scripts the entire reality. It is a 'causal intervention' that ripples through the lives of every individual the recipient will ever touch—and just as importantly, those they will now *never* meet due to the alteration of reality. It is the instantiating power behind reality, acting by select instantiations to edit the master-timeline of our universe in real-time.

When it comes to the instantiating power of a 'discrete miracle event', we must discard the delusion of locality. In a reality where gravity is infinite and the Evaluative Field is omnipresent; there is no such thing as a 'discrete' event. Just as moving a single rock from Mars to Jupiter recalculates the gravitational tension of the entire solar system, a miracle is a universal re-configuration. To heal a single individual or forestall a single death is to edit the spatiotemporal fabric of the whole. Every instantiation is therefore a total instantiation.

Parsimony in Miracles

The documentation of miracle events within the medical histories of select NDErs provides the ontological receipt that reorients the sprawling inefficiency of strange materialist's concepts such as *the Everett Interpretation* into its only workable corollary. The link between miracle events and The Everett Interpretation may not be obvious on the face, but it is a deep and incompatible collision.

The Everett Interpretation (Many-Worlds) is the ultimate monument to materialist waste; to avoid the elegant reality of the quantum world, it forces the universe to undergo a massive, infinite division' at every Planck second. It is a theory that burns the entire house down just to avoid admitting someone turned on the light.

The special case of miracles within our ontology accounts for two fundamental shifts:

(1) The objective collapse of the wave function within our specific realm.

(2) The instantiating power of a teleological, omnipotent, omniscient, immaterial, non-local, creative, Instantiator.

Under this power, reality has no need to create a trillion redundant worlds to account for the bouncing of a dropped paperclip; it simply updates the code for everyone, everywhere, at once, within the single universe in which we stand. This is the power of instantiation—Parsimony. Complete, total, efficient, and perfected parsimony. Something that cannot be achieved within an atheistic, ungoverned materialist reality.

In the instance of miracles, specifically in the example of a divine healing resulting from an NDE, we witness a truth that incinerates the materialist construction: the individual human, far from being an insignificant byproduct of cosmic dust, is revealed as the most cherished and causative factor behind the universe. The fact that the instantiating power of the entire cosmos would intercede in the capacity of a miracle for a single soul is dispositive proof of the nature of our reality. We are not lost in the machine; we are the reason the machine is running. This is a conclusion that no miracle is needed to tall. The message of the NDE itself states this same principle—the existence of living beings such as humans, as the end purpose and sole reason for our reality.

To clarify, in the context of miracles in our *Third Book, Transcendence,* we are using a more focused and specific definition of miracles than in our first two Books, *Book One: Instantiation,* and *Book Two: Emergence &Transformation.*

In the first Two Books where we were discussing scientific phenomena, we established the benchmark for what qualified as a miracle in our investigation. This benchmark was any event with a probability of less than 1 chance in 10 raised to the 140th power.

With this benchmark we discovered in Section 1 that virtual "zero-chance" miracles have taken place in our reality routinely, and that quadrillions of quadrillions of miracles take place every second just on Earth. (for instance, every imaginable event that takes place within the individual cell, be it mammal or bacteria, is literally a miracle from a material standpoint). We also found a single thread running through all of these miracles that bound them all together under one principle. All of these miracles have in common that they all are required if there is to be any biological life within our universe whatsoever. More specifically, all of the miracles in the first Two Books are all in service to the existence of intelligent observers and to the ability of intelligent observers to inspect, observe, experience, contemplate, deduce and perform science upon our reality.

We will provide three examples of the type of miracle we speak of from dozens that have been documented and are available in the universe of reported NDEs.

Miraculous NDEs

Donna Rebodow had a near-death experience while water skiing. A rope tangled around her leg, and she was unable to free herself prior to the boat taking off. She was pulled forcefully and the rope tightened perilously around her leg dragging her under the water. She was unable to free herself and subsequently drowned. This is where her NDE began.

When she returned to her body, she was still tangled in the rope, and it had wrapped around her leg so tightly that it was coiled down to the femur bone nearly severing her leg. We begin her story without inclusion of her NDE, rather after she returns:

"I heard the boat turn-off. The next physically impossible thing is, I looked down at my leg and rope was embedded in my leg. It strangled my leg to the femur; I had no space. It was actually strangled. They don't understand why I

didn't rupture my femoral artery. We have no idea. And I thought, 'how are we going to get this rope. (it) can't come undone'. I started lifting my hands to swim to the surface and human hands came under my armpit and the rope came undone. I got to the surface; there was no one there. I don't know who drug me up. I know what I physically felt. Then I had to get rid of all the water that was in my lungs. And my sister was watching this[1]."

In 1982, Mellen-Thomas had terminal brain cancer and died. His body was observed to remain without vital signs for an hour and a half after he died. Somehow, he returned to life and recounted having an NDE in the intervening period. After returning to life, he rapidly experienced a complete remission of his terminal cancer. We begin his story in the midst of his NDE.

"The next question I asked was why would you create humanity? Why would humanity be created when they are such dark and evil beings? And at that moment, the light turned into a mandala, like a big, round, stained glass window that was alive and I was breathed into the center of it. In that center it was like I could look into every human soul, including my own. And I could see no evil at all. No darkness in any human that has ever lived. And that may shock you. It shocked me. But I was there in the nexus, the mandala of human souls. There was something in us all and in all of nature and all of Gaia that's incorruptible, no matter what you've ever done–it's incorruptible. And that is that source in you that is perfect, that is there already. And in that moment, I heard the light say, 'Oh beautiful human.' It was a blessing. And in that moment, I fell in love with humanity again. And I think that's when my cancer was cured[2]."

Our final NDE was recorded by the famous NDE researcher, P.M.H. Atwater. She recounts a well-documented and astounding NDE involving Bruna Reyn, which eventually ended up being declared a miracle by the Catholic Church:

My name is Bruna Reyn and I live in Canada... from the age of two to seven I was a ward of the court in the care of The Sisters of Mercy of the Holy Cross... In 1994 I received a strange phone call from my mother. My mother said the sisters of the holy cross had contacted her to say that they had been looking for me. The reason: on a hot August day, when I was 3 years old, I had been found drowned in a cistern in a small yard at Laxenburg... They knew I was underwater for at least 6 hours. Attempts at resuscitation by the nuns who found me, as well as the doctor in attendance, failed. The doctor pronounced me dead and left. But, following passionate prayers directed towards Sr. Maria Theresia, the founder [of the Order], I opened my eyes. I had a very high fever for 3 days. I was told that everyone who was present for this event, including the doctor, was asked to provide a written description of what they had witnessed... This event was documented and made its way through various levels until it reached the Vatican. In 1994 the Vatican declared this a miracle. I was invited, together with my husband, to attend the 1995 Beatification Mass for Sr. Maria Theresia Scherer, and received a medal[3].

David Hume Says Miracles Are Not To Be Believed, Don't Believe Him.

Scottish philosopher David Hume's Of Miracles—found in chapter X of his *An Enquiry Concerning Human Understanding*—remains the constitutional document of materialist denialism. Hume argues that miracles are inherently unprovable, framing his skepticism around the psychology of the wise man, who must reject any novel instantiation of reality that defies uniform experience (what we commonly know and should expect) By centering the argument on the mind of the observer rather than the phenomenon itself, Hume constructed an impenetrable philosophical firewall, effectively reducing miracles to mere fantastical belief[4].

Hume's criteria for rejection are rigid: he claims miracles are against reason, lack reputable witnesses, and contradict the history of cause and effect. He asserts that the reputation of those reporting miracles is inherently impeached by the very nature of their claims. Borrowing from Dr. John Tillotson, Hume argues that since our belief in miracles relies on the reliability of the senses, and miracles themselves violate the implicate order those senses observe, the two are mutually exclusive. According to Tillotson, a belief in doctrines like transubstantiation would "call into question anything and everything that we think we might know on the basis of sense experience[5]." echoes this, stating that our evidence for religion "is less than the evidence for the truth of our senses[6]."

This is a profoundly flawed argument. As John Earman notes, if followed strictly, Hume's method would necessitate denying vast swaths of human history, including the beginning of the universe and the fundamental reality of quantum mechanics—events that appear miraculous and defy ordinary cause and effect[7]. If we used Hume's reasoning in our investigation, every one of the eighteen scientifically verified events in The Divine Algorithm would be denied simply because they are not possible under materialist "laws."

The irony, however, is that Hume's logic is self-defeating. In his own "Problem of Inductive Reasoning," Hume asserts that neither deductive nor inductive reasoning can establish universal facts. He acknowledges that what we call laws of physics are merely patterns we have noticed in the past—mental habits—not ironclad rules that dictate the future. Since we cannot logically prove that tomorrow must behave exactly like yesterday, the door is always open for a unique event to occur without violating logic. Hume essentially cuts the legs out from his own argument in *Of Miracles.*

In reality, our investigation suggests the opposite: miracles are not system glitches; they are the system glue and the fundamental principle our reality has been ordered around. They are, in effect, the foundational updates within our universe upon which reality depends.

Today's evidence makes the materialist denial of these events an invalidation of the spirit of science itself. The code has been updated, and the bell of the miraculous cannot be unrung.

A Miracle Testimony

Miracles do happen. I have experienced one. Had it not occurred, my life would have been unalterably changed.

Mount Si is located in the state of Washington just off the town of North Bend. Hiking up it is a 4-mile trip up with a number of switchbacks, and a number of beautiful vantage points. The top affords a view of Snoqualmie falls and parts west towards Seattle. At the very top there is a rock outcropping called "the haystack." It's near vertical and you need to use your hands and feet to scramble up to the top.

My friend and I decided to climb with him about 15 feet ahead of me.

When I got to the top, he was nowhere to be seen. The haystack has a taller portion to the left that's about 15 feet of additional rock that's higher than the rest. To the right is a series of jagged rocks. The area I had reached was a flat portion, about 8′ by 8′ feet. On the other side of what we'd climbed was an approximate 1,000 rock fall.

Standing at the top, there wasn't anywhere he could logically be. I thought, "Where is he?"

I looked around and then I leaned out to the western edge of the 8′ by 8′ space and looked down and to the left around the upper haystack portion, to my horror my friend was about 25 feet out on a near vertical face with the 1,000 foot drop off below him. I looked at him in disbelief, and then down at the possible fall. I was trying to understand the path he had taken, and it was strange. For the first 3-4 feet it almost looks like there is a trail around this edge. It starts out about 2 feet wide, but by 4 feet in it pinches off to nothing. This is the 'trail' my friend had somehow been convinced existed.

I tried to remain calm to not infect him with my panic and I asked him what he was doing and said he was "I'm following the trail". I looked at where he had gone again and back out to him literally clinging onto a near sheer portion of cliff. I said to him, trying to be calm, "there is no trail" so I wouldn't break whatever spell he was under, and he could come back as confidently as he had gone out. But I was firm. I needed him to stop. It was surreal to find ourselves in one of those situations where a recreational situation is horrifically miscalculated and possible death is suddenly at hand.

My friend looked around. Saw the huge drop off behind him, re-evaluated his position, realized there was no trail and then began to shift his position. Realizing there was no safe way back, I saw his body language assume the condition of someone in a precarious situation. His relaxed climb turned stiff.

I said, "can you make it back?" and he said, "I think". I could see that he was stuck, tentative and unable to move. As the horror sank in, I then realized that he was well aware that I was watching. In my mind, he may have been choosing his actions based upon my observation and selecting options that would not alarm me. For some reason, instead of staying with him to help guide him, my instinct was to tell him I would not look and go around the corner. This decision may have been key. I told him, "I don't want you to have to worry about me and what I think.

I went to the other side and sat with a sense of unreality.

After waiting no longer than 90 seconds my friend climbed down from the top part of the haystack and sat down beside me. I was shocked. I asked him how he ended up on top. He told me a few seconds after I had turned the corner, he looked up and saw a large root protruding from the rock above his head. He thought that if he could hoist himself up, he could make it to the top of the haystack about ten feet above. He grabbed the root, tugged at it to test that it was sturdy, and then when he put his full weight on it to make his move the root gave way in his hands and he began to free fall backwards. Then, suddenly he was on top safe, about 10-15 feet above where he had been.

I was certain I didn't understand, so I asked, "After the root came free you were able to get a hold of the rocks and climb up?" He replied, "No. I was falling backwards, then I was on top." I couldn't make sense of what he was saying, so I asked again, "You blacked out and scrambled up?" He said, "no. I fell then I was on top." We looked at each other, both disbelieving what happened, and slowly coming to terms with the fact that he had fallen to his death but somehow was put in a safe location.

It took me over twenty years of puzzling over this before we both realized, within the same week, that my instinct to not watch was perhaps one of the keys. What would have happened had I continued to watch? Would he have survived? Would I have been allowed to watch him fall backward to his death, and witnessed him being snatched from the sky, and then wondering where he was, before he climbed down beside me. Our instinct was "no." that would not have been allowed. My instinct may have been a message to not watch.

The only thing I'll add is that we were not happy or joyful about this event. It was disturbing. We didn't know if he was 'safe'. If perhaps the universe had made a mistake. If the world had glitched and allowed him to live, or we had both been transferred to an alternative world where he lived, while the one he died in was split off from us. Honestly, then and now, I think a miracle happened. But at the time, we didn't know what exactly had happened.

As it turns out this kind of teleportation/instantaneous relocation is more common in the literature than most realize. These incidents have been documented under the topic of *Bilocation*.

In P.M.H. Atwater's *Beyond the Light* (1994), Dr. Atwater documents multiple accounts of highway miracles where drivers are saved and relocated just before a horrific crash. In one case, a driver who saw a head-on collision experienced a "blip" in reality and instantly found himself 100 yards further down the road, in the correct lane, with the impending collision safely behind him[8].

In P.M.H. Atwater's *The Big Book of Near-Death Experiences* (2007), Atwater and other researchers detail accounts where survivors describe the sensation of freefall, followed by a sudden, seamless "edit", similar to a film cut, where they find themselves standing safely on the ground or back on the ledge[9].

Padre Pio and the Allied Pilots (1943): One of the most famous cases of bilocation occurred during WWII in San Giovanni Rotondo, Italy here several Allied pilots, including General Bernardo Rosini of the Italian Air Force, reported a "flying friar" in the sky, blocking their bomber flight paths or causing their bombs to drop harmlessly. The man they witnessed, Padre Pio was documented by the brothers in the monastery as being in deep prayer. General Rosini visited the monastery after the war and identified Padre Pio as the man he had seen[10].

St. Martin de Porres (1579–1639): In the official Vatican Canonization Records witnesses have provided sworn testimony that St. Martin de Porres was seen in Mexico, North Africa, and the Philippines, tending to the sick and providing provisions, yet during these times he was verified to have never left his monastery in Lima, Peru[11].

Gil Pérez (1593): as documented in the 17th-century spanish colonial records, *Sucesos de las islas Filipinas* by Antonio de Morga, On October 24, 1593, Gil Pérez, a soldier at the Governor's Palace in Manila, Philippines found himself in the Plaza Mayor of Mexico City. At the time he knew of the assassination of the Philippine Governor, news that would not reach Mexico by ship for several months. His presence in Manila in the morning was verified by passengers on the first ship to arrive from the Philippines[12].

In missionary literature, such as A.E. Glover's *A Thousand Miles of Miracle in China* (1904), there are recurring reports of *'Translation'* where within missionary accounts individuals have been somehow translated across flooded rivers or through miles of impassable jungle in seconds to reach a needed duty, such as a sick or dying person or to escape imminent death. How this took place is unknown to the recipients of the *translation*[13].

A Case For Miracles

> "Fiction is obliged to stick to possibilities. Truth isn't."
> -Mark Twain

In the book, *A Case For Miracles* by best-selling author and former atheist Lee Strobel, Strobel holds a discussion on miracles with Craig S. Keener, another former atheist, historian and author of the 1,200 page, 2-volume set, *Miracles: The Credibility of the New Testament Accounts.* In their discussion Keener listed examples of miracles, he had documented:

> "'Cataracts and goiters–instantly and visibly healed' he said, 'Paralytics suddenly able to walk. Multiple Sclerosis radically cured. Broken bones suddenly mended. Hearing for the deaf. Sight for the blind. Voices restored. Burns disappearing. Massive hemorrhaging stopped. Failing kidneys cured. Rheumatoid arthritis and osteoporosis–gone. Life given back to the dead, even after several hours. I have accounts from around the world–China, Mozambique, the Philippines, Nigeria, Argentina, Brazil, Cuba, Ecuador, Indonesia, South Korea, and other countries. Multiple and independent eyewitnesses with reputations for integrity, including physicians, names, dates, medical documentation in many cases. There's even a peer reviewed scientific study confirming the healing of the deaf[14]."

And Dr. Keener's assessment?

"Apart from some sort of divine intervention, many of these phenomena seem inexplicable. In other words, a lot of these cases better fit a supernatural explanation than a natural one[15]."

In opposition to Dr. Keener's assessment, we see the ghost of Hume haunting modern inquiry. This is the lingering insistence that miracles must be dismissed outright, an a priori rejection that forbids the conclusion before the investigation. By tethering their logic to this bias, they ensure that no amount of evidence can ever be 'enough.'

"'Belief in Miracles is unjustified' scoffed Larry Shapiro, professor of philosophy at the University of Wisconsin-Madison, in his book The Miracle Myth. 'No one has ever had or currently has good reasons for believing in miracles. The reasons people give for believing in miracles... are bad[16].'"

For those who have experienced or witnessed miracles, the issue isn't a quest for "justified belief"; it is the confrontation with an unavoidable event. The issue is that a miracle *has* happened, and it has altered the trajectory of their lives. This is the one-way effect in action: once the impossible becomes an empirical fact of one's own history, the world of possibility and the guard rails promised by the physicalist ontology are gone forever. The true nature of our reality is no longer possible to deny.

---

1 (Youtube.com Jeff Mara Podcast #480 She Had A Conversation With God During Her Near Death Experience - Donna Rebadow #480)
2 (Youtube.com, We Don't Die Podcast Episode 39 Inventor + NDEr Mellen Thomas Benedict on We Don't Die Radio)
3 (P.M.H. Atwater, Near-Death Experiences: The Rest of the Story, HRPUB, 2011, pg. 97-98)
4 (D. Hume, An Enquiry Concerning Human Understanding, 2007).
5 (M.P. Levine, Hume and the Problem of Miracles, 1989). Hume
6 (D. Hume, An Enquiry concerning Human Understanding, 2007).
7 (J. Earman, Hume's Abject Failure: The Argument Against Miracles, 2000).
8 (P.M.H. Atwater, Beyond the Light: What Isn't Being Said About Near-Death Experience, 1994, Carol Publishing Group)
9 (P.M.H. Atwater, The Big Book of Near-Death Experiences: The Ultimate Guide to The NDE and its Aftereffects, 2007, Hampton Roads Publishing)
10 (F. M. Rega, Amazing Miracles of Padre Pio: And The Stories Behind Them, 2019, Independently Published)
11 (G. Cavallini, St. Martin De Porres: Apostle of Charity, 1942, TAN Books) (Reverend Scott A. Haynes, Mystical Theology of the Mass Online, St. Martin de Porres, August 22, 2021 (as of May 2, 2026),
12 (A. De Morga, Sucesos de las Islas Filipinas, 1609, Casa de Geronimo Balli)
13 (A.E. Glover, A Thousand Miles of Miracle, 2001, Christian Focus Publisher)

[14] (Lee Strobel, The Case For Miracles, 2018, Zondervan, Pg. 97)150
[15] (Ibid, Lee Strobel, The Case For Miracles, 2018, Zondervan)151
[16] (Lee Strobel, The Case For Miracles, 2018, Zondervan, Pg. 87)152

9 Is the Creator in Book 3, the Instantiator in Books 1 & 2?

Assessing God

In Books One and Two, we examined the 'How': the mathematical and systemic fingerprints of an intended and instantiated environment. Here, we encounter the 'Who.' The following testimonial snippets from the humane empirical data points from the frontiers of human consciousness. They describe encounters with an entity that meets our definition for the instantiator in Books One and Two, that is, one that has:

(1) the requisite power to make instantiations across our universe.

(2) knows what everything must be to 'fit' together and make our universe wholly coherent

(3) is virtually eternal with regards to the universe, having made the universe and being responsible for instantiations in the present

(4) being of an immaterial, non-local and pure consciousness-based substance.

All of the descriptions we have found that speak of this entity proclaim that it is precisely the entity required to have made our universe. This entity is the Source of all we have investigated, and the substance that energizes the Maya of our universe. The citations below are collected from NDEs used in this section, books used as references in this section and the NDERF website.

-Source is the source, the ONE infinite consciousness. Source has one desire, to know itself.

-Yes, the almighty is real and controls the heavens, planets, galaxies, it is the infinity of everything. The beginning and end in Earthly words.

-He "spoke" to me through the LIGHT... I can feel his omnipotent presence.

-The light and the presence of the entity. It was clearly God. It was present everywhere and a part of everything. There were no limits on the power of God.

-The goldeness was like all knowing and wisdom…

-God (who is indescribable, unimaginable and not human) is definitely the creator of the universe and everything within it.

-The one reality includes all beingness or consciousness. It is the endlessly unknowable infinity of creativity and an apparent paradox of infinite numbers of unique individuals that are simultaneously one.

- ...my puny human faculties were unable to capture it all, as if a computer runs out of memory. I recall "knowing" everything while in this state but lost most of it when I came back.

-I knew that all knowledge was in the light and that once there I would be and know everything and was anxious to enter it. It was like going home only magnified a billion times.

-The collective consciousness that I became a part of, is God. Or that is the feeling I got. It is a state of pure energy and pure love.

-I experienced what God is. Different from all other entities I encountered. God is the Alpha and Omega and so perfect and saturated in LIGHT and JOY and LOVE…

- ... such sacred holiness and power... as such I can only say I have witnessed the ultimate of all that is, was and ever shall be. Yet, I cannot name that which cannot be described.

-The entire encounter was about God, the ultimate power of God, and God's forgiveness. The message was, "Love is the greatest power in the universe."

-I did not believe in God before my experience. The presence of God, the forms in which God presented itself to me were undeniable. I wasn't just given information I experienced the information so I would absolutely know it to be true.

-Direct experience and immersion in this presence with the gift of omniscience while I was there.

-I felt the existence of love and belonging and welcoming home. Yes, I didn't feel or see any shape but the feeling was so real and felt so true, I'm sorry I don't know what to say or describe other than I felt I was with God or the creator.

-I felt the touch of something that I instinctively knew as being the touch of God, or the creator of everything that can possibly exist... the fabric of existence itself.

-I knew because I felt the overwhelming presence of intelligent energy everywhere.

-I felt like I "knew" everything for just an instant. I realized then that God is "knowing". That's the only way I could have described it at the time, but I knew with my whole being that a supreme force of some kind existed.

-Everything is God. I would not use the word God, but it answers your question.

-I could see and feel and became one with a higher energy that is much greater than anything ever experienced. It taught me everything there was to know about existence. It did not call itself God or a higher being. It did not really name itself

at all. But I am sure it was comparable to God - though it was not a man and it was somehow one thing and thousands (maybe millions) of things all at once.

-God is complete unconditional love. God is pure light. God is "larger than religion". God is "larger than Earth and Earth's culture." God is science, all knowledge/intelligence and beyond description. God loves us more than we can imagine.

-I have only awareness that God was the creator of all that I was encountering.

-Yes I felt that face... it is everything... the force behind everything.

-"...he just said to me, 'I'm the one that you call God.' ... I could very clearly feel his thoughts and his experience... It was as if he knew everything throughout time, everything throughout space. He was aware of the entirety of the multiverse; absolutely everything was there.. He had this incredible awareness of all things simultaneously which I still have trouble explaining but that was what I experienced when I was there and he was joined to me."

10 For The Future

One NDEr asked the entity he encountered to see the beginning of the Universe. He then related this tale[1]:

> *" 'You're home now,' seemingly thousands of voices said to me. All were speaking at the same time as one voice. But it wasn't like I heard them with my ears. It's like they communicated directly with my soul. I replied that I don't know where I am, even though I did know, I didn't want to admit to myself where I was. They replied, 'This is where you came from and this is where you'll end,' they kept repeating it, again and again. The colors pulsated with their voice. As they spoke, I could feel and see a timeline of the whole of existence. It was like they were showing it to me. I could see everything, not just our galaxy. I saw the whole universe. I could see the big bang. Although it wasn't a bang, it was more like a slow drip. It was like a material being stretched and slowly a hole formed in the middle of it. We are in that gap. I begin to understand it was this 'God/energy' place turning itself into existence. It made a rule; as if it made a game and set wheels in motion. Everything started from a small point and grew from there slowly. The universe is a thing that created itself and is watching itself. We're all part of it. This 'place/thing/existence' that created us and is us, doesn't know where life will lead. That is what it enjoys[2].*

What else will humanity be allowed to witness once we embrace the reality of a creative consciousness behind existence that can be accessed and communicated with? Imagine what we may be able to learn, in science, in our life's, about spirituality, if we embrace and begin to utilize the power of the inexhaustible evaluative field along with the individual gifts and talents we have been bestowed.

The Divine Algorithm is many things, one of which is an invitation to experience a greater truth. Each individual need not wait for "humanity" to decide. We can choose for ourselves. The decision to embrace a new level of reality is for each individual to make. There are millions of individuals who have realized in a moment of insight, that they have been issued this same invitation.

Final Statement

Some of my earliest memories come from before I was able to speak. The moments I came alive were when my mother entered my field of view. But soon another idea began forming in my mind. As I moved through the world, I wondered what it was. What was actually going on? Why was I here? What were we supposed to be doing?

When I got older, I began to feel it might be possible to glimpse the boundary of our experiment—and perhaps what lay beyond it.

As my understanding of the world increased, I could see that the edge of our experiment extended far beyond anything physically reachable. Yet I intuitively knew this world could be transcended, because there was clearly more occurring than what was plainly visible within it.

When I was in my early twenties, I resolved to penetrate into the deeper levels of reality. I already knew the method I would use, and if I was fortunate, my spirit would be allowed to break free and descend into the deepest layers of existence.

My attempt was successful. My mind could no longer tell time, read clocks, or make sense of words. Numbers and letters shifted constantly, making any interpretation of symbols or spatial relationships impossible. I left the back of the house and became aware that every step I took would determine my reality and future possibilities until I reached the cold, moonlit field behind the house.

As I breathed in the quiet night air, I became more lucid with each step. Then I noticed the light of the moon. It shone like a beacon across the grass. I became aware the grass itself was making a sound—a high-pitched resonance, deep ringing, and what seemed like a chorus of living beings singing in unison. I realized the grass was praying to God because the Sun's light was being reflected to them by the Moon.

I realized I was in the presence of God.

And this is where something crucial took place.

As I became aware of His presence, my mind cleared and my spirit swelled with perfection—perfect space, perfect ideas, expansiveness, love, and divine grace. I was astonished. I thought to myself, "This can't be happening," because what I was experiencing was far beyond anything I could have imagined. It felt like I had entered the reality that mystics and visionaries only speak about in fragments.

Then suddenly, it was as if a vacuum began pulling away all of the beauty and perfection. It began receding instantly. And because it was leaving, I realized with complete certainty that it had been real. I had arrived. The plan had worked. I was standing within the divine.

Then the movement reversed, and the presence returned again, filling my entire soul and mind.

I had nearly lost the most precious moment of my life because I could not believe it. My disbelief itself had become a form of denial, and the moment it receded, I understood with absolute certainty that it had been real.

While in the presence of God, something unusual happened. God allowed me to borrow His intellect. Whatever question I asked, I instantly understood the answer. Many people who have had near-death experiences describe "information downloads," but this was different. It felt as though God had joined with me or momentarily allowed me access to a deeper mode of knowing.

I reveled in the clarity. We were all one. There was no separation, no evil, no imperfection—only unity beneath the illusion of division. All the hardships of my life dissolved. We all love one another, even when we cannot perceive it. The thing we seek is already here, immediate and ever-present.

I distinctly remember feeling like "the master quizmaster." No answer seemed beyond me.

Then I remembered the great question and asked,

"What is the meaning of life?"

The answer I received was: "Love."

And strangely, despite the clarity I had been given, I could not fully comprehend the answer. I received it completely, yet I was still left to grow into its meaning.

Today, many people who report near death experiences either ask this same question or are directly told that the reason for existence itself—for life, time, and reality—is love. Hearing this now is deeply validating. But my experience took place in the late 1980's, long before I had ever encountered such accounts. At the time, the answer perplexed me profoundly, likely because I still believed there was s separation between the spiritual and the physical.

How could a spiritual answer be the explanation for physical existence? It would take me many years to understand that the separate itself was the illusion. But having been told the truth, I knew love was the answer, even though I didn't understand exactly how it resulted in a physical existence.

And to clarify, I do not mean that I became equal to God or possessed infinite knowledge in any literal sense. I was not suddenly aware of every hidden process in nature, every star, every cell, or every mind. Rather, I was allowed to experience what was most essential—what was most true and most relevant for me to understand: the perfection of God, the perfection of creation, and the unity underlying life itself.

For years afterward, I struggled to live up to what had been revealed to me. I searched through philosophy, science, religion, psychology, cosmology, and consciousness research, trying to reassemble the framework I thought I had already understood. I already suspected that many of our dominant assumptions about reality were fundamentally incomplete, but I was unprepared for the depth to which they were wrong.

Like many others, I believed materialism described one category of existence while spirituality belonged to another entirely. But now I knew the wall between them had never truly existed. The immaterial did not merely interact with the material—it permeated it, sustained it, and gave rise to it.

I can only say that I could not continue as I was, because my entire system of understanding had been overturned. When belief becomes fused to the ego and forms the foundation of your identity, having it forcibly removed can feel as though the person you thought you were no longer exists.

This is what some of us mean when we speak about the destruction of the ego. It is not merely games, pride or self-importance that is destroyed, but the hidden assumptions upon which we unknowingly build our reality. We unknowingly build ourselves upon what we believe is bedrock, piece by piece, from our first moments. In truth, much of our foundation is based upon assumptions, dogma, or falsity.

There are people who have no need for belief because they feel no need for God. There are also those who have no need for belief because they know God. But many of us exist somewhere in between. We know God, we feel God, and some of us have encountered Him directly, yet we still carry a desperate longing for Him deep within our souls—a longing like a child reaching for its mother.

Having witnessed even a fragment of the beauty, grace, and unimaginable perfection of God, I also glimpsed something terrifying: the depth required for God to know all things completely, across all times, in all possible ways. There are no adequate words for this. There are only approximations.

Many additional things happened that night, but the last I will speak of is this: I understood that I had been asked to speak about God, and I accepted without hesitation. At the time, I did not know what that would mean. So, I continued living my life, waiting and watching, never forgetting.

And there was something else that happened that night. Like many others, I felt I had been asked by God to tell others about Him. At the time, I had no idea what that meant, but I knew it was not going to involve standing on a street corner holding a sign.

It was not until decades later, when I finally recognized the underlying algorithm, that I began to understand what had been meant.

That is why this book exists, despite the fact that I am neither a cosmologist, biologist, philosopher or guru. When I look back upon my life—meeting God, leaving my body, the sweat lodge, lucid dreams, miraculous events, impossible coincidences, and countless events that defied conventional explanation—combined with a lifelong career working as a therapist, trainer, mental health administrator and student of the deepest operations of the human mind, in all manner of conditions, I can now see the preparation that was taking place.

Over time, I came to understand not only the algorithm itself, but also the powerful forces that prevent people from seeing it clearly: belief fixation, social conditioning, peer pressure, institutional self-interest, undue influence, and the invisible gravity of culture. These forces continuously generate protective hypotheses that preserve a fundamentally false materialist ontology, even when its contradictions have become impossible to ignore.

I believed that God had already told me what my role would be. I simply did not understand it yet.

Final

What has emerged from this journey is the foundation for a new ontology of reality—one that includes the immaterial, the non-local, the eternal, and the conscious force underlying existence itself.

And yet all of this pales beside the final realization:

Our existence does not end with the failure of the biological body. This physical realm was not constructed as a prison, but as a stage of development meant to be fulfilled and ultimately transcended.

This earthly life is neither the beginning nor the end. It is a sojourn—a necessary immersion into the odyssey of existence, granting a perspective that enriches the soul.

Acknowledgments

Gratitude to the following people for their support, companionship, and goodwill during my life and the time I was writing this book: Daren Chamberlin, Dr. Douglas Axe, Ian Hough, Ben Guterson, Joseph Trager, Jason Motlagh, Michael McGregor, Soren Mitchell, Julie Maxson, Pablo Manga, Todd Hanson, Colin E. Waterton, Jr. and my sons Beau Eaves and Baptiste Eaves. My family, Honey, Gaga, Nathaniel, Trent, Sierra, Nigel, Kendelle, Alice, Nicholas, and all of my brothers and sisters and nieces, nephews and relatives in Philadelphia, New York City and in and around Nashville Tennessee. Last but not least, to Marla, Malissa, Megan, and Mary Angel, and far too many to be named Roughriders and Bulldogs from Seattle, WA.

---

[1] (A man during an NDE is shown the beginning of our universe as it happened, Aaron M. nderf.org, NDE 8350):
[2] (NDERF.org, 4379, Aaron M. NDE 8350)154

www.ingramcontent.com/pod-product-compliance
Ingram Content Group UK Ltd.
Pitfield, Milton Keynes, MK11 3LW, UK
UKHW020202290926
15471UKWH00005B/1020

9 798996 333